THE SYNAPSE:
STRUCTURE AND FUNCTION

THE SYNAPSE:
STRUCTURE AND FUNCTION

Edited by

VIRGINIA PICKEL AND MENAHEM SEGAL

ELSEVIER

Amsterdam • Boston • Heidelberg • London
New York • Oxford • Paris • San Diego
San Francisco • Sydney • Tokyo
Academic Press is an imprint of Elsevier

Academic Press is an imprint of Elsevier
The Boulevard, Langford Lane, Kidlington, Oxford OX5 1GB, UK
Radarweg 29, PO Box 211, 1000 AE Amsterdam, The Netherlands
225 Wyman Street, Waltham, MA 02451, USA
525 B Street, Suite 1800, San Diego, CA 92101-4495, USA

First edition 2014

Library of Congress Cataloging-in-Publication Data
Synapse (Pickel)
The synapse : structure and function / edited by Virginia Pickel, Menahem Segal. -- First edition.
p. ; cm.
Includes bibliographical references and index.
ISBN 978-0-12-418675-0
I. Pickel, Virginia, 1943- editor of compilation. II. Segal, Menahem, 1944- editor of compilation.
III. Title.
[DNLM: 1. Synapses--physiology. 2. Synaptic Transmission. WL 102.8]
QP364
612.8--dc23
2013039439

British Library Cataloguing in Publication Data
A catalogue record for this book is available from the British Library

ISBN: 978-0-12-418675-0

For information on all Academic Press publications visit our web site at store.elsevier.com

CONTENTS

LIST OF CONTRIBUTORS

Chiye Aoki
Center for Neural Science, New York University, New York, NY, USA

Uri Ashery
Department of Neurobiology, Tel-Aviv University, Tel-Aviv, Israel; Sagol School of Neuroscience, Tel-Aviv University, Tel-Aviv, Israel

Boaz Barak
Department of Neurobiology, Tel-Aviv University, Tel-Aviv, Israel; Sagol School of Neuroscience, Tel-Aviv University, Tel-Aviv, Israel

Dana Bar-On
Department of Neurobiology, Tel-Aviv University, Tel-Aviv, Israel

Yoav Ben-Simon
Department of Neurobiology, Tel-Aviv University, Tel-Aviv, Israel; Sagol School of Neuroscience, Tel-Aviv University, Tel-Aviv, Israel

Noa Bielopolski
Department of Neurobiology, Tel-Aviv University, Tel-Aviv, Israel

Tamara Blutstein
Department of Neuroscience, Tufts University School of Medicine, Boston, MA, USA

Michael E. Cahill
Department of Physiology, Northwestern University Feinberg School of Medicine, Chicago, IL, USA

Juliette E. Cheyne
Department of Synapse and Network Development, Netherlands Institute for Neuroscience, Amsterdam, The Netherlands

Heather A. Davies
Laboratory of Neurocytology, Deparment of Life Sciences, The Open University, Milton Keynes, UK

Mathias De Roo
Department and Center of Neuroscience, University of Geneva, Geneva, Switzerland

Laurent Descarries
Departments of Pathology & Cell Biology and of Physiology, and Groupe de recherche sur le système nerveux central, Université de Montréal, Montreal, QC, Canada

Joseph Dynes
Department of Physiology, University of California Irvine, Irvine, CA, USA

Alev Erisir
Department of Psychology, University of Virginia, Charlottesville, VA, USA

Shannon Farris
Reeve-Irvine Research Center, University of California Irvine, Irvine, CA, USA; Center for the Neurobiology of Learning and Memory, University of California Irvine, Irvine, CA, USA; Departments of Anatomy and Neurobiology, Neurobiology and Behavior, and Neurosurgery, University of California Irvine, Irvine, CA, USA

Irit Gottfried
Department of Neurobiology, Tel-Aviv University, Tel-Aviv, Israel

Yuko Hara
Friedman Brain Institute, Mount Sinai School of Medicine, New York, NY, USA; Fishberg Department of Neuroscience and Kastor Neurobiology of Aging Laboratories, Mount Sinai School of Medicine, New York, NY, USA

Philip G. Haydon
Department of Neuroscience, Tufts University School of Medicine, Boston, MA, USA

Christopher Heise
Department of Pharmacology, University of Milan, Milan, Italy

Martin Horak
Institute of Physiology AS CR, v.v.i., Videnska, Czech Republic

Kelly A. Jones
Department of Physiology, Northwestern University Feinberg School of Medicine, Chicago, IL, USA

Igor V. Kraev
Laboratory of Neurocytology, Deparment of Life Sciences, The Open University, Milton Keynes, UK

Ayal Lavi
Department of Neurobiology, Tel-Aviv University, Tel-Aviv, Israel; Sagol School of Neuroscience, Tel-Aviv University, Tel-Aviv, Israel

Christian Lohmann
Department of Synapse and Network Development, Netherlands Institute for Neuroscience, Amsterdam, The Netherlands

Bruce S. McEwen
Harold and Margaret Milliken Hatch Laboratory of Neuroendocrinology, The Rockefeller University, New York, NY, USA

Nikolay Medvedev
Laboratory of Neurocytology, Deparment of Life Sciences, The Open University, Milton Keynes, UK

Lirin Michaeli
Department of Neurobiology, Tel-Aviv University, Tel-Aviv, Israel

Teresa A. Milner
Department of Neurology and Neuroscience, Brain and Mind Research Institute, Weill Cornell Medical College, New York, NY, USA; Harold and Margaret Milliken Hatch Laboratory of Neuroendocrinology, The Rockefeller University, New York, NY, USA

John H. Morrison
Friedman Brain Institute, Mount Sinai School of Medicine, New York, NY, USA; Fishberg Department of Neuroscience and Kastor Neurobiology of Aging Laboratories, Mount Sinai School of Medicine, New York, NY, USA; Department of Geriatrics and Palliative Medicine, Mount Sinai School of Medicine, New York, NY, USA; Computational Neurobiology and Imaging Center, Mount Sinai School of Medicine, New York, NY, USA

Dominique Muller
Department and Center of Neuroscience, University of Geneva, Geneva, Switzerland

Irina Nikonenko
Department and Center of Neuroscience, University of Geneva, Geneva, Switzerland

Martin Parent
Department of Psychiatry and Neuroscience, Université Laval & Centre de recherche de l'Institut universitaire en santé mentale de Québec, Quebec City, QC, Canada

Peter Penzes
Department of Physiology, Northwestern University Feinberg School of Medicine, Chicago, IL, USA; Department of Psychiatry and Behavioral Sciences, Northwestern University Feinberg School of Medicine, Chicago, IL, USA

Ronald S. Petralia
NIDCD/NIH, Bethesda, MD

Victor I. Popov
Institute of Cell Biophysics, Russian Academy of Sciences, Pushchino, Russia

Christine Remmers
Department of Physiology, Northwestern University Feinberg School of Medicine, Chicago, IL, USA

Carlo Sala
Department of Pharmacology, University of Milan, Milan, Italy; Neuromuscular Diseases and Neuroimmunology, Neurological Institute Foundation Carlo Besta, Milan, Italy

Gail K. Seabold
NIDCD/NIH, Bethesda, MD

Susan R. Sesack
Departments of Neuroscience and Psychiatry, University of Pittsburgh, Pittsburgh, PA, USA

Zehavit Shapira
Department of Neurobiology, Tel-Aviv University, Tel-Aviv, Israel

Anton Sheinin
Department of Neurobiology, Tel-Aviv University, Tel-Aviv, Israel; Sagol School of Neuroscience, Tel-Aviv University, Tel-Aviv, Israel

Oswald Steward
Reeve-Irvine Research Center, University of California Irvine, Irvine, CA, USA; Center for the Neurobiology of Learning and Memory, University of California Irvine, Irvine, CA, USA; Departments of Anatomy and Neurobiology, Neurobiology and Behavior, and Neurosurgery, University of California Irvine, Irvine, CA, USA

Michael G. Stewart
Laboratory of Neurocytology, Deparment of Life Sciences, The Open University, Milton Keynes, UK

Jung-Hwa Tao-Cheng
National Institute of Neurological Disorders and Stroke, NIH, Bethesda, MD, USA

Jon-Eric VanLeeuwen
Department of Physiology, Northwestern University Feinberg School of Medicine, Chicago, IL, USA

Chiara Verpelli
Department of Pharmacology, University of Milan, Milan, Italy; Neuromuscular Diseases and Neuroimmunology, Neurological Institute Foundation Carlo Besta, Milan, Italy

Elizabeth M. Waters
Harold and Margaret Milliken Hatch Laboratory of Neuroendocrinology, The Rockefeller University, New York, NY, USA

Kevin M. Woolfrey
Department of Physiology, Northwestern University Feinberg School of Medicine, Chicago, IL, USA

CHAPTER ONE

Structure and Complexity of the Synapse and Dendritic Spine

Michael G. Stewart[1], Victor I. Popov[2], Igor V. Kraev[1], Nikolay Medvedev[1] and Heather A. Davies[1]

[1]Laboratory of Neurocytology, Deparment of Life Sciences, The Open University, Milton Keynes, UK
[2]Institute of Cell Biophysics, Russian Academy of Sciences, Pushchino, Russia

1. INTRODUCTION

Synapses are the point at which an impulse is transmitted from one nerve cell to another. Some signalling is purely electrical, where transmission occurs via electrical coupling of ion movements at gap junctions. However, most transmission occurs at chemical synapses where an action potential causes release of neurotransmitter from the presynaptic component (the bouton), which acts on receptors on the postsynaptic cell. Knowledge of the structure of synapse and their postsynaptic contacts is the key to understanding synaptic function. In this chapter, discussion will be restricted to the hippocampus of central nervous system (Figure 1) and will focus on a description of synapses and their dendritic spines contacts.

2.1. SYNAPSES AND DENDRITIC SPINES

2.1.1. Synapses

Synapses are axonal swellings, found either at the end of an axon, or formed as en passant contacts—made on a dendritic component (spine or shaft) as an axon courses on its pathway. Synapses are not visible at light microscope level but in a Golgi impregnated granule cell in the dentate gyrus of the hippocampus (Figure 2(A)), the dendritic spines are visible as small "bumps" and more clearly in Figure 2(B), as twig-like structures. Camillo Golgi first recognized spines in 1873 using a silver nitrate impregnation method of neurons, and these were later described in detail by Santiago Ramon y Cajal in 1893 (DeFelipe, 2006). However, to see spines in detail, electron microscopy (EM) is necessary, and Figure 2(C) is an electron micrograph from a semi thin section (300 nm) of a portion of the granule cell in Figure 2(B) where the silver from the Golgi technique has been replaced by gold (gold toning). Spines can be seen extending from either side of the dendrite and can be up to 3 μm in length. A limitation, however, of the Golgi technique is that even with gold toning, it is difficult to see contacts between synapses and spines.

The Synapse
http://dx.doi.org/10.1016/B978-0-12-418675-0.00001-8

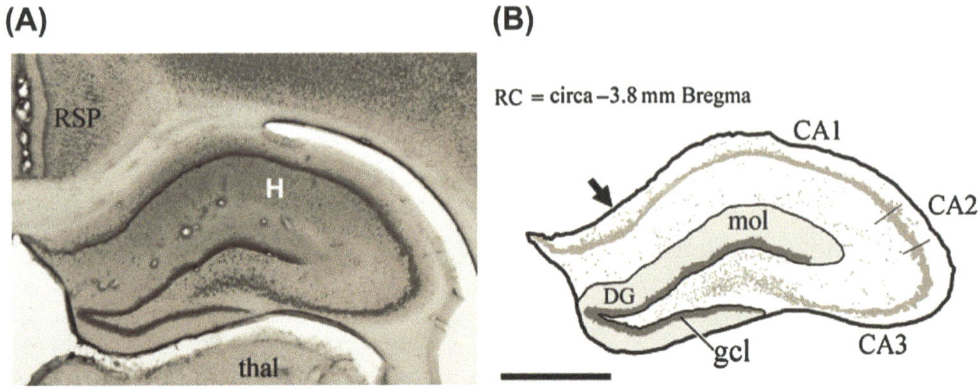

Figure 1 The hippocampus is shown in section of tissue (A) and reference diagram (B).

To see synapses, EM with "normal" staining—using lead and uranium salts (Davies and Stewart, 1998)—is necessary. While confocal microscopy and newer methods including stimulated emission depletion microscopy (Wilt et al., 2009; Nagerl et al., 2008, 2010) enable quantification of factors such as Ca^{2+} entry into spines, they do not yet enable visualization of synaptic boutons or postsynaptic density (PSD) parameters.

2.1.2. Postsynaptic Density

Synapses can be identified at the electron microscope (EM) level by the presence of a PSD and presynaptic vesicles close to the presynaptic membrane, as in Figure 1.2(D–H).

Figure 2 (A) Golgi impregnated granule cell in dentate gyrus of rat, arrow points to an apical dendritic branch, scale bar = 30 μm; (B) higher power image of dendrite shown by arrow in (A), scale bar = 10 μm. Note the presence of dendritic spines which cover the entire dendrite, some show by arrows; (C) electron micrograph of a portion of dendrite in (B) in which the gold has been replaced by silver. The dendritic spines (Sp) are now clearly visible but the synapses cannot be seen with this method of staining; (D) a small portion of dendrite as in (C) but with the tissue now stained with lead and uranium salts. Not only is the dendrite (den) and spine (sp neck) clearly visible but also the PSD (postsynaptic density), this is an asymmetric synapse (see below for definition). The dendrite contains smooth endoplasmic reticulum (sER) and there is a spine apparatus (SA) in the spine head, scale bar = 5 μm (E) Synapse at higher magnification with bouton containing vesicles (ves) and a mitochondrion (M). The bouton makes contact with a spine and the PSD is perforated (perf) (scale bar = 0.5 μm); (F) Electron micrograph of an asymmetric axo-spinous synaptic junction from stratum lacunosum-moleculare of dorsal anterior CA1 hippocampus. Spherical synaptic vesicles (ves) are present in the presynaptic axon terminal (also termed presynaptic bouton) (at). The PSD is indicated by the arrow. The post-synaptic spine head (Sp) arises from a dendritic shaft (DS). Mitochondria (mt) are visible in the axon terminal and the dendritic shaft. Scale bar: 0.5 μm (G) Schematic representation of the asymmetric axo-spinous synapse seen in H. The profiles of synaptic vesicles (ves) and mitochondria (mt) are illustrated. (H) A synapse with presynaptic bouton PrB makes contact with a dendritic spine (Sp) (purple color). The PSD is arrowed. Most notable is the fact that the synapse is surrounded by an astrocyte (Ast) (yellow color) scale bar = 0.5 μm. *Originals except (F) and (G), with permission from Donohue et al., 2006.*

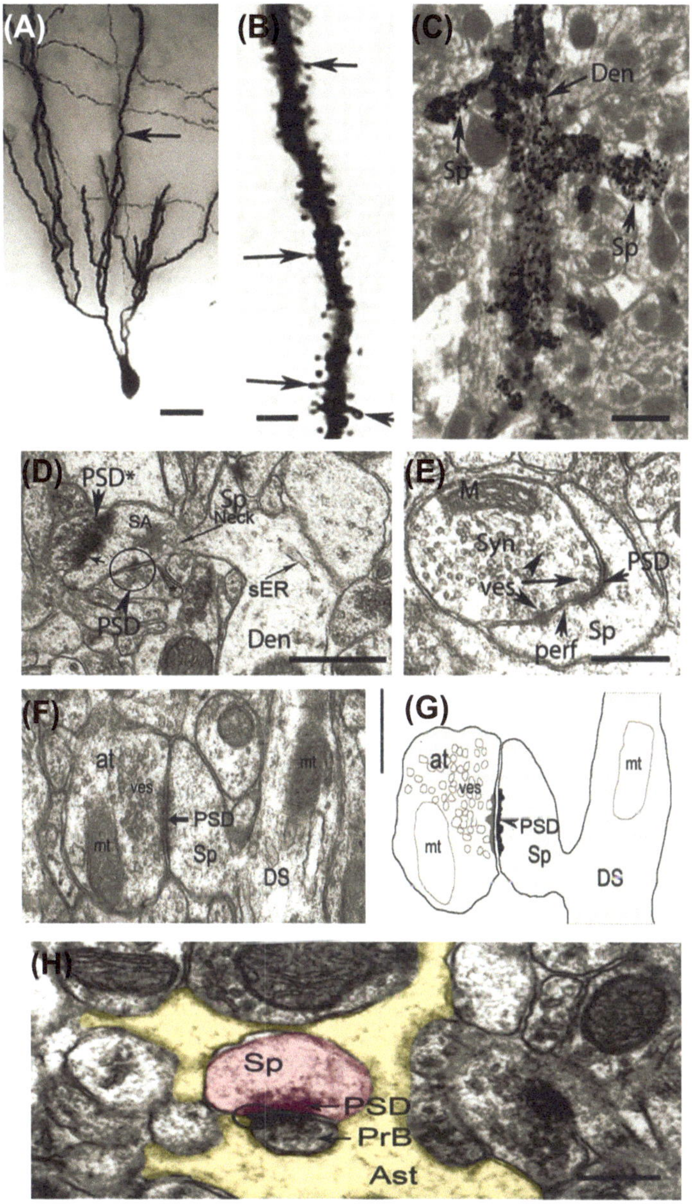
(A)
(B)
(C)
Den
Sp
Sp
(D)
PSD*
SA
Sp
Neck
sER
PSD
Den
(E)
M
Syn
ves
PSD
perf
Sp
(F)
at
ves
PSD
mt
Sp
mt
DS
(G)
at
ves
PSD
mt
Sp
mt
DS
(H)
Sp
PSD
PrB
Ast

Synaptic vesicle distribution appears to be related to function activity with redistribution of synaptic vesicles in the synaptic bouton toward the synaptic active zone following long-term potentiation (LTP) (Applegate et al., 1987; Applegate and Landfield, 1988; Lushnikova et al., 2008). PSDs are found typically on dendritic spines which play primarily connective, electrical, and biochemical roles in neuronal physiology (DeFelipe, 2006; Harris, 1999; Popov et al., 2004, 2005; Segal, 2005). The PSD contains receptors, scaffold, and signalling proteins (Gu and Zeng, 2009). There is only a fine synaptic cleft (~20 nm) separating the pre- and postsynaptic components. In Figure 1(D) and (F), the spine is attached to the dendrite by a thin spine (Tsp) neck and at the head of the spine is a PSD in apposition to a synaptic bouton, while at the side of the spine there is another synapse with a very fine PSD. The structure and form of spines is due a cytoskeleton, which is primarily made of filamentous actin (F-actin), but this cytoskeleton is not visible with normal staining in the electron microscope, immunocytochemistry would be required.

2.1.3. Three-Dimensional Reconstructions

To determine the structural organization of PSDs and dendritic spines most clearly, three-dimensional (3-D) reconstructions from ultrathin serial sections are necessary. Typically up to 150 serial ultrathin sections, from fixed and resin embedded material, are cut at ~50–60 nm and imaged in an electron microscope at ~10 K. The images are reconstructed in 3-D using software developed by Drs John Fiala and Kristen Harris (http://synapses.clm.utexas.edu/). The availability of these programmes has made the ability to carry out reconstructions of 2-D sections accessible to electron microscopists but the procedures require patience and experience. However, features of synapses and spines become readily apparent in 3-D reconstructions and these are described below.

2.1.4. Synapses and Spinules

Spinules are prominent structures arising from discontinuities in the PSD on the spine head, which evaginate into the presynaptic bouton. Figure 3(A) and (B) shows two images of a series of 100 sections that were used to reconstruct the spine and PSD which are presented in Figure 3(C).

This example is of a synaptic contact on a *mushroom spine* (see definitions below) and there are two spinules from this mushroom spine. Figure 3(D) and (E) are two serial electron micrographs from a *Tsp* in which two spinules project into an axon, and these are seen clearly in the 3-D reconstruction of the spine in Figure 3(F). Spinules from Tsp heads are less frequent than on mushroom spines. Spinules are suggested to play a role in exchange of material at synaptic junctions and may also affect the electrical coupling of the pre-/postsynaptic structures (Spacek and Harris, 2004). Interestingly the number of spinules per spine increases significantly on mushroom spines in the middle molecular layer of the dentate gyrus after LTP. The function of the spinules is unclear, but may

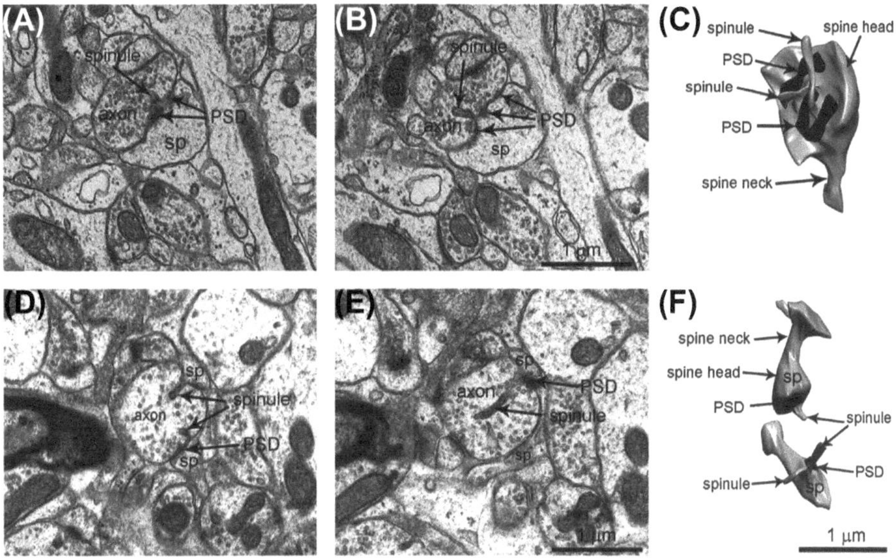

Figure 3 Spinules on Mushroom and Thin Spines. Two electron micrographs (from a series of 100) containing identified synapses and spines. Reconstructions of 100 serial sections of the synapse shown in (A) and (B), and presented in (C). This synaptic contact is on a mushroom spine and two spinules from the spine arise as discontinuities in the PSD on the spine head and evaginate into the presynaptic bouton. (D,E) Two serial electron micrographs through a thin spine from which two spinules project into an axon, as shown in the 3-D reconstruction in (F). Spinules from thin spine heads are less frequent than on mushroom spines (see below); abbreviations: sp, spine; PSD, postsynaptic density. *Reproduced with permission from Medvedev et al., 2010.*

be related to growth and remodeling of synaptic membranes or an increased motility of spines (Matus, 2000). Spacek and Harris (2004) suggested that trans-endocytosis of spinules removes excess pre- and postsynaptic membrane after activation resulting in transient or segmented synapses on mushroom spines.

2.1.5. Endosomes

Endosomes are involved with plasticity after synaptic stimulation and play a major role in the endocytotic route sending proteins and lipids to multiple destinations including the cell surface, Golgi complex, and lysosomes (Cooney et al., 2002; Murk et al., 2003; Popov et al., 2008).

2.1.6. Mitochondria

Note that in Figure 4(B) and (C), there is a spinule (sp) protruding from the bouton into the dendrite (D), both of which also contains a mitochondrion (MT). MT normally do not penetrate into dendritic spines, but there is an exception of thorny excrescences in stratum lucidum of hippocampal CA3. MT are normally depicted as being discrete

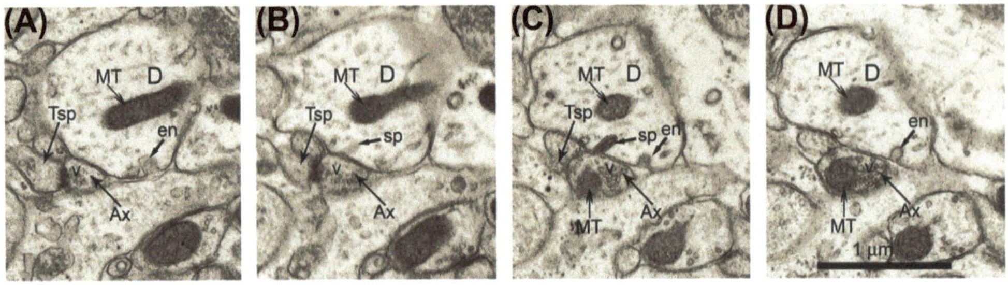

Figure 4 Electron micrographs from four serial sections in dentate gyrus showing a dendrite (D) from a granule cell contacted by a small synapse (Ax—or axon terminal). A thin spine (Tsp) is also labeled. An endosome is present in (A), (C) and (D), and a spinule (sp) begins to appear in Figure 3(B) and is more clearly present in (C). Scale bar = 1 µm.

entities in dendrites and axons but there is evidence that in dendrites in the CNS, MT actually form a syncytium when 3-D reconstructions are examined (Figure 5(A) and (B)).

2.1.7. Excitatory and Inhibitory Synapses

Synapses are normally classified as either excitatory or inhibitory. Excitatory synapses (Figure 2(D–F), Figure 3(A) and (B)) have an asymmetrical synaptic junction with a prominent PSD, and synaptic vesicles of spheroid character (in the hippocampus normally containing glutamate), whereas inhibitory synapses (one is encircled in Figure 2(D)) are typically GABA-ergic, have a synaptic junction that is symmetrical in form, do not contain a prominent PSD, and possess ellipsoidal synaptic vesicles. The synapse on a dendritic spine shown in Figure 2(E) has a perforated (also termed segmented) PSD, but when they are unperforated, they are termed macular as in Figure 2(D). Glial coverage of synapses is a notable feature in the hippocampus. Figure 2(H) shows an asymmetric synapse with a prominent PSD (between arrows) contacting a large spine. The spine and synapse are surrounded by an astrocyte; typically processes of this class of glial cell appose a large percentage of synapses in the hippocampus; in CA1 this may be up to 57% (Ventura and Harris, 1999). One of the most important aspects of knowledge of the structure of synapses and spines stems from the fact that the basis of learning and memory formation is considered to result from a change in the efficacy of synaptic transmission. This is believed to be reflected in changes in the morphology of spines and synapses, expressed as structural plasticity, due to strengthening and growth of existing synapses and/or development of new synapses.

2.2. SYNAPSE: SPINE RELATIONSHIP

The relationship between a spine and synapse is not necessarily one-to one (Fiala et al., 1998; Harris, 1999; Geinisman et al., 2001; Popov and Stewart, 2009). The

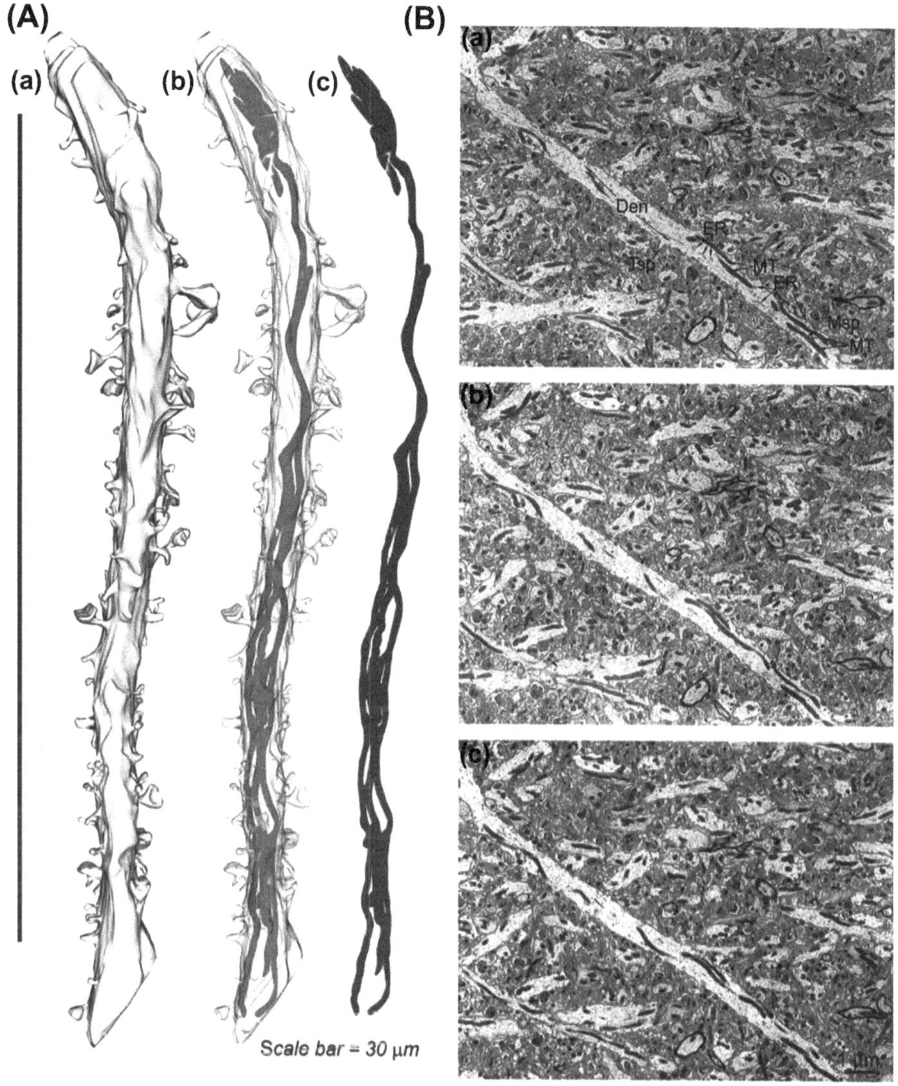

Figure 5 Reconstructed Dendritic Segment (~30 mm in Length) from Rat Dentate Gyrus Granule Cell. A(a): the complete reconstruction with dendritic spines. A(b): the internal structure with branched filamentous mitochondria. A(c): the filamentous mitochondria alone. B(a–c): three consecutive images (section thickness 60 nm) from a series in which the mitochondrial filaments reconstructed in A are present. MT, mitochondrion; Msp, mushroom spine; Tsp, thin space; ER, endoplasmic reticulum. Scale bars = 30 μm in A; 1 μm in c (applies to a–c). *Reproduced with permission from Popov et al., 2005.*

complexity of the relationship between spines and synapse can be shown by examination of Figure 1.6(A) where three synapses (termed axonal (Ax) boutons) contact a large spine and the shape of the PDSs varies from almost flat (Ax1 and Ax3) to U-shaped (Ax2).

2.2.1. CA3 Mossy Fibers and Thorny Excrescences

The greatest complexity of synapse—spine shape and interactions can be seen in CA3 of the hippocampus where the mossy fibers from the dentate granule cells contact the thorny excrescences of stratum lucidum of the giant pyramidal neurons (Figure 6(B)). The contacts take the form of giant boutons (GBs) (also named detonator synapses because of their size) and there is an abundance of spheroid vesicles (glutamate containing) in the GBs which make contact on the thorns that are unique to CA3 in the hippocampus.

2.2.2. Spine Apparatus and Smooth Endoplasmic Reticulum

Large spines in other areas of the hippocampus such as dentate gyrus can often show the presence of a spine apparatus, a type of endoplasmic reticulum (Figure 6(C)). The spine apparatus is shown in finer detail in two serial sections in Figures 6(D) and (E), where smooth endoplasmic reticulum is also present. The spine apparatus forms a lamellar type structure, penetrating into the large mushroom spines reaching even to the PSD (Spacek and Harris, 1997). The absence of spine apparatuses (as in synaptopodin-deficient mice) results in deficits in LTP and impaired spatial learning supporting the idea that the spine apparatus is involved in synaptic plasticity (Jedlicka et al., 2008, 2009). The large spines containing spine apparatus are termed *mushroom spines* (Msp) as opposed to Tsp, which will be discussed in more detail below.

2.2.3. Perforated PSDs

Figure 7 is from a 10 mm dendritic segment of a granule cell in the dentate gyrus. This shows 12 serial ultrathin sections (Figure 7A(a–l)), with a synaptic bouton contacting a spine; one of these (Figure 7A(f)), was shown in Figure 2(E). When the sections are combined in a 3-D reconstruction, the spine head and the PSD appear as in Figure 5B(a–c). Note that the PSD is perforated, which is now quite obvious in the reconstruction whereas it was only "apparent" in 2E. The role and importance of these perforations is not entirely clear but more complex synapses (such as the large synapses on mushroom

→

dendrite and on this asymmetric synapse makes contact (PrB, presynaptic bouton). Note the presence of smooth endoplasmic reticulum (sER) in the dendrite and also a mitochondrion (MT). Also present Tsp-thin spine, scale bar = 1 μm. (D) and (E) Two serial sections containing a mushroom spine protruding from a dendrite with a spine apparatus penetrating into the spine. The asymmetric synaptic apposition zone has a large PSD (indicated by an arrow). In the dendrite (E), sER is presented (indicated by branched arrows). Scale bars = 0.5 μm.

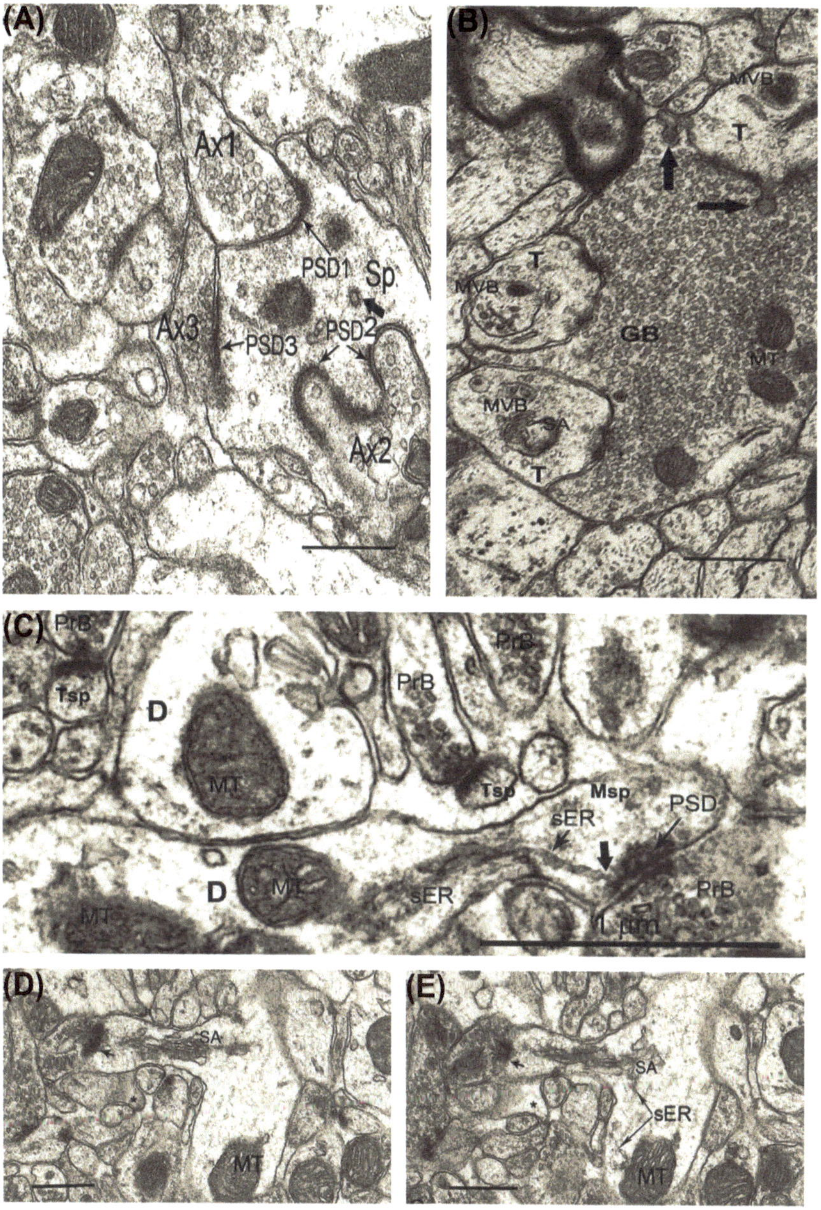

Figure 6 (A) Three axon terminals, or synaptic boutons (Ax1, 2, and 3) make contact with a large dendritic spine (Sp), the PDSs are of different forms. PSD1 shows curvature with concavity of the PSD, while PSD2 is U-shaped and PSD3 is flat. The functional significance of these different shapes is unclear (Medvedev et al., 2010). Scale bar = 1 μm. (B) Giant bouton (GB) from mossy fiber making contact on three thorns (T) from thorny excrescence in hippocampal CA3. Also multivesicular bodies (endosomes) and part of a spine apparatus (SA) are presented. Note the numerous spherical vesicles in the GB. (C) Part of a dendrite (D) in longitudinal section with a mushroom spine (Msp) emanating from the

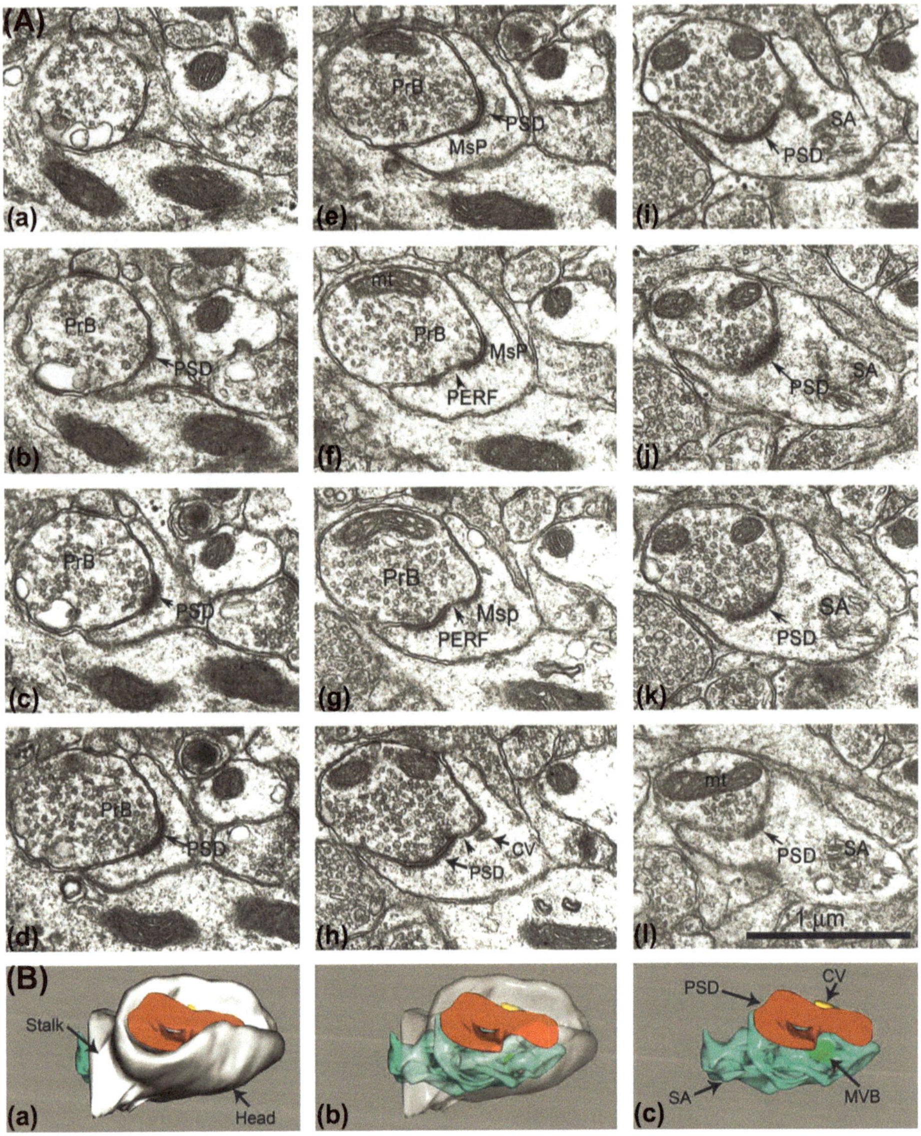

Figure 7 Series of 12 Ultrathin Sections through a Synapse with Presynaptic Bouton (PrB) and Mushroom Spine (Msp). A spine apparatus (SA) is visible in A(k). Note how the appearance of the PSD changes when the series is examined, from barely visible in A(a) to prominence with a single unperforated PSD, but perforated (PERF) in A(f–h) and then the PSD almost disappears in A(l). The true shape of the PSD and spine can only be fully appreciated in the reconstructions in B(a–c). B(b) is the spine and PSD showing then extensive spine apparatus (SA) and in B(c), the spine membrane is stripped off and shows the spine apparatus, PSD, a coated vesicle (CV) and a multivesicular body (MVB). None of this fine detail would be visible without 3-D reconstruction. Scale bar = 1 μm.

spines (Figure 3(A) and (B); Figures 7 and 8), have one or more perforations, and these may be related to enhancement of transmission of the signal across the synapse to the dendritic spine.

2.3. DENDRITIC SPINE CLASSIFICATIONS

There is a great variability in the morphological characteristics of spines types, including their size, the shape of the spine head and neck, and the size and distribution of the PSD (or densities where more than one synapse makes input to a spine). Whether this reflects entirely separate phenotypes, or simply spines that can change shape depending on physiological circumstances, is unclear.

2.3.1. Spine Shapes

These vary in proportions, from the large mushroom (so-called learning spines, Figure 8(A–E)) which in adult rat, dentate gyrus may comprise ~10% of the total spine number; to Tsps which make up ~75% (Figure 8(A–D)). Stubby spines are ~10% (not shown in this Figure—see below in Figure 9), with those found directly on dendritic shafts (Sh Sp) ~5%. Figure 8C(a and b) shows that the mushroom spines in Figure 8A(a and b) and Ba,b are actually branched mushroom spines where the heads with PSD contacts presynaptic boutons. Figure 8(D) is a 3-D reconstruction from ~100 serial sections in dentate gyrus of rat hippocampus, approximately 10 μm in length, and details in 3-D of a large mushroom spine is shown in Figure 8(E). Spine and synapse forms are similar throughout the hippocampus except in CA3 in stratum lucidum where the dendrites are covered with thorny excrescences which are contacted by the GBs of the mossy fibers (as seen in Figure 6(B)). When the thorny excrescences and GBs in CA3 are serially sectioned and reconstructed in 3D (Figure 8(F)), the complexity of the synapse–spine relationship becomes very clear.

Fluctuations in spine volume and thus enlargement and shrinkage of spines might explain synapse maintenance over long periods (Kasai et al., 2010(a)) and this would suggest that the spines types shown in Figure 8(A–E) are a continuum of forms. This is discussed in more detail in 2.3.2 below.

Spine morphology influences synaptic input. Thus the length of the spine neck is likely to affect the efficiency of the molecular mechanisms involved in the modification of glutamate receptors, or the accessibility of cytoskeletal elements associated with changes in glutamate-mediated function (van Rossum and Hanisch, 1999). Alteration of spine neck length acts as a regulatory mechanism for continuous fine-tuning of synaptic alterations. A constricted neck separates signals in the spine head, and protects the parent dendrite from excitotoxicity (Bourne and Harris, 2008). Dendritic spines protect neurons from synaptic activity-induced rises in intracellular calcium concentrations, thus spine density increases with enhanced synaptic activity, including LTP generation, and shrinks or is replaced by

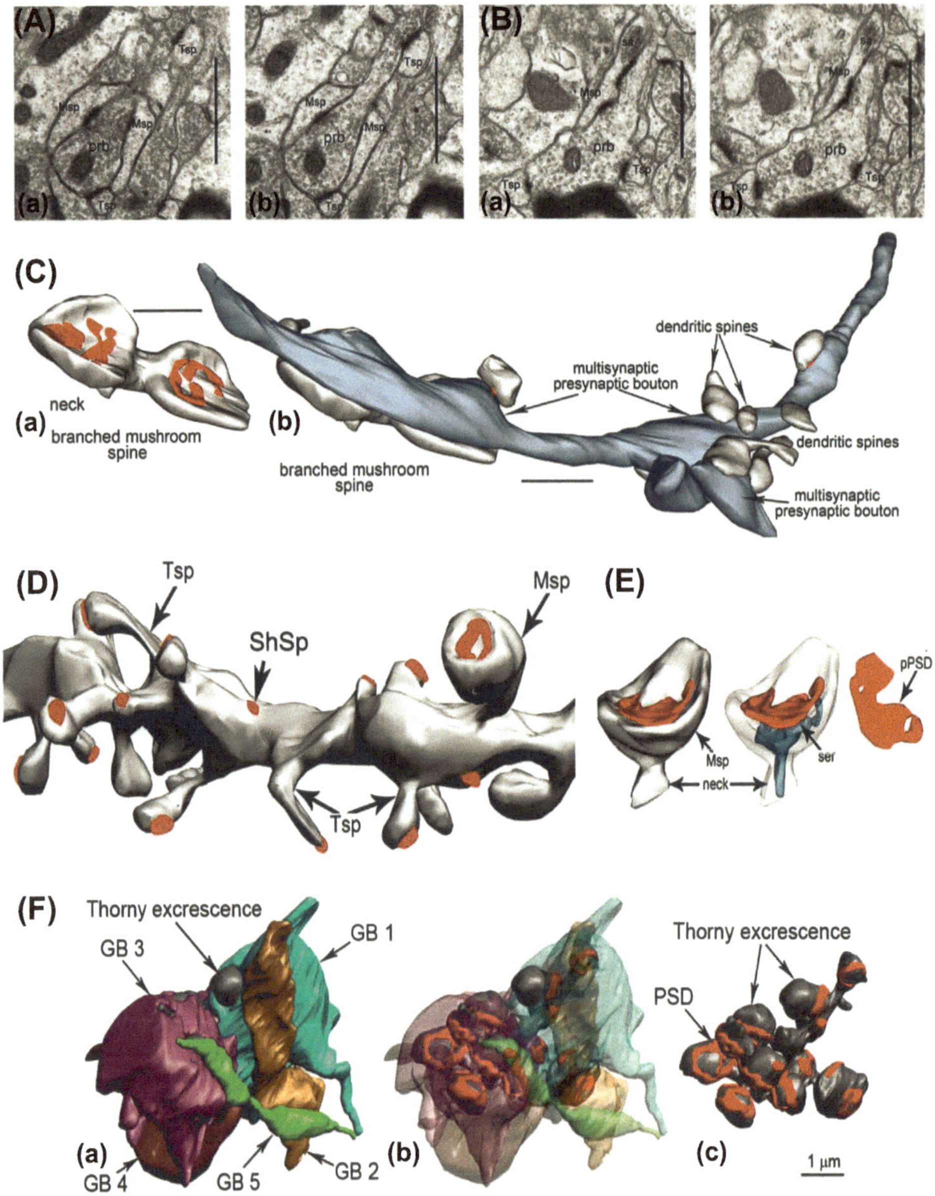

Figure 8 Four Electron Micrographs from Serial Sections of a Mushroom Spine (Msp) in Adult Rat Dentate Gyrus. C(a and b) shows that the mushroom spine shown in both A(a,b) and B(a,b) is branched mushroom spine where heads and PSD contacting same presynaptic bouton. D is a 3-D reconstruction from ~100 serial sections of a segment of dendrite in dentate gyrus of rat hippocampus, approximately 10 μm in length; E is a 3-D reconstruction of a large mushroom spine. Spine and synapse forms are similar throughout the hippocampus except in CA3 in stratum lucidum where the dendrites are covered with thorny excrescences which are contacted by the giant boutons of the mossy fibers as seen in (B) and shown in the 3-D reconstructions in (F). F(a), a complete reconstruction; F(b), with boutons translucent; F(c) without boutons, showing only the spiny thorny excrescence. All scale bars = 1 μm.

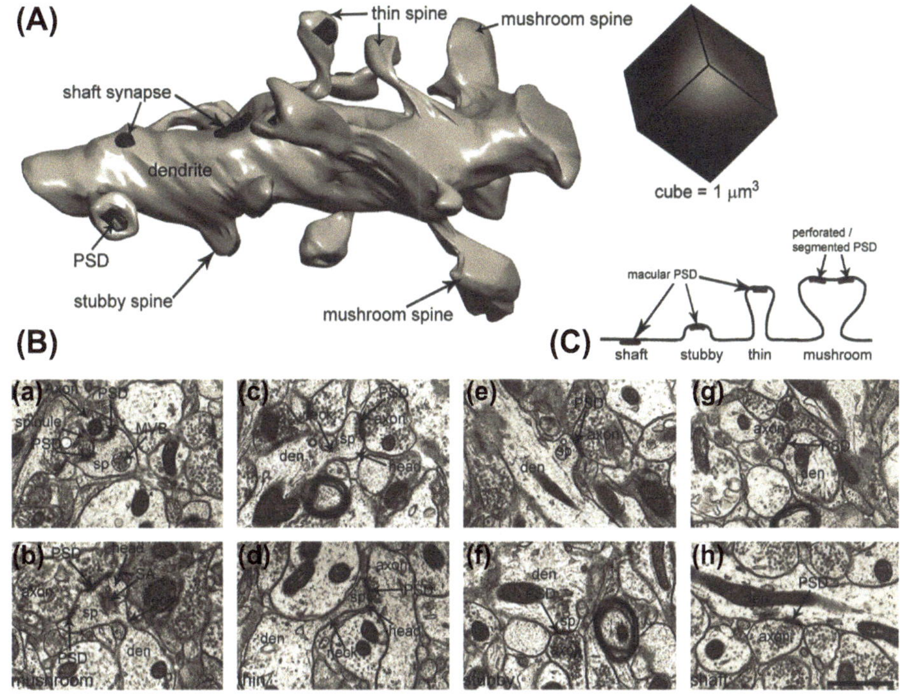

Figure 9 Categories of Dendritic Spines and Synapses. (A) 3-D reconstruction of a short (~10 mm) segment of dendrite with three spine categories distinguished (mushroom, thin, and stubby), and another category—synapses directly on the shaft, shaft synapses. This reconstruction was taken from over 100 serial ultrathin sections, eight of which are shown in electron micrographs in B(a–h) with individual elements of types of spines and synapses labeled—mushroom spine (a,b); thin spine (c,d); stubby spine (e,f) and shaft synapses (g,h). Abbreviations: PSD, postsynaptic density; sp, spine; den, dendrite; SA, spine apparatus; MVB, multivesicular body. A schematic of the spine categories is shown in C and can be compared with the reconstructions in A. These categories, as we discuss in the text, most likely represent a continuum where stubby spines become thin and then become thick (mushroom) following neural activation, and vice versa when activity downregulation occurs. The categories we show are thin—where the height is usually several times in excess of the width; stubby—where the spine protrudes only slightly from the dendritic shaft; mushroom—where the spine head is large and considerably in excess of the spine neck diameter, shaft synapses are a fourth category where the synapse contacts directly the dendritic shaft (A, C). Scale bar = 1 μm 3 (A); 1 μm (B). *Reproduced with permission from Medvedev et al., 2010.*

filopodia when synaptic activity is low, as is the case with the deafferented neuron (Segal, 2010). When spines are lost, neurons are more subject to large synaptic inputs impinging on their dendritic shafts, and these inputs may lead to their eventual death.

2.3.2. Spine Classifications

A gradation/continuum of spine forms from shaft to stubby to thin to mushroom may be seen following LTP of the perforant path in the hippocampus (schematic in Figure 9(C)).

After LTP, a proportion of stubby spines in the dentate gyrus appears to be transformed into the Tsps, which may change to mushroom types while existing mushroom spines enlarge (Popov et al., 2004). Mushroom spines have been termed memory spines (Matsuzaki et al., 2004; Bourne and Harris, 2007). The classification scheme used for spines is based subjectively according to their morphology and the nature of their synaptic contact (Peters and Kaiserman-Abramof, 1970). These spine and synapse forms are shown in more detail in Figure 9(A) which is a 3-D reconstructed segment of dendrite in the dentate gyrus of the rat hippocampus, approximately 10 μm in length. A series of 2-D electron micrographs (from which the 3-D reconstruction was made) is shown in Figure 9(B) indicating mushroom (Figure 9(a and b)), thin (Figure 9(c and d)), and stubby spines (Figure 9(e and f)), and synaptic contacts on dendritic shafts (Figure 9(g and h)). The different types of spines are shown diagrammatically in Figure 9(C).

2.3.3. Relationships between Axons and Dendrites, and Spines and Synapses

The intricacy of the interactions and relationships between spines and synapses can only be fully appreciated in 3-D reconstructions of which the images in Figure 8 and 9 are

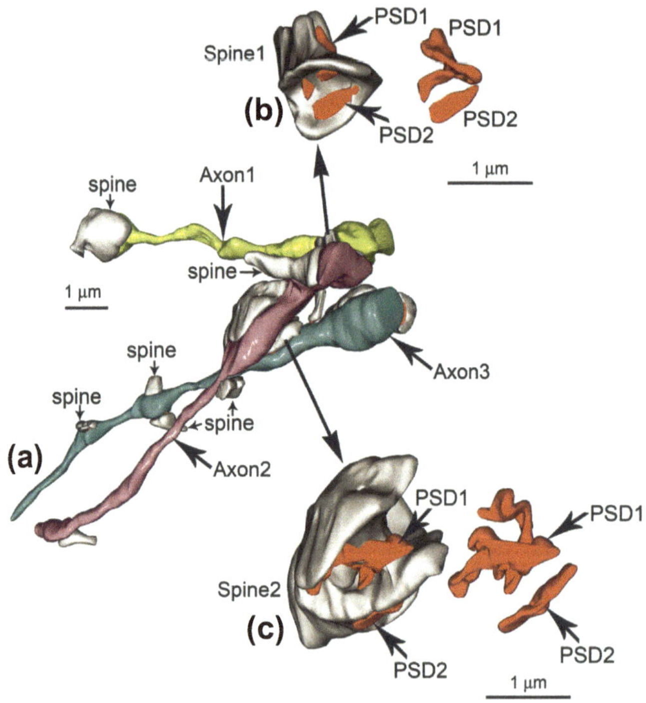

Figure 10 Three-Dimensional Reconstructions of Three Axons (1–3) and Their Associated Dendritic Spines. The complete reconstruction is presented in (a) and the separate spines and PSDs in (b,c). Scale bars = 1 μm.

a part. Figure 10 shows 3-D reconstructions of axonal segments and contacting them dendritic spines in dentate gyrus of rat. These reconstructions are composed of ~100 serial ultrathin sections in the dentate gyrus. Figure 10(a) shows the three axons (axons 1, 2, and 3), each of which makes en-passant contacts with many dendritic spines, among which are two dendritic spines that make contact on two axonal segment at the same time: spine 1 makes contact to axons 1 and 2 (Figure 10(b)); spine 2, with axons 2 and 3 (Figure 10(c)).

Note how the spine heads engulf the axon terminals and when the spines are shown separately, this engulfing nature means that the PSDs are not visible. However, it is not only the complexity of dendrite and axons that is shown more cogently in 3-D reconstructions.

2.3.4. Astroglia

These glial cells have an intimate relationship with synapses, forming a network around and between them. Astrocytes interact with neurons to regulate synaptic transmission; it is believed that they control synaptic strength by modulating extracellular homeostasis (Pannasch et al, 2011). The relationship with synapses is most clearly demonstrated in Figure 11. Figure 11(A) shows the astrocytic network around dendrites and axons, whereas when the astrocyte network is stripped away in Figure 10(B), the axons and dendrites can be clearly seen.

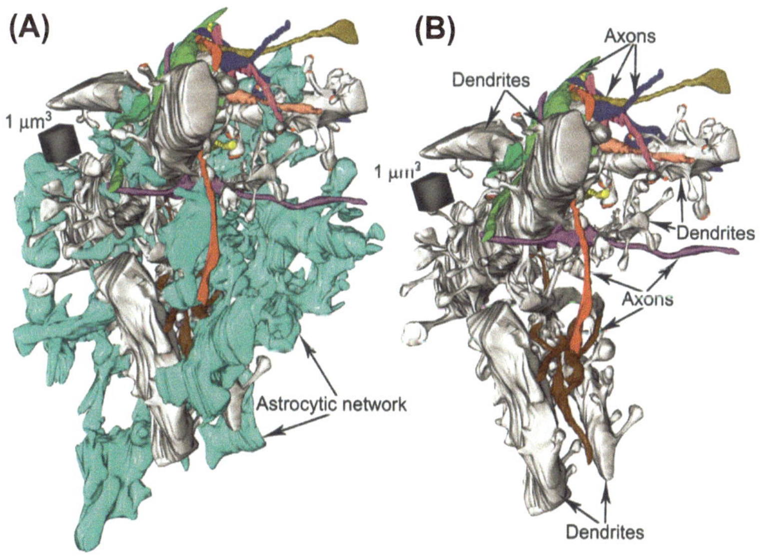

Figure 11 Example of Spatial Geometry of 3-D Reconstructions of Astrocytic Network, Dendritic, and Axonal Segments. (A) Six dendritic segments and 10 axonal segments surrounded by astrocytic network in ground squirrel. (B) As (A) but with the astrocyte stripped off leaving dendrites and axons. In normal temperature conditions of this hibernating mammal, the astrocytic network comprises ~12% of the tissue space. Scale = 1 μm^3. *Reproduced with permission from Popov et al., 2007.*

2.3.5. Multisynaptic Boutons

Synapses do not necessarily form a one-to-one relationship with their contacts, and may make several contacts termed *multisynaptic boutons*. The complexity of multisynaptic synaptic boutons can be seen when three spines from the same dendrite can wrap themselves in basket fashion around an *en passant* axonal bouton. This has three PSDs, as shown in Figure 12. The four electron micrographs (Figure 12A(1–4)) are from a series

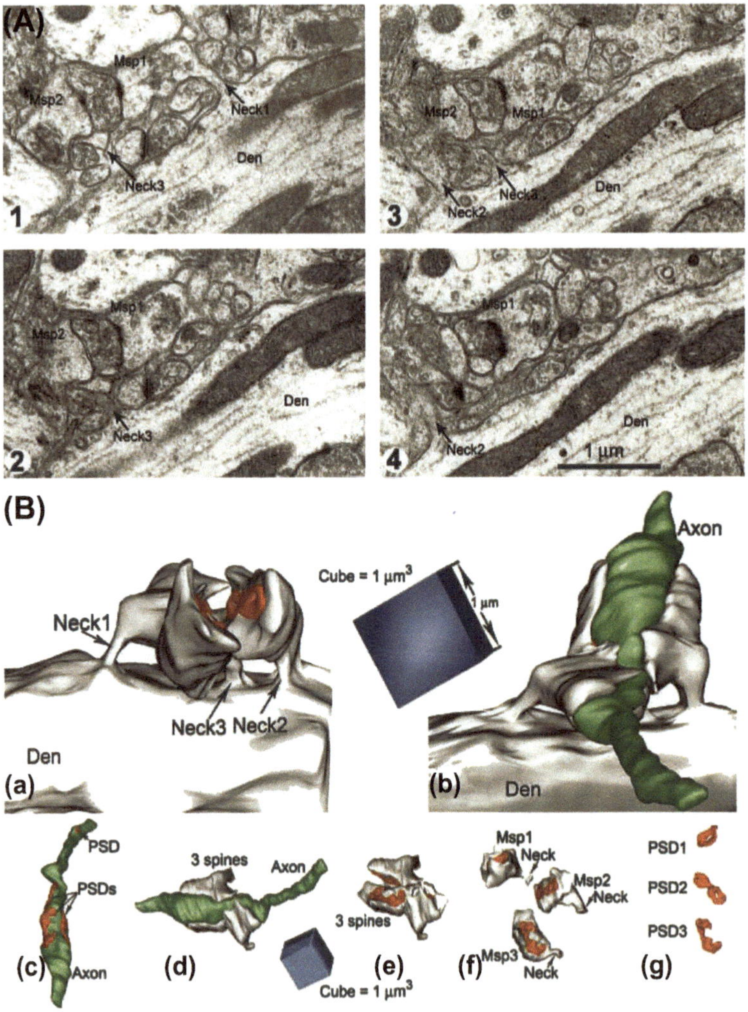

Figure 12 Multisynaptic Presynaptic Bouton. Four electron micrographs from four serial sections show three spine necks (1–3) belonging to the same dendrite and giving rise to three mushroom spines Msp 1–3. B(a–g) 3-D reconstructions from 150 sections which made up the series are presented. The three spines B(a, b) wrap around boutons of a multisynaptic axonal segment which is shown in B(c, d). The PSDs of the three spines have two perforated PSDs (PSD1 and 2) and one U-shaped PSD (PSD3) B(e–g). Scale = 1 μm³. *Reproduced with permission from Popov and Stewart, 2009.*

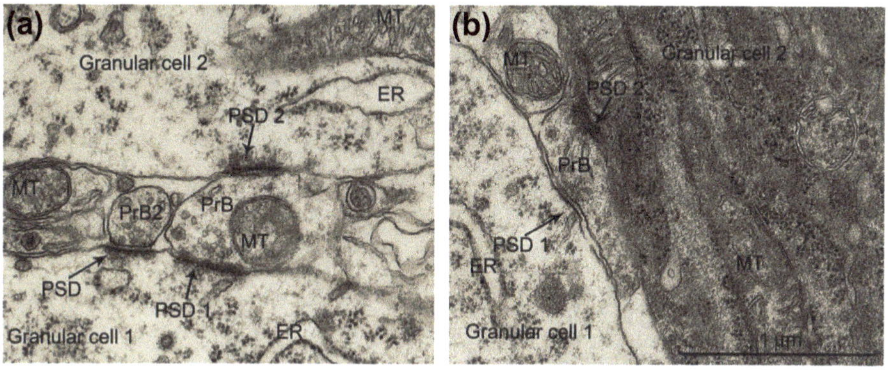

Figure 13 Electron Micrographs of Two Multisynaptic Boutons with Direct Contact on Two Granular Cell Bodies. These boutons (PrB) (a, b) form synapses (asymmetrical) with neuronal bodies of neighboring granular cells in the dentate gyrus of a young rat. Abbreviations: ER, endoplasmic reticulum; PSD, postsynaptic density; MT, mitochondrion. Scale bar = 1 μm. *Reproduced with permission from Popov and Stewart, 2009.*

of 150 sections used to reconstruct the dendrite, spines, and axons; three necks (1–3) can be seen emanating from the same dendritic shaft (Den) and Figure 12B(a and b) shows how the spines are wrapped around an axon while the spines are shown separated from the axon, in Figure 12B(c–g).

A multisynaptic bouton can also contact directly two granular cell bodies, as shown in Figure 13(a and b); these are two separate granular cells and the axonal bouton forms two contacts on the cell soma of each granular neuron.

2.3.6. Multi-Innervated Spines

The reverse situation occurs when dendritic spines are contacted by more than one synapse; and when this happens, it is termed a multi-innervated spine (MIS). Indeed there are dendritic spines that may be contacted by two or more axonal boutons (Nikonenko et al., 2003; Genoud et al., 2004; Stewart et al., 2005; Popov and Stewart, 2009; Radwanska et al., 2011), as in the 3D reconstructions in Figure 14(a–c). Note that in Figure 14(a and b), there are a total of four PSDs, and these four come from the four axons (1–4) shown in Figure 14(c), with the spine almost hidden by the axons.

3. CONCLUSIONS

Synapses and dendritic spines occur in a myriad of forms. Although transmission EM does not allow study of living tissue and dynamic cellular events, it does allow examination of the finest ultrastructural details and this is necessary to allow interpretation of dynamic events. EM also enables the complex relationships between synapses and dendritic components, especially spines, to be determined fully, though only when these

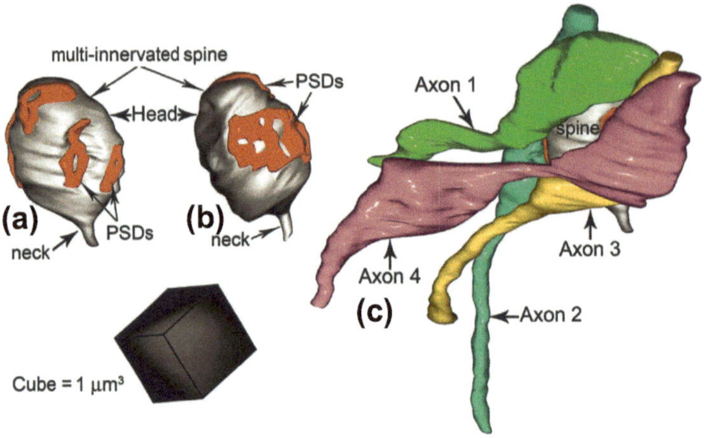

Figure 14 Multi-innervated spine (a,b) in which four axons (1–3) (c) make contact with four PSDs on the same spine. Scale = 1 μm^3.

are reconstructed in three dimensions is the relationship fully apparent. This chapter has demonstrated the great diversity of synaptic and dendritic forms and shown in detail a variety of two-dimensional electron microscope images and 3-D reconstructions of synapses and spines, in the mammalian hippocampus. It is important to note that synapses do not necessarily form a one-to-one relationship with their contacts, but a single synapse may make several contacts on dendritic spines and such synapses are termed *multisynaptic boutons*. In contrast, when dendritic spines are contacted by more than one synapse; and when this happens, it is termed an MIS.

REFERENCES

Applegate, M.D., Kerr, D.S., Landfield, P.W., 1987. Redistribution of synaptic vesicles during long-term potentiation in the hippocampus. Brain Res. 401 (2), 401–406.

Applegate, M.D., Landfield, P.W., 1988. Synaptic vesicle redistribution during hippocampal frequency potentiation and depression in young and aged rats. J. Neurosci. 8, 1096–1111.

Bourne, J., Harris, K.M., 2007. Do thin spines learn to be mushroom spines that remember? Curr. Opin. Neurobiol. 17, 381–386.

Bourne, J., Harris, K.M., 2008. Balancing structure and function at hippocampal dendritic spines. Annu. Rev. Neurosci. 31, 47–67.

Cooney, J.R., Hurlburt, J.L., Selig, D.K., Harris, K.M., Fiala, J.C., 2002. Endosomal compartments serve multiple hippocampal dendritic spines from a widespread rather than a local store of recycling membrane. J. Neurosci. 22 (6), 2215–2224.

Davies, H.A., Stewart, M.G., 1997. Preparation of tissue for electron microscopy. In: Lynch, M.A., O'Mara, S. (Eds.), Neuroscience Labfax. Academic press, London, pp. 47–66. Chapter 4.

DeFelipe, J., 2006. Brain plasticity and mental processes: Cajal again. Nat. Rev. Neurosci. 10, 811–817.

Donohue, H.S., Gabbott, P.L.A., Davies, H.A., Rodriguez, J.J., Cordero, M.I., Sandi, C., Medvedev, N.I., Popov, V.I., Colyer, F.M., Peddie, C.J., Stewart, M.G., 2006. Chronic restraint stress induces changes in synapse morphology in stratum lacunosum-moleculare CA1 rat hippocampus: a stereological and three-dimensional ultrastructural study. Neuroscience 140, 597–606.

Fiala, J.C., Feinberg, M., Popov, V., Harris, K.M., 1998. Synaptogenesis via dendritic filopodia in developing hippocampal area CA1. J. Neurosci. 18, 8900–8911.
Geinisman, Y., Berry, R.W., Disterhoft, John F., Power, J.M., Van der Zee, E.A., 2001. Associative learning elicits the formation of multiple-synapse boutons. J. Neurosci. 21 (15), 5568–5573.
Genoud, C., Knott, G.W., Sakata, K., Lu, B., Welker, E., 2004. Altered synapse formation in the adult somatosensory cortex of brain-derived neurotrophic factor heterozygote mice. J. Neurosci. 24, 2394–23400.
Gu, J., Aheng, J.Q., 2009. Microtubules in dendritic spine development and plasticity. Open Neurosci. J. 3, 128–133.
Harris, K.M., 1999. Calcium from internal stores modifies dendritic spine shape. Proc. Natl. Acad. Sci. USA 96 (22), 12213–12215.
Jedlicka, P., Vlachos, A., Schwarzacher, S.W., Deller, T., 2008. A role for the spine apparatus in LTP and spatial learning. Behav. Brain Res. 192 (1), 12–19.
Jedlicka, P., Schwarzacher, S.W., Winkels, R., Kienzler, F., Frotscher, M., Bramham, C.R., Schultz, C., Bas Orth, C., Deller, T., 2009. Impairment of in vivo theta-burst long-term potentiation and network excitability in the dentate gyrus of synaptopodin-deficient mice lacking the spine apparatus and the cisternal organelle. Hippocampus 19 (2), 130–140.
Kasai, H., Hayama, T., Ishikawa, M., Watanabe, S., Yagishita, S., Noguchi, J., 2010. Learning rules and persistence of dendritic spines. Eur. J. Neurosci. 32, 241–249.
Lushnikova, I.V., Nikonenko, I.R., Nikonenko, O.H., Skibo, H.H., 2008. Spatial distribution of synaptic vesicles in CA1 hippocampal synapses under conditions of induced long-term. Fiziol. Zhivotn. 54 (1), 35–44.
Matsuzaki, M., Honkura, N., Ellis-Davies, G.C., Kasai, H., 2004. Structural basis of long-term potentiation in single dendritic spines. Nature 429 (6993), 761–766.
Matus, A., 2000. Actin-based plasticity in dendritic spines. Science 290, 754–758.
Medvedev, N.I., Popov, V.I., Rodriguez Arellano, J.J., Dallérac, G., Davies, H.A., Gabbott, P.L., Laroche, S., Kraev, I.V., Doyère, V., Stewart, M.G., 2010. The NMDA receptor antagonist CPP alters synapse and spine structure and impairs LTP and LTD induced morphological plasticity in dentate gyrus of the awake rat. Neuroscience 165 (4), 1170–1181.
Murk, J.L., Humbel, B.M., Ziese, U., Griffith, J.M., Posthuma, G., Slot, J.W., Koster, A.J., Verkleij, A.J., Geuze, H.J., Kleijmeer, M.J., 2003. Endosomal compartmentalization in three dimensions: implications for membrane fusion. Proc. Natl. Acad. Sci. USA 100, 13332–13337.
Nägerl, U.V., Willig, K.I., Hein, B., Hell, S.W., Bonhoeffer, T., 2008. Live-cell imaging of dendritic spines by STED microscopy. Proc. Natl. Acad. Sci. USA 105 (48), 18982–18987.
Nägerl, U.V., Bonhoeffer, T., 2010. Imaging living synapses at the nanoscale by STED microscopy. J. Neurosci. 30 (28), 9341–9346.
Nikonenko, I., Jourdain, P., Muller, D., 2003. Presynaptic remodeling contributes to activity-dependent synaptogenesis. J. Neurosci. 23 (24), 8498–8505.
Pannasch, U., Vargová, L., Reingruber, J., Ezan, P., Holcman, D., Giaume, C., Syková, E., Rouach, N., 2011. Astroglial networks scale synaptic activity and plasticity. Proc. Natl. Acad. Sci. USA 108 (20), 8467–8472.
Peters, A., Kaiserman-Abramof, I.R., 1970. The small pyramidal neuron of the rat cerebral cortex: the perikaryon, dendrites and spines. Am. J. Anat. 127, 321–356.
Popov, V.I., Medvedev, N.I., Patrushev, I., Ignat'ev, D.A., Morenkov, E.D., Stewart, M.G., 2007. Reversible reduction in dendritic spines in CA1 of rat and ground squirrel subjected to hypothermia-normothermia in vivo: a 3-dimensional electron microscope study. Neuroscience 149, 549–560.
Popov, V.I., Davies, H.A., Rogachevskii, V.V., Errington, M.L., Gabbott, P.L.A., Bliss, T.V.P., Stewart, M.G., 2004. Remodelling of synaptic morphology but unchanged synaptic density during late phase LTP: a serial section EM study of the dentate gyrus in the anaesthetised rat. Neuroscience 28 (2), 251–262.
Popov, V.I., Medvedev, N.,I., Davies, H.A., Stewart, M.G., 2005. Mitochondria form a filamentous reticular network in hippocampal dendrites but are present as discrete bodies in axons: a three-dimensional ultrastructural study. J. Comp. Neurol. 492, 50–65.
Popov, V.I., Medvedev, N.I., Kraev, I.V., Gabbott, P.L., Davies, H.A., Lynch, M., Cowley, T.R., Berezin, V., Bock, E., Stewart, M.G., 2008. A cell adhesion molecule mimetic, FGL peptide, induces alterations in synapse and dendritic spine structure in the dentate gyrus of aged rats: a three-dimensional ultrastructural study. Eur. J. Neurosci. 2, 301–314.

Popov, V.I., Stewart, M.G., 2009. Complexity of contacts between synaptic boutons and dendritic spines in adult mammalian hippocampus: three-dimensional reconstructions from serial ultrathin sections *in vivo*. Synapse 63, 369–377.
Radwanska, K., Medvedev, N.I., Pereira, G.S., Engmann, O., Thiede, N., Moraes, M.F.D., Villiers, A., Irvine, E.E., Maunganidze, N.S., Pyza, E.M., Ris, L., Szymańska, M., Lipiński, M., Kaczmarek, L., Stewart, M.G., Giese, K.P., 2011. Mechanism for long-term memory formation when synaptic strengthening is impaired. Proc. Natl. Acad. Sci. USA 108 (45), 18471–18475.
Segal, M., 2005. Dendritic spines and long-term plasticity. Nat. Rev. Neurosci. 6, 277–284.
Segal, M., Vlachos, A., Korkotian, E., 2010. The spine apparatus, synaptopodin, and dendritic spine plasticity. Neuroscientist 16 (2), 125–131.
Spacek, J., Harris, K.M., 1997. Three-dimensional organization of smooth endoplasmic reticulum in hippocampal CA1 dendrites and dendritic spines of the immature and mature rat. J. Neurosci. 17, 190–203.
Spacek, J., Harris, K.M., 2004. Trans-endocytosis via spinules in adult rat hippocampus. J. Neurosci. 24, 4233–4241.
Stewart, M.G., Davies, H.A., Sandi, C., Kraev, I.V., Rogachevsky, V.V., Peddie, C.J., Rodriguez, J.J., Cordero, M.I., Donohue, H.S., Gabbott, P.L., Popov, V.I., 2005. Stress suppresses and learning induces plasticity in CA3 of rat hippocampus: a three-dimensional ultrastructural study of thorny excrescences and their postsynaptic densities. Neuroscience 13, 43–54.
Van Rossum, D., Hanisch, U.K., 1999. Cytoskeletal dynamics in dendritic spines: direct modulation by glutamate receptors? Trends Neurosci. 22 (7), 290–295.
Ventura, R., Harris, K.M., 1999. Three-dimensional relationships between hippocampal synapses and astrocytes. J. Neurosci. 19 (16), 6897–6906.
Wilt, B.A., Burns, L.D., Wei Ho, E.T., Ghosh, K.K., Mukamel, E.A., Schnitzer, M.J., 2009. Advances in light microscopy for neuroscience. Annu. Rev. Neurosci. 32, 435–506.

CHAPTER TWO

The Molecular Mechanisms Underlying Synaptic Transmission: A View of the Presynaptic Terminal

Uri Ashery[1,2], Noa Bielopolski[1], Ayal Lavi[1,2], Boaz Barak[1,2], Lirin Michaeli[1], Yoav Ben-Simon[1,2], Anton Sheinin[1,2], Dana Bar-On[1], Zehavit Shapira[1] and Irit Gottfried[1]
[1]Department of Neurobiology, Tel-Aviv University, Tel-Aviv, Israel
[2]Sagol School of Neuroscience, Tel-Aviv University, Tel-Aviv, Israel

1. INTRODUCTION

Structurally, the synapse is an asymmetric organization that includes the presynaptic terminal, which contains several hundreds to thousands of vesicles (filled with neurotransmitter), the synaptic cleft, and the postsynaptic cell containing the neurotransmitter–receptor scaffold and signaling machinery. For the purpose of this chapter, we will concentrate almost exclusively on the presynaptic terminal (Figure 1). The main processes in the presynaptic terminal are coordinated by a large number of synaptic proteins (Table 1), and can be summarized as follows: following the arrival of an action potential to the nerve terminal, the terminal plasma membrane (PM) is depolarized and voltage-gated Ca^{2+} channels are activated (Barnett and Larkman, 2007; Catterall, 2011; Mahapatra et al., 2012) allowing Ca^{2+} influx into the nerve terminal. Calcium ions then bind with specific synaptic proteins, each of which drives different processes in the terminal-like vesicle fusion, recycling, and translocation. Most models assume a series of sequential steps occurring in the synapse: synaptic vesicles translocate and dock in release sites where they undergo a priming reaction that renders them fusion-competent. Only such primed vesicles, forming the readily releasable pool (RRP) of synaptic vesicles, can fuse with the presynaptic PM in the active zone (AZ) upon arrival of an action potential and the concomitant increase in presynaptic Ca^{2+} concentration (Burgalossi et al., 2010;Verhage and Sorensen, 2008). Following fusion, the vesicle undergoes endocytosis to recycle its membrane and membrane proteins. The recycled vesicles then undergo neurotransmitter uptake and preparation for the next cycle of exocytosis. This chapter provides a detailed description of the molecular mechanisms underlying synaptic transmission and plasticity.

The Synapse
http://dx.doi.org/10.1016/B978-0-12-418675-0.00002-X

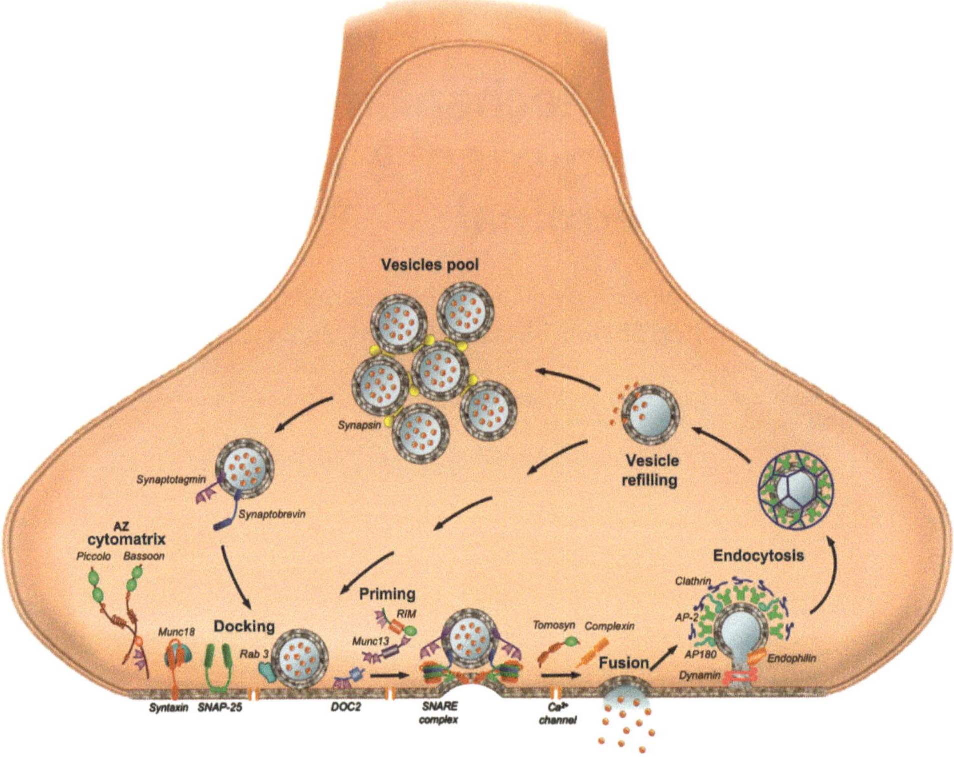

Figure 1 A schematic representation of the presynaptic terminal. The main steps of the synaptic vesicle cycle are depicted and representative synaptic proteins are indicated according to their function. Drawing is not to scale.

Table 1 A List of the Synaptic Proteins with a Short Description of Proteins' Function and Domains

Protein Name	Short Description of Protein Function and Domains
α-SNAP (soluble *N*-ethylmaleimide-sensitive factor (NSF) attachment protein)	A small protein that binds an SNARE complex and together with NSF, disassembles SNARE complexes.
α-Synuclein	Synucleins are found primarily in neural tissue. α-Synuclein is associated with Parkinson's disease. Synucleins bind to negatively charged phospholipids and pay a role in SNARE-complex formation.
Amphiphysin	Involved in clathrin-mediated endocytosis (CME; membrane curvature), actin modulation, and signaling pathways. Interacts with the PM through its N-BAR domain, and with dynamin and synaptojanin 1 through its SH3 domain. The proline–arginine-rich region binds endophilin and an additional domain enables binding to clathrin and AP-2.

Table 1 A List of the Synaptic Proteins with a Short Description of Proteins' Function and Domains—cont'd

Protein Name	Short Description of Protein Function and Domains
AP180/CALM	Involved in CME (coat assembly). Possesses an ANTH domain for $PI_{(4,5)}P2$-binding, and additional binding sites for AP-2 and clathrin. AP180 is exclusively expressed in neurons. CALM is a ubiquitous protein found in both neurons and glia and in non-neuronal tissues.
AP-2	The main adaptor complex of CME (clathrin recruitment and assembly, and cargo recognition). Built from four subunits. Expressed in both neuronal and nonneuronal cells.
ARP2/3	Actin-related protein complex, built from seven subunits. Promotes actin nucleation and polymerization.
Auxilin	Involved in CME (coat disassembly). Cofactor of the ATPase HSC70 involved in endocytic vesicle uncoating.
Bassoon	A scaffold protein of the presynaptic active zone (AZ). Part of a large multiprotein complex that includes CAST, RIM1 and Munc13-1. Bassoon is involved in the organization of Ca^{2+} channels and vesicles in AZ.
Bruchpilot (BRP)	The *Drosophila* homolog of CAST. A scaffold protein with the presynaptic T-bar structure in the NMJ AZ, involved in assembly of the T-bar complex and localization of AZ components with Ca^{2+} channels.
Calcineurin	Calcium-dependent serine–threonine phosphatase. During prolonged depolarization of central nerve terminals, it dephosphorylates seven endocytic proteins, thought to trigger bulk endocytosis. Also thought to regulate exchange of vesicles between the recycling pool and the RP.
CaMKII	Serine–threonine kinase activated by the Ca^{2+}–calmodulin complex during intense synaptic activity. Upon synaptic activation, it phosphorylates synapsin I and enables vesicle translocation to the AZ.
CAPS (calcium-dependent activated protein for secretion)	A large phosphoprotein essential for vesicle priming.
CDC42	Rho GTPase protein, involved in actin polymerization through the actin-related protein 2/3 (ARP2/3) complex. A positive regulator of secretion.
Clathrin	The key protein of CME. Three clathrin heavy chains and three clathrin light chains build a triskelial unit which polymerizes into pentagons and hexagons that cover membranal invaginations which mature into vesicles.
Complexin	A small SNARE-complex-binding protein that is essential for fast synchronous neurotransmitter release and is also suggested to clamp SNARE complexes for asynchronous and spontaneous release.
CSP (cysteine string protein)	Molecular chaperones composed of an N-terminal and J-domain and a central palmitoylated cysteine string. CSPs are associated with vesicles and interact with SNARE proteins and with HSC70.
DOC2	DOC2 proteins have two C2 domains (C2A and C2B) and a munc-13-binding domain. They bind calcium in the submicromolar range, regulate spontaneous and asynchronous release in neurons, and promote vesicle priming and refilling in chromaffin cells.

Continued

Table 1 A List of the Synaptic Proteins with a Short Description of Proteins' Function and Domains—cont'd

Protein Name	Short Description of Protein Function and Domains
Dynamin	Involved in endocytosis (vesicle fission). A large GTPase that is recruited to mature late-stage vesicles and upon GTP hydrolysis, mediates fission of the vesicle from the PM.
ELKS(ERC/ CAST)	A scaffold protein of the presynaptic AZ. Part of a large multiprotein complex that includes bassoon/piccolo, RIM1, and Munc13-1.
Endophilin	Involved in CME. Contains an SH3 domain that can bind dynamin and synaptojanin, and an N-BAR domain with a role in membrane-curvature induction of sensing. Might also be involved in vesicle uncoating.
Exocyst	An octameric complex that controls the docking and insertion of exocytic vesicles in growing areas of the PM in yeast, and mediates the addition of membrane during postnatal presynaptic maturation in the mammalian CNS.
Fodrin	Regulator of activity-dependent actin remodeling (the neural form of spectrin). Cross-links F-actin at low intracellular calcium concentrations, and dissociates from actin and enables local dissolution of the actin cortex upon calcium elevation.
HSC70	Involved in CME (vesicle uncoating). An ATPase, recruited by auxilin, binds clathrin and initiates uncoating of the vesicle.
Liprin-α	Involved in the formation and maintenance of the AZ. Interacts directly with RIM and ELKS.
MARCKS	Regulator of the activity-dependent actin remodeling. It binds and stabilizes actin filaments. MARCKS is displaced from actin upon phosphorylation, encouraging actin depolymerization and vesicle mobilization.
MINT (X11-like protein)	Adaptor protein that binds Munc18-1. Involved in cytoskeleton organization and protein scaffolding, found in the AZ.
Munc13	Essential presynaptic AZ protein, part of a large multiprotein complex that includes bassoon/piccolo, CAST and RIM1. Involved in vesicle priming by partially "opening" syntaxin-1.
Munc18	An arch-shaped protein involved in vesicle docking and priming. Belongs to the family of SM proteins (Sec1 and Munc18). Binds syntaxin alone (in its "closed" conformation) or the assembled SNARE complex containing syntaxin-1 (in an "open" conformation). Essential for synaptic transmission.
Neurexin	Presynaptic cell-adhesion molecule interacts with the postsynaptic neuroligin to connect the pre- and postsynaptic membranes. Known as receptor for α-latrotoxin.
NSF	Hexameric ATPase that functions together with SNAPs (soluble NSF attachment proteins) and disassembles SNARE complexes.
N-WASP	Involved in actin polymerization through the actin-related protein 2/3 (ARP2/3) complex, which promotes actin nucleation and polymerization.
Piccolo	A scaffold protein of the presynaptic AZ. Part of a large multiprotein complex that includes CAST, RIM1, and Munc13-1 and involved in the organization of the presynaptic AZ and in synaptic vesicle clustering.

Table 1 A List of the Synaptic Proteins with a Short Description of Proteins' Function and Domains—cont'd

Protein Name	Short Description of Protein Function and Domains
PKA	A kinase that phosphorylates synaptic proteins, such as synapsin, rabphilin, tamosyn, and RIM1, thereby regulating their function. Involved in long-term potentiation of the synapse.
PKC	A kinase that phosphorylates synaptic proteins, such as Munc13, SNAP-25, Munc18, and MARCKS, thereby causing potentiation of release.
PLD	A phospholipase that produces phosphatidic acid via hydrolysis of phosphatidylcholine, thus facilitating vesicle priming and fusion.
PRA1	Prenylated Rab acceptor 1, associates with piccolo and synaptobrevin-2 suggesting a function in synaptic vesicle trafficking and exocytosis.
Rab3	Small GTPase localized to vesicles and secretory granules. Functions as a mediator of membrane trafficking and vesicle docking.
Rab27	Small GTPase localized to vesicles and secretory granules. Functions as a mediator of membrane trafficking and vesicle docking.
Rabphilin	Rab3-dependent synaptic protein that is involved in docking of dense-core vesicles probably by simultaneously interaction with Rab27A on the vesicle and with SNAP-25 on the PM.
RIM1	A component of the AZ cycomatrix. Part of a large multiprotein complex that includes bassoon/piccolo, CAST, and Mun13-1. Involved in priming through interactions with Munc13 and Rab3.
RIBEYE	A scaffold protein exclusive to the photoreceptor ribbon AZ. Interacts with bassoon and connects the two parts of the ribbon.
Scinderin	An activity-dependent F-actin regulator. Disassembles cortical actin just below the PM, possibly allowing better access of vesicles to the PM.
SNAP-25 (synaptosome-associated protein of 25 kDa)	SNAP-25 is a Q-SNARE protein and a component of the neuronal SNARE complex. It is anchored to the membrane via palmitoyl side chains and has α-helices that form the SNARE complex together with synaptobrevin-2 and syntaxin-1. It plays a central role in vesicle docking, priming and fusion.
Snapin	Snapin contains heptad repeats in its C-terminal part. It is an SNAP-25-interacting protein that enhances the binding of synaptotagmin-1 to SNAREs in a phosphorylation-dependent manner.
SNARE (SNAP (soluble NSF attachment protein) receptor)	The core complex involved in synaptic vesicle docking, priming and fusion. Consists of the vesicular protein, synaptobrevin-2 (also known as VAMP-2), the presynaptic membrane protein syntaxin-1A, and the presynaptic membrane SNAP-25. The interaction between the SNARE motifs results in a four-helix coiled-coil structure, connecting the vesicle and the PM.
Stonin	Involved in CME (selective sorting of the endocytosed cargo). Thought to be the adaptor for synaptotagmin internalization.
Synapsins	A key family of proteins involved in vesicle organization and mobilization. Activity-induced synapsin phosphorylation decreases its affinity toward both vesicles and actin, thereby allowing vesicle exchange between pools and vesicle translocation to the AZ.

Continued

Table 1 A List of the Synaptic Proteins with a Short Description of Proteins' Function and Domains—cont'd

Protein Name	Short Description of Protein Function and Domains
Synaptobrevin	The vesicular R-SNARE complex protein (also known as VAMP-2). Contains a transmembrane domain and α-helix that forms the SNARE complex together with SNAP-25 and syntaxin-1. The complex plays a central role in synaptic vesicle docking, priming, and fusion machinery.
Synaptojanin	PIP2-phosphatase, involved in CME (veside uncoating). Probably recruited by endophilin. It causes a change in phospholipids that destabilizes clathrin coats and recruits auxilin. It is also a substrate for calcineuring.
Synaptophysin	A vesicular glycoprotein of unknown function. Contains four transmembrane domains. Interacts with the SNARE protein synaptobrevin.
Synaptotagmin	Synaptotagmin-1 belongs to a family of proteins that consists of 15 isoforms and is thought to be the calcium sensor for synchronous release. It is a calcium-binding protein anchored to synaptic vesicles at N-terminus by a transmembrane domain and contains two cytosolic C-terminal C2 domains. The C2 domains of synaptotagmin-1 are characterized by Ca^{2+}-dependent phospholipid and SNARE binding.
Syndapin	Involved in bulk endocytosis. Following intense stimulation, it interacts with dephosphorylated dynamin 1.
Syntaxin-1	Syntaxin-1a is a Q-SNARE protein and a component of the neuronal SNARE complex. It is anchored to the membrane via a transmembrane domain and has one α-helix that forms the SNARE complex together with synaptobrevin-2 and SNAP-25. It plays a central role in vesicle docking, priming, and fusion.
Tomosyn	Tomosyn inhibits vesicle priming. Binds syntaxin and SNAP-25 and creates "dead-end" SNARE complexes, excluding synaptobrevin.
Vacuolar proton pump	The vacuolar proton pump is involved in the acidification of synaptic vesicles.
VGAT	The vesicular GABA transporter is responsible for uptake of GABA by synaptic vesicles.
VGLUT1	The vesicular glutamate transporter is responsible for uptake of glutamate by synaptic vesicles.
Voltage-gated Ca^{2+} channel	Voltage-gated calcium channels (N, P/Q, T, and L type) mediate the influx of Ca^{2+} ions into the synapse and trigger synaptic vesicle fusion and neurotransmitter release.

2.1. SYNAPTIC STRUCTURE

2.1.1. Synaptic Vesicle Pools

The synaptic vesicle goes through cycles of fusion and recycling, which allow its repeated use during sustained activity. In the conventional synapse at rest, a very small percentage of the vesicles are docked to the membrane in the AZ; the remaining vesicles are stored remotely from the PM. Under the electron microscope (EM), all the synaptic vesicles look alike and the division into three distinct pools of vesicles is based mainly on the location of the vesicles at the synapse (Sudhof, 2004) and their availability for release. The main three pools of vesicles in the presynaptic terminal are the following:

1. *RRP*: These vesicles are docked at the synaptic AZ membrane and are available for immediate release. The vesicles of the RRP constitute 0.5–5% of all vesicles, and can be depleted rapidly by 5–15 pulses of high-frequency electrical stimulation or a few milliseconds of depolarization (Elmqvist and Quastel, 1965; Mennerick and Matthews, 1996; Schneggenburger et al., 1999). The RRP is also part of the recycling pool.
2. *The recycling pool*: These vesicles are released during moderate (physiological) stimulation, and constitute about 5–20% of all vesicles (Harata et al., 2001; Richards et al., 2003). Vesicles from this pool enter the exocytosis–endocytosis cycle under high-frequency physiological conditions (de Lange et al., 2003), so that they can be used repeatedly during sustained activity (Sudhof, 2004).
3. *The reserve/resting pool* (RP): The RP vesicles represent about 80–90% of all vesicles in the presynaptic terminal (Delgado et al., 2000; Heuser and Reese, 1973). Accepted dogma suggests that these vesicles are only seldomly, if ever, recruited during physiological activity, and are only released by intensive stimulation or when the recycling pool is depleted (Ikeda and Bekkers, 2009; Kuromi and Kidokoro, 2000). Interestingly, in the neuromuscular junction (NMJ), intense stimulation results in fusion of all the vesicles in the terminal (Heuser and Reese, 1973), whereas in hippocampal neurons, a substantial proportion of the vesicles remain unused, even during strong nonphysiological stimulation (Rizzoli and Betz, 2005). A recent study shows that in central nervous system (CNS) synapses and the NMJ, only 1–5% of all vesicles are recycled (Denker et al., 2011). Hence, the proportion of the RP vesicles that participate in synaptic transmission is still a matter of debate and await further investigations.

2.1.2. Organization of Synapses and the Active Zone

In the presynaptic terminal, a substructure termed active zone plays a pivotal role in the exocytic machinery (Landis et al., 1988) by being a "hot spot" for vesicle fusion and cell–cell contact. The AZ cytoskeleton matrix appears as an electron-dense region containing various proteins that are considered to be involved in signal transduction, cytoskeleton organization, cell–cell adhesion, neurotransmitter release (docking, priming and fusion of synaptic vesicles), and regulation of the changes in neurotransmitter release during short-term and long-term synaptic plasticity (Dresbach et al., 2001; Gundelfinger and tom Dieck, 2000).

Several types of synapses exist that differ in location, structure, and function. In this section, we focus on three types that have unique structures and are further referred to throughout this chapter (Figure 2).

1. *The conventional CNS synapse.* The AZ of the most common chemical synapse in the CNS is closely and precisely aligned with the postsynaptic density and appears as an electron-dense structure containing the AZ proteins and synaptic vesicles. This tight structure and its molecular organization are extremely important for signaling

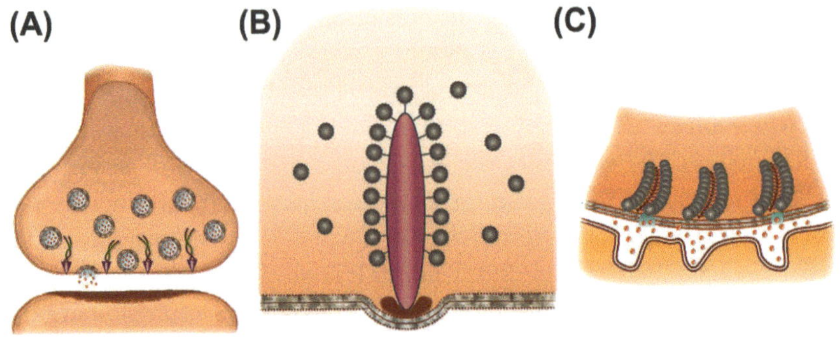

Figure 2 Illustration of vesicles and active zone organization in three types of synapses. (A) The conventional small central nervous system synapse, where vesicles are docked at the plasma membrane, between cone-shaped particles of the active zone. (B) The ribbon synapse, where vesicles cover a ribbon-shaped structure that enables efficient release. (C) The neuromuscular junction, where the vesicles are held in two parallel rows connected by an array-like structure. Drawing is not to scale.

precision. Electron micrographs of chemical synapses of the CNS show that the AZ contains a regular array of electron-dense cone-shaped particles, which extend approximately 50 nm into the presynaptic cytoplasm (Landis et al., 1988; Pfenninger et al., 1972). These particles are connected by cytoskeletal filaments, and long filamentous strands extend from them deep into the presynaptic zone (Landis et al., 1988; Phillips et al., 2001), while the docked vesicles are linked to the PM by dense material at the AZ (Siksou et al., 2009a). These long filamentous strands might represent key proteins of the AZ, which will be further discussed, although the exact composition of these structures is still not known.

2. *The ribbon structured synapse.* This unique type of chemical synapse exists in retinal photoreceptor cells and in the inner hair cells (Sobkowicz et al., 1982), and enables continuous release of large amounts of vesicles (Levic et al., 2011). The ribbon extends from the terminal membrane into the presynaptic cytoplasm and is covered by hundreds of vesicles (Rao-Mirotznik et al., 1995), so that membrane fusion and transmitter release occur focally at the ribbon base (Lenzi and von Gersdorff, 2001). This unique structure makes the photoreceptor ribbon synapse an efficient relay structure in sensory systems (Zhai and Bellen, 2004).

More specifically, the ribbon is structurally composed of two groups of proteins that are segregated morphologically. The assumption is that these two groups reflect segregation of AZ function. The AZ proteins are located at the base of the ribbon. These proteins are essential for docking, priming, and the connection to the second group of proteins (i.e. the ribbon). The second group is distinct and relatively remote from the PM and extends into the presynaptic zone. This group of proteins is thought to be involved in tethering synaptic vesicles to the ribbon and preparing them for exocytosis (tom Dieck et al., 2005).

3. *The NMJ synapse.* This chemical synapse connects the motor nerve with the muscle fiber, and enables rapid neurotransmitter release. Early freeze-fracture studies of the NMJ revealed two rows of prominent intramembranous structures that include the Ca^{2+} channels with the synaptic vesicle alongside (Heuser et al., 1974). EM tomography images revealed a 3D view of the macromolecules near the AZ that connect the vesicles to the PM and Ca^{2+} channels, suggesting that the AZ material is involved in both vesicle docking and anchoring of the Ca^{2+} channels. The NMJ AZ structure consists of two parallel rows of docked synaptic vesicles connected by an array-like structure built from "pegs," "ribs," and "beams" (Harlow et al., 2001). The *Drosophila* NMJ has a different structure, often displayed as electron-dense T-shaped bodies (T-bars) (Kittel et al., 2006).

2.1.2.1. The AZ Proteins: Molecular Composition

Most, if not all, of the regulated synaptic vesicle fusion events occur in the AZ. There are several proteins that are known as AZ cytomatrix components: Munc13, Rab3-interacting molecule (RIM), Bassoon, Piccolo, and ELKS/ERC/CAST. Both ribbon and conventional synapses possess protein complexes that include bassoon, piccolo, ELKS/ERC/CAST, RIM1, and Munc13-1 in their AZ (reviewed in Schoch and Gundelfinger (2006)). Therefore, these complexes are most probably important for maintaining the structure of the AZ. Despite detailed knowledge of the functions of many of the synaptic proteins in the different steps of the synaptic vesicle cycle, knowledge of the roles of some of the AZ proteins like Bassoon and Piccolo is rather incomplete, in part because of the difficulty to study their functions due to their large size and the redundancy of protein functions. Therefore, in the next section, we focus mainly on structures and interactions between these proteins, as well as putative functions.

Bassoon and piccolo are large multidomain proteins (420 and 530 kDa, respectively) that share structural similarities, and are the largest AZ-specific proteins identified to date (Wang et al., 1999). They are present in all vertebrate synapses, but are absent in invertebrates (Schoch and Gundelfinger, 2006), and they are found at nascent synapses in cultured neurons where they might play an important role in AZ assembly (Ahmari et al., 2000; Shapira et al., 2003).

These proteins share 10 regions of high sequence similarity, including highly homologous zinc-finger and coiled-coil (cc) domains (Cases-Langhoff et al., 1996; tom Dieck et al., 1998). Two zinc-finger regions of bassoon and piccolo show a small percentage of homology to α-RIMs and rabphilin (Wang et al., 1999). Piccolo has an additional unique PDZ domain and two C2 domains that are structurally related to RIMs (Cases-Langhoff et al., 1996; Fenster et al., 2000). Through their third cc domain, bassoon and piccolo bind the second cc of CAST and ELKS in a competitive manner (Takao-Rikitsu et al., 2004), as part of a large multiprotein complex that also includes RIM1 and Munc13-1 (Figure 3). Interfering with this complex's formation results in

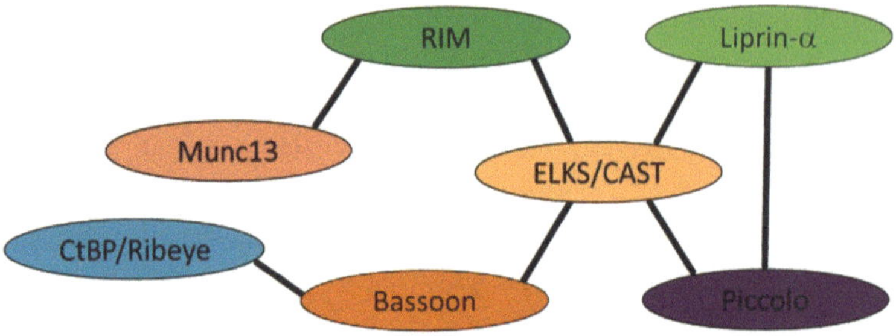

Figure 3 Active zone proteins interactions. Schematic representation of the interaction between the active zone proteins bassoon, piccolo, RIM, Munc13, Ribeye, liprin-a, and ELKS/CAST.

impaired synaptic transmission, indicating that it is involved in neurotransmitter release (Takao-Rikitsu et al., 2004).

In bassoon-mutant mice, synapse formation was not damaged but there was a reduction in normal synaptic transmission (Altrock et al., 2003). Vesicles were clustered and docked in normal numbers but were unable to fuse. Phenotypically, the loss of normal bassoon activity causes epileptic seizures (Altrock et al., 2003), although Mukherjee et al. (2010) could not find any deficits in the synaptic transmission of either GABAergic or glutamatergic synapses from piccolo-knockout/bassoon-knockdown mice (Mukherjee et al., 2010). The authors made several electrophysiological measurements, including spontaneous "mini" response, short-term plasticity (STP), evoked release and RRP, but none were affected. However, EM showed a significant reduction in synaptic vesicle clustering. Thus, we can assume that bassoon and piccolo do not have a direct impact on neurotransmitter release but may play a role in clustering the vesicles or a structural or regulatory role that remains to be elucidated (Mukherjee et al., 2010).

Evidence of an important role for bassoon in the establishment of functional release sites comes from the photoreceptor synapse. In this synapse, the presynaptic region has two molecular components: the ribbon and the ribbon base. A strong physical interaction between bassoon and ribeye (a scaffold protein exclusive to the photoreceptor ribbon) links the two components (tom Dieck et al., 2005). In retinae of bassoon-mutant mice, the connection between the two components is lost and as a result, the ribbons are not anchored to the presynaptic AZ, but are free-floating in the cytoplasm (tom Dieck et al., 2005). A milder phenotype is seen in the inner hair cells of bassoon-mutant synapses; these show an intermediate phenotype in which the ribbon is partially anchored to the membrane (Frank et al., 2010). Interestingly, in this phenotype, there was also a decrease in the number of Ca^{2+} channels and their clustering. These morphological changes were accompanied by a reduction in synaptic transmission and vesicle replenishment. Thus it seems that the mutation in bassoon and the associated ribbon loss reduce the number of functional release sites and impair their refilling, and consequently lower the RRP

(Frank et al., 2010). Hence, bassoon and the ribbon are important to the establishment of a large number of release sites via the organization of Ca^{2+} channels and vesicles, similar to bassoon role in conventional CNS synapses (Mukherjee et al., 2010), and they promote vesicle replenishment in the photoreceptor ribbon synapse.

Another assumption is that bassoon and piccolo are major components of the precursor vesicles that assemble at the AZ, and that the arrival and interaction of those proteins might regulate trafficking of vesicles in early synapse formation (Ahmari et al., 2000; Shapira et al., 2003). Bassoon was shown to interact with dynein light chain and travel along the axon in both anterograde and retrograde transport (Fejtova et al., 2009).

CAST (which is also called ERC2 or ELKS2) is a vertebrate protein belonging to a protein family that includes ELKS (which are also called RAB6IP2A, ERC1 or CAST2) (Wang et al., 2002). CAST and ELKS both weigh 120 kDa, and both have four cc regions and a unique C terminal amino acid motif—IWA (Takao-Rikitsu et al., 2004). ELKS's isoform, ELKSα, shows ~70% homology to CAST. Both CAST and ELKS are expressed mainly in the brain. In the cerebellum tissue, CAST, ELKSα, and bassoon localize in proximity to the presynaptic membrane (Deguchi-Tawarada et al., 2004). In ribbon synapses, CAST is at the base of the ribbon, whereas ELKSα is located on the ribbon itself (Deguchi-Tawarada et al., 2004). This segregation of localization may indicate different functions for these proteins, despite their similarity. ELKS is regulated by Rab6 and is therefore suggested to be involved in vesicle transport (Takao-Rikitsu et al., 2004). CAST and ELKS bind RIM1 via the IWA motif and thus indirectly bind Munc13-1 (Takao-Rikitsu et al., 2004) and may be indirectly involved in vesicle priming. Interfering with the ternary complex of bassoon/piccolo–CAST–RIM1 does not alter the localization of bassoon/piccolo, but mislocalizes RIM1 (Takao-Rikitsu et al., 2004). Eliminating the IWA motif of CAST or expressing RIM1 without the PDZ domain that binds CAST results in diffuse distribution of RIM1 and its mislocalization at the AZ (Takao-Rikitsu et al., 2004). However, eliminating the third cc of bassoon or piccolo or the second cc of CAST does not affect their localization at the AZ. This implies that CAST is not responsible for bassoon or piccolo anchoring at the presynaptic membrane.

Another interactor of ELKS is liprin-α (Ko et al., 2003), which is also suspected of being involved in the formation and maintenance of the AZ (Kaufmann et al., 2002). Liprin-α interacts directly with RIM and ELKS and indirectly with piccolo via GIT (Kim et al., 2003; Schoch et al., 2002). Vertebrates have four genes homologous to liprin-α (Serra-Pages et al., 1998), whereas invertebrates have only a single homologous gene—*syd-2* (van Roessel et al., 2004). Liprin-α2 and liprin-α3 are specifically expressed in the brain, whereas liprin-α1 and liprin-α4 are also found outside the nervous system. Damaged SYD-2 leads to an increase in AZ length and in the number of vesicles (Zhen and Jin, 1999), as well as increased synapse size and abnormal AZ structure (van Roessel et al., 2004). SYD-2 mutants display signs of impaired synaptic transmission as they suffer from reduced locomotion and defective egg laying (Zhen and Jin, 1999).

Other homologous genes of ELKS/CAST are the *Caenorhabditis elegans* CeCast (Deguchi-Tawarada et al., 2004; Kittel et al., 2006; Wagh et al., 2006) and the *Drosophila* bruchpilot (BRP) (Kittel et al., 2006). EM and stimulated emission depletion (STED) microscopy studies of the *Drosophila* NMJ in BRP-mutant flies showed defects in T-bar structure in the AZ (Kittel et al., 2006). The T-bars were mislocalized and the density of presynaptic Ca^{2+} channel clusters was decreased. This led to reduced probability of vesicle release. In addition, during low-frequency stimulation, postsynaptic currents were reduced, but this effect subsided as the frequency increased. Spontaneous release of neurotransmitter was not affected. BRP surrounds the T-bar, and probably plays a role in the assembly of the T-bar complex and the localization of AZ components with Ca^{2+} channels, and is also involved in vesicle release (Kittel et al., 2006).

An additional cytomatrix component of the *Drosophila* AZ is the RIM-binding protein (DRBP). DRBP is essential for neurotransmitter release and for AZ organization. In DRBP mutants, Ca^{2+} channel clustering and Ca^{2+} influx are impaired, and synaptic release probability is drastically reduced (Liu et al., 2011). The DRBP mutant still forms T-shaped structures but they reside on thinner pedestals. In a DRBP-null mutant, abnormally shaped electron-dense material was found and the T-shaped structures were missing. Hence proteins of the RBP family are important for AZ organization and coupling of synaptic vesicles, Ca^{2+} channels, and the fusion machinery (Liu et al., 2011).

As already noted, some proteins are highly enriched in the AZ, such as Munc13, RIM, bassoon, piccolo, and ELKS (ERC/CAST), and might serve as its core. However, additional proteins are associated with, albeit not restricted to, the AZ. These proteins play additional roles in the cell, such as (1) cytoskeleton proteins (actin, tubulin, myosin, spectrin, and β-catenin) which are the fundamental proteins of the AZ cytomatrix framework but are also abundant throughout the cell, (2) cell-adhesion molecules (neurexins, cadherins, integrins, sidekicks, and liprin-α), (3) proteins involved in synaptic vesicle fusion (syntaxin, SNAP-25, and Munc18), and (4) voltage-gated Ca^{2+} channels (reviewed in Schoch and Gundelfinger (2006); Zhai and Bellen (2004)). Some of these proteins are discussed later in this chapter.

2.2. THE SYNAPTIC VESICLE CYCLE

2.2.1. Overview

As already mentioned, vesicles undergo a series of sequential steps in the synapses called the synaptic vesicle cycle. Synaptic vesicles translocate and dock at the AZ. They then undergo priming that makes them fusion-competent, and following Ca^{2+} influx, they undergo fusion and recycling. Although these steps are described as separate processes, there is ample of evidence that they are not completely separate and that the same proteins participate in or contribute to multiple steps, suggesting these steps' continuity.

Nevertheless, in the following section, each of these steps is discussed separately, with an indication of the processes that converge.

2.2.2. Vesicle Mobilization

Vesicles mobilization between the different pools in the terminal is essential to ensuring their availability during periods of intense synaptic activity. A key family of proteins in vesicle organization and mobilization is the synapsins.

The synapsins are abundant phosphoproteins present in most conventional chemical synapses. The family consists of synapsins I and II which are essential for the clustering of synaptic vesicles near the AZs and for the organization of the RPs of synaptic vesicles, as well as of the developmentally regulated synapsin III (Ferreira et al., 2000). The synapsins exhibit several phosphorylation sites. These include a well-conserved protein kinase A (PKA) site that controls the affinity of synapsin to vesicles and regulates its function (Chi et al., 2001, 2003; Hosaka et al., 1999; Menegon et al., 2006; Orenbuch et al., 2012b) and two phosphorylation sites for Ca^{2+}-calmodulin kinase II (CaMKII), a serine–threonine protein kinase that is activated by the Ca^{2+}–calmodulin complex during intense synaptic activity (Benfenati et al., 1990; Tao-Cheng et al., 2006). Synapsin I is also a substrate for MAP kinase, CDK5, and others (Jovanovic et al., 2001; Matsubara et al., 1996). In a dephosphorylated state, synapsin bundles actin (Bahler and Greengard, 1987; Nielander et al., 1997) and associates with vesicles (Benfenati et al., 1993; Hosaka et al., 1999), suggesting that it maintains vesicle clusters by cross-linking or tethering vesicles to actin. Furthermore, following synaptic activity and CaMKII phosphorylation, its affinity toward both vesicles and actin decreases, which has been suggested to allow vesicle translocation to the AZ (for review, see Bykhovskaia (2011)). Hence, activity-induced synapsin phosphorylation may allow vesicle exchange between the pools as part of the mobilization of synaptic vesicles from the RP to the RRP (for review, see Bykhovskaia (2011)).

It should be noted that other proteins in addition to synapsin, and most intriguingly those of the endocytotic cycle, were proposed to cluster synaptic vesicles (Shupliakov et al., 2011). In agreement with these collective observations, microinjection of dephosphorylated synapsin inhibits neurotransmitter release (Coleman and Bykhovskaia, 2009; Llinas et al., 1985), whereas synapsin depletion or inactivation results in loss of clustered vesicles located distal from AZs (i.e. the recycling pool and RP) (Baldelli et al., 2007; Gitler et al., 2004; Hilfiker et al., 1998, 2005; Orenbuch et al., 2012a; Pieribone et al., 1995). Moreover, following repeated activity, these synapses undergo faster synaptic depression. Thus, it seems that by clustering vesicles to the RP and then releasing them during stimulation, synapsins are required for sustained high-frequency transmission (Cesca et al., 2010; Gitler et al., 2004, 2008; Li et al., 1995; Rosahl et al., 1995). It is of note that synapsins have been suggested to participate also in the endocytic segment of the vesicle cycle (Bloom et al., 2003), a notion that was recently expanded by the observation that endocytosis is critical for the activity-dependent redistribution of synapsin during intense synaptic activity (Orenbuch et al., 2012b).

The division of vesicles into pools is not static; vesicles undergo continuous, albeit slow, exchange between the RRP and recycling pool at rest (Pyle et al., 2000) as well as between the recycling pool and the RP (Kamin et al., 2010), in a manner that may be regulated bidirectionally by CDK5 and calcineurin activity (Kim and Ryan, 2010). Moreover, vesicles move between synapses (Darcy et al., 2006; Staras et al., 2010) in a synapsin-dependent manner (Orenbuch et al., 2012a). A physiological role for the exchange of vesicles between pools is supported by evidence implicating its participation in long-term synaptic plasticity (Ratnayaka et al., 2012).

The mechanism of vesicle trafficking between the different pools is not clear, but is thought to be carried out at least in part by the actin cytoskeleton through protein–protein interactions, under the regulation of Ca^{2+} (Doussau and Augustine, 2000; Giner et al., 2005; Lang et al., 2000; Neco et al., 2003; Oheim and Stuhmer, 2000, but see Sankaranarayanan et al., 2003). The actin network is continuously rearranged during the exocytic cycle. One of the key regulators of actin-activity-dependent remodeling is the myristoylated alanine-rich C kinase substrate (MARCKS) protein. Under resting conditions, unphosphorylated MARCKS binds and stabilizes actin filaments by cross-linking them to a dense, intricate actin meshwork (Cuchillo-Ibanez et al., 2004). This actin meshwork maintains the RP by physically barring vesicles from approaching the PM (Aunis and Bader, 1988; Burgoyne and Cheek, 1987; Trifaro et al., 1985; Vitale et al., 1995). Following intense stimulation and high cytosolic Ca^{2+} concentrations, PKC is activated and phosphorylates MARCKS. This phosphorylation displaces MARCKS from actin, encouraging actin depolymerization and allowing vesicle mobilization and recruitment to docking sites at the PM (Cuchillo-Ibanez et al., 2004; Doreian et al., 2009; Hartwig et al., 1992; Trifaro et al., 2002).

Fodrin, similar to MARCKS in terms of activity, is a neural form of spectrin, which has long been known to be a key factor in cortical actin fragmentation (Perrin and Aunis, 1985). It was found that in chromaffin cells, this protein cross-links F-actin at low concentrations of intracellular calcium, but following stimulation and Ca^{2+} elevation, fodrin dissociates from actin and there is a local dissolution of the actin cortex. The role of this protein in exocytosis was demonstrated when calcium-induced secretion from digitonin-permeabilized cells was partially inhibited by purified antifodrin antibodies (Perrin et al., 1987). Furthermore, alpha-fodrin antibodies were suggested as markers for common secretory disorders in humans (Szanto et al., 2008). Further evidence for the involvement of fodrin in exocytosis comes from the observation that it interacts with the soluble NSF attachment protein receptor (SNARE) protein syntaxin-1 (Nakano et al., 2001) and that it might constitute part of the fusion-pore complex (Jena et al., 2003).

Another key activity-dependent F-actin regulator is the scinderin protein, which can bind and sever actin (Rodriguez Del Castillo et al., 1990). Scinderin is a Ca^{2+}-dependent protein containing two Ca^{2+}-, two actin-, and two phosphatidylinositol-4,5-bisphosphate

(PIP2)-binding sites (Marcu et al., 1994). PIP2 competes with actin for interaction with scinderin. Indeed, it was found that PIP2 inhibits the actin-severing activity of scinderin (Maekawa and Sakai, 1990; Rodriguez Del Castillo et al., 1992; Zhang et al., 1996). This suggests that PIP2 levels and therefore PIP2 regulators, such as the small molecular weight G-proteins Rho and Rac, modulate scinderin activity and may indirectly regulate actin fragmentation and secretion (Hartwig et al., 1995). Under resting conditions, scinderin is localized to the PM via interaction with PIP2. However, following cell stimulation and a rise in intracellular Ca^{2+}, the scinderin–actin interaction becomes more favorable (Rodriguez Del Castillo et al., 1992), suggesting that under these conditions, scinderin detaches from its binding sites at the PM to interact and sever cortical actin, facilitating access of the vesicle to the PM. Indeed, it was found that in chromaffin cells, the severing activity of scinderin on actin results in enhanced secretion (Dumitrescu Pene et al., 2005; Lejen et al., 2002; Trifaro et al., 2008).

Several other actin-polymerization proteins are involved in the exocytic cycle. CDC42 is a Rho GTPase protein and among its effectors is the Wiskott–Aldrich syndrome protein, N-WASP. N-WASP links CDC42 to actin polymerization through the actin-related protein 2/3 (ARP2/3) complex which promotes actin nucleation and polymerization (Rohatgi et al., 1999). It was surprising to find that overexpression or activation of CDC42 or N-WASP increased secretion in neuroendocrine cells, and that inhibition or knockdown of CDC42 decreased it (Gasman et al., 2004). Consequently, the actin-physical-barrier model was put into question and some studies hypothesized that the CDC42-induced increase in release was due to optimization of the actin network to make release more efficient. However, other studies emphasized the diversity of roles that CDC42 and other Rho GTPases perform and suggested that the enhancement effect on secretion stems from acceleration of priming and of the transition through the ER–Golgi path, rather than from actin polymerization (Daniel et al., 2002; Nevins and Thurmond, 2005; Sato et al., 2012).

Proteins of the myosin family have also been suggested to play an active role in vesicle trafficking in the synapse (Peng et al., 2012). Generally, vesicle transport on actin filaments is carried out by the actin motor protein myosinVa which is associated to vesicles by protein–protein interaction with the vesicular GTP-Rab3A, Rab27 GTPase, and MyRIP proteins (Desnos et al., 2003). When the vesicle approaches the cell cortex, where the actin network is highly branched, their transport becomes dependent on myosin II (reviewed in Mooseker and Cheney, 1995; Villanueva et al., 2012). Myosin II is linked to cortical F-actin, and therefore accumulates in the cell periphery. There is abundant evidence supporting a role for myosin II in the transport of vesicles through the cortical region of neuroendocrine cells (Neco et al., 2002; Rose et al., 2003). For example, inhibitors of myosin II impair granule motion in cortical regions (Berberian et al., 2009; Neco et al., 2002). In addition, myosin II was shown to be involved in both vesicle motion and mobilization during neuronal activity and in controlling evoked neurotransmitter release (Peng et al., 2012).

2.2.3. SNARE Proteins: The Core Proteins in the Synaptic Vesicle Cycle

The core complex in the synaptic vesicle fusion machinery is formed by SNARE proteins and consists of the vesicular SNARE protein, synaptobrevin-2 (also known as VAMP-2), the presynaptic membrane SNARE syntaxin-1A, and the presynaptic membrane SNARE SNAP-25 (synaptosome-associated protein of 25 kDa) (Sollner et al., 1993). SNAREs belong to a small well-conserved group of membrane-anchored proteins that are characterized by conserved sequences of 60–70 amino acids called SNARE motifs (Weimbs et al., 1997). The interaction between SNARE motifs results in a four-helix coiled-coil structure called the SNARE complex (Sutton et al., 1998), which plays a central role in synaptic vesicle docking, priming, and fusion machinery (Sinha et al., 2011).

Some of the initial evidence demonstrating that these three proteins are essential components of the membrane-fusion machinery came from experiments with botulinum and tetanus toxins. SNARE proteins are specifically targeted and cleaved by clostridial and tetanus neurotoxins—potent inhibitors of neurotransmitter release (Blasi et al., 1993; Link et al., 1992; Schiavo et al., 1992, 2000). These findings provided one of the first molecular links to the *C. elegans* and yeast homologs of these proteins which had been shown to be essential for nervous system function in *C. elegans* (Brenner, 1974) and membrane trafficking in the secretory pathway in yeast (Novick et al., 1980). Later studies demonstrated that null mutants of any of the three SNARE proteins exhibit strong inhibition of neurotransmitter release at various synapses (Borisovska et al., 2005; Deitcher et al., 1998; Schoch et al., 2001; Schulze et al., 1995; Washbourne et al., 2002).

It is of interest to note that the SNARE proteins are involved in nearly every step of the synaptic vesicle cycle in the terminal, including vesicle docking, priming, fusion, and recycling. Their involvement in these steps is discussed in the following sections.

2.2.4. Vesicle Docking

Upon arrival at the AZ, vesicles detach from the cytoskeleton and undergo a process called docking. Docked vesicles have been defined by EM and total internal reflection fluorescence (TIRF) microscopy. On electron micrographs, docked vesicles are traditionally defined as those vesicles that appear closely apposed to the PM. Some studies score vesicles whose membrane is within a certain distance of the PM (for instance <30 nm) as being docked, and some score for the presence of a "contact patch" between the vesicle and the PM (Verhage and Sorensen, 2008). Today, the preferred fixation method for EM is high-pressure freezing as it immobilizes synaptic components on a millisecond timescale while avoiding artifacts associated with chemical fixation such as the collapse of presynaptic terminals and the redistribution of synaptic vesicles (Rostaing et al., 2006; Siksou et al., 2007; Weimer et al., 2006).

TIRF microscopy enables selective visualization of synaptic vesicle or large dense-core vesicle (LDCV) dynamics close to the PM in the evanescent field (~200 nm from the PM), where vesicle docking, priming, and fusion take place (Becherer et al., 2007; Toonen et al., 2006a). Toonen et al. (2006) used TIRF in chromaffin cells to find that most vesicles move in and out of the TIRF plane apparently without being retained, but some vesicles are retained for shorter or longer times by local tethering forces at the target; those latter vesicles were defined as docked. Accordingly, vesicle docking is suggested to correspond to changes in the axial mobility and resident time of vesicles near the membrane (Toonen et al., 2006a). In addition, vesicle docking in chromaffin cells was suggested to be calcium-dependent and show maximal docking at 500 nM calcium (Pasche et al., 2012), although the nature of this calcium dependency is still not fully known.

The molecular mechanism governing vesicle docking is not fully understood but several key synaptic proteins, including Munc18, have been shown to be essential. Munc18 belongs to the family of SM proteins (Sec1 and Munc18) that play an essential role in vesicle docking. SM proteins are cytosolic proteins of 600–700 residues that directly or indirectly bind to SNARE and have remained highly conserved throughout evolution (reviewed in Rizo and Sudhof (2002)). Munc18-1 folds into an arch-shaped structure (Misura et al., 2000) that binds one of the SNARE proteins, syntaxin, in a regulatory manner (Pevsner et al., 1994), and is essential for synaptic transmission (Verhage et al., 2000). In contrast to the many SNARE isoforms in vertebrates and yeast, only a few SM protein isoforms are expressed in vertebrates and in yeast.

In the absence of Munc18-1, LDCVs in mouse chromaffin cells fail to dock at the PM (Voets et al., 2001). This phenotype was detected as a significant reduction in the number of vesicles that are attached to the PM in electron micrographs and as a decrease in vesicle delivery rate and long-retained vesicle tethering as viewed in TIRF measurements (Toonen et al., 2006a). It is important to note that the absence of Munc18 did not reduce the level of synaptic vesicles docking in neurons (Verhage et al., 2000). This suggests that the mechanisms of vesicle docking are different between these two systems and that additional factors control vesicle docking in neurons.

Indeed, in the few years before 2013, it has been shown that Munc18 does not operate solely in vesicle docking; Munc18-1 is known to interact with neuronal SNARE proteins in two distinct modes: with isolated syntaxin-1 alone in a "closed"-syntaxin-1 conformation and with the assembled SNARE complexes containing syntaxin-1 in an "open"-syntaxin-1 conformation together with SNAP-25 and synaptobrevin-2 (Toonen and Verhage, 2007). Some studies suggested that Munc18 induces docking by trafficking syntaxin-1 to the PM (Arunachalam et al., 2008), but several findings argue against this: Toonen et al. (2005) argued that Munc18-1 stabilizes syntaxin-1, but is not essential for syntaxin-1 targeting to the PM and SNARE complex formation (Toonen et al., 2005). In addition, Munc-18 mutants with impaired docking phenotypes showed normal syntaxin-1 localization, which means that Munc18-1 stabilizes syntaxin-1 through a mechanism separate from the interaction that affects docking (Gulyas-Kovacs et al., 2007). In general, it is suggested that binding of Munc18 to the closed conformation of syntaxin-1 creates a complex that is considered a good candidate docking platform for regulated secretion (Gerber et al., 2008; Gulyas-Kovacs et al., 2007; Toonen et al., 2005). However, de Wit et al. (2009) argued that Munc18–syntaxin-1 does not function alone in the docking process, but alongside synaptotagmin and SNAP-25 (de Wit et al., 2009).

According to this and other studies, it was postulated that an intermediate step in the docking process is the formation of a 1:1 Syntaxin-1–SNAP-25 heterodimer (Fasshauer and Margittai, 2004; Zilly et al., 2006). Furthermore, it was suggested that Munc18-1 controls the assembly of syntaxin-1–SNAP-25 heterodimers and may even promote the formation of this "acceptor" complex relative to a proposed nonproductive syntaxin-1–SNAP-25 (2:1) complex (Fasshauer and Margittai, 2004; Xiao et al., 2001; Zilly et al., 2006). de Wit et al. (2009) found that synaptotagmin and SNAP-25 are required for docking of secretory vesicles as null mutations of either SNAP-25 or synaptotagmin-1 abolish docking. By using synaptotagmin-1 and SNAP-25 mutations that affect their interaction, they confirmed that both proteins tightly cooperate for correct anchoring of secretory vesicles to fusion sites. While docking was established between syntaxin-1–SNAP-25 acceptor complexes at the target membrane and synaptotagmin-1 on the vesicle membrane, they claimed that Munc18-1 promotes the formation or stability of the correct acceptor SNARE complexes.

Hence, SNARE and SM protein complexes might stabilize the attachment of vesicles to the target membrane and thereby participate in docking (de Wit et al., 2009). More specifically, their model (Figure 4) included the following steps. First, Munc18-1 binds the closed conformation of syntaxin-1. Second, SNAP-25 binds the syntaxin-1–Munc18-1 heterodimer (Burkhardt et al., 2008; Zilly et al., 2006). Third, secretory vesicles reach the target membrane area and associate via synaptotagmin-1 to this trimeric syntaxin-1–Munc18-1–SNAP-25 complex, which effectuates docking. Fourth, synaptobrevin-2 binds to the synaptotagmin-1–syntaxin-1–Munc18-1–SNAP-25 complex, resulting in formation of the SNARE bundle, and the beginning of the priming process (de Wit et al., 2009). Therefore, the docking process involves Munc18-1, syntaxin-1, SNAP-25, and synaptotagmin-1.

Another protein complex that is suggested to be linked with docking is the exocyst. The exocyst is an octameric complex initially identified through the isolation of yeast's temperature-sensitive mutants that were defective in secretion (Hsu et al., 2004; Munson and Novick, 2006). Six of its eight subunits are associated to vesicles (Sec5, Sec6, Sec8, Sec10, Sec15, and Exo84) and two (Exo70 and Sec3) to the PM. This octameric complex controls the docking and insertion of exocytic vesicles in growing areas of the PM in yeast (Hertzog and Chavrier, 2011). According to the exocyst's function in targeting vesicles to the tip of the budding daughter cell in yeast, it was reasonable to assume that the exocyst is involved in exocytosis and synaptic transmission, possibly as a main docking complex.

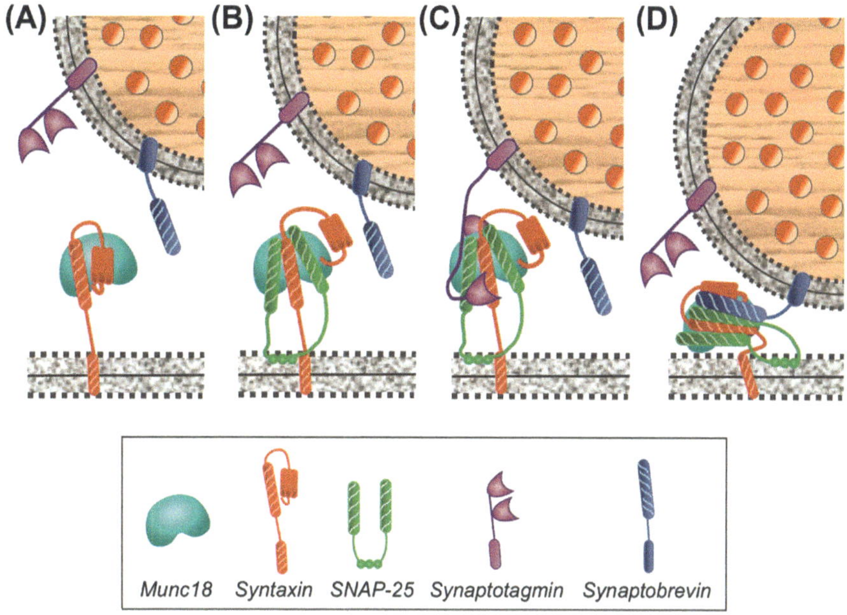

Figure 4 Molecular mechanism of vesicle docking. (A) Munc18-1 binds the closed conformation of syntaxin-1. (B) SNAP-25 binds the syntaxin-1–Munc18-1 heterodimer. (C) The vesicle associates via synaptotagmin-1 to the trimeric syntaxin-1–Munc18-1–SNAP-25 complex, which effectuates docking. (D) Synaptobrevin-2 binds to the synaptotagmin-1–syntaxin-1–Munc18-1–SNAP-25 complex, resulting in formation of the SNARE complex. It is unknown whether synaptotagmin-1 dissociates after the formation of the SNARE complex.

In animal cells, the role of the exocyst in exocytosis and cell-surface expansion has been demonstrated in a variety of cell types, such as protein targeting in epithelial cells (Grindstaff et al., 1998), dendritic delivery of *N*-methyl-D-aspartate (NMDA) receptors at postsynaptic membranes (Sans et al., 2003), and insulin-induced exocytosis of glucose transporter in adipocytes (Inoue et al., 2003). However, the exocyst is not required for all types of exocytic events in animal cells as it does not seem to play a major role during neurotransmitter release at mature synapses (He and Guo, 2009; Murthy et al., 2003).

As to docking of synaptic vesicles in central mammalian synapses, it was found that the exocyst complex mediates the addition of membrane during postnatal presynaptic maturation, but does not function as a tethering complex in the local recycling of vesicles within the synaptic vesicle cycle (Schwenger and Kuner, 2010). Schwenger and Kuner (2010) expressed Exo70 or its dominant-negative truncation construct Exo70ΔC in globular bushy cells, and their calyces were analyzed for changes in maturation and synaptic transmission. Their results suggested that the exocyst complex is involved in the formation of synaptic terminals during the early phases of synapse maturation but does not contribute to the synaptic vesicle cycle in mammals (Schwenger and Kuner, 2010).

Additional docking mechanism was suggested in neuroendocrine cells, involving two small GTPases, Rab3A, and Rab27A. Rab proteins belong to the Ras-related superfamily of small monomeric GTPases and function as mediators of membrane trafficking in all eukaryotic cells (Merrins and Stuenkel, 2008). The isoforms of Rab3 and Rab27 are localized to vesicles and secretory granules (Handley et al., 2007) and have been directly implicated in regulated exocytosis. Overexpression of Rab3A inhibited regulated exocytosis (Holz et al., 1994; Johannes et al., 1998; Regazzi et al., 1996), while Rab3A knockout mice showed an increased Ca^{2+}-triggered exocytosis that led to faster synaptic depression in cultured hippocampal neurons (Geppert et al., 1994a, 1997) and reduced spontaneous neurotransmitter release at the neuromuscular synapse (Sons and Plomp, 2006). On the other hand, overexpression of Rab27A inhibited exocytosis in some studies but enhanced exocytosis in other studies (Desnos et al., 2003; Johnson et al., 2005; Menasche et al., 2003; Yi et al., 2002).

Silencing either Rab3A or Rab27A in PC12 cells significantly decreased the number of LDCVs docked to the PM without altering the kinetics of individual exocytic events. Endogenous Rab27A (or Rab3A) was unable to fully compensate the function of Rab3A (or Rab27A), suggesting that the functions of Rab3A and Rab27A are not completely redundant. Simultaneous silencing of Rab3A and Rab27A caused a significantly greater decrease in the number of vesicles docked to the PM, suggesting that these GTPases cooperatively regulate the docking step(s) of LDCVs to the PM (Tsuboi and Fukuda, 2006). A possible mechanism for Rab3A and Rab27A function comes from the finding that Slp4-a (Synaptotagmin-like protein4-a/granuphilin-a) promotes the docking of dense-core vesicles to the PM by simultaneously interacting with Rab27A on the vesicle and with syntaxin-1a/Munc18-1 on the PM (Coppola et al., 2002; Fukuda, 2003; Fukuda et al., 2005; Torii et al., 2004). In addition, it was proposed that rabphilin controls the docking step of dense-core vesicle exocytosis by simultaneously interacting with Rab27A on the vesicle and with SNAP-25 on the PM (Fukuda, 2007; Tsuboi and Fukuda, 2005). On the other hand, in mouse hippocampal neurons, it was suggested that rabphilin and SNAP-25 regulate priming activity after intense synaptic stimulation as knockout of rabphilin resulted in accelerated recovery following synaptic depression (Deak et al., 2006).

2.2.5. Vesicle Priming

Priming renders the vesicle release-competent so that it can fuse with the PM when the trigger (Ca^{2+}) arrives (Verhage and Sorensen, 2008). In electrophysiological measurements, *priming* is often defined as the process by which the RRP is refilled (von Ruden and Neher, 1993). The size of the primed pool of vesicles can be measured by either electrophysiological or fluorescence means and it is defined as the number of vesicles that are released following a depleting/emptying stimulus (Pyle et al., 2000; Rosenmund and Stevens, 1996; Verhage and Sorensen, 2008). These stimulations consist of either a train of action potentials or application of hypertonic sucrose solution that are thought to trigger the fusion of readily releasable vesicles (Rosenmund and Stevens, 1996). Further characterization of the primed vesicle by TIRF microscopy suggested that priming was associated with restricted lateral mobility of the vesicle (Nofal et al., 2007; Yizhar and Ashery, 2008, but see Degtyar et al. (2007)).

As noted, the core of the synaptic vesicle priming and fusion machinery is formed by three members of the SNARE protein family: the vesicle SNARE protein, synaptobrevin-2, and the presynaptic membrane SNAREs syntaxin-1A and SNAP-25 (Sinha et al., 2011). Complementary sets of those SNAREs are present in both of the membranes undergoing fusion (van den Bogaart and Jahn, 2011). The four SNARE motifs of the synaptic SNARE proteins assemble into the four-helix coiled-coil structure of the SNARE complex (Poirier et al., 1998; Sudhof and Rizo, 2011; Sutton et al., 1998), with one SNARE motif from synaptobrevin, one from syntaxin-1, and two from SNAP-25 (Sudhof and Rizo, 2011).

Mechanistically, the SNARE complex contributes to priming in its "trans" form as it brings the synaptic vesicle membrane into close proximity to the PM, a process which is thought to be essential for their fusion (Baumert et al., 1989; Bennett et al., 1992; Hanson et al., 1997; Lin and Scheller, 1997; Sutton et al., 1998, reviewed in Weimer and Jorgensen (2003)). Consistent with this notion, the size of the RRP (with presumably primed vesicles) correlates with the level of SNARE complex assembly in vivo (Lonart and Sudhof, 2000, reviewed in Weimer and Jorgensen (2003)), and a high number of SNARE complexes is thought to increase the priming rate and fusion competence of the vesicle (Mohrmann and Sorensen, 2012). However, SNARE proteins do not function alone; several other proteins regulate and control SNARE complex formation.

A simplified sequence of the events leading to the formation of SNARE complexes suggests that the first step is the formation of a binary complex between syntaxin-1 and SNAP-25 on the PM. Later, this binary complex interacts with synaptobrevin-2 that resides on the synaptic vesicle and a trans-SNARE complex is formed, leading to vesicle priming (Figure 5). As already mentioned, before the assembly of SNARE complexes, syntaxin-1 molecules are found in closed conformation in close association with Munc18-1, and the acceptor complex is in association with synaptotagmin.

Munc18-1 is thought to promote the assembly of syntaxin-1–SNAP-25 heterodimers (Zilly et al., 2006) and provide the platform for further interaction and formation of trans-SNARE complexes with synaptobrevin. Accordingly it was shown that the interaction of Munc18 with syntaxin-1 facilitates the assembly of SNARE complexes between synaptobrevin-2 and binary syntaxin-1–SNAP-25 complex (Shen et al., 2007; Zilly et al., 2006) and thus facilitates vesicle priming and fusion. However, as the binding

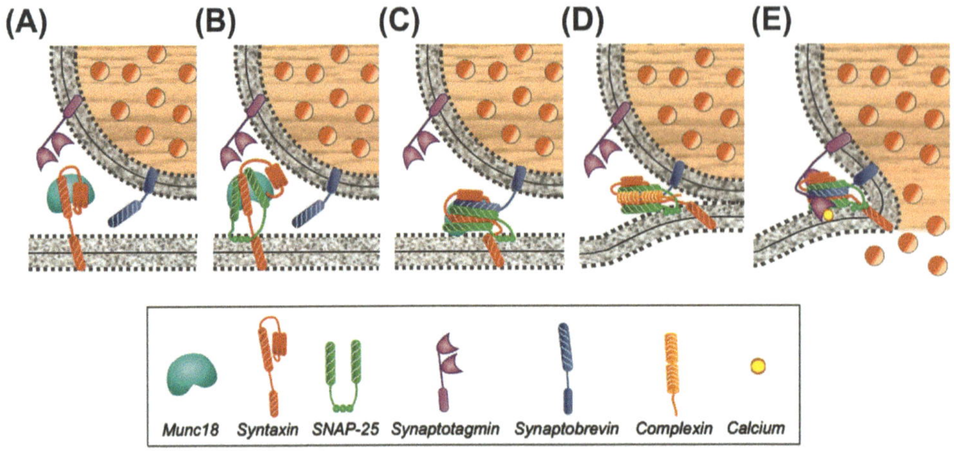

Figure 5 Molecular mechanism of vesicle priming and fusion. (A) Munc18-1 binds the closed conformation of syntaxin-1. (B) SNAP-25 binds the syntaxin-1–Munc18-1 heterodimer. (C) Synaptobrevin-2 binds to the syntaxin-1–Munc18-1–SNAP-25 complex, resulting in formation of the SNARE complex. (D) Binding of complexin further tightens the SNARE bundle and causes maximal proximation between the plasma membrane and the vesicle. Additional proteins such as Munc13, RIM, CAPS, and tomosyn, which also affect priming, are not represented in this scheme. (E) Increase in calcium concentration causes displacement of complexin from the SNARE complex and binding of synaptotagmin to the complex. This in turn initiates vesicle fusion.

of Munc18 to closed syntaxin-1 was thought to prevent formation of SNARE complexes, the molecular nature of this interaction was unclear for many years, as was the question of whether and how formation of Munc18-1–SNARE complexes is essential for exocytosis (Khvotchev et al., 2007).

Answers were revealed by a series of studies that described another mode of interaction between syntaxin-1 and Munc18, through the N-terminal peptide of syntaxin. Several findings suggested that this N-terminal peptide is essential for Munc18 binding to the fully assembled SNARE complex and activation of fusion (Burkhardt et al., 2008; Gerber et al., 2008; Johnson et al., 2009; Khvotchev et al., 2007; McEwen and Kaplan, 2008; Munson and Bryant, 2009; Rathore et al., 2010; Shen et al., 2007). It was suggested that the N-terminus of syntaxin-1 physically recruits Munc18 to the SNARE complex, facilitating the latter's assembly (Rathore et al., 2010). Point mutations in Munc18-1 that preserved tight binding to closed syntaxin-1 but markedly disrupted Munc18-1 binding to SNARE complexes containing open syntaxin-1 impaired synaptic vesicle priming but not Ca^{2+}-triggered fusion of primed vesicles (Deak et al., 2009).

It has been recently suggested that Munc18 interaction with the Habc domain of syntaxin-1 in its closed conformation may regulate priming while Munc18 interaction with the N-terminal syntaxin-1 peptide regulates the early initiation stage of fusion (Rathore et al., 2010; Sudhof and Rizo, 2011). In addition, it was suggested that after association, Munc18-1 and the SNARE bundle drive the merging of membrane bilayers without further involvement of the N-terminal peptide. Hence, the N-terminal peptide acts as an initiation factor for the assembly of the fusion–competent complex (Rathore et al., 2010). It can be assumed that the closed conformation of syntaxin-1 gates the initiation of the synaptic vesicle priming reaction, which is then mediated by SNARE complex–Munc18-1 assemblies. It is clear that the interaction of the N-terminal peptide of syntaxin-1 with Munc18 plays an essential role at the border between vesicle priming and initiation of fusion.

It should be noted that for many years, the interaction of Munc18 with the closed conformation of syntaxin-1 was suggested to inhibit SNARE complex formation and fusion (Dulubova et al., 1999; Wu et al., 2001). However, the recent evidence presented above clearly place Munc18 as a positive regulator of synaptic transmission. These results are also supported by the finding that Munc18-1 deficiency caused a loss of synaptic transmission rather than an increase in synaptic transmission one would expect after deletion of a negative regulator of release (Verhage et al., 2000), and overexpression of Munc18 enhanced both vesicle docking and exocytosis in chromaffin cells (Nili et al., 2006).

Although syntaxin-1 must interact with SNAP-25 and later with synaptobrevin-2 to prime vesicles for release, in its closed conformation, its N-terminus folds over and occludes the SNARE-binding domain (Calakos et al., 1994; Dulubova et al., 1999, reviewed in Weimer and Jorgensen (2003)). Thus, syntaxin-1 needs to be "opened" for priming to occur. Two key synaptic proteins that might be involved in this process are Munc13 and RIM.

Hence, in addition to SNARE proteins and Munc18, two large presynaptic AZ proteins, Unc13/Munc13s, and Unc10/a-RIMs, are known to have key roles in synaptic vesicle priming. Priming is completely abolished upon genetic ablation of Unc13 in invertebrates or of its homologs Munc13-1 and Munc13-2 in mice (Aravamudan et al., 1999; Augustin et al., 1999b; Richmond et al., 1999; Varoqueaux et al., 2002, reviewed in Ma et al. (2011)), whereas strong priming defects are also observed in the absence of *C. elegans* Unc10 or its mammalian homologs RIM1a and RIM2a (Koushika et al., 2001; Schoch et al., 2002, 2006, reviewed in Ma et al. (2011)). Notably, these priming defects were partially rescued by overexpression of the open syntaxin-1 mutant in *C. elegans* unc-13 or unc-10 null mutants (Koushika et al., 2001; Mennerick and Matthews, 1996; Richmond et al., 2001, reviewed in Ma et al. (2011)), suggesting that Unc13 and Unc10 are somehow involved in opening syntaxin-1. Indeed, Ma et al. (2011) found that transition from the closed syntaxin-1–Munc18-1 complex to the SNARE complex is dramatically accelerated by the MUN domain of Munc13 (the minimal domain necessary for priming). This activity critically depends on the weak interaction of the MUN domain with the syntaxin-1 SNARE motif and can be bypassed by an open form of syntaxin. The finding that the rescue is only partial suggests that Munc13 plays a role(s) in the priming process additional to physically opening syntaxin-1 (Mennerick and Matthews, 1996).

As to RIM's contribution to the priming process, the Zn^{2+} finger domain of RIM binds to the N-terminal C2A domain of Munc13, which forms a homodimer in the absence of RIM but a heterodimer with it (Dulubova et al., 2005; Lu et al., 2006). Munc13 homodimerization is suggested to inhibit its priming activity. Deng et al. (2011) found that the isolated Zn^{2+} finger domain of RIM autonomously promotes vesicle priming by binding to Munc13, thereby relieving Munc13 homodimerization. Constitutively monomeric mutants of Munc13 rescued priming in RIM-deficient synapses, whereas wild-type Munc13 that created homodimers did not. Thus, homodimerization of Munc13 inhibits its priming function, and RIMs activate priming, at least partially, by disrupting Munc13 homodimerization (Deng et al., 2011).

Calcium-dependent activator protein for secretion (CAPS) is a large phosphoprotein that is regulated by Ca^{2+} and protein kinase CK2 phosphorylation (Nojiri et al., 2009). In its phosphorylated state, it is essential for priming of both LDCVs and synaptic vesicles and dephosphorylation terminates its activity (Jockusch et al., 2007; Nojiri et al., 2009; Stevens and Rettig, 2009). There are two CAPS isoforms, and both contain a Munc13 homology domain (Speidel et al., 2003). Double knockout of CAPS1 and 2 results in reduction of the RRP size in several organisms as well as decreased sustained release, which is an indicator of priming deficiencies (Liu et al., 2008). In addition, CAPS1 knockout causes disruptions in loading of LDCVs with catecholamine and expression of CAPS1 in the CAPS-knockout background restores all properties damaged by its absence (Liu et al., 2008; Speidel et al., 2005). The open form of syntaxin-1 can also overcome the loss of CAPS, similar to the rescue of Unc-13 knockout (Richmond et al., 2001), suggesting that CAPS is a key priming factor which works on syntaxin-1 similar to Munc13s (Liu et al., 2010). However, Munc13-1 overexpression in the absence of CAPS did not enhance priming, implying that Munc13 and CAPS cannot compensate one for each other despite their similarities (Liu et al., 2010).

Another synaptic protein involved in priming regulation is tomosyn, which has been suggested to compete with Munc18 for syntaxin-1 binding both in vitro (Fujita et al., 1998) and in vivo (Gladycheva et al., 2007). Through a C terminal SNARE-binding motif, tomosyn also forms a dead-end SNARE complex with syntaxin-1 and SNAP-25 that excludes synaptobrevin-2 (Hatsuzawa et al., 2003; Pobbati et al., 2004). Based on in vitro experiments, it was hypothesized that these complexes reduce the formation of fusogenic trans-SNARE complexes and inhibit priming and fusion. Accordingly, tomosyn was shown to inhibit vesicle priming and negatively regulate neurosecretion in a variety of preparations (Ashery et al., 2009; Gracheva et al., 2006, 2007, 2010; McEwen et al., 2006; Sakisaka et al., 2008; Yizhar et al., 2004). Consequently, loss of tomosyn in *C. elegans* (TOM-1) results in increased vesicle priming (Gracheva et al., 2006).

It is interesting to note that the tomosyn SNARE motif makes up only 10% of the protein. The larger part of the protein is suggested to fold into two β-propeller WD-40 domains (Williams et al., 2011) and intact WD-40 domains are important for its inhibitory activity (Yizhar et al., 2007). Tomosyn has recently been found to interact with the calcium sensor synaptotagmin-1. Yamamoto et al. (2010) demonstrated that formation of the Ca^{2+}-dependent synaptotagmin-1–tomosyn–SNARE complex through tomosyn's WD-40 domain ensures inactivation of the fusion machineries and thereby inhibits priming of synaptic vesicles (Yamamoto et al., 2010). It is therefore clear that both the tomosyn SNARE motif and its WD-40 domains are important for its inhibitory effects, although the exact contribution of each domain remains unclear.

While tomosyn is classified as a protein that inhibits priming and synaptic transmission, several studies have shown an additional effect on vesicle docking, with changes observed upon tomosyn deletion or overexpression. In *C. elegans* TOM-1 mutant synapses, in addition to the effect on synaptic transmission, a concomitant increase in the number of docked vesicles was detected with EM, suggesting an effect on vesicle docking. Interestingly, in wild-type synapses, contacting vesicles were preferentially localized within ~150 nm of the presynaptic density, whereas in TOM-1 mutants, the increase in membrane-contacting vesicles was distributed throughout the terminal (Gracheva et al., 2006). In addition, TIRF experiments showed that tomosyn overexpression decreases the vesicles' ability to undergo tethering and docking to the PM. Vesicles under tomosyn overexpression had a higher turnover rate—the rate of vesicle arrival to the PM (initial docking phase) and disappearance into the cytosol (undocking) was enhanced although the total number at the PM was not affected (Yizhar and Ashery, 2008)—leading to a reduction in the time spent by the vesicle on the PM. Morphologically, the number of visible vesicle was not altered but one can suggest that the number of "functionally docked" vesicle was decreased. These data suggest a first link between docking and priming. Interestingly, similar effect on docking was demonstrated for Munc13, a key protein involved in vesicle priming. In slice cultures from mice deficient in Munc13, vesicles were not docked to the AZ, suggesting that Munc13 manipulation also alters docking (Siksou et al., 2009b).

At the molecular level, it is possible that both tomosyn and Munc13 act on a single protein, syntaxin, but affect synaptic transmission in opposite ways. As syntaxin-1 is a key protein in both vesicle docking and priming, we can envision the following scenario: the interaction of tomosyn with syntaxin-1 reduces the formation of syntaxin-1–SNAP-25 complexes that are involved in docking, hence reducing the number of docking sites and later, also reducing the probability of forming SNARE complexes, thereby leading to reduced priming. On the other hand, Munc13 might allow more open syntaxin-1 to form binary SNARE complexes that serve as acceptor complex for vesicle docking and subsequently it lead to formation of more SNARE complexes, enhancing vesicle priming. Reduction of syntaxin-1 availability by Munc13 knockout or tomosyn overexpression reduces both docking and priming. These results indicate that docking and priming are molecularly interlinked (Ashery et al., 2009; Siksou et al., 2007) and represent continual steps in the protein–protein interactions involving

SNARE complexes, Munc13, and tomosyn and lead to the formation of assembled SNARE complexes and fusion.

The last protein family to be described in this section acts at the border between priming and fusion. Complexins are small SNARE-complex-binding proteins that have shown to compete with a-SNAP for binding to the SNARE complex and potentially modulate neurotransmitter release (McMahon et al., 1995; Pabst et al., 2000). Complexins are essential for fast synchronous neurotransmitter release triggered by Ca^{2+}-binding to synaptotagmin (Huntwork and Littleton, 2007; Maximov et al., 2009; Reim et al., 2001; Tang et al., 2006; Xue et al., 2007; Yang et al., 2010) (Figure 5). Initial findings using complexin-knockout mice suggest that neurons lacking complexins show a dramatic reduction in transmitter-release efficiency due to decreased Ca^{2+} sensitivity of the synaptic secretion process (Huntwork and Littleton, 2007; Maximov et al., 2009; Reim et al., 2001; Tang et al., 2006; Xue et al., 2007), as well as reduced spontaneous release (Xue et al., 2008). The interaction between the complexin N-terminus and the SNARE complex C terminus was shown to be important to the facilitating function of complexins in mouse neurons (Xue et al., 2010). Nevertheless, in addition to findings showing that complexins activate SNARE complexes for Ca^{2+}-triggered synchronous release, it is now also suggested that complexins clamp SNARE complexes for asynchronous and spontaneous release.

Yang et al. (2010) showed that complexins activate synchronous release, at least in part, by promoting vesicle priming, and that they clamp asynchronous and spontaneous release by blocking a secondary Ca^{2+} sensor (Yang et al., 2010). Complexins' dual functions, activation and clamping, are encoded by separate, autonomously acting sequences, and provide evidence that the accessory α-helix of complexions—previously shown to mediate their clamping function (Giraudo et al., 2009; Maximov et al., 2009; Xue et al., 2009)—likely blocks asynchronous and spontaneous release via insertion into nascent SNARE complexes, thereby preventing completion of SNARE complex assembly. Support for this conclusion comes from the finding that in complexin knockdown, asynchronous release is enhanced (Yang et al., 2010). Complexins function together with synaptotagmins as molecular switches in activating and clamping release, with functionally overlapping, but mechanistically distinct modes of action. Kummel et al. (2011) suggest that the assembled clamp itself might promote fusion by simply setting the stage: multiple SNARE complexes are gathered in orientations close to that required for fusion-pore formation even as their cross-linking impedes it, and they are already half-zippered (Kummel et al., 2011). This alone will allow for fast, efficient fusion as soon as the clamp is released upon stimulus. According to the complexin clamp hyphotesis, as soon as calcium binds to syntaptotagmin, syntaptotagmin displaces complexin from the SNARE complex, releasing complexin's clamping inhibition and triggers fusion (Figure 5). Nevertheless, in light of the positive role suggested in some studies and the clamping role suggested in other studies, further studies will be needed to elucidate the exact function of complexin.

2.2.6. Vesicle Fusion

2.2.6.1. Overview

The final zippering of the SNARE complex plays an important role in vesicle fusion. The four-helix bundle is formed primarily by hydrophobic interactions. However, it contains a conserved central hydrophilic layer of interacting amino acids composed of three glutamine residues and one arginine residue (Fasshauer et al., 1998). Accordingly, it was suggested that SNARE motifs can be classified into four groups that are referred to as Qa-, Qb-, Qc-, and R-SNAREs based on this central layer residue (Bock et al., 2001). All in vivo SNARE complexes are composed of one of each of the four SNARE groups (Fasshauer et al., 1998; Yang et al., 1999). In most cases, an R-SNARE is localized to a transport vesicle that fuses with an acceptor membrane containing all three Q-SNAREs; hence the R-SNARE is also called a v-SNARE (for "vesicular SNARE"), and the three Q-SNAREs are referred to as t-SNAREs (for "target-membrane SNARE").

Zipper-like assembly of the SNARE motifs from their N-terminal ends toward their membrane-proximal C termini results in the formation of a highly stable heterotrimeric "trans-SNARE complex" (also called "SNAREpin"), consisting of four parallel a-helices, which bring the two membranes into close apposition for fusion (Hanson et al., 1997; Hu et al., 2003; Lin and Scheller, 1997; Sinha et al., 2011; Weber et al., 1998). Full zippering of trans-SNARE complexes likely produces fusion-pore opening. It is suggested that nucleation and zippering of the highly stable SNARE core complex generate the energy required to bring opposing membranes to a state of proximity that initiates the fusion event (Lam et al., 2008); hence SNARE complex formation may promote membrane fusion by simple mechanical force (Sudhof and Rizo, 2011). A recent finding suggests that the full zippering only stresses the membranes and that fusion-pore opening then occurs, mediated by Munc18 (Jorgacevski and Zorec, 2012). After fusion-pore opening, the two membranes merge completely, and trans-SNARE complexes are converted into cis-SNARE complexes (SNARE complexes on a single membrane), which are dissociated into monomers by an α/βSNAP–NSF complex (Sudhof and Rizo, 2011), and synaptic vesicle membrane and protein components are recovered by endocytosis (Burgalossi et al., 2010).

t-SNAREs, syntaxin-1, and SNAP-25 are typically arranged in clusters on the membrane (Bar-On et al., 2012; Lang et al., 2001; Ohara-Imaizumi et al., 2004, reviewed in Mohrmann and Sorensen (2012)). The assembly of the two PM-localized t-SNAREs into a heterodimer, which can serve as a docking/priming acceptor complex, is a prerequisite for fusion. It is hypothesized that docking and fusion of vesicles occur with larger probability at or near clusters in neurosecretory cells (Lang et al., 2001; Ohara-Imaizumi et al., 2004, reviewed in Mohrmann and Sorensen (2012)). Accordingly, it has been proposed that dimer formation occurs at the edge of syntaxin-1 clusters, where syntaxin-1 adopts a half-opened form (Bar-On et al., 2012; Sieber et al., 2007). These syntaxin-1 clusters can therefore serve to enhance the chance for productive SNARE complex formation, accelerate docking and priming or serve as a reservoir of SNAREs to be used during and following periods of intense vesicle fusion (Mohrmann and Sorensen, 2012). However, more direct evidence is necessary to support this hypothesis.

The precise number of SNARE complexes minimally required to drive membrane fusion is highly debated—current estimates range between 1 and 15 (reviewed in Sinha et al. (2011)). Montecucco et al. (2005) suggested that different systems may require different numbers of SNARE complexes for fusion. They suggested that different secretory organelles (synaptic vesicles and LDCVs) might have different requirements in terms of size and rate of opening of the membrane-fusion complex (Montecucco et al., 2005). High-order multimers of SNARE complex could be required in tightly regulated fast exocytosis from synaptic vesicles, whereas lower order multimers might be sufficient for the slower exocytosis of dense-core granules of chromaffin and PC12 cells (Karatekin et al., 2010; Montecucco et al., 2005). In an in vitro system, van den Bogaart et al. (2010) created liposomes containing only a single copy of v- and t-SNARE proteins and showed that a single SNARE complex is sufficient for membrane fusion of small artificial vesicles (van den Bogaart et al., 2010). However by using in vitro assays, it is impossible to completely mimic the reaction conditions of the cell. According to studies in PC12 cells and in neurons, the number of SNARE complexes catalyzing membrane fusion was estimated to range from 3 to 15 (Han et al., 2004; Hua and Scheller, 2001; Keller et al., 2004, reviewed in van den Bogaart and Jahn (2011)).

Mohrmann et al. (2010) employed an expression system in chromaffin cells that resulted in different levels of active versus inactive SNAP-25 per cell. To determine these ratios, they fused both proteins to spectrally separate fluorescent proteins, expressed them at different ratios using various internal ribosome entry site sequences and examined them by fluorescence microscopy. Subsequently, granule release from each cell was characterized by electrophysiology, allowing high-resolution measurements of the exocytic process (Haucke et al., 2011; Mohrmann et al., 2010). They were able to accurately determine the effect on exocytosis while reducing the number of available active SNAREs. For fast synchronous release, they obtained a third-power dependency on the number of SNAREs, whereas slower overall release showed a linear dependency. This indicated that three SNARE complexes are required for fast evoked release, whereas a single complex is sufficient for the slower phase of exocytic release (Mohrmann et al., 2010). This is in agreement with Domanska et al. (2010), who suggested the requirement for approximately three SNARE complexes for efficient fusion.

Another in vivo experiment studied neurotransmitter release from neurons of synaptobrevin-2/cellubrevin double-knockout mice. Using rescue experiments with Synapto-pHluorin, the authors concluded that at least two synaptobrevin-2 molecules are required for fusion (Sinha et al., 2011). It is worth mentioning that as the number of synaptobrevins on the synaptic vesicle is estimated to be 70 (Takamori et al., 2006) and t-SNAREs are abundant on the PM (Bar-On et al., 2008; Lang et al., 2002), it is possible that in vivo, more than the minimal two to three SNARE complexes are formed, accelerating priming and fusion. Hence the experimental data available to date might be underestimating this value. Taken together, these findings suggest that the number of SNARE complexes required for fusion might depend on the biological requirements in the same tissue and moreover, would explain why the SNARE concentration varies significantly among different trafficking organelles (van den Bogaart et al., 2010).

The final step of fusion is triggered by calcium ions. It has long been suggested that synaptotagmin-1 is the calcium sensor for synchronous release. Synaptotagmin-1 belongs to a large family of proteins consisting of 15 isoforms in mammals. It is anchored to synaptic vesicles at its N-terminus by a transmembrane domain and contains two cytosolic C terminal C2 domains. The C2 domains of synaptotagmin-1 are characterized by Ca^{2+}-dependent phospholipid and SNARE binding (Arac et al., 2006; Bhalla et al., 2006; Geppert et al., 1994b; Lynch et al., 2007), and the SNARE-binding site for synaptotagmin-1 overlaps with the one for complexin. Ca^{2+} binding to synaptotagmin-1 triggers the displacement of complexin (Tang et al., 2006). Thus, binding of Ca^{2+} to synaptotagmin-1 allows it to displace complexin and facilitates SNARE complex fusion of the vesicle membrane with the PM.

In addition to its role in releasing the SNARE complex from its clamps, synaptotagmin-1 plays another facilitating role in primed vesicle fusion by inserting its C2 domains into the PM to the depth of the lipid glycerol backbone, thereby changing membrane curvature (Martens et al., 2007; Martens and McMahon, 2008). It has recently been suggested that synaptotagmin-1 triggers vesicle fusion by the Ca^{2+}-dependent induction of membrane curvature (Hui et al., 2009; Martens et al., 2007, reviewed in McMahon et al. (2010)). Calcium binding by the two C2 domains results in an interaction with and insertion of C2 domain residues into the PM (Campelo et al., 2008; Chapman, 2008; Herrick et al., 2006; Hui et al., 2006). As a result of membrane insertion, the C2 domains of synaptotagmin-1 buckle the PM toward the vesicle and the lipids in the protein-free end cap of the buckled membrane are under curvature stress. This curvature stress is released during the fusion process. Both the decrease in intermembrane distance and the curvature stress dramatically increase the probability of fusion by decreasing the energy barrier needed for fusion, possibly even creating a partial mixing of the synaptic and PMs. Thus synaptotagmin-1, and by analogy other proteins with double C2 domains, play a central role during membrane fusion, acting as fusogens that reduce the energy barrier needed for fusion alongside the SNARE complex (Martens, 2010).

2.2.6.2. Regulation of Vesicle Fusion by Lipids and Cytoskeleton

Membrane curvature is also affected by the lipid composition in the membrane. Lipids that have a small head group and relatively large hydrophobic tail form concave structures and, by convention, are denoted as having intrinsic negative curvature, whereas lipids that have a large head group and relatively small hydrophobic tail form convex structures and are denoted as having intrinsic positive curvature (Furber et al., 2009). Thus, lipid compositions that facilitate the formation of these various stalk-pore intermediates decrease the energy requirement for lipid merger and enable membrane fusion to occur within the constraints of characterized biological systems (Furber et al., 2009). For example, cholesterol, an abundant component of native membranes, is substantially enriched in secretory vesicle membranes (Churchward et al., 2005;Takamori et al., 2006, reviewed in Furber et al. (2009)). As a molecule of intrinsic negative curvature (Chen and Rand, 1997), cholesterol supports fusion by contributing to the formation of high-curvature stalk-pore intermediates. Depletion of cholesterol using the chelating reagent methyl-β-cyclodextrin (mβcd) results in dose-dependent inhibition of the rate, Ca^{2+} sensitivity, and kinetics of fusion of isolated sea urchin egg cortical vesicles (Churchward et al., 2005). However, depletion of cholesterol also affects the membrane organization and clustering of protein that are important for fusion.

Phosphatidic acid is a cone-shaped lipid that is predicted to promote negative curvature in the cytoplasmic leaflet of the PM, thereby facilitating the formation of the hemifusion intermediates required for the fusion of two membranes (Chernomordik and Kozlov, 2005). One phosphatidic acid-producing pathway is via hydrolysis of phosphatidylcholine by the enzyme phospholipase D (PLD) (Athenstaedt and Daum, 1999). This occurs on a relatively fast timescale, and is thought to underlie the dynamic regulation of phosphatidic acid that allows it to function as a signaling lipid in agonist-stimulated cell biological responses such as secretion and changes in cellular morphology. This can be also used to facilitate vesicle priming following intense stimulation. Indeed, reduced PLD activity results in reduce vesicle priming and fusion (Zeniou-Meyer et al., 2007) and this can be reversed by adding a lipid such as lysophosphatidylcholine to the outer leaflet of the PM that favors positive membrane curvature (Zeniou-Meyer et al., 2007). This suggests that phosphatidic acid triggers or facilitates hemifusion intermediates between docked granules and the PM, thereby directly promoting the formation of exocytic fusion pores.

In addition to lipids, the cytoskeleton has also been shown to affect vesicle fusion. There is substantial evidence that myosin II affects the fusion process. For instance, myosin II inhibitors affect the fast phases of exocytosis associated with the fusion of primed vesicles (Kumakura et al., 1994). Furthermore, myosin II was directly implicated in vesicle fusion when overexpression of a nonphosphorylatable form of the regulatory light chain of myosin II was found to alter single-vesicle-fusion kinetics, as studied by amperometry (Neco et al., 2004). A subsequent patch amperometry study directly implicated myosin II phosphorylation in the control of fusion-pore expansion (Neco et al., 2008). F-actin/myosin II's role in the cortical cytoskeleton in controlling the final phases of membrane fusion was demonstrated when kiss-and-run (KR) and full collapse events were distinguished by altering the frequency of electrical stimulation of mouse chromaffin cells (Doreian et al., 2008). Furthermore, it was found that disruption of the F-actin cortex by myosin II phosphorylation favors full collapse of chromaffin granules, while stabilization of F-actin structures increases KR events. Hence, the fusion process itself is affected by synaptic proteins, phospholipids, and cytoskeletal proteins that can control the site, the mode, and the kinetics of vesicle fusion.

In this respect, a new model suggest that following arrival to the PM, the vesicles undergo a process called "positional priming" that might involve better coupling of the vesicle and its fusion machinery to Ca^{2+} channels (Cohen and Atlas, 2004; Neher and Sakaba, 2008), thereby converting a slowly releasing vesicle into a rapidly releasing one (Neher and Sakaba, 2008). This might involve some movement and rearrangement of the vesicle's position near the release site and would explain vesicle movement shown to occur just before fusion (Degtyar et al., 2007). A connection between vesicle position's in the fusion site and its mode of fusion was suggested by Park et al. (2012). In that study, determination of the time, position, and mode of fusion in living hippocampal nerve terminals revealed that vesicles starting close to their ultimate fusion sites tend to fuse earlier than those positioned further away. The mode of fusion depended on the prior motion of the vesicles, with long-dwelling vesicles preferring KR over full-collapse fusion. KR fusion events were concentrated near the center of the synapse, whereas full-collapse fusion events were broadly dispersed (Park et al., 2012).

2.2.7. Recycling of Synaptic Vesicles

As thoroughly explained above, synaptic activity is based on the fusion of synaptic vesicles with the PM in the AZ. This fusion results in the addition of synaptic vesicle proteins and lipids to the PM, and requires a mechanism for retrieval. The retrieval of synaptic vesicles from the PM is crucial for keeping the size of the cell/terminal constant by removing the excess membrane, refilling vesicle pools to prevent pool depletion, and local recycling of synaptic proteins that are needed for the next round of synaptic transmission.

The recycling of synaptic vesicles was first evidenced in EM sections of stimulated frog NMJs (Heuser and Reese, 1973). It was shown that following stimulation, synaptic vesicle numbers are reduced, a change compensated for by the appearance of cisternal membrane invaginations and endocytic vesicles. In recent years, the use of capacitance measurements and optical markers, such as FM dyes and Green fluorescent protein (GFP) or pHluorin-fused proteins, has enabled dynamic recordings of endocytosis and further investigation of the process. In parallel, identification of the molecules involved in endocytosis has allowed further refinement of the molecular mechanism of vesicle recycling.

Experimental evidence shows three alternative albeit not necessarily exclusive pathways for vesicle recycling that likely also depend on the type of synapse and stimulation: (1) the classic, clathrin-mediated endocytosis (CME), (2) KR endocytosis, and (3) bulk endocytosis (Figure 6). These pathways differ in their timescales and molecular mechanisms.

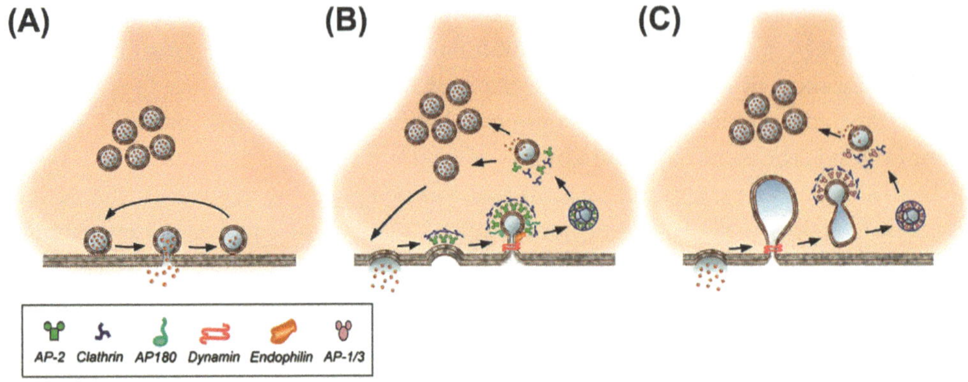

Figure 6 Pathways of synaptic vesicles recycling. (a) Kiss-and-run. Upon stimulation, a transient opening of a fusion pore occurs, enabling the release of neurotransmitter without full collapse of the vesicle into the PM. The pore is then closed, and the vesicle only requires only neurotransmitter refilling to be ready for reuse. (B) Clathrin-mediated endocytosis. Upon stimulation, the synaptic vesicle fully collapses into the plasma membrane. Following, vesicle's membrane and proteins are recaptured and internalized by the clathrin-mediated endocytosis machinery. The recovered vesicle can join the different vesicular pools. Key proteins of the process are represented. (C) Bulk endocytosis. Upon intense stimulation, vesicles collapse into the plasma membrane, and recaptured by the internalization of a large amount of membrane into the cell. This creates a vacuole from which synaptic vesicles eventually bud. Vesicles recovered by this mechanism will probably repopulate the reserve/resting pool of vesicles. Drawing is not to scale.

2.2.7.1. Clathrin-Mediated Endocytosis

CME is a well-studied mechanism involved in many important nonneuronal cell functions, such as the continuous uptake of essential nutrients (Jones et al., 1984; Pearse, 1982) and modulation of signal transduction, by both controlling the levels of surface receptors and serving as a mechanism for signal compartmentalization (Huang et al., 1995; McMahon and Boucrot, 2011; Vieira et al., 1996). In neuronal cells, it has additional roles, most notably the recycling of synaptic vesicles.

The key protein in this process is clathrin, a triskelial assembly unit built from three clathrin heavy chains and three clathrin light chains (Kirchhausen and Harrison, 1981). Triskelia with the aid of endocytic adaptors polymerize into pentagons and hexagons that cover a membrane pit that by progressive membrane deformation driven by accessory endocytic proteins matures into a coated vesicle (Fotin et al., 2004b). The vesicle detaches from the membrane, undergoes uncoating and reacidification, and further processing for reuse. This process is complex and requires a large and still growing number of accessory proteins (Schmid et al., 2006). According to the leading paradigm (mostly studied in nonneuronal cells), initiation of this process depends on FCHo proteins that together with Eps15/Eps15R and intersectins form a complex that stabilizes adaptor protein (AP-2) bound to PIP2 and to cargo molecules at the PM (Cocucci et al., 2012;

Henne et al., 2010). AP-2 and its partners recruit clathrin from the cytoplasm, and assembly of the clathrin lattice at the membrane begins.

Coat assembly is supported by additional proteins, such as AP180, CALM, and epsins, which also affect membrane curvature and deepen the pit invagination (Ford et al., 2002). The large GTPase dynamin (Faelber et al., 2011; Ford et al., 2011) is recruited to mature late-stage CCPs by SH3 domain-containing proteins, many of which also contain membrane-curvature inducing/sensing BAR domains that may also regulate dynamin self-assembly. These factors include amphiphysin, endophilin, SNX9, intersectin, and syndapin. Upon GTP hydrolysis, dynamin mediates fission of the vesicle from the PM (Praefcke and McMahon, 2004), although the precise mechanism remains a matter of debate. The released vesicle is stripped of clathrin and other accessory proteins with the involvement of the PIP2-phosphatase synaptojanin, HSC70, and auxilin (Umeda et al., 2000). The uncoated vesicle is then ready for further processing and reuse.

The involvement of clathrin in synaptic vesicle recycling has been shown by several means, from the early EM sections of Heuser and his colleagues (Heuser and Reese, 1973; Miller and Heuser, 1984), to expression of clathrin light chain fused to GFP which showed recruitment of clathrin to synapses of hippocampal neurons after stimulation (Mueller et al., 2004). Reducing clathrin levels at the synapse using small interfering RNA, or interfering with coat assembly or dynamics, blocked compensatory endocytosis in primary hippocampal neurons (Granseth et al., 2006) and at lamprey reticulospinal synapses (von Kleist et al., 2011).

The assembly of clathrin coat is dependent on several adaptor proteins. The major one, AP-2, is a four-subunit complex that has a role in clathrin recruitment and assembly (Robinson, 2004) as well as in cargo recognition. Interestingly, nematodes lacking the AP-2 subunit μ2 showed no dramatic effect on locomotion (Gu et al., 2008) despite reduction of endocytic efficiency, perhaps due to compensation by other proteins. Similarly, the use of μ2 short hairpin RNA (shRNA) in rat primary CNS neurons resulted in impaired kinetics but not a complete block of endocytosis, though some compensation by AP-1 was detected (Kim and Ryan, 2009). Other accessory proteins might play a role in selective sorting of the endocytosed cargo (i.e. synaptic vesicle proteins). Stonin2, for example, is thought to be the adaptor for synaptotagmin since it binds it directly and facilitates its internalization from the PM (Diril et al., 2006). CALM and AP180 are thought to serve as specific cargo adaptors for the SNARE protein synaptobrevin-2. In their absence, synaptobrevin-2 recycling is impaired and it accumulates on the cell surface. The interaction between synaptobrevin-2 and the adaptor proteins occurs via its SNARE motif, and therefore internalization most likely occurs after fusion and disassembly of the SNARE complex on the PM (Koo et al., 2011). Supporting this specificity, the recycling of another vesicular protein, vGLUT1, was not affected by the elimination of AP180 or CALM (Koo et al., 2011).

Endophilin has been suggested to participate in several points along the process. Its structure contains an SH3 domain that can bind dynamin and synaptojanin (Cestra et al., 1999; Ringstad et al., 1997). Microinjection of endophilin antibodies into lamprey giant axons caused stimulus-dependent depletion of synaptic vesicles and accumulation of shallow clathrin-coated pits (Ringstad et al., 1999). In flies and worms, lack of endophilin caused mislocalization of synaptojanin (Schuske et al., 2003; Verstreken et al., 2003). Therefore, endophilin has been suggested to function as a molecular scaffold. In addition, endophilin has an N-BAR domain with a role in membrane curvature induction or sensing (Farsad et al., 2001; Gallop et al., 2006). In endophilin-null mutants of *C. elegans*, synaptic vesicle endocytosis could be rescued by endophilin lacking the SH3 domain, but not by a mutant in the N-BAR domain (Bai et al., 2010), suggesting that the membrane-bending activity is most important.

Despite all these findings, a study performed on mice knocked out in all three endophilin genes suggests a predominant role in vesicle uncoating. Endophilin knockout mice died shortly after birth, and the main cellular effect was the accumulation of clathrin-coated vesicles, a phenotype similar to that of synaptojanin-1 knockout mice (Cremona et al., 1999), a PIP2-phosphatase recruited by endophilin to late-stage CCPs (Milosevic et al., 2011). The predominant phenotype of loss of endophilin may thus differ somewhat between species and/or the type of synapse studied, likely because of different degrees of functional overlap with other BAR-SH3 domain proteins. Taken together, endophilin may be involved in several steps of the endocytic process.

Dynamin is a GTPase that plays a key role in endocytosis. Its monomers assemble into helical polymers around the necks of the budding vesicles; GTP hydrolysis then causes a conformational change that promotes fission of the membrane and generates a free endocytic vesicle (Hinshaw and Schmid, 1995; Takei et al., 1995). This function is well demonstrated by the temperature-sensitive *Drosophila* mutant, shibire, which shows depletion of synaptic vesicles and accumulation of collared pits in synapses after acute stimulation at nonpermissive temperature (Chen et al., 1991; Koenig and Ikeda, 1989). Accordingly, mice lacking dynamin-1 show limited viability, poor vesicle retrieval, and a marked accumulation of clathrin-coated structures (Ferguson et al., 2007).

The main proteins involved in stripping the internalized vesicle of the clathrin lattice and accessory proteins are the ATPase HSC70 and its cofactor, auxilin (Schlossman et al., 1984; Ungewickell et al., 1995). Auxilin associates with the clathrin terminal domain of clathrin triskelia (Fotin et al., 2004a; Scheele et al., 2001) after budding of the clathrin-coated vesicle (Massol et al., 2006). Then, auxilin recruits HSC70 to a specific motif in the clathrin heavy chain below the clathrin hub (Rapoport et al., 2008; Xing et al., 2010), where the uncoating reaction is initiated. Accordingly, auxilin-knockout mice show reduced viability, and those that survive have an increased number of clathrin-coated vesicles, despite partial compensation by GAK (auxilin's ubiquitously expressed homolog) (Yim et al., 2010). Another protein involved in this step is the phosphatase synaptojanin, presumably by hydrolysis of PIP2 (Cremona et al., 1999; Verstreken et al., 2003), a key phosphoinositide required for stabilizing clathrin coats at membranes. In addition, the change in phosphoinositide composition facilitates auxilin recruitment (Massol et al., 2006).

The order of recruitment of major proteins involved in CME was demonstrated in an elaborate study utilizing TIRF microscopy (Taylor et al., 2011). Today, superresolution imaging methods, such as stochastic optical reconstruction microscopy (STORM), STED, and photoactivation localization microscopy (PALM) (Betzig et al., 2006; Klar et al., 2000; Rust et al., 2006; van de Linde et al., 2011), are becoming more common. A pioneering study of live-cell STORM following clathrin pits and cargo molecules (transferrin) achieved a spatial resolution of 20–50 nm and a temporal resolution of 0.5–2 s (Jones et al., 2011). These methods are expected to provide a better understanding of the organization of the proteins as the endocytic process progresses.

The time needed for vesicle internalization by CME in nonneuronal cells is in the tens of seconds range (10–60 s) (Miller and Heuser, 1984; Mueller et al., 2004; Ryan et al., 1996; Sankaranarayanan and Ryan, 2000). Whether this timeframe is fast enough to sustain secretion during high-frequency stimulation is still debated. Hence, it is thought that other forms of endocytosis coexist and are activated under certain conditions (Pyle et al., 2000; Sara et al., 2002). Whether or to which degree these pathways are clathrin-dependent is also unclear and will be discussed in more detail below. Interestingly, the existence of a pool of membrane regions with a preassembled endocytic machinery was recently observed, which would enable rapid retrieval of vesicles upon stimulation (Hua et al., 2011). This so-called readily retrievable pool of SVs surrounds AZs and may constitute a preferred substrate for activity-dependent endocytosis that equals the size of the RRP.

The process of KR describes the transient opening of a fusion pore that enables the release of neurotransmitter without full collapse of the vesicle into the PM. This flickering is very fast, ranging from several milliseconds up to 1 s, and is therefore considered the mechanism governing rapid endocytosis. This mechanism was first suggested by Ceccarelli et al. (1972, 1973) who, using EM, detected the existence of uncoated omega-shaped membranes connected to the PM by a narrow neck in the frog NMJ AZ (Ceccarelli et al., 1972, 1973).

Stevens and Williams (2000) estimated that KR accounts for 20% of the exocytic events in hippocampal neurons, based on their comparison of glutamate and dye release. Earlier works used the uptake and release of different FM dyes in hippocampal synapses (Klingauf et al., 1998) to show the existence of kinetically different populations of vesicles. One of the most compelling pieces of evidence for this process was provided by He et al. (2006) who, using cell-attached measurements on the calyx of held, detected fast, transient, and reversible capacitance changes suitable for KR in 3% of the fusion events. A more recent work followed the release of Q-dot internalized to vesicles, and suggested that KR is favored by RRP vesicles, and is more prevalent at the beginning of the stimulus (63% of events) than at the end (<10% of events) (Zhang et al., 2009a). These findings were later challenged by Granseth et al. (2009) who used synaptophysin–pHluorin dynamics and a mathematical model to show that KR is not a dominant mechanism of recycling at any stage of the stimulus.

While KR exoendocytosis in neurons has been a subject to controversy, this mechanism undoubtedly is used in neuroendocrine cells. Evidence is accumulating for the notion that different proteins are involved in, or can affect the process, but no complete mechanism has been established. Two of the leading assumptions are that (1) KR involves some of the endocytic proteins involved in CME, but without the coating process, thus enabling the faster kinetics, and (2) KR is an incomplete opening of the fusion pore, and therefore it is regulated by exocytic proteins related to pore dynamics.

Supporting the involvement of endocytic proteins, Artalejo et al. (2002) defined two types of endocytosis in chromaffin cells: slow endocytosis that is clathrin- and dynamin-2-dependent, and rapid endocytosis that is assumed to be KR (Artalejo et al., 2002). Rapid endocytosis was clathrin-independent, but dynamin-1- and calcium-dependent. Accordingly, the use of Dynasore, a reversible inhibitor of both dynamins, blocked all types of endocytosis in rat hippocampal neurons (Newton et al., 2006). Similarly, in the *Drosophila* NMJ, high-frequency stimulation caused complete synaptic depression due to the fusion of all synaptic vesicles in dynamin mutants while in endophilin mutants, a portion of the vesicles was always recycled, allowing sustained neurotransmitter release under high-frequency stimulation as well (Verstreken et al., 2002). This pool of vesicles might represent those vesicles that undergo KR and whose recycling is dependent on dynamin and not on coat or curvature proteins such as endophilin.

In support of incomplete opening of the fusion pore, Archer et al. (2002) demonstrated that overexpression of complexin-2 causes a change in the release kinetics in chromaffin cells that is consistent with KR (Archer et al., 2002). This effect was abolished when a mutated version of complexin, incapable of SNARE binding, was overexpressed. Therefore, the authors suggested that the interaction of complexin-2 with the assembled SNARE complex regulates the fast closure of the fusion pore required for KR. In PC12 cells, different synaptotagmins were shown to play a role in the different endocytic pathways. Overexpression of synaptotagmin-1 promoted full fusion, whereas overexpression of synaptotagmin-4 promoted KR events of norepinephrine-containing vesicles, as measured by amperometry. Each pathway was dependent on a different domain in the protein, and required different calcium levels (Wang et al., 2003). Capacitance measurements did not detect a difference in the prevalence of KR or full fusion in synaptotagmin-4-overexpressing cells, but showed that the kinetics of the pore changed for LDCVs (but not for small vesicles) (Zhang et al., 2010). Furthermore, another group showed that synaptotagmin-1 and synaptotagmin-9 function redundantly in exocytosis in PC12 cells, and that downregulation of both results in an increase in KR events, raising the possibility that synaptotagmins play a role in pore dilation and can regulate the balance between the endocytic pathways (Zhu et al., 2007).

The cytoskeleton also participates in the process, through stabilization of the fusion pore and prevention of its dilation. It was shown that under low-frequency stimulation (0.5 Hz), the F-actin cell cortex is intact and KR fusion events occur. However, under increased stimulation (15 Hz), the actin cortex was disrupted and full fusion events occurred (Doreian et al., 2008). Moreover, treatment with cytochalasin D, which disrupts the F-actin cortex, resulted in a shift from KR events to full-collapse exocytosis under low-frequency stimulation, whereas stabilizing F-actin caused KR events under 15 Hz stimulation. In addition, these authors demonstrated that inhibition of myosin II or myosin light-chain kinase under elevated stimulation frequencies inhibits fusion-pore dilation and maintains the vesicles in KR exocytosis (Doreian et al., 2008).

The balance between full fusion events and KR was shown to change following treatment with staurosporine (a kinase inhibitor) in favor of KR, suggesting that protein phosphorylation state can regulate this balance, but the target in this process was not found (Henkel et al., 2001). Similarly, use of the SRC tyrosine kinase inhibitor PP1 on rat brain synaptosomes enhanced release in the KR pathway (Baldwin et al., 2006), again with no specific protein being identified as the regulator.

2.2.7.2. Bulk Endocytosis

Bulk endocytosis refers to the internalization of a large amount of membrane into the cell and the formation of a vacuole from which synaptic vesicles eventually bud. This process has been demonstrated in central nerve terminals using EM (Takei et al., 1996), FM dyes (Richards et al., 2000), and dextrans and HRP labeling (Clayton et al., 2008).

One of the key proteins responsible for bulk endocytosis seems to be dynamin-1. This protein goes through a cycle of phosphorylations and dephosphorylations that are essential for bulk endocytosis. In its proline-rich domain, two serine residues are phosphorylated by the kinases GSK3 and Cdk5 (Clayton et al., 2010). This phosphorylation is required for bulk endocytosis but not for CME (Clayton et al., 2010). These residues are dephosphorylated by the protein phosphatase calcineurin. During prolonged depolarization of central nerve terminals, calcineurin dephosphorylates seven substrates, called the dephosphins (reviewed in Cousin and Robinson (2001)). These proteins all have an essential role in endocytosis: the GTPase dynamin-1, the scaffold proteins amphiphysins 1 and 2, the lipid phosphatase synaptojanin, and the adapter proteins epsin, Eps15 and AP180.

The strongest evidence that calcineurin triggers bulk endocytosis and not other types of endocytosis came from a study showing that dephosphorylation of dynamin-1 in nerve terminals only occurs above the activity threshold that triggers bulk endocytosis (Clayton et al., 2009). This dephosphorylation is of a specific splice variant, dynamin-1xb, and it stimulates a specific dynamin-1–syndapin-1 interaction (Xue et al., 2011). Blocking this interaction or knockdown of syndapin-1 partially inhibits bulk endocytosis during strong stimulation, but does not affect CME (Clayton et al., 2009). Syndapin-1 itself has been shown to be important for bulk endocytosis. Microinjection of syndapin antibodies into lamprey reticulospinal synapse inhibited vesicle recycling evoked by intense (5 Hz or more), but not by mild (0.2 Hz) stimulation. In addition, it resulted in massive accumulation of membranous cisternae and invaginations, which contained vesicle membrane, around release sites (Andersson et al., 2008).

Glyvuk et al. (2010) found accumulation of large endosomal intermediates in >σ-adaptin-knockout mice. These intermediates might have originated from bulk endocytosis and therefore suggest a role for AP-1 complex in the recycling of synaptic vesicles through bulk endocytosis (Glyvuk et al., 2010). Recently it was demonstrated that both AP-1 and AP-3 are essential for bulk endocytosis. However, their role is not at the PM, but at the later stages of vesicle budding from the endosomes (Cheung and Cousin, 2012).

The model emerging from these studies suggests that bulk endocytosis involves the dynamin-mediated fission of PM-derived BAR-domain protein-coated vacuoles from which SVs reform by AP-1- and/or AP-3-dependent budding.

A recent study suggests that vesicles that originate from each recycling pathway are destined to specific vesicle pools (Cheung et al., 2010). Using dyes with different kinetics, the authors could label vesicles under different stimulation protocols and follow their fate. They concluded that vesicles originated from CME refill the RRP, while vesicles originated from bulk endocytosis join the RP (Cheung et al., 2010).

A key unsolved problem remains the question as to how a tight coupling between exocytosis and endocytosis at the synapse is achieved. For example, it is unknown how the numbers of synaptic vesicles undergoing fusion is matched to those retrieved by endocytosis. Moreover, it appears likely that structural scaffolds exist that spatially organize exoendocytic proteins at or around release sites. Several synaptic vesicle proteins have been proposed to regulate the kinetics of synaptic vesicle membrane retrieval (i.e. synaptophysin (Gordon et al., 2011; Kwon and Chapman, 2011), SV2 (Yao et al., 2010), and synaptobrevin-2 (Deak et al., 2004)) and by extension participate in exoendocytic coupling. One of them is synaptotagmin-1, a major calcium sensor for exocytosis (see above) that binds the SNARE complex and phospholipids (Chapman, 2008; Koh and Bellen, 2003). In addition, it can associate with the endocytic adaptor proteins AP-2 and stonin2 (Haucke et al., 2000; Walther et al., 2004; Zhang et al., 1994). As with other exocytic proteins, interfering with its function results in defective exocytosis, making effects on endocytosis difficult to distinguish.

Yao et al. (2012) have recently demonstrated a role for synaptotagmin-1 in endocytosis, independent of its role in exocytosis (Yao et al., 2012). They showed that either of the C2 domains of synaptotagmin-1, C2A or C2B, is able to function as a Ca^{2+} sensor for endocytosis. These data are consistent with genetic data in *C. elegans* (Jorgensen et al., 1995) and with acute photoinactivation of synaptotagmin-1 at *Drosophila* NMJs (Poskanzer et al., 2003) that have revealed an important role of this protein in the endocytic retrieval of synaptic vesicle membranes. Thus, synaptotagmin-1 functions as a dual Ca^{2+} sensor for both endo- and exocytosis, potentially coupling these two components of the vesicle cycle (Yao et al., 2012). Synaptotagmins may also affect the balance between different endocytic pathways (elaborated in the KR section).

Another major exocytic protein that has been suggested to function in endocytosis is the SNARE protein synaptobrevin-2. Neurons from synaptobrevin-2 knockout mice show delayed pool replenishment after hypertonic-sucrose-induced RRP depletion. Therefore, the authors suggested that synaptobrevin-2 is essential for two fast synapse-specific processes: fast exocytosis for neurotransmitter release and fast endocytosis which mediates rapid reuse of synaptic vesicles (Deak et al., 2004). Consistent with a role in exoendocytic coupling, Hosoi et al. (2009) observed a pronounced short-term depression of neurotransmitter release following microinjection of a synaptobrevin-derived proline-rich peptide into the calyx of Held, indicative of a possible defect in clearance of release sites (Hosoi et al., 2009). Possible mechanisms of exoendocytic coupling have been reviewed recently by Haucke et al. (2011).

2.3. A GLIMPSE INTO THE NANOMETRIC MOLECULAR ORGANIZATION OF THE SYNAPSE

2.3.1. Overview

Resolving the nanoscale organization and distribution of proteins in synapses holds the key to understanding their function, characterizing plasticity, and identifying synaptic malfunction. Our knowledge of protein organization and function in the synapse comes from a combination of physiological, imaging, biochemical, and genetic, structural, and morphological experiments. However, the exact molecular composition, as well as the relative distribution of the proteins on the nanoscale level, is still missing. Imaging synaptic processes was limited up until 2010 by the size of the synaptic entities (vesicles, proteins, AZ, etc.), which is well below the diffraction limit of conventional light/fluorescence microscopy (Sigrist and Sabatini, 2012). To observe those elements, EM was required. Even though the resolution of EM and of EM tomography (Harlow et al., 2001) is extremely high, it is still difficult to precisely define the molecular composition of the synapse using this method due to difficulties with the labeling procedures. Moreover, EM is limited as it provides only static, rather than dynamic aspects of this organization (McEvoy et al., 2010).

2.3.2. Superresolution Imaging as an Instrumental Tool for Exploring Synaptic Protein Architecture

From 2001 to 2010, the diffraction barrier, limiting the resolution of conventional light microscopy to ~200 nm or more, has been overcome by the development of several optical superresolution methods that can achieve resolutions of up to 10 nm (Galbraith and Galbraith, 2011). Nanoscopic techniques such as STED (Hell, 2007), stochastic optical reconstruction microscopy (STORM) (Zhuang, 2009), direct STORM ($_d$storm) (van de Linde et al., 2011), PALM (Betzig et al., 2006), and structured illumination microscopy (Gustafsson, 2000) are based on breaching the diffraction limit by temporally or spatially modulating the excitation or activation light (Galbraith and Galbraith, 2011).

In STED microscopy, for example, improved resolution is achieved by optically reducing the diameter of the point spread function (PSF) using an aligned pair of scanning beams: one for excitation and the second for high-intensity doughnut-shaped depletion laser beam. The smaller PSF is then scanned over the sample in a confocal manner (Hell, 2007). On the other hand, in single-molecule localization-based approaches such as PALM and STORM/$_d$storm, a small subset of fluorescent molecules are randomly switched on and imaged. Precise position determination (localization) of individual fluorophores is achieved by approximating their fluorescence-emission pattern with a two-dimensional Gaussian function. These pointillistic approaches make use of standard fluorophores (e.g. Alexa Fluor and ATTO dyes) and modified photoactivatable fluorescent probes (Bates et al., 2007).

2.3.3. Resolving Synaptic Architecture and Dynamics

STED has been successfully applied from 2001 to 2010 to characterize the organization of several proteins at the synaptic AZ membranes. Kittel et al. (2006) showed that the *Drosophila* ELKS-family member BRP adopts an elongated conformation of around 100 nm, with the N-terminus superimposed on Ca^{2+}-channel clusters centered in the AZs. Another STED-based study showed that synaptotagmin-1, a vesicle-membrane protein, remains clustered on the presynaptic membrane even after intensive stimulation, suggesting that this vesicle constituent remains closely packed until a certain period after exocytosis when it is endocytosed (Willig et al., 2006). This study does not coincide with previous studies that showed that synaptic vesicles exchange their protein contents with PM proteins (Dittman and Kaplan, 2006; Fernandez-Alfonso et al., 2006; Sankaranarayanan and Ryan, 2000; Wienisch and Klingauf, 2006, but see Ceridono et al. (2011)). STED has also been shown to be a powerful approach for tracking fast dynamics; it was used to noninvasively image activity-dependent morphological plasticity of dendritic spines and allowed more accurate quantification of morphological parameters such as the neck, width, and curvature of the spines' heads (Nagerl et al., 2008). Synaptic vesicle mobility was monitored by video-rate STED at a speed of 28 frame/s, showing that vesicles within boutons tended to be much less mobile than those outside the boutons (Westphal et al., 2008). These data are consistent with previous studies in which vesicles have been shown to continuously transit through the axons (Lemke and Klingauf, 2005).

Another study (Hua et al., 2011) concentrated on monitoring and characterizing the so-called "readily retrievable pool," which is hypothesized to support fast CME. Using dual color isotropic STED (Schmidt et al., 2008), they revealed the three-dimensional spatial organization of synaptotagmin-1 surface pool that organizes in doughnut-shaped patches around the AZ and the PSD complexes that represent preassembled readily retrievable pool.

SNARE proteins are known to be nonhomogeneously distributed on the PM, forming nanoclusters containing dozens of protein copies (Lang, 2007; Lang et al., 2001, 2002; Murray and Tamm, 2009; Rickman et al., 2010; Sieber et al., 2006). The most characterized SNARE member in this respect is syntaxin-1. An STED-based study investigated the distribution of syntaxin-1 on the membrane and showed its partitioning into 30- to 50-nm diameter clusters. Sieber et al. (2007) suggested the existence of two pools of syntaxin-1: clustered and single freely diffusing molecules (Sieber et al., 2007). In a recently published $_d$storm-based study, it was possible to distinguish the single-molecule pool of syntaxin-1 from the clustered pool and characterize an internal density gradient in the clusters that reflected the inner organization of the proteins within them. SNAP-25 clusters were characterized as well and were shown to be significantly larger than syntaxin-1 clusters. Moreover, they differed in their shape and density and formed more contacts with each other (Bar-On et al., 2012).

Several STORM-based studies have been published in the past few years focusing on synaptic protein organization. Multicolor three-dimensional STORM was recently applied to accurately trace neuronal connectivity in cultured hippocampal neurons using "brainbow" labeling (Lakadamyali et al., 2012). This approach is expected to be applicable in the near future to brain tissues. An impressive systematic multicolor three-dimensional STORM study characterized the molecular architecture of synapses in brain tissues at nanometer-scale precision (Dani et al., 2010). The distribution of 10 proteins located at the presynaptic AZ or at the postsynaptic density was characterized and changes in synapse morphology and distribution were shown in different synapses and in different brain regions (Dani et al., 2010). On the presynaptic side, bassoon and piccolo were shown to be organized in an extended and oriented manner in the AZ by labeling certain epitopes located in different regions of these proteins. A map consisting of the axial positions of a large set of synaptic proteins was assembled and the mean distance between the proteins was determined, reflecting the overall crowdedness and symmetry of pre- and postsynaptic regions. Bassoon and piccolo N-terminals were found to be located 80–90 nm from the synaptic cleft, while RIM1 was found to be much closer to the presynaptic membrane, consistent with its role in vesicle priming (Dani et al., 2010).

Superresolution studies over 2010 have contributed significantly to research into synaptic organization and dynamics and they are expected to become common practice once the technical limitations are overcome. Advances in the field are expected to be achieved by combining and correlating superresolution electron and light microscopy, the latter potentially allowing direct identification of protein localizations. Another future challenge is to use the pointillistic approaches for quantitative analysis of proteins in a reliable manner, allowing exploration of multiheteroprotein complexes such as the

AZ and homoprotein clusters such as SNARE clusters. The extension of the method to living cells and complex thick brain tissues will definitely offer a wide range of possibilities for investigation in the field.

2.4. SYNCHRONOUS, A-SYNCHRONOUS AND SPONTANEOUS RELEASE: PHYSIOLOGICAL AND MOLECULAR PERSPECTIVES

2.4.1. Overview

Following the arrival of an action potential to the nerve terminal, the terminal PM is depolarized, causing voltage-gated Ca^{2+} channels to open (Barnett and Larkman, 2007; Catterall, 2011; Mahapatra et al., 2012). Ca^{2+} is then driven into the nerve terminal by the chemical-electrical force, generating a transient concentration gradient, which decreases with increasing distance from the channel and then diffuses throughout the cytosol. The transient Ca^{2+} gradients can reach concentrations of tens of micromoles within microseconds in the AZ area and trigger the fusion of primed vesicles with high Ca^{2+} cooperativity (reviewed by Neher and Sakaba (2008)).

Several kinetic and molecular models for calcium-dependent synaptic transmission have been suggested over the past two decades to explain the various synaptic release types. Some assign them to different releasable vesicle pools, each with its own synaptic machinery (multiple releasable pool models; Fredj and Burrone, 2009; Mezer et al., 2004; Sara et al., 2005; Zikich et al., 2008), whereas others describe multiple synaptic proteins competing for the release of the same releasable pool (single releasable pool model; Lou et al., 2005; Sun et al., 2007). We present several models, which relate experimental data on synaptic proteins to the various types of synaptic release.

There are three kinetically distinct types of synaptic release, defined by their spatiotemporal association with the action potential: evoked or synchronous release, asynchronous release, and spontaneous release. These types rely on the high dynamic range of Ca^{2+} concentration under physiological conditions and possibly on complementary sets of synaptic proteins. The proportion between them determines, to a large extent, the release properties of various types of neurons and presynaptic terminals (Hagler and Goda, 2001; Neher and Sakaba, 2008).

Evoked synaptic activity induces synchronized (phasic) release of vesicles from the presynaptic terminal. This synchronous synaptic release is time-locked to the presynaptic action-potential (delay of <2 ms) (Hagler and Goda, 2001; Kochubey et al., 2011; Schneggenburger and Neher, 2000) and driven by short-lived increases in Ca^{2+} concentration in the submicrometer vicinity of the Ca^{2+} channel. This high-$[Ca^{2+}]$ environment is associated with an area of ~0.1 μm termed the "release AZ" (Schneggenburger and Neher, 2005) where a high density of Ca^{2+} channels and docked vesicles exist (Schneggenburger et al., 2012). Hence, synchronous release is driven by the spatial proximity between the Ca^{2+} entry site and the readily releasable vesicles. This spatial association synchronizes the time of the action potential and the time of vesicle fusion (Schneggenburger and Neher, 2005; Stanley, 1997).

The second type of release follows the synchronous release and thus begins >2 ms after an action potential (Kochubey et al., 2011). This asynchronous (tonic) release phase is driven by the low concentration of residual calcium (up to ~1 μM [Ca^{2+}] outside the AZ domain), which accumulates in the presynaptic terminal (Hagler and Goda, 2001). The vesicle release rate in this phase is two orders of magnitude slower than that of the synchronized release (Hagler and Goda, 2001). Furthermore, this phase shows lower calcium cooperativity than synchronous release (Kochubey et al., 2011; Sun et al., 2007, but see Goda and Stevens (1994)) and lasts up to hundreds of milliseconds after an action potential; hence asynchronous release can therefore accumulate throughout neuronal activity (Kochubey et al., 2011).

The third type of presynaptic release is spontaneous release, which is not time-locked to an action potential and can occur without association to previous neuronal activity. It can be readily recorded in the presence of tetrodotoxin which blocks action potential-induced presynaptic membrane depolarization and thus avoids the intervention of ongoing activity. Therefore, it is assumed to occur under baseline Ca^{2+} levels (30–50 nM) and shows no cooperativity. However, spontaneous release rate is facilitated by increased Ca^{2+} concentrations (Kochubey et al., 2011; Lou et al., 2005; Neher and Sakaba, 2008).

It should be noted that synaptic release is a combination of resource availability (vesicles and neurotransmitters) and release-driving forces (synaptic proteins and Ca^{2+}). The origin of synchronous release is the RRP, but the source and physiological relevance of asynchronous and spontaneous release are still under debate (Neher and Sakaba, 2008; Otsu and Murphy, 2003; Pare et al., 1997; Ramirez and Kavalali, 2011; Sun et al., 2007). It is possible that they originate from the same vesicle pool (Pang et al., 2006; Sun et al., 2007) or from different pools of primed vesicles (Sara et al., 2005). Interestingly, upon uniform increases in Ca^{2+} concentration (with calcium-caged compounds) in the presynaptic terminal, it is still possible to observe the slow (asynchronous) release component (Sun et al., 2007). This means that asynchronous release does not merely stem from the distance between the Ca^{2+} channel and the vesicles but involves other molecular mechanisms that are discussed below.

2.4.2. Kinetic Models for Synaptic Transmission

Over the years, several kinetic models have been developed to describe and quantify the process of synaptic transmission. Here, we present three of these models: two models present presynaptic proteins which are competing for the same vesicles while in another model, simultaneous release can occur from several vesicular pools. Furthermore, two models use a set of kinetic equations to describe the binding of Ca^{2+} to a calcium sensor and uncover the connection between intracellular Ca^{2+} concentration and vesicle release rate. The other model simulates a serial set of kinetic binding equations between synaptic proteins by taking into account the intracellular Ca^{2+} concentration and the concentrations of key synaptic proteins.

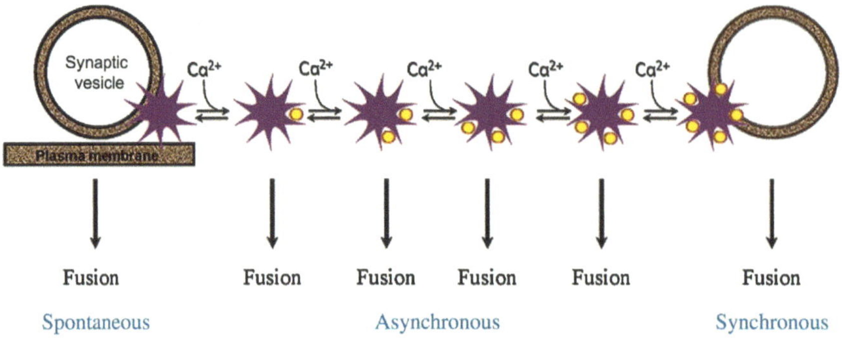

Figure 7 The allosteric model represents Ca^{2+}-dependent release over a wide range of intracellular calcium. Fusion can occur from all states including the partially bound states. The model allows low rates of vesicle fusion at the Ca^{2+}-free state (representing spontaneous release) and each Ca^{2+} ion increases the fusion probability of the vesicles (asynchronous release) reaching fast fusion when all five sites are occupied (synchronous release). Calcium ions represented by yellow circles.

The first model is the five-site model, which was suggested by Schneggenburger and Neher (2000) and was later extended to the allosteric model (Lou et al., 2005; Schneggenburger and Neher, 2000) (Figure 7). The five-site model is based on the dose–response curve of neurotransmitter release versus Ca^{2+} concentration from caged-calcium experiments in the calyx of held synapses. In these experiments, Ca^{2+} levels are manipulated by a combination of calcium buffers and calcium-caged compounds with various affinities. Ca^{2+} concentration is elevated by a combination of calcium-caged compounds and various UV flash intensities and Ca^{2+} levels are recorded with a range of fluorescent calcium indicators, which match the affinity of the measured intracellular Ca^{2+} levels. This model best simulates the high calcium cooperativity in fast synaptic release and hypothesizes that five Ca^{2+} ions are required to achieve this release. Its extension, the allosteric model, better accounts for slow synaptic release resulting from variations in the submicromolar $[Ca^{2+}]$ range (see also Wolfel et al. (2007)). By fitting the experimental data, both models predict a cooperativity of 4.2 for synchronous release. While the five-site model requires that all five ions bind to a putative calcium sensor to trigger release, the allosteric model allows synaptic release to occur when less than five Ca^{2+} ions are bound. It was hypothesized that increasing the number of bound Ca^{2+} ions drives the system from spontaneous release (no bound Ca^{2+} ions) to asynchronous release (one to four bound Ca^{2+} ions) and synchronous release (all five sites are occupied).

It is interesting to note that although a single synaptotagmin protein can bind five Ca^{2+} ions (via its two C2 domains (Kochubey et al., 2011)) and plays a key role in calcium-dependent release, this model does not attempt to describe synaptotagmin-1 or synaptotagmin-2 as the putative calcium sensor but rather describes the required calcium cooperativity to drive the various types of synaptic release. Nevertheless, it is assumed that synaptotagmin-1/2 can play this role. In addition, the allosteric model explains the PKC/Munc13 potentiation of release as an increase in the fusion-willingness of the release mechanism (see also Basu et al. (2007)). It demonstrates, both experimentally and computationally, how phorbol-ester potentiation of PKC/Munc13 activity by PDBU shifts the calcium-dependent release curve to the left (making it more sensitive) and thus the same release rate is achieved at lower Ca^{2+} levels (Lou et al., 2005). This simplified model refers to most experimental data in the framework of increasing or decreasing the energy barrier for synaptic release. Its simplistic nature is appealing but it does not identify the proteins that mediate all types of synaptic release. Rather, it groups all the interactions between the various synaptic proteins under the vague term of energy barrier for synaptic release and the various manipulations simply shift the calcium-dependent release rate curve.

The second model is the two-Ca^{2+} sensor model, suggested by Goda and Stevens (1994) after observing two components of synaptic release in their recordings. In that study, the ratio of Ca^{2+} to Sr^{2+} in the extracellular medium was changed and a change in the ratio of immediate (synchronous) to delayed (asynchronous) release was measured (Goda and Stevens, 1994). Thus, the authors concluded that the sensor for synchronous release is synaptotagmin-1 and the other sensor should have higher affinity to Ca^{2+} as it should act at the low concentration of residual Ca^{2+}. This model was later extended by Sun et al. (2007) who suggested that the fast calcium sensor requires five Ca^{2+} ions to induce synchronous release and independently, the slow sensor requires two Ca^{2+} ions to induce asynchronous release (Figure 8). Hence, the model allows the combination of all Ca^{2+}-binding states for the two sensors (overall 3×6 states and a state for spontaneous release) and suggests that spontaneous release occurs when none of the sensors bind Ca^{2+} (a total of 19 combinations of states) (Sun et al., 2007). The model is based on calcium uncaging measurements over an extended range of $[Ca^{2+}]i$ (Lou et al., 2005) for wild-type and synaptotagmin-2-knockout mice (the fast calcium sensor in these cells) (Kochubey et al., 2011; Sun et al., 2007; Xu et al., 2009). The synaptotagmin-2-knockout synapses show an acute loss of synchronous release and an increase in asynchronous and spontaneous release.

This phenomenon is similar to the effect of synaptotagmin-1 knockout in hippocampal neurons (Geppert et al., 1994b). The measurements in the absence of synaptotagmin-2 revealed the activity of a secondary calcium sensor and showed that the slow sensor has lower cooperativity (~2; although a recent study reported that the asynchronous sensor has no cooperativity and that its calcium-dependent release is near-linear (Kochubey and Schneggenburger, 2011)). Since synaptotagmin-1/2 binds the SNARE complex, it was suggested that the fast sensor clamps the SNAREs and competes with the activity of the slow calcium sensor (Sun et al., 2007). Therefore, one interpretation of the model (together with (Kochubey and Schneggenburger, 2011)) is that the increased asynchronous release experiments in the absence of synaptotagmin-1/2 (Maximov et al.,

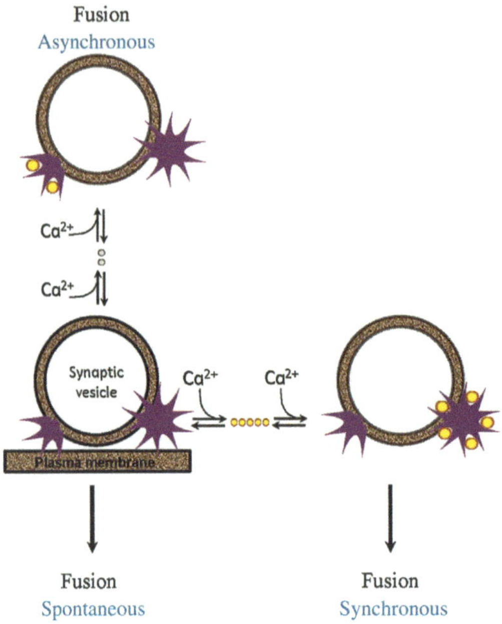

Figure 8 The two-calcium sensor model represents Ca^{2+}-dependent release from two independent calcium sensors. Release can occur only in the fully bound state of the fast sensor (synchronous) or the slow sensor (asynchronous). The two sensors compete for the binding of Ca^{2+} ion to induce asynchronous release at low $[Ca^{2+}]i$ by the slow sensor and synchronous release at high $[Ca^{2+}]i$ by the fast sensor. Calcium ions are represented by yellow circles.

2009; Yang et al., 2010) comes from unclamping of the slow sensor. As complexin-1 may work upstream of synaptotagmin-1/2, its absence also produces the same increase in asynchronous release (Maximov et al., 2009).

This model provides a mechanistic approach to explain the experimental results described above and the synaptotagmin-1/2 knockout mouse data unravel the specific calcium sensors for fast synchronous release. However, it has some pitfalls. First, it does not identify the slow calcium sensor (DOC2 proteins are the leading candidates suggested in Yao et al. (2011) and will be discussed below). Furthermore, it provides a complicated picture in which the SNARE complex is created, clamped by complexin-1 and synaptotagmin-1/2, and then requires unclamping to induce fast release. Moreover, the activity of the second slow sensor is detected only in the absence of the fast sensor; this raises doubt as to the relevance of this sensor under physiological conditions in the presence of the fast sensor (see also Kochubey and Schneggenburger (2011)) or suggests that the release kinetics will be determined from the ratio between the two sensors. It is also important to note that the allosteric model can explain the same experimental results by an increase or decrease in the energy barrier for release; in that sense, complexin-1 acts as an adaptor which decreases the energy barrier for synchronous release by synaptotagmin-1/2 (Neher, 2010). Finally, the model allows many degrees of freedom (up to 19 possible states); the question whether all of them exist in each synapse and what are the limiting factors remains open.

It is important to note that the Allosteric model and two-Ca^{2+} sensor model are not mutually exclusive. While the first takes a more general approach, the second takes a more mechanistic one. It is possible that for some types of neurons, one of the models would be able to describe several types of synaptic release, but give an incomplete picture of other types. Furthermore, these two models (Kochubey et al., 2011; Sun et al., 2007) suggest the existence of a calcium sensor for slow synaptic release. To date, the leading candidates for this role have been members of the DOC2 protein family, which are important for the regulation of asynchronous release and whose absence decreases the frequency of spontaneous release (Groffen et al., 2010; Yao et al., 2011).

The third model was suggested by Mezer et al. (2004), and was later extended by Zikich et al. (2008). It describes exocytosis as a sequence of differential rate equations. The model is based on a sequential set of reversible binary equations expressed by chemical differential rate equations that describe synaptic protein interactions (Mezer et al., 2004; Zikich et al., 2008). It takes into account the assembly of the SNARE complex, the involvement of Munc13 and calmodulin, and the Ca^{2+}-dependent priming of the vesicles, up until the Ca^{2+}-dependent release from the readily and slowly releasable vesicle pools in chromaffin cells (RRP and SRP, respectively). This model successfully recreates the secretion profiles measured from wild-type, synaptotagmin-1-knockout or Munc13-1-overexpressing chromaffin cells. It predicts that the two types of primed vesicle pools, SRP and RRP, work independently but are interconnected; the fusion from each pool is controlled by a different isoform of synaptotagmin or another calcium sensor. Thus, it predicts a calcium sensor for fast synaptic release (synaptotagmin-1/2) and a second, yet to be identified, calcium sensor for slow synaptic release. However, this model does not directly explain how spontaneous release might take place. Furthermore, since synaptic release in chromaffin cells comes from LDCVs, it might be regulated differently than the synaptic release mechanism in the CNS, which comprises the smaller synaptic vesicles.

In general, all three models agree on the existence of a fast calcium sensor for synchronous release (presumably synaptotagmin-1 or synaptotagmin-2) and offer three options for the origin of asynchronous release (see also Maximov and Sudhof (2005)). Asynchronous release might come from vesicles, which reside on the PM and are ready for release, but reside far from Ca^{2+} channels, which delays their fusion. This hypothesis is countered by the fact that calcium-caged experiments, in which Ca^{2+} concentration is homogeneously raised across the terminal, also show delayed asynchronous release (Sun et al., 2007). Therefore, the delayed release might be explained by vesicles that are not docked to the PM and undergo translocation upon elevation in Ca^{2+} concentration (Mennerick and Matthews, 1996; Pasche et al., 2012; Yizhar and Ashery, 2008).

Alternatively, the molecular machinery of the vesicles undergoing asynchronous and synchronous release might be different, and each release mode might exploit different sets of proteins (Young and Neher, 2009). For example, it was shown that DOC2 and synaptobrevin-4 enhance asynchronous release, and therefore the ratio of synaptotagmin-1 to DOC2 as the calcium sensors or synaptobrevin-2 to synaptobrevin-4 on the synaptic vesicle would determine the degree of synchronous versus asynchronous release, respectively (Raingo et al., 2012; Yao et al., 2011). This molecular explanation can also be extended by assuming different clamping proteins for asynchronous and synchronous release. In this case, the delay in asynchronous release might come from a slower release of the clamp. It is reasonable to assume that combinations of the above scenarios exist in different synapses.

2.4.3. DOC2 Proteins and Nonsynchronous Release

As already mentioned, DOC2 proteins have been suggested to play a key role in calcium-dependent asynchronous release. The DOC2 protein family has three isoforms which are expressed in the brain and in secretory cells (Verhage et al., 1997). DOC2 proteins have two C2 domains (C2A and C2B) which bind calcium in the submicromolar range (Groffen et al., 2006) and in addition, a Munc13-binding domain. DOC2A and DOC2B translocate to the PM upon elevation of calcium concentration to the submicromolar range and promote the priming and vesicle refilling of LDCVs in chromaffin cells (Friedrich et al., 2008; Groffen et al., 2006; McMahon et al., 2010). Although several contradictory results undermine the role of DOC2 proteins in synaptic release, they have been suggested as the leading candidates for spontaneous and asynchronous release (Groffen et al., 2010; Yao et al., 2011). Hippocampal autapses—synapses between a single neuron and itself—in DOC2A- and DOC2B-double-knockout mice show reduced frequency of spontaneous miniature neurotransmitter release, which is regained upon reintroduction of DOC2B (Groffen et al., 2010).

Moreover, DOC2B's effect on spontaneous release depends on Ca^{2+} concentration, suggesting that its function depends on the activity of the neuron (Groffen et al., 2010; but see Pang et al. (2011) for alternative explanation). Furthermore, the importance of DOC2 proteins in the regulation of asynchronous release was described by Yao et al. (2011). DOC2A-knockdown in synaptotagmin-1-knockout neurons decreased the asynchronous release seen in synaptotagmin-1 knockout, an effect that was rescued by DOC2B overexpression. Finally, DOC2A-knockdown neuronal cultures showed less reverberatory activity associated with asynchronous release (Yao et al., 2011). Therefore, it was concluded that DOC2 proteins are important for the regulation of asynchronous synaptic release. Interestingly, the properties of DOC2 proteins are better revealed in the absence of synaptotagmin proteins, suggesting DOC/synaptotagmin interplay in the same reservoir of Ca^{2+} and vesicles. The above studies strongly suggest that DOC2 proteins have an important role as calcium sensors in asynchronous and spontaneous synaptic release, perhaps through competition with the calcium sensor for synchronous release, synaptotagmin-1/2. It is possible that synapses with different proportions of DOC2A/B to synaptotagmin-1/2 display different ratio of synchronous to asynchronous synaptic release.

2.4.4. Alternative Modes of Synaptic Transmission

The rapid timescale of synchronous release (a few hundred microseconds and even <100 μs in some model systems; Sabatini and Regehr, 1996) matches that of Ca^{2+}-channel opening. This led to the suggestion that the Ca^{2+} channel itself is not only the Ca^{2+} source but also a calcium sensor that is part of the exocytic machinery (Lerner et al., 2006; Marom et al., 2007; Trus et al., 2007). Indeed, the voltage-gated Ca^{2+} channels interact physically and functionally with synaptotagmin-1 and the SNARE complex membrane-anchored proteins SNAP-25 and syntaxin-1A (Cohen and Atlas, 2004; Tobi et al., 1998; Wiser et al., 1997, 1999). This notion gained further support when secretion was monitored in the presence of lanthanides such as La3+ in Ca^{2+}-free solutions. La3+ ions are similar to Ca^{2+} in terms of radius and charge and are able to bind to the pore of the Ca^{2+} channel, but they fail to flow through it; these ions were able to induce catecholamine release from chromaffin cells (Lerner et al., 2006; Marom et al., 2007; Trus et al., 2007). A mutant Ca^{2+}-impermeable channel also succeeded in inducing secretion following membrane depolarization (Hagalili et al., 2008; Hohaus et al., 2005). These findings led to the hypothesis that upon Ca^{2+} binding to the channel, the Ca^{2+} channel undergoes conformational changes, which in turn trigger release (reviewed by Atlas (2010)). Nevertheless, it is not clear if this mechanism does function under physiological conditions and what is its overall contribution.

2.5. MOLECULAR ASPECTS OF SYNAPTIC PLASTICITY

2.5.1. Overview

One of the most fascinating characteristics of the synapse is its plasticity, defined as "activity-dependent modification of the strength or efficacy of synaptic transmission at synapses" (Citri and Malenka, 2008). Synaptic plasticity is the postulated mechanism through which past experiences can influence future actions (Hebb, 1949; Huang et al., 1996; Malenka, 1994). At the basis of this process are myriad synaptic proteins that are able to recognize changes in the pattern of trans-synaptic activity and respond by modifying synapse structure and function (O'Rourke et al., 2012). The ubiquity and diversity of plasticity processes across the various phyla, along with the multiple types of synapses they have evolved, make the individual synaptic terminal much more than a mere conveyor of information (Abbott and Regehr, 2004; Zador, 2001). Of the two principal forms of synaptic plasticity, namely pre- and postsynaptic plasticity (Salin et al., 1996b), in this segment we discuss only the former, elaborating on the various proteins and mechanisms that contribute to the formation of short- and long-term changes to its efficacy. In addition, we will only relate to changes in synaptic proteins and second messengers, and neglect changes in ion channels that lead to changes in excitability.

The basic mechanism of calcium-dependent vesicle fusion is very well conserved and shared by all neuronal synapses, but its level of fidelity varies greatly between synapses of different neurons, and even between synapses belonging to the same axon (Branco et al., 2008; Citri and Malenka, 2008; Dobrunz, 2002; Dobrunz and Stevens, 1997; Rosenmund et al., 1993). The term "release probability", or P_r, was coined to denote the probability of the release of a single vesicle upon arrival of an action potential to the terminal (Hessler et al., 1993; Rosenmund et al., 1993). Calculation of this value is most readily achieved by either electrophysiological techniques or the use of FM dyes, in which the fusion of a single vesicle can be measured and divided by the size of the RRP. Although the exact interplay between factors that determine P_r—such as Ca^{2+} levels, protein composition, and availability of vesicles and release sites—is not fully known, plasticity of the presynaptic terminal can be described as the change in the baseline P_r value of a given synapse.

By far the most widely used paradigm for measuring changes in P_r is the paired-pulse ratio (PPR) (Zucker and Regehr, 2002). In this paradigm, two consecutive stimulations are given in brief succession, usually several tens of milliseconds apart, while the postsynaptic response to these stimulations is measured extracellularly, intracellularly or by imaging methods. Although the P_r of a given synapse cannot be calculated directly from the PPR value, the latter is often used as an index to test for changes in basal P_r following a manipulation of the synapse, such as a genetic modification or application of a pharmacological agent. In synapses in which this ratio changes following the manipulation, plasticity of the presynaptic terminal can be said to have occurred.

Although Ca^{2+} kinetics and SNARE complex activity do not vary much from synapse to synapse, large heterogeneity in CNS synapses is often observed when comparing parameters such as P_r and STP (Branco et al., 2008; Dobrunz, 2002; Dobrunz and Stevens, 1997; Hawkins et al., 1993; Hessler et al., 1993). The reason for this heterogeneity lies in the unique composite of proteins such as kinases, autoreceptors, and synaptic proteins that are not necessarily required for proper function of the SNARE mechanism itself, but regulate its activity by mitigating or augmenting its efficacy in response to incoming action potentials. The combined activity of these proteins can be said to set the basal P_r at a given synapse. As a result, small perturbations in the level or activity of any of these proteins can cause the shifts in the synapse's P_r that underlie plasticity, the time course of which will depend on the protein affected.

While paired-pulse facilitation (or depression) is used as an indicator of changes in P_r (Zucker and Regehr, 2002), it is also a plastic phenomenon in its own right, part of a group of phenomena termed STP. This group consists of short-lived changes to P_r that rely heavily on intrasynaptic Ca^{2+} levels and last for at most several minutes (Kamiya and Zucker, 1994; Pan and Zucker, 2009; Zucker, 1999; Zucker and Regehr, 2002). In contrast, long-term changes to P_r, which persist long after Ca^{2+} levels have returned to baseline, can also be elicited, and unlike STP, these changes can persist for hours, days, and possibly even years (Huang et al., 1996; Malenka, 1994; Nicoll and Malenka, 1999). This group is collectively referred to as long-term plasticity (LTP).

2.5.2. Short-Term Plasticity

Plasticity changes that occur within the timeframe of several milliseconds to several minutes fall under the rubric of STP, a process that originates predominantly in the presynaptic terminal (Pan and Zucker, 2009; Powell, 2006). Short-term enhancement of release is composed of several overlapping processes with different kinetics and duration: facilitation is responsible for the rapid increase in discharge that occurs in the timeframe of several tens of milliseconds, while the slower acting augmentation and posttetanic potentiation are slow to rise but can last for several minutes (Fisher et al., 1997). The same process, however, can also produce synaptic depression as the rate of vesicle release begins to exceed that of vesicle replenishment (Hessler et al., 1993; Leao and von Gersdorff, 2009; Neher and Sakaba, 2008). This depression is commonly measured in synapses that have high initial P_r.

Changes in initial P_r shape the mode of STP in a given synapse. A good example of such differences can be found in the synapses formed on cerebellar Purkinje cells by the climbing and parallel fibers: while the former have a relatively high P_r, their efficacy is reduced as the stimulation frequency increases creating synaptic depression; the latter do just the opposite: with low initial P_r, the response of Purkinje cells to parallel fiber stimulation can increase four- to fivefold as stimulation frequency increases (Dittman et al., 2000). Many studies have focused on the different parameters that can correlate with initial P_r of a given synapse and have attempted to account for the large variance observed in such measurements. For example, while it stands to reason that synapses of the same axon will share many similarities in their properties, it was found that variance in initial P_r is larger between terminals of the same axon than between different axons synapsing on the same dendritic branch (Branco et al., 2008). Moreover, a strong correlation was established between the initial P_r and the distance of the synapse from the soma of the postsynaptic neuron (Branco and Hausser, 2011).

A different contributing parameter is the type of cell that a given neuron forms synapses on: in the hippocampal mossy fiber (MF) pathway, dentate gyrus granule cells form large excitatory synapses on CA3 pyramidal cells called thorny excrescences. From these large synaptic complexes, small filopodium-like extensions branch out to synapses of neighboring interneurons. While the MF synapse on the pyramidal cells is known to be one of the strongest facilitating synapses in the CNS, the terminals of the pathway's filopodia are strongly depressed at high frequencies (McBain, 2008; Nicoll and Schmitz, 2005). This property is thought to enhance the pathway's ability to filter out low-activity frequencies. As for the synapses described above, as well as many others, it is presumed that the different organization and composition of synaptic proteins causes the observed variations in release patterns; we will discuss possible contributions of some of these proteins in the next section.

2.5.3. Long-Term Plasticity

The synapse's ability to change its efficacy and strength for relatively long periods of time was first discovered in the hippocampus (Bliss and Lomo, 1973). LTP changes could take place in the form of either enhancement (or long-term potentiation) or reduction (long-term depression (LTD)). For many years, the issue of whether the origin of LTP is presynaptic or postsynaptic was a matter of debate (Nicoll and Malenka, 1995). It is generally accepted that the most studied form of plasticity, NMDA receptor-dependent LTP, as measured in the CA1 region of the hippocampus, is expressed at the postsynaptic level (Zalutsky and Nicoll, 1990). However, LTP can be also dependent on mechanisms that control and modulate neurotransmitter release and are independent of postsynaptic NMDA activity. The best-studied examples of presynaptically expressed NMDA-independent LTP are those observed in hippocampal MFs (Harris and Cotman, 1986; Zalutsky and Nicoll, 1990), the cerebellar parallel fiber LTP (Sakurai, 1987; Salin et al., 1996a), and the LTP of the corticothalamic terminals (Castro-Alamancos and Calcagnotto, 1999). It has been suggested that the above forms of presynaptic LTP share a common mechanism of induction and expression: the presynaptic Ca^{2+} influx activates a calcium/calmodulin-regulated adenylyl cyclase which increases the intracellular cAMP level and activates PKA (Nguyen and Woo, 2003; Wang and Storm, 2003). Potentiation of neurotransmitter release occurs upon PKA-mediated phosphorylation of particular proteins in the presynaptic terminal.

2.5.4. Molecular Changes Associated with Synaptic Plasticity

In the next section, we review the array of proteins that modulate synaptic P_r and modify it in an activity-dependent manner. These proteins can be divided into three groups: (1) synaptic and AZ proteins which act on the SNARE mechanism, hindering or facilitating its formation, (2) protein kinases which regulate entire subsets of phosphorylatable proteins, and (3) G-protein-coupled receptors (GPCRs) located on the terminal's membrane. The GPCRs that have been found to influence P_r are all $G_{i/o}$ family members which regulate the levels of cAMP via their a subunit and can also mitigate membrane potential and Ca^{2+} influx by affecting voltage-gated channels and inward rectifying G-protein coupled receptors (GIRK) channels via their βγ subunit (Niswender et al., 2008; Peters and Scott, 2009).

While many proteins that contribute to the plasticity of the presynaptic terminal have been identified, elucidation of their role in the mechanism has proved a daunting task. The reason for this is mainly the differential and sometimes opposite effects that many of these proteins have in different types of synapses. This heterogeneity probably results from the unique and still unknown composition of the proteins in different synapses. The main two principal types of synapses discussed herein are Schaffer's collateral synapses which display NMDA-dependent LTP (Nicoll and Malenka, 1999), and hippocampal MF synapses which display NMDA-independent LTP (McBain, 2008; Nicoll and Schmitz, 2005; Salin et al., 1996b). While almost all the discussed proteins are expressed in both types of synapses, little or no overlap can be found in their activity, in either the direction or duration of their effect. In addition, this will be complemented by studies performed at the level of cultured neurons.

2.5.4.1. Involvement of Synaptic Proteins

Rab3A: Experiments performed with various Rab3 isoforms have demonstrated a crucial role for RAB3A in determining synaptic P_r but interestingly, different preparations yield very different effects. Measurements from dissociated cultures taken from Rab3A-knockout mice showed that it plays an active role in priming vesicles for release by reducing the amount of Ca^{2+} required for vesicle fusion (Lledo et al., 1993; Schluter et al., 2004, 2006). However, while recordings from dissociated hippocampal culture revealed a decrease in STP in knocked out animals, in measurements taken from MF terminals of acute hippocampal slices, there was a drastic reduction in LTP, but no detectable change in STP (Castillo et al., 1997). A different study in the CA1 subregion found no change in STP or LTP following deletion, but it did find a reduction in endocannabinoid-dependent LTD (Tsetsenis et al., 2011). Since none of the studies performed on Rab3-knockout mice showed impairment in the frequency or amplitude of spontaneous release, it is generally assumed that RAB3A is required for priming of the vesicle rather than for the release itself (Lonart et al., 1998; Muller et al., 2011).

RIM1α: RIM1α is considered one of the key elements in control of vesicle priming and release, mainly through its interactions with Rab3A and Munc13 (Betz et al., 2001; Calakos et al., 2004; Deng et al., 2011; Han et al., 2011; Liu et al., 2011; Schoch et al., 2002). Being a Rab3 effector, RIM1α deletions share many of the phenotypes exhibited by deletions of the Rab3A gene (Schoch et al., 2002). RIM1α knockout has been shown both in autaptic neurons and in the Schaffer's collateral pathway of the hippocampus to reduce basal P_r in the synapse (Calakos et al., 2004; Schoch et al., 2002). However, in the MF pathway, no such reduction was detected (Castillo et al., 2002; Yang and Calakos, 2010). Both of these results are almost identical to measurements taken from Rab3A-knockout mice. Differences between the two knockout phenotypes emerged when P_r measurements were taken from inhibitory synapses of the hippocampus, cerebellum, and lateral amygdala: a reduction in P_r was recorded in RIM1α-knockout slices but not in Rab3A-knockout slices (Fourcaudot et al., 2008; Lachamp et al., 2009; Tsetsenis et al., 2011).

RIM1α has been shown to be required for MF (Castillo et al., 2002) and cerebellar parallel fiber presynaptic LTP (Lonart et al., 2003). However, normal LTP was observed in the hippocampal CA1 region of RIM1α-deficient mice (Schoch et al., 2002), suggesting that RIM is needed for presynaptic LTP while the postsynaptic, NMDA-dependent LTP does not require RIM1α. In addition, Lonart et al. (2003) showed that a prerequisite for presynaptic LTP is the phosphorylation of RIM1α at a single N-terminal site (serine 413) by PKA. However, later studies demonstrated that the phosphorylation at serine 413 is not necessary for either parallel fiber or MF LTP (Kaeser et al., 2008; Yang and Calakos, 2010), leaving the mechanism for PKA and RIM1α regulation of the presynaptic LTP unknown.

Munc13: Because of its importance in vesicle priming as well as its interaction with Rab3A and RIM1α in the AZ, Munc13 has also been a target in research into the mechanisms of STP (Bello et al., 2012; Betz et al., 2001; Dulubova et al., 2005; Huang et al., 2011). Unlike RIM1α and Rab3A, Munc13 is essential for vesicle exocytosis and in synapses lacking both isoforms of the protein—munc13-1/2, release is critically impaired and the mice die (Augustin et al., 1999a,b). Munc13 has been shown to influence release probability through interactions with calcium, calmodulin, PKC, and PIP_2 in various binding sites (Junge et al., 2004; Lou et al., 2008; Shin et al., 2010). However, it is worth noting that manipulations of the individual isoforms have a differential effect on P_r and that this effect varies among different types of synapses: while Munc13-1 is essential for release in most hippocampal glutamatergic synapses (Augustin et al., 1999b), it has been shown that some excitatory, as well as inhibitory synapses utilize Munc13-2 as a priming factor instead (Augustin et al., 1999a; Breustedt et al., 2010).

Accordingly, Munc13-1 was shown to govern release in ~90% of cultured hippocampal synapses, leaving release in the other 10% unaffected (Rosenmund et al., 2002; Varoqueaux et al., 2002). In addition, the different isoforms contribute to different forms of STP: rescue of synaptic transmission, in neurons derived from Munc13-1/2 double knockout mice, leads to synaptic depression when Munc13-1 is expressed while rescue with Munc13-2 leads to facilitation and synaptic augmentation (Rosenmund et al., 2002). Therefore, the mode of STP can be regulated by changes in the expression level and activity of different Munc13 isoforms. Deletion of each isoform is deleterious to exocytosis in the synapses that utilize it, an effect that is most readily manifested as a marked decrease in P_r. A different study, performed with hippocampal slices, complemented these results by finding that deletion of Munc13-2 affects STP in MF synapses only, sparing the synapses of other hippocampal pathways (Breustedt et al., 2010).

In contrast, interfering with the interaction between RIM and Munc13-1 reduced LTP in the MF terminal, suggesting that Munc13-2 plays an important role in STP, and Munc13-1 in LTP under these conditions.

Munc18: The Sec1/Munc18 (SM) proteins regulate membrane fusion through SNARE complex formation (Toonen and Verhage, 2007). Although Munc18-1 knockout completely abolishes synaptic transmission, autaptic GABA- and glutamatergic neurons from heterozygous Munc18-1+/− mice displayed increased depression during intense stimulation (Toonen et al., 2006b). In addition, Munc18-1 heterozygous autapses had a reduced vesicles' RRP (Toonen et al., 2006b). Conversely, Munc-18-1 overexpression increased the size of the RRP and accelerated the rate of RRP replenishment (Nili et al., 2006; Toonen et al., 2006b). Overall, these results suggest a regulatory role for Munc18-1 in determining the number of readily released vesicles and their replenishment.

Bassoon and Piccolo: Spontaneous EPSCs and excitatory postsynaptic current (EPSCs) evoked at low frequencies were unaffected in cerebellar MF to granule cells synapses of bassoon-deficient mice (Hallermann et al., 2010). However, the short-term depression was enhanced during sustained high-frequency trains. In addition, the recovery from depression was significantly slower in the absence of Bassoon. The above results suggest that Bassoon speeds up the reloading of vesicles at release sites and speeds up the vesicle replenishment process (Hallermann et al., 2010).

Complexin: Complexin I and II were initially identified as SNARE complex-associated proteins (Ishizuka et al., 1995). Complexin-II-deficient mice are viable and appear normal. In addition, the basal synaptic neurotransmission and STP are not affected. However, the magnitude of LTP in both CA1 and CA3 hippocampal regions is significantly reduced (Takahashi et al., 1999). In another study utilizing triple-deletion (complexin I, II, and III) mice, complexin deficiency in both excitatory and inhibitory neurons caused a decrease in spontaneous and Ca^{2+}-triggered neurotransmitter release, probably indicating a facilitative role for complexin in neurotransmission (Xue et al., 2008).

Synaptotagmin: Synaptotagmins belong to a large family of proteins with similar structure which are regarded as the Ca^{2+} sensors for vesicle fusion (Pang and Sudhof, 2010; Sudhof, 2004). In neurons obtained from syntaptotagmin-1 knockout mice, the lack of synaptotagmin-1 blocked synchronous GABA and glutamate release, when tested at low (>0.1 Hz) stimulation frequencies (Maximov and Sudhof, 2005). At higher stimulation frequencies, deletion of synaptotagmin-1 altered synchronous release and dramatically enhanced "delayed asynchronous release" during and following the stimulus train (Maximov and Sudhof, 2005). As a result, synaptotagmin-1 knockout only modestly reduced synaptic charge transfer at high-stimulation frequencies (Maximov and Sudhof, 2005; Xu et al., 2012). This might explain the lack of effect of the synaptotagmin-1 knockout in the hippocampal CA1 region on acquisition of recent contextual memories (Xu et al., 2012) (see next section for more details).

SV2: Synaptic vesicle protein 2 is a vesicle-specific glycoprotein that is present in neurons and neuroendocrine cells (Bajjalieh et al., 1992; Buckley and Kelly, 1985). In mammals, three isoforms of SV2 have been identified: SV2A, SV2B, and SV2C (Bajjalieh et al., 1993, 1994; Janz and Sudhof, 1999). SV2 has been suggested to act as a positive modulator of calcium-dependent exocytosis. Thus, neurons lacking SV2A and SV2A + B display reduced excitatory and inhibitory evoked synaptic responses and decreased initial release probability, but show intact spontaneous neurotransmission (Chang and Sudhof, 2009; Crowder et al., 1999; Nowack et al., 2010; Schivell et al., 2005; Yao et al., 2010). Surprisingly, overexpression of SV2A in excitatory neurons also resulted in reduced release probability and accordingly, it has been suggested that changes in SV2 expression in either direction can reduce release probability (Nowack et al., 2011) calling for further investigation of SV2's activity.

2.5.4.2. Protein Kinases

PKA: Long-term, PKA-dependent synaptic plasticity occurs at certain synapses of the CNS. In the hippocampal MF pathway, synaptic facilitation produced by forskolin-induced activation of cAMP/PKA signaling occluded expression of tetanus-induced LTP. In addition, tetanus-induced LTP could also be blocked by application of PKA inhibitors (Huang and Kandel, 1994) suggesting the involvement of PKA in this form of presynaptic LTP. Theoretically, at the presynaptic level, PKA may exert its effects via two different mechanisms: modulation of P_r and recruitment of the proposed "silent" release sites. PKA can influence P_r via different pathways; while some are responsible for fast, short-lived increases, other pathways, which involve activation of downstream transcription factors in the nucleus, could be responsible for stable, long-term modifications (Bolshakov et al., 1997).

Similarly, in monosynaptic CA3–CA1 synapses, application of the PKA activator Sp-cAMP induced the late phase of LTP that was associated with a reduction in paired-pulse facilitation (Yu et al., 2001), indicating an increase in P_r. In addition to the increase in P_r, data from imaging studies confirmed an increase in the number of functional release sites (Bozdagi et al., 2000; Cousin and Evans, 2011; Kim and Thayer, 2001; Ma et al., 1999). Thus, for example, using FM dyes to label the pool of recycling synaptic vesicles in cultured cerebellar neurons, Cousin and Evans (2011) found that the "silent" and weak release sites could be activated by PKA; conversely, the active synapses could be silenced by PKA inhibition (Cousin and Evans, 2011). Although, several presynaptic proteins mentioned above including synapsin, RIM, rabphillin, and tomosyn are known as PKA targets, the exact molecular mechanism of PKA regulation of presynaptic plasticity remains unclear.

CaMKII and Synapsin: Although it is mainly associated with postsynaptic LTP, at many synapses of the CNS, this calmodulin-dependent kinase has also been shown to partake in short-term processes (Chapman et al., 1995; Hojjati et al., 2007; Salin et al., 1996b). However, much like the aforementioned proteins, this influence varies greatly among the different types of synapses (Salin et al., 1996b). Activation of

CaMKII is triggered by binding of calmodulin in its calcium-bound conformation (Bronstein et al., 1993; DeRiemer et al., 1984). This attribute links CaMKII activity with local Ca^{2+} elevations that follow action potentials, conferring its role as an activity-dependent protein (Benfenati et al., 1989; Chi et al., 2001; Hosaka et al., 1999; Scott et al., 2010; Sihra et al., 1989). Following its activation, CaMKII phosphorylates synapsins which consequently release cytoskeleton-bound vesicles, enabling them to join the RRP and undergo fusion (Benfenati et al., 1989, 1990; Sihra et al., 1989). However, knockout of both CaMKII and synapsin, as well as pharmacological blockers, has failed to exhibit changes in STP across different synaptic models (Chapman et al., 1995; Hojjati et al., 2007; Rosahl et al., 1995; Silva et al., 1996). Exceptions to this are the hippocampal MFs, *Drosophila* sensory neurons and synapses of the lateral-line lobe of the electric fish brain, in which these manipulations caused a significant reduction in different forms of short-term enhancement (Harvey-Girard et al., 2010; Jin et al., 1998; Salin et al., 1996b), and αCaMKII$^{+/-}$ heterozygous mutant mice showed impaired STP in the CA1 (Silva et al., 1996).

2.5.4.3. G-Protein-Coupled Receptors

mGluR2/3: This family of mGluRs is most tightly linked to regulation of presynaptic activity. These autoreceptors are stereotypically located at the extrasynaptic membrane and are activated when excess glutamate spills over from the synaptic cleft (Anwyl, 1992, 1999). Activation of these receptors has a strong inhibitory effect on the synapses in which they are expressed; in MF terminals, their effect is so dominant that agonists of the receptor will completely abolish all activity in the synapse (Cosgrove et al., 2011; Kwon and Castillo, 2008; Scanziani et al., 1997; Toth et al., 2000), while in perforant path synapses, it reduces the rate of synaptic depression which follows high-frequency stimulation (Kilbride et al., 2001; Rush et al., 2002). The most readily observed result of mGluR2/3 agonists on slice electrophysiology is a shift from depression to facilitation. This transition is thought to be the consequence of a drop in P_r following the receptor's activation (Anwyl, 1999; Bellone et al., 2008).

mAChRs: The muscarinic ACh receptors M2 and M4 belong to the $G_{i/o}$ group (Wess et al., 2007) and are strongly involved in suppression of neurotransmitter release from presynaptic terminals (Vannucchi and Pepeu, 1995; Volpicelli and Levey, 2004). These receptors are highly likely to function mainly as autoreceptors, mostly in cholinergic terminals, moderating their synaptic activity (Zheng et al., 2012). In addition to the above mechanism, which is very similar in nature to the effects exerted by other GPCRs of the $G_{i/o}$ group, M2 and M4 have been shown to influence release in a voltage-, as well as ligand-, dependent manner (Ben-Chaim et al., 2006). This mechanism is quite distinct from the activities of the other autoreceptors discussed in this section; it allows for rapid control, in the millisecond time range, of neurotransmitter release, which can balance the effects of the rapid calcium influxes that follow an action potential (Parnas and Parnas, 2007).

This finding is the basis for the "voltage-calcium hypothesis" which suggests that in addition to calcium, voltage has a crucial role in regulating the time course of transmitter release (Parnas and Parnas, 2007).

GABAb receptors: As GABAergic interneurons do not tend to synapse directly on the terminals of excitatory synapses (the cerebellar rosettes and the olfactory bulb glomeruli are two notable exceptions), GABAb receptor activation has been suggested to be tonic in nature and strongest in synapses close to the neuron's soma, where inhibitory afferents are most abundant. Supporting this notion is the finding that activation of this receptor is inversely correlated with initial P_r (Laviv et al., 2010, 2011). This mechanism might therefore be responsible for some of the variance observed in P_r values as synapses of the same type might contact the postsynaptic cell at various points along its dendritic arbor, a parameter which is seldom taken into account.

Endocannabinoid receptors: Recent studies unveiled a novel form of synaptic plasticity mediated by endocannabinoids (Wilson and Nicoll, 2001). Endocannabinoids are substances of lipid origin and generated on demand by cleavage from membrane lipids (Cadas et al., 1996; Stella et al., 1997). The endocannabinoids are retrograde messengers: they are released to the synaptic cleft from the postsynaptic membrane and activate their receptors on the presynaptic membrane. The endocannabinoid receptor CB1 is found in many brain regions (hippocampus, striatum, prefrontal cortex, amygdala, and nucleus accumbens), and to a lower extent, in several peripheral tissues (Galiegue et al., 1995; Munro et al., 1993). It is coupled to $G_{i/o}$ protein and its activation leads to reducing of adenylyl cyclase activity, reducing of Ca^{2+} influx through the N-, P/Q-, and L-type channels (Gebremedhin et al., 1999; Mackie and Hille, 1992; Twitchell et al., 1997), and increase in K^+ influx through the inwardly rectifying potassium channels (Mackie et al., 1995). These effects, in their turn, reduce the neurotransmitter release. In addition, the endocannabinoids may exert their effects on neurotransmitter release through the presynaptic cAMP/PKA signaling and relevant synaptic proteins like synapsin, rabphillin, tomosyn, and RIM1α, since disruption of these cascades leads to impairment of LTD (Chevaleyre et al., 2007).

Activation of CB1 receptors contributes to different forms of short-term synaptic plasticity in both excitatory and inhibitory synapses. Thus, the endocannabinoid release following the postsynaptic depolarization could lead to reducing of GABA or glutamate release (depolarization-induced suppression of inhibition (DSI) (Ohno-Shosaku et al., 2001; Wilson et al., 2001), and depolarization-induced suppression of excitation (DSE) (Kreitzer and Regehr, 2001)). In addition to modulation of STP, endocannabinoids may originate more persistent forms of synaptic plasticity. The endocannabinoid-dependent form of LTD was observed in many brain regions, e.g. striatum, prefrontal cortex, amygdala, nucleus accumbens, and cerebellum (Auclair et al., 2000; Chevaleyre and Castillo, 2003; Chevaleyre et al., 2007; Gerdeman et al., 2002; Robbe et al., 2003; Safo and Regehr, 2005). As in the case of DSI, the endocannabinoid-dependent LTD is initiated on the presynaptic membrane and involves rise in postsynaptic Ca^{2+} concentration (Alger, 2002; Freund et al., 2003), although Ca^{2+}-independent mechanisms of endocannabinoid release have been reported (Kim et al., 2002; Maejima et al., 2001).

2.6. GENETIC MANIPULATIONS OF SYNAPTIC PROTEINS AND BEHAVIORAL CONSEQUENCES

The well-characterized and documented synaptic proteins have been mainly studied in vitro. Nevertheless, increasing interest in the interplay between synaptic transmission and behavior between 1990 and 2010 has encouraged researchers to genetically manipulate synaptic proteins in vivo to reveal their effects on behavior and cognitive phenomena, mainly in the field of learning and memory. The interaction between synaptic properties and behavior is probably most associated with Hebb's postulation, stating that during learning, memory is being stored in synaptic connections that are being strengthened and/or formed (Hebb, 1949). It is now recognized that synaptic plasticity is one of the main processes enabling such synaptic modifications. To explore this field more accurately, transgenic manipulations and genetic approaches enable localizing, as well as determining the temporal specificity of the manipulated protein. These genetic tools involve knocking-in, knocking-out, overexpressing, or downregulating synaptic proteins, which may take place at the whole organism level, or be a local transgenic manipulation affecting a brain region or a specific neuronal circuit, for example, a specific cell subtype, or even only a few specific cells. Here, we discuss some of the knockout and genetic manipulations that have been performed on synaptic proteins to demonstrate their involvement in behavior. Early studies on natural and induced mutations in nematodes and *Drosophila* first linked mutations or deletions of genes to animal physiology and behavior. Starting in the 1990s, knockout techniques, performed mainly in mice, were the first to allow neuroscientists to explore the interaction between synaptic proteins and behavior and cognition.

Synaptotagmin-4 is a synaptic vesicle protein that is highly expressed in the cerebellum (Berton et al., 1997), in Golgi-derived dense-core vesicles and microvesicles in the posterior pituitary nerve terminals (Zhang et al., 2009b), and in BDNF-containing vesicles in hippocampal neurons (Dean et al., 2009). Synaptotagmin-4 is known to bind calcium ions through its C2 domains (Davletov and Sudhof, 1993) and to interact with other presynaptic proteins, thus regulating the synaptic transmission process. In synaptotagmin-4 knockout mice, deficits in motor control on the rotarod as well as hippocampal-dependent memory abilities were found, probably due to alteration of synaptic regulation in the cerebellum that coordinates movement, and in the hippocampus that is responsible for learning and memory (Ferguson et al., 2000). In addition, synaptotagmin-4-knockout mice showed deficits in contextual fear conditioning and the social transmission of food preference, both hippocampus-dependent tasks. Although synaptotagmin-4 is not the main synaptotagmin isoform in the hippocampus, it is reasonable to assume that faulty regulation of synaptic transmission in the hippocampus was responsible for these deficits.

Similarly, Munc13-3, which is cerebellum-specific (Augustin et al., 2001), was shown to be involved in motor learning: munc13-3 knockout mice showed normal spontaneous motor activity, but were impaired in motor learning of complex tasks. As Munc13 proteins are essential for vesicle priming, the impaired motor learning was probably due to a reduction in P_r at the parallel fiber-Purkinje cell synapses due to interference with vesicle priming. As a result, these synapses showed increased paired-pulse facilitation that contributed to the observed deficit in motor learning (Augustin et al., 2001).

In addition, the linkage between impaired STP and behavioral deficits was also demonstrated by knocking out key presynaptic proteins. In αCaMKII$^{+/-}$, synapsin-II$^{-/-}$ and synapsin-I/II$^{-/-}$ mutant mice, characterized by decreased paired-pulse facilitation or posttetanic potentiation, profound impairments were measured in learning tasks (Silva et al., 1996). On the other hand, no such behavioral deficits were measured in synapsin-I$^{-/-}$ mice, characterized by increased paired-pulse facilitation (Silva et al., 1996). These experiments suggest that changes in STP can also lead to alterations in animal behavior and that these are not always necessarily linked to alterations in LTP.

Rab3a, a synaptic vesicle-associated protein, takes part in the process of synaptic vesicle fusion to the PM (Schluter et al., 2002). Surprisingly, in Rab3a-knockout mice, no deficits in contextual, spatial, or working memory were measured (Hensbroek et al., 2003), although the absence of both MF LTP and LTD were measured physiologically.

Some of the synaptic proteins have also been shown to be involved in psychiatric disorders. For example, synapsin II knockout mice were shown to have behavioral deficits related to schizophrenia, such as deficits in prepulse inhibition, decreased social behavior, and locomotor hyperactivity (Dyck et al., 2009). More specifically, these schizophrenia-related behavioral deficits were noticed when synapsin II levels were downregulated specifically in the medial-prefrontal cortex of wild-type rats (Dyck et al., 2011). Another synaptic protein shown to be involved in schizophrenia is RIM1α, known to mediate presynaptic plasticity and cognitive function: RIM1α-knockout mice exhibited abnormalities in the same schizophrenia-relevant behavioral tasks (Blundell et al., 2010) but in addition, mice deficient in RIM1α were also shown to be impaired in learning and memory abilities, deficits that were attributed to the absence of MF–CA3 LTP, and decreased P_r in the CA1 area of the mouse hippocampus (Powell et al., 2004).

Tomosyn is a syntaxin-binding protein that inhibits vesicle priming and synaptic transmission via a mechanism attributed to its interference with the formation of SNARE complexes (for review, see Ashery et al. (2009)). Interestingly, knocking down tomosyn levels affected odor learning in *Drosophila* by enhancing synaptic strength, resulting in substantially reduced late, but not short, aversive odor memory (Chen et al., 2011).

Although many researchers have used knocked-out mice to elucidate the interplay between a manipulated protein and behavior and physiology, substantial problems are involved with such genetic manipulations. For example, many of the knocked-out mice that lack expression of a synaptic protein are not fertile or viable, making it impossible to study the mature animal. Moreover, knocking out a protein will result in its deficiency in all types of cells, entailing developmental complications that may complicate the interpretation of the behavioral and physiological findings. Local genetic manipulation, on the other hand, is a rapid and an efficient technique for manipulating the protein level in individual or multiple brain regions. This can be done on animals with different genetic backgrounds, lasts for a long time, and is cell-specific. Therefore, although used in only a few studies to date, local genetic manipulation makes it easier to identify the role and physiology of the synaptic protein manipulated in the specific brain region in adult mice, an insight that is hard to achieve using knockout manipulation.

One of the main studies in recent years that made use of local genetic manipulation of a synaptic protein blocked synaptic transmission by cell type-restricted and temporally controlled expression of the tetanus toxin (TeTX) light chain (Yamamoto et al., 2003), which specifically cleaves synaptobrevin-2 (Schiavo et al., 1992). In doing so, a significant reduction in synaptobrevin-2 expression levels was noted in CA1 and CA3, and in part of the dentate gyrus, resulting in TeTX-mediated blockade of synaptic transmission (Nakashiba et al., 2008). Measuring the behavioral aspect of this manipulation, it was shown that the CA3 output in the trisynaptic pathway is dispensable for both acquisition and recall of incremental spatial learning and memory recall in the Morris water maze task. Moreover, when tested in contextual fear conditioning, it was demonstrated that the CA3 output is crucial for rapid one-trial learning in a novel context.

In a recent study (Xu et al., 2012), researchers knocked down synaptotagmin-1 levels in the hippocampal CA1 region or in the prefrontal cortex by stereotaxic injection of lentivirus expressing synaptotagmin-1-short hairpin. Knockdown of synaptotagmin-1 blocked the information transfer carried by a single action potential (as it blocked synchronous release) but allowed information transfer carried out during bursts of action potentials (due to access to excess of asynchronous release in synaptotagmin-1 knockdown that accumulates during bursts). Interestingly, knocking down synaptotagmin-1 in the hippocampus did not impede acquisition of recent contextual fear memories; however, it did impair their precision. When synaptotagmin-1 levels were knocked down in the prefrontal cortex, remote fear memories were impaired. Moreover, when synaptic transmission was blocked in the hippocampus by TeTX, recent, but not remote contextual fear memory was impaired, while such blockage in the prefrontal cortex abrogated remote, but not recent fear memory. The conclusions from these experiments enabled a correlation between physiological synaptic activity (isolated spikes versus bursting activity) and information transfer: the hippocampus can rely solely on spike bursts to transfer information downstream, while isolated spike transmission in the hippocampus is not required for contextual fear memory. In contrast, isolated spike transmission and/or precise timing of spikes are pivotal for the proper function of the prefrontal cortex.

Overall, these studies demonstrate the diverse and complex involvement of synaptic proteins in synaptic plasticity and behavioral aspects such as learning and memory, motor learning, schizophrenia-associated behavior, and more. With the great improvement in local genetic manipulations, allowing rapid and efficient manipulation of protein level in single or multiple brain regions, a more precise and informative characterization of the molecular mechanism responsible for the different behavioral alterations will be revealed. Hence by specific manipulation of discrete steps in synaptic transmission, we can learn about the contribution of a defined pathway or synapse to defined behavior.

3. CONCLUSIONS

In the past few decades, the main molecular players in synaptic transmission have been identified and characterized. However, the precise sequence of protein–protein interactions and the nature of the protein complexes that drive specific steps in this process are not fully understood. Nevertheless, it has become clear that the relative expression of these proteins, or their activity levels, determine some of the key properties of the synaptic transmission and plasticity of a given synapse. Therefore, one of the challenges in the coming years will be to understand how a specific combination of proteins affects synapse function, and its consequent effect at the network level and on animal behavior. Furthermore, dynamic information on the nanoscale distribution of synaptic proteins and the sequence of their interactions with other proteins will be critical to achieving a precise understanding of the molecular synaptic transmission machinery. Such information may be achieved in the near future via the development and application of new molecular and genetic tools, allowing simultaneous detection of single molecules and correlation with direct physiological measurements of synaptic activity in those synapses. These findings will provide mechanistic insights into the molecular regulation of synaptic function in health, during learning and ageing, and in disease.

ACKNOWLEDGMENTS

We are very thankful to Profs Jens Rettig, Volker Haucke, Thomas Dresbach, Ralf Schneggenburger, and Daniel Gitler for their comments and suggestions on early versions of this chapter and for Varda Wexler for graphic assistance.

REFERENCES

Abbott, L.F., Regehr, W.G., 2004. Synaptic computation. Nature 431, 796–803.

Ahmari, S.E., Buchanan, J., Smith, S.J., 2000. Assembly of presynaptic active zones from cytoplasmic transport packets. Nat. Neurosci. 3, 445–451.

Alger, B.E., 2002. Retrograde signaling in the regulation of synaptic transmission: focus on endocannabinoids. Prog. Neurobiol. 68, 247–286.

Altrock, W.D., tom Dieck, S., Sokolov, M., Meyer, A.C., Sigler, A., Brakebusch, C., Fassler, R., Richter, K., Boeckers, T.M., Potschka, H., Brandt, C., Loscher, W., Grimberg, D., Dresbach, T., Hempelmann, A., Hassan, H., Balschun, D., Frey, J.U., Brandstatter, J.H., Garner, C.C., Rosenmund, C., Gundelfinger, E.D., 2003. Functional inactivation of a fraction of excitatory synapses in mice deficient for the active zone protein bassoon. Neuron 37, 787–800.

Andersson, F., Jakobsson, J., Low, P., Shupliakov, O., Brodin, L., 2008. Perturbation of syndapin/PACSIN impairs synaptic vesicle recycling evoked by intense stimulation. J. Neurosci. 28, 3925–3933.

Anwyl, R., 1992. Metabotropic glutamate receptors: electrophysiological properties and role in plasticity. Rev. Neurosci. 3, 217–231.

Anwyl, R., 1999. Metabotropic glutamate receptors: electrophysiological properties and role in plasticity. Brain Res. 29, 83–120.

Arac, D., Chen, X., Khant, H.A., Ubach, J., Ludtke, S.J., Kikkawa, M., Johnson, A.E., Chiu, W., Sudhof, T.C., Rizo, J., 2006. Close membrane-membrane proximity induced by Ca(2+)-dependent multivalent binding of synaptotagmin-1 to phospholipids. Nat. Struct. Mol. Biol. 13, 209–217.

Aravamudan, B., Fergestad, T., Davis, W.S., Rodesch, C.K., Broadie, K., 1999. Drosophila UNC-13 is essential for synaptic transmission. Nat. Neurosci. 2, 965–971.

Archer, D.A., Graham, M.E., Burgoyne, R.D., 2002. Complexin regulates the closure of the fusion pore during regulated vesicle exocytosis. J. Biol. Chem. 277, 18249–18252.

Artalejo, C.R., Elhamdani, A., Palfrey, H.C., 2002. Sustained stimulation shifts the mechanism of endocytosis from dynamin-1-dependent rapid endocytosis to clathrin- and dynamin-2-mediated slow endocytosis in chromaffin cells. Proc. Natl. Acad. Sci. USA 99, 6358–6363.

Arunachalam, L., Han, L., Tassew, N.G., He, Y., Wang, L., Xie, L., Fujita, Y., Kwan, E., Davletov, B., Monnier, P.P., Gaisano, H.Y., Sugita, S., 2008. Munc18-1 is critical for plasma membrane localization of syntaxin1 but not of SNAP-25 in PC12 cells. Mol. Biol. Cell 19, 722–734.

Ashery, U., Bielopolski, N., Barak, B., Yizhar, O., 2009. Friends and foes in synaptic transmission: the role of tomosyn in vesicle priming. Trends Neurosci. 32, 275–282.

Athenstaedt, K., Daum, G., 1999. Phosphatidic acid, a key intermediate in lipid metabolism. Eur. J. Biochem./FEBS 266, 1–16.

Atlas, D., 2010. Signaling role of the voltage-gated calcium channel as the molecular on/off-switch of secretion. Cell. Signal. 22, 1597–1603.

Auclair, N., Otani, S., Soubrie, P., Crepel, F., 2000. Cannabinoids modulate synaptic strength and plasticity at glutamatergic synapses of rat prefrontal cortex pyramidal neurons. J. Neurophysiol. 83, 3287–3293.

Augustin, I., Betz, A., Herrmann, C., Jo, T., Brose, N., 1999a. Differential expression of two novel Munc13 proteins in rat brain. Biochem. J. 337 (Pt 3), 363–371.

Augustin, I., Korte, S., Rickmann, M., Kretzschmar, H.A., Sudhof, T.C., Herms, J.W., Brose, N., 2001. The cerebellum-specific Munc13 isoform Munc13-3 regulates cerebellar synaptic transmission and motor learning in mice. J. Neurosci. 21, 10–17.

Augustin, I., Rosenmund, C., Sudhof, T.C., Brose, N., 1999b. Munc13-1 is essential for fusion competence of glutamatergic synaptic vesicles. Nature 400, 457–461.

Aunis, D., Bader, M.F., 1988. The cytoskeleton as a barrier to exocytosis in secretory cells. J. Exp. Biol. 139, 253–266.

Bahler, M., Greengard, P., 1987. Synapsin I bundles F-actin in a phosphorylation-dependent manner. Nature 326, 704–707.

Bai, J., Hu, Z., Dittman, J.S., Pym, E.C., Kaplan, J.M., 2010. Endophilin functions as a membrane-bending molecule and is delivered to endocytic zones by exocytosis. Cell 143, 430–441.

Bajjalieh, S.M., Frantz, G.D., Weimann, J.M., McConnell, S.K., Scheller, R.H., 1994. Differential expression of synaptic vesicle protein 2 (SV2) isoforms. J. Neurosci. 14, 5223–5235.

Bajjalieh, S.M., Peterson, K., Linial, M., Scheller, R.H., 1993. Brain contains two forms of synaptic vesicle protein 2. Proc. Natl. Acad. Sci. USA 90, 2150–2154.

Bajjalieh, S.M., Peterson, K., Shinghal, R., Scheller, R.H., 1992. SV2, a brain synaptic vesicle protein homologous to bacterial transporters. Science 257, 1271–1273.

Baldelli, P., Fassio, A., Valtorta, F., Benfenati, F., 2007. Lack of synapsin I reduces the readily releasable pool of synaptic vesicles at central inhibitory synapses. J. Neurosci. 27, 13520–13531.

Baldwin, M.L., Cammarota, M., Sim, A.T., Rostas, J.A., 2006. Src family tyrosine kinases differentially modulate exocytosis from rat brain nerve terminals. Neurochem. Int. 49, 80–86.
Bar-On, D., Winter, U., Nachliel, E., Gutman, M., Fasshauer, D., Lang, T., Ashery, U., 2008. Imaging the assembly and disassembly kinetics of cis-SNARE complexes on native plasma membranes. FEBS Lett. 582, 3563–3568.
Bar-On, D., Wolter, S., van de Linde, S., Heilemann, M., Nudelman, G., Nachliel, E., Gutman, M., Sauer, M., Ashery, U., 2012. Super-resolution imaging reveals the internal architecture of nano-sized syntaxin clusters. J. Biol. Chem. 287, 27158–27167.
Barnett, M.W., Larkman, P.M., 2007. The action potential. Pract. Neurol. 7, 192–197.
Basu, J., Betz, A., Brose, N., Rosenmund, C., 2007. Munc13-1 C1 domain activation lowers the energy barrier for synaptic vesicle fusion. J. Neurosci. 27, 1200–1210.
Bates, M., Huang, B., Dempsey, G.T., Zhuang, X., 2007. Multicolor super-resolution imaging with photo-switchable fluorescent probes. Science 317, 1749–1753.
Baumert, M., Maycox, P.R., Navone, F., De Camilli, P., Jahn, R., 1989. Synaptobrevin: an integral membrane protein of 18,000 daltons present in small synaptic vesicles of rat brain. EMBO J. 8, 379–384.
Becherer, U., Pasche, M., Nofal, S., Hof, D., Matti, U., Rettig, J., 2007. Quantifying exocytosis by combination of membrane capacitance measurements and total internal reflection fluorescence microscopy in chromaffin cells. PloS One 2, e505.
Bello, O.D., Zanetti, M.N., Mayorga, L.S., Michaut, M.A., 2012. RIM, Munc13, and Rab3A interplay in acrosomal exocytosis. Exp. Cell Res. 318, 478–488.
Bellone, C., Luscher, C., Mameli, M., 2008. Mechanisms of synaptic depression triggered by metabotropic glutamate receptors. Cell. Mol. Life Sci. 65, 2913–2923.
Ben-Chaim, Y., Chanda, B., Dascal, N., Bezanilla, F., Parnas, I., Parnas, H., 2006. Movement of 'gating charge' is coupled to ligand binding in a G-protein-coupled receptor. Nature 444, 106–109.
Benfenati, F., Bahler, M., Jahn, R., Greengard, P., 1989. Interactions of synapsin I with small synaptic vesicles: distinct sites in synapsin I bind to vesicle phospholipids and vesicle proteins. J. Cell Biol. 108, 1863–1872.
Benfenati, F., Neyroz, P., Bahler, M., Masotti, L., Greengard, P., 1990. Time-resolved fluorescence study of the neuron-specific phosphoprotein synapsin I. Evidence for phosphorylation-dependent conformational changes. J. Biol. Chem. 265, 12584–12595.
Benfenati, F., Valtorta, F., Rossi, M.C., Onofri, F., Sihra, T., Greengard, P., 1993. Interactions of synapsin I with phospholipids: possible role in synaptic vesicle clustering and in the maintenance of bilayer structures. J. Cell Biol. 123, 1845–1855.
Bennett, M.K., Calakos, N., Scheller, R.H., 1992. Syntaxin: a synaptic protein implicated in docking of synaptic vesicles at presynaptic active zones. Science, New York, 257, 255–259.
Berberian, K., Torres, A.J., Fang, Q., Kisler, K., Lindau, M., 2009. F-actin and myosin II accelerate catecholamine release from chromaffin granules. J. Neurosci. 29, 863–870.
Berton, F., Iborra, C., Boudier, J.A., Seagar, M.J., Marqueze, B., 1997. Developmental regulation of synaptotagmin I, II, III, and IV mRNAs in the rat CNS. J. Neurosci. 17, 1206–1216.
Betz, A., Thakur, P., Junge, H.J., Ashery, U., Rhee, J.S., Scheuss, V., Rosenmund, C., Rettig, J., Brose, N., 2001. Functional interaction of the active zone proteins Munc13-1 and RIM1 in synaptic vesicle priming. Neuron 30, 183–196.
Betzig, E., Patterson, G.H., Sougrat, R., Lindwasser, O.W., Olenych, S., Bonifacino, J.S., Davidson, M.W., Lippincott-Schwartz, J., Hess, H.F., 2006. Imaging intracellular fluorescent proteins at nanometer resolution. Science 313, 1642–1645.
Bhalla, A., Chicka, M.C., Tucker, W.C., Chapman, E.R., 2006. Ca(2+)-synaptotagmin directly regulates t-SNARE function during reconstituted membrane fusion. Nat. Struct. Mol. Biol. 13, 323–330.
Blasi, J., Chapman, E.R., Link, E., Binz, T., Yamasaki, S., De Camilli, P., Sudhof, T.C., Niemann, H., Jahn, R., 1993. Botulinum neurotoxin A selectively cleaves the synaptic protein SNAP-25. Nature 365, 160–163.
Bliss, T.V., Lomo, T., 1973. Long-lasting potentiation of synaptic transmission in the dentate area of the anaesthetized rabbit following stimulation of the perforant path. J. Physiol. 232, 331–356.
Bloom, O., Evergren, E., Tomilin, N., Kjaerulff, O., Low, P., Brodin, L., Pieribone, V.A., Greengard, P., Shupliakov, O., 2003. Colocalization of synapsin and actin during synaptic vesicle recycling. J. Cell Biol. 161, 737–747.

Blundell, J., Kaeser, P.S., Sudhof, T.C., Powell, C.M., 2010. RIM1alpha and interacting proteins involved in presynaptic plasticity mediate prepulse inhibition and additional behaviors linked to schizophrenia. J. Neurosci. 30, 5326–5333.
Bock, J.B., Matern, H.T., Peden, A.A., Scheller, R.H., 2001. A genomic perspective on membrane compartment organization. Nature 409, 839–841.
Bolshakov, V.Y., Golan, H., Kandel, E.R., Siegelbaum, S.A., 1997. Recruitment of new sites of synaptic transmission during the cAMP-dependent late phase of LTP at CA3–CA1 synapses in the hippocampus. Neuron 19, 635–651.
Borisovska, M., Zhao, Y., Tsytsyura, Y., Glyvuk, N., Takamori, S., Matti, U., Rettig, J., Sudhof, T., Bruns, D., 2005. v-SNAREs control exocytosis of vesicles from priming to fusion. EMBO J. 24, 2114–2126.
Bozdagi, O., Shan, W., Tanaka, H., Benson, D.L., Huntley, G.W., 2000. Increasing numbers of synaptic puncta during late-phase LTP: N-cadherin is synthesized, recruited to synaptic sites, and required for potentiation. Neuron 28, 245–259.
Branco, T., Hausser, M., 2011. Synaptic integration gradients in single cortical pyramidal cell dendrites. Neuron 69, 885–892.
Branco, T., Staras, K., Darcy, K.J., Goda, Y., 2008. Local dendritic activity sets release probability at hippocampal synapses. Neuron 59, 475–485.
Brenner, S., 1974. The genetics of *Caenorhabditis elegans*. Genetics 77, 71–94.
Breustedt, J., Gundlfinger, A., Varoqueaux, F., Reim, K., Brose, N., Schmitz, D., 2010. Munc13-2 differentially affects hippocampal synaptic transmission and plasticity. Cereb. Cortex 20, 1109–1120.
Bronstein, J.M., Farber, D.B., Wasterlain, C.G., 1993. Regulation of type-II calmodulin kinase: functional implications. Brain Res. Brain Res. Rev. 18, 135–147.
Buckley, K., Kelly, R.B., 1985. Identification of a transmembrane glycoprotein specific for secretory vesicles of neural and endocrine cells. J. Cell Biol. 100, 1284–1294.
Burgalossi, A., Jung, S., Meyer, G., Jockusch, W.J., Jahn, O., Taschenberger, H., O'Connor, V.M., Nishiki, T., Takahashi, M., Brose, N., Rhee, J.S., 2010. SNARE protein recycling by alpha SNAP and beta SNAP supports synaptic vesicle priming. Neuron 68, 473–487.
Burgoyne, R.D., Cheek, T.R., 1987. Reorganisation of peripheral actin filaments as a prelude to exocytosis. Biosci. Rep. 7, 281–288.
Burkhardt, P., Hattendorf, D.A., Weis, W.I., Fasshauer, D., 2008. Munc18a controls SNARE assembly through its interaction with the syntaxin N-peptide. EMBO J. 27, 923–933.
Bykhovskaia, M., 2011. Synapsin regulation of vesicle organization and functional pools. Semin. Cell Dev. Biol. 22, 387–392.
Cadas, H., Gaillet, S., Beltramo, M., Venance, L., Piomelli, D., 1996. Biosynthesis of an endogenous cannabinoid precursor in neurons and its control by calcium and cAMP. J. Neurosci. 16, 3934–3942.
Calakos, N., Bennett, M.K., Peterson, K.E., Scheller, R.H., 1994. Protein–protein interactions contributing to the specificity of intracellular vesicular trafficking. Science, New York, 263, 1146–1149.
Calakos, N., Schoch, S., Sudhof, T.C., Malenka, R.C., 2004. Multiple roles for the active zone protein RIM1alpha in late stages of neurotransmitter release. Neuron 42, 889–896.
Campelo, F., McMahon, H.T., Kozlov, M.M., 2008. The hydrophobic insertion mechanism of membrane curvature generation by proteins. Biophys. J. 95, 2325–2339.
Cases-Langhoff, C., Voss, B., Garner, A.M., Appeltauer, U., Takei, K., Kindler, S., Veh, R.W., De Camilli, P., Gundelfinger, E.D., Garner, C.C., 1996. Piccolo, a novel 420 kDa protein associated with the presynaptic cytomatrix. Eur. J. Cell Biol. 69, 214–223.
Castillo, P.E., Janz, R., Sudhof, T.C., Tzounopoulos, T., Malenka, R.C., Nicoll, R.A., 1997. Rab3A is essential for mossy fibre long-term potentiation in the hippocampus. Nature 388, 590–593.
Castillo, P.E., Schoch, S., Schmitz, F., Sudhof, T.C., Malenka, R.C., 2002. RIM1alpha is required for presynaptic long-term potentiation. Nature 415, 327–330.
Castro-Alamancos, M.A., Calcagnotto, M.E., 1999. Presynaptic long-term potentiation in corticothalamic synapses. J. Neurosci. 19, 9090–9097.
Catterall, W.A., 2011. Voltage-gated calcium channels. Cold Spring Harbor Perspect. Biol. 3, a003947.
Ceccarelli, B., Hurlbut, W.P., Mauro, A., 1972. Depletion of vesicles from frog neuromuscular junctions by prolonged tetanic stimulation. J. Cell Biol. 54, 30–38.

Ceccarelli, B., Hurlbut, W.P., Mauro, A., 1973. Turnover of transmitter and synaptic vesicles at the frog neuromuscular junction. J. Cell Biol. 57, 499–524.

Ceridono, M., Ory, S., Momboisse, F., Chasserot-Golaz, S., Houy, S., Calco, V., Haeberle, A.M., Demais, V., Bailly, Y., Bader, M.F., Gasman, S., 2011. Selective recapture of secretory granule components after full collapse exocytosis in neuroendocrine chromaffin cells. Traffic 12, 72–88.

Cesca, F., Baldelli, P., Valtorta, F., Benfenati, F., 2010. The synapsins: key actors of synapse function and plasticity. Prog. Neurobiol. 91, 313–348.

Cestra, G., Castagnoli, L., Dente, L., Minenkova, O., Petrelli, A., Migone, N., Hoffmuller, U., Schneider-Mergener, J., Cesareni, G., 1999. The SH3 domains of endophilin and amphiphysin bind to the proline-rich region of synaptojanin 1 at distinct sites that display an unconventional binding specificity. J. Biol. Chem. 274, 32001–32007.

Chang, W.P., Sudhof, T.C., 2009. SV2 renders primed synaptic vesicles competent for Ca^{2+}-induced exocytosis. J. Neurosci. 29, 883–897.

Chapman, E.R., 2008. How does synaptotagmin trigger neurotransmitter release? Annu. Rev. Biochem. 77, 615–641.

Chapman, P.F., Frenguelli, B.G., Smith, A., Chen, C.M., Silva, A.J., 1995. The alpha-Ca^{2+}/calmodulin kinase II: a bidirectional modulator of presynaptic plasticity. Neuron 14, 591–597.

Chen, K., Richlitzki, A., Featherstone, D.E., Schwarzel, M., Richmond, J.E., 2011. Tomosyn-dependent regulation of synaptic transmission is required for a late phase of associative odor memory. Proc. Natl. Acad. Sci. USA 108, 18482–18487.

Chen, M.S., Obar, R.A., Schroeder, C.C., Austin, T.W., Poodry, C.A., Wadsworth, S.C., Vallee, R.B., 1991. Multiple forms of dynamin are encoded by shibire, a Drosophila gene involved in endocytosis. Nature 351, 583–586.

Chen, Z., Rand, R.P., 1997. The influence of cholesterol on phospholipid membrane curvature and bending elasticity. Biophys. J. 73, 267–276.

Chernomordik, L.V., Kozlov, M.M., 2005. Membrane hemifusion: crossing a chasm in two leaps. Cell 123, 375–382.

Cheung, G., Cousin, M.A., 2012. Adaptor protein complexes 1 and 3 are essential for generation of synaptic vesicles from activity-dependent bulk endosomes. J. Neurosci. 32, 6014–6023.

Cheung, G., Jupp, O.J., Cousin, M.A., 2010. Activity-dependent bulk endocytosis and clathrin-dependent endocytosis replenish specific synaptic vesicle pools in central nerve terminals. J. Neurosci. 30, 8151–8161.

Chevaleyre, V., Castillo, P.E., 2003. Heterosynaptic LTD of hippocampal GABAergic synapses: a novel role of endocannabinoids in regulating excitability. Neuron 38, 461–472.

Chevaleyre, V., Heifets, B.D., Kaeser, P.S., Sudhof, T.C., Castillo, P.E., 2007. Endocannabinoid-mediated long-term plasticity requires cAMP/PKA signaling and RIM1alpha. Neuron 54, 801–812.

Chi, P., Greengard, P., Ryan, T.A., 2001. Synapsin dispersion and reclustering during synaptic activity. Nat. Neurosci. 4, 1187–1193.

Chi, P., Greengard, P., Ryan, T.A., 2003. Synaptic vesicle mobilization is regulated by distinct synapsin I phosphorylation pathways at different frequencies. Neuron 38, 69–78.

Churchward, M.A., Rogasevskaia, T., Hofgen, J., Bau, J., Coorssen, J.R., 2005. Cholesterol facilitates the native mechanism of Ca^{2+}-triggered membrane fusion. J. Cell Sci. 118, 4833–4848.

Citri, A., Malenka, R.C., 2008. Synaptic plasticity: multiple forms, functions, and mechanisms. Neuropsychopharmacology 33, 18–41.

Clayton, E.L., Anggono, V., Smillie, K.J., Chau, N., Robinson, P.J., Cousin, M.A., 2009. The phospho-dependent dynamin-syndapin interaction triggers activity-dependent bulk endocytosis of synaptic vesicles. J. Neurosci. 29, 7706–7717.

Clayton, E.L., Evans, G.J., Cousin, M.A., 2008. Bulk synaptic vesicle endocytosis is rapidly triggered during strong stimulation. J. Neurosci. 28, 6627–6632.

Clayton, E.L., Sue, N., Smillie, K.J., O'Leary, T., Bache, N., Cheung, G., Cole, A.R., Wyllie, D.J., Sutherland, C., Robinson, P.J., Cousin, M.A., 2010. Dynamin I phosphorylation by GSK3 controls activity-dependent bulk endocytosis of synaptic vesicles. Nat. Neurosci. 13, 845–851.

Cocucci, E., Aguet, F., Boulant, S., Kirchhausen, T., 2012. The first five seconds in the life of a clathrin-coated pit. Cell 150, 495–507.

Cohen, R., Atlas, D., 2004. R-type voltage-gated Ca(2+) channel interacts with synaptic proteins and recruits synaptotagmin to the plasma membrane of Xenopus oocytes. Neuroscience 128, 831–841.
Coleman, W.L., Bykhovskaia, M., 2009. Synapsin I accelerates the kinetics of neurotransmitter release in mouse motor terminals. Synapse 63, 531–533.
Coppola, T., Frantz, C., Perret-Menoud, V., Gattesco, S., Hirling, H., Regazzi, R., 2002. Pancreatic beta-cell protein granuphilin binds Rab3 and Munc-18 and controls exocytosis. Mol. Biol. Cell 13, 1906–1915.
Cosgrove, K.E., Galvan, E.J., Barrionuevo, G., Meriney, S.D., 2011. mGluRs modulate strength and timing of excitatory transmission in hippocampal area CA3. Mol. Neurobiol. 44, 93–101.
Cousin, M.A., Evans, G.J., 2011. Activation of silent and weak synapses by cAMP-dependent protein kinase in cultured cerebellar granule neurons. J. Physiol. 589, 1943–1955.
Cousin, M.A., Robinson, P.J., 2001. The dephosphins: dephosphorylation by calcineurin triggers synaptic vesicle endocytosis. Trends Neurosci. 24, 659–665.
Cremona, O., Di Paolo, G., Wenk, M.R., Luthi, A., Kim, W.T., Takei, K., Daniell, L., Nemoto, Y., Shears, S.B., Flavell, R.A., McCormick, D.A., De Camilli, P., 1999. Essential role of phosphoinositide metabolism in synaptic vesicle recycling. Cell 99, 179–188.
Crowder, K.M., Gunther, J.M., Jones, T.A., Hale, B.D., Zhang, H.Z., Peterson, M.R., Scheller, R.H., Chavkin, C., Bajjalieh, S.M., 1999. Abnormal neurotransmission in mice lacking synaptic vesicle protein 2A (SV2A). Proc. Natl. Acad. Sci. USA 96, 15268–15273.
Cuchillo-Ibanez, I., Lejen, T., Albillos, A., Rose, S.D., Olivares, R., Villarroya, M., Garcia, A.G., Trifaro, J.M., 2004. Mitochondrial calcium sequestration and protein kinase C cooperate in the regulation of cortical F-actin disassembly and secretion in bovine chromaffin cells. J. Physiol. 560, 63–76.
Dani, A., Huang, B., Bergan, J., Dulac, C., Zhuang, X., 2010. Superresolution imaging of chemical synapses in the brain. Neuron 68, 843–856.
Daniel, S., Noda, M., Cerione, R.A., Sharp, G.W., 2002. A link between Cdc42 and syntaxin is involved in mastoparan-stimulated insulin release. Biochemistry 41, 9663–9671.
Darcy, K.J., Staras, K., Collinson, L.M., Goda, Y., 2006. Constitutive sharing of recycling synaptic vesicles between presynaptic boutons. Nat. Neurosci. 9, 315–321.
Davletov, B.A., Sudhof, T.C., 1993. A single C2 domain from synaptotagmin I is sufficient for high affinity Ca^{2+}/phospholipid binding. J. Biol. Chem. 268, 26386–26390.
de Lange, R.P., de Roos, A.D., Borst, J.G., 2003. Two modes of vesicle recycling in the rat calyx of Held. J. Neurosci. 23, 10164–10173.
de Wit, H., Walter, A.M., Milosevic, I., Gulyas-Kovacs, A., Riedel, D., Sorensen, J.B., Verhage, M., 2009. Synaptotagmin-1 docks secretory vesicles to syntaxin-1/SNAP-25 acceptor complexes. Cell 138, 935–946.
Deak, F., Schoch, S., Liu, X., Sudhof, T.C., Kavalali, E.T., 2004. Synaptobrevin is essential for fast synaptic-vesicle endocytosis. Nat. Cell Biol. 6, 1102–1108.
Deak, F., Shin, O.H., Tang, J., Hanson, P., Ubach, J., Jahn, R., Rizo, J., Kavalali, E.T., Sudhof, T.C., 2006. Rabphilin regulates SNARE-dependent re-priming of synaptic vesicles for fusion. EMBO J. 25, 2856–2866.
Deak, F., Xu, Y., Chang, W.P., Dulubova, I., Khvotchev, M., Liu, X., Sudhof, T.C., Rizo, J., 2009. Munc18-1 binding to the neuronal SNARE complex controls synaptic vesicle priming. J. Cell Biol. 184, 751–764.
Dean, C., Liu, H., Dunning, F.M., Chang, P.Y., Jackson, M.B., Chapman, E.R., 2009. Synaptotagmin-IV modulates synaptic function and long-term potentiation by regulating BDNF release. Nat. Neurosci. 12, 767–776.
Degtyar, V.E., Allersma, M.W., Axelrod, D., Holz, R.W., 2007. Increased motion and travel, rather than stable docking, characterize the last moments before secretory granule fusion. Proc. Natl. Acad. Sci. USA 104, 15929–15934.
Deguchi-Tawarada, M., Inoue, E., Takao-Rikitsu, E., Inoue, M., Ohtsuka, T., Takai, Y., 2004. CAST2: identification and characterization of a protein structurally related to the presynaptic cytomatrix protein CAST. Genes Cells 9, 15–23.
Deitcher, D.L., Ueda, A., Stewart, B.A., Burgess, R.W., Kidokoro, Y., Schwarz, T.L., 1998. Distinct requirements for evoked and spontaneous release of neurotransmitter are revealed by mutations in the Drosophila gene neuronal-synaptobrevin. J. Neurosci. 18, 2028–2039.

Delgado, R., Maureira, C., Oliva, C., Kidokoro, Y., Labarca, P., 2000. Size of vesicle pools, rates of mobilization, and recycling at neuromuscular synapses of a Drosophila mutant, shibire. Neuron 28, 941–953.
Deng, L., Kaeser, P.S., Xu, W., Sudhof, T.C., 2011. RIM proteins activate vesicle priming by reversing autoinhibitory homodimerization of Munc13. Neuron 69, 317–331.
Denker, A., Bethani, I., Krohnert, K., Korber, C., Horstmann, H., Wilhelm, B.G., Barysch, S.V., Kuner, T., Neher, E., Rizzoli, S.O., 2011. A small pool of vesicles maintains synaptic activity in vivo. Proc. Natl. Acad. Sci. USA 108, 17177–17182.
DeRiemer, S.A., Kaczmarek, L.K., Lai, Y., McGuinness, T.L., Greengard, P., 1984. Calcium/calmodulin-dependent protein phosphorylation in the nervous system of Aplysia. J. Neurosci. 4, 1618–1625.
Desnos, C., Schonn, J.S., Huet, S., Tran, V.S., El-Amraoui, A., Raposo, G., Fanget, I., Chapuis, C., Menasche, G., de Saint Basile, G., Petit, C., Cribier, S., Henry, J.P., Darchen, F., 2003. Rab27A and its effector MyRIP link secretory granules to F-actin and control their motion towards release sites. J. Cell Biol. 163, 559–570.
Diril, M.K., Wienisch, M., Jung, N., Klingauf, J., Haucke, V., 2006. Stonin 2 is an AP-2-dependent endocytic sorting adaptor for synaptotagmin internalization and recycling. Dev. Cell 10, 233–244.
Dittman, J.S., Kaplan, J.M., 2006. Factors regulating the abundance and localization of synaptobrevin in the plasma membrane. Proc. Natl. Acad. Sci. USA 103, 11399–11404.
Dittman, J.S., Kreitzer, A.C., Regehr, W.G., 2000. Interplay between facilitation, depression, and residual calcium at three presynaptic terminals. J. Neurosci. 20, 1374–1385.
Dobrunz, L.E., 2002. Release probability is regulated by the size of the readily releasable vesicle pool at excitatory synapses in hippocampus. Int. J. Dev. Neurosci. 20, 225–236.
Dobrunz, L.E., Stevens, C.F., 1997. Heterogeneity of release probability, facilitation, and depletion at central synapses. Neuron 18, 995–1008.
Domanska, M.K., Kiessling, V., Tamm, L.K., 2010. Docking and fast fusion of synaptobrevin vesicles depends on the lipid compositions of the vesicle and the acceptor SNARE complex-containing target membrane. Biophys. J. 99, 2936–2946.
Doreian, B.W., Fulop, T.G., Meklemburg, R.L., Smith, C.B., 2009. Cortical F-actin, the exocytic mode, and neuropeptide release in mouse chromaffin cells is regulated by myristoylated alanine-rich C-kinase substrate and myosin II. Mol. Biol. Cell 20, 3142–3154.
Doreian, B.W., Fulop, T.G., Smith, C.B., 2008. Myosin II activation and actin reorganization regulate the mode of quantal exocytosis in mouse adrenal chromaffin cells. J. Neurosci. 28, 4470–4478.
Doussau, F., Augustine, G.J., 2000. The actin cytoskeleton and neurotransmitter release: an overview. Biochimie 82, 353–363.
Dresbach, T., Qualmann, B., Kessels, M.M., Garner, C.C., Gundelfinger, E.D., 2001. The presynaptic cytomatrix of brain synapses. Cell. Mol. Life Sci. 58, 94–116.
Dulubova, I., Lou, X., Lu, J., Huryeva, I., Alam, A., Schneggenburger, R., Sudhof, T.C., Rizo, J., 2005. A Munc13/RIM/Rab3 tripartite complex: from priming to plasticity? EMBO J. 24, 2839–2850.
Dulubova, I., Sugita, S., Hill, S., Hosaka, M., Fernandez, I., Sudhof, T.C., Rizo, J., 1999. A conformational switch in syntaxin during exocytosis: role of munc18. EMBO J. 18, 4372–4382.
Dumitrescu Pene, T., Rose, S.D., Lejen, T., Marcu, M.G., Trifaro, J.M., 2005. Expression of various scinderin domains in chromaffin cells indicates that this protein acts as a molecular switch in the control of actin filament dynamics and exocytosis. J. Neurochem. 92, 780–789.
Dyck, B.A., Beyaert, M.G., Ferro, M.A., Mishra, R.K., 2011. Medial prefrontal cortical synapsin II knock-down induces behavioral abnormalities in the rat: examining synapsin II in the pathophysiology of schizophrenia. Schizophr. Res. 130, 250–259.
Dyck, B.A., Skoblenick, K.J., Castellano, J.M., Ki, K., Thomas, N., Mishra, R.K., 2009. Behavioral abnormalities in synapsin II knockout mice implicate a causal factor in schizophrenia. Synapse 63, 662–672.
Elmqvist, D., Quastel, D.M., 1965. A quantitative study of end-plate potentials in isolated human muscle. J. Physiol. 178, 505–529.
Faelber, K., Posor, Y., Gao, S., Held, M., Roske, Y., Schulze, D., Haucke, V., Noe, F., Daumke, O., 2011. Crystal structure of nucleotide-free dynamin. Nature 477, 556–560.
Farsad, K., Ringstad, N., Takei, K., Floyd, S.R., Rose, K., De Camilli, P., 2001. Generation of high curvature membranes mediated by direct endophilin bilayer interactions. J. Cell Biol. 155, 193–200.

Fasshauer, D., Margittai, M., 2004. A transient N-terminal interaction of SNAP-25 and syntaxin nucleates SNARE assembly. J. Biol. Chem. 279, 7613–7621.
Fasshauer, D., Sutton, R.B., Brunger, A.T., Jahn, R., 1998. Conserved structural features of the synaptic fusion complex: SNARE proteins reclassified as Q- and R-SNAREs. Proc. Natl. Acad. Sci. USA 95, 15781–15786.
Fejtova, A., Davydova, D., Bischof, F., Lazarevic, V., Altrock, W.D., Romorini, S., Schone, C., Zuschratter, W., Kreutz, M.R., Garner, C.C., Ziv, N.E., Gundelfinger, E.D., 2009. Dynein light chain regulates axonal trafficking and synaptic levels of Bassoon. J. Cell Biol. 185, 341–355.
Fenster, S.D., Chung, W.J., Zhai, R., Cases-Langhoff, C., Voss, B., Garner, A.M., Kaempf, U., Kindler, S., Gundelfinger, E.D., Garner, C.C., 2000. Piccolo, a presynaptic zinc finger protein structurally related to bassoon. Neuron 25, 203–214.
Ferguson, G.D., Anagnostaras, S.G., Silva, A.J., Herschman, H.R., 2000. Deficits in memory and motor performance in synaptotagmin IV mutant mice. Proc. Natl. Acad. Sci. USA 97, 5598–5603.
Ferguson, S.M., Brasnjo, G., Hayashi, M., Wolfel, M., Collesi, C., Giovedi, S., Raimondi, A., Gong, L.W., Ariel, P., Paradise, S., O'Toole, E., Flavell, R., Cremona, O., Miesenbock, G., Ryan, T.A., De Camilli, P., 2007. A selective activity-dependent requirement for dynamin 1 in synaptic vesicle endocytosis. Science, New York, 316, 570–574.
Fernandez-Alfonso, T., Kwan, R., Ryan, T.A., 2006. Synaptic vesicles interchange their membrane proteins with a large surface reservoir during recycling. Neuron 51, 179–186.
Ferreira, A., Kao, H.T., Feng, J., Rapoport, M., Greengard, P., 2000. Synapsin III: developmental expression, subcellular localization, and role in axon formation. J. Neurosci. 20, 3736–3744.
Fisher, S.A., Fischer, T.M., Carew, T.J., 1997. Multiple overlapping processes underlying short-term synaptic enhancement. Trends Neurosci. 20, 170–177.
Ford, M.G., Jenni, S., Nunnari, J., 2011. The crystal structure of dynamin. Nature 477, 561–566.
Ford, M.G., Mills, I.G., Peter, B.J., Vallis, Y., Praefcke, G.J., Evans, P.R., McMahon, H.T., 2002. Curvature of clathrin-coated pits driven by epsin. Nature 419, 361–366.
Fotin, A., Cheng, Y., Grigorieff, N., Walz, T., Harrison, S.C., Kirchhausen, T., 2004a. Structure of an auxilin-bound clathrin coat and its implications for the mechanism of uncoating. Nature 432, 649–653.
Fotin, A., Cheng, Y., Sliz, P., Grigorieff, N., Harrison, S.C., Kirchhausen, T., Walz, T., 2004b. Molecular model for a complete clathrin lattice from electron cryomicroscopy. Nature 432, 573–579.
Fourcaudot, E., Gambino, F., Humeau, Y., Casassus, G., Shaban, H., Poulain, B., Luthi, A., 2008. cAMP/PKA signaling and RIM1alpha mediate presynaptic LTP in the lateral amygdala. Proc. Natl. Acad. Sci. USA 105, 15130–15135.
Frank, T., Rutherford, M.A., Strenzke, N., Neef, A., Pangrsic, T., Khimich, D., Fejtova, A., Gundelfinger, E.D., Liberman, M.C., Harke, B., Bryan, K.E., Lee, A., Egner, A., Riedel, D., Moser, T., 2010. Bassoon and the synaptic ribbon organize Ca(2)+ channels and vesicles to add release sites and promote refilling. Neuron 68, 724–738.
Fredj, N.B., Burrone, J., 2009. A resting pool of vesicles is responsible for spontaneous vesicle fusion at the synapse. Nat. Neurosci. 12, 751–758.
Freund, T.F., Katona, I., Piomelli, D., 2003. Role of endogenous cannabinoids in synaptic signaling. Physiol. Rev. 83, 1017–1066.
Friedrich, R., Groffen, A.J., Connell, E., van Weering, J.R., Gutman, O., Henis, Y.I., Davletov, B., Ashery, U., 2008. DOC2B acts as a calcium switch and enhances vesicle fusion. J. Neurosci. 28, 6794–6806.
Fujita, Y., Shirataki, H., Sakisaka, T., Asakura, T., Ohya, T., Kotani, H., Yokoyama, S., Nishioka, H., Matsuura, Y., Mizoguchi, A., Scheller, R.H., Takai, Y., 1998. Tomosyn: a syntaxin-1-binding protein that forms a novel complex in the neurotransmitter release process. Neuron 20, 905–915.
Fukuda, M., 2003. Slp4-a/granuphilin-a inhibits dense-core vesicle exocytosis through interaction with the GDP-bound form of Rab27A in PC12 cells. J. Biol. Chem. 278, 15390–15396.
Fukuda, M., 2007. The role of synaptotagmin and synaptotagmin-like protein (Slp) in regulated exocytosis. In: Regazzi, R. (Ed.), Molecular Mechanisms of Exocytosis. Landes Bioscience, Austin, TX, USA, pp. 42–61.
Fukuda, M., Imai, A., Nashida, T., Shimomura, H., 2005. Slp4-a/granuphilin-a interacts with syntaxin-2/3 in a Munc18-2-dependent manner. J. Biol. Chem. 280, 39175–39184.

Furber, K.L., Churchward, M.A., Rogasevskaia, T.P., Coorssen, J.R., 2009. Identifying critical components of native Ca^{2+}-triggered membrane fusion. Integrating studies of proteins and lipids. Ann. N.Y. Acad. Sci. 1152, 121–134.
Galbraith, C.G., Galbraith, J.A., 2011. Super-resolution microscopy at a glance. J. Cell Sci. 124, 1607–1611.
Galiegue, S., Mary, S., Marchand, J., Dussossoy, D., Carriere, D., Carayon, P., Bouaboula, M., Shire, D., Le Fur, G., Casellas, P., 1995. Expression of central and peripheral cannabinoid receptors in human immune tissues and leukocyte subpopulations. Eur. J. Biochem. 232, 54–61.
Gallop, J.L., Jao, C.C., Kent, H.M., Butler, P.J., Evans, P.R., Langen, R., McMahon, H.T., 2006. Mechanism of endophilin N-BAR domain-mediated membrane curvature. EMBO J. 25, 2898–2910.
Gasman, S., Chasserot-Golaz, S., Malacombe, M., Way, M., Bader, M.F., 2004. Regulated exocytosis in neuroendocrine cells: a role for subplasmalemmal Cdc42/N-WASP-induced actin filaments. Mol. Biol. Cell 15, 520–531.
Gebremedhin, D., Lange, A.R., Campbell, W.B., Hillard, C.J., Harder, D.R., 1999. Cannabinoid CB1 receptor of cat cerebral arterial muscle functions to inhibit L-type Ca^{2+} channel current. Am. J. Physiol. 276, H2085–H2093.
Geppert, M., Bolshakov, V.Y., Siegelbaum, S.A., Takei, K., De Camilli, P., Hammer, R.E., Sudhof, T.C., 1994a. The role of Rab3A in neurotransmitter release. Nature 369, 493–497.
Geppert, M., Goda, Y., Hammer, R.E., Li, C., Rosahl, T.W., Stevens, C.F., Sudhof, T.C., 1994b. Synaptotagmin I: a major Ca^{2+} sensor for transmitter release at a central synapse. Cell 79, 717–727.
Geppert, M., Goda, Y., Stevens, C.F., Sudhof, T.C., 1997. The small GTP-binding protein Rab3A regulates a late step in synaptic vesicle fusion. Nature 387, 810–814.
Gerber, S.H., Rah, J.C., Min, S.W., Liu, X., de Wit, H., Dulubova, I., Meyer, A.C., Rizo, J., Arancillo, M., Hammer, R.E., Verhage, M., Rosenmund, C., Sudhof, T.C., 2008. Conformational switch of syntaxin-1 controls synaptic vesicle fusion. Science, New York, 321, 1507–1510.
Gerdeman, G.L., Ronesi, J., Lovinger, D.M., 2002. Postsynaptic endocannabinoid release is critical to long-term depression in the striatum. Nat. Neurosci. 5, 446–451.
Giner, D., Neco, P., Frances Mdel, M., Lopez, I., Viniegra, S., Gutierrez, L.M., 2005. Real-time dynamics of the F-actin cytoskeleton during secretion from chromaffin cells. J. Cell Sci. 118, 2871–2880.
Giraudo, C.G., Garcia-Diaz, A., Eng, W.S., Chen, Y., Hendrickson, W.A., Melia, T.J., Rothman, J.E., 2009. Alternative zippering as an on–off switch for SNARE-mediated fusion. Science, New York, 323, 512–516.
Gitler, D., Cheng, Q., Greengard, P., Augustine, G.J., 2008. Synapsin IIa controls the reserve pool of glutamatergic synaptic vesicles. J. Neurosci. 28, 10835–10843.
Gitler, D., Takagishi, Y., Feng, J., Ren, Y., Rodriguiz, R.M., Wetsel, W.C., Greengard, P., Augustine, G.J., 2004. Different presynaptic roles of synapsins at excitatory and inhibitory synapses. J. Neurosci. 24, 11368–11380.
Gladycheva, S.E., Lam, A.D., Liu, J., D'Andrea-Merrins, M., Yizhar, O., Lentz, S.I., Ashery, U., Ernst, S.A., Stuenkel, E.L., 2007. Receptor-mediated regulation of tomosyn-syntaxin 1A interactions in bovine adrenal chromaffin cells. J. Biol. Chem. 282, 22887–22899.
Glyvuk, N., Tsytsyura, Y., Geumann, C., D'Hooge, R., Huve, J., Kratzke, M., Baltes, J., Boening, D., Klingauf, J., Schu, P., 2010. AP-1/sigma1B-adaptin mediates endosomal synaptic vesicle recycling, learning and memory. EMBO J. 29, 1318–1330.
Goda, Y., Stevens, C.F., 1994. Two components of transmitter release at a central synapse. Proc. Natl. Acad. Sci. USA 91, 12942–12946.
Gordon, S.L., Leube, R.E., Cousin, M.A., 2011. Synaptophysin is required for synaptobrevin retrieval during synaptic vesicle endocytosis. J. Neurosci. 31, 14032–14036.
Gracheva, E.O., Burdina, A.O., Holgado, A.M., Berthelot-Grosjean, M., Ackley, B.D., Hadwiger, G., Nonet, M.L., Weimer, R.M., Richmond, J.E., 2006. Tomosyn inhibits synaptic vesicle priming in *Caenorhabditis elegans*. PLoS Biol. 4, e261.
Gracheva, E.O., Burdina, A.O., Touroutine, D., Berthelot-Grosjean, M., Parekh, H., Richmond, J.E., 2007. Tomosyn negatively regulates CAPS-dependent peptide release at *Caenorhabditis elegans* synapses. J. Neurosci. 27, 10176–10184.
Gracheva, E.O., Maryon, E.B., Berthelot-Grosjean, M., Richmond, J.E., 2010. Differential regulation of synaptic vesicle tethering and docking by UNC-18 and TOM-1. Front. Synaptic Neurosci. 2, 141.

Granseth, B., Odermatt, B., Royle, S.J., Lagnado, L., 2006. Clathrin-mediated endocytosis is the dominant mechanism of vesicle retrieval at hippocampal synapses. Neuron 51, 773–786.

Granseth, B., Odermatt, B., Royle, S.J., Lagnado, L., 2009. Comment on "The dynamic control of kiss-and-run and vesicular reuse probed with single nanoparticles". Science, New York, 325, 1499 author reply 1499.

Grindstaff, K.K., Yeaman, C., Anandasabapathy, N., Hsu, S.C., Rodriguez-Boulan, E., Scheller, R.H., Nelson, W.J., 1998. Sec6/8 complex is recruited to cell–cell contacts and specifies transport vesicle delivery to the basal-lateral membrane in epithelial cells. Cell 93, 731–740.

Groffen, A.J., Friedrich, R., Brian, E.C., Ashery, U., Verhage, M., 2006. DOC2A and DOC2B are sensors for neuronal activity with unique calcium-dependent and kinetic properties. J. Neurochem. 97, 818–833.

Groffen, A.J., Martens, S., Diez Arazola, R., Cornelisse, L.N., Lozovaya, N., de Jong, A.P., Goriounova, N.A., Habets, R.L., Takai, Y., Borst, J.G., Brose, N., McMahon, H.T., Verhage, M., 2010. Doc2b is a high-affinity Ca^{2+} sensor for spontaneous neurotransmitter release. Science, New York, 327, 1614–1618.

Gu, M., Schuske, K., Watanabe, S., Liu, Q., Baum, P., Garriga, G., Jorgensen, E.M., 2008. Mu2 adaptin facilitates but is not essential for synaptic vesicle recycling in *Caenorhabditis elegans*. J. Cell Biol. 183, 881–892.

Gulyas-Kovacs, A., de Wit, H., Milosevic, I., Kochubey, O., Toonen, R., Klingauf, J., Verhage, M., Sorensen, J.B., 2007. Munc18-1: sequential interactions with the fusion machinery stimulate vesicle docking and priming. J. Neurosci. 27, 8676–8686.

Gundelfinger, E.D., tom Dieck, S., 2000. Molecular organization of excitatory chemical synapses in the mammalian brain. Naturwissenschaften 87, 513–523.

Gustafsson, M.G., 2000. Surpassing the lateral resolution limit by a factor of two using structured illumination microscopy. J. Microsc. 198, 82–87.

Hagalili, Y., Bachnoff, N., Atlas, D., 2008. The voltage-gated Ca(2+) channel is the Ca(2+) sensor protein of secretion. Biochemistry 47, 13822–13830.

Hagler Jr., D.J., Goda, Y., 2001. Properties of synchronous and asynchronous release during pulse train depression in cultured hippocampal neurons. J. Neurophysiol. 85, 2324–2334.

Hallermann, S., Fejtova, A., Schmidt, H., Weyhersmuller, A., Silver, R.A., Gundelfinger, E.D., Eilers, J., 2010. Bassoon speeds vesicle reloading at a central excitatory synapse. Neuron 68, 710–723.

Han, X., Wang, C.T., Bai, J., Chapman, E.R., Jackson, M.B., 2004. Transmembrane segments of syntaxin line the fusion pore of Ca^{2+}-triggered exocytosis. Science, New York, 304, 289–292.

Han, Y., Kaeser, P.S., Sudhof, T.C., Schneggenburger, R., 2011. RIM determines Ca(2)+ channel density and vesicle docking at the presynaptic active zone. Neuron 69, 304–316.

Handley, M.T., Haynes, L.P., Burgoyne, R.D., 2007. Differential dynamics of Rab3A and Rab27A on secretory granules. J. Cell Sci. 120, 973–984.

Hanson, P.I., Roth, R., Morisaki, H., Jahn, R., Heuser, J.E., 1997. Structure and conformational changes in NSF and its membrane receptor complexes visualized by quick-freeze/deep-etch electron microscopy. Cell 90, 523–535.

Harata, N., Pyle, J.L., Aravanis, A.M., Mozhayeva, M., Kavalali, E.T., Tsien, R.W., 2001. Limited numbers of recycling vesicles in small CNS nerve terminals: implications for neural signaling and vesicular cycling. Trends Neurosci. 24, 637–643.

Harlow, M.L., Ress, D., Stoschek, A., Marshall, R.M., McMahan, U.J., 2001. The architecture of active zone material at the frog's neuromuscular junction. Nature 409, 479–484.

Harris, E.W., Cotman, C.W., 1986. Long-term potentiation of guinea pig mossy fiber responses is not blocked by N-methyl D-aspartate antagonists. Neurosci. Lett. 70, 132–137.

Hartwig, J.H., Bokoch, G.M., Carpenter, C.L., Janmey, P.A., Taylor, L.A., Toker, A., Stossel, T.P., 1995. Thrombin receptor ligation and activated Rac uncap actin filament barbed ends through phosphoinositide synthesis in permeabilized human platelets. Cell 82, 643–653.

Hartwig, J.H., Thelen, M., Rosen, A., Janmey, P.A., Nairn, A.C., Aderem, A., 1992. MARCKS is an actin filament cross-linking protein regulated by protein kinase C and calcium-calmodulin. Nature 356, 618–622.

Harvey-Girard, E., Lewis, J., Maler, L., 2010. Burst-induced anti-Hebbian depression acts through short-term synaptic dynamics to cancel redundant sensory signals. J. Neurosci. 30, 6152–6169.

Hatsuzawa, K., Lang, T., Fasshauer, D., Bruns, D., Jahn, R., 2003. The R-SNARE motif of tomosyn forms SNARE core complexes with syntaxin 1 and SNAP-25 and down-regulates exocytosis. J. Biol. Chem. 278, 31159–31166.

Haucke, V., Neher, E., Sigrist, S.J., 2011. Protein scaffolds in the coupling of synaptic exocytosis and endocytosis. Nat. Rev. Neurosci. 12, 127–138.

Haucke, V., Wenk, M.R., Chapman, E.R., Farsad, K., De Camilli, P., 2000. Dual interaction of synaptotagmin with mu2- and alpha-adaptin facilitates clathrin-coated pit nucleation. EMBO J. 19, 6011–6019.

Hawkins, R.D., Kandel, E.R., Siegelbaum, S.A., 1993. Learning to modulate transmitter release: themes and variations in synaptic plasticity. Annu. Rev. Neurosci. 16, 625–665.

He, B., Guo, W., 2009. The exocyst complex in polarized exocytosis. Curr. Opin. Cell Biol. 21, 537–542.

He, L., Wu, X.S., Mohan, R., Wu, L.G., 2006. Two modes of fusion pore opening revealed by cell-attached recordings at a synapse. Nature 444, 102–105.

Hebb, D.O., 1949. The Organization of Behavior. Wiley & Sons, New York.

Hell, S.W., 2007. Far-field optical nanoscopy. Science 316, 1153–1158.

Henkel, A.W., Kang, G., Kornhuber, J., 2001. A common molecular machinery for exocytosis and the 'kiss-and-run' mechanism in chromaffin cells is controlled by phosphorylation. J. Cell Sci. 114, 4613–4620.

Henne, W.M., Boucrot, E., Meinecke, M., Evergren, E., Vallis, Y., Mittal, R., McMahon, H.T., 2010. FCHo proteins are nucleators of clathrin-mediated endocytosis. Science 328, 1281–1284.

Hensbroek, R.A., Kamal, A., Baars, A.M., Verhage, M., Spruijt, B.M., 2003. Spatial, contextual and working memory are not affected by the absence of mossy fiber long-term potentiation and depression. Behav. Brain Res. 138, 215–223.

Herrick, D.Z., Sterbling, S., Rasch, K.A., Hinderliter, A., Cafiso, D.S., 2006. Position of synaptotagmin I at the membrane interface: cooperative interactions of tandem C2 domains. Biochemistry 45, 9668–9674.

Hertzog, M., Chavrier, P., 2011. Cell polarity during motile processes: keeping on track with the exocyst complex. Biochem. J. 433, 403–409.

Hessler, N.A., Shirke, A.M., Malinow, R., 1993. The probability of transmitter release at a mammalian central synapse. Nature 366, 569–572.

Heuser, J.E., Reese, T.S., 1973. Evidence for recycling of synaptic vesicle membrane during transmitter release at the frog neuromuscular junction. J. Cell Biol. 57, 315–344.

Heuser, J.E., Reese, T.S., Landis, D.M., 1974. Functional changes in frog neuromuscular junctions studied with freeze-fracture. J. Neurocytol. 3, 109–131.

Hilfiker, S., Benfenati, F., Doussau, F., Nairn, A.C., Czernik, A.J., Augustine, G.J., Greengard, P., 2005. Structural domains involved in the regulation of transmitter release by synapsins. J. Neurosci. 25, 2658–2669.

Hilfiker, S., Schweizer, F.E., Kao, H.T., Czernik, A.J., Greengard, P., Augustine, G.J., 1998. Two sites of action for synapsin domain E in regulating neurotransmitter release. Nat. Neurosci. 1, 29–35.

Hinshaw, J.E., Schmid, S.L., 1995. Dynamin self-assembles into rings suggesting a mechanism for coated vesicle budding. Nature 374, 190–192.

Hohaus, A., Beyl, S., Kudrnac, M., Berjukow, S., Timin, E.N., Marksteiner, R., Maw, M.A., Hering, S., 2005. Structural determinants of L-type channel activation in segment IIS6 revealed by a retinal disorder. J. Biol. Chem. 280, 38471–38477.

Hojjati, M.R., van Woerden, G.M., Tyler, W.J., Giese, K.P., Silva, A.J., Pozzo-Miller, L., Elgersma, Y., 2007. Kinase activity is not required for alpha-CaMKII-dependent presynaptic plasticity at CA3–CA1 synapses. Nat. Neurosci. 10, 1125–1127.

Holz, R.W., Brondyk, W.H., Senter, R.A., Kuizon, L., Macara, I.G., 1994. Evidence for the involvement of Rab3A in Ca(2+)-dependent exocytosis from adrenal chromaffin cells. J. Biol. Chem. 269, 10229–10234.

Hosaka, M., Hammer, R.E., Sudhof, T.C., 1999. A phospho-switch controls the dynamic association of synapsins with synaptic vesicles. Neuron 24, 377–387.

Hosoi, N., Holt, M., Sakaba, T., 2009. Calcium dependence of exo- and endocytotic coupling at a glutamatergic synapse. Neuron 63, 216–229.

Hsu, S.C., TerBush, D., Abraham, M., Guo, W., 2004. The exocyst complex in polarized exocytosis. Int. Rev. Cytol. 233, 243–265.

Hu, C., Ahmed, M., Melia, T.J., Sollner, T.H., Mayer, T., Rothman, J.E., 2003. Fusion of cells by flipped SNAREs. Science, New York, 300, 1745–1749.
Hua, Y., Scheller, R.H., 2001. Three SNARE complexes cooperate to mediate membrane fusion. Proc. Natl. Acad. Sci. USA 98, 8065–8070.
Hua, Y., Sinha, R., Thiel, C.S., Schmidt, R., Huve, J., Martens, H., Hell, S.W., Egner, A., Klingauf, J., 2011. A readily retrievable pool of synaptic vesicles. Nat. Neurosci. 14, 833–839.
Huang, C.C., Yang, D.M., Lin, C.C., Kao, L.S., 2011. Involvement of Rab3A in vesicle priming during exocytosis: interaction with Munc13-1 and Munc18-1. Traffic 12, 1356–1370.
Huang, Y.Y., Kandel, E.R., 1994. Recruitment of long-lasting and protein kinase A-dependent long-term potentiation in the CA1 region of hippocampus requires repeated tetanization. Learn. Mem. 1, 74–82.
Huang, Y.Y., Nguyen, P.V., Abel, T., Kandel, E.R., 1996. Long-lasting forms of synaptic potentiation in the mammalian hippocampus. Learn. Mem. 3, 74–85.
Huang, Z., Chen, Y., Nissenson, R.A., 1995. The cytoplasmic tail of the G-protein-coupled receptor for parathyroid hormone and parathyroid hormone-related protein contains positive and negative signals for endocytosis. J. Biol. Chem. 270, 151–156.
Hui, E., Bai, J., Chapman, E.R., 2006. Ca2+-triggered simultaneous membrane penetration of the tandem C2-domains of synaptotagmin I. Biophys. J. 91, 1767–1777.
Hui, E., Johnson, C.P., Yao, J., Dunning, F.M., Chapman, E.R., 2009. Synaptotagmin-mediated bending of the target membrane is a critical step in Ca(2+)-regulated fusion. Cell 138, 709–721.
Huntwork, S., Littleton, J.T., 2007. A complexin fusion clamp regulates spontaneous neurotransmitter release and synaptic growth. Nat. Neurosci. 10, 1235–1237.
Ikeda, K., Bekkers, J.M., 2009. Counting the number of releasable synaptic vesicles in a presynaptic terminal. Proc. Natl. Acad. Sci. USA 106, 2945–2950.
Inoue, M., Chang, L., Hwang, J., Chiang, S.H., Saltiel, A.R., 2003. The exocyst complex is required for targeting of Glut4 to the plasma membrane by insulin. Nature 422, 629–633.
Ishizuka, T., Saisu, H., Odani, S., Abe, T., 1995. Synaphin: a protein associated with the docking/fusion complex in presynaptic terminals. Biochem. Biophys. Res. Commun. 213, 1107–1114.
Janz, R., Sudhof, T.C., 1999. SV2C is a synaptic vesicle protein with an unusually restricted localization: anatomy of a synaptic vesicle protein family. Neuroscience 94, 1279–1290.
Jena, B.P., Cho, S.J., Jeremic, A., Stromer, M.H., Abu-Hamdah, R., 2003. Structure and composition of the fusion pore. Biophys. J. 84, 1337–1343.
Jin, P., Griffith, L.C., Murphey, R.K., 1998. Presynaptic calcium/calmodulin-dependent protein kinase II regulates habituation of a simple reflex in adult Drosophila. J. Neurosci. 18, 8955–8964.
Jockusch, W.J., Speidel, D., Sigler, A., Sorensen, J.B., Varoqueaux, F., Rhee, J.S., Brose, N., 2007. CAPS-1 and CAPS-2 are essential synaptic vesicle priming proteins. Cell 131, 796–808.
Johannes, L., Lledo, P.M., Chameau, P., Vincent, J.D., Henry, J.P., Darchen, F., 1998. Regulation of the Ca^{2+} sensitivity of exocytosis by Rab3a. J. Neurochem. 71, 1127–1133.
Johnson, J.L., Ellis, B.A., Noack, D., Seabra, M.C., Catz, S.D., 2005. The Rab27a-binding protein, JFC1, regulates androgen-dependent secretion of prostate-specific antigen and prostatic-specific acid phosphatase. Biochem. J. 391, 699–710.
Johnson, J.R., Ferdek, P., Lian, L.Y., Barclay, J.W., Burgoyne, R.D., Morgan, A., 2009. Binding of UNC-18 to the N-terminus of syntaxin is essential for neurotransmission in *Caenorhabditis elegans*. Biochem. J. 418, 73–80.
Jones, A.L., Hradek, G.T., Hornick, C., Renaud, G., Windler, E.E., Havel, R.J., 1984. Uptake and processing of remnants of chylomicrons and very low density lipoproteins by rat liver. J. Lipid Res. 25, 1151–1158.
Jones, S.A., Shim, S.H., He, J., Zhuang, X., 2011. Fast, three-dimensional super-resolution imaging of live cells. Nat. Methods 8, 499–508.
Jorgacevski, J., Zorec, R., 2012. Munc18-1, exocytotic fusion pore regulation and local membrane anisotropy. Commun. Integr. Biol. 5, 74–77.
Jorgensen, E.M., Hartwieg, E., Schuske, K., Nonet, M.L., Jin, Y., Horvitz, H.R., 1995. Defective recycling of synaptic vesicles in synaptotagmin mutants of *Caenorhabditis elegans*. Nature 378, 196–199.
Jovanovic, J.N., Sihra, T.S., Nairn, A.C., Hemmings Jr., H.C., Greengard, P., Czernik, A.J., 2001. Opposing changes in phosphorylation of specific sites in synapsin I during Ca^{2+}-dependent glutamate release in isolated nerve terminals. J. Neurosci. 21, 7944–7953.

Junge, H.J., Rhee, J.S., Jahn, O., Varoqueaux, F., Spiess, J., Waxham, M.N., Rosenmund, C., Brose, N., 2004. Calmodulin and Munc13 form a Ca^{2+} sensor/effector complex that controls short-term synaptic plasticity. Cell 118, 389–401.

Kaeser, P.S., Kwon, H.B., Blundell, J., Chevaleyre, V., Morishita, W., Malenka, R.C., Powell, C.M., Castillo, P.E., Sudhof, T.C., 2008. RIM1alpha phosphorylation at serine-413 by protein kinase A is not required for presynaptic long-term plasticity or learning. Proc. Natl. Acad. Sci. USA 105, 14680–14685.

Kamin, D., Lauterbach, M.A., Westphal, V., Keller, J., Schonle, A., Hell, S.W., Rizzoli, S.O., 2010. High- and low-mobility stages in the synaptic vesicle cycle. Biophys. J. 99, 675–684.

Kamiya, H., Zucker, R.S., 1994. Residual Ca^{2+} and short-term synaptic plasticity. Nature 371, 603–606.

Karatekin, E., Di Giovanni, J., Iborra, C., Coleman, J., O'Shaughnessy, B., Seagar, M., Rothman, J.E., 2010. A fast, single-vesicle fusion assay mimics physiological SNARE requirements. Proc. Natl. Acad. Sci. USA 107, 3517–3521.

Kaufmann, N., DeProto, J., Ranjan, R., Wan, H., Van Vactor, D., 2002. Drosophila liprin-alpha and the receptor phosphatase Dlar control synapse morphogenesis. Neuron 34, 27–38.

Keller, J.E., Cai, F., Neale, E.A., 2004. Uptake of botulinum neurotoxin into cultured neurons. Biochemistry 43, 526–532.

Khvotchev, M., Dulubova, I., Sun, J., Dai, H., Rizo, J., Sudhof, T.C., 2007. Dual modes of Munc18-1/SNARE interactions are coupled by functionally critical binding to syntaxin-1 N terminus. J. Neurosci. 27, 12147–12155.

Kilbride, J., Rush, A.M., Rowan, M.J., Anwyl, R., 2001. Presynaptic group II mGluR inhibition of short-term depression in the medial perforant path of the dentate gyrus in vitro. J. Neurophysiol. 85, 2509–2515.

Kim, D., Thayer, S.A., 2001. Cannabinoids inhibit the formation of new synapses between hippocampal neurons in culture. J. Neurosci. 21, RC146.

Kim, J., Isokawa, M., Ledent, C., Alger, B.E., 2002. Activation of muscarinic acetylcholine receptors enhances the release of endogenous cannabinoids in the hippocampus. J. Neurosci. 22, 10182–10191.

Kim, S., Ko, J., Shin, H., Lee, J.R., Lim, C., Han, J.H., Altrock, W.D., Garner, C.C., Gundelfinger, E.D., Premont, R.T., Kaang, B.K., Kim, E., 2003. The GIT family of proteins forms multimers and associates with the presynaptic cytomatrix protein Piccolo. J. Biol. Chem. 278, 6291–6300.

Kim, S.H., Ryan, T.A., 2009. Synaptic vesicle recycling at CNS synapses without AP-2. J. Neurosci. 29, 3865–3874.

Kim, S.H., Ryan, T.A., 2010. CDK5 serves as a major control point in neurotransmitter release. Neuron 67, 797–809.

Kirchhausen, T., Harrison, S.C., 1981. Protein organization in clathrin trimers. Cell 23, 755–761.

Kittel, R.J., Wichmann, C., Rasse, T.M., Fouquet, W., Schmidt, M., Schmid, A., Wagh, D.A., Pawlu, C., Kellner, R.R., Willig, K.I., Hell, S.W., Buchner, E., Heckmann, M., Sigrist, S.J., 2006. Bruchpilot promotes active zone assembly, Ca^{2+} channel clustering, and vesicle release. Science 312, 1051–1054.

Klar, T.A., Jakobs, S., Dyba, M., Egner, A., Hell, S.W., 2000. Fluorescence microscopy with diffraction resolution barrier broken by stimulated emission. Proc. Natl. Acad. Sci. USA 97, 8206–8210.

Klingauf, J., Kavalali, E.T., Tsien, R.W., 1998. Kinetics and regulation of fast endocytosis at hippocampal synapses. Nature 394, 581–585.

Ko, J., Na, M., Kim, S., Lee, J.R., Kim, E., 2003. Interaction of the ERC family of RIM-binding proteins with the liprin-alpha family of multidomain proteins. J. Biol. Chem. 278, 42377–42385.

Kochubey, O., Lou, X., Schneggenburger, R., 2011. Regulation of transmitter release by Ca(2+) and synaptotagmin: insights from a large CNS synapse. Trends Neurosci. 34, 237–246.

Kochubey, O., Schneggenburger, R., 2011. Synaptotagmin increases the dynamic range of synapses by driving Ca(2)+-evoked release and by clamping a near-linear remaining Ca(2)+ sensor. Neuron 69, 736–748.

Koenig, J.H., Ikeda, K., 1989. Disappearance and reformation of synaptic vesicle membrane upon transmitter release observed under reversible blockage of membrane retrieval. J. Neurosci. 9, 3844–3860.

Koh, T.W., Bellen, H.J., 2003. Synaptotagmin I, a Ca^{2+} sensor for neurotransmitter release. Trends Neurosci. 26, 413–422.

Koo, S.J., Markovic, S., Puchkov, D., Mahrenholz, C.C., Beceren-Braun, F., Maritzen, T., Dernedde, J., Volkmer, R., Oschkinat, H., Haucke, V., 2011. SNARE motif-mediated sorting of synaptobrevin by the endocytic adaptors clathrin assembly lymphoid myeloid leukemia (CALM) and AP180 at synapses. Proc. Natl. Acad. Sci. USA 108, 13540–13545.

Koushika, S.P., Richmond, J.E., Hadwiger, G., Weimer, R.M., Jorgensen, E.M., Nonet, M.L., 2001. A post-docking role for active zone protein Rim. Nat. Neurosci. 4, 997–1005.

Kreitzer, A.C., Regehr, W.G., 2001. Retrograde inhibition of presynaptic calcium influx by endogenous cannabinoids at excitatory synapses onto Purkinje cells. Neuron 29, 717–727.

Kumakura, K., Sasaki, K., Sakurai, T., Ohara-Imaizumi, M., Misonou, H., Nakamura, S., Matsuda, Y., Nonomura, Y., 1994. Essential role of myosin light chain kinase in the mechanism for MgATP-dependent priming of exocytosis in adrenal chromaffin cells. J. Neurosci. 14, 7695–7703.

Kummel, D., Krishnakumar, S.S., Radoff, D.T., Li, F., Giraudo, C.G., Pincet, F., Rothman, J.E., Reinisch, K.M., 2011. Complexin cross-links prefusion SNAREs into a zigzag array. Nat. Struct. Mol. Biol. 18, 927–933.

Kuromi, H., Kidokoro, Y., 2000. Tetanic stimulation recruits vesicles from reserve pool via a cAMP-mediated process in Drosophila synapses. Neuron 27, 133–143.

Kwon, H.B., Castillo, P.E., 2008. Role of glutamate autoreceptors at hippocampal mossy fiber synapses. Neuron 60, 1082–1094.

Kwon, S.E., Chapman, E.R., 2011. Synaptophysin regulates the kinetics of synaptic vesicle endocytosis in central neurons. Neuron 70, 847–854.

Lachamp, P.M., Liu, Y., Liu, S.J., 2009. Glutamatergic modulation of cerebellar interneuron activity is mediated by an enhancement of GABA release and requires protein kinase A/RIM1alpha signaling. J. Neurosci. 29, 381–392.

Lakadamyali, M., Babcock, H., Bates, M., Zhuang, X., Lichtman, J., 2012. 3D multicolor super-resolution imaging offers improved accuracy in neuron tracing. PloS One 7, e30826.

Lam, A.D., Tryoen-Toth, P., Tsai, B., Vitale, N., Stuenkel, E.L., 2008. SNARE-catalyzed fusion events are regulated by Syntaxin1A-lipid interactions. Mol. Biol. Cell 19, 485–497.

Landis, D.M., Hall, A.K., Weinstein, L.A., Reese, T.S., 1988. The organization of cytoplasm at the presynaptic active zone of a central nervous system synapse. Neuron 1, 201–209.

Lang, T., 2007. SNARE proteins and 'membrane rafts'. J. Physiol. 585, 693–698.

Lang, T., Bruns, D., Wenzel, D., Riedel, D., Holroyd, P., Thiele, C., Jahn, R., 2001. SNAREs are concentrated in cholesterol-dependent clusters that define docking and fusion sites for exocytosis. EMBO J. 20, 2202–2213.

Lang, T., Margittai, M., Holzler, H., Jahn, R., 2002. SNAREs in native plasma membranes are active and readily form core complexes with endogenous and exogenous SNAREs. J. Cell Biol. 158, 751–760.

Lang, T., Wacker, I., Wunderlich, I., Rohrbach, A., Giese, G., Soldati, T., Almers, W., 2000. Role of actin cortex in the subplasmalemmal transport of secretory granules in PC-12 cells. Biophys. J. 78, 2863–2877.

Laviv, T., Riven, I., Dolev, I., Vertkin, I., Balana, B., Slesinger, P.A., Slutsky, I., 2010. Basal GABA regulates GABA(B)R conformation and release probability at single hippocampal synapses. Neuron 67, 253–267.

Laviv, T., Vertkin, I., Berdichevsky, Y., Fogel, H., Riven, I., Bettler, B., Slesinger, P.A., Slutsky, I., 2011. Compartmentalization of the GABAB receptor signaling complex is required for presynaptic inhibition at hippocampal synapses. J. Neurosci. 31, 12523–12532.

Leao, R.M., von Gersdorff, H., 2009. Synaptic vesicle pool size, release probability and synaptic depression are sensitive to Ca^{2+} buffering capacity in the developing rat calyx of Held. Braz. J. Med. Biol. Res. 42, 94–104.

Lejen, T., Pene, T.D., Rose, S.D., Trifaro, J.M., 2002. The role of different Scinderin domains in the control of F-actin cytoskeleton during exocytosis. Ann. N.Y. Acad. Sci. 971, 248–250.

Lemke, E.A., Klingauf, J., 2005. Single synaptic vesicle tracking in individual hippocampal boutons at rest and during synaptic activity. J. Neurosci. 25, 11034–11044.

Lenzi, D., von Gersdorff, H., 2001. Structure suggests function: the case for synaptic ribbons as exocytotic nanomachines. Bioessays 23, 831–840.

Lerner, I., Trus, M., Cohen, R., Yizhar, O., Nussinovitch, I., Atlas, D., 2006. Ion interaction at the pore of Lc-type Ca^{2+} channel is sufficient to mediate depolarization-induced exocytosis. J. Neurochem. 97, 116–127.

Levic, S., Bouleau, Y., Dulon, D., 2011. Developmental acquisition of a rapid calcium-regulated vesicle supply allows sustained high rates of exocytosis in auditory hair cells. PloS One 6, e25714.

Li, L., Chin, L.S., Shupliakov, O., Brodin, L., Sihra, T.S., Hvalby, O., Jensen, V., Zheng, D., McNamara, J.O., Greengard, P., et al., 1995. Impairment of synaptic vesicle clustering and of synaptic transmission, and increased seizure propensity, in synapsin I-deficient mice. Proc. Natl. Acad. Sci. USA 92, 9235–9239.

Lin, R.C., Scheller, R.H., 1997. Structural organization of the synaptic exocytosis core complex. Neuron 19, 1087–1094.
Link, E., Edelmann, L., Chou, J.H., Binz, T., Yamasaki, S., Eisel, U., Baumert, M., Sudhof, T.C., Niemann, H., Jahn, R., 1992. Tetanus toxin action: inhibition of neurotransmitter release linked to synaptobrevin proteolysis. Biochem. Biophys. Res. Commun. 189, 1017–1023.
Liu, K.S., Siebert, M., Mertel, S., Knoche, E., Wegener, S., Wichmann, C., Matkovic, T., Muhammad, K., Depner, H., Mettke, C., Buckers, J., Hell, S.W., Muller, M., Davis, G.W., Schmitz, D., Sigrist, S.J., 2011. RIM-binding protein, a central part of the active zone, is essential for neurotransmitter release. Science 334, 1565–1569.
Liu, Y., Schirra, C., Edelmann, L., Matti, U., Rhee, J., Hof, D., Bruns, D., Brose, N., Rieger, H., Stevens, D.R., Rettig, J., 2010. Two distinct secretory vesicle-priming steps in adrenal chromaffin cells. J. Cell Biol. 190, 1067–1077.
Liu, Y., Schirra, C., Stevens, D.R., Matti, U., Speidel, D., Hof, D., Bruns, D., Brose, N., Rettig, J., 2008. CAPS facilitates filling of the rapidly releasable pool of large dense-core vesicles. J. Neurosci. 28, 5594–5601.
Lledo, P.M., Vernier, P., Vincent, J.D., Mason, W.T., Zorec, R., 1993. Inhibition of Rab3B expression attenuates Ca(2+)-dependent exocytosis in rat anterior pituitary cells. Nature 364, 540–544.
Llinas, R., McGuinness, T.L., Leonard, C.S., Sugimori, M., Greengard, P., 1985. Intraterminal injection of synapsin I or calcium/calmodulin-dependent protein kinase II alters neurotransmitter release at the squid giant synapse. Proc. Natl. Acad. Sci. USA 82, 3035–3039.
Lonart, G., Janz, R., Johnson, K.M., Sudhof, T.C., 1998. Mechanism of action of rab3A in mossy fiber LTP. Neuron 21, 1141–1150.
Lonart, G., Schoch, S., Kaeser, P.S., Larkin, C.J., Sudhof, T.C., Linden, D.J., 2003. Phosphorylation of RIM1alpha by PKA triggers presynaptic long-term potentiation at cerebellar parallel fiber synapses. Cell 115, 49–60.
Lonart, G., Sudhof, T.C., 2000. Assembly of SNARE core complexes prior to neurotransmitter release sets the readily releasable pool of synaptic vesicles. J. Biol. Chem. 275, 27703–27707.
Lou, X., Korogod, N., Brose, N., Schneggenburger, R., 2008. Phorbol esters modulate spontaneous and Ca^{2+}-evoked transmitter release via acting on both Munc13 and protein kinase C. J. Neurosci. 28, 8257–8267.
Lou, X., Scheuss, V., Schneggenburger, R., 2005. Allosteric modulation of the presynaptic Ca^{2+} sensor for vesicle fusion. Nature 435, 497–501.
Lu, J., Machius, M., Dulubova, I., Dai, H., Sudhof, T.C., Tomchick, D.R., Rizo, J., 2006. Structural basis for a Munc13-1 homodimer to Munc13-1/RIM heterodimer switch. PLoS Biol. 4, e192.
Lynch, K.L., Gerona, R.R., Larsen, E.C., Marcia, R.F., Mitchell, J.C., Martin, T.F., 2007. Synaptotagmin C2A loop 2 mediates Ca^{2+}-dependent SNARE interactions essential for Ca^{2+}-triggered vesicle exocytosis. Mol. Biol. Cell 18, 4957–4968.
Ma, C., Li, W., Xu, Y., Rizo, J., 2011. Munc13 mediates the transition from the closed syntaxin–Munc18 complex to the SNARE complex. Nat. Struct. Mol. Biol. 18, 542–549.
Ma, L., Zablow, L., Kandel, E.R., Siegelbaum, S.A., 1999. Cyclic AMP induces functional presynaptic boutons in hippocampal CA3–CA1 neuronal cultures. Nat. Neurosci. 2, 24–30.
Mackie, K., Hille, B., 1992. Cannabinoids inhibit N-type calcium channels in neuroblastoma–glioma cells. Proc. Natl. Acad. Sci. USA 89, 3825–3829.
Mackie, K., Lai, Y., Westenbroek, R., Mitchell, R., 1995. Cannabinoids activate an inwardly rectifying potassium conductance and inhibit Q-type calcium currents in AtT20 cells transfected with rat brain cannabinoid receptor. J. Neurosci. 15, 6552–6561.
Maejima, T., Hashimoto, K., Yoshida, T., Aiba, A., Kano, M., 2001. Presynaptic inhibition caused by retrograde signal from metabotropic glutamate to cannabinoid receptors. Neuron 31, 463–475.
Maekawa, S., Sakai, H., 1990. Inhibition of actin regulatory activity of the 74-kDa protein from bovine adrenal medulla (adseverin) by some phospholipids. J. Biol. Chem. 265, 10940–10942.
Mahapatra, S., Calorio, C., Vandael, D.H., Marcantoni, A., Carabelli, V., Carbone, E., 2012. Calcium channel types contributing to chromaffin cell excitability, exocytosis and endocytosis. Cell Calcium 51, 321–330.
Malenka, R.C., 1994. Synaptic plasticity in the hippocampus: LTP and LTD. Cell 78, 535–538.
Marcu, M.G., Rodriguez del Castillo, A., Vitale, M.L., Trifaro, J.M., 1994. Molecular cloning and functional expression of chromaffin cell scinderin indicates that it belongs to the family of Ca(2+)-dependent F-actin severing proteins. Mol. Cell. Biochem. 141, 153–165.

Marom, M., Sebag, A., Atlas, D., 2007. Cations residing at the selectivity filter of the voltage-gated Ca^{2+}-channel modify fusion-pore kinetics. Channels (Austin) 1, 377–386.
Martens, S., 2010. Role of C2 domain proteins during synaptic vesicle exocytosis. Biochem. Soc. Trans. 38, 213–216.
Martens, S., Kozlov, M.M., McMahon, H.T., 2007. How synaptotagmin promotes membrane fusion. Science, New York, 316, 1205–1208.
Martens, S., McMahon, H.T., 2008. Mechanisms of membrane fusion: disparate players and common principles. Nat. Rev. 9, 543–556.
Massol, R.H., Boll, W., Griffin, A.M., Kirchhausen, T., 2006. A burst of auxilin recruitment determines the onset of clathrin-coated vesicle uncoating. Proc. Natl. Acad. Sci. USA 103, 10265–10270.
Matsubara, M., Kusubata, M., Ishiguro, K., Uchida, T., Titani, K., Taniguchi, H., 1996. Site-specific phosphorylation of synapsin I by mitogen-activated protein kinase and Cdk5 and its effects on physiological functions. J. Biol. Chem. 271, 21108–21113.
Maximov, A., Sudhof, T.C., 2005. Autonomous function of synaptotagmin 1 in triggering synchronous release independent of asynchronous release. Neuron 48, 547–554.
Maximov, A., Tang, J., Yang, X., Pang, Z.P., Sudhof, T.C., 2009. Complexin controls the force transfer from SNARE complexes to membranes in fusion. Science, New York, 323, 516–521.
McBain, C.J., 2008. Differential mechanisms of transmission and plasticity at mossy fiber synapses. Prog. Brain Res. 169, 225–240.
McEvoy, A.L., Greenfield, D., Bates, M., Liphardt, J., 2010. Q&A: single-molecule localization microscopy for biological imaging. BMC Biol. 8, 106.
McEwen, J.M., Kaplan, J.M., 2008. UNC-18 promotes both the anterograde trafficking and synaptic function of syntaxin. Mol. Biol. Cell 19, 3836–3846.
McEwen, J.M., Madison, J.M., Dybbs, M., Kaplan, J.M., 2006. Antagonistic regulation of synaptic vesicle priming by Tomosyn and UNC-13. Neuron 51, 303–315.
McMahon, H.T., Boucrot, E., 2011. Molecular mechanism and physiological functions of clathrin-mediated endocytosis. Nat. Rev. 12, 517–533.
McMahon, H.T., Kozlov, M.M., Martens, S., 2010. Membrane curvature in synaptic vesicle fusion and beyond. Cell 140, 601–605.
McMahon, H.T., Missler, M., Li, C., Sudhof, T.C., 1995. Complexins: cytosolic proteins that regulate SNAP receptor function. Cell 83, 111–119.
Menasche, G., Feldmann, J., Houdusse, A., Desaymard, C., Fischer, A., Goud, B., de Saint Basile, G., 2003. Biochemical and functional characterization of Rab27a mutations occurring in Griscelli syndrome patients. Blood 101, 2736–2742.
Menegon, A., Bonanomi, D., Albertinazzi, C., Lotti, F., Ferrari, G., Kao, H.T., Benfenati, F., Baldelli, P., Valtorta, F., 2006. Protein kinase A-mediated synapsin I phosphorylation is a central modulator of Ca^{2+}-dependent synaptic activity. J. Neurosci. 26, 11670–11681.
Mennerick, S., Matthews, G., 1996. Ultrafast exocytosis elicited by calcium current in synaptic terminals of retinal bipolar neurons. Neuron 17, 1241–1249.
Merrins, M.J., Stuenkel, E.L., 2008. Kinetics of Rab27a-dependent actions on vesicle docking and priming in pancreatic beta-cells. J. Physiol. 586, 5367–5381.
Mezer, A., Nachliel, E., Gutman, M., Ashery, U., 2004. A new platform to study the molecular mechanisms of exocytosis. J. Neurosci. 24, 8838–8846.
Miller, T.M., Heuser, J.E., 1984. Endocytosis of synaptic vesicle membrane at the frog neuromuscular junction. J. Cell Biol. 98, 685–698.
Milosevic, I., Giovedi, S., Lou, X., Raimondi, A., Collesi, C., Shen, H., Paradise, S., O'Toole, E., Ferguson, S., Cremona, O., De Camilli, P., 2011. Recruitment of endophilin to clathrin-coated pit necks is required for efficient vesicle uncoating after fission. Neuron 72, 587–601.
Misura, K.M., Scheller, R.H., Weis, W.I., 2000. Three-dimensional structure of the neuronal-Sec1–syntaxin 1a complex. Nature 404, 355–362.
Mohrmann, R., de Wit, H., Verhage, M., Neher, E., Sorensen, J.B., 2010. Fast vesicle fusion in living cells requires at least three SNARE complexes. Science, New York, 330, 502–505.
Mohrmann, R., Sorensen, J.B., 2012. SNARE requirements en route to exocytosis: from many to few. J. Mol. Neurosci.

Montecucco, C., Schiavo, G., Pantano, S., 2005. SNARE complexes and neuroexocytosis: how many, how close? Trends Biochem Sci. 30, 367–372.
Mooseker, M.S., Cheney, R.E., 1995. Unconventional myosins. Annu. Rev. Cell Dev. Biol. 11, 633–675.
Mueller, V.J., Wienisch, M., Nehring, R.B., Klingauf, J., 2004. Monitoring clathrin-mediated endocytosis during synaptic activity. J. Neurosci. 24, 2004–2012.
Mukherjee, K., Yang, X., Gerber, S.H., Kwon, H.B., Ho, A., Castillo, P.E., Liu, X., Sudhof, T.C., 2010. Piccolo and bassoon maintain synaptic vesicle clustering without directly participating in vesicle exocytosis. Proc. Natl. Acad. Sci. USA 107, 6504–6509.
Muller, M., Pym, E.C., Tong, A., Davis, G.W., 2011. Rab3-GAP controls the progression of synaptic homeostasis at a late stage of vesicle release. Neuron 69, 749–762.
Munro, S., Thomas, K.L., Abu-Shaar, M., 1993. Molecular characterization of a peripheral receptor for cannabinoids. Nature 365, 61–65.
Munson, M., Bryant, N.J., 2009. A role for the syntaxin N-terminus. Biochem. J. 418, e1–e3.
Munson, M., Novick, P., 2006. The exocyst defrocked, a framework of rods revealed. Nat. Struct. Mol. Biol. 13, 577–581.
Murray, D.H., Tamm, L.K., 2009. Clustering of syntaxin-1A in model membranes is modulated by phosphatidylinositol 4,5-bisphosphate and cholesterol. Biochemistry 48, 4617–4625.
Murthy, M., Garza, D., Scheller, R.H., Schwarz, T.L., 2003. Mutations in the exocyst component Sec5 disrupt neuronal membrane traffic, but neurotransmitter release persists. Neuron 37, 433–447.
Nagerl, U.V., Willig, K.I., Hein, B., Hell, S.W., Bonhoeffer, T., 2008. Live-cell imaging of dendritic spines by STED microscopy. Proc. Natl. Acad. Sci. USA 105, 18982–18987.
Nakano, M., Nogami, S., Sato, S., Terano, A., Shirataki, H., 2001. Interaction of syntaxin with alpha-fodrin, a major component of the submembranous cytoskeleton. Biochem. Biophys. Res. Commun. 288, 468–475.
Nakashiba, T., Young, J.Z., McHugh, T.J., Buhl, D.L., Tonegawa, S., 2008. Transgenic inhibition of synaptic transmission reveals role of CA3 output in hippocampal learning. Science 319, 1260–1264.
Neco, P., Fernandez-Peruchena, C., Navas, S., Gutierrez, L.M., de Toledo, G.A., Ales, E., 2008. Myosin II contributes to fusion pore expansion during exocytosis. J. Biol. Chem. 283, 10949–10957.
Neco, P., Gil, A., Del Mar Frances, M., Viniegra, S., Gutierrez, L.M., 2002. The role of myosin in vesicle transport during bovine chromaffin cell secretion. Biochem. J. 368, 405–413.
Neco, P., Giner, D., del Mar Frances, M., Viniegra, S., Gutierrez, L.M., 2003. Differential participation of actin- and tubulin-based vesicle transport systems during secretion in bovine chromaffin cells. Eur. J. Neurosci. 18, 733–742.
Neco, P., Giner, D., Viniegra, S., Borges, R., Villarroel, A., Gutierrez, L.M., 2004. New roles of myosin II during vesicle transport and fusion in chromaffin cells. J. Biol. Chem. 279, 27450–27457.
Neher, E., 2010. Complexin: does it deserve its name? Neuron 68, 803–806.
Neher, E., Sakaba, T., 2008. Multiple roles of calcium ions in the regulation of neurotransmitter release. Neuron 59, 861–872.
Nevins, A.K., Thurmond, D.C., 2005. A direct interaction between Cdc42 and vesicle-associated membrane protein 2 regulates SNARE-dependent insulin exocytosis. J. Biol. Chem. 280, 1944–1952.
Newton, A.J., Kirchhausen, T., Murthy, V.N., 2006. Inhibition of dynamin completely blocks compensatory synaptic vesicle endocytosis. Proc. Natl. Acad. Sci. USA 103, 17955–17960.
Nguyen, P.V., Woo, N.H., 2003. Regulation of hippocampal synaptic plasticity by cyclic AMP-dependent protein kinases. Prog. Neurobiol. 71, 401–437.
Nicoll, R.A., Malenka, R.C., 1995. Contrasting properties of two forms of long-term potentiation in the hippocampus. Nature 377, 115–118.
Nicoll, R.A., Malenka, R.C., 1999. Expression mechanisms underlying NMDA receptor-dependent long-term potentiation. Ann. N.Y. Acad. Sci. 868, 515–525.
Nicoll, R.A., Schmitz, D., 2005. Synaptic plasticity at hippocampal mossy fibre synapses. Nat. Rev. 6, 863–876.
Nielander, H.B., Onofri, F., Schaeffer, E., Menegon, A., Fesce, R., Valtorta, F., Greengard, P., Benfenati, F., 1997. Phosphorylation-dependent effects of synapsin IIa on actin polymerization and network formation. Eur. J. Neurosci. 9, 2712–2722.
Nili, U., de Wit, H., Gulyas-Kovacs, A., Toonen, R.F., Sorensen, J.B., Verhage, M., Ashery, U., 2006. Munc18-1 phosphorylation by protein kinase C potentiates vesicle pool replenishment in bovine chromaffin cells. Neuroscience 143, 487–500.

Niswender, C.M., Johnson, K.A., Luo, Q., Ayala, J.E., Kim, C., Conn, P.J., Weaver, C.D., 2008. A novel assay of Gi/o-linked G protein-coupled receptor coupling to potassium channels provides new insights into the pharmacology of the group III metabotropic glutamate receptors. Mol. Pharmacol. 73, 1213–1224.
Nofal, S., Becherer, U., Hof, D., Matti, U., Rettig, J., 2007. Primed vesicles can be distinguished from docked vesicles by analyzing their mobility. J. Neurosci. 27, 1386–1395.
Nojiri, M., Loyet, K.M., Klenchin, V.A., Kabachinski, G., Martin, T.F., 2009. CAPS activity in priming vesicle exocytosis requires CK2 phosphorylation. J. Biol. Chem. 284, 18707–18714.
Novick, P., Field, C., Schekman, R., 1980. Identification of 23 complementation groups required for post-translational events in the yeast secretory pathway. Cell 21, 205–215.
Nowack, A., Malarkey, E.B., Yao, J., Bleckert, A., Hill, J., Bajjalieh, S.M., 2011. Levetiracetam reverses synaptic deficits produced by overexpression of SV2A. PloS One 6, e29560.
Nowack, A., Yao, J., Custer, K.L., Bajjalieh, S.M., 2010. SV2 regulates neurotransmitter release via multiple mechanisms. Am. J. Physiol. 299, C960–C967.
O'Rourke, N.A., Weiler, N.C., Micheva, K.D., Smith, S.J., 2012. Deep molecular diversity of mammalian synapses: why it matters and how to measure it. Nat. Rev. 13, 365–379.
Ohara-Imaizumi, M., Nishiwaki, C., Kikuta, T., Kumakura, K., Nakamichi, Y., Nagamatsu, S., 2004. Site of docking and fusion of insulin secretory granules in live MIN6 beta cells analyzed by TAT-conjugated anti-syntaxin 1 antibody and total internal reflection fluorescence microscopy. J. Biol. Chem. 279, 8403–8408.
Oheim, M., Stuhmer, W., 2000. Tracking chromaffin granules on their way through the actin cortex. Eur. Biophys. J. 29, 67–89.
Ohno-Shosaku, T., Maejima, T., Kano, M., 2001. Endogenous cannabinoids mediate retrograde signals from depolarized postsynaptic neurons to presynaptic terminals. Neuron 29, 729–738.
Orenbuch, A., Shalev, L., Marra, V., Sinai, I., Lavy, Y., Kahn, J., Burden, J.J., Staras, K., Gitler, D., 2012a. Synapsin selectively controls the mobility of resting pool vesicles at hippocampal terminals. J. Neurosci. 32, 3969–3980.
Orenbuch, A., Shulman, Y., Lipstein, N., Bechar, A., Lavy, Y., Brumer, E., Vasileva, M., Kahn, J., Barki-Harrington, L., Kuner, T., Gitler, D., 2012b. Inhibition of exocytosis or endocytosis blocks activity-dependent redistribution of synapsin. J. Neurochem. 120, 248–258.
Otsu, Y., Murphy, T.H., 2003. Miniature transmitter release: accident of nature or careful design? Sci. STKE 2003, pe54.
Pabst, S., Hazzard, J.W., Antonin, W., Sudhof, T.C., Jahn, R., Rizo, J., Fasshauer, D., 2000. Selective interaction of complexin with the neuronal SNARE complex. Determination of the binding regions. J. Biol. Chem. 275, 19808–19818.
Pan, B., Zucker, R.S., 2009. A general model of synaptic transmission and short-term plasticity. Neuron 62, 539–554.
Pang, Z.P., Bacaj, T., Yang, X., Zhou, P., Xu, W., Sudhof, T.C., 2011. Doc2 supports spontaneous synaptic transmission by a Ca(2+)-independent mechanism. Neuron 70, 244–251.
Pang, Z.P., Sudhof, T.C., 2010. Cell biology of Ca^{2+}-triggered exocytosis. Curr. Opin. Cell Biol. 22, 496–505.
Pang, Z.P., Sun, J., Rizo, J., Maximov, A., Sudhof, T.C., 2006. Genetic analysis of synaptotagmin 2 in spontaneous and Ca^{2+}-triggered neurotransmitter release. EMBO J. 25, 2039–2050.
Pare, D., Lebel, E., Lang, E.J., 1997. Differential impact of miniature synaptic potentials on the soma and dendrites of pyramidal neurons in vivo. J. Neurophysiol. 78, 1735–1739.
Park, H., Li, Y., Tsien, R.W., 2012. Influence of synaptic vesicle position on release probability and exocytotic fusion mode. Science, New York.
Parnas, H., Parnas, I., 2007. The chemical synapse goes electric: Ca^{2+}- and voltage-sensitive GPCRs control neurotransmitter release. Trends Neurosci. 30, 54–61.
Pasche, M., Matti, U., Hof, D., Rettig, J., Becherer, U., 2012. Docking of LDCVs is modulated by lower intracellular [Ca^{2+}] than priming. PloS One 7, e36416.
Pearse, B.M., 1982. Coated vesicles from human placenta carry ferritin, transferrin, and immunoglobulin G. Proc. Natl. Acad. Sci. USA 79, 451–455.
Peng, A., Rotman, Z., Deng, P.Y., Klyachko, V.A., 2012. Differential motion dynamics of synaptic vesicles undergoing spontaneous and activity-evoked endocytosis. Neuron 73, 1108–1115.
Perrin, D., Aunis, D., 1985. Reorganization of alpha-fodrin induced by stimulation in secretory cells. Nature 315, 589–592.

Perrin, D., Langley, O.K., Aunis, D., 1987. Anti-alpha-fodrin inhibits secretion from permeabilized chromaffin cells. Nature 326, 498–501.

Peters, M.F., Scott, C.W., 2009. Evaluating cellular impedance assays for detection of GPCR pleiotropic signaling and functional selectivity. J. Biomol. Screen. 14, 246–255.

Pevsner, J., Hsu, S.C., Braun, J.E., Calakos, N., Ting, A.E., Bennett, M.K., Scheller, R.H., 1994. Specificity and regulation of a synaptic vesicle docking complex. Neuron 13, 353–361.

Pfenninger, K., Akert, K., Moor, H., Sandri, C., 1972. The fine structure of freeze-fractured presynaptic membranes. J. Neurocytol. 1, 129–149.

Phillips, G.R., Huang, J.K., Wang, Y., Tanaka, H., Shapiro, L., Zhang, W., Shan, W.S., Arndt, K., Frank, M., Gordon, R.E., Gawinowicz, M.A., Zhao, Y., Colman, D.R., 2001. The presynaptic particle web: ultrastructure, composition, dissolution, and reconstitution. Neuron 32, 63–77.

Pieribone, V.A., Shupliakov, O., Brodin, L., Hilfiker-Rothenfluh, S., Czernik, A.J., Greengard, P., 1995. Distinct pools of synaptic vesicles in neurotransmitter release. Nature 375, 493–497.

Pobbati, A.V., Razeto, A., Boddener, M., Becker, S., Fasshauer, D., 2004. Structural basis for the inhibitory role of tomosyn in exocytosis. J. Biol. Chem. 279, 47192–47200.

Poirier, M.A., Xiao, W., Macosko, J.C., Chan, C., Shin, Y.K., Bennett, M.K., 1998. The synaptic SNARE complex is a parallel four-stranded helical bundle. Nat. Struct. Biol. 5, 765–769.

Poskanzer, K.E., Marek, K.W., Sweeney, S.T., Davis, G.W., 2003. Synaptotagmin I is necessary for compensatory synaptic vesicle endocytosis in vivo. Nature 426, 559–563.

Powell, C.M., 2006. Gene targeting of presynaptic proteins in synaptic plasticity and memory: across the great divide. Neurobiol. Learn. Mem. 85, 2–15.

Powell, C.M., Schoch, S., Monteggia, L., Barrot, M., Matos, M.F., Feldmann, N., Sudhof, T.C., Nestler, E.J., 2004. The presynaptic active zone protein RIM1alpha is critical for normal learning and memory. Neuron 42, 143–153.

Praefcke, G.J., McMahon, H.T., 2004. The dynamin superfamily: universal membrane tubulation and fission molecules? Nat. Rev. Mol. Cell Biol. 5, 133–147.

Pyle, J.L., Kavalali, E.T., Piedras-Renteria, E.S., Tsien, R.W., 2000. Rapid reuse of readily releasable pool vesicles at hippocampal synapses. Neuron 28, 221–231.

Raingo, J., Khvotchev, M., Liu, P., Darios, F., Li, Y.C., Ramirez, D.M., Adachi, M., Lemieux, P., Toth, K., Davletov, B., Kavalali, E.T., 2012. VAMP4 directs synaptic vesicles to a pool that selectively maintains asynchronous neurotransmission. Nat. Neurosci. 15, 738–745.

Ramirez, D.M., Kavalali, E.T., 2011. Differential regulation of spontaneous and evoked neurotransmitter release at central synapses. Curr. Opin. Neurobiol. 21, 275–282.

Rao-Mirotznik, R., Harkins, A.B., Buchsbaum, G., Sterling, P., 1995. Mammalian rod terminal: architecture of a binary synapse. Neuron 14, 561–569.

Rapoport, I., Boll, W., Yu, A., Bocking, T., Kirchhausen, T., 2008. A motif in the clathrin heavy chain required for the Hsc70/auxilin uncoating reaction. Mol. Biol. Cell 19, 405–413.

Rathore, S.S., Bend, E.G., Yu, H., Hammarlund, M., Jorgensen, E.M., Shen, J., 2010. Syntaxin N-terminal peptide motif is an initiation factor for the assembly of the SNARE-Sec1/Munc18 membrane fusion complex. Proc. Natl. Acad. Sci. USA 107, 22399–22406.

Ratnayaka, A., Marra, V., Bush, D., Burden, J.J., Branco, T., Staras, K., 2012. Recruitment of resting vesicles into recycling pools supports NMDA receptor-dependent synaptic potentiation in cultured hippocampal neurons. J. Physiol. 590, 1585–1597.

Regazzi, R., Ravazzola, M., Iezzi, M., Lang, J., Zahraoui, A., Andereggen, E., Morel, P., Takai, Y., Wollheim, C.B., 1996. Expression, localization and functional role of small GTPases of the Rab3 family in insulin-secreting cells. J. Cell Sci. 109 (Pt 9), 2265–2273.

Reim, K., Mansour, M., Varoqueaux, F., McMahon, H.T., Sudhof, T.C., Brose, N., Rosenmund, C., 2001. Complexins regulate a late step in Ca2+-dependent neurotransmitter release. Cell 104, 71–81.

Richards, D.A., Guatimosim, C., Betz, W.J., 2000. Two endocytic recycling routes selectively fill two vesicle pools in frog motor nerve terminals. Neuron 27, 551–559.

Richards, D.A., Guatimosim, C., Rizzoli, S.O., Betz, W.J., 2003. Synaptic vesicle pools at the frog neuromuscular junction. Neuron 39, 529–541.

Richmond, J.E., Davis, W.S., Jorgensen, E.M., 1999. UNC-13 is required for synaptic vesicle fusion in *C. elegans*. Nat. Neurosci. 2, 959–964.

Richmond, J.E., Weimer, R.M., Jorgensen, E.M., 2001. An open form of syntaxin bypasses the requirement for UNC-13 in vesicle priming. Nature 412, 338–341.
Rickman, C., Medine, C.N., Dun, A.R., Moulton, D.J., Mandula, O., Halemani, N.D., Rizzoli, S.O., Chamberlain, L.H., Duncan, R.R., 2010. t-SNARE protein conformations patterned by the lipid microenvironment. J. Biol. Chem. 285, 13535–13541.
Ringstad, N., Gad, H., Low, P., Di Paolo, G., Brodin, L., Shupliakov, O., De Camilli, P., 1999. Endophilin/SH3p4 is required for the transition from early to late stages in clathrin-mediated synaptic vesicle endocytosis. Neuron 24, 143–154.
Ringstad, N., Nemoto, Y., De Camilli, P., 1997. The SH3p4/Sh3p8/SH3p13 protein family: binding partners for synaptojanin and dynamin via a Grb2-like Src homology 3 domain. Proc. Natl. Acad. Sci. USA 94, 8569–8574.
Rizo, J., Sudhof, T.C., 2002. Snares and Munc18 in synaptic vesicle fusion. Nat. Rev. Neurosci. 3, 641–653.
Rizzoli, S.O., Betz, W.J., 2005. Synaptic vesicle pools. Nat. Rev. Neurosci. 6, 57–69.
Robbe, D., Alonso, G., Manzoni, O.J., 2003. Exogenous and endogenous cannabinoids control synaptic transmission in mice nucleus accumbens. Ann. N.Y. Acad. Sci. 1003, 212–225.
Robinson, M.S., 2004. Adaptable adaptors for coated vesicles. Trends Cell Biol. 14, 167–174.
Rodriguez Del Castillo, A., Lemaire, S., Tchakarov, L., Jeyapragasan, M., Doucet, J.P., Vitale, M.L., Trifaro, J.M., 1990. Chromaffin cell scinderin, a novel calcium-dependent actin filament-severing protein. EMBO J. 9, 43–52.
Rodriguez Del Castillo, A., Vitale, M.L., Trifaro, J.M., 1992. Ca^{2+} and pH determine the interaction of chromaffin cell scinderin with phosphatidylserine and phosphatidylinositol 4,5,-biphosphate and its cellular distribution during nicotinic-receptor stimulation and protein kinase C activation. J. Cell Biol. 119, 797–810.
Rohatgi, R., Ma, L., Miki, H., Lopez, M., Kirchhausen, T., Takenawa, T., Kirschner, M.W., 1999. The interaction between N-WASP and the Arp2/3 complex links Cdc42-dependent signals to actin assembly. Cell 97, 221–231.
Rosahl, T.W., Spillane, D., Missler, M., Herz, J., Selig, D.K., Wolff, J.R., Hammer, R.E., Malenka, R.C., Sudhof, T.C., 1995. Essential functions of synapsins I and II in synaptic vesicle regulation. Nature 375, 488–493.
Rose, S.D., Lejen, T., Casaletti, L., Larson, R.E., Pene, T.D., Trifaro, J.M., 2003. Myosins II and V in chromaffin cells: myosin V is a chromaffin vesicle molecular motor involved in secretion. J. Neurochem. 85, 287–298.
Rosenmund, C., Clements, J.D., Westbrook, G.L., 1993. Nonuniform probability of glutamate release at a hippocampal synapse. Science, New York, 262, 754–757.
Rosenmund, C., Sigler, A., Augustin, I., Reim, K., Brose, N., Rhee, J.S., 2002. Differential control of vesicle priming and short-term plasticity by Munc13 isoforms. Neuron 33, 411–424.
Rosenmund, C., Stevens, C.F., 1996. Definition of the readily releasable pool of vesicles at hippocampal synapses. Neuron 16, 1197–1207.
Rostaing, P., Real, E., Siksou, L., Lechaire, J.P., Boudier, T., Boeckers, T.M., Gertler, F., Gundelfinger, E.D., Triller, A., Marty, S., 2006. Analysis of synaptic ultrastructure without fixative using high-pressure freezing and tomography. Eur. J. Neurosci. 24, 3463–3474.
Rush, A.M., Kilbride, J., Rowan, M.J., Anwyl, R., 2002. Presynaptic group III mGluR modulation of short-term plasticity in the lateral perforant path of the dentate gyrus in vitro. Brain Res. 952, 38–43.
Rust, M.J., Bates, M., Zhuang, X., 2006. Sub-diffraction-limit imaging by stochastic optical reconstruction microscopy (STORM). Nat. Methods 3, 793–795.
Ryan, T.A., Smith, S.J., Reuter, H., 1996. The timing of synaptic vesicle endocytosis. Proc. Natl. Acad. Sci. USA 93, 5567–5571.
Sabatini, B.L., Regehr, W.G., 1996. Timing of neurotransmission at fast synapses in the mammalian brain. Nature 384, 170–172.
Safo, P.K., Regehr, W.G., 2005. Endocannabinoids control the induction of cerebellar LTD. Neuron 48, 647–659.
Sakisaka, T., Yamamoto, Y., Mochida, S., Nakamura, M., Nishikawa, K., Ishizaki, H., Okamoto-Tanaka, M., Miyoshi, J., Fujiyoshi, Y., Manabe, T., Takai, Y., 2008. Dual inhibition of SNARE complex formation by tomosyn ensures controlled neurotransmitter release. J. Cell Biol. 183, 323–337.

Sakurai, M., 1987. Synaptic modification of parallel fibre-Purkinje cell transmission in in vitro guinea-pig cerebellar slices. J. Physiol. 394, 463–480.
Salin, P.A., Malenka, R.C., Nicoll, R.A., 1996a. Cyclic AMP mediates a presynaptic form of LTP at cerebellar parallel fiber synapses. Neuron 16, 797–803.
Salin, P.A., Scanziani, M., Malenka, R.C., Nicoll, R.A., 1996b. Distinct short-term plasticity at two excitatory synapses in the hippocampus. Proc. Natl. Acad. Sci. USA 93, 13304–13309.
Sankaranarayanan, S., Atluri, P.P., Ryan, T.A., 2003. Actin has a molecular scaffolding, not propulsive, role in presynaptic function. Nat. Neurosci. 6, 127–135.
Sankaranarayanan, S., Ryan, T.A., 2000. Real-time measurements of vesicle-SNARE recycling in synapses of the central nervous system. Nat. Cell Biol. 2, 197–204.
Sans, N., Prybylowski, K., Petralia, R.S., Chang, K., Wang, Y.X., Racca, C., Vicini, S., Wenthold, R.J., 2003. NMDA receptor trafficking through an interaction between PDZ proteins and the exocyst complex. Nat. Cell Biol. 5, 520–530.
Sara, Y., Mozhayeva, M.G., Liu, X., Kavalali, E.T., 2002. Fast vesicle recycling supports neurotransmission during sustained stimulation at hippocampal synapses. J. Neurosci. 22, 1608–1617.
Sara, Y., Virmani, T., Deak, F., Liu, X., Kavalali, E.T., 2005. An isolated pool of vesicles recycles at rest and drives spontaneous neurotransmission. Neuron 45, 563–573.
Sato, M., Kitaguchi, T., Numano, R., Ikematsu, K., Kakeyama, M., Murata, M., Sato, K., Tsuboi, T., 2012. The small GTPase Cdc42 modulates the number of exocytosis-competent dense-core vesicles in PC12 cells. Biochem. Biophys. Res. Commun. 420, 417–421.
Scanziani, M., Salin, P.A., Vogt, K.E., Malenka, R.C., Nicoll, R.A., 1997. Use-dependent increases in glutamate concentration activate presynaptic metabotropic glutamate receptors. Nature 385, 630–634.
Scheele, U., Kalthoff, C., Ungewickell, E., 2001. Multiple interactions of auxilin 1 with clathrin and the AP-2 adaptor complex. J. Biol. Chem. 276, 36131–36138.
Schiavo, G., Benfenati, F., Poulain, B., Rossetto, O., Polverino de Laureto, P., DasGupta, B.R., Montecucco, C., 1992. Tetanus and botulinum-B neurotoxins block neurotransmitter release by proteolytic cleavage of synaptobrevin. Nature 359, 832–835.
Schiavo, G., Matteoli, M., Montecucco, C., 2000. Neurotoxins affecting neuroexocytosis. Physiol. Rev. 80, 717–766.
Schivell, A.E., Mochida, S., Kensel-Hammes, P., Custer, K.L., Bajjalieh, S.M., 2005. SV2A and SV2C contain a unique synaptotagmin-binding site. Mol. Cell. Neurosci. 29, 56–64.
Schlossman, D.M., Schmid, S.L., Braell, W.A., Rothman, J.E., 1984. An enzyme that removes clathrin coats: purification of an uncoating ATPase. J. Cell Biol. 99, 723–733.
Schluter, O.M., Basu, J., Sudhof, T.C., Rosenmund, C., 2006. Rab3 superprimes synaptic vesicles for release: implications for short-term synaptic plasticity. J. Neurosci. 26, 1239–1246.
Schluter, O.M., Khvotchev, M., Jahn, R., Sudhof, T.C., 2002. Localization versus function of Rab3 proteins. Evidence for a common regulatory role in controlling fusion. J. Biol. Chem. 277, 40919–40929.
Schluter, O.M., Schmitz, F., Jahn, R., Rosenmund, C., Sudhof, T.C., 2004. A complete genetic analysis of neuronal Rab3 function. J. Neurosci. 24, 6629–6637.
Schmid, E.M., Ford, M.G., Burtey, A., Praefcke, G.J., Peak-Chew, S.Y., Mills, I.G., Benmerah, A., McMahon, H.T., 2006. Role of the AP2 beta-appendage hub in recruiting partners for clathrin-coated vesicle assembly. PLoS Biol. 4, e262.
Schmidt, R., Wurm, C.A., Jakobs, S., Engelhardt, J., Egner, A., Hell, S.W., 2008. Spherical nanosized focal spot unravels the interior of cells. Nat. Methods 5, 539–544.
Schneggenburger, R., Han, Y., Kochubey, O., 2012. Ca(2+) channels and transmitter release at the active zone. Cell Calcium.
Schneggenburger, R., Meyer, A.C., Neher, E., 1999. Released fraction and total size of a pool of immediately available transmitter quanta at a calyx synapse. Neuron 23, 399–409.
Schneggenburger, R., Neher, E., 2000. Intracellular calcium dependence of transmitter release rates at a fast central synapse. Nature 406, 889–893.
Schneggenburger, R., Neher, E., 2005. Presynaptic calcium and control of vesicle fusion. Curr. Opin. Neurobiol. 15, 266–274.

Schoch, S., Castillo, P.E., Jo, T., Mukherjee, K., Geppert, M., Wang, Y., Schmitz, F., Malenka, R.C., Sudhof, T.C., 2002. RIM1alpha forms a protein scaffold for regulating neurotransmitter release at the active zone. Nature 415, 321–326.
Schoch, S., Deak, F., Konigstorfer, A., Mozhayeva, M., Sara, Y., Sudhof, T.C., Kavalali, E.T., 2001. SNARE function analyzed in synaptobrevin/VAMP knockout mice. Science, New York, 294, 1117–1122.
Schoch, S., Gundelfinger, E.D., 2006. Molecular organization of the presynaptic active zone. Cell Tissue Res. 326, 379–391.
Schoch, S., Mittelstaedt, T., Kaeser, P.S., Padgett, D., Feldmann, N., Chevaleyre, V., Castillo, P.E., Hammer, R.E., Han, W., Schmitz, F., Lin, W., Sudhof, T.C., 2006. Redundant functions of RIM1alpha and RIM2alpha in Ca(2+)-triggered neurotransmitter release. EMBO J. 25, 5852–5863.
Schulze, K.L., Broadie, K., Perin, M.S., Bellen, H.J., 1995. Genetic and electrophysiological studies of Drosophila syntaxin-1A demonstrate its role in nonneuronal secretion and neurotransmission. Cell 80, 311–320.
Schuske, K.R., Richmond, J.E., Matthies, D.S., Davis, W.S., Runz, S., Rube, D.A., van der Bliek, A.M., Jorgensen, E.M., 2003. Endophilin is required for synaptic vesicle endocytosis by localizing synaptojanin. Neuron 40, 749–762.
Schwenger, D.B., Kuner, T., 2010. Acute genetic perturbation of exocyst function in the rat calyx of Held impedes structural maturation, but spares synaptic transmission. Eur. J. Neurosci. 32, 974–984.
Scott, L.L., Kogan, D., Shamma, A.A., Quinlan, E.M., 2010. Differential regulation of synapsin phosphorylation by monocular deprivation in juveniles and adults. Neuroscience 166, 539–550.
Serra-Pages, C., Medley, Q.G., Tang, M., Hart, A., Streuli, M., 1998. Liprins, a family of LAR transmembrane protein-tyrosine phosphatase-interacting proteins. J. Biol. Chem. 273, 15611–15620.
Shapira, M., Zhai, R.G., Dresbach, T., Bresler, T., Torres, V.I., Gundelfinger, E.D., Ziv, N.E., Garner, C.C., 2003. Unitary assembly of presynaptic active zones from Piccolo–Bassoon transport vesicles. Neuron 38, 237–252.
Shen, J., Tareste, D.C., Paumet, F., Rothman, J.E., Melia, T.J., 2007. Selective activation of cognate SNAREpins by Sec1/Munc18 proteins. Cell 128, 183–195.
Shin, O.H., Lu, J., Rhee, J.S., Tomchick, D.R., Pang, Z.P., Wojcik, S.M., Camacho-Perez, M., Brose, N., Machius, M., Rizo, J., Rosenmund, C., Sudhof, T.C., 2010. Munc13 C2B domain is an activity-dependent Ca^{2+} regulator of synaptic exocytosis. Nat. Struct. Mol. Biol. 17, 280–288.
Shupliakov, O., Haucke, V., Pechstein, A., 2011. How synapsin I may cluster synaptic vesicles. Semin. Cell Dev. Biol. 22, 393–399.
Sieber, J.J., Willig, K.I., Heintzmann, R., Hell, S.W., Lang, T., 2006. The SNARE motif is essential for the formation of syntaxin clusters in the plasma membrane. Biophys. J. 90, 2843–2851.
Sieber, J.J., Willig, K.I., Kutzner, C., Gerding-Reimers, C., Harke, B., Donnert, G., Rammner, B., Eggeling, C., Hell, S.W., Grubmuller, H., Lang, T., 2007. Anatomy and dynamics of a supramolecular membrane protein cluster. Science, New York, 317, 1072–1076.
Sigrist, S.J., Sabatini, B.L., 2012. Optical super-resolution microscopy in neurobiology. Curr. Opin. Neurobiol. 22, 86–93.
Sihra, T.S., Wang, J.K., Gorelick, F.S., Greengard, P., 1989. Translocation of synapsin I in response to depolarization of isolated nerve terminals. Proc. Natl. Acad. Sci. USA 86, 8108–8112.
Siksou, L., Rostaing, P., Lechaire, J.P., Boudier, T., Ohtsuka, T., Fejtova, A., Kao, H.T., Greengard, P., Gundelfinger, E.D., Triller, A., Marty, S., 2007. Three-dimensional architecture of presynaptic terminal cytomatrix. J. Neurosci. 27, 6868–6877.
Siksou, L., Triller, A., Marty, S., 2009a. An emerging view of presynaptic structure from electron microscopic studies. J. Neurochem. 108, 1336–1342.
Siksou, L., Varoqueaux, F., Pascual, O., Triller, A., Brose, N., Marty, S., 2009b. A common molecular basis for membrane docking and functional priming of synaptic vesicles. Eur. J. Neurosci. 30, 49–56.
Silva, A.J., Rosahl, T.W., Chapman, P.F., Marowitz, Z., Friedman, E., Frankland, P.W., Cestari, V., Cioffi, D., Sudhof, T.C., Bourtchuladze, R., 1996. Impaired learning in mice with abnormal short-lived plasticity. Curr. Biol. 6, 1509–1518.
Sinha, R., Ahmed, S., Jahn, R., Klingauf, J., 2011. Two synaptobrevin molecules are sufficient for vesicle fusion in central nervous system synapses. Proc. Natl. Acad. Sci. USA 108, 14318–14323.

Sobkowicz, H.M., Rose, J.E., Scott, G.E., Slapnick, S.M., 1982. Ribbon synapses in the developing intact and cultured organ of Corti in the mouse. J. Neurosci. 2, 942–957.
Sollner, T., Bennett, M.K., Whiteheart, S.W., Scheller, R.H., Rothman, J.E., 1993. A protein assembly–disassembly pathway in vitro that may correspond to sequential steps of synaptic vesicle docking, activation, and fusion. Cell 75, 409–418.
Sons, M.S., Plomp, J.J., 2006. Rab3A deletion selectively reduces spontaneous neurotransmitter release at the mouse neuromuscular synapse. Brain Res. 1089, 126–134.
Speidel, D., Bruederle, C.E., Enk, C., Voets, T., Varoqueaux, F., Reim, K., Becherer, U., Fornai, F., Ruggieri, S., Holighaus, Y., Weihe, E., Bruns, D., Brose, N., Rettig, J., 2005. CAPS1 regulates catecholamine loading of large dense-core vesicles. Neuron 46, 75–88.
Speidel, D., Varoqueaux, F., Enk, C., Nojiri, M., Grishanin, R.N., Martin, T.F., Hofmann, K., Brose, N., Reim, K., 2003. A family of Ca^{2+}-dependent activator proteins for secretion: comparative analysis of structure, expression, localization, and function. J. Biol. Chem. 278, 52802–52809.
Stanley, E.F., 1997. The calcium channel and the organization of the presynaptic transmitter release face. Trends Neurosci. 20, 404–409.
Staras, K., Branco, T., Burden, J.J., Pozo, K., Darcy, K., Marra, V., Ratnayaka, A., Goda, Y., 2010. A vesicle superpool spans multiple presynaptic terminals in hippocampal neurons. Neuron 66, 37–44.
Stella, N., Schweitzer, P., Piomelli, D., 1997. A second endogenous cannabinoid that modulates long-term potentiation. Nature 388, 773–778.
Stevens, C.F., Williams, J.H., 2000. "Kiss and run" exocytosis at hippocampal synapses. Proc. Natl. Acad. Sci. USA 97, 12828–12833.
Stevens, D.R., Rettig, J., 2009. The Ca(2+)-dependent activator protein for secretion CAPS: do I dock or do I prime? Mol. Neurobiol. 39, 62–72.
Sudhof, T.C., 2004. The synaptic vesicle cycle. Annu. Rev. Neurosci. 27, 509–547.
Sudhof, T.C., Rizo, J., 2011. Synaptic vesicle exocytosis. Cold Spring Harbor Perspect. Biol. 3.
Sun, J., Pang, Z.P., Qin, D., Fahim, A.T., Adachi, R., Sudhof, T.C., 2007. A dual-Ca^{2+}-sensor model for neurotransmitter release in a central synapse. Nature 450, 676–682.
Sutton, R.B., Fasshauer, D., Jahn, R., Brunger, A.T., 1998. Crystal structure of a SNARE complex involved in synaptic exocytosis at 2.4 A resolution. Nature 395, 347–353.
Szanto, A., Csipo, I., Horvath, I., Biro, E., Szodoray, P., Zeher, M., 2008. Autoantibodies to alfa-fodrin in patients with Hashimoto thyroiditis and Sjogren's syndrome: possible markers for a common secretory disorder. Rheumatol. Int. 28, 1169–1172.
Takahashi, S., Ujihara, H., Huang, G.Z., Yagyu, K.I., Sanbo, M., Kaba, H., Yagi, T., 1999. Reduced hippocampal LTP in mice lacking a presynaptic protein: complexin II. Eur. J. Neurosci. 11, 2359–2366.
Takamori, S., Holt, M., Stenius, K., Lemke, E.A., Gronborg, M., Riedel, D., Urlaub, H., Schenck, S., Brugger, B., Ringler, P., Muller, S.A., Rammner, B., Grater, F., Hub, J.S., De Groot, B.L., Mieskes, G., Moriyama, Y., Klingauf, J., Grubmuller, H., Heuser, J., Wieland, F., Jahn, R., 2006. Molecular anatomy of a trafficking organelle. Cell 127, 831–846.
Takao-Rikitsu, E., Mochida, S., Inoue, E., Deguchi-Tawarada, M., Inoue, M., Ohtsuka, T., Takai, Y., 2004. Physical and functional interaction of the active zone proteins, CAST, RIM1, and Bassoon, in neurotransmitter release. J. Cell Biol. 164, 301–311.
Takei, K., McPherson, P.S., Schmid, S.L., De Camilli, P., 1995. Tubular membrane invaginations coated by dynamin rings are induced by GTP-gamma S in nerve terminals. Nature 374, 186–190.
Takei, K., Mundigl, O., Daniell, L., De Camilli, P., 1996. The synaptic vesicle cycle: a single vesicle budding step involving clathrin and dynamin. J. Cell. Biol. 133, 1237–1250.
Tang, J., Maximov, A., Shin, O.H., Dai, H., Rizo, J., Sudhof, T.C., 2006. A complexin/synaptotagmin 1 switch controls fast synaptic vesicle exocytosis. Cell 126, 1175–1187.
Tao-Cheng, J.H., Dosemeci, A., Winters, C.A., Reese, T.S., 2006. Changes in the distribution of calcium calmodulin-dependent protein kinase II at the presynaptic bouton after depolarization. Brain Cell Biol. 35, 117–124.
Taylor, M.J., Perrais, D., Merrifield, C.J., 2011. A high precision survey of the molecular dynamics of mammalian clathrin-mediated endocytosis. PLoS Biol. 9, e1000604.

Tobi, D., Wiser, O., Trus, M., Atlas, D., 1998. N-type voltage-sensitive calcium channel interacts with syntaxin, synaptotagmin and SNAP-25 in a multiprotein complex. Receptors Channels 6, 89–98.
tom Dieck, S., Altrock, W.D., Kessels, M.M., Qualmann, B., Regus, H., Brauner, D., Fejtova, A., Bracko, O., Gundelfinger, E.D., Brandstatter, J.H., 2005. Molecular dissection of the photoreceptor ribbon synapse: physical interaction of Bassoon and RIBEYE is essential for the assembly of the ribbon complex. J. Cell Biol. 168, 825–836.
tom Dieck, S., Sanmarti-Vila, L., Langnaese, K., Richter, K., Kindler, S., Soyke, A., Wex, H., Smalla, K.H., Kampf, U., Franzer, J.T., Stumm, M., Garner, C.C., Gundelfinger, E.D., 1998. Bassoon, a novel zinc-finger CAG/glutamine-repeat protein selectively localized at the active zone of presynaptic nerve terminals. J. Cell Biol. 142, 499–509.
Toonen, R.F., de Vries, K.J., Zalm, R., Sudhof, T.C., Verhage, M., 2005. Munc18-1 stabilizes syntaxin 1, but is not essential for syntaxin 1 targeting and SNARE complex formation. J. Neurochem. 93, 1393–1400.
Toonen, R.F., Kochubey, O., de Wit, H., Gulyas-Kovacs, A., Konijnenburg, B., Sorensen, J.B., Klingauf, J., Verhage, M., 2006a. Dissecting docking and tethering of secretory vesicles at the target membrane. EMBO J. 25, 3725–3737.
Toonen, R.F., Wierda, K., Sons, M.S., de Wit, H., Cornelisse, L.N., Brussaard, A., Plomp, J.J., Verhage, M., 2006b. Munc18-1 expression levels control synapse recovery by regulating readily releasable pool size. Proc. Natl. Acad. Sci. USA 103, 18332–18337.
Toonen, R.F., Verhage, M., 2007. Munc18-1 in secretion: lonely Munc joins SNARE team and takes control. Trends Neurosci. 30, 564–572.
Torii, S., Takeuchi, T., Nagamatsu, S., Izumi, T., 2004. Rab27 effector granuphilin promotes the plasma membrane targeting of insulin granules via interaction with syntaxin 1a. J. Biol. Chem. 279, 22532–22538.
Toth, K., Suares, G., Lawrence, J.J., Philips-Tansey, E., McBain, C.J., 2000. Differential mechanisms of transmission at three types of mossy fiber synapse. J. Neurosci. 20, 8279–8289.
Trifaro, J.M., Bader, M.F., Doucet, J.P., 1985. Chromaffin cell cytoskeleton: its possible role in secretion. Can. J. Biochem. Cell Biol. 63, 661–679.
Trifaro, J.M., Gasman, S., Gutierrez, L.M., 2008. Cytoskeletal control of vesicle transport and exocytosis in chromaffin cells. Acta Physiol. (Oxf.) 192, 165–172.
Trifaro, J.M., Lejen, T., Rose, S.D., Pene, T.D., Barkar, N.D., Seward, F. P., 2002. Pathways that control cortical F-actin dynamics during secretion. Neurochem. Res. 27, 1371–1385.
Trus, M., Corkey, R.F., Nesher, R., Richard, A.M., Deeney, J.T., Corkey, B.E., Atlas, D., 2007. The L-type voltage-gated Ca^{2+} channel is the Ca^{2+} sensor protein of stimulus-secretion coupling in pancreatic beta cells. Biochemistry 46, 14461–14467.
Tsetsenis, T., Younts, T.J., Chiu, C.Q., Kaeser, P.S., Castillo, P.E., Sudhof, T.C., 2011. Rab3B protein is required for long-term depression of hippocampal inhibitory synapses and for normal reversal learning. Proc. Natl. Acad. Sci. USA 108, 14300–14305.
Tsuboi, T., Fukuda, M., 2005. The C2B domain of rabphilin directly interacts with SNAP-25 and regulates the docking step of dense core vesicle exocytosis in PC12 cells. J. Biol. Chem. 280, 39253–39259.
Tsuboi, T., Fukuda, M., 2006. Rab3A and Rab27A cooperatively regulate the docking step of dense-core vesicle exocytosis in PC12 cells. J. Cell Sci. 119, 2196–2203.
Twitchell, W., Brown, S., Mackie, K., 1997. Cannabinoids inhibit N- and P/Q-type calcium channels in cultured rat hippocampal neurons. J. Neurophysiol. 78, 43–50.
Umeda, A., Meyerholz, A., Ungewickell, E., 2000. Identification of the universal cofactor (auxilin 2) in clathrin coat dissociation. Eur. J. Cell Biol. 79, 336–342.
Ungewickell, E., Ungewickell, H., Holstein, S.E., Lindner, R., Prasad, K., Barouch, W., Martin, B., Greene, L.E., Eisenberg, E., 1995. Role of auxilin in uncoating clathrin-coated vesicles. Nature 378, 632–635.
van de Linde, S., Loschberger, A., Klein, T., Heidbreder, M., Wolter, S., Heilemann, M., Sauer, M., 2011. Direct stochastic optical reconstruction microscopy with standard fluorescent probes. Nat. Protoc. 6, 991–1009.
van den Bogaart, G., Holt, M.G., Bunt, G., Riedel, D., Wouters, F.S., Jahn, R., 2010. One SNARE complex is sufficient for membrane fusion. Nat. Struct. Mol. Biol. 17, 358–364.

van den Bogaart, G., Jahn, R., 2011. Inside insight to membrane fusion. Proc. Natl. Acad. Sci. USA 108, 11729–11730.

van Roessel, P., Elliott, D.A., Robinson, I.M., Prokop, A., Brand, A.H., 2004. Independent regulation of synaptic size and activity by the anaphase-promoting complex. Cell 119, 707–718.

Vannucchi, M.G., Pepeu, G., 1995. Muscarinic receptor modulation of acetylcholine release from rat cerebral cortex and hippocampus. Neurosci. Lett. 190, 53–56.

Varoqueaux, F., Sigler, A., Rhee, J.S., Brose, N., Enk, C., Reim, K., Rosenmund, C., 2002. Total arrest of spontaneous and evoked synaptic transmission but normal synaptogenesis in the absence of Munc13-mediated vesicle priming. Proc. Natl. Acad. Sci. USA 99, 9037–9042.

Verhage, M., de Vries, K.J., Roshol, H., Burbach, J.P., Gispen, W.H., Sudhof, T.C., 1997. DOC2 proteins in rat brain: complementary distribution and proposed function as vesicular adapter proteins in early stages of secretion. Neuron 18, 453–461.

Verhage, M., Maia, A.S., Plomp, J.J., Brussaard, A.B., Heeroma, J.H., Vermeer, H., Toonen, R.F., Hammer, R.E., van den Berg, T.K., Missler, M., Geuze, H.J., Sudhof, T.C., 2000. Synaptic assembly of the brain in the absence of neurotransmitter secretion. Science, New York, 287, 864–869.

Verhage, M., Sorensen, J.B., 2008. Vesicle docking in regulated exocytosis. Traffic 9, 1414–1424.

Verstreken, P., Kjaerulff, O., Lloyd, T.E., Atkinson, R., Zhou, Y., Meinertzhagen, I.A., Bellen, H.J., 2002. Endophilin mutations block clathrin-mediated endocytosis but not neurotransmitter release. Cell 109, 101–112.

Verstreken, P., Koh, T.W., Schulze, K.L., Zhai, R.G., Hiesinger, P.R., Zhou, Y., Mehta, S.Q., Cao, Y., Roos, J., Bellen, H.J., 2003. Synaptojanin is recruited by endophilin to promote synaptic vesicle uncoating. Neuron 40, 733–748.

Vieira, A.V., Lamaze, C., Schmid, S.L., 1996. Control of EGF receptor signaling by clathrin-mediated endocytosis. Science, New York, 274, 2086–2089.

Villanueva, J., Torregrosa-Hetland, C.J., Garcia-Martinez, V., Del Mar Frances, M., Viniegra, S., Gutierrez, L.M., 2012. The F-actin cortex in chromaffin granule dynamics and fusion: a mini review. J. Mol. Neurosci.

Vitale, M.L., Seward, E.P., Trifaro, J.M., 1995. Chromaffin cell cortical actin network dynamics control the size of the release-ready vesicle pool and the initial rate of exocytosis. Neuron 14, 353–363.

Voets, T., Toonen, R.F., Brian, E.C., de Wit, H., Moser, T., Rettig, J., Sudhof, T.C., Neher, E., Verhage, M., 2001. Munc18-1 promotes large dense-core vesicle docking. Neuron 31, 581–591.

Volpicelli, L.A., Levey, A.I., 2004. Muscarinic acetylcholine receptor subtypes in cerebral cortex and hippocampus. Prog. Brain Res. 145, 59–66.

von Kleist, L., Stahlschmidt, W., Bulut, H., Gromova, K., Puchkov, D., Robertson, M.J., MacGregor, K.A., Tomilin, N., Pechstein, A., Chau, N., Chircop, M., Sakoff, J., von Kries, J.P., Saenger, W., Krausslich, H.G., Shupliakov, O., Robinson, P.J., McCluskey, A., Haucke, V., 2011. Role of the clathrin terminal domain in regulating coated pit dynamics revealed by small molecule inhibition. Cell 146, 471–484.

von Ruden, L., Neher, E., 1993. A Ca-dependent early step in the release of catecholamines from adrenal chromaffin cells. Science, New York, 262, 1061–1065.

Wagh, D.A., Rasse, T.M., Asan, E., Hofbauer, A., Schwenkert, I., Durrbeck, H., Buchner, S., Dabauvalle, M.C., Schmidt, M., Qin, G., Wichmann, C., Kittel, R., Sigrist, S.J., Buchner, E., 2006. Bruchpilot, a protein with homology to ELKS/CAST, is required for structural integrity and function of synaptic active zones in Drosophila. Neuron 49, 833–844.

Walther, K., Diril, M.K., Jung, N., Haucke, V., 2004. Functional dissection of the interactions of stonin 2 with the adaptor complex AP-2 and synaptotagmin. Proc. Natl. Acad. Sci. USA 101, 964–969.

Wang, C.T., Lu, J.C., Bai, J., Chang, P.Y., Martin, T.F., Chapman, E.R., Jackson, M.B., 2003. Different domains of synaptotagmin control the choice between kiss-and-run and full fusion. Nature 424, 943–947.

Wang, H., Storm, D.R., 2003. Calmodulin-regulated adenylyl cyclases: cross-talk and plasticity in the central nervous system. Mol. Pharmacol. 63, 463–468.

Wang, X., Kibschull, M., Laue, M.M., Lichte, B., Petrasch-Parwez, E., Kilimann, M.W., 1999. Aczonin, a 550-kD putative scaffolding protein of presynaptic active zones, shares homology regions with Rim and Bassoon and binds profilin. J. Cell Biol. 147, 151–162.

Wang, Y., Liu, X., Biederer, T., Sudhof, T.C., 2002. A family of RIM-binding proteins regulated by alternative splicing: implications for the genesis of synaptic active zones. Proc. Natl. Acad. Sci. USA 99, 14464–14469.

Washbourne, P., Thompson, P.M., Carta, M., Costa, E.T., Mathews, J.R., Lopez-Bendito, G., Molnar, Z., Becher, M.W., Valenzuela, C.F., Partridge, L.D., Wilson, M.C., 2002. Genetic ablation of the t-SNARE SNAP-25 distinguishes mechanisms of neuroexocytosis. Nat. Neurosci. 5, 19–26.

Weber, T., Zemelman, B.V., McNew, J.A., Westermann, B., Gmachl, M., Parlati, F., Sollner, T.H., Rothman, J.E., 1998. SNAREpins: minimal machinery for membrane fusion. Cell 92, 759–772.

Weimbs, T., Low, S.H., Chapin, S.J., Mostov, K.E., Bucher, P., Hofmann, K., 1997. A conserved domain is present in different families of vesicular fusion proteins: a new superfamily. Proc. Natl. Acad. Sci. USA 94, 3046–3051.

Weimer, R.M., Gracheva, E.O., Meyrignac, O., Miller, K.G., Richmond, J.E., Bessereau, J.L., 2006. UNC-13 and UNC-10/rim localize synaptic vesicles to specific membrane domains. J. Neurosci. 26, 8040–8047.

Weimer, R.M., Jorgensen, E.M., 2003. Controversies in synaptic vesicle exocytosis. J. Cell Sci. 116, 3661–3666.

Wess, J., Eglen, R.M., Gautam, D., 2007. Muscarinic acetylcholine receptors: mutant mice provide new insights for drug development. Nat. Rev. 6, 721–733.

Westphal, V., Rizzoli, S.O., Lauterbach, M.A., Kamin, D., Jahn, R., Hell, S.W., 2008. Video-rate far-field optical nanoscopy dissects synaptic vesicle movement. Science 320, 246–249.

Wienisch, M., Klingauf, J., 2006. Vesicular proteins exocytosed and subsequently retrieved by compensatory endocytosis are nonidentical. Nat. Neurosci. 9, 1019–1027.

Williams, A.L., Bielopolski, N., Meroz, D., Lam, A.D., Passmore, D.R., Ben-Tal, N., Ernst, S.A., Ashery, U., Stuenkel, E.L., 2011. Structural and functional analysis of tomosyn identifies domains important in exocytotic regulation. J. Biol. Chem. 286, 14542–14553.

Willig, K.I., Rizzoli, S.O., Westphal, V., Jahn, R., Hell, S.W., 2006. STED microscopy reveals that synaptotagmin remains clustered after synaptic vesicle exocytosis. Nature 440, 935–939.

Wilson, R.I., Kunos, G., Nicoll, R.A., 2001. Presynaptic specificity of endocannabinoid signaling in the hippocampus. Neuron 31, 453–462.

Wilson, R.I., Nicoll, R.A., 2001. Endogenous cannabinoids mediate retrograde signalling at hippocampal synapses. Nature 410, 588–592.

Wiser, O., Tobi, D., Trus, M., Atlas, D., 1997. Synaptotagmin restores kinetic properties of a syntaxin-associated N-type voltage sensitive calcium channel. FEBS Lett. 404, 203–207.

Wiser, O., Trus, M., Hernandez, A., Renstrom, E., Barg, S., Rorsman, P., Atlas, D., 1999. The voltage sensitive Lc-type Ca^{2+} channel is functionally coupled to the exocytotic machinery. Proc. Natl. Acad. Sci. USA 96, 248–253.

Wolfel, M., Lou, X., Schneggenburger, R., 2007. A mechanism intrinsic to the vesicle fusion machinery determines fast and slow transmitter release at a large CNS synapse. J. Neurosci. 27, 3198–3210.

Wu, M.N., Schulze, K.L., Lloyd, T.E., Bellen, H.J., 2001. The ROP-syntaxin interaction inhibits neurotransmitter release. Eur. J. Cell Biol. 80, 196–199.

Xiao, W., Poirier, M.A., Bennett, M.K., Shin, Y.K., 2001. The neuronal t-SNARE complex is a parallel four-helix bundle. Nat. Struct. Biol. 8, 308–311.

Xing, Y., Bocking, T., Wolf, M., Grigorieff, N., Kirchhausen, T., Harrison, S.C., 2010. Structure of clathrin coat with bound Hsc70 and auxilin: mechanism of Hsc70-facilitated disassembly. EMBO J. 29, 655–665.

Xu, J., Pang, Z.P., Shin, O.H., Sudhof, T.C., 2009. Synaptotagmin-1 functions as a Ca^{2+} sensor for spontaneous release. Nat. Neurosci. 12, 759–766.

Xu, W., Morishita, W., Buckmaster, P.S., Pang, Z.P., Malenka, R.C., Sudhof, T.C., 2012. Distinct neuronal coding schemes in memory revealed by selective erasure of fast synchronous synaptic transmission. Neuron 73, 990–1001.

Xue, J., Graham, M.E., Novelle, A.E., Sue, N., Gray, N., McNiven, M.A., Smillie, K.J., Cousin, M.A., Robinson, P.J., 2011. Calcineurin selectively docks with the dynamin Ixb splice variant to regulate activity-dependent bulk endocytosis. J. Biol. Chem. 286, 30295–30303.

Xue, M., Craig, T.K., Xu, J., Chao, H.T., Rizo, J., Rosenmund, C., 2010. Binding of the complexin N terminus to the SNARE complex potentiates synaptic-vesicle fusogenicity. Nat. Struct. Mol. Biol. 17, 568–575.

Xue, M., Lin, Y.Q., Pan, H., Reim, K., Deng, H., Bellen, H.J., Rosenmund, C., 2009. Tilting the balance between facilitatory and inhibitory functions of mammalian and Drosophila Complexins orchestrates synaptic vesicle exocytosis. Neuron 64, 367–380.

Xue, M., Reim, K., Chen, X., Chao, H.T., Deng, H., Rizo, J., Brose, N., Rosenmund, C., 2007. Distinct domains of complexin I differentially regulate neurotransmitter release. Nat. Struct. Mol. Biol. 14, 949–958.
Xue, M., Stradomska, A., Chen, H., Brose, N., Zhang, W., Rosenmund, C., Reim, K., 2008. Complexins facilitate neurotransmitter release at excitatory and inhibitory synapses in mammalian central nervous system. Proc. Natl. Acad. Sci. USA 105, 7875–7880.
Yamamoto, M., Wada, N., Kitabatake, Y., Watanabe, D., Anzai, M., Yokoyama, M., Teranishi, Y., Nakanishi, S., 2003. Reversible suppression of glutamatergic neurotransmission of cerebellar granule cells in vivo by genetically manipulated expression of tetanus neurotoxin light chain. J. Neurosci. 23, 6759–6767.
Yamamoto, Y., Mochida, S., Miyazaki, N., Kawai, K., Fujikura, K., Kurooka, T., Iwasaki, K., Sakisaka, T., 2010. Tomosyn inhibits synaptotagmin-1-mediated step of Ca^{2+}-dependent neurotransmitter release through its N-terminal WD40 repeats. J. Biol. Chem. 285, 40943–40955.
Yang, B., Gonzalez Jr., L., Prekeris, R., Steegmaier, M., Advani, R.J., Scheller, R.H., 1999. SNARE interactions are not selective. Implications for membrane fusion specificity. J. Biol. Chem. 274, 5649–5653.
Yang, X., Kaeser-Woo, Y.J., Pang, Z.P., Xu, W., Sudhof, T.C., 2010. Complexin clamps asynchronous release by blocking a secondary Ca(2+) sensor via its accessory alpha helix. Neuron 68, 907–920.
Yang, Y., Calakos, N., 2010. Acute in vivo genetic rescue demonstrates that phosphorylation of RIM1alpha serine 413 is not required for mossy fiber long-term potentiation. J. Neurosci. 30, 2542–2546.
Yao, J., Gaffaney, J.D., Kwon, S.E., Chapman, E.R., 2011. Doc2 is a Ca^{2+} sensor required for asynchronous neurotransmitter release. Cell 147, 666–677.
Yao, J., Kwon, S.E., Gaffaney, J.D., Dunning, F.M., Chapman, E.R., 2012. Uncoupling the roles of synaptotagmin I during endo- and exocytosis of synaptic vesicles. Nat. Neurosci. 15, 243–249.
Yao, J., Nowack, A., Kensel-Hammes, P., Gardner, R.G., Bajjalieh, S.M., 2010. Cotrafficking of SV2 and synaptotagmin at the synapse. J. Neurosci. 30, 5569–5578.
Yi, Z., Yokota, H., Torii, S., Aoki, T., Hosaka, M., Zhao, S., Takata, K., Takeuchi, T., Izumi, T., 2002. The Rab27a/granuphilin complex regulates the exocytosis of insulin-containing dense-core granules. Mol. Cell. Biol. 22, 1858–1867.
Yim, Y.I., Sun, T., Wu, L.G., Raimondi, A., De Camilli, P., Eisenberg, E., Greene, L.E., 2010. Endocytosis and clathrin-uncoating defects at synapses of auxilin knockout mice. Proc. Natl. Acad. Sci. USA 107, 4412–4417.
Yizhar, O., Ashery, U., 2008. Modulating vesicle priming reveals that vesicle immobilization is necessary but not sufficient for fusion-competence. PloS One 3, e2694.
Yizhar, O., Lipstein, N., Gladycheva, S.E., Matti, U., Ernst, S.A., Rettig, J., Stuenkel, E.L., Ashery, U., 2007. Multiple functional domains are involved in tomosyn regulation of exocytosis. J. Neurochem. 103, 604–616.
Yizhar, O., Matti, U., Melamed, R., Hagalili, Y., Bruns, D., Rettig, J., Ashery, U., 2004. Tomosyn inhibits priming of large dense-core vesicles in a calcium-dependent manner. Proc. Natl. Acad. Sci. USA 101, 2578–2583.
Young Jr., S.M., Neher, E., 2009. Synaptotagmin has an essential function in synaptic vesicle positioning for synchronous release in addition to its role as a calcium sensor. Neuron 63, 482–496.
Yu, T.P., McKinney, S., Lester, H.A., Davidson, N., 2001. Gamma-aminobutyric acid type A receptors modulate cAMP-mediated long-term potentiation and long-term depression at monosynaptic CA3–CA1 synapses. Proc. Natl. Acad. Sci. USA 98, 5264–5269.
Zador, A.M., 2001. Synaptic connectivity and computation. Nat. Neurosci. 4, 1157–1158.
Zalutsky, R.A., Nicoll, R.A., 1990. Comparison of two forms of long-term potentiation in single hippocampal neurons. Science 248, 1619–1624.
Zeniou-Meyer, M., Zabari, N., Ashery, U., Chasserot-Golaz, S., Haeberle, A.M., Demais, V., Bailly, Y., Gottfried, I., Nakanishi, H., Neiman, A.M., Du, G., Frohman, M.A., Bader, M.F., Vitale, N., 2007. Phospholipase D1 production of phosphatidic acid at the plasma membrane promotes exocytosis of large dense-core granules at a late stage. J. Biol. Chem. 282, 21746–21757.
Zhai, R.G., Bellen, H.J., 2004. The architecture of the active zone in the presynaptic nerve terminal. Physiology (Bethesda) 19, 262–270.
Zhang, J.Z., Davletov, B.A., Sudhof, T.C., Anderson, R.G., 1994. Synaptotagmin I is a high affinity receptor for clathrin AP-2: implications for membrane recycling. Cell 78, 751–760.
Zhang, L., Marcu, M.G., Nau-Staudt, K., Trifaro, J.M., 1996. Recombinant scinderin enhances exocytosis, an effect blocked by two scinderin-derived actin-binding peptides and PIP2. Neuron 17, 287–296.

Zhang, Q., Li, Y., Tsien, R.W., 2009a. The dynamic control of kiss-and-run and vesicular reuse probed with single nanoparticles. Science, New York, 323, 1448–1453.
Zhang, Z., Bhalla, A., Dean, C., Chapman, E.R., Jackson, M.B., 2009b. Synaptotagmin IV: a multifunctional regulator of peptidergic nerve terminals. Nat. Neurosci. 12, 163–171.
Zhang, Z., Zhang, Z., Jackson, M.B., 2010. Synaptotagmin IV modulation of vesicle size and fusion pores in PC12 cells. Biophys. J. 98, 968–978.
Zhen, M., Jin, Y., 1999. The liprin protein SYD-2 regulates the differentiation of presynaptic termini in *C. elegans*. Nature 401, 371–375.
Zheng, F., Wess, J., Alzheimer, C., 2012. M2 muscarinic acetylcholine receptors regulate long-term potentiation at hippocampal CA3 pyramidal cell synapses in an input-specific fashion. J. Neurophysiol. 108, 91–100.
Zhu, D., Zhou, W., Liang, T., Yang, F., Zhang, R.Y., Wu, Z.X., Xu, T., 2007. Synaptotagmin I and IX function redundantly in controlling fusion pore of large dense core vesicles. Biochem. Biophys. Res. Commun. 361, 922–927.
Zhuang, X., 2009. Nano-imaging with Storm. Nat. Photonics 3, 365–367.
Zikich, D., Mezer, A., Varoqueaux, F., Sheinin, A., Junge, H.J., Nachliel, E., Melamed, R., Brose, N., Gutman, M., Ashery, U., 2008. Vesicle priming and recruitment by ubMunc13-2 are differentially regulated by calcium and calmodulin. J. Neurosci. 28, 1949–1960.
Zilly, F.E., Sorensen, J.B., Jahn, R., Lang, T., 2006. Munc18-bound syntaxin readily forms SNARE complexes with synaptobrevin in native plasma membranes. PLoS Biol. 4, e330.
Zucker, R.S., 1999. Calcium- and activity-dependent synaptic plasticity. Curr. Opin. Neurobiol. 9, 305–313.
Zucker, R.S., Regehr, W.G., 2002. Short-term synaptic plasticity. Annu. Rev. Physiol. 64, 355–405.

CHAPTER THREE

The First Hour in the Life of a Synapse: Contact Formation, Partner Selection, and Onset of Function

Juliette E. Cheyne and Christian Lohmann
Department of Synapse and Network Development, Netherlands Institute for Neuroscience, Amsterdam, The Netherlands

1. INTRODUCTION

The brain contains a vast number of different cell types which are connected in precise ways. The specificity of these connections is crucial for the correct functioning of the brain. Despite decades of research into the processes involved in synapse formation, we still do not have a complete understanding of how synapses are formed and how specificity is established. However, much progress has been made and we now know a lot about how synapse formation precedes between certain cell types. In this chapter, we will focus on the three different phases in the process of synapse formation: contact formation, partner selection, and onset of function.

Synapse formation begins with the contact of axons and dendrites but there are different ways that this can occur. Are synapses formed by the concerted growth of both axons and dendrites or does one side of the synapse play a larger role? Do presynaptic axons initiate contact or do postsynaptic filopodia reach out to find appropriate partners? Are there differences in contact formation at glutamatergic versus GABAergic synapses? Often different cell types are present in the same region yet the particular connections are formed, with cellular and sometimes even subcellular specificity. Do synapses only form when the appropriate synaptic partners are contacted or are inappropriate synapses formed but then later eliminated? How is the formation of a correct synapse signaled within neurons? Are cell adhesion molecules involved in partner selection? Once a contact has been formed, there are many steps involved in making a functional synapse. These steps include the recruitment of presynaptic release machinery, postsynaptic receptors, and scaffolding proteins. Are the mechanisms for protein recruitment similar on both sides of the synapse? How long does it take for a synapse to become fully functional? In addition to answering these questions, we will address what role activity plays in each step of this process.

The Synapse
http://dx.doi.org/10.1016/B978-0-12-418675-0.00003-1

2.1. CONTACT FORMATION

Contact between two neurons is the first necessary step for synapse formation. However, before contact can occur, neurons need to differentiate and form specific neuronal processes. Both the structure and function of neurons are highly polarized, with axons that send and dendrites that receive synaptic inputs. Such polarity is established independent of contact with other neurons, as axons and dendrites differentiate in the absence of interaction with other neurons (Bartlett and Banker, 1984a). The different subcellular domains of neurons are further specialized by the formation of different types of synapses at precise locations. Inhibitory synapses (symmetrical in electron micrographs) are mostly located on cell bodies and dendritic shafts. In contrast, excitatory synapses (asymmetrical in electron micrographs) form on dendritic spines and dendritic shafts (Bartlett and Banker, 1984b). Synaptogenesis research has focused primarily on excitatory glutamatergic synapses. Interestingly, recent research has suggested that the formation of inhibitory GABAergic synapses is fundamentally different to the formation of excitatory glutamatergic synapses (Figure 1).

Glutamate is the neurotransmitter at a majority of the excitatory synapses in the brain. In the mature brain excitatory synapses largely occur on dendritic spines, which are small (usually $<1\ \mu m^3$) protrusions in the postsynaptic membrane opposed to the presynaptic active zone (Ehlers, 2002; Hering and Sheng, 2001). However, during early development, the majority of glutamatergic synapses are present on dendritic shafts, with spine synapses first appearing during the second postnatal week (Harris et al., 1992; Fiala et al., 1998). Different models have been proposed for the formation of excitatory synapses (reviewed by Yuste and Bonhoeffer (2004)); however, the filopodia model predominates during development. Filopodia are small dynamic protrusions from dendrites, which play a primary role in searching the local environment and finding a synaptic partner (Ziv and Smith, 1996). In the filopodia model, the synapse is formed after the filopodium has contacted an axon and developed into a dendritic spine (Figure 1(A): Ziv and Smith, 1996; Dailey and Smith, 1996; Fiala et al., 1998) or a new branch (see below). It should be noted that spines can also grow directly, without the prior need for a filopodium; however, this direct spinogenesis appears to be more prominent in the adult brain (Section 2.4.3).

Live cell imaging of dendrites has revealed that dynamic filopodia are abundant during early development, at the peak of synaptogenesis (Ziv and Smith, 1996; Lohmann and Bonhoeffer, 2008; Dailey and Smith, 1996). Dendritic filopodia are very motile structures that appear, grow, and retract rapidly. The average half lives of filopodia are short, only a few minutes. Filopodia search their environment for possible synaptic partners and thus initiate contact with axons. Contacts are frequently observed and while most are short-lived (seconds to a few minutes) (Ziv and Smith, 1996; Lohmann and

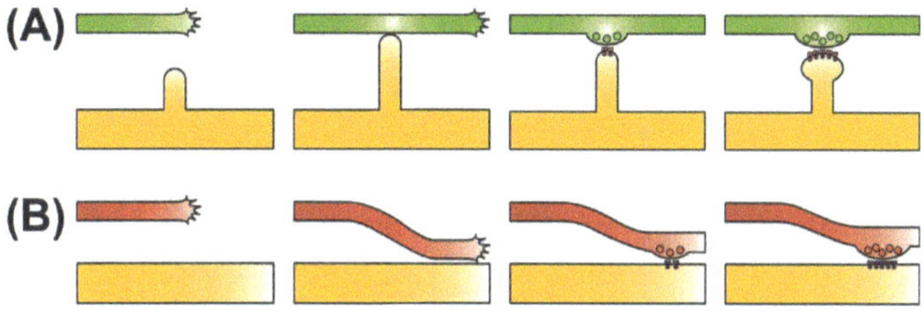

Figure 1 Different Modes of Synapse Formation at Glutamatergic and GABAergic Synapses. (A) At excitatory synapses, filopodia grow from the dendrite (yellow) to seek glutamatergic axons (green). After contact a synapse forms on the filopodium, which later matures into a dendritic spine. (B) In contrast, at GABAergic synapses, filopodia are not involved. Instead, synapses form at sites of contact between the dendrite and the GABAergic axon (red).

Bonhoeffer, 2008), some become stabilized and trigger the formation of functional presynaptic boutons (Ziv and Smith, 1996; De Roo et al., 2008). Electron microscopy studies have confirmed that early in development approximately one-quarter of synapses occur onto filopodia (Fiala et al., 1998). Synapses can be found anywhere over the filopodial surface and multiple synapses can be present on a filopodium. Filopodia may also recruit shaft synapses and give rise to spines (Fiala et al., 1998; Marrs et al., 2001). Either way there is strong evidence that during development filopodia initiate contact which sometimes leads to synapse formation.

To examine whether the identity of the contacted axon may determine why some contacts are stabilized while others are abandoned, different axons were selectively visualized (Lohmann and Bonhoeffer, 2008). When a mixed population of axons was labeled, the duration of contacts varied considerably, some last only seconds whereas others were maintained for longer periods of time. When GABAergic axons were specifically labeled, all contacts were short-lived (Lohmann and Bonhoeffer, 2008). Thus the identity of the presynaptic axon is an important factor in filopodial partner selection (Section 3.3).

Although dendritic filopodia select which axons to form connections with, it appears that axons also play an important role in synapse formation. Even in the absence of contact with dendrites, presynaptic terminals are formed at distinct locations along the axon (Sabo et al., 2006). Filopodia appear to slide along target axons in search of these regions to settle upon (Lohmann and Bonhoeffer, 2008). Therefore, axons direct the specific locations for synapse formation. This occurs independent of neurotransmitter release as blocking ionotropic glutamate receptors does not alter contact lifetimes or stabilization (Lohmann and Bonhoeffer, 2008).

Although activity induced by neurotransmitter release does not appear to play a role in initial contact formation, the growth of axons and dendrites is controlled by local calcium signaling. Axonal growth cones are involved in making choices about where to grow and this process is regulated by local calcium signals (Rehder and Cheng, 1998; Gomez et al., 2001; Zheng, 2000). In addition, local calcium transients regulate the growth of both dendrites and filopodia by triggering stabilization (Lohmann et al., 2002, 2005).

The formation of synapses may contribute to the arborization of dendrites. The synaptotropic hypothesis suggests that dendrites and axons grow toward areas where synaptic connections are made (Vaughn, 1989). Live imaging studies in zebrafish larvae have provided experimental evidence that this hypothesis is correct (Niell et al., 2004). Newly formed synapses were observed to appear at the tips of newly extended filopodia, some of which became stabilized and subsequently matured into dendritic branches. Activity plays an important role in this process as reduced AMPA (α-amino-3-hydroxy-5-methyl-4-isoxazolepropionic acid) receptor (AMPAR) expression prevented the formation of complex arbors in the *Xenopus* retinotectal system (Haas et al., 2006). On the contrary, in the mouse hippocampus, synapse formation and the development of dendritic arbors preceded normally when glutamatergic receptor expression was completely absent. Taken together this suggests that there are species differences in the instructive role of glutamatergic synapse formation and/or activity in directing dendritic arbor growth.

In contrast to glutamatergic synapses, filopodia do not play a role in the formation of GABAergic synapses during development. Instead, GABAergic synapses on dendritic shafts are formed by new presynaptic boutons which appear at preexisting sites of axonal and dendritic contact (Figure 1(B): Wierenga et al., 2008). GABAergic interneurons often contact excitatory neurons with subcellular precision. For example, fast spiking parvalbumin-positive interneurons form synapses on the cell bodies of pyramidal neurons in the cortex, in contrast to somatostatin-positive interneurons, which form synapses on the dendrites of cortical pyramidal neurons (Fino et al., 2012). Several studies have investigated whether activity plays a role in mediating the subcellular precision of GABAergic inputs. In the visual cortex, GABAergic synapses still form at precise subcellular locations in the absence of experience-driven activity (Di Cristo et al., 2004; Chattopadhyaya et al., 2004). Therefore, at both glutamatergic and GABAergic synapses, contact formation is independent of activity.

2.2. PARTNER SELECTION

Not every contact between a dendrite and an axon will develop a synapse but how are some contacts selected while others are abandoned? The identity of the contacted axon is important in this decision-making process, as are the events at filopodia, which are initiated rapidly after appropriate contacts are formed (Figure 2, Lohmann and Bonhoeffer, 2008). When filopodia contact an appropriate synaptic partner, local dendritic calcium transients are observed within 1 min (Lohmann and Bonhoeffer, 2008). The contacts which display calcium transients are selectively maintained (Figure 2(A). In contrast, calcium transients are not seen at the contacts that are lost (Lohmann and Bonhoeffer, 2008). Contacts with GABAergic axons never initiate calcium transients

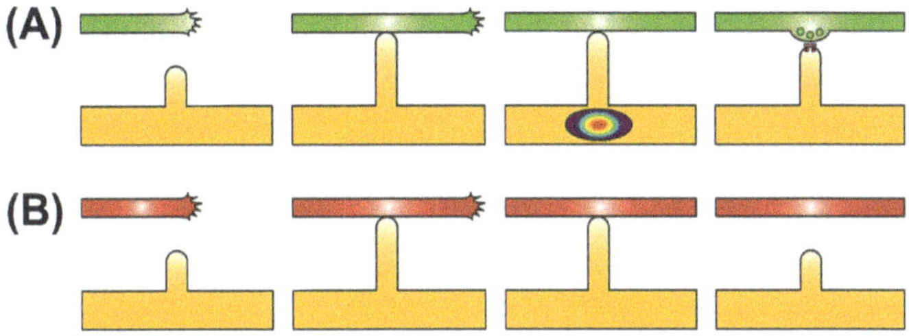

Figure 2 Partner Selection at Glutamatergic Synapses is Controlled by Local Calcium Signaling. (A) When filopodia extending from dendrites (yellow) make contact with glutamatergic axons (green), calcium transients (rainbow) are sometimes induced. If calcium transients occur, filopodia become stabilized and synapse formation proceeds. (B) In contrast, when filopodia contact GABAergic axons (red), no calcium transients occur and filopodia retract.

and thus are always lost (Figure 2(B)). Therefore, local calcium transients at filopodia indicate the formation of an appropriate contact. Given the differences in the formation of GABAergic synapses, there may also be differences in the mechanism of partner selection; however, this currently remains an open question.

Pharmacological experiments suggest that calcium signaling in filopodia during partner selection does not require glutamate release (Lohmann and Bonhoeffer, 2008). Contact-induced calcium transients are not mediated by NMDA (*N*-methyl-D-aspartate) receptors (NMDARs) or AMPARs, as they are still observed when ionotropic glutamate receptors are blocked (Lohmann and Bonhoeffer, 2008). Although the involvement of metabotropic glutamate receptors in calcium signaling upon contact has not been specifically addressed, it is known that metabotropic glutamate receptor antagonists do not affect the frequency of calcium transients at this stage of development, when filopodia are abundant (Lohmann et al., 2005).

The source of the calcium transients during partner selection is currently unknown; however, transsynaptic adhesion molecules are likely candidates (Section 3.3.1) for mediating these calcium signals, either alone or via tension-induced signaling (Heiman and Shaham, 2009b). It has been suggested that tension at the tip of filopodia upon binding to an axon leads to the opening of mechanosensitive channels, such as TRPM7, which generate calcium signals (Heiman and Shaham, 2009a,b; Wei et al., 2009; Jacques-Fricke et al., 2006). Such a tension-induced mechanism could elegantly generate the same type of signal in filopodia, independent of the exact identities of adhesion molecules present at the membrane, as any tight binding could induce local calcium transients and trigger stabilization of the filopodia.

2.2.1. Role of Adhesion Molecules in Synapse Specificity

In the brain, different cell types form specific connections; however, little is known about how selection is determined and signaled within neurons. It is known that the generation of synapse specificity is mediated by a number of different processes including the guidance of axons to correct locations and the selection of correct targets following contact (Sanes and Yamagata, 2009; Shen and Scheiffele, 2010). Because adhesion molecules span the gap between the synaptic cleft, they are perfectly located to mediate specificity at the level of individual synapses. Many different families of transsynaptic adhesion molecules have been identified that can induce the formation of pre- or postsynaptic terminals, mediate matching of both sides of the synapse by homophilic or heterophilic interactions and provide molecular diversity for synapse specificity.

One theory suggests that synapse specificity could be generated by different pre- and postsynaptic terminals having a sort of molecular tag that makes them distinct from other terminals (Sperry, 1963). Such tags could be generated by alternate splicing of adhesion molecules such as neurexin or the protocadherin family (Chih et al., 2006; Rowen et al., 2002; Tabuchi and Sudhof, 2002; Ullrich et al., 1995; Ushkaryov and Sudhof, 1993; Frank and Kemler, 2002; Washbourne et al., 2004; Zipursky and Sanes, 2010; Chen et al., 2012; Iijima et al., 2011). Matching of pre- and postsynaptic molecular tags could either trigger the formation of only correct synapses or act post-hoc such that synapses are formed in a promiscuous manner and later eliminated. In this model, appropriate adhesion molecules would instead be important for stabilization processes after synapse formation, without which inappropriate synapses would be unstable and thus eliminated.

A detailed discussion of all of the families of adhesion molecules is outside the scope of this chapter. However, here we will highlight just a few examples that may be important for partner selection and synapse formation. Other adhesion molecules that may also be important for partner selection and synapse formation include syncams (Biederer et al., 2002), integrins (Coppolino et al., 1997; Shi and Ethell, 2006), protocadherins (Washbourne et al., 2004; Zipursky and Sanes, 2010; Chen et al., 2012), and ephrin (Nolt et al., 2011; Kayser et al., 2008).

2.2.1.1. Neuroligins and Neurexins

Neuroligins and neurexins are involved in the formation of synaptic specializations and are important for the trafficking of synaptic proteins to newly generated synapses. Neurexins are expressed presynaptically, in contrast to neuroligins, which are found postsynaptically, with different isoforms at glutamatergic (neuroligin-1, -3, and -4) and GABAergic synapses (neuroligin-2: Song et al., 1999; Graf et al., 2004). Neurexins are sufficient to induce differentiation of both glutamatergic and GABAergic postsynaptic specializations (Dean and Dresbach, 2006; Kang et al., 2008; Nam and Chen, 2005; Chih et al., 2005). Conversely, presynaptic differentiation in both glutamatergic and GABAergic axons is induced by neuroligins (Scheiffele et al., 2000; Graf et al., 2004; Dean and Dresbach, 2006). Furthermore, the expression of neuroligins regulates synapse formation and density in an input-specific manner. Neuroligin-1 expression bidirectionally regulates the formation of stable contacts and therefore modulates synapse density (Craig and Kang, 2007; Krueger et al., 2012; Arstikaitis et al., 2011). This regulation of synapse density occurs in a competitive manner due to the differential expression level of neuroligin-1 between neurons and is not observed for when neuroligin-1 is knocked out globally (Kwon et al., 2012). Interestingly, in CA1 hippocampal neurons neuroligin-2 expression regulates synapse formation in an input-specific manner as neuroligin-2 knockout alters only synapses in the stratum radiatum but not stratum lacunosum-moleculare (Denardo et al., 2012).

Localization of neuroligins to postsynaptic sites triggers the recruitment of other synaptic proteins, through both direct and indirect protein interactions. Neuroligin-1 directly binds postsynaptic density (PSD) protein 95 kDa (PSD-95) and NMDARs and independently regulates their recruitment to synapses (Dresbach et al., 2004; Barrow et al., 2009). The postsynaptic specializations induced by neurexins become functional within 2 h with the recruitment of GluR2-containing AMPARs (Heine et al., 2008). Taken together, this research suggests that the formation of functional synapses can be triggered by neuroligins and neurexins in an input-specific manner; however, knowledge is still lacking concerning the first moments in a synapse's lifetime. It is known that neuroligin can be present in filopodia (Barrow et al., 2009), but, whether its presence is correlated with calcium transients during partner selection is currently unknown.

2.2.1.2. Cadherins

Cadherins are a large family of adhesion proteins (more than 100 members) that play roles in synapse formation and are important mediators of synapse specificity (Salinas and Price, 2005). Cadherins are expressed pre- and postsynaptically and form homophilic interactions across the synapse. An elegant example of the role of a specific cadherin in generating synapse specificity was recently revealed (Williams et al., 2011). Surprisingly, specificity of mossy fiber synapses, which connect dentate gyrus granule cells with CA3 pyramidal neurons, is maintained in dissociated hippocampal cultures. Cadherin-9 was identified as a candidate for mediating this synapse's specificity by screening adhesion molecules that are selectively expressed in these two neuronal populations. Knockdown of cadherin-9 specifically impaired the formation of mossy fiber synapses onto CA3 neurons both in vitro and in vivo. This shows that transsynaptic homophilic interactions between pre- and postsynaptic cadherin-9 are important for the development of this synapse (Williams et al., 2011). In summary, adhesion molecules likely play an important role in mediating partner selection.

2.3. ONSET OF FUNCTION

After an appropriate contact has formed and been selected, there are still several steps to complete before a synapse becomes functional. This includes the formation of the pre- and postsynaptic specializations via recruitment of a vast array of synaptic proteins over the course of about 1 h (Waites et al., 2005). Numerous experiments with in vitro hippocampal culture preparations have shown that assembly of the presynaptic active zone is fundamentally different to formation of the PSD. Presynaptic differentiation occurs before postsynaptic differentiation at glutamatergic synapses with presynaptic terminals becoming fully functional only 30 min after the initial contact, while postsynaptic expression of glutamate receptors (GluRs) takes 35–45 min (Friedman et al., 2000; Zito et al., 2009). Following the initial calcium transients induced by appropriate contact formation, synaptic calcium transients appear within 12–72 min (Lohmann and Bonhoeffer, 2008). These transients are longer lasting than the initial transients (0.7 vs 0.5 s: Lohmann and Bonhoeffer, 2008).

2.3.1. Development of the Presynaptic Terminal

Active zone proteins and synaptic vesicle proteins are recruited to presynaptic terminals separately (Figure 3.3: Ziv and Garner, 2004). Active zone protein containing vesicles are known as Piccolo–Bassoon transport vesicles (PTVs), which are 80 nm dense core granulated vesicles containing Piccolo, Bassoon, Syntaxin, synaptosomal-associated protein 25 kDa (SNAP-25), and N-cadherin and chromogranin-B (Zhai et al., 2001; Shapira et al., 2003). Sorting of such proteins into these active zone precursor vesicles occurs within the trans-Golgi compartment (Dresbach et al., 2006). Active zone assembly occurs rapidly via the insertion of two or three PTVs into the synaptic membrane (Shapira et al., 2003).

Synaptic vesicle proteins such as vesicle-associated membrane protein (VAMP-2/synaptobrevin-II), synaptophysin, synaptotagmin, and synapsin are not contained within PTVs nor is GABA transporter 1 (GAT-1) which resides near but not within the active zone (Shapira et al., 2003; Zhai et al., 2001). Synaptic vesicle proteins, such as synaptobrevin (VAMP) and synaptotagmins, are however united early in the synthesis pathway and trafficked along axons in discrete vesicles called synaptic vesicle protein transport vesicles (STVs) (Ahmari et al., 2000). The recruitment of PTVs and STVs is coordinated which may be important for rapid assembly of the presynaptic terminal (Bury and Sabo, 2011). Synapsin is not contained within either STVs or PTVs but is recruited in a separate transport packet (Easley-Neal et al., 2013). Use of the dye FM1-43 to visualize release of synaptic vesicles has shown that release sites can be functional in the absence of postsynaptic specializations (Krueger et al., 2003). However, clustering of presynaptic vesicle proteins such as synapsin-I and synaptophysin have been reported to only occur upon contact with dendrites (Fletcher et al., 1991).

2.3.2. Development of the Postsynaptic Terminal

In contrast to formation of the presynaptic active zone, the PSD is formed gradually by protein recruitment (Figure 3; Bresler et al., 2004). Time lapse imaging of PSD-95

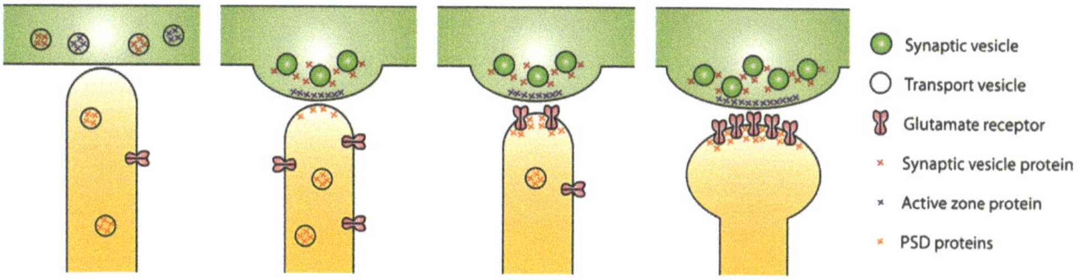

Figure 3 Presynaptic Terminals Form More Rapidly than Postsynaptic Terminals. The presynaptic terminal forms quickly by the fusion of PTVs which contain active zone proteins (blue) and STVs which contain synaptic vesicle proteins (red). The postsynaptic terminal forms slower by gradual recruitment of receptors (pink) and different postsynaptic proteins (orange).

has shown that PSDs can form and disappear rapidly in <15 min (Marrs et al., 2001). Sometimes synapses form via the recruitment of PSD proteins into a filopodium, whereas others form by the extension of spines carrying preformed PSD (Marrs et al., 2001). PSD-95 is recruited early during synaptogenesis and can be present within 20 min of contact (Bresler et al., 2001; Okabe et al., 2001). NMDARs, Shank2, and Shank3 are recruited gradually (within 30–40 min) to PSD with very similar kinetics suggesting that a common upstream rate limiting process restricts their recruitment (Bresler et al., 2004). PSD-95, guanylate kinase-associated protein, neuroligin-1, and Shank have been reported to exist in a mobile cluster that is delivered to synapses via actin transport (Gerrow et al., 2006). Glutamate receptors can either arrive in transport vesicles (Washbourne et al., 2002) or diffuse into laterally synapses from the extrasynaptic sites (Tovar and Westbrook, 2002). AMPARs and NMDARs are recruited to synapses independently (Section 3.4.3; Jeyifous et al., 2009; Washbourne et al., 2002).

2.3.3. Role of Activity in Onset of Function

Activity is required for the maturation of synapses and refinement of neuronal circuits; however, synapses can still form in the absence of activity. A range of different methods have been used to investigate the role that activity plays in synapse formation including blocking presynaptic neurotransmitter release, inhibiting expression of functional NMDARs, and use of autaptic hippocampal cultures. Synapses form normally when neurotransmitter release is prevented, either by the removal of proteins that are essential for vesicle priming (Varoqueaux et al., 2002; Verhage et al., 2000) or by inhibition of SNARE-mediated release in cultured hippocampal neurons (Harms and Craig, 2005). In hippocampal cultures grown from mouse embryos lacking the NR1 subunit of the NMDAR (which is required for formation of functional NMDARs), synapses still form and spine density is similar to that seen for wild-type neurons (Okabe et al., 1998). Microisland cultures, which contain isolated neurons grown on small islands, have provided a useful model system to probe the necessity of neurotransmitter release in the development of glutamatergic and GABAergic synapses (Rao et al., 2000). In this model system, neurons form autaptic synapses, and are thus only exposed to the type of neurotransmitter that they release. In these cultures, synapses form in the absence of both action potentials and activation of GluRs (Gomperts et al., 2000). For pyramidal neurons in microisland cultures, GABA input is not necessary for $GABA_A$ receptor or gephyrin clustering. For GABAergic neurons on a microisland, glutamate input is not necessary for PSD-95 or NMDAR clustering but is required for AMPAR clustering (Rao et al., 2000). Similarly, transient connections can form independent of NMDAR activity in vivo in *Xenopus*; however, NMDAR activation is required for persistent synapse formation and maturation (Chen et al., 2010). Taken together, these results suggest that synapses can form and express NMDA receptors without the need for activity.

Activity is, however, important for the regulation of AMPAR expression at synapses, with NMDAR activity bidirectionally regulating AMPAR expression. There is some evidence that new spines can contain AMPARs rapidly after their formation, however, the prominence of synapses lacking AMPARs early in development has lead to the assumption that synapses lack AMPARs initially. To date few studies have followed individual synapses from the time of their creation to becoming functional and as a result whether newly formed synapses contain AMPARs initially is still a matter of debate which is currently unsolved. Here we will outline the evidence from both sides of the debate and discuss new possibilities that suggest that neighboring synapses may be an important regulator of synapse function.

Glutamatergic synapses always contain NMDARs; however, they can exist in different states depending on whether or not they contain AMPARs in the postsynaptic membrane. NMDAR only synapses are called silent synapses as they have no synaptic responses at normal postsynaptic membrane potentials due to the voltage-dependent magnesium (Mg^{2+}) block on NMDARs (Isaac et al., 1995; Monyer et al., 1992). The presence of silent synapses was initially suggested after individual analysis of the variability of the peak amplitude of synaptic AMPAR and NMDAR responses before and after long-term potentiation (LTP) induction (Kullmann, 1994). Peak amplitudes vary based on the amount of neurotransmitter release which is analyzed by calculating the coefficient of variation. The coefficient of variation was consistently higher for AMPARs than NMDARs prior to LTP suggesting that a proportion of synapses lack functional AMPARs (Kullmann, 1994). Subsequently, synapses containing only NMDARs were demonstrated using electrophysiology in CA1 (Isaac et al., 1995; Liao et al., 1995; Durand et al., 1996), CA3 of hippocampal slices (Montgomery et al., 2001), and in the visual cortex (Rumpel et al., 1998). After LTP, these silent synapses gained AMPAR responses (Isaac et al., 1995; Liao et al., 1995; Montgomery et al., 2001; Rumpel et al., 1998). In addition, silent synapses can be created through the induction of NMDAR-dependent long-term depression (Montgomery and Madison, 2002) which leads to AMPAR removal from synapses (Dudek and Bear, 1992; Beattie et al., 2000; Carroll et al., 1999a,b). Taken together, this shows that NMDARs bidirectionally regulate AMPAR expression which can lead to the activation or creation of silent synapses.

Results from initial studies which showed that AMPARs are inserted into synapses in response to LTP have been further supported by evidence that these receptors are also regulated by both spontaneous activity and sensory experience. In the early postnatal hippocampus, spontaneous activity leads to recruitment of GluR4-containing AMPARs to synapses in an NMDAR-dependent manner (Zhu et al., 2000). In the barrel cortex, synaptic potentiation induced by experience was found to drive GluR1-containing AMPARs into synapses (Takahashi et al., 2003). Blocking the delivery of GluR1-containing AMPARs blocked the experience-dependent potentiation (Takahashi et al., 2003). In addition, unsilencing of synapses in the barrel cortex is driven by sensory experience (Ashby and Isaac, 2011). The connections between neurons in slices from normal developing mice and deprived mice that had their whiskers trimmed were probed by utilizing two-photon glutamate uncaging. The development of spines did not require sensory experience, however, the development of functional connections containing AMPARs did. In the absence of sensory activity, neurons were still connected but the majority of connections were silent. Taken together, this suggests that neurons intrinsically form a network of silent connections but experience is required to unsilence these synapses (Ashby and Isaac, 2011).

The prominence of silent synapses during early development has been revealed using a wide range of different techniques (Liao et al., 1999; Petralia et al., 1999; Busetto et al., 2008). Use of fluorescent labeling in dissociated hippocampal cultures revealed that the percentage of silent synapses (with NMDAR labeling but no AMPAR labeling) is very high at first but decreases with time such that ~90% are silent at 7 days in vitro (DIV), ~70% are silent at 14 DIV and ~40% are silent at 21 DIV (Liao et al., 1999). Similar findings have been seen using immunogold electron microscopy of the hippocampus (Petralia et al., 1999). Electrophysiology combined with two-photon uncaging of glutamate in cortical slices revealed that silent synapses are commonly found on layer 2/3 pyramidal neurons, early in development but later disappear (Busetto et al., 2008). In addition, differences in the developmental regulation of silent synapses have been reported for different layers of the cortex (Rumpel et al., 2004).

This reduction in the prominence of silent synapses with development has lead to the assumption that synapses begin silent and are unsilenced by activity; however, this may not be the case as there is evidence that newly formed synapses can express functional AMPARs. The detailed analysis of newly generated spines relative to their more mature neighbors suggests that spines may not start silent (Zito et al., 2009). New spines were observed to have a mature complement of AMPARs within 35 min of their generation. NMDAR-mediated calcium currents were reduced at young spines, which may be due to altered subunit composition or phosphorylation state (Zito et al., 2009). This suggests that the presence of silent synapses is not correlated with the age of a spine.

AMPARs may be initially incorporated into synapses but later removed through activity-dependent mechanisms. Newly formed synapses contain NR2B-NMDARs that negatively regulate AMPARs recruitment by blocking expression of AMPAR scaffolding proteins (Hall et al., 2007). When NMDARs are knocked down before or during synaptogenesis, the number of functional synapses is increased suggesting that basal NMDAR activity actually blocks AMPAR recruitment to silent synapses (Adesnik et al., 2008). NMDAR activity can also destabilize synapses during development leading to synapse elimination (Colonnese et al., 2005). Therefore, NMDARs may trigger removal of AMPARs out of synapses after their initial incorporation leading to abundant silent synapses. In addition, activity can create silent synapses by removal of AMPARs, independent of NMDAR activation (Xiao et al., 2004; Wasling et al., 2012).

During development, the generation of spines can occur independently of activity; however, glutamate can also induce the formation of new spines. It has been shown that spine growth, near an existing spine, can be induced by uncaging of glutamate (Kwon and Sabatini, 2011). In this situation, spine growth is rapid (<20 s) and requires NMDARs and PKA. New spines show similar AMPAR excitatory postsynaptic currents (EPSCs) as mature spines but have reduced NMDA EPSCs (Kwon and Sabatini, 2011). However, this direct spinogenesis was induced extrinsically by glutamate application and it is currently unknown whether this mechanism for glutamate-driven synapse formation occurs naturally in the brain. However, the fact that spine growth could only be induced near existing spines agrees with other studies showing the importance of nearby synapses to the formation, activity, and plasticity of synapses.

Recently it has been shown that neighboring synapses are more often coactive than distant synapses in cultured hippocampal slices during early (Kleindienst et al., 2011) and late development as well as in vivo (Takahashi et al., 2012). Importantly, this clustering of inputs is activity-dependent (Kleindienst et al., 2011; Takahashi et al., 2012). Learning paradigms can drive both the clustered expression of AMPARs (Makino and Malinow, 2011) and the formation of clustered spines (Fu et al., 2012). Taken together these results suggest that the formation and function of synapses is dynamically regulated based on the activity experienced by the synapse and the neighborhood of the synapse.

3. CONCLUSIONS

In this chapter, we have outlined what happens during three phases of synaptogenesis: contact formation, partner selection, and onset of function. During development, filopodia play a prominent role in seeking and selecting axonal partners for the formation of glutamatergic synapses. This process occurs independent of activity but it is regulated by local calcium activity. In contrast, GABAergic synapses form at points of contact between axons and dendrites without the need for filopodia. Details about how partner selection occurs at GABAergic synapses are still lacking. Adhesion molecules are however likely to be important mediators of synapse specificity at both glutamatergic and GABAergic synapses. While some specific examples of synapse specificity mediated by adhesion proteins have been demonstrated, more research is needed to understand the roles of all the different adhesion proteins. Some of these adhesion proteins are also involved in the protein recruitment which is required for the onset of synapse function. Such protein recruitment occurs rapidly for presynaptic terminals (~30 min) and more gradually for postsynaptic terminals (~45 min–1 h). Synapses can form in the absence of activity; however, activity is very important for the dynamic regulation of AMPAR presence at synapses.

REFERENCES

Adesnik, H., Li, G., During, M.J., Pleasure, S.J., Nicoll, R.A., 2008. NMDA receptors inhibit synapse unsilencing during brain development. Proc. Natl. Acad. Sci. USA 105, 5597–5602.

Ahmari, S.E., Buchanan, J., Smith, S.J., 2000. Assembly of presynaptic active zones from cytoplasmic transport packets. Nat. Neurosci. 3, 445–451.

Arstikaitis, P., Gauthier-Campbell, C., Huang, K., El-Husseini, A., Murphy, T.H., 2011. Proteins that promote filopodia stability, but not number, lead to more axonal-dendritic contacts. PLoS One 6, e16998.

Ashby, M.C., Isaac, J.T., 2011. Maturation of a recurrent excitatory neocortical circuit by experience-dependent unsilencing of newly formed dendritic spines. Neuron 70, 510–521.

Barrow, S.L., Constable, J.R., Clark, E., El-Sabeawy, F., McAllister, A.K., Washbourne, P., 2009. Neuroligin1: a cell adhesion molecule that recruits PSD-95 and NMDA receptors by distinct mechanisms during synaptogenesis. Neural Dev. 4, 17.

Bartlett, W.P., Banker, G.A., 1984a. An electron microscopic study of the development of axons and dendrites by hippocampal neurons in culture. I. Cells which develop without intercellular contacts. J. Neurosci. 4, 1944–1953.

Bartlett, W.P., Banker, G.A., 1984b. An electron microscopic study of the development of axons and dendrites by hippocampal neurons in culture. II. Synaptic relationships. J. Neurosci. 4, 1954–1965.

Beattie, E.C., Carroll, R.C., Yu, X., Morishita, W., Yasuda, H., von Zastrow, M., Malenka, R.C., 2000. Regulation of AMPA receptor endocytosis by a signaling mechanism shared with LTD. Nat. Neurosci. 3, 1291–1300.

Biederer, T., Sara, Y., Mozhayeva, M., Atasoy, D., Liu, X., Kavalali, E.T., Sudhof, T.C., 2002. SynCAM, a synaptic adhesion molecule that drives synapse assembly. Science 297, 1525–1531.

Bresler, T., Ramati, Y., Zamorano, P.L., Zhai, R., Garner, C.C., Ziv, N.E., 2001. The dynamics of SAP90/PSD-95 recruitment to new synaptic junctions. Mol. Cell. Neurosci. 18, 149–167.

Bresler, T., Shapira, M., Boeckers, T., Dresbach, T., Futter, M., Garner, C.C., Rosenblum, K., Gundelfinger, E.D., Ziv, N.E., 2004. Postsynaptic density assembly is fundamentally different from presynaptic active zone assembly. J. Neurosci. 24, 1507–1520.

Bury, L.A., Sabo, S.L., 2011. Coordinated trafficking of synaptic vesicle and active zone proteins prior to synapse formation. Neural Dev. 6, 24.

Busetto, G., Higley, M.J., Sabatini, B.L., 2008. Developmental presence and disappearance of postsynaptically silent synapses on dendritic spines of rat layer 2/3 pyramidal neurons. J. Physiol. 586, 1519–1527.

Carroll, R.C., Beattie, E.C., Xia, H., Luscher, C., Altschuler, Y., Nicoll, R.A., Malenka, R.C., von Zastrow, M., 1999a. Dynamin-dependent endocytosis of ionotropic glutamate receptors. Proc. Natl. Acad. Sci. USA 96, 14112–14117.

Carroll, R.C., Lissin, D.V., von Zastrow, M., Nicoll, R.A., Malenka, R.C., 1999b. Rapid redistribution of glutamate receptors contributes to long-term depression in hippocampal cultures. Nat. Neurosci. 2, 454–460.

Chattopadhyaya, B., Di Cristo, G., Higashiyama, H., Knott, G.W., Kuhlman, S.J., Welker, E., Huang, Z.J., 2004. Experience and activity-dependent maturation of perisomatic GABAergic innervation in primary visual cortex during a postnatal critical period. J. Neurosci. 24, 9598–9611.

Chen, S.X., Tari, P.K., She, K., Haas, K., 2010. Neurexin–neuroligin cell adhesion complexes contribute to synaptotropic dendritogenesis via growth stabilization mechanisms in vivo. Neuron 67, 967–983.
Chen, W.V., Alvarez, F.J., Lefebvre, J.L., Friedman, B., Nwakeze, C., Geiman, E., Smith, C., Thu, C.A., Tapia, J.C., Tasic, B., Sanes, J.R., Maniatis, T., 2012. Functional significance of isoform diversification in the protocadherin gamma gene cluster. Neuron 75, 402–409.
Chih, B., Gollan, L., Scheiffele, P., 2006. Alternative splicing controls selective trans-synaptic interactions of the neuroligin–neurexin complex. Neuron 51, 171–178.
Chih, B., Engelman, H., Scheiffele, P., 2005. Control of excitatory and inhibitory synapse formation by neuroligins. Science 307, 1324–1328.
Colonnese, M.T., Zhao, J.P., Constantine-Paton, M., 2005. NMDA receptor currents suppress synapse formation on sprouting axons in vivo. J. Neurosci. 25, 1291–1303.
Coppolino, M.G., Woodside, M.J., Demaurex, N., Grinstein, S., St Arnaud, R., Dedhar, S., 1997. Calreticulin is essential for integrin-mediated calcium signalling and cell adhesion. Nature 386, 843–847.
Craig, A.M., Kang, Y., 2007. Neurexin–neuroligin signaling in synapse development. Curr. Opin. Neurobiol. 17, 43–52.
Dailey, M.E., Smith, S.J., 1996. The dynamics of dendritic structure in developing hippocampal slices. J. Neurosci. 16, 2983–2994.
De Roo, M., Klauser, P., Mendez, P., Poglia, L., Muller, D., 2008. Activity-dependent PSD formation and stabilization of newly formed spines in hippocampal slice cultures. Cereb. Cortex 18, 151–161.
Dean, C., Dresbach, T., 2006. Neuroligins and neurexins: linking cell adhesion, synapse formation and cognitive function. Trends Neurosci. 29, 21–29.
Denardo, L.A., de Wit, J., Otto-Hitt, S., Ghosh, A., 2012. NGL-2 regulates input-specific synapse development in CA1 pyramidal neurons. Neuron 76, 762–775.
Di Cristo, G., Wu, C., Chattopadhyaya, B., Ango, F., Knott, G., Welker, E., Svoboda, K., Huang, Z.J., 2004. Subcellular domain-restricted GABAergic innervation in primary visual cortex in the absence of sensory and thalamic inputs. Nat. Neurosci. 7, 1184–1186.
Dresbach, T., Neeb, A., Meyer, G., Gundelfinger, E.D., Brose, N., 2004. Synaptic targeting of neuroligin is independent of neurexin and SAP90/PSD95 binding. Mol. Cell. Neurosci. 27, 227–235.
Dresbach, T., Torres, V., Wittenmayer, N., Altrock, W.D., Zamorano, P., Zuschratter, W., Nawrotzki, R., Ziv, N.E., Garner, C.C., Gundelfinger, E.D., 2006. Assembly of active zone precursor vesicles: obligatory trafficking of presynaptic cytomatrix proteins Bassoon and Piccolo via a trans-Golgi compartment. J. Biol. Chem. 281, 6038–6047.
Dudek, S.M., Bear, M.F., 1992. Homosynaptic long-term depression in area CA1 of hippocampus and effects of N-methyl-D-aspartate receptor blockade. Proc. Natl. Acad. Sci. USA 89, 4363–4367.
Durand, G.M., Kovalchuk, Y., Konnerth, A., 1996. Long-term potentiation and functional synapse induction in developing hippocampus. Nature 381, 71–75.
Ehlers, M.D., 2002. Molecular morphogens for dendritic spines. Trends Neurosci. 25, 64–67.
Easley-Neal, C., Fierro, J., Buchanan, J., Washbourne, P., 2013. Late recruitment of synapsin to nascent synapses is regulated by Cdk5. Cell Rep. 3 (4), 1199–1212.
Fiala, J.C., Feinberg, M., Popov, V., Harris, K.M., 1998. Synaptogenesis via dendritic filopodia in developing hippocampal area CA1. J. Neurosci. 18, 8900–8911.
Fino, E., Packer, A.M., Yuste, R., 2012. The logic of inhibitory connectivity in the neocortex. Neuroscientist 19 (3), 228–237.
Fletcher, T.L., Cameron, P., De Camilli, P., Banker, G., 1991. The distribution of synapsin-I and synaptophysin in hippocampal neurons developing in culture. J. Neurosci. 11, 1617–1626.
Frank, M., Kemler, R., 2002. Protocadherins. Curr. Opin. Cell Biol. 14, 557.
Friedman, H.V., Bresler, T., Garner, C.C., Ziv, N.E., 2000. Assembly of new individual excitatory synapses: time course and temporal order of synaptic molecule recruitment. Neuron 27, 57–69.
Fu, M., Yu, X., Lu, J., Zuo, Y., 2012. Repetitive motor learning induces coordinated formation of clustered dendritic spines in vivo. Nature 483, 92–95.
Gerrow, K., Romorini, S., Nabi, S.M., Colicos, M.A., Sala, C., El Husseini, A., 2006. A preformed complex of postsynaptic proteins is involved in excitatory synapse development. Neuron 49, 547–562.
Gomez, T.M., Robles, E., Poo, M., Spitzer, N.C., 2001. Filopodial calcium transients promote substrate-dependent growth cone turning. Science 291, 1983–1987.

Gomperts, S.N., Carroll, R., Malenka, R.C., Nicoll, R.A., 2000. Distinct roles for ionotropic and metabotropic glutamate receptors in the maturation of excitatory synapses. J. Neurosci. 20, 2229–2237.
Graf, E.R., Zhang, X., Jin, S.X., Linhoff, M.W., Craig, A.M., 2004. Neurexins induce differentiation of GABA and glutamate postsynaptic specializations via neuroligins. Cell 119, 1013–1026.
Haas, K., Li, J., Cline, H.T., 2006. AMPA receptors regulate experience-dependent dendritic arbor growth in vivo. Proc. Natl. Acad. Sci. USA 103, 12127–12131.
Hall, B.J., Ripley, B., Ghosh, A., 2007. NR2B signaling regulates the development of synaptic AMPA receptor current. J. Neurosci. 27, 13446–13456.
Harms, K.J., Craig, A.M., 2005. Synapse composition and organization following chronic activity blockade in cultured hippocampal neurons. J. Comp. Neurol. 490, 72–84.
Harris, K.M., Jensen, F.E., Tsao, B., 1992. Three-dimensional structure of dendritic spines and synapses in rat hippocampus (CA1) at postnatal day 15 and adult ages: implications for the maturation of synaptic physiology and long-term potentiation. J. Neurosci. 12, 2685–2705.
Heiman, M.G., Shaham, S., 2009a. DEX-1 and DYF-7 establish sensory dendrite length by anchoring dendritic tips during cell migration. Cell 137, 344–355.
Heiman, M.G., Shaham, S., 2009b. Twigs into branches: how a filopodium becomes a dendrite. Curr. Opin. Neurobiol. 20, 86–91.
Heine, M., Thoumine, O., Mondin, M., Tessier, B., Giannone, G., Choquet, D., 2008. Activity-independent and subunit-specific recruitment of functional AMPA receptors at neurexin/neuroligin contacts. Proc. Natl. Acad. Sci. USA 105, 20947–20952.
Hering, H., Sheng, M., 2001. Dendritic spines: structure, dynamics and regulation. Nat. Rev. Neurosci. 2, 880–888.
Iijima, T., Wu, K., Witte, H., Hanno-Iijima, Y., Glatter, T., Richard, S., Scheiffele, P., 2011. SAM68 regulates neuronal activity-dependent alternative splicing of neurexin-1. Cell 147, 1601–1614.
Isaac, J.T., Nicoll, R.A., Malenka, R.C., 1995. Evidence for silent synapses: implications for the expression of LTP. Neuron 15, 427–434.
Jacques-Fricke, B.T., Seow, Y., Gottlieb, P.A., Sachs, F., Gomez, T.M., 2006. Ca2+ influx through mechanosensitive channels inhibits neurite outgrowth in opposition to other influx pathways and release from intracellular stores. J. Neurosci. 26, 5656–5664.
Jeyifous, O., Waites, C.L., Specht, C.G., Fujisawa, S., Schubert, M., Lin, E.I., Marshall, J., Aoki, C., de, S.T., Montgomery, J.M., Garner, C.C., Green, W.N., 2009. SAP97 and CASK mediate sorting of NMDA receptors through a previously unknown secretory pathway. Nat. Neurosci. 12, 1011–1019.
Kang, Y., Zhang, X., Dobie, F., Wu, H., Craig, A.M., 2008. Induction of GABAergic postsynaptic differentiation by alpha-neurexins. J. Biol. Chem. 283, 2323–2334.
Kayser, M.S., Nolt, M.J., Dalva, M.B., 2008. EphB receptors couple dendritic filopodia motility to synapse formation. Neuron 59, 56–69.
Kleindienst, T., Winnubst, J., Roth-Alpermann, C., Bonhoeffer, T., Lohmann, C., 2011. Activity-dependent clustering of functional synaptic inputs on developing hippocampal dendrites. Neuron 72, 1012–1024.
Krueger, D.D., Tuffy, L.P., Papadopoulos, T., Brose, N., 2012. The role of neurexins and neuroligins in the formation, maturation, and function of vertebrate synapses. Curr. Opin. Neurobiol. 22, 412–422.
Krueger, S.R., Kolar, A., Fitzsimonds, R.M., 2003. The presynaptic release apparatus is functional in the absence of dendritic contact and highly mobile within isolated axons. Neuron 40, 945–957.
Kullmann, D.M., 1994. Amplitude fluctuations of dual-component EPSCs in hippocampal pyramidal cells: implications for long-term potentiation. Neuron 12, 1111–1120.
Kwon, H.B., Sabatini, B.L., 2011. Glutamate induces de novo growth of functional spines in developing cortex. Nature.
Kwon, H.B., Kozorovitskiy, Y., Oh, W.J., Peixoto, R.T., Akhtar, N., Saulnier, J.L., Gu, C., Sabatini, B.L., 2012. Neuroligin-1-dependent competition regulates cortical synaptogenesis and synapse number. Nat. Neurosci. 15 (12), 1667–1674.
Liao, D., Hessler, N.A., Malinow, R., 1995. Activation of postsynaptically silent synapses during pairing-induced LTP in CA1 region of hippocampal slice. Nature 375, 400–404.
Liao, D., Zhang, X., O'Brien, R., Ehlers, M.D., Huganir, R.L., 1999. Regulation of morphological postsynaptic silent synapses in developing hippocampal neurons. Nat. Neurosci. 2, 37–43.

Lohmann, C., Bonhoeffer, T., 2008. A role for local calcium signaling in rapid synaptic partner selection by dendritic filopodia. Neuron 59, 253–260.
Lohmann, C., Finski, A., Bonhoeffer, T., 2005. Local calcium transients regulate the spontaneous motility of dendritic filopodia. Nat. Neurosci. 8, 305–312.
Lohmann, C., Myhr, K.L., Wong, R.O., 2002. Transmitter-evoked local calcium release stabilizes developing dendrites. Nature 418, 177–181.
Makino, H., Malinow, R., 2011. Compartmentalized versus global synaptic plasticity on dendrites controlled by experience. Neuron 72, 1001–1011.
Marrs, G.S., Green, S.H., Dailey, M.E., 2001. Rapid formation and remodeling of postsynaptic densities in developing dendrites. Nat. Neurosci. 4, 1006–1013.
Montgomery, J.M., Madison, D.V., 2002. State-dependent heterogeneity in synaptic depression between pyramidal cell pairs. Neuron 33, 765–777.
Montgomery, J.M., Pavlidis, P., Madison, D.V., 2001. Pair recordings reveal all-silent synaptic connections and the postsynaptic expression of long-term potentiation. Neuron 29, 691–701.
Monyer, H., Sprengel, R., Schoepfer, R., Herb, A., Higuchi, M., Lomeli, H., Burnashev, N., Sakmann, B., Seeburg, P.H., 1992. Heteromeric NMDA receptors: molecular and functional distinction of subtypes. Science 256, 1217–1221.
Nam, C.I., Chen, L., 2005. Postsynaptic assembly induced by neurexin–neuroligin interaction and neurotransmitter. Proc. Natl. Acad. Sci. 102, 6137–6142.
Niell, C.M., Meyer, M.P., Smith, S.J., 2004. In vivo imaging of synapse formation on a growing dendritic arbor. Nat. Neurosci. 7, 254–260.
Nolt, M.J., Lin, Y., Hruska, M., Murphy, J., Sheffler-Colins, S.I., Kayser, M.S., Passer, J., Bennett, M.V.L., Zukin, R.S., Dalva, M.B., 2011. EphB controls NMDA receptor function and synaptic targeting in a subunit-specific manner. J. Neurosci. 31, 5353–5364.
Okabe, S., Miwa, A., Okado, H., 2001. Spine formation and correlated assembly of presynaptic and postsynaptic molecules. J. Neurosci. 21, 6105–6114.
Okabe, S., Vicario-Abejon, C., Segal, M., McKay, R.D., 1998. Survival and synaptogenesis of hippocampal neurons without NMDA receptor function in culture. Eur. J. Neurosci. 10, 2192–2198.
Petralia, R.S., Esteban, J.A., Wang, Y.X., Partridge, J.G., Zhao, H.M., Wenthold, R.J., Malinow, R., 1999. Selective acquisition of AMPA receptors over postnatal development suggests a molecular basis for silent synapses. Nat. Neurosci. 2, 31–36.
Rao, A., Cha, E.M., Craig, A.M., 2000. Mismatched appositions of presynaptic and postsynaptic components in isolated hippocampal neurons. J. Neurosci. 20, 8344–8353.
Rehder, V., Cheng, S., 1998. Autonomous regulation of growth cone filopodia. J. Neurobiol. 34, 179–192.
Rowen, L., Young, J., Birditt, B., Kaur, A., Madan, A., Philipps, D.L., Qin, S., Minx, P., Wilson, R.K., Hood, L., Graveley, B.R., 2002. Analysis of the human neurexin genes: alternative splicing and the generation of protein diversity. Genomics 79, 587–597.
Rumpel, S., Hatt, H., Gottmann, K., 1998. Silent synapses in the developing rat visual cortex: evidence for postsynaptic expression of synaptic plasticity. J. Neurosci. 18, 8863–8874.
Rumpel, S., Kattenstroth, G., Gottmann, K., 2004. Silent synapses in the immature visual cortex: layer-specific developmental regulation. J. Neurophysiol. 91, 1097–1101.
Sabo, S.L., Gomes, R.A., McAllister, A.K., 2006. Formation of presynaptic terminals at predefined sites along axons. J. Neurosci. 26, 10813–10825.
Salinas, P.C., Price, S.R., 2005. Cadherins and catenins in synapse development. Curr. Opin. Neurobiol. 15, 73–80.
Sanes, J.R., Yamagata, M., 2009. Many paths to synaptic specificity. Annu. Rev. Cell Dev. Biol. 25, 161–195.
Scheiffele, P., Fan, J., Choih, J., Fetter, R., Serafini, T., 2000. Neuroligin expressed in nonneuronal cells triggers presynaptic development in contacting axons. Cell 101, 657–669.
Shapira, M., Zhai, R.G., Dresbach, T., Bresler, T., Torres, V.I., Gundelfinger, E.D., Ziv, N.E., Garner, C.C., 2003. Unitary assembly of presynaptic active zones from Piccolo–Bassoon transport vesicles. Neuron 38, 237–252.
Shen, K., Scheiffele, P., 2010. Genetics and cell biology of building specific synapse connectivity. Annu. Rev. Neurosci. 33, 473–507.

Shi, Y., Ethell, I.M., 2006. Integrins control dendritic spine plasticity in hippocampal neurons through NMDA receptor and Ca2+/calmodulin-dependent protein kinase II-mediated actin reorganization. J. Neurosci. 26, 1813–1822.

Song, J.Y., Ichtchenko, K., Sudhof, T.C., Brose, N., 1999. Neuroligin 1 is a postsynaptic cell-adhesion molecule of excitatory synapses. Proc. Natl. Acad. Sci. USA 96, 1100–1105.

Sperry, R.W., 1963. Chemoaffinity in the orderly growth of nerve fiber patterns and connections. Proc. Natl. Acad. Sci. USA 50, 703–710.

Tabuchi, K., Sudhof, T.C., 2002. Structure and evolution of neurexin genes: insight into the mechanism of alternative splicing. Genomics 79, 849–859.

Takahashi, N., Kitamura, K., Matsuo, N., Mayford, M., Kano, M., Matsuki, N., Ikegaya, Y., 2012. Locally synchronized synaptic inputs. Science 335, 353–356.

Takahashi, T., Svoboda, K., Malinow, R., 2003. Experience strengthening transmission by driving AMPA receptors into synapses. Science 299, 1585–1588.

Tovar, K.R., Westbrook, G.L., 2002. Mobile NMDA receptors at hippocampal synapses. Neuron 34, 255–264.

Ullrich, B., Ushkaryov, Y.A., Sudhof, T.C., 1995. Cartography of neurexins: more than 1000 isoforms generated by alternative splicing and expressed in distinct subsets of neurons. Neuron 14, 497–507.

Ushkaryov, Y.A., Sudhof, T.C., 1993. Neurexin III alpha: extensive alternative splicing generates membrane-bound and soluble forms. Proc. Natl. Acad. Sci. USA 90, 6410–6414.

Varoqueaux, F., Sigler, A., Rhee, J.S., Brose, N., Enk, C., Reim, K., Rosenmund, C., 2002. Total arrest of spontaneous and evoked synaptic transmission but normal synaptogenesis in the absence of Munc13-mediated vesicle priming. Proc. Natl. Acad. Sci. USA 99, 9037–9042.

Vaughn, J.E., 1989. Fine structure of synaptogenesis in the vertebrate central nervous system. Synapse 3, 255–285.

Verhage, M., Maia, A.S., Plomp, J.J., Brussaard, A.B., Heeroma, J.H., Vermeer, H., Toonen, R.F., Hammer, R.E., van den Berg, T.K., Missler, M., Geuze, H.J., Sudhof, T.C., 2000. Synaptic assembly of the brain in the absence of neurotransmitter secretion. Science 287, 864–869.

Waites, C.L., Craig, A.M., Garner, C.C., 2005. Mechanisms of vertebrate synaptogenesis. Annu. Rev. Neurosci. 28, 251–274.

Washbourne, P., Bennett, J.E., McAllister, A.K., 2002. Rapid recruitment of NMDA receptor transport packets to nascent synapses. Nat. Neurosci. 5, 751–759.

Washbourne, P., Dityatev, A., Scheiffele, P., Biederer, T., Weiner, J.A., Christopherson, K.S., El Husseini, A., 2004. Cell adhesion molecules in synapse formation. J. Neurosci. 24, 9244–9249.

Wasling, P., Strandberg, J., Hanse, E., 2012. AMPA receptor activation causes silencing of AMPA receptor-mediated synaptic transmission in the developing hippocampus. PLoS One 7, e34474

Wei, C., Wang, X., Chen, M., Ouyang, K., Song, L.S., Cheng, H., 2009. Calcium flickers steer cell migration. Nature 457, 901–905.

Wierenga, C.J., Becker, N., Bonhoeffer, T., 2008. GABAergic synapses are formed without the involvement of dendritic protrusions. Nat. Neurosci. 11, 1044–1052.

Williams, M.E., Wilke, S.A., Daggett, A., Davis, E., Otto, S., Ravi, D., Ripley, B., Bushong, E.A., Ellisman, M.H., Klein, G., Ghosh, A., 2011. Cadherin-9 regulates synapse-specific differentiation in the developing hippocampus. Neuron 71, 640–655.

Xiao, M.Y., Wasling, P., Hanse, E., Gustafsson, B., 2004. Creation of AMPA-silent synapses in the neonatal hippocampus. Nat. Neurosci. 7, 236–243.

Yuste, R., Bonhoeffer, T., 2004. Genesis of dendritic spines: insights from ultrastructural and imaging studies. Nat. Rev. Neurosci. 5, 24–34.

Zhai, R.G., Vardinon-Friedman, H., Cases-Langhoff, C., Becker, B., Gundelfinger, E.D., Ziv, N.E., Garner, C.C., 2001. Assembling the presynaptic active zone: a characterization of an active one precursor vesicle. Neuron 29, 131–143.

Zheng, J.Q., 2000. Turning of nerve growth cones induced by localized increases in intracellular calcium ions. Nature 403, 89–93.

Zhu, J.J., Esteban, J.A., Hayashi, Y., Malinow, R., 2000. Postnatal synaptic potentiation: delivery of GluR4-containing AMPA receptors by spontaneous activity. Nat. Neurosci. 3, 1098–1106.

Zipursky, S.L., Sanes, J.R., 2010. Chemoaffinity revisited: dscams, protocadherins, and neural circuit assembly. Cell 143, 343–353.

Zito, K., Scheuss, V., Knott, G., Hill, T., Svoboda, K., 2009. Rapid functional maturation of nascent dendritic spines. Neuron 61, 247–258.

Ziv, N.E., Garner, C.C., 2004. Cellular and molecular mechanisms of presynaptic assembly. Nat. Rev. Neurosci. 5, 385–399.

Ziv, N.E., Smith, S.J., 1996. Evidence for a role of dendritic filopodia in synaptogenesis and spine formation. Neuron 17, 91–102.

CHAPTER FOUR

Structural and Functional Organization of the Postsynaptic Density

Chiara Verpelli[1,2], Christopher Heise[1] and Carlo Sala[1,2]
[1]Department of Pharmacology, University of Milan, Milan, Italy
[2]Neuromuscular Diseases and Neuroimmunology, Neurological Institute Foundation Carlo Besta, Milan, Italy

1. INTRODUCTION

Communication between neurons occur through the synapses which are specialized contacts. Assembly and formation of synapses takes place in various stages involving the accumulation of specific pre- and postsynaptic proteins at the contact site. Several biochemical, genetic, and imaging studies have elucidated the molecular mechanisms underlying synapse formation, growth, and maturation. It is clear that understanding the development and organization of neuron/neuron synapses is crucial to our comprehension of the development of the nervous system. But it is now clear that even the adult brain is constantly reorganizing neuronal connections in response to experience and that synaptic developmental molecular mechanisms are utilized in the mature brain during synaptic plasticity. More importantly, alterations of these processes contribute to several developmental and degenerative neurological and behavioral disorders. Chemical synapses, where neurotransmitters are released by the axon of the presynaptic neuron and diffuse across a narrow cleft in order to act on corresponding receptors on the postsynaptic neuron, are mostly used by the neurons to communicate. Depending on the type of neurotransmitter, the ability to depolarize or hyperpolarize the postsynaptic membrane, and the morphological differences, synapses are divided into excitatory and inhibitory synapses and most of the excitatory synapses use glutamate as neurotransmitter in the mammalian brain.

The postsynaptic density (PSD) is a specific electron microscopic (EM) electron-dense organelle localized at the postsynaptic sites of excitatory synapses where the two main types of glutamate-receptor channels, N-methyl-D-aspartate (NMDA), and α-amino-3-hydroxy-5-methyl-4-isoxazole propionic acid (AMPA) receptors (NMDARs and AMPARs, respectively) are clustered. Several published works have demonstrated that the PSD regulates synapse function. On the contrary, the inhibitory synapses in the brain are mainly mediated by γ-amino-butyric (GABA) neurotransmitters, whose receptor channels are localized on the postsynaptic membrane where a clear PSD is not present

The Synapse
http://dx.doi.org/10.1016/B978-0-12-418675-0.00004-3

because the molecular composition of the inhibitory postsynaptic compartment is much more simplified and less enriched in proteins. Thus, excitatory and inhibitory synapses are very clearly distinguishable on the basis of both morphology and molecular composition of the respective postsynaptic compartments.

Interestingly, each neuron in the brain exhibits both excitatory and inhibitory synapses, which are located in cross-proximity to the same dendrites in strictly controlled numbers and ratios to provide the correct functioning of brain circuits.

This chapter will concentrate on describing the development, structure, and function of the brain excitatory synapses. Emphasis will be placed on their molecular and structural composition and on the alterations of PSD proteins that cause synapse dysfunctions associated to genetic-based cognitive disabilities in humans.

2.1. EXCITATORY SYNAPSES: POSTSYNAPTIC ORGANIZATION

The PSD is a specialized organelle formed by a protein complex that adheres to the postsynaptic membrane. For this organelle, multiple functions have been proposed including the regulation of adhesion, the control of receptor clustering, and the regulation of synapse functions. Morphologically, the PSD is a disc-like structure that is 30–40 nm thick and a few hundred nanometers wide. The proteins that form the PSD are neurotransmitter and transmembrane receptors, various scaffold proteins, cytoskeletal elements, and regulatory enzymes. In recent years, the protein composition of the PSD has been extensively studied because of its central role in synaptic transmission and plasticity.

Indeed, PSD molecular architecture is designed to specialize in signal transduction, and its protein composition is highly modifiable by activity in order to allow for the strengthening or weakening of glutamatergic synapses. Thus, neural activity—involving mechanisms such as protein phosphorylation, local translation, ubiquitination, degradation, and subcellular redistribution—dynamically regulates both the function and composition of the PSD. PSD protein components have been gradually discovered over the years using two-hybrid yeast and biochemical approaches. Biochemically, the PSD is typically purified by means of differential centrifugation, sucrose gradient sedimentation, and detergent extraction because of the peculiar PSD properties that cannot be solubilized using mild nonionic detergents such as Triton X-100 (Hunt et al., 1996).

The laboratory of Mary Kennedy first revealed several PSD proteins (e.g. PSD-95, densin-180) (Apperson et al., 1996; Cho et al., 1992) by sequencing protein bands from one-dimensional gels of PSD preparations. However, in the 1990s, many other PSD proteins were identified as binding four partners of known postsynaptic proteins using the two-hybrid yeast system. For example, the PSD-95 family of PDZ-containing scaffolds was also found to interact with the cytoplasmic tails of glutamate NMDARs (Kornau et al., 1995; Niethammer et al., 1996). More recently, modern proteomic and imaging techniques have probably identified all the proteins of the PSD, and partially quantified the relative stoichiometry of several of them (Cheng et al., 2006; Sugiyama et al., 2005). On the basis of their functions, PSD proteins can be divided into four subgroups: scaffold proteins, receptor transmembrane proteins, cytoskeletal proteins, and signaling proteins. In the following sections of this chapter, we will discuss what we consider to be the major PSD components in each subgroup (Figure 1).

2.1.1. The PSD-95 Family and Other Major PSD Scaffold Proteins

Without doubt, PSD-95 is the most widely studied PSD scaffold protein. It belongs to the membrane-associated guanylate kinase (MAGUK) superfamily, characterized by the presence of at least one PDZ and one GK domain. The PSD-95 family by itself is formed by four genes: PSD-95/SAP90 (synapse-associated protein 90), PSD-93/chapsyn-110, SAP102, and SAP97. Each member is characterized by the assembly of three PDZ (PSD-95, Dlg, and ZO-1 homology) domains, an SRC homology (SH3) domain, and a guanylate kinase-like (GK) domain. Only SAP97 also has a LIN2/LIN7 (L27) domain located at the N-terminus (Elias and Nicoll, 2007; Sheng and Sala, 2001).

PSD-95 is localized very close to the postsynaptic membrane (a mean distance of 12 nm), a good position for interacting with postsynaptic membrane proteins such as receptors, ion channels, and cell-adhesion molecules, as well as with cytoplasmic proteins. This has been clearly demonstrated by a number of immuno-EM and tomography studies (Chen et al., 2008; Petersen et al., 2003; Valtschanoff and Weinberg, 2001).

There is still disagreement as to whether PSD-95 is essential for the synaptic clustering of NMDARs in mammals, for sure they are not eliminated by the genetic disruption of PSD-95 (Migaud et al., 1998; Sprengel et al., 1998). However, consistent data suggest that PSD-95 clusters are regulated by interacting synaptic adhesion molecules (Irie et al., 1997; Kim et al., 2006; Mah et al., 2010; Woo et al., 2009) and the size of PSD-95 clusters is proportional to the number of synaptic AMPARs (El-Husseini et al., 2000; Elias et al., 2006).

Indeed the most important biochemical function of PSD-95 is to organize the signaling complexes at the postsynaptic membrane. This is because PSD-95 interacts with a wide variety of cytoplasmic signaling molecules that are physically linked to surface receptors and therefore facilitate signal coupling in the PSD. It has been demonstrated that the overexpression of PSD-95 in hippocampal neurons increases the number of dendritic spines, the maturation of excitatory synapses, and the accumulation of AMPARs (Ehrlich et al., 2007; El-Husseini et al., 2000). Instead, PSD-95 knock-down (using RNA interference) in hippocampal neurons during development modifies the excitatory/inhibitory synapse ratio in favor of inhibitory synapses (Gerrow et al., 2006). PSD-95 is directly associated with proteins that regulate actin polymerization and remodeling such as kalirin-7, a guanine nucleotide exchange factor (GEF) for RAC1 (Penzes et al., 2001). GEF promotes spine formation, possibly through downstream effectors such as the EphB

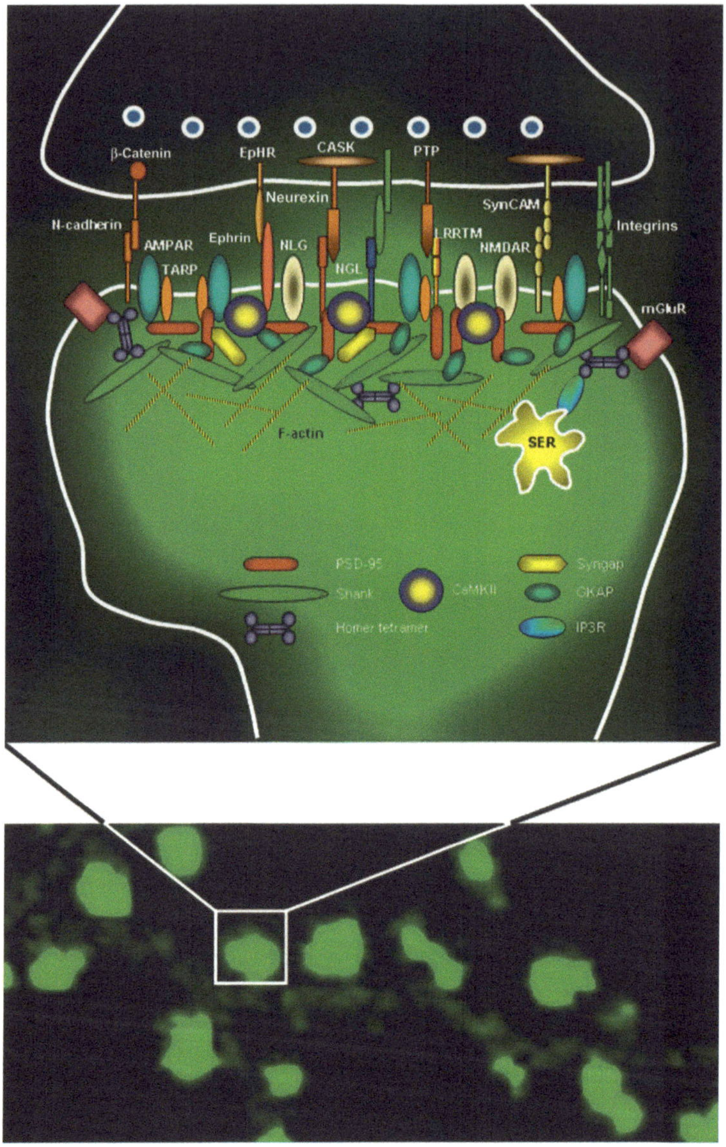

Figure 1 Top, the image of a dendritic spine has been enlarged and overlapped with a diagram showing a schematic organization of the network of proteins in the PSD. Major families and certain classes of PSD proteins are designed. Some interactions between them are schematically indicated. Bottom, the image shows a dendrite of hippocampal neuron transfected with GFPShank1. Dendritic spines are filled by GFPShank1.

receptor, SPAR, an inhibitory GTPase-activating protein (GAP) for Ras-related proteins (RAP) that promotes the growth of dendritic spines, and IRSp53, a downstream effector of Cdc42 small GTPase (Choi et al., 2005; Pak et al., 2001; Penzes et al., 2003).

Instead, the ability of PSD-95 to recruit AMPARs seems to be mediated by its binding to the tetra-spanning membrane proteins Stargazin and its relatives TARP, which are essential for AMPAR surface expression, synaptic accumulation, and function (Chen et al., 2000; Nicoll et al., 2006; Schnell et al., 2002).

In the PSD, there are three other major scaffold components that directly or indirectly interact with PSD-95 and each other: Shank, GKAP, and Homer protein families.

Shank, which is also known as proline-rich synapse-associated protein (ProSAP), is a large scaffold protein whose multidomain organization consists of an ankyrin repeat near the N-terminal, followed by an SH3 and a PDZ domain, a long proline-rich region, and a sterile alpha motif domain (SAM) at the C-terminus (Boeckers et al., 1999; Naisbitt et al., 1999; Tu et al., 1999).

Shank proteins (encoded by the three genes Shank1-3) molecularly link two glutamate receptor (GluR) subtypes: NMDARs and type-I metabotropic GluRs (mGluRs). The Shank PDZ domain binds to the C-terminus of GKAP. Homer interaction with the proline-rich domain ensures the association of Shank with type-I mGluRs. Shank also interacts with actin-binding proteins such as cortactin, Abp1, fodrin, the Rac1 and Cdc42 exchange factor βPIX, and the Cdc42-binding protein IRSp53 (Bockers et al., 2001; Bockmann et al., 2002; Park et al., 2003; Qualmann et al., 2004; Soltau et al., 2002). Other proteins that regulate dendritic morphology and formation, such as Dasm1 and Densin-180, are Shank interactors and their activity seems to be regulated by Shank (Quitsch et al., 2005).

The importance of Shank in regulating synapse function is underlined by the fact that several mutations in the Shank genes are associated with neurological dysfunctions (see later). Another major scaffold protein family at the PSD is called GKAP/SAPAP whose four members were originally identified as proteins interacting with the GK domain of PSD-95. These proteins have five repeats of 14 amino acids at the N-terminal domain involved in interacting with PSD-95 while the C-terminal tail binds to the PDZ domain of Shank. They bind to S-SCAM, nArgBP2, Dynein light chain, and Shank and may therefore function as scaffolding proteins that link PSD protein complexes to motor proteins (Hirao et al., 1998; Kawabe et al., 1999; Kim et al., 1997; Naisbitt et al., 2000).

The other major scaffold family of proteins is that formed by the Homer proteins encoded by three genes (Homer 1–3). Their structure consists of an N-terminal Ena/VASP homology (EVH1) domain followed by a coiled-coil domain that mediates tetramerization with other Homer proteins (Brakeman et al., 1997; Xiao et al., 2000). The EVH1 domain of Homer1 binds to a PPXXF sequence motif of Shank, to the mGluR1/5, inositol-1,4,5-trisphosphate (IP3) receptor, the ryanodine receptor and to members of the TRPC family of ion channels (Sala et al., 2005; Tu et al., 1998, 1999; Yuan et al., 2003).

Through the ability of the coiled-coil domain (called "CC-Homer" or Homer1b in the case of the Homer1 gene) to self-associate, Homer can physically and functionally link the proteins and receptors that bind to the EVH1 domain (Fagni et al., 2002; Hayashi et al., 2009; Huang et al., 2007; Kammermeier et al., 2000; Xiao et al., 2000). Instead, Homer1a, the short splice-variant of Homer1 that lacks the coiled-coil domain, can function as a natural dominant negative. Interestingly, synaptic activity induces Homer1a expression at the mRNA level (Ango et al., 2001; Hu et al., 2010; Ronesi et al., 2012; Sala et al., 2003).

A different set of scaffolding proteins are linked to AMPARs: GRIP/ABP (encoded by the two distinct genes GRIP1 and ABP/GRIP2), and the protein interacting with C kinase 1 (PICK1) (Dong et al., 1997; Hirbec et al., 2003; Kim et al., 2001; Takamiya et al., 2008). Both GRIP/ABP and PICK1 proteins are important regulators of AMPAR trafficking to and from the postsynaptic membrane (Malinow and Malenka, 2002).

GRIP/ABP proteins have up to seven PDZ domains used to interact with a number of proteins, including GluA2/3, Eph receptors and their ephrin ligands, an RAS guanine nucleotide exchange factor (RasGEF), liprin-α, the transmembrane protein Fraser syndrome 1 (FRAS1), and the metabotropic kainate-type GluRs. Thus, these scaffold proteins not only regulate synaptic trafficking and/or stabilization of AMPARs but also can participate in synaptic function by associating with Eph receptors and their ephrin ligands, which have been found to be involved in dendritic spine morphogenesis and hippocampal synaptic plasticity (Hirbec et al., 2003; Hoogenraad et al., 2005; Shin et al., 2003; Silverman et al., 2007; Takamiya et al., 2004; Wyszynski et al., 2002).

Interestingly, the widespread cellular distribution of GRIP and its interaction with motor proteins (directly with conventional kinesin KIF5 (Setou et al., 2002) or indirectly with KIF1A via liprin-α), suggest that GRIP/ABP might link motor proteins to AMPARs and could contribute to the transport of AMPARs (Hoogenraad et al., 2005).

The unique PDZ domain of PICK1 has relatively promiscuous binding properties; in fact, both pre- and postsynaptic partners have been found: PKCα, GluR2/3, the netrin receptor UNC5H, various mGluR subtypes, the dopamine plasma-membrane transporter, the erythroblastic leukemia viral oncogene homolog 2 (ErbB2) receptor tyrosine kinase (RTK), and TSPAN7 (Bassani et al., 2012; Hanley, 2008; Xia et al., 1999). Most of these interactors are regulated in their subcellular localization and/or surface expression by PICK1, but PICK1 is also involved in regulating dendritic spines cytoskeleton (Nakamura et al., 2011; Rocca et al., 2008).

2.1.2. Receptor and Transmembrane Proteins

Of course the GluRs are the major and most abundant receptors at the PSD membrane. As mentioned before, at the PSD these receptors are organized into supramolecular signaling complexes by interacting with specific PDZ domain-containing scaffold proteins, such as PSD-95/SAP90 and their associated proteins of the NMDARs, and

GRIP/ABP, PICK, and TARP for the AMPA-type receptors. During development, the synaptic GluR distribution and the synaptic content change in a way that immature synapses preferentially contain NMDARs and gradually acquire AMPARs (Matsuzaki et al., 2001; Takumi et al., 1999). This observation has led to the discovery of so-called "silent synapses" whose lack of AMPARs makes them less responsive to glutamate (Kerchner and Nicoll, 2008; Petralia et al., 1999; Shi et al., 1999).

Finally, while AMPA and NMDA receptors seem to be almost homogenously mixed and distributed inside the PSD membrane, the mGluRs are mainly concentrated at the border of the synapse (Lujan et al., 1997).

The second major classes of PSD membrane proteins are all the families of transmembrane adhesion molecules that regulate synapse formation and structures. The prototype is Neuroligin, the first identified synaptic adhesion molecule that plays an important role in synapse formation. Neuroligin proteins (codified by the four genes Neuroligin1–4) interact with PSD-95 and trans-synaptically with the neurexin family of proteins, which in turn bind to the PDZ domain of CASK/LIN2 (another scaffold of the MAGUK superfamily of proteins), which is enriched on both sides of the synapse and interacts with other synaptic membrane proteins such as syndecan and SynCAM (Biederer et al., 2002; Han and Kim, 2008; Hsueh et al., 1998; Irie et al., 1997; Prange et al., 2004; Yamagata et al., 2003).

Thus, the trans-synaptic neuroligin/neurexin interaction is important for inducing pre- and postsynaptic synaptic differentiation and the amount of PSDS-95 regulates the balance between the number of inhibitory and excitatory synapses (Levinson et al., 2005; Prange et al., 2004). The activity of the neuroligin/neurexin complex was nicely demonstrated by coculturing heterologous cells overexpressing neuroligin with neurons; transfected cells are able to induce morphological and functional presynaptic specializations in contacting axons. A significant number of families of synaptic adhesion molecules, including SALMs, Netrin-G ligand, ADAM22, and LRRTM2 associate with PSD-95. This suggests that they may act in concert to couple trans-synaptic adhesion to molecular organization of synaptic proteins (de Wit et al., 2009; Kwon et al., 2010; Mah et al., 2010). Thus, PSD-95 functions as one of the central organizers of synaptic adhesion by recruiting diverse proteins to sites of synaptic adhesion, by promoting trans-synaptic signaling, and by coupling neuronal activity with changes in synaptic adhesion (Han and Kim, 2008; Margeta and Shen, 2010).

However, other synaptic cell-adhesion molecules like SynCAM have a cytoplasmic tail that binds to PDZ proteins such as CASK but not PSD-95 (Biederer et al., 2002). SynCAM is an immunoglobulin domain-containing protein that is found in synaptic membranes and can also induce presynaptic differentiation in contacting axons through homophilic cell-adhesion activity (Fogel et al., 2011; Robbins et al., 2010).

Cadherins are also present on both pre- and postsynaptic membranes of synapses and are important for synapse integrity and the morphological maturation of dendritic spines (Malinverno et al., 2010; Monea et al., 2006; Saglietti et al., 2007; Togashi et al., 2002). It also seems clear that cadherins play both an initiating role in synapse formation and a supporting role in synapse growth and plasticity (Saglietti et al., 2007; Togashi et al., 2002). With their cytoplasmic tails, cadherins bind α- and β-catenins, which in turn bind and regulate the actin cytoskeleton that can consolidate the cadherin cell-adhesion complex and synapses (Abe et al., 2004).

Another important family of membrane proteins associated with synapses is the integrins which are formed by the heterodimeric glycoprotein α- and β-subunits. The expression of many of the 16 α- and eight β-subunits has been identified in different spatial patterns in the brain (Pinkstaff et al., 1999). Integrins function as cell-to-matrix or cell-to-cell adhesion molecules by respectively binding to extracellular matrix proteins or immunoglobulin-superfamily receptors on other cells. Their cytoplasmic domains interact with actin-binding proteins, such as talin and α-actinin, and also activate nonreceptor protein tyrosine kinases, such as the focal-adhesion kinase that affects cytoskeletal organization and intracellular signal transduction (Denda and Reichardt, 2007). Integrins also interact with AMPARs and regulate their function and trafficking to synapses and also appear to be modulators of homeostatic plasticity (Chavis and Westbrook, 2001; Cingolani et al., 2008).

Finally, an important tyrosine kinase EphB (ephrinB) family of RTK and their cognate ephrinB ligands are localized at synapses (Bruckner et al., 1999; Contractor et al., 2002). These proteins play an important role in axon guidance and topographic map formation in the nervous system but are also involved in the development of excitatory synapses and dendritic spines by means of phosphorylating cell surface heparan/sulphate proteoglycan Syndecan and by recruiting Kalirin (Ethell et al., 2001; Penzes et al., 2003). Also EphB2 interacts physically with NMDARs, enhances their channel activity, and can induce their clustering upon interaction with ephrinB ligands (Dalva et al., 2000; Nolt et al., 2011; Takasu et al., 2002).

2.1.3. Signaling Proteins

In the PSD, a number of crucial signaling proteins are associated with PSD-95. The PDZ domain of the nitric oxide synthase (nNOS)—a Ca^{2+}/calmodulin-activated enzyme that produces the nitric oxide involved in regulating neurotransmission and excitotoxicity—binds specifically to the second PDZ domain of PSD-95 (Hillier et al., 1999). It has been shown that the ternary NMDAR/PSD-95/nNOS complex may functionally couple NMDAR gating to nNOS activation, because disrupting the NMDAR/PSD-95 interaction with synthetic peptides that mimic the last nine residues of NR2B reduces NMDAR-induced excitotoxicity without affecting NMDAR function (Aarts et al., 2002; Cook et al., 2012).

However, the most abundant signaling protein in the PSD is the kinase CaMKII, whose multiple functions include regulating receptor targeting and function at synapses. CaMKII activation is essential for inducing synaptic plasticity and its phosphor-substrates, some of which have not yet been completely identified, for sure include several PSD proteins (Colbran and Brown, 2004; Hayashi et al., 2000). Because CaMKII is quantitatively the most abundant protein at the PSD, it also has a scaffold function, e.g. by recruiting proteasomes to synapses (Bingol et al., 2010). Ras GTPase-activating protein (SynGAP) is another abundant PSD signaling molecule that binds to PSD-95 (Chen et al., 1998; Kim et al., 1998). SynGAP regulates synaptic plasticity by suppressing the ERK pathway when it becomes activated by CaMKII (Kim et al., 2005). PSD-95 also associates with the nonreceptor tyrosine kinases of the Src family and their upstream activator, proline-rich tyrosine kinase 2 (Pyk2), both of which are thought to be important for synaptic plasticity (Seabold et al., 2003). Thus, the Pyk2–Src signaling cascade is linked closely to NMDARs by PSD-95 although the importance of PSD-95 scaffolds in synaptic regulation by tyrosine phosphorylation has not been directly investigated (Bartos et al., 2010). Finally, smaller amounts of several other signaling proteins are associated with the PSD, including a number of kinases and phosphatases whose functions have not been completely clarified (Bhattacharyya et al., 2009; Jurado et al., 2010).

2.1.4. Cytoskeletal Proteins

The actin cytoskeleton in dendritic spines and excitatory synapses is regulated by a number of actin-binding proteins (usually Rho and Rac GTPase effectors regulated by Ca^{2+}) and most of them are shaped by the concatenation of different protein interaction domains and thus function as scaffolds assembled in multimolecular complexes that are enriched in the PSD (Sekino et al., 2007; Tada and Sheng, 2006).

Drebrin, an abundant cytoskeletal protein, is an F-actin-binding protein that is mainly expressed in neurons and highly concentrated in the PSD.

The overexpression of drebrin in hippocampal neurons increases the length of dendritic spines and promotes actin assembly and the synaptic clustering of PSD-95 in the PSD (Hayashi and Shirao, 1999). It was also nicely shown that drebrin (but also profilin II and αN-catenin) are redistributed to dendritic spines with an increased F-actin content after the induction of LTP in the dentate gyrus or after synaptic activation and NMDA receptor activation (Abe et al., 2004; Ackermann and Matus, 2003). Translocation of these proteins to synapses increases F-actin polymerization and/or stabilization, a phenomenon important for synapse stabilization taking place during LTP (Bosch and Hayashi, 2011; Fukazawa et al., 2003).

Two other F-actin-binding proteins present in the PSD are spinophilin (or neurabin II) and neurabin I, both of which have similar domain structures containing a PDZ and a coiled-coil domain that form homo- and heterodimers. Spinophilin is predominantly localized through its actin-binding domain on the dendritic spines of pyramidal neurons (as its name suggests). Knock-out mice for spinophilin have more filopodia as well as immature spines and show altered glutamatergic transmission (Allen et al., 1997; Feng et al., 2000; Ryan et al., 2005). In contrast, neurabin I overexpression in immature cultured hippocampal neurons induces the formation of filopodia while promoting the enlargement of dendritic spines in older neurons. Thus spinophilin and neurabin I have opposite effect on spines (Ryan et al., 2005; Satoh et al., 1998).

α-actinin-2 and IRSp53 are two other abundant actin-binding proteins localized at the PSD. α-actinin is a member of the spectrin/dystrophin family of actin-binding proteins that binds also to the NR1 and NR2B subunits of NMDARs and provides a cytoskeletal bridge between NMDARs and actin (Schulz et al., 2004; Wyszynski et al., 1997).

IRSp53 interacts with actin but also with Shank, PSD-95, and PSD-93. Its overexpression in cultured neurons increases the density of dendritic spines, while its siRNA-mediated knock-down reduces spine density, length, and width (Choi et al., 2005). Thus, IRSp53 plays a role in linking PSD-95 to activated Rac1/Cdc42 and effectors of actin polymerization in dendritic spines (Bockmann et al., 2002; Soltau et al., 2004).

2.2. SYNAPTIC ADHESION MOLECULES GENE MUTATIONS

The correct development of synaptic specializations and the establishment of appropriate connectivity patterns are crucial events for the assembly of functional neuronal connections. In humans, deletions or mutations in neuroligins and neurexin (their synaptic functions is reviewed in Dalva et al. (2007), Dean and Dresbach (2006), Südhof (2008)) are associated with several cognitive disorders, including autism spectrum disorders (ASDs), mental retardation, and schizophrenia, although the exact connection between genetic background and biochemical mechanisms by which these mutations contribute to the diseases are still unknown.

A strong association with mutations in the genes encoding Nrxn1, Nlgn1, Nlgn3, and Nlgn4 has been described in patients affected by autism. For the Nrxn1 gene, seven point mutations, two distinct translocations, and four different large-scale deletions have been identified (Feng et al., 2006; Kim et al., 2008; Marshall et al., 2008; Szatmari et al., 2007; Yan et al., 2008b; Zahir et al., 2008). For the Nlgn4 gene, at least 10 different mutations have been observed (2 frameshifts, 5 missense mutations, and 3 internal deletions), while for Nlgn3, a single point mutation has been identified (the R451C substitution) (Jamain et al., 2003; Laumonnier et al., 2004; Talebizadeh et al., 2006; Yan et al., 2005, 2008a). In addition, deletions of X-chromosomal DNA that includes the Nlgn4 locus were detected in patients affected by autism (Chocholska et al., 2006; Lawson-Yuen et al., 2008; Macarov et al., 2007; Marshall et al., 2008).

However, it's quite important to note that often the clinical manifestations in patients do not correlate with the genetic alterations. For example, identical mutations can be carried by symptomatic patients and nonsymptomatic relatives or can cause a completely different phenotype in different individuals. For example, Lawson-Yuen at al. described a family where the same Nlgn4 microdeletion caused a severe autism in one patient and Tourette's syndrome in his brother (Lawson-Yuen et al., 2008).

Interestingly, mutations in the Nrxn1α gene have been found in individuals affected by pure schizophrenia (Kirov et al., 2008; Walsh et al., 2008) while Nrxn3 alterations have been connected to different types of addiction (Hishimoto et al., 2007; Lachman et al., 2007). Thus, dysfunctions in synaptic adhesion molecules seem to be characterized by the manifestation of a continuum of intellectual disabilities that include autism, mental retardation, and schizophrenia (Morrow et al., 2008).

In parallel with the human mutations also some autism-like phenotypes have been observed in Nlgn1, Nlgn3, and Nlgn4 mutant mice (Blundell et al., 2010; Comoletti et al., 2004; Jamain et al., 2008; Tabuchi et al., 2007).

Of course, the best way to understand whether Nrxns and Nlgns mutations found in humans are directly correlated to the manifestation of Alzheimer's Disease (ADS) is to develop animal models for the same mutations. But in some cases, there is not a linear relationship between the genetic alteration in mice and the phenotype manifestation in humans.

A number of mouse models for three Nlgn mutations have been produced, the Nlgn3 R451C substitution and the Nlgn1 and Nlgn4 loss-of-function mutation (Blundell et al., 2010; Jamain et al., 2008; Tabuchi et al., 2007).

The R451C knock-in mice showed normal motor and anxiety behaviors, but the social interaction was partially impaired and the spatial learning capability was greatly increased. However, this behavioral phenotype does not recapitulate completely the human phenotype because the R415C substitution does not impair cognitive function in the mice while in humans, it causes severe intellectual disability (Jamain et al., 2003). Consistent with the idea that autism might be due to the alteration between excitatory/inhibitory balance, the R451C mutant mice display an increase in inhibitory synaptic transmission in the somatosensory cortex.

On the other hand, Chadman et al. (2008) reported that the humanized R451C mutation in mice does not result in apparent autism-like phenotypes but rather leads to detectable functional consequences that may be interpreted in terms of problems of physical development and/or reduced sensitivity to stimuli. Further studies should clarify the discrepancy between the two studies. Finally, both Nlgn4 and Nlgn1 deletions cause an autism-like phenotype in animal models (Blundell et al., 2010; Jamain et al., 2008).

All these studies indicate that single mutations or deletions of one of the Nlgns or Nrxns do not perturb the overall synapse structure and formation, suggesting that these proteins are not only implicated in building the synapses, but small changes in their functions induce important changes in the neural network leading to cognitive impairments.

Thus, it would not be surprising if mutations of other synaptic adhesion molecules or synaptogenic proteins will be discovered in the context of ADS and intellectual disability. An example is the interleukin-1 receptor accessory protein-like 1 (IL1RAPL1) whose mutations are associated with cognitive impairment ranging from nonsyndromic X-linked (the gene is on the X chromosome) intellectual disability to autism

(Carrie et al., 1999). Recently, we have shown that IL1RAPL1 can bind to PSD-95 and regulate its phosphorylation and synaptic association by activating c-Jun terminal kinase (JNK) (Pavlowsky et al., 2010). But interestingly we also found that the extracellular domain of IL1RAPL1 and IL1RAPL2 (the paralog) are able to induce presynaptic differentiation by binding to PTPδ (Valnegri et al., 2011; Yoshida et al., 2011). Thus, our data suggest that the IL1RAPL family of proteins have the function of a classical synaptic adhesion molecule.

Therefore, genetic and functional data indicate that mutations in one of the multiple synaptic adhesion proteins is always associated with intellectual disabilities suggesting that all these molecules play an important role in regulating synapse structure and function.

2.3. POSTSYNAPTIC SCAFFOLDS AND THEIR RELATION TO NEUROPSYCHIATRIC DISORDERS

Mislocalization and dysregulation of postsynaptic scaffold proteins are crucial events during the pathophysiological course of several so-called "synaptopathies" like distinct forms of autism, schizophrenia, or Alzheimer's disease. Indeed, in the recent years, several neuropsychiatric disorders have been directly linked to an alteration in synaptic and dendritic spine morphology and function and potential new therapeutic approaches will require a deeper understanding of the molecular mechanisms underlying the disruption of subcellular structures of the synapse like the PSD or the active zone at the presynaptic buttons.

2.3.1. The ProSAP/Shank Family

The ProSAP2/Shank3 gene haploinsufficiency, which underlies the cause for the 22q13.3 deletion syndrome (now called Phelan–McDermid syndrome, PMS), is the most direct link between the absence of a postsynaptic scaffold protein and a syndromic disorder whose major clinical features include intellectual disability, global developmental delay, absent or severely delayed speech, muscular hypotonia, and "autistic-like" behavior (Bonaglia et al., 2001; Manning et al., 2004; Phelan et al., 2001; Wilson et al., 2003).

The description of the minimal critical region deleting ProSAP2/Shank3 and of a recurrent breakpoint within intron 8 of the ProSAP2/Shank3 gene (Bonaglia et al., 2001, 2006, 2011; Delahaye et al., 2009; Wilson et al., 2003) has excluded the possibility that the deletion of other genes, also located on 22q13.3, are responsible of the PMS neuropsychiatric symptoms. On the other hand, the identification of de novo mutations in the ProSAP2/Shank3 (Durand et al., 2007; Gauthier et al., 2009; Moessner et al., 2007) and ProSAP1/Shank2 genes in individuals diagnosed with ASD (Berkel et al., 2010; Pinto et al., 2010), strongly support the concept that Shank family of proteins are essential for the correct functioning of excitatory synapses.

When transfected in cultured neurons, Shank3 mutants are not localized to synapses and appear to reduce the number excitatory synapses (Durand et al., 2007, 2011). Thus, these mutations probably have a severe impact on the molecular organization of the whole postsynaptic protein complex in the affected patients by disrupting its physiological properties which are essential for normal synaptic homeostasis and the balance between excitatory and inhibitory synapses (Verpelli et al., 2011).

Indeed, as mentioned before, alterations in the synaptic adhesion proteins like the presynaptic neurexins (Ching et al., 2010; Kim et al., 2008) or postsynaptic neuroligins (Jamain et al., 2003) have also been identified in autistic individuals and the synaptic Neurexin–Neuroligin–ProSAP/SHANK pathway has been proposed as one of the major altered synaptic core complexes that causes development defects of ASD patients (Bourgeron, 2009).

Four recent in vivo studies emphasize the importance of the Shank3 protein at the molecular and behavioral level. Shank3 heterozygous and homozygous mice display abnormal social behavior, communication patterns, and learning and memory. These behavioral alterations are associated with a strong impairment of basal synaptic transmission in CA3–CA1 connections, a reduction in GluA1 clusters, and an alteration of activity-dependent AMPA receptor-mediated synaptic plasticity (Bozdagi et al., 2010; Wang et al., 2011).

In a publication of Peça et al. (2011), mice with genetic deletions of two major Shank3 splice variants exhibit self-injurious repetitive grooming and deficits in social interaction and these behavioral defects are associated with a major alterations at striatal synapses and corticostriatal circuits but not in the hippocampus suggesting that the remaining Shank3 splice variant might be sufficient to maintain a normal synapse function and structure in the hippocampus.

On the contrary, Bangash et al. (2011) showed that a Shank3 protein missing the C-terminal fragment has a gain of function phenotype: it specifically reduces NR1 at synapses but does not alter synaptic AMPAR function and composition. Instead, we recently showed that in neuronal cultures, a knock-down of all the major Shank3 splice variants mainly has an effect on the expression of mGluR5 receptors (Verpelli et al., 2011).

Interestingly, de novo mutations in the ProSAP2/Shank3 have also been found in patients ascertained for schizophrenia (Gauthier et al., 2010), thus suggesting that also schizophrenia is another neuropsychiatric disorder directly related to ProSAP/Shank failure. Finally, additional studies have also implicated ProSAP/Shank complex disassembly in neurodegeneration pathologies. A reduction of Shank1 and Shank3 proteins has been described in brains of Alzheimer Disease patients and in brains of transgenic Alzheimer Disease mouse models and is induced by a beta-amyloid oligomer accumulation in rat frontocortical cell cultures (Pham et al., 2010; Roselli et al., 2009). On the contrary, Shank2 protein levels seem higher in similar conditions (Gong et al., 2009).

It is relevant to underline the fact that both Shank2 and 3 bind and are assembled and stabilized at synapses by Zinc ions (Baron et al., 2006; Grabrucker et al., 2011a; Gundelfinger et al., 2006). Interestingly, because Zinc ions are able to bind beta-Amyloid (Matsubara et al., 2003) and physiological levels of Zn^{2+} are known to cause beta-amyloid aggregation (Friedlich et al., 2004), it was suggested that beta-amyloid accumulation within the synaptic cleft might cause cognitive impairment by trapping synaptic zinc rather than through direct neuronal toxicity (Adlard et al., 2010; Deshpande et al., 2009). Indeed, Grabrucker et al., 2011b have recently demonstrated that beta-amyloid treatment of primary hippocampal cultures causes a dysregulation of the ProSAP/Shank scaffold followed by a loss of synapses because zinc ions fail to reach their postsynaptic targets.

2.3.2. The MAGUK Family and Other Scaffold Proteins

The DLG4 gene that codifies for PSD-95, the most abundant scaffold protein in the PSD, was extensively studied for the presence of specific polymorphisms and mutations which might be associated with mental diseases. Indeed, a preliminary study suggests the association between DLG4 gene variations and ASDs, Williams' syndrome (Feyder et al., 2010), and perhaps a polymorphism in the promoter gene is associated with schizophrenia (Cheng et al., 2010). Instead, alterations in the expression of PSD-95 have been found in patients with Alzheimer's disease and in mouse models of Fragile-X (Leuba et al., 2008; Zalfa et al., 2007; Zhu et al., 2010). Regarding the other members of the MAGUK family, a clear association of mutations with mental retardation was found in the DLG3 gene that codifies for SAP102 (Tarpey et al., 2004; Zanni et al., 2010). The mutations described by Tarpey introduces a stop codon within or before the third PDZ domain, and apparently this mutation impairs the ability of SAP102 to interact with the NMDA receptor and/or other proteins involved in downstream NMDA receptor signaling pathways.

For the GKAP/SAPAP family, a clear evidence for a major role of SAPAP3 in brain function was demonstrated by Feng's lab. Mice deficient for SAPAP3 develop an obsessive–compulsive disorder (OCD)-like phenotype, which includes compulsive grooming and increased anxiety (Welch et al., 2007). In humans, multiple rare SAPAP3 missense variants have been found to be associated with trichotillomania (TTM) and OCD (Züchner et al., 2009). Thus, further genetic studies could provide more clear associations between SAPA3 mutations in humans and TTM or OCD diseases. Similarly, additional studies could decipher whether there is an association between the mutation of the HOMER2 gene and schizophrenia or the maintenance of alcohol and substance-use disorders which have been found in HOMER2 knock-out mice (Szumlinski et al., 2004) but not in humans (Szumlinski et al., 2006). Interestingly all, this data suggest that most of the major synaptic scaffold proteins have specific functions that cannot be rescued by other members of the family.

3. CONCLUSIONS

Over the past few years, our knowledge of the molecules and molecular mechanisms involved in the formation and function of synapses and dendritic spines has considerably increased and protein complexes and signaling pathways forming the PSD have been independently and extensively identified by different laboratories. All this information has demonstrated the ability of major PSD scaffold and synaptic adhesion proteins to assemble different protein complexes and to functionally link different signaling pathways important for synapse and spine formation and plasticity. Interestingly, the search for genetic causes of intellectual disability and autism has identified numerous mutated genes that code for proteins involved in synapse formation and function. Several studies indicate that mild alterations in synapse morphology and function can cause severe neuronal pathologies including intellectual disabilities, schizophrenia, and neurodegenerative diseases. These studies have opened the possibility to test pharmacological agents that might be able to counteract these morphological and functional synaptic anomalies and ameliorate the symptoms of some of these pathological conditions. In future years, we will see how many of these therapeutic possibilities will become reality.

REFERENCES

Aarts, M., Liu, Y., Liu, L., Besshoh, S., Arundine, M., Gurd, J.W., Wang, Y.T., Salter, M.W., Tymianski, M., 2002. Treatment of ischemic brain damage by perturbing NMDA receptor- PSD-95 protein interactions. Science 298, 846–850.

Abe, K., Chisaka, O., Van Roy, F., Takeichi, M., 2004. Stability of dendritic spines and synaptic contacts is controlled by alpha N-catenin. Nat. Neurosci. 7, 357–363.

Ackermann, M., Matus, A., 2003. Activity-induced targeting of profilin and stabilization of dendritic spine morphology. Nat. Neurosci. 6, 1194–1200.

Adlard, P.A., Parncutt, J.M., Finkelstein, D.I., Bush, A.I., 2010. Cognitive loss in zinc transporter-3 knock-out mice: a phenocopy for the synaptic and memory deficits of Alzheimer's disease? J. Neurosci. 30, 1631–1636.

Allen, P.B., Ouimet, C.C., Greengard, P., 1997. Spinophilin, a novel protein phosphatase 1 binding protein localized to dendritic spines. Proc. Natl. Acad. Sci. 94, 9956–9961.

Ango, F., Prezeau, L., Muller, T., Tu, J.C., Xiao, B., Worley, P.F., Pin, J.P., Bockaert, J., Fagni, L., 2001. Agonist-independent activation of metabotropic glutamate receptors by the intracellular protein Homer. Nature 411, 962–965.

Apperson, M.L., Moon, I.S., Kennedy, M.B., 1996. Characterization of densin-180, a new brain-specific synaptic protein of the O-sialoglycoprotein family. J. Neurosci. 16, 6839–6852.

Bangash, M.A., Park, J.M., Melnikova, T., Wang, D., Jeon, S.K., Lee, D., Syeda, S., Kim, J., Kouser, M., Schwartz, J., et al., 2011. Enhanced polyubiquitination of Shank3 and NMDA receptor in a mouse model of autism. Cell 145, 758–772.

Baron, M.K., Boeckers, T.M., Vaida, B., Faham, S., Gingery, M., Sawaya, M.R., Salyer, D., Gundelfinger, E.D., Bowie, J.U., 2006. An architectural framework that may lie at the core of the postsynaptic density. Science 311, 531–535.

Bartos, J.A., Ulrich, J.D., Li, H., Beazely, M.A., Chen, Y., Macdonald, J.F., Hell, J.W., 2010. Postsynaptic clustering and activation of Pyk2 by PSD-95. J. Neurosci. 30, 449–463.

Bassani, S., Cingolani, L.A., Valnegri, P., Folci, A., Zapata, J., Gianfelice, A., Sala, C., Goda, Y., Passafaro, M., 2012. The X-linked intellectual disability protein TSPAN7 regulates excitatory synapse development and AMPAR trafficking. Neuron 73, 1143–1158.

Berkel, S., Marshall, C.R., Weiss, B., Howe, J., Roeth, R., Moog, U., Endris, V., Roberts, W., Szatmari, P., Pinto, D., et al., 2010. Mutations in the SHANK2 synaptic scaffolding gene in autism spectrum disorder and mental retardation. Nat. Genet. 42, 489–491.
Bhattacharyya, S., Biou, V., Xu, W., Schlüter, O., Malenka, R.C., 2009. A critical role for PSD-95/AKAP interactions in endocytosis of synaptic AMPA receptors. Nat. Neurosci. 12, 172–181.
Biederer, T., Sara, Y., Mozhayeva, M., Atasoy, D., Liu, X., Kavalali, E.T., Sudhof, T.C., 2002. SynCAM, a synaptic adhesion molecule that drives synapse assembly. Science 297, 1525–1531.
Bingol, B., Wang, C.F., Arnott, D., Cheng, D., Peng, J., Sheng, M., 2010. Autophosphorylated CaMKIIalpha acts as a scaffold to recruit proteasomes to dendritic spines. Cell 140, 567–578.
Blundell, J., Blaiss, C.A., Etherton, M.R., Espinosa, F., Tabuchi, K., Walz, C., Bolliger, M.F., Südhof, T.C., Powell, C.M., 2010. Neuroligin-1 deletion results in impaired spatial memory and increased repetitive behavior. J. Neurosci. 30, 2115–2129.
Bockers, T.M., Mameza, M.G., Kreutz, M.R., Bockmann, J., Weise, C., Buck, F., Richter, D., Gundelfinger, E.D., Kreienkamp, H.J., 2001. Synaptic scaffolding proteins in rat brain. Ankyrin repeats of the multidomain Shank protein family interact with the cytoskeletal protein alpha-fodrin. J. Biol. Chem. 276, 40104–40112.
Bockmann, J., Kreutz, M.R., Gundelfinger, E.D., Bockers, T.M., 2002. ProSAP/Shank postsynaptic density proteins interact with insulin receptor tyrosine kinase substrate IRSp53. J. Neurochem. 83, 1013–1017.
Boeckers, T.M., Kreutz, M.R., Winter, C., Zuschratter, W., Smalla, K.H., Sanmarti-Vila, L., Wex, H., Langnaese, K., Bockmann, J., Garner, C.C., Gundelfinger, E.D., 1999. Proline-rich synapse-associated protein-1/cortactin binding protein 1 (ProSAP1/CortBP1) is a PDZ-domain protein highly enriched in the postsynaptic density. J. Neurosci. 19, 6506–6518.
Bonaglia, M.C., Giorda, R., Beri, S., De Agostini, C., Novara, F., Fichera, M., Grillo, L., Galesi, O., Vetro, A., Ciccone, R., et al., 2011. Molecular mechanisms generating and stabilizing terminal 22q13 deletions in 44 subjects with Phelan/McDermid Syndrome. PLoS Genet. 7, e1002173.
Bonaglia, M.C., Giorda, R., Borgatti, R., Felisari, G., Gagliardi, C., Selicorni, A., Zuffardi, O., 2001. Disruption of the ProSAP2 gene in a t(12;22)(q24.1;q13.3) is associated with the 22q13.3 deletion syndrome. Am. J. Hum. Genet. 69, 261–268.
Bonaglia, M.C., Giorda, R., Mani, E., Aceti, G., Anderlid, B.M., Baroncini, A., Pramparo, T., Zuffardi, O., 2006. Identification of a recurrent breakpoint within the SHANK3 gene in the 22q13.3 deletion syndrome. J. Med. Genet. 43, 822–828.
Bosch, M., Hayashi, Y., 2011. Structural plasticity of dendritic spines. Curr. Opin. Neurobiol.
Bourgeron, T., 2009. A synaptic trek to autism. Curr. Opin. Neurobiol. 19, 231–234.
Bozdagi, O., Sakurai, T., Papapetrou, D., Wang, X., Dickstein, D.L., Takahashi, N., Kajiwara, Y., Yang, M., Katz, A.M., Scattoni, M.L., et al., 2010. Haploinsufficiency of the autism associated Shank3 gene leads to deficits in synaptic function, social interaction, and social communication. Mol. Autism 1, 15.
Brakeman, P.R., Lanahan, A.A., O'Brien, R., Roche, K., Barnes, C.A., Huganir, R.L., Worley, P.F., 1997. Homer: a protein that selectively binds metabotropic glutamate receptors. Nature 386, 284–288.
Bruckner, K., Pablo Labrador, J., Scheiffele, P., Herb, A., Seeburg, P.H., Klein, R., 1999. EphrinB ligands recruit GRIP family PDZ adaptor proteins into raft membrane microdomains. Neuron 22, 511–524.
Carrie, A., Jun, L., Bienvenu, T., Vinet, M.C., McDonell, N., Couvert, P., Zemni, R., Cardona, A., Van Buggenhout, G., Frints, S., et al., 1999. A new member of the IL-1 receptor family highly expressed in hippocampus and involved in X-linked mental retardation. Nat. Genet. 23, 25–31.
Chadman, K.K., Gong, S., Scattoni, M.L., Boltuck, S.E., Gandhy, S.U., Heintz, N., Crawley, J.N., 2008. Minimal aberrant behavioral phenotypes of neuroligin-3 R451C knockin mice. Autism Res. 1, 147–158.
Chavis, P., Westbrook, G., 2001. Integrins mediate functional pre- and postsynaptic maturation at a hippocampal synapse. Nature 411, 317–321.
Chen, H.J., Rojas-Soto, M., Oguni, A., Kennedy, M.B., 1998. A synaptic Ras-GTPase activating protein (p135 SynGAP) inhibited by CaM kinase II. Neuron 20, 895–904.
Chen, L., Chetkovich, D.M., Petralia, R.S., Sweeney, N.T., Kawasaki, Y., Wenthold, R.J., Bredt, D.S., Nicoll, R.A., 2000. Stargazin regulates synaptic targeting of AMPA receptors by two distinct mechanisms. Nature 408, 936–943.

Chen, X., Winters, C., Azzam, R., Li, X., Galbraith, J.A., Leapman, R.D., Reese, T.S., 2008. Organization of the core structure of the postsynaptic density. Proc. Natl. Acad. Sci. USA 105, 4453–4458.

Cheng, D., Hoogenraad, C.C., Rush, J., Ramm, E., Schlager, M.A., Duong, D.M., Xu, P., Rukshan, S., Hanfelt, J., Nakagawa, T., et al., 2006. Relative and absolute quantification of postsynaptic density proteome isolated from rat forebrain and cerebellum. Mol. Cell. Proteomics.

Cheng, M.C., Lu, C.L., Luu, S.U., Tsai, H.M., Hsu, S.H., Chen, T.T., Chen, C.H., 2010. Genetic and functional analysis of the DLG4 gene encoding the post-synaptic density protein 95 in schizophrenia. PLoS One 5, e15107.

Ching, M.S., Shen, Y., Tan, W.H., Jeste, S.S., Morrow, E.M., Chen, X., Mukaddes, N.M., Yoo, S.Y., Hanson, E., Hundley, R., et al., 2010. Deletions of NRXN1 (neurexin-1) predispose to a wide spectrum of developmental disorders. Am. J. Med. Genet. B Neuropsychiatr. Genet. 153B, 937–947.

Cho, K.-O., Hunt, C.A., Kennedy, M.B., 1992. The rat brain postsynaptic density fraction contains a homolog of the Drosophila discs-large tumor suppressor protein. Neuron 9, 929–942.

Chocholska, S., Rossier, E., Barbi, G., Kehrer-Sawatzki, H., 2006. Molecular cytogenetic analysis of a familial interstitial deletion Xp22.2–22.3 with a highly variable phenotype in female carriers. Am. J. Med. Genet. A 140, 604–610.

Choi, J., Ko, J., Racz, B., Burette, A., Lee, J.R., Kim, S., Na, M., Lee, H.W., Kim, K., Weinberg, R.J., Kim, E., 2005. Regulation of dendritic spine morphogenesis by insulin receptor substrate 53, a downstream effector of Rac1 and Cdc42 small GTPases. J. Neurosci. 25, 869–879.

Cingolani, L.A., Thalhammer, A., Yu, L.M., Catalano, M., Ramos, T., Colicos, M.A., Goda, Y., 2008. Activity-dependent regulation of synaptic AMPA receptor composition and abundance by beta3 integrins. Neuron 58, 749–762.

Colbran, R.J., Brown, A.M., 2004. Calcium/calmodulin-dependent protein kinase II and synaptic plasticity. Curr. Opin. Neurobiol. 14, 318–327.

Comoletti, D., De Jaco, A., Jennings, L.L., Flynn, R.E., Gaietta, G., Tsigelny, I., Ellisman, M.H., Taylor, P., 2004. The Arg451Cys-neuroligin-3 mutation associated with autism reveals a defect in protein processing. J. Neurosci. 24, 4889–4893.

Contractor, A., Rogers, C., Maron, C., Henkemeyer, M., Swanson, G.T., Heinemann, S.F., 2002. Trans-synaptic Eph receptor-ephrin signaling in hippocampal mossy fiber LTP. Science 296, 1864–1869.

Cook, D.J., Teves, L., Tymianski, M., 2012. Treatment of stroke with a PSD-95 inhibitor in the gyrencephalic primate brain. Nature 483, 213–217.

Dalva, M.B., McClelland, A.C., Kayser, M.S., 2007. Cell adhesion molecules: signalling functions at the synapse. Nat. Rev. Neurosci. 8, 206–220.

Dalva, M.B., Takasu, M.A., Lin, M.Z., Shamah, S.M., Hu, L., Gale, N.W., Greenberg, M.E., 2000. EphB receptors interact with NMDA receptors and regulate excitatory synapse formation. Cell 103, 945–956.

de Wit, J., Sylwestrak, E., O'Sullivan, M.L., Otto, S., Tiglio, K., Savas, J.N., Yates, J.R., Comoletti, D., Taylor, P., Ghosh, A., 2009. LRRTM2 interacts with Neurexin1 and regulates excitatory synapse formation. Neuron 64, 799–806.

Dean, C., Dresbach, T., 2006. Neuroligins and neurexins: linking cell adhesion, synapse formation and cognitive function. Trends Neurosci. 29, 21–29.

Delahaye, A., Toutain, A., Aboura, A., Dupont, C., Tabet, A.C., Benzacken, B., Elion, J., Verloes, A., Pipiras, E., Drunat, S., 2009. Chromosome 22q13.3 deletion syndrome with a de novo interstitial 22q13.3 cryptic deletion disrupting SHANK3. Eur. J. Med. Genet. 52, 328–332.

Denda, S., Reichardt, L.F., 2007. Studies on integrins in the nervous system. Methods Enzymol. 426, 203–221.

Deshpande, A., Kawai, H., Metherate, R., Glabe, C.G., Busciglio, J., 2009. A role for synaptic zinc in activity-dependent Abeta oligomer formation and accumulation at excitatory synapses. J. Neurosci. 29, 4004–4015.

Dong, H., O'Brien, R.J., Fung, E.T., Lanahan, A.A., Worley, P.F., Huganir, R.L., 1997. GRIP: a synaptic PDZ domain-containing protein that interacts with AMPA receptors. Nature 386, 279–284.

Durand, C.M., Betancur, C., Boeckers, T.M., Bockmann, J., Chaste, P., Fauchereau, F., Nygren, G., Rastam, M., Gillberg, I.C., Anckarsater, H., et al., 2007. Mutations in the gene encoding the synaptic scaffolding protein SHANK3 are associated with autism spectrum disorders. Nat. Genet. 39, 25–27.

Durand, C.M., Perroy, J., Loll, F., Perrais, D., Fagni, L., Bourgeron, T., Montcouquiol, M., Sans, N., 2011. SHANK3 mutations identified in autism lead to modification of dendritic spine morphology via an actin-dependent mechanism. Mol. Psychiatry.

Ehrlich, I., Klein, M., Rumpel, S., Malinow, R., 2007. PSD-95 is required for activity driven synapse stabilization. Proc. Natl. Acad. Sci. USA 104, 4176–4181.

El-Husseini, A.E., Schnell, E., Chetkovich, D.M., Nicoll, R.A., Bredt, D.S., 2000. PSD-95 involvement in maturation of excitatory synapses. Science 290, 1364–1368.

Elias, G.M., Funke, L., Stein, V., Grant, S.G., Bredt, D.S., Nicoll, R.A., 2006. Synapse specific and developmentally regulated targeting of AMPA receptors by a family of MAGUK scaffolding proteins. Neuron 52, 307–320.

Elias, G.M., Nicoll, R.A., 2007. Synaptic trafficking of glutamate receptors by MAGUK scaffolding proteins. Trends Cell Biol. 17, 343–352.

Ethell, I.M., Irie, F., Kalo, M.S., Couchman, J.R., Pasquale, E.B., Yamaguchi, Y., 2001. EphB/syndecan-2 signaling in dendritic spine morphogenesis. Neuron 31, 1001–1013.

Fagni, L., Worley, P.F., Ango, F., 2002. Homer as both a scaffold and transduction molecule. Sci. STKE 2002, RE8.

Feng, J., Schroer, R., Yan, J., Song, W., Yang, C., Bockholt, A., Cook, E.H., Skinner, C., Schwartz, C.E., Sommer, S.S., 2006. High frequency of neurexin 1beta signal peptide structural variants in patients with autism. Neurosci. Lett. 409, 10–13.

Feng, J., Yan, Z., Ferreira, A., Tomizawa, K., Liauw, J.A., Zhuo, M., Allen, P.B., Ouimet, C.C., Greengard, P., 2000. Spinophilin regulates the formation and function of dendritic spines. Proc. Natl. Acad. Sci. USA 97, 9287–9292.

Feyder, M., Karlsson, R.M., Mathur, P., Lyman, M., Bock, R., Momenan, R., Munasinghe, J., Scattoni, M.L., Ihne, J., Camp, M., et al., 2010. Association of mouse Dlg4 (PSD-95) gene deletion and human DLG4 gene variation with phenotypes relevant to autism spectrum disorders and Williams' syndrome. Am. J. Psychiatry 167, 1508–1517.

Fogel, A.I., Stagi, M., Perez de Arce, K., Biederer, T., 2011. Lateral assembly of the immunoglobulin protein SynCAM 1 controls its adhesive function and instructs synapse formation. EMBO J. 30, 4728–4738.

Friedlich, A.L., Lee, J.Y., van Groen, T., Cherny, R.A., Volitakis, I., Cole, T.B., Palmiter, R.D., Koh, J.Y., Bush, A.I., 2004. Neuronal zinc exchange with the blood vessel wall promotes cerebral amyloid angiopathy in an animal model of Alzheimer's disease. J. Neurosci. 24, 3453–3459.

Fukazawa, Y., Saitoh, Y., Ozawa, F., Ohta, Y., Mizuno, K., Inokuchi, K., 2003. Hippocampal LTP is accompanied by enhanced F-actin content within the dendritic spine that is essential for late LTP maintenance in vivo. Neuron 38, 447–460.

Gauthier, J., Champagne, N., Lafrenière, R.G., Xiong, L., Spiegelman, D., Brustein, E., Lapointe, M., Peng, H., Côté, M., Noreau, A., et al., 2010. De novo mutations in the gene encoding the synaptic scaffolding protein SHANK3 in patients ascertained for schizophrenia. Proc. Natl. Acad. Sci. USA 107, 7863–7868.

Gauthier, J., Spiegelman, D., Piton, A., Lafreniere, R.G., Laurent, S., St-Onge, J., Lapointe, L., Hamdan, F.F., Cossette, P., Mottron, L., et al., 2009. Novel de novo SHANK3 mutation in autistic patients. Am. J. Med. Genet. B Neuropsychiatr. Genet. 150B, 421–424.

Gerrow, K., Romorini, S., Nabi, S.M., Colicos, M.A., Sala, C., El-Husseini, A., 2006. A preformed complex of postsynaptic proteins is involved in excitatory synapse development. Neuron 49, 547–562.

Gong, Y., Lippa, C.F., Zhu, J., Lin, Q., Rosso, A.L., 2009. Disruption of glutamate receptors at Shank-postsynaptic platform in Alzheimer's disease. Brain Res. 1292, 191–198.

Grabrucker, A.M., Knight, M.J., Proepper, C., Bockmann, J., Joubert, M., Rowan, M., Nienhaus, G.U., Garner, C.C., Bowie, J.U., Kreutz, M.R., et al., 2011a. Concerted action of zinc and ProSAP/Shank in synaptogenesis and synapse maturation. EMBO J. 30, 569–581.

Grabrucker, A.M., Schmeisser, M.J., Udvardi, P.T., Arons, M., Schoen, M., Woodling, N.S., Andreasson, K.I., Hof, P.R., Buxbaum, J.D., Garner, C.C., Boeckers, T.M., 2011b. Amyloid beta protein-induced zinc sequestration leads to synaptic loss via dysregulation of the ProSAP2/Shank3 scaffold. Mol. Neurodegener. 6, 65.

Gundelfinger, E.D., Boeckers, T.M., Baron, M.K., Bowie, J.U., 2006. A role for zinc in postsynaptic density asSAMbly and plasticity? Trends Biochem. Sci. 31, 366–373.

Han, K., Kim, E., 2008. Synaptic adhesion molecules and PSD-95. Prog. Neurobiol. 84, 263–283.
Hanley, J.G., 2008. PICK1: a multi-talented modulator of AMPA receptor trafficking. Pharmacol. Ther. 118, 152–160.
Hayashi, K., Shirao, T., 1999. Change in the shape of dendritic spines caused by overexpression of drebrin in cultured cortical neurons. J. Neurosci. 19, 3918–3925.
Hayashi, M.K., Tang, C., Verpelli, C., Narayanan, R., Stearns, M.H., Xu, R.M., Li, H., Sala, C., Hayashi, Y., 2009. The postsynaptic density proteins Homer and Shank form a polymeric network structure. Cell 137, 159–171.
Hayashi, Y., Shi, S.H., Esteban, J.A., Piccini, A., Poncer, J.C., Malinow, R., 2000. Driving AMPA receptors into synapses by LTP and CaMKII: requirement for GluR1 and PDZ domain interaction. Science 287, 2262–2267.
Hillier, B.J., Christopherson, K.S., Prehoda, K.E., Bredt, D.S., Lim, W.A., 1999. Unexpected modes of PDZ domain scaffolding revealed by structure of nNOS-syntrophin complex. Science 284, 812–815.
Hirao, K., Hata, Y., Ide, N., Takeuchi, M., Irie, M., Yao, I., Deguchi, M., Toyoda, A., Sudhof, T.C., Takai, Y., 1998. A novel multiple PDZ domain-containing molecule interacting with N-methyl-D-aspartate receptors and neuronal cell adhesion proteins. J. Biol. Chem. 273, 21105–21110.
Hirbec, H., Francis, J.C., Lauri, S.E., Braithwaite, S.P., Coussen, F., Mulle, C., Dev, K.K., Coutinho, V., Meyer, G., Isaac, J.T., et al., 2003. Rapid and differential regulation of AMPA and kainate receptors at hippocampal mossy fibre synapses by PICK1 and GRIP. Neuron 37, 625–638.
Hishimoto, A., Liu, Q.R., Drgon, T., Pletnikova, O., Walther, D., Zhu, X.G., Troncoso, J.C., Uhl, G.R., 2007. Neurexin 3 polymorphisms are associated with alcohol dependence and altered expression of specific isoforms. Hum. Mol. Genet. 16, 2880–2891.
Hoogenraad, C.C., Milstein, A.D., Ethell, I.M., Henkemeyer, M., Sheng, M., 2005. GRIP1 controls dendrite morphogenesis by regulating EphB receptor trafficking. Nat. Neurosci.
Hsueh, Y.-P., Yang, F.-C., Kharazia, V., Naisbitt, S., Cohen, A.R., Weinberg, R.J., Sheng, M., 1998. Direct interaction of CASK/LIN-2 and syndecan heparan sulfate proteoglycan and their overlapping distribution in neuronal synapses. J. Cell Biol. 142, 139–151.
Hu, J.H., Park, J.M., Park, S., Xiao, B., Dehoff, M.H., Kim, S., Hayashi, T., Schwarz, M.K., Huganir, R.L., Seeburg, P.H., et al., 2010. Homeostatic scaling requires group I mGluR activation mediated by Homer1a. Neuron 68, 1128–1142.
Huang, G., Kim, J.Y., Dehoff, M., Mizuno, Y., Kamm, K.E., Worley, P.F., Muallem, S., Zeng, W., 2007. Ca2+ signaling in microdomains: Homer1 mediates the interaction between RyR2 and Cav1.2 to regulate excitation–contraction coupling. J. Biol. Chem. 282, 14283–14290.
Hunt, C.A., Schenker, L.J., Kennedy, M.B., 1996. PSD-95 is associated with the postsynaptic density and not with the presynaptic membrane at forebrain synapses. J. Neurosci. 16, 1380–1388.
Irie, M., Hata, Y., Takeuchi, M., Ichtchenko, K., Toyoda, A., Hirao, K., Takai, Y., Rosahl, T.W., Südhof, T.C., 1997. Binding of neuroligins to PSD-95. Science 277, 1511–1515.
Jamain, S., Quach, H., Betancur, C., Rastam, M., Colineaux, C., Gillberg, I.C., Soderstrom, H., Giros, B., Leboyer, M., Gillberg, C., Bourgeron, T., 2003. Mutations of the X-linked genes encoding neuroligins NLGN3 and NLGN4 are associated with autism. Nat. Genet. 34, 27–29.
Jamain, S., Radyushkin, K., Hammerschmidt, K., Granon, S., Boretius, S., Varoqueaux, F., Ramanantsoa, N., Gallego, J., Ronnenberg, A., Winter, D., et al., 2008. Reduced social interaction and ultrasonic communication in a mouse model of monogenic heritable autism. Proc. Natl. Acad. Sci. USA 105, 1710–1715.
Jurado, S., Benoist, M., Lario, A., Knafo, S., Petrok, C.N., Esteban, J.A., 2010. PTEN is recruited to the postsynaptic terminal for NMDA receptor-dependent long-term depression. EMBO J. 29, 2827–2840.
Kammermeier, P.J., Xiao, B., Tu, J.C., Worley, P.F., Ikeda, S.R., 2000. Homer proteins regulate coupling of group I metabotropic glutamate receptors to N-type calcium and M-type potassium channels. J. Neurosci. 20, 7238–7245.
Kawabe, H., Hata, Y., Takeuchi, M., Ide, N., Mizoguchi, A., Takai, Y., 1999. nArgBP2, a novel neural member of ponsin/ArgBP2/vinexin family that interacts with synapse associated protein 90/postsynaptic density-95-associated protein (SAPAP). J. Biol. Chem. 274, 30914–30918.
Kerchner, G.A., Nicoll, R.A., 2008. Silent synapses and the emergence of a postsynaptic mechanism for LTP. Nat. Rev. Neurosci. 9, 813–825.

Kim, C.H., Chung, H.J., Lee, H.K., Huganir, R.L., 2001. Interaction of the AMPA receptor subunit GluR2/3 with PDZ domains regulates hippocampal long-term depression. Proc. Natl. Acad. Sci. USA 98, 11725–11730.

Kim, E., Naisbitt, S., Hsueh, Y.-P., Rao, A., Rothschild, A., Craig, A.M., Sheng, M., 1997. GKAP, a novel synaptic protein that interacts with the guanylate kinase-like domain of the PSD-95/SAP90 family of channel clustering molecules. J. Cell Biol. 136, 669–678.

Kim, H.G., Kishikawa, S., Higgins, A.W., Seong, I.S., Donovan, D.J., Shen, Y., Lally, E., Weiss, L.A., Najm, J., Kutsche, K., et al., 2008. Disruption of neurexin 1 associated with autism spectrum disorder. Am. J. Hum. Genet. 82, 199–207.

Kim, J.H., Liao, D., Lau, L.F., Huganir, R.L., 1998. SynGAP: a synaptic RasGAP that associates with the PSD-95/SAP90 protein family. Neuron 20, 683–691.

Kim, M.J., Dunah, A.W., Wang, Y.T., Sheng, M., 2005. Differential roles of NR2A- and NR2B-containing NMDA receptors in Ras-ERK signaling and AMPA receptor trafficking. Neuron 46, 745–760.

Kim, S., Burette, A., Chung, H.S., Kwon, S.K., Woo, J., Lee, H.W., Kim, K., Kim, H., Weinberg, R.J., Kim, E., 2006. NGL family PSD-95-interacting adhesion molecules regulate excitatory synapse formation. Nat. Neurosci. 9, 1294–1301.

Kirov, G., Gumus, D., Chen, W., Norton, N., Georgieva, L., Sari, M., O'Donovan, M.C., Erdogan, F., Owen, M.J., Ropers, H.H., Ullmann, R., 2008. Comparative genome hybridization suggests a role for NRXN1 and APBA2 in schizophrenia. Hum. Mol. Genet. 17, 458–465.

Kornau, H.-C., Schenker, L.T., Kennedy, M.B., Seeburg, P.H., 1995. Domain interaction between NMDA receptor subunits and the postsynaptic density protein PSD-95. Science 269, 1737–1740.

Kwon, S.K., Woo, J., Kim, S.Y., Kim, H., Kim, E., 2010. Trans-synaptic adhesions between netrin-G ligand-3 (NGL-3) and receptor tyrosine phosphatases LAR, protein tyrosine phosphatase delta (PTPdelta), and PTPsigma via specific domains regulate excitatory synapse formation. J. Biol. Chem. 285, 13966–13978.

Lachman, H.M., Fann, C.S., Bartzis, M., Evgrafov, O.V., Rosenthal, R.N., Nunes, E.V., Miner, C., Santana, M., Gaffney, J., Riddick, A., et al., 2007. Genomewide suggestive linkage of opioid dependence to chromosome 14q. Hum. Mol. Genet. 16, 1327–1334.

Laumonnier, F., Bonnet-Brilhault, F., Gomot, M., Blanc, R., David, A., Moizard, M.P., Raynaud, M., Ronce, N., Lemonnier, E., Calvas, P., et al., 2004. X-linked mental retardation and autism are associated with a mutation in the NLGN4 gene, a member of the neuroligin family. Am. J. Hum. Genet. 74, 552–557.

Lawson-Yuen, A., Saldivar, J.S., Sommer, S., Picker, J., 2008. Familial deletion within NLGN4 associated with autism and Tourette syndrome. Eur. J. Hum. Genet. 16, 614–618.

Leuba, G., Savioz, A., Vernay, A., Carnal, B., Kraftsik, R., Tardif, E., Riederer, I., Riederer, B.M., 2008. Differential changes in synaptic proteins in the Alzheimer frontal cortex with marked increase in PSD-95 postsynaptic protein. J. Alzheimer's Dis. 15, 139–151.

Levinson, J.N., Chéry, N., Huang, K., Wong, T.P., Gerrow, K., Kang, R., Prange, O., Wang, Y.T., El-Husseini, A., 2005. Neuroligins mediate excitatory and inhibitory synapse formation: involvement of PSD-95 and neurexin-1beta in neuroligin-induced synaptic specificity. J. Biol. Chem. 280, 17312–17319.

Lujan, R., Roberts, J.D., Shigemoto, R., Ohishi, H., Somogyi, P., 1997. Differential plasma membrane distribution of metabotropic glutamate receptors mGluR1 alpha, mGluR2 and mGluR5, relative to neurotransmitter release sites. J. Chem. Neuroanat. 13, 219–241.

Macarov, M., Zeigler, M., Newman, J.P., Strich, D., Sury, V., Tennenbaum, A., Meiner, V., 2007. Deletions of VCX-A and NLGN4: a variable phenotype including normal intellect. J. Intellect. Disabil. Res. 51, 329–333.

Mah, W., Ko, J., Nam, J., Han, K., Chung, W.S., Kim, E., 2010. Selected SALM (synaptic adhesion-like molecule) family proteins regulate synapse formation. J. Neurosci. 30, 5559–5568.

Malinow, R., Malenka, R.C., 2002. AMPA receptor trafficking and synaptic plasticity. Annu. Rev. Neurosci. 25, 103–126.

Malinverno, M., Carta, M., Epis, R., Marcello, E., Verpelli, C., Cattabeni, F., Sala, C., Mulle, C., Di Luca, M., Gardoni, F., 2010. Synaptic localization and activity of ADAM10 regulate excitatory synapses through N-cadherin cleavage. J. Neurosci. 30, 16343–16355.

Manning, M.A., Cassidy, S.B., Clericuzio, C., Cherry, A.M., Schwartz, S., Hudgins, L., Enns, G.M., Hoyme, H.E., 2004. Terminal 22q deletion syndrome: a newly recognized cause of speech and language disability in the autism spectrum. Pediatrics 114, 451–457.

Margeta, M.A., Shen, K., 2010. Molecular mechanisms of synaptic specificity. Mol. Cell. Neurosci. 43, 261–267.

Marshall, C.R., Noor, A., Vincent, J.B., Lionel, A.C., Feuk, L., Skaug, J., Shago, M., Moessner, R., Pinto, D., Ren, Y., et al., 2008. Structural variation of chromosomes in autism spectrum disorder. Am. J. Hum. Genet. 82, 477–488.

Matsubara, E., Bryant-Thomas, T., Pacheco Quinto, J., Henry, T.L., Poeggeler, B., Herbert, D., Cruz-Sanchez, F., Chyan, Y.J., Smith, M.A., Perry, G., et al., 2003. Melatonin increases survival and inhibits oxidative and amyloid pathology in a transgenic model of Alzheimer's disease. J. Neurochem. 85, 1101–1108.

Matsuzaki, M., Ellis-Davies, G.C., Nemoto, T., Miyashita, Y., Iino, M., Kasai, H., 2001. Dendritic spine geometry is critical for AMPA receptor expression in hippocampal CA1 pyramidal neurons. Nat. Neurosci. 4, 1086–1092.

Migaud, M., Charlesworth, P., Dempster, M., Webster, L.C., Watabe, A.M., Makhinson, M., He, Y., Ramsay, M.F., Morris, R.G., Morrison, J.H., et al., 1998. Enhanced long-term potentiation and impaired learning in mice with mutant postsynaptic density-95 protein. Nature 396, 433–439.

Moessner, R., Marshall, C.R., Sutcliffe, J.S., Skaug, J., Pinto, D., Vincent, J., Zwaigenbaum, L., Fernandez, B., Roberts, W., Szatmari, P., Scherer, S.W., 2007. Contribution of SHANK3 mutations to autism spectrum disorder. Am. J. Hum. Genet. 81, 1289–1297.

Monea, S., Jordan, B.A., Srivastava, S., DeSouza, S., Ziff, E.B., 2006. Membrane localization of membrane type 5 matrix metalloproteinase by AMPA receptor binding protein and cleavage of cadherins. J. Neurosci. 26, 2300–2312.

Morrow, E.M., Yoo, S.Y., Flavell, S.W., Kim, T.K., Lin, Y., Hill, R.S., Mukaddes, N.M., Balkhy, S., Gascon, G., Hashmi, A., et al., 2008. Identifying autism loci and genes by tracing recent shared ancestry. Science 321, 218–223.

Naisbitt, S., Kim, E., Tu, J.C., Xiao, B., Sala, C., Valtschanoff, J., Weinberg, R.J., Worley, P.F., Sheng, M., 1999. Shank, a novel family of postsynaptic density proteins that binds to the NMDA receptor/PSD-95/GKAP complex and cortactin. Neuron 23, 569–582.

Naisbitt, S., Valtschanoff, J., Allison, D., Sala, C., Kim, E., Craig, A., Weinberg, R., Sheng, M., 2000. Interaction of the postsynaptic density-95/guanylate kinase domain-associated protein complex with a light chain of myosin-V and dynein. J. Neurosci. 20, 4524–4534.

Nakamura, Y., Wood, C.L., Patton, A.P., Jaafari, N., Henley, J.M., Mellor, J.R., Hanley, J.G., 2011. PICK1 inhibition of the Arp2/3 complex controls dendritic spine size and synaptic plasticity. EMBO J. 30, 719–730.

Nicoll, R.A., Tomita, S., Bredt, D.S., 2006. Auxiliary subunits assist AMPA-type glutamate receptors. Science 311, 1253–1256.

Niethammer, M., Kim, E., Sheng, M., 1996. Interaction between the C terminus of NMDA receptor subunits and multiple members of the PSD-95 family of membrane-associated guanylate kinases. J. Neurosci. 16, 2157–2163.

Nolt, M.J., Lin, Y., Hruska, M., Murphy, J., Sheffler-Colins, S.I., Kayser, M.S., Passer, J., Bennett, M.V., Zukin, R.S., Dalva, M.B., 2011. EphB controls NMDA receptor function and synaptic targeting in a subunit-specific manner. J. Neurosci. 31, 5353–5364.

Pak, D.T., Yang, S., Rudolph-Correia, S., Kim, E., Sheng, M., 2001. Regulation of dendritic spine morphology by SPAR, a PSD-95-associated RapGAP. Neuron 31, 289–303.

Park, E., Na, M., Choi, J., Kim, S., Lee, J.R., Yoon, J., Park, D., Sheng, M., Kim, E., 2003. The Shank family of postsynaptic density proteins interacts with and promotes synaptic accumulation of the beta PIX guanine nucleotide exchange factor for Rac1 and Cdc42. J. Biol. Chem. 278, 19220–19229.

Pavlowsky, A., Gianfelice, A., Pallotto, M., Zanchi, A., Vara, H., Khelfaoui, M., Valnegri, P., Rezai, X., Bassani, S., Brambilla, D., et al., 2010. A postsynaptic signaling pathway that may account for the cognitive defect due to IL1RAPL1 mutation. Curr. Biol. 20, 103–115.

Penzes, P., Beeser, A., Chernoff, J., Schiller, M.R., Eipper, B.A., Mains, R.E., Huganir, R.L., 2003. Rapid induction of dendritic spine morphogenesis by trans-synaptic ephrinB-EphB receptor activation of the Rho-GEF kalirin. Neuron 37, 263–274.

Penzes, P., Johnson, R.C., Sattler, R., Zhang, X., Huganir, R.L., Kambampati, V., Mains, R.E., Eipper, B.A., 2001. The neuronal Rho-GEF Kalirin-7 interacts with PDZ domain containing proteins and regulates dendritic morphogenesis. Neuron 29, 229–242.

Petersen, J.D., Chen, X., Vinade, L., Dosemeci, A., Lisman, J.E., Reese, T.S., 2003. Distribution of postsynaptic density (PSD)-95 and Ca2+/calmodulin-dependent protein kinase II at the PSD. J. Neurosci. 23, 11270–11278.

Petralia, R.S., Esteban, J.A., Wang, Y.-X., Partridge, J.G., Zhao, H.-M., Wenthold, R.J., Malinow, R., 1999. Selective acquisition of AMPA receptors over postnatal development suggests a molecular basis for silent synapses. Nat. Neurosci. 2, 31–36.

Peça, J., Feliciano, C., Ting, J.T., Wang, W., Wells, M.F., Venkatraman, T.N., Lascola, C.D., Fu, Z., Feng, G., 2011. Shank3 mutant mice display autistic-like behaviours and striatal dysfunction. Nature.

Pham, E., Crews, L., Ubhi, K., Hansen, L., Adame, A., Cartier, A., Salmon, D., Galasko, D., Michael, S., Savas, J.N., et al., 2010. Progressive accumulation of amyloid-beta oligomers in Alzheimer's disease and in amyloid precursor protein transgenic mice is accompanied by selective alterations in synaptic scaffold proteins. FEBS J. 277, 3051–3067.

Phelan, M.C., Rogers, R.C., Saul, R.A., Stapleton, G.A., Sweet, K., McDermid, H., Shaw, S.R., Claytor, J., Willis, J., Kelly, D.P., 2001. 22q13 deletion syndrome. Am. J. Med. Genet. 101, 91–99.

Pinkstaff, J.K., Detterich, J., Lynch, G., Gall, C., 1999. Integrin subunit gene expression is regionally differentiated in adult brain. J. Neurosci. 19, 1541–1556.

Pinto, D., Pagnamenta, A.T., Klei, L., Anney, R., Merico, D., Regan, R., Conroy, J., Magalhaes, T.R., Correia, C., Abrahams, B.S., et al., 2010. Functional impact of global rare copy number variation in autism spectrum disorders. Nature 466, 368–372.

Prange, O., Wong, T.P., Gerrow, K., Wang, Y.T., El-Husseini, A., 2004. A balance between excitatory and inhibitory synapses is controlled by PSD-95 and neuroligin. Proc. Natl. Acad. Sci. USA 101, 13915–13920.

Qualmann, B., Boeckers, T.M., Jeromin, M., Gundelfinger, E.D., Kessels, M.M., 2004. Linkage of the actin cytoskeleton to the postsynaptic density via direct interactions of Abp1 with the ProSAP/Shank family. J. Neurosci. 24, 2481–2495.

Quitsch, A., Berhorster, K., Liew, C.W., Richter, D., Kreienkamp, H.J., 2005. Postsynaptic shank antagonizes dendrite branching induced by the leucine-rich repeat protein Densin-180. J. Neurosci. 25, 479–487.

Robbins, E.M., Krupp, A.J., Perez de Arce, K., Ghosh, A.K., Fogel, A.I., Boucard, A., Südhof, T.C., Stein, V., Biederer, T., 2010. SynCAM 1 adhesion dynamically regulates synapse number and impacts plasticity and learning. Neuron 68, 894–906.

Rocca, D.L., Martin, S., Jenkins, E.L., Hanley, J.G., 2008. Inhibition of Arp2/3-mediated actin polymerization by PICK1 regulates neuronal morphology and AMPA receptor endocytosis. Nat. Cell Biol. 10, 259–271.

Ronesi, J.A., Collins, K.A., Hays, S.A., Tsai, N.P., Guo, W., Birnbaum, S.G., Hu, J.H., Worley, P.F., Gibson, J.R., Huber, K.M., 2012. Disrupted Homer scaffolds mediate abnormal mGluR5 function in a mouse model of fragile X syndrome. Nat. Neurosci.

Roselli, F., Hutzler, P., Wegerich, Y., Livrea, P., Almeida, O.F., 2009. Disassembly of shank and homer synaptic clusters is driven by soluble beta-amyloid(1–40) through divergent NMDAR-dependent signalling pathways. PLoS One 4, e6011.

Ryan, X.P., Alldritt, J., Svenningsson, P., Allen, P.B., Wu, G.Y., Nairn, A.C., Greengard, P., 2005. The Rho-specific GEF Lfc interacts with neurabin and spinophilin to regulate dendritic spine morphology. Neuron 47, 85–100.

Saglietti, L., Dequidt, C., Kamieniarz, K., Rousset, M.C., Valnegri, P., Thoumine, O., Beretta, F., Fagni, L., Choquet, D., Sala, C., et al., 2007. Extracellular interactions between GluR2 and N-cadherin in spine regulation. Neuron 54, 461–477.

Sala, C., Futai, K., Yamamoto, K., Worley, P.F., Hayashi, Y., Sheng, M., 2003. Inhibition of dendritic spine morphogenesis and synaptic transmission by activity-inducible protein Homer1a. J. Neurosci. 23, 6327–6337.

Sala, C., Roussignol, C., Meldolesi, J., Fagni, L., 2005. Key role of the postsynaptic density scaffold proteins Shank and Homer in the functional architecture of Ca2+ homeostasis at dendritic spines in hippocampal neurons. J. Neurosci. 25, 4587–4592.

Satoh, A., Nakanishi, H., Obaishi, H., Wada, M., Takahashi, K., Satoh, K., Hirao, K., Nishioka, H., Hata, Y., Mizoguchi, A., Takai, Y., 1998. Neurabin-II/spinophilin. J. Biol. Chem. 273, 3470–3475.

Schnell, E., Sizemore, M., Karimzadegan, S., Chen, L., Bredt, D.S., Nicoll, R.A., 2002. Direct interactions between PSD-95 and stargazin control synaptic AMPA receptor number. Proc. Natl. Acad. Sci. USA 99, 13902–13907.

Schulz, T.W., Nakagawa, T., Licznerski, P., Pawlak, V., Kolleker, A., Rozov, A., Kim, J., Dittgen, T., Kohr, G., Sheng, M., et al., 2004. Actin/alpha-actinin-dependent transport of AMPA receptors in dendritic spines: role of the PDZ–LIM protein RIL. J. Neurosci. 24, 8584–8594.

Seabold, G.K., Burette, A., Lim, I.A., Weinberg, R.J., Hell, J.W., 2003. Interaction of the tyrosine kinase Pyk2 with the N-methyl-D-aspartate receptor complex via the Src homology 3 domains of PSD-95 and SAP102. J. Biol. Chem. 278, 15040–15048.

Sekino, Y., Kojima, N., Shirao, T., 2007. Role of actin cytoskeleton in dendritic spine morphogenesis. Neurochem. Int. 51, 92–104.

Setou, M., Seog, D.H., Tanaka, Y., Kanai, Y., Takei, Y., Kawagishi, M., Hirokawa, N., 2002. Glutamate-receptor-interacting protein GRIP1 directly steers kinesin to dendrites. Nature 417, 83–87.

Sheng, M., Sala, C., 2001. PDZ domains and the organization of supramolecular complexes. Annu. Rev. Neurosci. 24, 1–29.

Shi, S.H., Hayashi, Y., Petralia, R.S., Zaman, S.H., Wenthold, R.J., Svoboda, K., Malinow, R., 1999. Rapid spine delivery and redistribution of AMPA receptors after synaptic NMDA receptor activation. Science 284, 1811–1816.

Shin, H., Wyszynski, M., Huh, K.H., Valtschanoff, J.G., Lee, J.R., Ko, J., Streuli, M., Weinberg, R.J., Sheng, M., Kim, E., 2003. Association of the kinesin motor KIF1A with the multimodular protein liprin-alpha. J. Biol. Chem. 278, 11393–11401.

Silverman, J.B., Restituito, S., Lu, W., Lee-Edwards, L., Khatri, L., Ziff, E.B., 2007. Synaptic anchorage of AMPA receptors by cadherins through neural plakophilin-related arm protein AMPA receptor-binding protein complexes. J. Neurosci. 27, 8505–8516.

Soltau, M., Berhörster, K., Kindler, S., Buck, F., Richter, D., Kreienkamp, H.J., 2004. Insulin receptor substrate of 53 kDa links postsynaptic shank to PSD-95. J. Neurochem. 90, 659–665.

Soltau, M., Richter, D., Kreienkamp, H.J., 2002. The insulin receptor substrate IRSp53 links postsynaptic shank1 to the small G-protein cdc42. Mol. Cell. Neurosci. 21, 575–583.

Sprengel, R., Suchanek, B., Amico, C., Brusa, R., Burnashev, N., Rozov, A., Hvalby, O., Jensen, V., Paulsen, O., Andersen, P., et al., 1998. Importance of the intracellular domain of NR2 subunits for NMDA receptor function in vivo. Cell 92, 279–289.

Sugiyama, Y., Kawabata, I., Sobue, K., Okabe, S., 2005. Determination of absolute protein numbers in single synapses by a GFP-based calibration technique. Nat. Methods 2, 677–684.

Szatmari, P., Paterson, A.D., Zwaigenbaum, L., Roberts, W., Brian, J., Liu, X.Q., Vincent, J.B., Skaug, J.L., Thompson, A.P., Senman, L., et al., 2007. Mapping autism risk loci using genetic linkage and chromosomal rearrangements. Nat. Genet. 39, 319–328.

Szumlinski, K.K., Dehoff, M.H., Kang, S.H., Frys, K.A., Lominac, K.D., Klugmann, M., Rohrer, J., Griffin, W., Toda, S., Champtiaux, N.P., et al., 2004. Homer proteins regulate sensitivity to cocaine. Neuron 43, 401–413.

Szumlinski, K.K., Kalivas, P.W., Worley, P.F., 2006. Homer proteins: implications for neuropsychiatric disorders. Curr. Opin. Neurobiol. 16, 251–257.

Südhof, T.C., 2008. Neuroligins and neurexins link synaptic function to cognitive disease. Nature 455, 903–911.

Tabuchi, K., Blundell, J., Etherton, M.R., Hammer, R.E., Liu, X., Powell, C.M., Südhof, T.C., 2007. A neuroligin-3 mutation implicated in autism increases inhibitory synaptic transmission in mice. Science 318, 71–76.

Tada, T., Sheng, M., 2006. Molecular mechanisms of dendritic spine morphogenesis. Curr. Opin. Neurobiol. 16, 95–101.

Takamiya, K., Kostourou, V., Adams, S., Jadeja, S., Chalepakis, G., Scambler, P.J., Huganir, R.L., Adams, R.H., 2004. A direct functional link between the multi-PDZ domain protein GRIP1 and the Fraser syndrome protein Fras1. Nat. Genet. 36, 172–177.

Takamiya, K., Mao, L., Huganir, R.L., Linden, D.J., 2008. The glutamate receptor interacting protein family of GluR2-binding proteins is required for long-term synaptic depression expression in cerebellar Purkinje cells. J. Neurosci. 28, 5752–5755.

Takasu, M.A., Dalva, M.B., Zigmond, R.E., Greenberg, M.E., 2002. Modulation of NMDA receptor-dependent calcium influx and gene expression through EphB receptors. Science 295, 491–495.

Takumi, Y., Ramirez-Leon, V., Laake, P., Rinvik, E., Ottersen, O.P., 1999. Different modes of expression of AMPA and NMDA receptors in hippocampal synapses. Nat. Neurosci. 2, 618–624.

Talebizadeh, Z., Lam, D.Y., Theodoro, M.F., Bittel, D.C., Lushington, G.H., Butler, M.G., 2006. Novel splice isoforms for NLGN3 and NLGN4 with possible implications in autism. J. Med. Genet. 43, e21.

Tarpey, P., Parnau, J., Blow, M., Woffendin, H., Bignell, G., Cox, C., Cox, J., Davies, H., Edkins, S., Holden, S., et al., 2004. Mutations in the DLG3 gene cause nonsyndromic X-linked mental retardation. Am. J. Hum. Genet. 75, 318–324.

Togashi, H., Abe, K., Mizoguchi, A., Takaoka, K., Chisaka, O., Takeichi, M., 2002. Cadherin regulates dendritic spine morphogenesis. Neuron 35, 77–89.

Tu, J.C., Xiao, B., Naisbitt, S., Yuan, J.P., Petralia, R.S., Brakeman, P., Doan, A., Aakalu, V.K., Lanahan, A.A., Sheng, M., Worley, P.F., 1999. Coupling of mGluR/Homer and PSD-95 complexes by the Shank family of postsynaptic density proteins. Neuron 23, 583–592.

Tu, J.C., Xiao, B., Yuan, J.P., Lanahan, A.A., Leoffert, K., Li, M., Linden, D.J., Worley, P.F., 1998. Homer binds a novel proline-rich motif and links group 1 metabotropic glutamate receptors with IP3 receptors. Neuron 21, 717–726.

Valnegri, P., Montrasio, C., Brambilla, D., Ko, J., Passafaro, M., Sala, C., 2011. The X-linked intellectual disability protein IL1RAPL1 regulates excitatory synapse formation by binding PTP{delta} and RhoGAP2. Hum. Mol. Genet.

Valtschanoff, J.G., Weinberg, R.J., 2001. Laminar organization of the NMDA receptor complex within the postsynaptic density. J. Neurosci. 21, 1211–1217.

Verpelli, C., Dvoretskova, E., Vicidomini, C., Rossi, F., Chiappalone, M., Schoen, M., Di Stefano, B., Mantegazza, R., Broccoli, V., Boeckers, T.M., et al., 2011. Importance of shank3 in regulating metabotropic glutamate receptor 5 (mGluR5) expression and signaling at synapses. J. Biol. Chem.

Walsh, T., McClellan, J.M., McCarthy, S.E., Addington, A.M., Pierce, S.B., Cooper, G.M., Nord, A.S., Kusenda, M., Malhotra, D., Bhandari, A., et al., 2008. Rare structural variants disrupt multiple genes in neurodevelopmental pathways in schizophrenia. Science 320, 539–543.

Wang, X., McCoy, P.A., Rodriguiz, R.M., Pan, Y., Je, H.S., Roberts, A.C., Kim, C.J., Berrios, J., Colvin, J.S., Bousquet-Moore, D., et al., 2011. Synaptic dysfunction and abnormal behaviors in mice lacking major isoforms of Shank3. Hum. Mol. Genet.

Welch, J.M., Lu, J., Rodriguiz, R.M., Trotta, N.C., Peca, J., Ding, J.D., Feliciano, C., Chen, M., Adams, J.P., Luo, J., et al., 2007. Cortico-striatal synaptic defects and OCD-like behaviours in Sapap3-mutant mice. Nature 448, 894–900.

Wilson, H.L., Wong, A.C., Shaw, S.R., Tse, W.Y., Stapleton, G.A., Phelan, M.C., Hu, S., Marshall, J., McDermid, H.E., 2003. Molecular characterisation of the 22q13 deletion syndrome supports the role of haploinsufficiency of SHANK3/PROSAP2 in the major neurological symptoms. J. Med. Genet. 40, 575–584.

Woo, J., Kwon, S.K., Choi, S., Kim, S., Lee, J.R., Dunah, A.W., Sheng, M., Kim, E., 2009. Trans-synaptic adhesion between NGL-3 and LAR regulates the formation of excitatory synapses. Nat. Neurosci. 12, 428–437.

Wyszynski, M., Kim, E., Dunah, A.W., Passafaro, M., Valtschanoff, J.G., Serra-Pages, C., Streuli, M., Weinberg, R.J., Sheng, M., 2002. Interaction between GRIP and Liprinalpha/SYD2 is required for AMPA receptor targeting. Neuron 34, 39–52.

Wyszynski, M., Lin, J., Rao, A., Nigh, E., Beggs, A.H., Craig, A.M., Sheng, M., 1997. Competitive binding of alpha-actinin and calmodulin to the NMDA receptor. Nature 385, 439–442.

Xia, J., Zhang, X., Staudinger, J., Huganir, R.L., 1999. Clustering of AMPA receptors by the synaptic PDZ domain-containing protein PICK1. Neuron 22, 179–187.

Xiao, B., Tu, J.C., Worley, P.F., 2000. Homer: a link between neural activity and glutamate receptor function. Curr. Opin. Neurobiol. 10, 370–374.

Yamagata, M., Sanes, J.R., Weiner, J.A., 2003. Synaptic adhesion molecules. Curr. Opin. Cell Biol. 15, 621–632.

Yan, J., Feng, J., Schroer, R., Li, W., Skinner, C., Schwartz, C.E., Cook, E.H., Sommer, S.S., 2008a. Analysis of the neuroligin 4Y gene in patients with autism. Psychiatr. Genet. 18, 204–207.

Yan, J., Noltner, K., Feng, J., Li, W., Schroer, R., Skinner, C., Zeng, W., Schwartz, C.E., Sommer, S.S., 2008b. Neurexin 1alpha structural variants associated with autism. Neurosci. Lett. 438, 368–370.

Yan, Y., Yang, D., Zarnowska, E.D., Du, Z., Werbel, B., Valliere, C., Pearce, R.A., Thomson, J.A., Zhang, S.C., 2005. Directed differentiation of dopaminergic neuronal subtypes from human embryonic stem cells. Stem Cells 23, 781–790.

Yoshida, T., Yasumura, M., Uemura, T., Lee, S.J., Ra, M., Taguchi, R., Iwakura, Y., Mishina, M., 2011. IL-1 receptor accessory protein-like 1 associated with mental retardation and autism mediates synapse formation by trans-synaptic interaction with protein tyrosine phosphatase {delta}. J. Neurosci. 31, 13485–13499.

Yuan, J.P., Kiselyov, K., Shin, D.M., Chen, J., Shcheynikov, N., Kang, S.H., Dehoff, M.H., Schwarz, M.K., Seeburg, P.H., Muallem, S., Worley, P.F., 2003. Homer binds TRPC family channels and is required for gating of TRPC1 by IP3 receptors. Cell 114, 777–789.

Zahir, F.R., Baross, A., Delaney, A.D., Eydoux, P., Fernandes, N.D., Pugh, T., Marra, M.A., Friedman, J.M., 2008. A patient with vertebral, cognitive and behavioural abnormalities and a de novo deletion of NRXN1alpha. J. Med. Genet. 45, 239–243.

Zalfa, F., Eleuteri, B., Dickson, K.S., Mercaldo, V., De Rubeis, S., di Penta, A., Tabolacci, E., Chiurazzi, P., Neri, G., Grant, S.G., Bagni, C., 2007. A new function for the fragile X mental retardation protein in regulation of PSD-95 mRNA stability. Nat. Neurosci. 10, 578–587.

Zanni, G., van Esch, H., Bensalem, A., Saillour, Y., Poirier, K., Castelnau, L., Ropers, H.H., de Brouwer, A.P., Laumonnier, F., Fryns, J.P., Chelly, J., 2010. A novel mutation in the DLG3 gene encoding the synapse-associated protein 102 (SAP102) causes non-syndromic mental retardation. Neurogenetics 11, 251–255.

Zhu, Z.W., Xu, Q., Zhao, Z.Y., Gu, W.Z., Wu, D.W., 2010. Spatiotemporal expression of PSD-95 in Fmr1 knockout mice brain. Neuropathology.

Züchner, S., Wendland, J.R., Ashley-Koch, A.E., Collins, A.L., Tran-Viet, K.N., Quinn, K., Timpano, K.C., Cuccaro, M.L., Pericak-Vance, M.A., Steffens, D.C., et al., 2009. Multiple rare SAPAP3 missense variants in trichotillomania and OCD. Mol. Psychiatry 14, 6–9.

CHAPTER FIVE

The Tripartite Synapse: A Role for Glial Cells in Modulating Synaptic Transmission

Tamara Blutstein and Philip G. Haydon
Department of Neuroscience, Tufts University School of Medicine, Boston, MA, USA

1. INTRODUCTION

1.1. Glial Cells

Over the past 20 years, there has been a growing appreciation for the role of glial cells as active participants in synaptic communication. Glial cells were first identified in 1846 by Virchow who coined the term glia, from the Greek for glue, because he believed that their purpose was to fasten the nerve cells together (Seth and Koul, 2008). By 1871 Camillo Golgi determined that glial cells were a distinct cellular population and proposed a role for them in providing metabolic support for neurons (Verkhratsky et al., 2011; Parpura et al., 2012)

Glial cells can be divided into two general classes, microglia and macroglia. Microglia are macrophage type cells that are essentially the immune cells of the brain (Kreutzberg, 1996). Macroglia can be further divided into three categories: (1) ependymal cells, (2) oligodendrocytes, and (3) astrocytes. Ependymal cells are specialized glial cells that line the ventricles, creating a barrier between the neuropil and the cerebrospinal fluid (Bruni, 1998; Kettenmann and Ransom, 2005). Oligodendrocytes, and their peripheral counterparts, Schwann cells, are the myelin forming cells of nervous system. This chapter will focus on the astrocyte, which is the most heterogeneous glial cell type in the brain and are named because of their star-like morphology (Kettenmann and Ransom, 2005). The astrocyte processes make contact with both the vasculature (Anderson and Nedergaard, 2003) as well as individual synapses (Araque et al., 1999). This allows the astrocyte to function as a critical integration centers and implicating them in a multitude of central nervous system (CNS) functions.

2.1. ASTROCYTE FUNCTIONS

2.1.1. Homeostasis

One of the major roles of astrocytes is to maintain homeostatic balance within the CNS, including regulating extracellular ion concentration, neurotransmitter levels, pH

The Synapse
http://dx.doi.org/10.1016/B978-0-12-418675-0.00005-5

and water homeostasis. Under resting conditions, neurons have a high intracellular K^+ concentration which helps maintain its resting membrane potential. Extracellular K^+ rises after neuronal activity and astrocytes play a critical role in removing this excess K^+ from the extracellular space. This is accomplished by inward rectifying K^+ channels, which maintain the astrocytic membrane potential close to the equilibrium potential for K^+, thus K^+ flows into the astrocyte as its concentrations rise in the extracellular space. Once inside the astrocyte, this K^+ is then spatial buffered throughout the glial network through K^+ permeable gap junctions (Simard and Nedergaard, 2004; Butt and Kalsi, 2006; Halassa and Haydon, 2010; Verkhratsky et al., 2011). Astrocytes also have the ability to sense changes in extracellular osmolarity and can regulate the exchange of water between the blood and the brain. In response to hypo-osmolarity, astrocytes swell and then initiate a corrective process that involves extrusion of ions such as K^+ and Cl^- (Simard and Nedergaard, 2004).

Astrocytes also play an essential role in regulating extracellular neurotransmitter concentration; this is specifically important for glutamate, which if it remains in the extracellular space for an extended period of time can become excitotoxic. This is accomplished through sodium-dependent glutamate transporters GLAST (EAAT1) and GLT-1 (EAAT2), which exchange glutamate, $2Na^+$ ions and $1H^+$ ion for $1K^+$ ion (Shigeri et al., 2004; Tzingounis and Wadiche, 2007). The dysregulation of glutamate transport is associated with a number of neurological disorders including amyotrophic lateral sclerosis, Alzheimer's disease, and Parkinson's disease, among others (Maragakis and Rothstein, 2004).

2.1.2. Metabolism

Astrocytes serve a metabolic function in the brain, regulating ammonium levels, scavenging free radicals and energy metabolism (Ransom and Ransom, 2012). The glutamate–glutamine cycle is an important example of neuronal glial cooperation in the nervous system. Once inside the astrocyte, the transported glutamate is converted by the glial enzyme glutamine synthetase into glutamine, the nonsignaling precursor to glutamate. The glutamine is transported back to the neuron where it is converted by glutaminase into glutamate. This is a major source of neuronal glutamate and gamma aminobutyric acid (GABA) as neurons lack the enzyme pyruvate carboxylase, which is required for de novo synthesis of glutamate from glucose (Hertz and Zielke, 2004).

Astrocytes also have the ability to take up glucose from the bloodstream and this glucose is transported through astrocytic gap junctions to areas of metabolic need. This glucose can be stored as glycogen or metabolized to lactate. Astrocytes show a preference for the production and release of lactate over the entry of pyruvate into the tricarboxylic acid cycle and this lactate can be released into the extracellular space and used by neurons as a substrate for adenosine triphosphate (ATP) production (Belanger et al., 2011).

2.1.3. Neurovascular Coupling

Astrocytes play an important role in integrating neuronal activation to cerebral blood flow through their structural association with both synapses and cerebral blood vessels (Haydon and Carmignoto, 2006; Halassa and Haydon, 2010). Increased neuronal activity in a specific brain region leads to increased blood flow in that area, a process known as functional hyperemia (Roy and Sherrington, 1890). This process requires neuron to astrocyte signaling, as glutamate-mediated Ca^{2+} oscillations in astrocyte leads to activity-dependent vasodilatation (Zonta et al., 2003). Astrocytes also produce a number of vasoactive substances including nitric oxide, cyclooxygenase products, and ATP (Haydon and Carmignoto, 2006).

2.1.4. Synaptic Transmission

Astrocytes play a crucial role in synaptic transmission, which will be the focus of the rest of this chapter. The concept of the tripartite synapse was introduced in 1999, and suggested that there are three essential components involved in synaptic transmission, the presynaptic neuron, the postsynaptic neuron, and the astrocyte (Araque et al., 1999). This was in response to the observations that glia cells could respond to neuronal activity and release transmitters themselves, which can act to modulate neuronal activity. Astrocytes are uniquely positioned to have a strong influence on neurotransmission. They occupy nonoverlapping spatial territories, with a single astrocyte contacting approximately 100,000 synapses in the hippocampus (Bushong et al., 2002) and approximately 300–600 dendrites in the cortex (Halassa et al., 2007).

2.2. CA^{2+} SIGNALING IN ASTROCYTES

2.2.1. Overview

Astrocytes are electrically inexcitable cells, which made studying their role in synaptic transmission challenging until the advent of new technologies for Ca^{2+} imaging in cells. Using this technology, Cornell-Bell et al. (1990) observed that activation of astrocytic glutamate receptors in cultured hippocampal astrocytes leads to an elevation of cytoplasmic Ca^{2+} which propagates in waves throughout the astrocyte and to neighboring astrocytes. Using organotypic hippocampal slice cultures it was later determined that these astrocytic Ca^{2+} waves occur in response to synaptically released glutamate (Dani et al., 1992). This led to the observation that a Ca^{2+} response in the astrocyte could trigger a subsequent increase in cytosolic Ca^{2+} in neurons (Nedergaard, 1994; Parpura et al., 1994) that was determined to be due to the astrocytic release of glutamate that acted on neighboring neurons to increase their cytosolic Ca^{2+} (Parpura et al., 1994). This was the start of a renaissance in the field of glial biology and jump started the exploration of Ca^{2+} dynamics in astrocytes and how this can lead to transmitter release from the astrocyte.

2.2.2. Sources of Astrocytic Ca^{2+}?

There are a variety of sources of Ca^{2+} in the astrocyte, a major source being the endoplasmic reticulum (ER). Ca^{2+} can be mobilized from the ER by activation of inositol (1,4,5)-triphosphate receptors (IP_3R) as well as ryanodine receptors. Ca^{2+} can also enter the astrocyte from the extracellular space through voltage-gated Ca^{2+} channels, store operated Ca^{2+} channels, and plasma membrane Na^+/Ca^{2+} exchangers (reviewed in Zorec et al. (2012)).

2.2.3. Activation of the Astrocytic Ca^{2+} Wave

The studies of the early 1990s established that glial cell excitability was mediated by Ca^{2+} waves propagating throughout the astrocyte and from one astrocyte to another. Further studies revealed the stimuli that lead to the generation of these Ca^{2+} waves. For example, astrocytes express metabotropic glutamate receptors, which are G-protein-coupled receptors (GPCR) (D'Antoni et al., 2008). In response to glutamate, the GPCR stimulates phospholipase C and the production of IP_3. IP_3 then binds to its receptor on the ER, which causes the release of Ca^{2+} from internal stores (Cai et al., 2000). Stimulation of hippocampal interneurons activates $GABA_B$ receptors on astrocytes that in turn stimulate astrocytic Ca^{2+} signaling (Kang et al., 1998). Also in the hippocampus, stimulation of the basal forebrain cholinergic inputs lead to a muscarinic acetylcholine receptor-mediated increase in astrocytic Ca^{2+} (Araque et al., 2002). Both ATP and adenosine can also induce increases in astrocytic Ca^{2+} (Porter and McCarthy, 1995).

In some cases, astrocytes can exhibit Ca^{2+} oscillations that are independent of neuronal activity. For example in the ventrobasal hypothalamus, astrocytes exhibit spontaneous Ca^{2+} oscillations of various frequencies and patterns. Within this population are a set of "pacemaker" astrocytes that exhibit a regular and robust pattern of Ca^{2+} oscillations (Parri and Crunelli, 2001) that are mediated by IP_3R activation (Parri and Crunelli, 2003).

Interestingly, recent work on human brain slices has shown that cortical and hippocampal astrocytes exhibit spontaneous Ca^{2+} oscillations as well as in response to ATP, glutamate, and the cannabinoid receptor 1 (CB1) agonist WIN (Navarrete et al., 2012).

2.2.4. Ca^{2+} Wave Heterogeneity

Ca^{2+} signaling in the astrocyte is not an all or none phenomenon but rather exhibits similar heterogeneity to neuronal signaling. In fact, astrocyte Ca^{2+} signaling is highly plastic, responding to changes in neuronal activity (Pasti et al., 1997). Some of this plasticity is dependent on the location of the astrocytes. In the hippocampus, the number of simultaneously activated synapses is correlated with the amplitude of the astrocytic Ca^{2+} elevation (Honsek et al., 2012). In the cerebellum, Bergmann glia, a specific type of astrocyte, exhibit multiple types of Ca^{2+} transients in response to different stimuli, like locomotion, and neuronal input dictates the extent of the Ca^{2+} wave propagation (Santello et al., 2012). Ca^{2+} waves can also vary in their temporal properties, with some responses occurring on slow, long duration time courses and others on a millisecond time scale (reviewed in Santello et al. (2012)).

2.3. GLIOTRANSMISSION

2.3.1. Overview

One of the primary functions of astrocytic Ca^{2+} signaling is to induce the release of chemical transmitters from the astrocyte in a process called gliotransmission.

2.3.2. Definition of a Gliotransmitter

Specific criteria have been defined in order to define a compound as a gliotransmitter (Parpura and Zorec, 2010). The compound must be synthesized or stored in glia, release must be regulated and triggered by physiological and/or pathological stimuli, this release should lead to the activation of rapid (milliseconds to seconds) responses in neighboring cells and it must play a role in physiological or pathological processes. To date a variety of compounds have been identified as gliotransmitters, including glutamate, D-serine, ATP/adenosine, and some peptides.

2.3.3. Gliotransmitter Release

Gliotransmitters can be released in three main ways. First, through plasma membrane channels which include anion channels (Pasantes Morales and Schousboe, 1988), hemichannels (Cotrina et al., 1998; Iglesias et al., 2009), and P2X7 receptors (Duan et al., 2003). They can also be released through plasma membrane transporters. This occurs through the reversal of the excitatory amino acid transporters (Szatkowski et al., 1990), exchange via the cysteine–glutamate exchanger (Warr et al., 1999), or organic anion transporters (Rosenberg et al., 1994). Finally, they can be released through Ca^{2+}-dependent exocytosis (Parpura et al., 1994), which will be described in detail below.

2.3.4. Ca^{2+}-Dependent Exocytosis in Astrocytes

There is specific machinery that must be present in the astrocyte in order for astrocytes to successfully release transmitter via Ca^{2+}-dependent exocytosis. The most critical of which is the soluble N-ethyl maleimide-sensitive fusion protein attachment protein receptor (SNARE) complex, which contains syntaxin, synaptobrevin 2 (or vesicle-associated membrane protein 2; VAMP2) and synaptosome-associated protein of 25 kDa (SNAP-25) or its homolog SNAP-23 (Montana et al., 2006). This complex is responsible for mediating the fusion of the vesicular and plasma membrane, which is necessary for exocytosis. Astrocytes express synaptobrevin 2, syntaxin 1, and SNAP-23 as well as other proteins involved in the exocytotic machinery including munc-18, and complexin 2 (Montana et al., 2006). Additionally, astrocytes also express the proteins necessary for vesicular filling, including the vacuolar-type H^+-ATPase (V-ATPase) which drives protons into the vesicular lumen creating the proton gradient necessary for transport of transmitter into the vesicle, as well as the vesicular glutamate transporters (VGLUT1, 2, and 3) and the vesicular nucleotide transporter (VNUT) which are responsible for transporting transmitters into vesicles (Montana et al., 2006; Parpura and Zorec, 2010; Zorec et al., 2012).

Further studies have revealed that this machinery is functional in the astrocyte. For example, clostridial toxin, tetanus toxin, and some types of botulinum toxins, which cleave SNARE proteins necessary for exocytosis cause a reduction in basal and evoked glutamate release from cultured astrocytes (Parpura et al., 1995; Jeftinija et al., 1997) and Ca^{2+}-dependent glutamate release from astrocytes (Parpura and Zorec, 2010). Similarly, the expression of the cytoplasmic tail of synaptobrevin 2, which contains the SNARE domain but lacks the ability to anchor to the vesicular membrane thus allowing it to act as a dominant negative SNARE (dnSNARE), resulted in the inhibition of evoked glutamate from astrocytes (Zhang et al., 2004a). Reducing synaptotagmin 4, the Ca^{2+} sensor in astrocytes, via RNA interference or expressing a mutant form of synaptotagmin 4 leads to a reduction in Ca^{2+}-dependent glutamate release from astrocytes (Zhang et al., 2004b). Rose Bengal, an inhibitor of VGLUTs, inhibited Ca^{2+}-dependent glutamate release from astrocytes (Montana et al., 2004).

2.3.5. Types of Gliotransmitters

2.3.5.1. Glutamate

The first demonstration of transmitter release from astrocytes was the Ca^{2+}-dependent release of glutamate (Parpura et al., 1994). This was shown in cultured astrocytes using high-pressure liquid chromatography analysis of culture superfusate. The astrocytes were incubated in a solution that depleted internal Ca^{2+} stores and prevented Ca^{2+} entry from the extracellular space or normal media. When stimulated with the Ca^{2+} ionophore ionomycin, there was an increase in the release of glutamate from the astrocytes incubated in normal media, this was absent in the astrocytes incubated in Ca^{2+}-free media (Parpura et al., 1994). This was the first demonstration that elevated intracellular Ca^{2+} was necessary and sufficient to stimulate glutamate release from astrocytes. Later it was shown that other stimuli that increased astrocytic Ca^{2+} also stimulated the release of glutamate from the astrocyte, including mechanical stimulation, photostimulation, and photolysis of caged Ca^{2+}. Similarly, the Ca^{2+} chelator 1,2-bis(2-aminophenoxy)ethane-*N*, *N*, *N*′, *N*′-tetraacetic acid (BAPTA) and thapsigargin (inhibitor of the store specific Ca^{2+} ATPase) also reduced evoked glutamate release from astrocytes (reviewed in Parpura and Zorec (2010)).

2.3.5.2. D-Serine

Astrocytes also have the ability to release D-serine, which is an amino acid that is synthesized from L-serine by the enzyme serine racemase, which is found predominantly in astrocytes (Wolosker et al., 1999). D-Serine can be released by astrocytes and bind to the glycine binding site of the n-methyl-D-aspartate (NMDA) receptor (NMDAR) (Schell et al., 1995). AMPA and metabotropic glutamate receptor stimulation leads to the Ca^{2+}-dependent release of D-serine from astrocytes and this can be blocked by disrupting Ca^{2+} homeostasis, inhibiting the V-ATPase or by tetanus toxin (Mothet et al., 2005). Confocal imaging has also shown the colocalization of D-serine with VAMP2 (Mothet et al., 2005).

2.3.5.3. ATP/Adenosine

ATP, a major energy source in the brain, also acts as a chemical transmitter either by acting directly on purinergic receptors or being hydrolyzed by membrane-bound ectonucleotidases to adenosine diphosphate, adenosine monophosphate (AMP), and adenosine. ATP was first identified as a gliotransmitter using luciferin/luciferase studies in cultured astrocytes, which demonstrated that Ca^{2+} waves triggered ATP release from astrocytes (Wang et al., 2000). Subsequently, ATP was shown to be released in response to NO in cultured cortical astrocytes and this could be blocked by BAPTA or by impairing vesicular release (Bal-Price et al., 2002). Mechanically stimulated astrocytes release ATP and this release can be blocked by the application of bafilomycin A1 or tetanus toxin (Coco et al., 2003). A VNUT has been identified and found to be enriched in astrocytes lending further support to the idea of vesicular release of ATP from astrocytes (Sawada et al., 2008).

Recently, action potential-dependent release of adenosine has been demonstrated in the cerebellum and shown to be mediated by a direct release pathway consistent with exocytosis of adenosine (Klyuch et al., 2012). The extracellular 5′ hydrolase, which is necessary for the hydrolysis of AMP to adenosine, was eliminated using $CD73^{-/-}$. Slices were taken from these mice and adenosine levels were measured with adenosine biosensors. Despite the absence of CD73, activity-dependent adenosine accumulation was detected and it was sensitive to bafilomycin A1, an inhibitor of the proton pump that is required to fill vesicles with transmitter. One caveat to this study is that there are multiple enzymes that can hydrolyze AMP to adenosine, thus a $CD73^{-/-}$ alone may be insufficient to conclude that adenosine is released directly. In support of their contention, the authors show that the hydrolysis of exogenous AMP was reduced by 91% in $CD73^{-/-}$ mice suggesting that the measured adenosine is released from adenosine-filled vesicles. The nature of the transporter responsible for filling the vesicles is still to be defined.

Under conditions of elevated metabolic demand, when intracellular adenosine levels rise, adenosine may also be directly released through the equilibrative nucleoside transporters (ENTs). The ENTs facilitate the transport of adenosine along concentration gradients (Parkinson et al., 2011). One condition under which this pathway probably contributes significantly to elevating extracellular adenosine is during hypoxia.

2.3.5.4. *Peptides*

The first evidence that astrocytes release peptides in a regulated manner came from EM studies, which showed the presence of large dense-core vesicles in astrocytes (Calegari et al., 1999). Dense-core vesicles typically contain peptidergic transmitters. Using secretogranin II as a marker, Calegari et al. (1999) showed that astrocytes release dense-core vesicles in response to a variety of stimuli, with the most effective being Ca^{2+}. One of the first peptides identified to be released from astrocytes was atrial natriuretic peptide (ANP), which is involved in vasodilatation. Cultured astrocytes were transfected with a construct that expressed a fluorescent ANP, which was observed as fluorescent puncta inside the astrocytes. Stimulation with ionomycin reduced the number of puncta, indicating Ca^{2+}-dependent release of ANP-containing vesicles, which was further confirmed using FM 4-64 dye (Krzan et al., 2003). Additionally, it has been shown that astrocytes are capable of endocytosing pro-BDNF and then routing it to an SNARE-dependent secretory pathway (Bergami et al., 2008). Since BDNF is de novo synthesized in neurons, this suggests a unique mechanism of neuronal glial communication.

2.4. ASTROCYTIC MODULATION OF SYNAPTIC TRANSMISSION

2.4.1. Overview

Astrocytes have now been shown to release a number of neuroactive substances including glutamate, D-serine, GABA, TNF-alpha, prostaglandins, and peptides, all of which have the potential to influence neuronal activity and synaptic communication (summarized in Figure 1).

2.4.2. Astrocytic Glutamate

The release of glutamate from astrocytes has been shown to cause an NMDAR-dependent elevation in neuronal Ca^{2+} in cortical cultures (Parpura et al., 1994). In mixed hippocampal cultures, mechanical or electrical stimulation of astrocytes to increase their Ca^{2+} led to a glutamate-dependent slow inward current (SIC) in the associated neurons. When BAPTA was used to prevent the increase in astrocytic Ca^{2+}, the glutamate-dependent SIC in the neurons was also prevented (Araque et al., 1998a, b). Similarly, Ca^{2+}-dependent glutamate release from astrocytes increased the frequency of excitatory and inhibitory miniature postsynaptic currents via the activation of extrasynaptic NMDAR (Araque et al., 1998a, b).

In the hippocampus, increases in astrocytic Ca^{2+} transiently increase the probability of transmitter release at the CA3–CA1 synapse. This is a result of Ca^{2+}-induced astrocytic SNARE-dependent release of glutamate acting on presynaptic mGluRs (Perea and Araque, 2007). One mechanism by which this glia glutamate can be released is through the activation of astrocytic CB1 receptors by neuronally released endocannabinoids (Navarrete and Araque, 2010). In the dentate gyrus, astrocytic glutamate acting on presynaptic NR2B-containing NMDARs facilitates excitatory transmission. The release

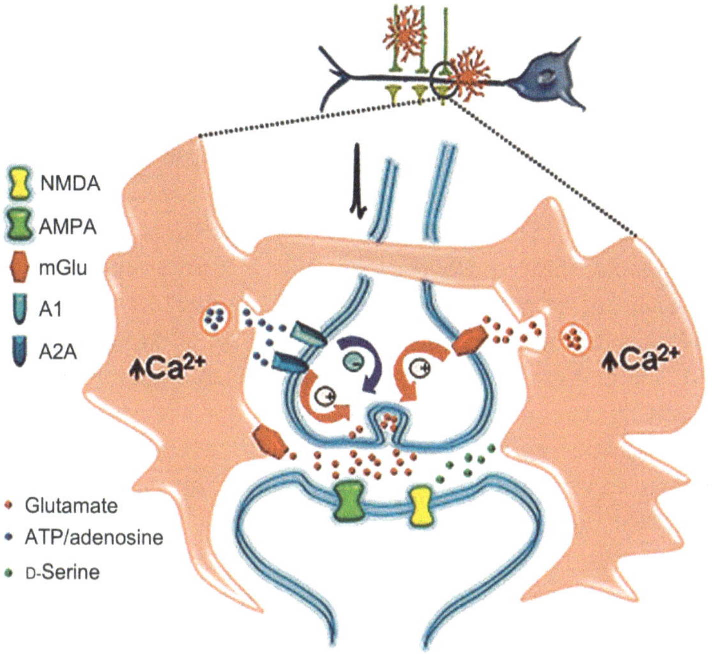

Figure 1 Astrocytes Modulate Synaptic Transmission. An astrocyte calcium wave is triggered by neuronal activity and stimulates the release of gliotransmitters, such as glutamate, ATP/adenosine, or D-serine. These gliotransmitters act on specific glutamate AMPA, NMDAR, and mGluR and adenosine (A1 and A2A) receptor types localized at the pre- or postsynaptic neuron. This then leads to depression or potentiation of synaptic transmission. *Adapted from Zorec et al. (2012).*

of glutamate from dentate molecular layer astrocytes is stimulated by neuronal activity-dependent activation of astrocytic P2Y1Rs (Jourdain et al., 2007). In the CA1 region of the hippocampus, astrocyte-derived glutamate activates extrasynaptic NMDARs leading to synchronous activation of groups of neurons (Fellin et al., 2004).

2.4.3. Astrocytic D-Serine

The most well studied form of synaptic plasticity is long-term potentiation (LTP) or long-term depression (LTD), which are mediated by NMDARs and involve the insertion or removal of AMPA receptors from the postsynaptic membrane (Malenka and Bear, 2004). NMDARs require the binding of a coagonist as well as glutamate. This coagonist was originally thought to be glycine (Johnson and Ascher, 1987). However, recently D-serine has also been shown to be a coagonist for the NMDAR (Mothet et al., 2000) and, as discussed above, is released by astrocytes (Mothet et al., 2005). In the retina, NMDAR currents are augmented by the addition of D-serine, while degradation of D-serine by D-amino acid oxygenase reduces the magnitude of these currents (Stevens et al., 2003). Hippocampal neurons, when cultured without astrocytes, needed D-serine to induce synaptic plasticity. Similarly in mixed astrocyte/neuron cultures and hippocampal slices, the degradation of D-serine blocked the induction of LTP (Yang et al., 2003). In slices from aged rats, there is an impairment of LTP that can be rescued by the addition of D-serine (Yang et al., 2005).

One of the most striking examples of the importance of astrocyte-derived D-serine is in the hypothalamic supraoptic nucleus (SON), which contains oxytocin and vasopressin neurons and highly plastic astrocytes. Under conditions such as lactation, parturition, and chronic dehydration, there is a pronounced reduction in the astrocytic coverage of magnocellular neurons (Theodosis, 2002). This changes the availability of gliotransmitters such as D-serine to these SON neurons. A comparison of virgin and lactating rats revealed a reduced NMDA component of the synaptic current in the SON that correlated with the reduced glial coverage of these synapses and exogenous addition of D-serine enhances the NMDA component in lactating animals compared to virgin rats (Panatier et al., 2006). This suggests that glial D-serine can locally regulate the degree to which synaptic glutamate can activate NMDARs and therefore the plasticity of that synapse. Whether a synapse exhibits LTP or LTD is dictated by the amount of glial coverage at that synapse and the availability of D-serine. LTD occurs when the astrocytic processes retract and there is a reduced level of synaptic D-serine, whereas the same stimulus induces LTP when there is a high degree of glial coverage.

A recent study in hippocampal slices demonstrated that LTP in area CA1 depends on the release of D-serine from astrocytes. When intracellular Ca^{2+} was clamped in individual hippocampal astrocytes, LTP induction was blocked at nearby excitatory synapses and this could be reversed by the application of exogenous D-serine or glycine. Depleting D-serine or disrupting exocytosis in a single astrocyte also blocked local LTP, suggesting that Ca^{2+}-dependent release of D-serine from astrocytes controls NMDAR-dependent plasticity in nearby excitatory synapses (Henneberger et al., 2010).

2.4.4. Astrocytic ATP/Adenosine

In the paraventricular nucleus of the hypothalamus, norepinephrine leads to an increase in the amplitude of miniature excitatory postsynaptic potentials. This is dependent on ATP released from astrocytes, which promotes the insertion of AMPA receptors via the activation of neuronal P2X7Rs (Gordon et al., 2005). This is an example of a direct action of glial-derived ATP on synaptic transmission; however, often the effect of glial-derived ATP is mediated by its metabolites, specifically adenosine, described in detail below. Interestingly, a recent study has raised the possibility that AMP could be a signal to activate the A1R in addition to adenosine. In this study, the authors tested the ability of AMP and nonhydrolyzable analogs of AMP to activate A1R and A2BR. While adenosine activated both receptors, addition of AMP selectively activated A1R. One potential confound with this study is the contamination of AMP with adenosine given that AMP could be hydrolyzed to adenosine which is the natural agonist. Therefore, trace contamination, or hydrolysis could lead to selective A1R activation based on a threshold effect mediated by the differential affinity of the receptors for adenosine. However, an elegant aspect of this study, which provided strong support for AMP being an agonist for the A1R, was the use of a single amino acid substitution that prevented AMP from activating the A1R without altering the EC_{50} for adenosine (Rittiner et al., 2012). Future studies will be needed to elucidate the physiological role for AMP activation of A1R but it does expand the potential role for glial-derived ATP.

As discussed above, glial-derived ATP can be converted extracellularly by ectonucleotidases into adenosine, which has a wide range of effects on synaptic transmission. In the retina, activation of astrocytes and Muller glial cells causes the release of ATP that is converted extracellularly to adenosine. This adenosine leads to the activation of potassium currents that hyperpolarize retinal ganglion neurons (Newman, 2003).

Heterosynaptic depression is a form of synaptic plasticity that occurs in synapses that were not active during the induction protocol (Chistiakova and Volgushev, 2009). Heterosynaptic depression in the CA1 region of the hippocampus is mediated by the activation of NMDARs, which cause release of adenosine that acts presynaptically to inhibit the release of glutamate from excitatory synapses (Manzoni et al., 1994). Further studies have determined that this adenosine is glial derived. Activation of glial $GABA_B$ receptors leads to the release of ATP from astrocytes; this ATP is then converted extracellularly to adenosine that acts on presynaptic A1R to inhibit excitatory transmission (Serrano et al., 2006). In addition, the expression of dnSNARE selectively in astrocytes increased the magnitude of fEPSPs and decreased the magnitude of LTP in hippocampus. Exogenously applied ATP or 2-chloro-*N* cyclopentyladenosine (CCPA) (A1R agonist) could restore the WT phenotype in dnSNARE mice suggesting that this effect is mediated by glial-derived ATP being converted to adenosine.

The extracellular tone of adenosine causes a persistent activation of A1R leading to presynaptic inhibition of excitatory synaptic transmission (Dunwiddie and Masino, 2001). In hippocampal slices isolated from mice in which dnSNARE was expressed selectively in astrocytes, there is an enhanced magnitude of synaptic transmission at the CA3–CA1 synapse that is a result of reduced A1R activation. The addition of A1R antagonists (either 8-cyclopentyl-1,3-dimethylxanthine (CPT) or 8-cyclopentyl-1,3-dipropylxanthine (DPCPX)) enhanced the magnitude of excitatory synaptic transmission but in dnSNARE mice, the relative enhancement was attenuated compared to littermate controls, suggesting that astrocytic expression of dnSNARE reduced the extracellular level of adenosine. Subsequent in vivo studies have shown a similar role for astrocytes in the control of cortical adenosine tone and the use of adenosine biosensors in hippocampal slices has supported the idea of an astrocytic control of adenosine by showing that astrocytic dnSNARE expression reduces the concentration of extracellular adenosine (Schmitt et al., 2012).

Adenosine acting through A2AR regulates dopamine receptor 2 (D2R) internalization and in the ventral striatum, there is a downregulation of D2/D3 receptors after sleep deprivation that may play a role in decreased wakefulness (Volkow et al., 2012). In the spinal cord, stimulation of the A1R has antinociceptive effects on baseline pain (Gong et al., 2010) and dnSNARE mice have a reduced threshold to mechanical nociception (Foley et al., 2011).

Astrocyte-derived adenosine can also regulate NMDARs. In dnSNARE mice, there is a selective decrease in synaptic NMDARs without a corresponding alteration in AMPA receptors. There is also an increased AMPA/NMDA ratio that is due to a reduction in the synaptic NMDA component. Surface biotinylation studies found that the surface expression of the NR2A and NR2B subunits of the NMDA receptor were reduced in dnSNARE mice. In agreement with the biochemistry, pharmacological studies indicated that this effect could not be accounted for solely by a reduction in coagonist availability, but rather there was a change in the distribution and trafficking of NMDARs (Deng et al., 2011).

Glial-derived adenosine also has the ability to modulate circuit function. During non-rapid eye movement (NREM) sleep cortical neurons exhibit slow oscillations (also termed up and down states) at a frequency of <1 Hz (Steriade et al., 1993a, b). These events are relatively synchronized among cortical pyramidal neurons allowing their detection with local field potential measurements (Steriade, 2006). When one compares slow oscillations in wild-type and dnSNARE mice, it is clear that astrocytic SNARE expression reduces the power of the slow oscillation (Fellin et al., 2009). This, together with electroencephalography (EEG) studies discussed beneath, provides the first evidence that astrocytes modulate the function of neural circuits.

Sleep homeostasis can be monitored using several measures. Central among these is the cortical EEG/electromyography (EMG) that is used to measure the amount of time that an animal is in three different vigilance states—wakefulness, NREM sleep, and REM sleep—as well as to measure the power of different frequency bands of the EEG in specific vigilance states. It is well known that in response to sleep deprivation, the sleep homeostat increases the pressure or the drive to sleep. Sleep pressure is measured by increased slow wave activity (SWA), the 0.5–4.0 Hz frequency range of the EEG during NREM sleep. SWA is elevated normally at ZT0 and decreases as the animal is allowed to sleep; it is also enhanced following a period of sleep deprivation. Increased sleep pressure also causes a compensatory increase in sleep time (Sinton and McCarley, 2004; Fuller et al., 2006).

EEG/EMG recordings were used to ask whether sleep homeostasis involves an astrocytic component. Recordings were made for an initial 24-h period (baseline) followed by a subsequent day in which mice were sleep deprived for 6 h between ZT0 and 6 which is equivalent to a sleep deprivation being performed for the first half of the subjective night time. Conditional expression of dnSNARE in astrocytes led to reduced sleep pressure. During the baseline day, the power of SWA during NREM sleep was reduced compared to littermate controls. More striking was that in dnSNARE mice, the increase in sleep pressure in response to sleep deprivation was significantly attenuated compared to littermate controls. In addition to these changes in EEG, dnSNARE expression prevented the compensatory increase in sleep time that normally follows sleep deprivation. Infusion of an A1R antagonist was able to phenocopy the dnSNARE mice (Halassa et al., 2009). The conditional knockout of forebrain A1R causes a similar phenotype to the dnSNARE mice (Bjorness et al., 2009). In addition, manipulation of the expression of adenosine kinase, the enzyme responsible for the phosphorylation of adenosine, and highly expressed in astrocytes, reduces extracellular adenosine and also causes a sleep homeostasis phenotype (Palchykova et al., 2010).

It is well known that sleep can be important for the consolidation of certain forms of memory and that sleep deprivation can prevent the consolidation of long-term memories (Kalia, 2006; Walker and Stickgold, 2006; Goel et al., 2009). The above studies implicating astrocytes in the homeostatic response to sleep deprivation led to the question of whether glial cells contribute to related phenotypes—memory consolidation and late-long-term potentiation (L-LTP). A period of training was performed for either novel object recognition or spatial object recognition tasks (at ZT = 0) and the animals were then tested for recognition memory 24 h later. When mice were left undisturbed following training and allowed to sleep ad libitum, recognition memory was intact 24 h later. However, when mice were subjected to sleep deprivation for 6 h following training, recognition memory was impaired. Expression of dnSNARE in astrocytes dramatically affected this phenotype. When mice were left undisturbed following training and allowed to sleep, recognition memory exhibited 24 h after the training period was equivalent to wild-type animals. In dnSNARE mice 6 h of sleep deprivation did not prevent the consolidation of recognition memory. The administration of the A1R antagonist CPT either in the ventricle or directly into the hippocampus phenocopied dnSNARE expression (Halassa et al., 2009).

Hippocampal L-LTP requires protein synthesis and is considered to be a cellular correlate of memory consolidation. Hippocampal slices were cut from wild-type and dnSNARE mice that either had been left undisturbed or were sleep deprived. Slices obtained from sleep-deprived mice failed to exhibit L-LTP whereas normal L-LTP was observed in undisturbed mice. Similar to the behavioral data, astrocytic dnSNARE expression or CPT application both rescued L-LTP following a period of sleep deprivation (Florian et al., 2011). One possible mechanism by which this could occur is through cAMP. It is known that a postsynaptic elevation of cAMP is required for L-LTP and memory consolidation (Vecsey et al., 2009). Since A1R can couple to G_i, which reduces cAMP, perhaps a glial source of adenosine during sleep deprivation prevents the neuronal elevation of cAMP that is required for memory consolidation.

3. CONCLUSIONS

The advancement of the field of glial biology, especially in the past 20 years has led to a greater appreciation for the role of the glial cell in synaptic transmission. One of the most significant developments has been the discovery of Ca^{2+}-dependent exocytosis from glial cells. Glial cells have the ability to respond to neuronal release of transmitter with Ca^{2+} waves, which cause the release of gliotransmitters that in turn act to influence neuronal activity. Gliotransmitters affect synaptic transmission on a time course of seconds, playing an important role in spike-timing-dependent plasticity on the time course of tens of milliseconds (Min and Nevian, 2012). At the other extreme, gliotransmission plays a critical role in sleep homeostasis and memory consolidation, processes that involve coordination of circuits and occur in a time course of hours (Halassa et al., 2009; Florian et al., 2011; Havekes et al., 2012). The ability of this single cell type to integrate information over a range of timescales that spans more than five orders of magnitude speaks to its diversity of function and suggests that further advancement in the field of glial biology may provide significant insights into brain function overall. Similarly, as we begin to understand the role of glial cells in neurodegenerative disorders, we may be able to identify novel therapeutic targets that are glial specific.

REFERENCES

Anderson, C.M., Nedergaard, M., 2003. Astrocyte-mediated control of cerebral microcirculation. Trends Neurosci. 26, 340–344 author reply 344–345.
Araque, A., Martin, E.D., Perea, G., Arellano, J.I., Buno, W., 2002. Synaptically released acetylcholine evokes Ca2+ elevations in astrocytes in hippocampal slices. J. Neurosci. 22, 2443–2450.
Araque, A., Parpura, V., Sanzgiri, R.P., Haydon, P.G., 1998a. Glutamate-dependent astrocyte modulation of synaptic transmission between cultured hippocampal neurons. Eur. J. Neurosci. 10, 2129–2142.
Araque, A., Sanzgiri, R.P., Parpura, V., Haydon, P.G., 1998b. Calcium elevation in astrocytes causes an NMDA receptor-dependent increase in the frequency of miniature synaptic currents in cultured hippocampal neurons. J. Neurosci. 18, 6822–6829.
Araque, A., Parpura, V., Sanzgiri, R.P., Haydon, P.G., 1999. Tripartite synapses: glia, the unacknowledged partner. Trends Neurosci. 22, 208–215.
Bal-Price, A., Moneer, Z., Brown, G.C., 2002. Nitric oxide induces rapid, calcium-dependent release of vesicular glutamate and ATP from cultured rat astrocytes. Glia 40, 312–323.
Belanger, M., Allaman, I., Magistretti, P.J., 2011. Brain energy metabolism: focus on astrocyte–neuron metabolic cooperation. Cell Metab. 14, 724–738.
Bergami, M., Santi, S., Formaggio, E., Cagnoli, C., Verderio, C., Blum, R., Berninger, B., Matteoli, M., Canossa, M., 2008. Uptake and recycling of pro-BDNF for transmitter-induced secretion by cortical astrocytes. J. Cell Biol. 183, 213–221.
Bjorness, T.E., Kelly, C.L., Gao, T., Poffenberger, V., Greene, R.W., 2009. Control and function of the homeostatic sleep response by adenosine A1 receptors. J. Neurosci. 29, 1267–1276.
Bruni, J.E., 1998. Ependymal development, proliferation, and functions: a review. Microsc. Res. Tech. 41, 2–13.
Bushong, E.A., Martone, M.E., Jones, Y.Z., Ellisman, M.H., 2002. Protoplasmic astrocytes in CA1 stratum radiatum occupy separate anatomical domains. J. Neurosci. 22, 183–192.
Butt, A.M., Kalsi, A., 2006. Inwardly rectifying potassium channels (Kir) in central nervous system glia: a special role for Kir4.1 in glial functions. J. Cell. Mol. Med. 10, 33–44.

Cai, Z., Schools, G.P., Kimelberg, H.K., 2000. Metabotropic glutamate receptors in acutely isolated hippocampal astrocytes: developmental changes of mGluR5 mRNA and functional expression. Glia 29, 70–80.

Calegari, F., Coco, S., Taverna, E., Bassetti, M., Verderio, C., Corradi, N., Matteoli, M., Rosa, P., 1999. A regulated secretory pathway in cultured hippocampal astrocytes. J. Biol. Chem. 274, 22539–22547.

Chistiakova, M., Volgushev, M., 2009. Heterosynaptic plasticity in the neocortex. Exp. Brain Res. 199, 377–390.

Coco, S., Calegari, F., Pravettoni, E., Pozzi, D., Taverna, E., Rosa, P., Matteoli, M., Verderio, C., 2003. Storage and release of ATP from astrocytes in culture. J. Biol. Chem. 278, 1354–1362.

Cornell-Bell, A.H., Finkbeiner, S.M., Cooper, M.S., Smith, S.J., 1990. Glutamate induces calcium waves in cultured astrocytes: long-range glial signaling. Science 247, 470–473.

Cotrina, M.L., Lin, J.H., Alves-Rodrigues, A., Liu, S., Li, J., Azmi-Ghadimi, H., Kang, J., Naus, C.C., Nedergaard, M., 1998. Connexins regulate calcium signaling by controlling ATP release. Proc. Natl. Acad. Sci. USA 95, 15735–15740.

D'Antoni, S., Berretta, A., Bonaccorso, C.M., Bruno, V., Aronica, E., Nicoletti, F., Catania, M.V., 2008. Metabotropic glutamate receptors in glial cells. Neurochem. Res. 33, 2436–2443.

Dani, J.W., Chernjavsky, A., Smith, S.J., 1992. Neuronal activity triggers calcium waves in hippocampal astrocyte networks. Neuron 8, 429–440.

Deng, Q., Terunuma, M., Fellin, T., Moss, S.J., Haydon, P.G., 2011. Astrocytic activation of A1 receptors regulates the surface expression of NMDA receptors through a Src kinase dependent pathway. Glia 59, 1084–1093.

Duan, S., Anderson, C.M., Keung, E.C., Chen, Y., Chen, Y., Swanson, R.A., 2003. P2X7 receptor-mediated release of excitatory amino acids from astrocytes. J. Neurosci. 23, 1320–1328.

Dunwiddie, T.V., Masino, S.A., 2001. The role and regulation of adenosine in the central nervous system. Annu. Rev. Neurosci. 24, 31–55.

Fellin, T., Halassa, M.M., Terunuma, M., Succol, F., Takano, H., Frank, M., Moss, S.J., Haydon, P.G., 2009. Endogenous nonneuronal modulators of synaptic transmission control cortical slow oscillations in vivo. Proc. Natl. Acad. Sci. USA 106, 15037–15042.

Fellin, T., Pascual, O., Gobbo, S., Pozzan, T., Haydon, P.G., Carmignoto, G., 2004. Neuronal synchrony mediated by astrocytic glutamate through activation of extrasynaptic NMDA receptors. Neuron 43, 729–743.

Florian, C., Vecsey, C.G., Halassa, M.M., Haydon, P.G., Abel, T., 2011. Astrocyte-derived adenosine and A1 receptor activity contribute to sleep loss-induced deficits in hippocampal synaptic plasticity and memory in mice. J. Neurosci. 31, 6956–6962.

Foley, J.C., McIver, S.R., Haydon, P.G., 2011. Gliotransmission modulates baseline mechanical nociception. Mol. Pain 7, 93.

Fuller, P.M., Gooley, J.J., Saper, C.B., 2006. Neurobiology of the sleep–wake cycle: sleep architecture, circadian regulation, and regulatory feedback. J. Biol. Rhythms 21, 482–493.

Goel, N., Rao, H., Durmer, J.S., Dinges, D.F., 2009. Neurocognitive consequences of sleep deprivation. Semin. Neurol. 29, 320–339.

Gong, Q.J., Li, Y.Y., Xin, W.J., Wei, X.H., Cui, Y., Wang, J., Liu, Y., Liu, C.C., Li, Y.Y., Liu, X.G., 2010. Differential effects of adenosine A1 receptor on pain-related behavior in normal and nerve-injured rats. Brain Res. 1361, 23–30.

Gordon, G.R., Baimoukhametova, D.V., Hewitt, S.A., Rajapaksha, W.R., Fisher, T.E., Bains, J.S., 2005. Norepinephrine triggers release of glial ATP to increase postsynaptic efficacy. Nat. Neurosci. 8, 1078–1086.

Halassa, M.M., Fellin, T., Takano, H., Dong, J.H., Haydon, P.G., 2007. Synaptic islands defined by the territory of a single astrocyte. J. Neurosci. 27, 6473–6477.

Halassa, M.M., Florian, C., Fellin, T., Munoz, J.R., Lee, S.Y., Abel, T., Haydon, P.G., Frank, M.G., 2009. Astrocytic modulation of sleep homeostasis and cognitive consequences of sleep loss. Neuron 61, 213–219.

Halassa, M.M., Haydon, P.G., 2010. Integrated brain circuits: astrocytic networks modulate neuronal activity and behavior. Annu. Rev. Physiol. 72, 335–355.

Havekes, R., Vecsey, C.G., Abel, T., 2012. The impact of sleep deprivation on neuronal and glial signaling pathways important for memory and synaptic plasticity. Cell Signal 24, 1251–1260.

Haydon, P.G., Carmignoto, G., 2006. Astrocyte control of synaptic transmission and neurovascular coupling. Physiol. Rev. 86, 1009–1031.

Henneberger, C., Papouin, T., Oliet, S.H., Rusakov, D.A., 2010. Long-term potentiation depends on release of D-serine from astrocytes. Nature 463, 232–236.
Hertz, L., Zielke, H.R., 2004. Astrocytic control of glutamatergic activity: astrocytes as stars of the show. Trends Neurosci. 27, 735.
Honsek, S.D., Walz, C., Kafitz, K.W., Rose, C.R., 2012. Astrocyte calcium signals at Schaffer collateral to CA1 pyramidal cell synapses correlate with the number of activated synapses but not with synaptic strength. Hippocampus 22, 29–42.
Iglesias, R., Dahl, G., Qiu, F., Spray, D.C., Scemes, E., 2009. Pannexin 1: the molecular substrate of astrocyte "hemichannels". J. Neurosci. 29, 7092–7097.
Jeftinija, S.D., Jeftinija, K.V., Stefanovic, G., 1997. Cultured astrocytes express proteins involved in vesicular glutamate release. Brain Res. 750, 41–47.
Johnson, J.W., Ascher, P., 1987. Glycine potentiates the NMDA response in cultured mouse brain neurons. Nature 325, 529–531.
Jourdain, P., Bergersen, L.H., Bhaukaurally, K., Bezzi, P., Santello, M., Domercq, M., Matute, C., Tonello, F., Gundersen, V., Volterra, A., 2007. Glutamate exocytosis from astrocytes controls synaptic strength. Nat. Neurosci. 10, 331–339.
Kalia, M., 2006. Neurobiology of sleep. Metabolism 55, S2–S6.
Kang, J., Jiang, L., Goldman, S.A., Nedergaard, M., 1998. Astrocyte-mediated potentiation of inhibitory synaptic transmission. Nat. Neurosci. 1, 683–692.
Kettenmann, H., Ransom, B.R. (Eds.), 2005. Neuroglia. Oxford University Press, New York.
Klyuch, B.P., Dale, N., Wall, M.J., 2012. Deletion of ecto-5′-nucleotidase (CD73) reveals direct action potential-dependent adenosine release. J. Neurosci. 32, 3842–3847.
Kreutzberg, G.W., 1996. Microglia: a sensor for pathological events in the CNS. Trends Neurosci. 19, 312–318.
Krzan, M., Stenovec, M., Kreft, M., Pangrsic, T., Grilc, S., Haydon, P.G., Zorec, R., 2003. Calcium-dependent exocytosis of atrial natriuretic peptide from astrocytes. J. Neurosci. 23, 1580–1583.
Malenka, R.C., Bear, M.F., 2004. LTP and LTD: an embarrassment of riches. Neuron 44, 5–21.
Manzoni, O.J., Manabe, T., Nicoll, R.A., 1994. Release of adenosine by activation of NMDA receptors in the hippocampus. Science 265, 2098–2101.
Maragakis, N.J., Rothstein, J.D., 2004. Glutamate transporters: animal models to neurologic disease. Neurobiol. Dis. 15, 461–473.
Min, R., Nevian, T., 2012. Astrocyte signaling controls spike timing-dependent depression at neocortical synapses. Nat. Neurosci. 15, 746–753.
Montana, V., Malarkey, E.B., Verderio, C., Matteoli, M., Parpura, V., 2006. Vesicular transmitter release from astrocytes. Glia 54, 700–715.
Montana, V., Ni, Y., Sunjara, V., Hua, X., Parpura, V., 2004. Vesicular glutamate transporter-dependent glutamate release from astrocytes. J. Neurosci. 24, 2633–2642.
Mothet, J.P., Parent, A.T., Wolosker, H., Brady Jr., R.O., Linden, D.J., Ferris, C.D., Rogawski, M.A., Snyder, S.H., 2000. D-serine is an endogenous ligand for the glycine site of the N-methyl-D-aspartate receptor. Proc. Natl. Acad. Sci. USA 97, 4926–4931.
Mothet, J.P., Pollegioni, L., Ouanounou, G., Martineau, M., Fossier, P., Baux, G., 2005. Glutamate receptor activation triggers a calcium-dependent and SNARE protein-dependent release of the gliotransmitter D-serine. Proc. Natl. Acad. Sci. USA 102, 5606–5611.
Navarrete, M., Araque, A., 2010. Endocannabinoids potentiate synaptic transmission through stimulation of astrocytes. Neuron 68, 113–126.
Navarrete, M., Perea, G., Maglio, L., Pastor, J., Garcia de Sola, R., Araque, A., 2012. Astrocyte calcium signal and gliotransmission in human brain tissue. Cereb. Cortex.
Nedergaard, M., 1994. Direct signaling from astrocytes to neurons in cultures of mammalian brain cells. Science 263, 1768–1771.
Newman, E.A., 2003. Glial cell inhibition of neurons by release of ATP. J. Neurosci. 23, 1659–1666.
Palchykova, S., Winsky-Sommerer, R., Shen, H.Y., Boison, D., Gerling, A., Tobler, I., 2010. Manipulation of adenosine kinase affects sleep regulation in mice. J. Neurosci. 30, 13157–13165.
Panatier, A., Theodosis, D.T., Mothet, J.P., Touquet, B., Pollegioni, L., Poulain, D.A., Oliet, S.H., 2006. Glia-derived D-serine controls NMDA receptor activity and synaptic memory. Cell 125, 775–784.

Parkinson, F.E., Damaraju, V.L., Graham, K., Yao, S.Y., Baldwin, S.A., Cass, C.E., Young, J.D., 2011. Molecular biology of nucleoside transporters and their distributions and functions in the brain. Curr. Top. Med. Chem. 11, 948–972.

Parpura, V., Basarsky, T.A., Liu, F., Jeftinija, K., Jeftinija, S., Haydon, P.G., 1994. Glutamate-mediated astrocyte–neuron signalling. Nature 369, 744–747.

Parpura, V., Heneka, M.T., Montana, V., Oliet, S.H., Schousboe, A., Haydon, P.G., Stout Jr., R.F., Spray, D.C., Reichenbach, A., Pannicke, T., Pekny, M., Pekna, M., Zorec, R., Verkhratsky, A., 2012. Glial cells in (patho)physiology. J. Neurochem. 121, 4–27.

Parpura, V., Liu, F., Brethorst, S., Jeftinija, K., Jeftinija, S., Haydon, P.G., 1995. Alpha-latrotoxin stimulates glutamate release from cortical astrocytes in cell culture. FEBS Lett. 360, 266–270.

Parpura, V., Zorec, R., 2010. Gliotransmission: exocytotic release from astrocytes. Brain Res. Rev. 63, 83–92.

Parri, H.R., Crunelli, V., 2001. Pacemaker calcium oscillations in thalamic astrocytes in situ. Neuroreport 12, 3897–3900.

Parri, H.R., Crunelli, V., 2003. The role of Ca2+ in the generation of spontaneous astrocytic Ca2+ oscillations. Neuroscience 120, 979–992.

Pasantes Morales, H., Schousboe, A., 1988. Volume regulation in astrocytes: a role for taurine as an osmoeffector. J. Neurosci. Res. 20, 503–509.

Pasti, L., Volterra, A., Pozzan, T., Carmignoto, G., 1997. Intracellular calcium oscillations in astrocytes: a highly plastic, bidirectional form of communication between neurons and astrocytes in situ. J. Neurosci. 17, 7817–7830.

Perea, G., Araque, A., 2007. Astrocytes potentiate transmitter release at single hippocampal synapses. Science 317, 1083–1086.

Porter, J.T., McCarthy, K.D., 1995. Adenosine receptors modulate [Ca2+]i in hippocampal astrocytes in situ. J. Neurochem. 65, 1515–1523.

Ransom, B.R., Ransom, C.B., 2012. Astrocytes: multitalented stars of the central nervous system. Methods Mol. Biol. 814, 3–7.

Rittiner, J.E., Korboukh, I., Hull-Ryde, E.A., Jin, J., Janzen, W.P., Frye, S.V., Zylka, M.J., 2012. AMP is an adenosine A1 receptor agonist. J. Biol. Chem. 287, 5301–5309.

Rosenberg, P.A., Knowles, R., Knowles, K.P., Li, Y., 1994. Beta-adrenergic receptor-mediated regulation of extracellular adenosine in cerebral cortex in culture. J. Neurosci. 14, 2953–2965.

Roy, C.S., Sherrington, C.S., 1890. On the regulation of the blood-supply of the brain. J. Physiol. 11, 85–158 117.

Santello, M., Cali, C., Bezzi, P., 2012. Gliotransmission and the tripartite synapse. Adv. Exp. Med. Biol. 970, 307–331.

Sawada, K., Echigo, N., Juge, N., Miyaji, T., Otsuka, M., Omote, H., Yamamoto, A., Moriyama, Y., 2008. Identification of a vesicular nucleotide transporter. Proc. Natl. Acad. Sci. USA 105, 5683–5686.

Schell, M.J., Molliver, M.E., Snyder, S.H., 1995. D-serine, an endogenous synaptic modulator: localization to astrocytes and glutamate-stimulated release. Proc. Natl. Acad. Sci. USA 92, 3948–3952.

Schmitt, L.I., Sims, R.E., Dale, N., Haydon, P.G., 2012. Wakefulness affects synaptic and network activity by increasing extracellular astrocyte-derived adenosine. J. Neurosci. 32, 4417–4425.

Serrano, A., Haddjeri, N., Lacaille, J.C., Robitaille, R., 2006. GABAergic network activation of glial cells underlies hippocampal heterosynaptic depression. J. Neurosci. 26, 5370–5382.

Seth, P., Koul, N., 2008. Astrocyte, the star avatar: redefined. J. Biosci. 33, 405–421.

Shigeri, Y., Seal, R.P., Shimamoto, K., 2004. Molecular pharmacology of glutamate transporters, EAATs and VGLUTs. Brain Res. Brain Res. Rev. 45, 250–265.

Simard, M., Nedergaard, M., 2004. The neurobiology of glia in the context of water and ion homeostasis. Neuroscience 129, 877–896.

Sinton, C.M., McCarley, R.W., 2004. Neurophysiological mechanisms of sleep and wakefulness: a question of balance. Semin. Neurol. 24, 211–223.

Steriade, M., 2006. Grouping of brain rhythms in corticothalamic systems. Neuroscience 137, 1087–1106.

Steriade, M., Contreras, D., Curro Dossi, R., Nunez, A., 1993a. The slow (<1 Hz) oscillation in reticular thalamic and thalamocortical neurons: scenario of sleep rhythm generation in interacting thalamic and neocortical networks. J. Neurosci. 13, 3284–3299.

Steriade, M., Nunez, A., Amzica, F., 1993b. Intracellular analysis of relations between the slow (<1 Hz) neocortical oscillation and other sleep rhythms of the electroencephalogram. J. Neurosci. 13, 3266–3283.

Stevens, E.R., Esguerra, M., Kim, P.M., Newman, E.A., Snyder, S.H., Zahs, K.R., Miller, R.F., 2003. D-serine and serine racemase are present in the vertebrate retina and contribute to the physiological activation of NMDA receptors. Proc. Natl. Acad. Sci. USA 100, 6789–6794.

Szatkowski, M., Barbour, B., Attwell, D., 1990. Non-vesicular release of glutamate from glial cells by reversed electrogenic glutamate uptake. Nature 348, 443–446.

Theodosis, D.T., 2002. Oxytocin-secreting neurons: a physiological model of morphological neuronal and glial plasticity in the adult hypothalamus. Front. Neuroendocrinol. 23, 101–135.

Tzingounis, A.V., Wadiche, J.I., 2007. Glutamate transporters: confining runaway excitation by shaping synaptic transmission. Nat. Rev. Neurosci. 8, 935–947.

Vecsey, C.G., Baillie, G.S., Jaganath, D., Havekes, R., Daniels, A., Wimmer, M., Huang, T., Brown, K.M., Li, X.Y., Descalzi, G., Kim, S.S., Chen, T., Shang, Y.Z., Zhuo, M., Houslay, M.D., Abel, T., 2009. Sleep deprivation impairs cAMP signalling in the hippocampus. Nature 461, 1122–1125.

Verkhratsky, A., Parpura, V., Rodriguez, J.J., 2011. Where the thoughts dwell: the physiology of neuronal-glial "diffuse neural net". Brain Res. Rev. 66, 133–151.

Volkow, N.D., Tomasi, D., Wang, G.J., Telang, F., Fowler, J.S., Logan, J., Benveniste, H., Kim, R., Thanos, P.K., Ferre, S., 2012. Evidence that sleep deprivation downregulates dopamine D2R in ventral striatum in the human brain. J. Neurosci. 32, 6711–6717.

Walker, M.P., Stickgold, R., 2006. Sleep, memory, and plasticity. Annu. Rev. Psychol. 57, 139–166.

Wang, Z., Haydon, P.G., Yeung, E.S., 2000. Direct observation of calcium-independent intercellular ATP signaling in astrocytes. Anal. Chem. 72, 2001–2007.

Warr, O., Takahashi, M., Attwell, D., 1999. Modulation of extracellular glutamate concentration in rat brain slices by cystine–glutamate exchange. J. Physiol. 514 (Pt 3), 783–793.

Wolosker, H., Blackshaw, S., Snyder, S.H., 1999. Serine racemase: a glial enzyme synthesizing D-serine to regulate glutamate-N-methyl-D-aspartate neurotransmission. Proc. Natl. Acad. Sci. USA 96, 13409–13414.

Yang, S., Qiao, H., Wen, L., Zhou, W., Zhang, Y., 2005. D-serine enhances impaired long-term potentiation in CA1 subfield of hippocampal slices from aged senescence-accelerated mouse prone/8. Neurosci. Lett. 379, 7–12.

Yang, Y., Ge, W., Chen, Y., Zhang, Z., Shen, W., Wu, C., Poo, M., Duan, S., 2003. Contribution of astrocytes to hippocampal long-term potentiation through release of D-serine. Proc. Natl. Acad. Sci. USA 100, 15194–15199.

Zhang, Q., Fukuda, M., Van Bockstaele, E., Pascual, O., Haydon, P.G., 2004a. Synaptotagmin IV regulates glial glutamate release. Proc. Natl. Acad. Sci. USA 101, 9441–9446.

Zhang, Q., Pangrsic, T., Kreft, M., Krzan, M., Li, N., Sul, J.Y., Halassa, M., Van Bockstaele, E., Zorec, R., Haydon, P.G., 2004b. Fusion-related release of glutamate from astrocytes. J. Biol. Chem. 279, 12724–12733.

Zonta, M., Angulo, M.C., Gobbo, S., Rosengarten, B., Hossmann, K.A., Pozzan, T., Carmignoto, G., 2003. Neuron-to-astrocyte signaling is central to the dynamic control of brain microcirculation. Nat. Neurosci. 6, 43–50.

Zorec, R., Araque, A., Carmignoto, G., Haydon, P.G., Verkhratsky, A., Parpura, V., 2012. Astroglial excitability and gliotransmission: an appraisal of Ca2+ as a signalling route. ASN Neuro 4.

CHAPTER SIX

Local Protein Synthesis at Synapses

Oswald Steward[1,2,3], Joseph Dynes[4] and Shannon Farris[1,2,3]
[1]Reeve-Irvine Research Center, University of California Irvine, Irvine, CA, USA
[2]Center for the Neurobiology of Learning and Memory, University of California Irvine, Irvine, CA, USA
[3]Departments of Anatomy and Neurobiology, Neurobiology and Behavior, and Neurosurgery, University of California Irvine, Irvine, CA, USA
[4]Department of Physiology, University of California Irvine, Irvine, CA, USA

1. INTRODUCTION

It has been 30 years since our discovery of the selective positioning of polyribosomes at spine synapses (Steward and Levy, 1982). In that article, we described what we now call "synapse-associated polyribosome complexes" (SPRCs) and proposed that they allowed local synthesis of key molecules of the synaptic junction, that local synthesis is regulated by synaptic activity, and that local synthesis of key proteins is important for synaptic plasticity. This discovery and accompanying ideas triggered a new line of thinking about how individual synapses could be modified in a protein synthesis-dependent fashion, specifically that synaptic activation could modulate the translation of mRNAs at individual synapses, and the newly synthesized proteins could then act locally to alter synaptic properties. Subsequent studies have confirmed these basic ideas, although the story has become much richer in detail. Findings from many labs have revealed that local protein synthesis plays an important role in synapse construction and plasticity. Although the core ideas about the importance of local protein synthesis in synaptic function are accepted, concepts continue to evolve.

This chapter presents a snapshot of current ideas about protein synthesis at synapses and mechanisms underlying dendritic transport and targeting of mRNA to synapses. Our studies on transport and targeting of mRNA were made possible because of the discovery of a unique immediate early gene (IEG) called activity-regulated cytoskeleton protein (Arc) (Lyford et al., 1995), also known as Arg3.1 (Link et al., 1995). Arc is unique among known IEGs because Arc mRNA is rapidly transported into dendrites and localizes selectively near active synapses (see below). In addition to revealing mechanisms of mRNA trafficking to synapses, studies of Arc have provided important insights into the role of local protein synthesis in synapse function, particularly in the forms of synaptic plasticity that underlie memory storage.

The Synapse
http://dx.doi.org/10.1016/B978-0-12-418675-0.00006-7

2.1. SPRCS: ULTRASTRUCTURE OF THE MACHINERY THAT UNDERLIES PROTEIN SYNTHESIS AT SYNAPSES

2.1.1. Overview

Figure 1 illustrates an example of SPRCs at a spine synapse. Our description of SPRCs came at a time when the accepted view was that protein synthesis occurred only in the cell body of neurons. Thus, our proposal for local protein synthesis at synapses was at first controversial. This was despite the fact that ribosomes were known to be present in dendrites, especially proximal dendrites (Peters et al., 1976). The localization of ribosomes at synapses had also been noted (Bodian, 1965; Palacios-Pru et al., 1981). However, the selectivity of the localization of polyribosomes had not been documented—viz., that in mature neurons, the majority of dendritic polyribosomes were selectively localized beneath postsynaptic sites (see below). It was the quantitative documentation of this selective positioning that inspired our hypotheses about the role of local protein synthesis in synaptic function.

Our initial quantitative analysis of SPRCs was carried out on spine synapses on granule cells in the hippocampal dentate gyrus (Steward and Levy, 1982). Later studies documented SPRCs in the subsynaptic cytoplasm of different classes of synapses and in different types of neurons (Steward, 1983). In addition to being localized at spine synapses, polyribosomes are positioned beneath the postsynaptic membrane specialization

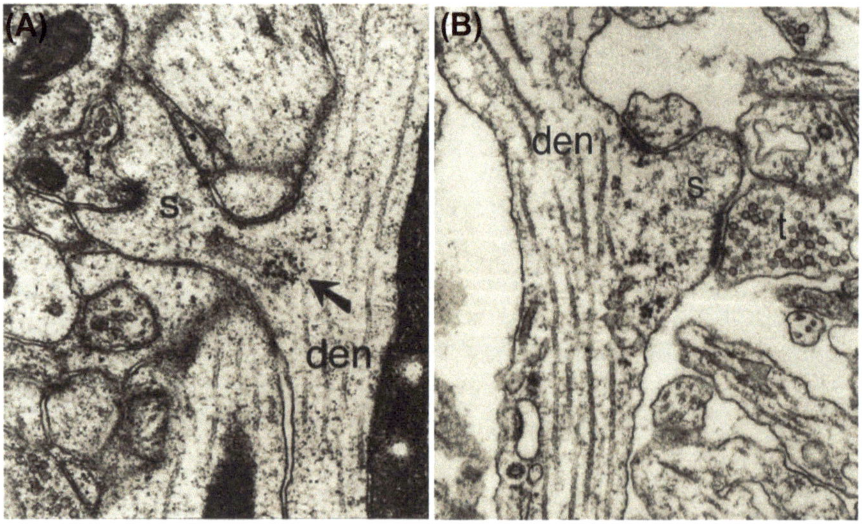

Figure 1 Synapse-associated polyribosome complexes (SPRCs) at spine synapses on the dendrites of CNS neurons. (A) Electron micrograph illustrating clusters of ribosomes (polyribosomes) selectively positioned at the base of a spine on a granule cell in the hippocampal dentate gyrus. (B) Electron micrograph illustrating polyribosomes beneath and within a developing spine. *From Steward and Falk (1986).*

of nonspine synapses including inhibitory synaptic connections on axon initial segments (Steward and Ribak, 1986). Thus, SPRCs are ubiquitous components of the subsynaptic cytoplasm.

Polyribosomes are not seen at every synapse in the mature brain, and even when present, are not seen in every section through the synapse. In fact, it is relatively rare to capture the profile of a spine perfectly in a single electron micrograph such as the one illustrated in Figure 1(A). This may be one reason that the selective localization of polyribosomes at synapses had not been noted prior to our study. Our initial study of serial section reconstructions of dendrites of dentate granule cells in adult rats revealed that about 25% of the spines on mid proximodistal dendrites have underlying polyribosomes (Steward and Levy, 1982). Incidence is higher on proximal dendrites. Other studies involving serial section analyses of pyramidal neurons in the cerebral cortex revealed that 82% of the reconstructed spines had ribosomes in the head, 42% had ribosomes in the neck, and 62% had ribosomes at the base (Spacek and Hartmann, 1983). In cerebellar Purkinje cells, 13% of the spines had ribosomes in the head and 22% had ribosomes at the base. The differences in estimates of incidence may be due in part to the criteria for counting. Spacek and Hartmann counted both singlet ribosomes and polyribosomes, whereas only polyribosomes with at least three singlet ribosomes were counted in Steward and Levy (1982). In any case, it is clear that at any given point in time, some mature synapses lack polyribosomes.

The fact that polyribosomes are present at some synapses but not others raises the question of whether SPRCs are more or less permanently stationed at particular synapses or whether they shuttle from synapse to synapse. If polyribosomes are permanently stationed, this would imply that local protein synthesis can occur at some synapses but not others. If polyribosomes can move from one synapse for another, then all synapses may have the same capacity for local protein synthesis. It will be of interest to use live-cell imaging techniques to test whether ribosomes can move from one synapse to another.

2.1.2. Selectivity of Polyribosome Localization beneath Synapses

At spine synapses, SPRCs are most often localized at the base of the spine (Figure 1). Selective localization was demonstrated through quantitative analyses of polyribosome distribution within dendrites of granule cells in the dentate gyrus (Steward and Levy, 1982). This study demonstrated that about 10% of the polyribosomes in dendrites were localized beneath or within "identified spine bases" (that is, profiles where neck of the spine could be seen to emerge from the dendrite); about 71% of the polyribosomes in dendrites were localized in "mounds", which serial section reconstruction revealed to be the base of spines that extend out of the plane of the single section. Thus, a reasonable estimate is that about 80% of the polyribosome clusters in dendrites are localized beneath or within spines.

SPRCs are also seen within the spine neck and head, and there is evidence that synaptic activation can cause ribosomes to move from the dendritic shaft into spines. For example, electron microscopic studies following tetanic stimulation of the Schaffer collateral system in hippocampal slices from 15-day-old rats revealed that an increased percentage of spines contained polyribosomes with a corresponding decrease in polyribosomes in dendritic shafts (Ostroff et al., 2002). The likely explanation is activity-dependent translocation of polyribosomes from the shaft into the spine proper. On the other hand, Desmond and Levy (1990) report a decrease in the number of polyribosomes in spines following induction of LTP in the perforant path. It is noteworthy that this occurs at the same time that spine heads undergo striking changes in shape, adopting a highly characteristic chalice-like morphology in which the spine head wraps around the presynaptic terminal (Desmond and Levy, 1983). The interesting observations regarding the disappearance of ribosomes with LTP have never been followed up. One speculation is that the ribosomes become closely associated with the postsynaptic density, which would make them invisible by electron microscopy.

SPRCs located within or near the portal between the spine neck and the shaft of the dendrite are ideally positioned to be influenced by electrical and/or chemical signals from the synapse as well as by events within the dendrite proper. As discussed further below, our recent studies demonstrate that one of the mRNAs that is delivered into dendrites (the mRNA for Arc) also localizes with a high degree of precision at the base of dendritic spines. This implies that the spine base is a specialized cytoplasmic micro-compartment where particular mRNAs and protein synthetic machinery dock.

Although most dendritic polyribosomes are localized beneath synapses, a few clusters of ribosomes are localized within the core of the dendritic shaft. Polyribosomes in the dendritic core may be associated with mRNAs that encode proteins that are not destined for synaptic sites that play some other role in dendritic function. This speculation is of particular interest given the diversity of the mRNAs that have been identified in dendrites (see below). Alternatively, clusters of polyribosomes in the core of the dendrite may be in transit from the cell body.

2.1.3. Association between SPRCs and Other Organelles

SPRCs are often associated with membranous organelles. The association can often be seen in single sections (Figure 1), but is more evident when synapses are evaluated using serial sections or with stereoviewing of relatively thick sections (Steward and Reeves, 1988). For example, serial section reconstructions of mid proximodistal dendrites of dentate granule cells and hippocampal pyramidal cells have revealed that about 50% of the polyribosomes are near tubular cisterns. A common configuration is one in which the ribosomes seem to surround a blind end of a cistern. The relationship between the ribosomes and the cisternal membrane is similar to what is seen in rough endoplasmic reticulum (RER), except that the dendritic cisterns are tubular rather than flat like the

RER cell bodies. One obvious hypothesis suggested by these observations is that the SPRC/cisternal complex is a form of RER, which is a potential site for the synthesis of integral membrane proteins or soluble proteins destined for release.

Interestingly, the cisterns with which SPRCs are associated are sometimes connected with a spine apparatus (Steward and Reeves, 1988). Sometimes this connection is prominent; in other cases, the connection is very tenuous. This association invites the speculation that the spine apparatus is involved in some aspects of posttranslational processing of proteins that are synthesized by the SPRCs (more on this below).

2.1.4. SPRCs at Nonspine Synapses

Although there have been no detailed quantitative evaluations of polyribosome distribution in the dendrites of nonspiny neurons, it is clear that the same basic relationships exist as in spiny dendrites. Indeed, the earliest observation of ribosomes at synapses was at nonspine synapses on motoneurons in the spinal cord (Bodian, 1965). Also polyribosomes are present beneath both the asymmetric and the symmetric synapses that are present on dendrites of cortical neurons (Steward et al., 1996). Polyribosomes are also localized beneath GABAergic synapses on axon initial segments (Steward and Ribak, 1986).

2.1.5. Prominence of SPRCs at Developing Synapses

Polyribosomes are especially abundant at developing synapses (Steward and Falk, 1985, 1986, 1991; Figure 1(B)). This suggests that local protein synthesis is especially important as synapses are being constructed. It is not known whether different types of proteins are synthesized at developing vs mature synapses; this question is of considerable importance because many studies seeking to identify mRNAs in dendrites have used young neurons growing in culture (see below).

2.2. PROTEIN SYNTHESIS IN DENDRITES

The demonstration of ribosomes at synapses raised the question of whether there was in fact ongoing protein synthesis. Initially, this was explored using subcellular fractions containing pinched-off dendrites called either synaptodendrosomes or synaptoneurosomes depending on the method of isolation. Synaptodendrosomes are isolated by differential centrifugation in sucrose gradients (Rao and Steward, 1991), whereas synaptoneurosomes are prepared by filtration (Weiler et al., 1991; Weiler and Greenough, 1991). Both types of fractions are enriched in pinched-off nerve terminals that are connected to small pinched-off dendritic fragments and/or spines, which appear as snowman figures at the electron microscopic level. An important limitation is that the fractions can be contaminated by resealed fragments of neuronal and glial cell bodies (Rao and Steward, 1993). Nevertheless, when care is taken to ensure minimal contamination by other elements (Bagni et al., 2000), the preparations can provide information about the mRNAs that are present in dendrites (Rao and Steward, 1993) and the proteins that are locally synthesized (Rao and Steward, 1991).

Studies involving synaptodendrosomes and synaptoneurosomes have also been useful for assessing how dendritic protein synthesis is influenced by depolarization (high potassium for example) and bath application of neurotransmitters and growth factors (Weiler et al., 1991; Weiler and Greenough, 1991, 1993). Measures include biochemical studies of metabolically labeled proteins (Leski and Steward, 1995, 1996; Villanueva and Steward, 2001a,b) and the degree of ribosome loading onto particular mRNAs (Zalfa et al., 2003).

Other studies demonstrated local protein synthesis by dendrites that had been mechanically separated from their cell bodies. One approach was to grow neurons on a sandwich made up of a millipore filter attached to a glass cover slip with Matrigel (Torre and Steward, 1992). Neurons growing on the filter extend their axons and dendrites into the Matrigel. The filter with cell bodies can then be pulled off, leaving the separated dendrites in the Matrigel. Dendrites isolated in this way exhibited local incorporation of radiolabeled amino acids into protein.

More recent studies used unique fluorescent reporters to demonstrate both the sites and dynamics of local protein synthesis in the dendrites of living neurons in culture (Aakalu et al., 2001). These experiments convincingly settle the question of whether dendrites are capable of local protein synthesis. Experiments involving mechanical separation of dendrites from cell bodies in hippocampal slices also provided strong evidence that dendritic protein synthesis is critical for changes in synaptic efficacy (Aakalu et al., 2001; Huber et al., 2000; Kang and Schuman, 1996) and for a review, see Steward and Schuman (2001).

2.3. POSTTRANSLATIONAL PROCESSING WITHIN DENDRITES

Evidence that components of the endoplasmic reticulum (ER) and Golgi complex are present in dendrites has come from studies that assessed the distribution of protein markers and enzyme activities characteristic of the RER and Golgi apparatus (GA). One approach has been to pulse-label neurons with the sugar precursors which are the substrates of various glycosyl transferases that are present in the RER and GA and assess sites of incorporation using autoradiography (Torre and Steward, 1996). This study revealed both mannosyl transferase as well as higher order (Golgi-like) glycosyltransferase activity in the dendrites of hippocampal neurons grown in culture. Immunocytochemistry further revealed that proteins that are considered markers of the endomembrane system are present in dendrites, including ribophorin I, which is a marker for the RER (Torre and Steward, 1996).

Subsequent studies used electron microscopic immunocytochemistry to assess how the different processing enzymes map onto membranous organelles at the spine. The membranous cisterns near spine synapses stain for Sec61alpha protein complex, which is part of the machinery for translocation of proteins through the RER during their synthesis (Pierce et al., 2000). The cisterns exhibiting labeling have the same appearance as the cisterns that represent the membranous component of SPRCs.

Immunostaining for the ribosomal protein S3 revealed labeling over the same membranous cisterns that were labeled for Sec61alpha. A follow-up study (Pierce et al., 2001) further mapped the distribution of key enzymes, showing that immunostaining for TGN38 and ERGIC53/58 was present in the dendritic shaft, membranous organelles at the base of spines stained for Rab6, spine apparatuses stained for mannosidase II and Giantin, and immunostaining for RAB1b was present in spine heads (Figure 2).

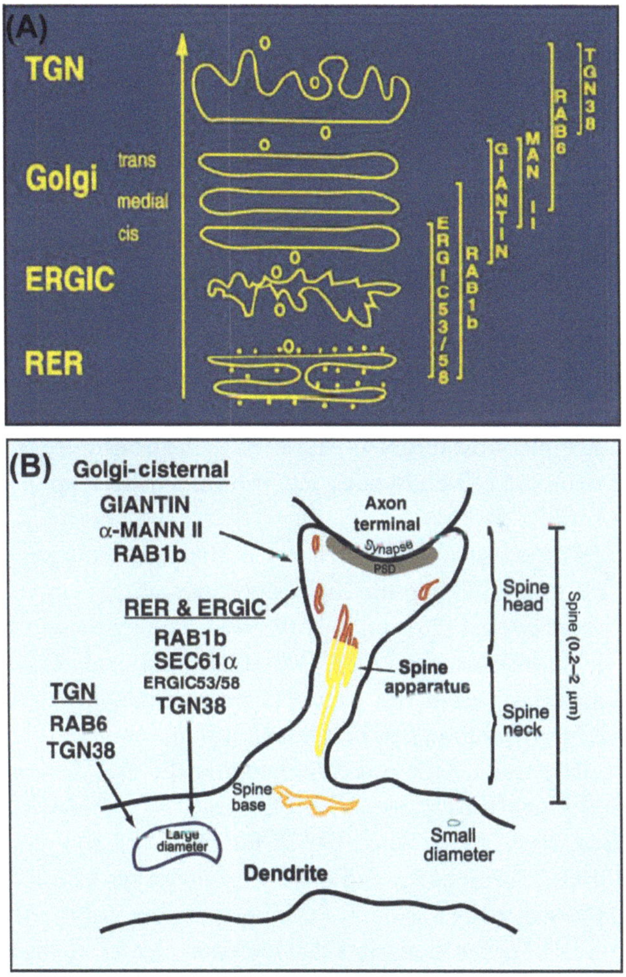

Figure 2 Schematic illustration of the distribution of elements of the endoplasmic reticulum (ER) and golgi apparatus at spine synapses. (A) Distribution of different proteins involved in posttranslational processing of proteins. RER, rough endoplasmic reticulum; ERGIC, endoplasmic reticulum to Golgi intermediate compartment; TGN, trans-Golgi network. (B) Schematic of the distribution of protein biosynthetic compartments in dendrites based on immunostaining. Early secretory pathway components (RER and ERGIC) are localized throughout spines and within dendrites. Middle compartments (Golgi cisternae) are primarily localized in spine heads. Late Golgi and trans-Golgi network (TGN) are predominantly found in dendrites. *From Pierce et al. (2001).*

These data indicate that all the machinery that is necessary to glycosylate newly synthesized protein is present at spines.

An interesting twist is that live-cell imaging studies reveal that Golgi vesicles in dendrites are mobile (Horton and Ehlers, 2003). Live-cell imaging using a green fluorescent protein (GFP)-tagged temperature-sensitive vesicular stomatitis virus glycoprotein (VSVG), and ER and Golgi apparatus markers, documented the presence of functional ER, post-ER carriers, Golgi Apparatus, and post-Golgi apparatus trafficking in dendrites. VSVG–GFP moves from ER elements in dendrites into tuberovesicular structures that move bidirectionally in dendrites, fusing with stationary structures that stained for Golgi apparatus markers (GM130). The accumulation of VSVG–GFP into tuberovesicular structures occurred at defined immobile foci that were positive for Sec13, a marker of ER exit sites in other cells. These data suggest the interesting possibility that elements of the Golgi can shuttle from spine to spine, sometimes moving tens of micrometers.

2.4. DENDRITIC mRNAs

The discovery of polyribosomes beneath synapses naturally raised the question of what mRNAs were present in dendrites. The development of cloning techniques and in situ hybridization in the late 1980s made it possible to address this question directly. Many early studies revealed that most mRNAs expressed by neurons were largely restricted to the region of the neuronal cell body, but a few mRNAs turned up that were present in dendrites at high levels.

The first two mRNAs that were found to be localized in dendrites by in situ hybridization were the mRNA encoding the high-molecular-weight microtubule-associated protein MAP2 (Garner et al., 1988) and the mRNA encoding the a-subunit of calcium/calmodulin-dependent protein kinase II (CaMKII) (Burgin et al., 1990). These mRNAs are present at high levels within the dendrites of forebrain neurons. Subsequently, a different set of mRNAs was found to be present at high levels in the dendrites of cerebellar Purkinje cells including the mRNA encoding the type I inositol trisphosphate receptor (InsP3 receptor) (Furuichi et al., 1993), and the mRNAs for proteins called L7 and PEP19 (Bian et al., 1996). The mRNA for the InsP3 receptor is present at the highest level in Purkinje cells, but is also present in dendrites of forebrain neurons. L7 and PEP19 are expressed at high levels in Purkinje cells, but not in most other neurons.

Later studies revealed other transcripts that are easily detectable by in situ hybridization in dendrites of neurons in vivo including the postsynaptic scaffold protein shank (Bockers et al., 2004; Kindler et al., 2004), dendrin (Herb et al., 1997), and elongation factor 1 alpha (Huang et al., 2005). When induced by activity, Arc mRNA is present at high levels in dendrites (more on this below), and BDNF mRNA appears in dendrites following seizures that lead to epilepsy (Tongiorgi et al., 2004). In addition to the mRNAs that are present in dendrites at high levels, there is increasing evidence that a large number of different mRNAs are present in dendrites at lower levels. This story line began with studies involving isolation and amplification of mRNAs from dendrites of young neurons growing in culture (Miyashiro et al., 1994). Many mRNAs were identified, but there were caveats because mRNAs were harvested from very young neurons growing in culture and there were no follow-up in situ hybridization analyses to verify dendritic localization in neurons in the CNS.

The question of how many mRNAs are in dendrites has been readdressed in recent studies that included many elegant controls (Cajigas et al., 2012). In this study, mRNA was isolated from dendritic laminae of hippocampal slices from adult rats and analyzed by deep sequencing. Because dendritic laminae of the hippocampus contain the cell bodies of glial cells and many types of interneurons, mRNAs expressed by glia and inhibitory interneurons were subtracted out. Even after the subtraction, 2550 candidate dendritic mRNAs were identified. High-resolution in situ hybridization analyses of a subset of the mRNAs revealed that a majority of mRNAs could in fact be detected in dendrites or axons. Interestingly, the mRNAs for a large proportion of known synaptic molecules were localized in dendrites including signaling molecules, scaffolds, and receptors.

One noteworthy finding was that the abundance of the different dendritically localized transcripts varied over a threefold range. The significance of having thousands of different mRNAs in dendrites remains to be explored. All proteins turn over, and synaptic synapses turn over with half-lives that are probably in the range of hours to days. The accepted view has been that most synaptic proteins are synthesized in the cell body and transported to synapses except for the few whose mRNAs are present in dendrites in abundance. The presence of thousands of different mRNAs in dendrites raises the alternative possibility that many synaptic proteins could be replaced via local synthesis. This idea raises many questions, however. If there are only a few copies of a particular transcript in the dendrites of a neuron, how could such a small complement replace proteins at the thousands of synapses on a dendrite? What determines where mRNAs will be translated, especially those present at very low abundance? Do the protein constituents of multiprotein signaling complexes turn over independently, and if so, how does a newly synthesized protein get assembled into a complex? If relative levels of the different mRNAs are the main determinant of the levels of synthesis of different proteins, can the nonabundant transcripts mediate synthesis of a meaningful percent of the total amount needed for ongoing turnover? During periods of synapse modification, which of the thousands of different mRNAs are actually needed?

Finally, there is the issue of the stoichiometric relationships between mRNAs and the translational machinery. The number of ribosomes may be rate limiting for translation (Schuman et al., 2006). At any given point in time, a single ribosome can only associate with a single mRNA, so even if each ribosome at a synapse was engaged with a different mRNA, only a limited number of different proteins could be produced at one time. If there are thousands of different transcripts available for translation, how are particular mRNAs selected for translation initiation? It remains possible that even though the mRNAs for many proteins are detectable in dendrites, it could still be that most of their protein products are synthesized in the cell body. Clearly, these new findings raise a myriad of new questions about how mRNA translation at synapses is regulated both during normal turnover and during periods of synapse modification.

2.5. DENDRITIC TRANSPORT OF RIBOSOMES AND mRNA

2.5.1. Overview

How are ribosomes and mRNAs delivered into dendrites? This question was initially addressed by assessing by pulse-labeling neurons in low-density cultures with 3H-uridine and then assessing the distribution of newly synthesized RNA with autoradiography (Davis et al., 1987, 1990). One hour after exposing neurons to 3H-uridine, newly synthesized RNA is restricted to the nucleus. Over the course of several hours, labeled RNA migrated into dendrites, reaching the distal tips of dendrites within about 12 h. Using mathematical modeling, the average rate of transport of labeled RNA was calculated to be 11–12 μm/h (Davis et al., 1990). Rates of transport were somewhat faster in longer dendrites. This study also documented transport of labeled RNA into axons, albeit at lower levels than dendrites.

When pulse-labeling neurons in culture with 3H-uridine, about 80% of the label incorporated into newly synthesized RNA is actually into ribosomal RNA (unpublished observations). Thus, in retrospect, it is likely that the slow transport documented by autoradiography mainly involved the transport of ribosomes. This is especially likely because subsequent studies revealed that mRNAs are transported much more rapidly (see below).

2.5.2. Dendritic Transport of mRNA; Lessons Learned from Studies of Arc

The initial hypotheses about protein synthesis at synapses focused on the translation of mRNAs that were present constitutively in dendrites. A new dimension of the story emerged with the discovery of a dendritic mRNA that is expressed as an IEG. Arc-associated protein (activity-regulated cytoskeleton-associated protein) (Lyford et al., 1995), also known as Arg3.1 (Link et al., 1995), is expressed at low levels under resting conditions (Figure 3(A) and (D)), but its transcription is strongly induced by physiological activity. For example, different types of sensory stimulation induce Arc in the activated portions of the sensory cortex, and learning experiences induce Arc in different neuron types depending on the nature of the experience. Seizures or intense synaptic activation sufficient to activate NMDA receptors trigger robust induction. For example, a single electroconvulsive seizure (ECS) activates Arc transcription in large numbers of forebrain neurons (Figure 3(B) and (E)).

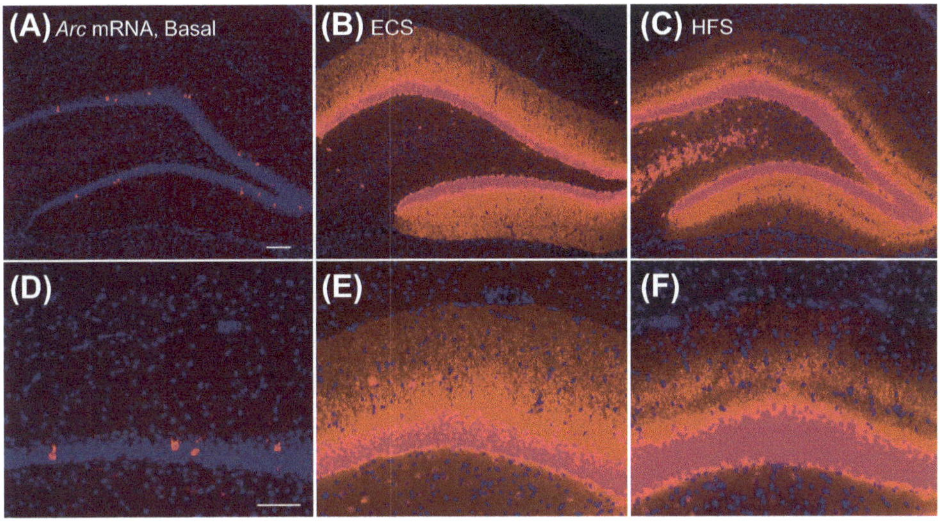

Figure 3 Activity-dependent induction and Arc mRNA and localization near active synapses as revealed by fluorescent In situ hybridization (FISH). (A) Arc mRNA is expressed at low levels under basal conditions. The image illustrates the distribution of Arc mRNA (red) on the nonstimulated side of an anesthetized rat. (B) Induction of Arc expression after an electroconvulsive seizure. Note that Arc is expressed by large numbers of dentate granule cells and Arc mRNA is transported throughout dendrites. (C) Selective targeting of Arc mRNA to active synapses. The panel illustrates the selective localization of Arc mRNA in the middle molecular layer of the dentate gyrus after high-frequency stimulation of the medial perforant path. (D, E, and F) High-magnification views of the fields shown in A, B, and C, respectively. Selective targeting of Arc mRNA to the activated portions of dendrites was initially described by Steward et al. (1998).

Because Arc is expressed as an IEG, it is possible to follow the life history of Arc mRNA from its transcription in the nucleus to its delivery into dendrites using in situ hybridization. Studies using high-resolution fluorescent in situ hybridization (FISH) reveal that a discrete behavioral experience induces Arc within minutes. Studies using a probe that recognizes the intron that is spliced out revealed two discrete foci of unspliced Arc mRNA in the nucleus of neurons in the CA1 region of the hippocampus 5 min after introducing a rat to a novel spatial environment. These represent transcription foci. Studies using probes for mature Arc mRNA revealed that newly synthesized Arc mRNA appears in the cytoplasm within 30 min (Guzowski et al., 1999, 2005). In CA1 neurons, epochs of transcription induced by experience are brief, lasting as little as 16 min, and levels of Arc mRNA then decrease rapidly. After ECS, Arc is activated in many forebrain neurons, but in most neurons, transcription turns off within a few minutes. An exception is the dentate gyrus, where transcription continues at high levels for hours in dentate granule cells (Lyford et al., 1995; Wallace et al., 1998).

2.5.3. Rapid Dendritic Transport of Arc mRNA

Once induced, Arc mRNA is rapidly transported throughout dendrites at a rate consistent with rapid microtubule-based transport. The original estimates of the rate of movement of Arc mRNA into dendrites were derived by inducing Arc expression with an ECS, and assessing the distribution of Arc mRNA over time using in situ hybridization (Wallace et al., 1998). These experiments indicated that Arc mRNA moved from the nucleus to the distal tips of dendrites of dentate granule cells (about 300 μm) within 60 min, indicating a minimal rate of 300 μm/h. It was recognized that this was an underestimate, however, because of the time required for Arc mRNA to be transcribed and undergo nucleocytoplasmic transport (approximately 20–30 min, see above). This study also revealed that Arc mRNA was delivered into dendrites in the presence of protein synthesis inhibitors, indicating that delivery into dendrites was determined by sequences within the Arc mRNA itself and not by the encoded protein. As discussed further below, the presence of the 3′-UTR of Arc in fusion transcripts is sufficient to lead to dendritic transport. Thus, signals that mark Arc mRNA for dendritic delivery are localized in the 3′-UTR, as is also true for most other mRNAs that are localized in different cell types.

2.5.4. Multiple Rates of Transport of Arc mRNA-Containing Particles

To directly visualize mRNA transport in living neurons, we used an exogenous expression approach (Dynes and Steward, 2007). Constructs containing the 3′-UTR of Arc mRNA and a sequence that binds the bacterial protein MS2 (Arc/MS2 mRNA) were coexpressed with transcripts encoding fusion proteins of MS2 and GFP. When expressed by neurons in culture, Arc/MS2 mRNA binds MS2/GFP protein, tagging the mRNA. The mRNA is visible in the cytoplasm in discrete fluorescent particles that extend throughout dendrites (Figure 4(A)).

Live-cell imaging of Arc/MS2 mRNA particle movement in unstimulated, cortical neurons in vitro revealed bidirectional dendritic transport at a wide range of velocities ranging from <1 μm/s to about 70 μm/s Figure 4(B) illustrates a kymograph of a single dendrite imaged for 500 s in which particles can be seen to move at different rates. In this segment, most particles move from proximal to distal, although long bouts of retrograde movement were seen in other segments. The fast and slow movements occurred at rates that are consistent with microtubule and actin/myosin-based transport, respectively, but there was also a distribution of velocities that were intermediate between what is considered typical for microtubule-based and actin/myosin-based mechanisms. At the fastest rate, Arc mRNA could move from the nucleus to a synapse hundreds of micrometers distant within minutes. Taken together, these results mean that synaptic activity can trigger transcription and newly synthesized gene transcripts can reach active synapses within minutes, which could allow translation-dependent synaptic modifications to occur much faster than previously imagined.

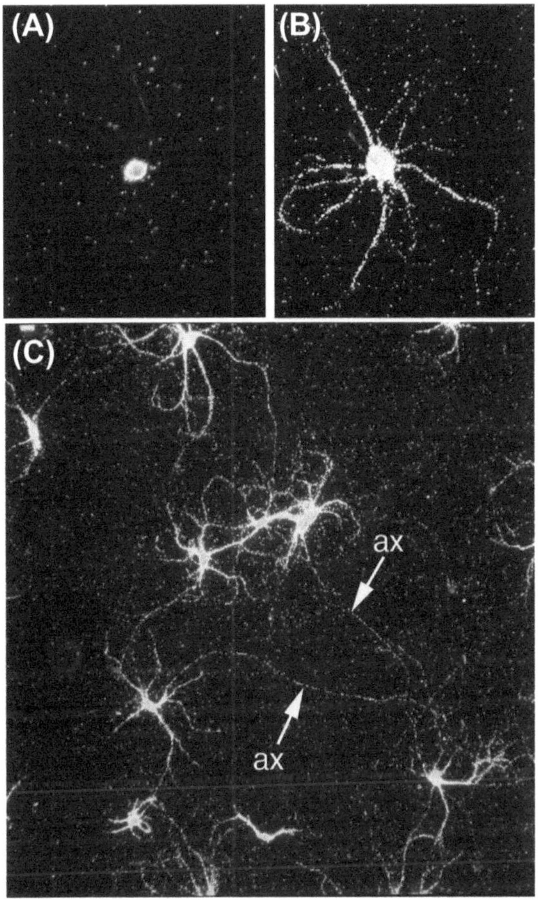

Figure 4 Slow dendritic transport of RNA. Hippocampal neurons in low-density culture were pulse-labeled with 3H-uridine for 1 h and then prepared for autoradiography at different times after the pulse. (A) At the end of the 1 h pulse, newly synthesized RNA has not yet left the nucleus. (B) After 12 h, labeled RNA extends throughout the dendrites. (C) A few axons (ax) also exhibit labeling, indicating RNA transport into axons, albeit at a lower level than into dendrites. *From Davis et al. (1990).*

Other studies have reported transport rates for other mRNAs or mRNA-binding proteins. The mean rate and maximal instantaneous velocity of Arc/MS2 is about 4–6 fold faster than reported for complexes containing CaMKII mRNA (Rook et al., 2000) and 4–6 fold slower than for the actin mRNA-binding protein ZBP1 (Tiruchinapalli et al., 2003). However, transport rates for Arc are not markedly different from those for complexes containing the mRNA-binding proteins Staufen (Kohrmann et al., 1999), Fragile-X Mental Retardation Protein (FMRP) (De Diego Otero et al., 2002), Cytoplasmic Polyadenylation Element Binding Protein (CPEB) (Huang et al., 2003), and RNA particles of unknown type labeled with syto 14 (Knowles et al., 1996). It is noteworthy that complexes that contain KIF5A and Pur ? also move more slowly (maximum of 7.4 μm/min) and for shorter distances (maximum of 8 μm) (Kanai et al., 2004) than the Arc mRNA containing particles. Possibly, the KIF5A motor may be used by Arc mRNA particles for the slow, local phase of movement (see below). It should be noted, however, that differences in experimental parameters (such as neuron type and maturity, image acquisition rate, and incubation temperature) make direct comparisons of data between studies problematic. Also, given the wide range of transport velocities, measures of central tendency (either mean or median) may not be very meaningful.

In our studies (Dynes and Steward, 2007), transcripts lacking the 3′-UTR of Arc were not efficiently delivered into dendrites, indicating that the presence of the 3′-UTR was sufficient for targeting fusion transcripts for dendritic delivery. Studies involving transfection of parts of the 3′-UTR followed by in situ hybridization indicate that two regions mapping between bases 2162–2513 and 2518–2889 are sufficient to confer some degree of dendritic localization (Kobayashi et al., 2005). This information has not yet been helpful in identifying unique mRNA-binding proteins involved in selective localization.

2.5.5. Behavior as Particles Stop Moving Suggests Docking

We noted that some Arc/MS2 particles moved at a characteristic rate and suddenly stopped during the recording period. Many of the particles that stopped moving at their characteristic velocity exhibited low velocity movements just before stopping completely. One of these is noted by the star in Figure (B) and another involving a particle moving in the retrograde direction is shown in Figure (A). Nearly half of these movements were in the opposite direction from the preceding bouts. These reverse slow movements were progressive, and thus are difficult to explain as being due to random dissociations from the transport machinery. None of the particles that reversed direction before stopping resumed rapid transport during the rest of the imaging session. This behavior suggests some final targeting based on local cues. An obvious speculation is that the particles stopped at synapses, but this was not assessed directly in our initial study.

The question of whether Arc/MS2 mRNA docked at synapses was assessed in a follow-up study involving somewhat older (17–20 div) neurons with well-developed dendritic spines (Dynes and Steward, 2012). In this experiment, Arc/MS2 particles were imaged 16–26 h after cotransfection when most particles are stationary. Arc/MS2 particles that remained stationary over the recording interval were identified by averaging consecutive images. Remarkably, many stationary particles were parked precisely at the base of dendritic spines (Figure 5). To quantify this association, we determined the position of every Arc/MS2 mRNA particle and spine base in a sample of dendritic segments with a combined length of 479 μm. This analysis revealed that 69/216 spines (32%) had an Arc/MS2 mRNA particle at or near the spine base (defined as within 0.35 μm), while 71/216 (33%) had no visible particle within 1 mm of the spine base.

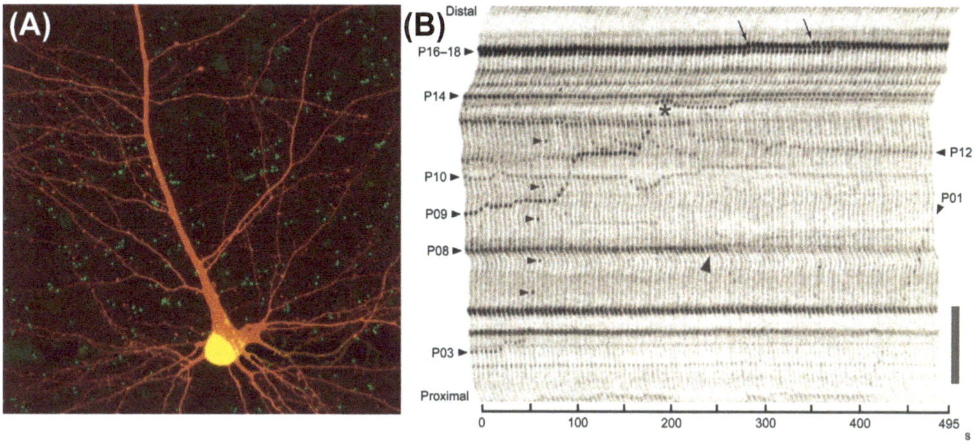

Figure 5 Rapid dendritic transport of Arc/MS2 mRNA. (A) The image illustrates a neuron in culture that had been biolistically transfected with constructs containing the 3′-UTR of Arc mRNA and a sequence that binds the bacterial protein MS2 (Arc/MS2 mRNA) and transcripts encoding fusion proteins of MS2 and GFP. The Arc/MS2 construct also contained the coding sequence for dsRed. Red fluorescence is present throughout the cell body, axons, and dendrites of transfected neurons. Arc/MS2 mRNA tagged with MS2/GFP protein is present in particles in the cell body and dendrites (yellow fluorescence). Larger green particles are the beads that were biolistically delivered to the cultures. (B) The image illustrates a kymograph of a single dendritic segment that was imaged for 500 s in which particles can be seen to move at different rates. Particles are numbered from 01 to 18. Particle #3 exhibits the fastest rate of movement approaching 70 μm/s. Particle number 09 exhibits slow movements just before stopping completely (asterisk). Large arrow head illustrates a particle that undergoes fragmentation. *From Dynes and Steward (2007).*

Conversely, of 235 particles analyzed, 71 (30%) were positioned within 0.35 μm of a spine base. This is likely to underestimate the number of particles at spines because some particles might be under spines that extend perpendicular to the dish, which may be invisible (Figure 6).

Although Arc/MS2 particles clearly localize with a high degree of precision at the base of spines, the results do not prove that the final movements of Arc/MS2 particles in the live-cell imaging experiments actually indicate docking at synapses. Also it remains to be established whether native Arc mRNA is transported in the same way as Arc/MS2. Fluorescent in situ hybridization analyses reveal that newly synthesized native Arc mRNA is present in particles that appear to be roughly the same size as Arc/MS2 particles (Figure 3). Nevertheless, native Arc mRNA undergoes nuclear processing including splicing, which fusion transcripts expressed by transfected DNAs do not, and proteins of the splice junction complex remain bound to the mRNA as it moves into dendrites (Giorgi et al., 2007). If proteins involved in nuclear processing are important for proper localization, then the movement and localization of fusion transcripts may not fully recapitulate what actually occurs with native mRNAs.

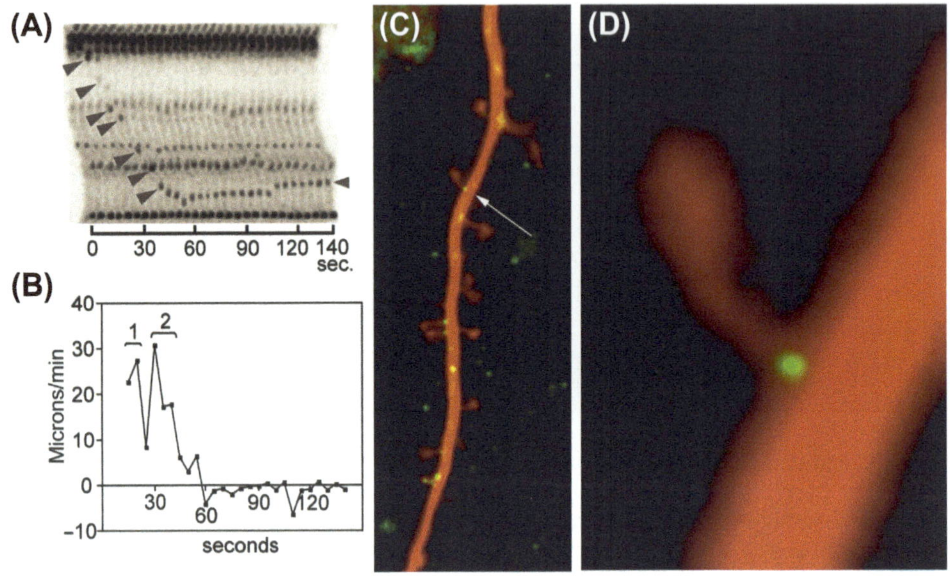

Figure 6 Precise localization of Arc/MS2 mRNA at the base of dendritic spines. (A) The kymograph illustrates the stopping behavior of a particle moving at a steady rate in a retrograde direction. Note slow movements in the reverse direction just before stopping completely. (B) Illustration of the instantaneous velocity of the article, determined by the distance moved from one image to the next. (C) Image of a dendrite in which immobile particles are identified by image averaging over time. (D) Higher power view of an Arc/MS2 particle parked at the base of a dendritic spine. The image was smoothed by Gaussian blurring. *A and B are from Dynes and Steward (2007). C and D are from Dynes and Steward (2012).*

2.5.6. Localization of Arc mRNA near Active Synaptic Sites

A striking feature of Arc mRNA is that it localizes selectively near synapses that have been recently activated. This was demonstrated in studies involving high-frequency stimulation of the pathway from the entorhinal cortex to the dentate gyrus (perforant path) in anesthetized rats (Steward et al., 1998). The perforant path terminates in a topographically organized fashion along the dendrites of dentate granule cells; projections from the medial entorhinal cortex terminate in the middle molecular layer on mid proximodistal dendrites and projections from the lateral entorhinal cortex terminate on distal dendrites (Steward, 1976). Delivery of HFS that induces perforant path LTP (8-pulse trains at 400 Hz at 1/10 s intervals) also induces Arc transcription in dentate granule cells. If stimulation is continued as Arc mRNA moves into dendrites, the newly synthesized mRNA localizes selectively in a discrete band in the activated portion of the dendrite (Steward et al., 1998). This selective localization is in striking contrast to the uniform distribution of Arc mRNA throughout dendrites after a seizure (Figure 3). Especially noteworthy is that if synapses on mid proximodistal dendrites are active as Arc mRNA moves into dendrites, Arc mRNA never reaches more distal dendritic laminae. This suggests that there is some active docking mechanism that captures Arc mRNA that is in transit.

2.5.7. Mechanisms Underlying the Localization of Arc mRNA near Active Synapses

One clue to the localization mechanism is that simple induction of LTP, which can be achieved by delivering thirty 400 Hz trains at 1/10 s intervals, induces Arc transcription, but when Arc mRNA reaches the dendrites 30–45 min later, there is little evidence for selective localization in the activated lamina. Instead, the mRNA is delivered throughout dendrites including the distal regions beyond the zone of synapses that had been active. There are two possible explanations for the failure to localize after simple LTP induction: (1) localization may require stimulation over long periods of time; (2) the signals generated by activity may be relatively short lived. Evidence against the first possibility came from experiments in which Arc expression is induced by an ECS, time was allowed for the mRNA to migrate into dendrites, and then stimulation was delivered to activate synapses terminating on particular dendritic domains. In this paradigm, 15 min of synaptic activation led to a selective localization of Arc mRNA in the activated dendritic lamina (Steward and Worley, 2001). Because 15 min of stimulation is sufficient to generate a localized distribution, the likely explanation of the failure of Arc mRNA to localize in a simple LTP paradigm is that by the time the Arc mRNA that is induced by the stimulation reaches the dendrite, the signal for localization has dissipated.

To define other properties of the localization mechanism, we used the paradigm described above, in which Arc is first induced by a generalized seizure and then targeted by synaptic stimulation (Steward and Worley, 2001). This approach was essential because some of the same intracellular signals are involved in both induction of transcription and localization. We first demonstrated that localization of Arc mRNA at active synapses was blocked by local infusion of NMDA receptor antagonists (Steward and Worley, 2001). Subsequent studies revealed that stimulation sufficient to cause localization triggers striking activation of the MAP kinase pathway as revealed using phosphospecific antibodies for ERK1/2 (Huang et al., 2007). Moreover, pharmacological blockade of ERK activation with U0126 completely blocked the localization of Arc mRNA that would otherwise occur. Also, stimulation that induced localization triggers polymerization of actin at active synapses and blockade of actin polymerization also blocks the localization of Arc mRNA that would otherwise occur during synaptic activation. These findings indicate that localization is driven by NMDA receptor activation and requires activation of MAP kinase and polymerization of actin at active synapses (Huang et al., 2007).

There is evidence that synaptic activation can cause mRNAs to move from the dendritic shaft into spines. For example, in one study, perforant path LTP was induced in unanesthetized adult rats, and subcellular fractionation techniques were then used to isolate synaptoneurosomes (Havik et al., 2003). Real-time PCR was then used to determine the levels of the mRNAs for alpha-CaMKII and Arc in the synaptoneurosomes and also in total homogenates. It was found that levels of aCaMKII mRNA increased about 2.5-fold in the synaptoneurosomes after inducing LTP, whereas there was no change in aCaMKII mRNA levels in total homogenates. These findings suggest that aCaMKII mRNA translocates into spines as a result of stimulation that induces perforant path LTP. There were also increases in the levels of Arc mRNA in synaptoneurosomes, but these were accompanied by increases in Arc mRNA levels in the total homogenates. Thus, the increases in synaptoneurosomes could reflect the overall increase in Arc mRNA levels, and not necessarily translocation of Arc mRNA into spines.

Although Arc mRNA is targeted to regions of dendrites contacted by active synapses, it remains to be established whether the mRNA localizes specifically at active synapses or spines. This is an important issue because of recent evidence that Arc protein plays a role in receptor endocytosis leading to synaptic depression. Induction of LTP increases synaptic efficacy, so if Arc mRNA localizes at active synapses, it would be expected that Arc protein might be delivered preferentially to the potentiated synapses. Alternatively, it is conceivable that Arc mRNA localizes selectively at inactive synapses on dendritic segments that are strongly activated by other synapses. Clearly, there is still much to learn about the mechanisms and functional significance of the remarkable localization of Arc mRNA induced by synaptic stimulation.

2.5.8. Multiple Patterns of mRNA Localization within Dendrites

It is common to talk of mRNAs as being dendritically localized or restricted to the region of the cell body, but there are actually a number of different localization patterns. For example, although the mRNAs for both MAP2 and the a-subunit of CaMKII are present in the dendrites of cortical neurons, hippocampal pyramidal cells, and dentate granule cells, the subcellular distributions of the two mRNAs are quite different (Paradies and Steward, 1997). The mRNA for aCaMKII is present at moderately high levels throughout the neuropil layers. In contrast, the levels of MAP2 mRNA are highest in the portion of the neuropil that contains proximal dendrites. Levels of labeling fall off dramatically in portions of the neuropil that contain distal dendrites.

Of the other dendritic RNAs, the mRNA for Den is present throughout dendrites; Arc (when induced) comes to have a distribution within granule cells of the dentate gyrus that is similar to aCaMKII unless there is ongoing synaptic activity. The RNA pol III transcript BC1 has a distribution that is similar to MAP2 in hippocampal neurons but is expressed at low levels by dentate granule cells. Some other mRNAs that are localized primarily in cell bodies also extend slightly into proximal dendrites. The mRNAs that are present in the dendrites of Purkinje cells also exhibit somewhat different patterns of distribution. The mRNA for the InsP3 receptor is present throughout dendrites but is concentrated in the proximal one-third of the dendrite. L7 mRNA appears to be more uniformly distributed in Purkinje cell dendrites (Bian et al., 1996). The fact that there are a number of different subcellular distribution patterns of the mRNAs localized in dendrites suggests that there will also be different localization signals that mark the mRNAs for dendritic delivery.

3. CONCLUSIONS

The goal of this chapter was to provide a snapshot of current ideas about the protein synthetic machinery that is present at postsynaptic sites on CNS neurons—machinery which provides the basis for gene expression at the synapse. Despite 30 years of research, there are still a number of unresolved questions. To note just a few are the following:

1. What are the address markers that control the dendritic distribution of different mRNAs?
2. Do ribosomes and mRNAs shuttle from one synapse to another so that all synapses are capable of local synthesis of proteins?
3. How is the translation of different mRNAs regulated? Is the synthesis of different proteins determined by the abundance of the different mRNAs or is there selection based on signals from the synapse?
4. What is the mechanism that mediates docking of mRNAs at spine bases, and how is their subsequent movement within the spine controlled?

ACKNOWLEDGMENTS

The authors' work on protein synthesis at synapses is supported by NIH grant NS12333.

REFERENCES

Aakalu, G., Smith, W.B., Nguyen, N., Jiang, C., Schuman, E.M., 2001. Dynamic visualization of local protein synthesis in hippocampal neurons. Neuron 30, 489–502.

Bagni, C., Mannucci, L., Dotti, C.G., Amaldi, F., 2000. Chemical stimulation of synaptosomes modulates alpha-Ca2+/calmodulin-dependent protein kinase II mRNA association to polysomes. J. Neurosci. 20, RC76 1–6.

Bian, F., Chu, T., Schilling, K., Oberdick, J., 1996. Differential mRNA transport and the regulation of protein synthesis: selective sensitivity of purkinje cell dendritic mRNAs to translational inhibition. Mol. Cell. Neurosci. 7, 116–133.

Bockers, T.M., Segger-Junius, M., Iglauer, P., Bockmann, J., Gundelfinger, E.D., et al., 2004. Differential expression and dendritic transcript localization of Shank family members: identification of a dendritic targeting element in the 3′ untranslated region of Shank1 mRNA. Mol. Cell. Neurosci. 26, 182–190.

Bodian, D., 1965. A suggestive relationship of nerve cell RNA with specific synaptic sites. Proc. Natl. Acad. Sci. 53, 418–425.

Burgin, K.E., Washam, M.N., Rickling, S., Westgate, S.A., Mobley, W.C., Kelly, P.T., 1990. In situ hybridization histochemistry of CA++/calmodulin-dependent protein kinase in developing rat brain. J. Neurosci. 10, 1788–1798.

Cajigas, I.J., Tushev, G., Will, T.J., tom Dieck, S., Fuerst, N., Schuman, E.M., 2012. The local transcriptome in the synaptic neuropil revealed by deep sequencing and high-resolution imaging. Neuron 74, 453–466.

Davis, L., Banker, G.A., Steward, O., 1987. Selective dendritic transport of RNA in hippocampal neurons in culture. Nature 330, 477–479.

Davis, L., Burger, B., Banker, G.A., Steward, O., 1990. Dendritic transport: quantitative analysis of the time course of somatodendritic transport of recently synthesized RNA. J. Neurosci. 10, 3056–3068.

De Diego Otero, Y., Severinjnen, L., van Cappellen, G., Schrier, M., Oostra, B., 2002. Transport of fragile X mental retardation protein via granules in neurites of PC12 cells. Mol. Cell. Biol. 22, 8332–8341.

Desmond, N.L., Levy, W.B., 1983. Synaptic correlates of associative potentiation/depression: an ultrastructural study in the hippocampus. Brain Res. 265, 21–30.

Desmond, N.L., Levy, W.B., 1990. Morphological correlates of long-term potentiation imply the modification of existing synapses, not synaptogenesis, in the hippocampal dentate gyrus. Synapse 5, 139–143.

Dynes, J.L., Steward, O., 2007. Dynamics of bi-directional transport of Arc mRNA in neuronal dendrites. J. Comp. Neurol. 500, 433–447.

Dynes, J.L., Steward, O., 2012. Arc mRNA docks precisely at the base of individual dendritic spines indicating the existence of a specialized microdomain for synapse-specific mRNA translation. J. Comp. Neurol. 520, 3105–3119.

Furuichi, T., Simon-Chazottes, D., Fujino, I., Yamada, N., Hasegawa, M., et al., 1993. Widespread expression of inositol 1,4,5-trisphosphate receptor type 1 gene (Insp3r1) in the mouse central nervous system. Recept. Channels 1, 11–24.

Garner, C.C., Tucker, R.P., Matus, A., 1988. Selective localization of messenger RNA for cytoskeletal protein MAP2 in dendrites. Nature 336, 674–677.

Giorgi, C., Yeo, G.W., Stone, M.E., Katz, D.B., Burge, C., et al., 2007. The EJC factor eIF4AIII modulates synaptic strength and neuronal protein expression. Cell 130, 179–191.

Guzowski, J.F., McNaughton, B.L., Barnes, C.A., Worley, P.F., 1999. Environment-specific induction of the immediate early gene Arc in hippocampal neuronal ensembles. Nat. Neurosci. 2, 1120–1124.

Guzowski, J.F., Timlin, J.A., Roysam, B., McNaughton, B.L., Worley, P.F., Barnes, C.A., 2005. Mapping behaviorally relevant neural circuits with immediate-early gene expression. Curr. Opin. Neurobiol. 15, 599–606.

Havik, B., Rokke, H., Bardsen, K., Davanger, S., Bramham, C., 2003. Bursts of high-frequency stimulation trigger rapid delivery of pre-existing alpha-CaMKII mRNA to synapses: a mechanism in dendritic protein synthesis during long-term potentiation in adult awake rats. Eur. J. Neurosci. 17, 2679–2689.

Herb, A., Wisden, W., Catania, M.V., Marechal, D., Dresse, A., Seeberg, P.H., 1997. Prominent dendritic localization in forebrain neurons of a novel mRNA and its product, dendrin. Mol. Cell. Neurosci. 8, 367–374.

Horton, A.C., Ehlers, M.D., 2003. Dual modes of endoplasmic reticulum-to-Golgi transport in dendrites revealed by live-cell imaging. J. Neurosci. 23, 6188–6199.

Huang, F., Chotiner, J.K., Steward, O., 2005. The mRNA for EF1alpha is localized in dendrites and translated in response to treatments that induce long-term depression (LTD). J. Neurosci. 25, 7199–7209.

Huang, F., Chotiner, J.K., Steward, O., 2007. Actin polymerization and ERK phosphorylation are required for Arc/Arg3.1 mRNA targeting to activated synaptic sites on dendrites. J. Neurosci. 27, 9054–9067.

Huang, Y.S., Carson, J.H., Barbarese, E., Richter, J.D., 2003. Facilitation of dendritic mRNA transport by CPEB. Genes Dev. 17, 638–653.

Huber, K.M., Kayser, M.S., Bear, M.F., 2000. Role for rapid dendritic protein synthesis in hippocampal mGluR-dependent long-term depression. Science 288, 1254–1257.

Kanai, Y., Dohmae, N., Hirokawa, N., 2004. Kinesin transports RNA: isolation and characterization of an RNA-transporting granule. Neuron 43, 513–525.

Kang, H., Schuman, E.M., 1996. A requirement for local protein synthesis in neurotrophin-induced hippocampal synaptic plasticity. Science 273, 1402–1406.

Kindler, S., Rehbein, M., Classen, B., Richter, D., Bockers, T.M., 2004. Distinct spatiotemporal expression of SAPAP transcripts in the developing rat brain: a novel dendritically localized mRNA. Brain Res. Mol. Brain Res. 126, 14–21.

Knowles, R.B., Sabry, J.H., Martone, M.E., Deerinck, T.J., Ellisman, M.H., et al., 1996. Translocation of RNA granules in living neurons. J. Neurosci. 16, 7812–7820.

Kobayashi, H., Yamamoto, S., Maruo, T., Murakami, F., 2005. Identification of a cis-acting element required for dendritic targeting of activity-regulated cytoskeleton-associated protein mRNA. Eur. J. Neurosci. 22, 2977–2984.

Kohrmann, M., Luo, M., Kaether, C., DesGroseillers, L., Dotti, C.G., Keibler, M.A., 1999. Microtubule-dependent recruitment of Staufen-green fluorescent protein into large RNA-containing granules and subsequent dendritic transport in living hippocampal neurons. Mol. Biol. Cell 10, 2945–2953.

Leski, M.L., Steward, O., 1995. Proteins that are synthesized at postsynaptic sites: a two-dimensional gel electrophoresis analysis of proteins synthesized within synaptodendrosomes. Soc. Neurosci. Abstr. 29, 20.

Leski, M.L., Steward, O., 1996. Synthesis of proteins within dendrites: ionic and neurotransmitter modulation of synthesis of particular polypeptides characterized by gel electrophoresis. Neurochem. Res. 21, 681–690.

Link, W., Konietzko, G., Kauselmann, G., Krug, M., Schwanke, B., et al., 1995. Somatodendritic expression of an immediate early gene is regulated by synaptic activity. Proc. Natl. Acad. Sci. 92, 5734–5738.

Lyford, G., Yamagata, K., Kaufmann, W., Barnes, C., Sanders, L., et al., 1995. Arc, a growth factor and activity-regulated gene, encodes a novel cytoskeleton-associated protein that is enriched in neuronal dendrites. Neuron 14, 433–445.

Miyashiro, K., Dichter, M., Eberwine, J., 1994. On the nature and differential distribution of mRNAs in hippocampal neurites: implications for neuronal functioning. Proc. Natl. Acad. Sci. 91, 10800–10804.

Ostroff, L.E., Fiala, J.C., Allwardt, B., Harris, K.M., 2002. Polyribosomes redistribute from dendritic shafts into spines with enlarged synapses during LTP in developing rat hippocampal slices. Neuron 35, 535–545.

Palacios-Pru, E.L., Palacios, L., Mendoza, R.V., 1981. Synaptogenetic mechanisms during chick cerebellar cortex development. J. Submicrosc. Cytol. 13, 145–167.

Paradies, M.A., Steward, O., 1997. Multiple subcellular mRNA distribution patterns in neurons: a nonisotopic in situ hybridization analysis. J. Neurobiol. 33, 473–493.

Peters, A., Palay, S.L., de Webster, H.F., 1976. The Fine Structure of the Nervous System: the Neurons and Supporting Cells. Saunders, Philadelphia.

Pierce, J.P., Mayer, T., McCarthy, J.B., 2001. Evidence for a satellite secretory pathway in neuronal dendritic spines. Curr. Biol. 11, 351–355.

Pierce, J.P., van Leyen, K., McCarthy, J.B., 2000. Translocation machinery for synthesis of integral membrane and secretory proteins in dendritic spines. Nat. Neurosci. 3, 311–313.

Rao, A., Steward, O., 1991. Evidence that protein constituents of postsynaptic membrane specializations are locally synthesized: analysis of proteins synthesized within synaptosomes. J. Neurosci. 11, 2881–2895.

Rao, A., Steward, O., 1993. Evaluation of RNAs present in synaptodendrosomes: dendritic, glial and neuronal cell body contribution. J. Neurochem. 61, 835–844.

Rook, M.S., Lu, M., Kosik, K.S., 2000. CaMKII alpha 3′ untranslated region-directed mRNA translocation in living neurons; visualization by GFP linkage. J. Neurosci. 20, 6385–6393.

Schuman, E.M., Dynes, J.L., Steward, O., 2006. Synaptic regulation of translation of dendritic mRNAs. J. Neurosci. 26, 7143–7146.

Spacek, J., Hartmann, M., 1983. Three-dimensional analysis of dendritic spines. I. Quantitative observations related to dendritic spine and synaptic morphology in cerebral and cerebellar cortices. Anat. Embryol. (Berl.) 167, 289–310.

Steward, O., 1976. Topographic organization of the projections from the entorhinal area to the hippocampal formation of the rat. J. Comp. Neurol. 167, 285–314.

Steward, O., 1983. Polyribosomes at the base of dendritic spines of CNS neurons: their possible role in synapse construction and modification. Cold Spring Harbor Symp. Quant. Biol. 48, 745–759.

Steward, O., Falk, P.M., 1985. Polyribosomes under developing spine synapses: growth specializations of dendrites at sites of synaptogenesis. J. Neurosci. Res. 13, 75–88.

Steward, O., Falk, P.M., 1986. Protein synthetic machinery at postsynaptic sites during synaptogenesis; a quantitative study of the association between polyribosomes and developing synapses. J. Neurosci. 6, 412–423.

Steward, O., Falk, P.M., 1991. Selective localization of polyribosomes beneath developing synapses: a quantitative analysis of the relationships between polyribosomes and developing synapses in the hippocampus and dentate gyrus. J. Comp. Neurol. 314, 545–557.

Steward, O., Falk, P.M., Torre, E.R., 1996. Ultrastructural basis for gene expression at the synapse: synapse-associated polyribosome complexes. J. Neurocytol. 25, 717–734.

Steward, O., Levy, W.B., 1982. Preferential localization of polyribosomes under the base of dendritic spines in granule cells of the dentate gyrus. J. Neurosci. 2, 284–291.

Steward, O., Reeves, T.M., 1988. Protein synthetic machinery beneath postsynaptic sites on CNS neurons: association between polyribosomes and other organelles at the synaptic site. J. Neurosci. 8, 176–184.

Steward, O., Ribak, C.E., 1986. Polyribosomes associated with synaptic sites on axon initial segments: localization of protein synthetic machinery at inhibitory synapses. J. Neurosci. 6, 3079–3085.

Steward, O., Schuman, E.M., 2001. Protein synthesis at synaptic sites on dendrites. Annu. Rev. Neurosci. 24, 299–325.

Steward, O., Wallace, C.S., Lyford, G.L., Worley, P.F., 1998. Synaptic activation causes the mRNA for the IEG Arc to localize selectively near activated postsynaptic sites on dendrites. Neuron 21, 741–751.

Steward, O., Worley, P.F., 2001. Selective targeting of newly synthesized Arc mRNA to active synapses requires NMDA receptor activation. Neuron 30, 227–240.

Tiruchinapalli, D.M., Oleynikov, Y., Kelic, S., Shenoy, S.M., Hartley, A., et al., 2003. Activity-dependent trafficking and dynamic localization of zipcode binding protein 1 and β-actin mRNA in dendrites and spines of hippocampal neurons. J. Neurosci. 23, 3251–3261.

Tongiorgi, E., Armellin, M., Giulianini, P.G., Bregola, G., Zucchini, S., et al., 2004. Brain-derived neurotrophic factor mRNA and protein are targeted to discrete dendritic laminas by events that trigger epileptogenesis. J. Neurosci. 24, 6842–6852.

Torre, E.R., Steward, O., 1992. Demonstration of local protein synthesis within dendrites using a new cell culture system that permits the isolation of living axons and dendrites from their cell bodies. J. Neurosci. 12, 762–772.

Torre, E.R., Steward, O., 1996. Protein synthesis within dendrites: glycosylation of newly synthesized proteins in dendrites of hippocampal neurons in culture. J. Neurosci. 16, 5967–5978.

Villanueva, S., Steward, O., 2001a. Glycoprotein synthesis at the synapse: fractionation of polypeptides synthesized within isolated dendritic fragments by concanavalin A affinity chromatography. Mol. Brain Res. 91, 137–147.

Villanueva, S., Steward, O., 2001b. Protein synthesis at the synapse: developmental changes, subcellular localization and regional distribution of polypeptides synthesized in isolated dendritic fragments. Mol. Brain Res. 91, 148–153.

Wallace, C.S., Lyford, G.L., Worley, P.F., Steward, O., 1998. Differential intracellular sorting of immediate early gene mRNAs depends on signals in the mRNA sequence. J. Neurosci. 18, 26–35.

Weiler, I.J., Davenport, W.J., Greenough, W.T., 1991. Depolarization-dependent stimulation of polyribosome loading in synaptoneurosomes. Soc. Neurosci. Abstr. 17, 1516.

Weiler, I.J., Greenough, W.T., 1991. Potassium ion stimulation triggers protein translation in synaptoneuronsomal polyribosomes. Mol. Cell. Neurosci. 2, 305–314.

Weiler, I.J., Greenough, W.T., 1993. Metabotropic glutamate receptors trigger postsynaptic protein synthesis. Proc. Natl. Acad. Sci. 90, 7168–7171.

Zalfa, F., Giorgi, M., Primerano, B., Reis, S., Oostra, B.A., Bagni, C., 2003. The fragile X syndrome protein FMRP associates with BC1 RNA and regulates the translation of specific mRNAs at synapses. Cell 112, 317–327.

CHAPTER SEVEN

Estrogen Effects on Hippocampal Synapses

Teresa A. Milner[1,2], Bruce S. McEwen[2] and Elizabeth M. Waters[2]
[1]Department of Neurology and Neuroscience, Brain and Mind Research Institute, Weill Cornell Medical College, New York, NY, USA
[2]Harold and Margaret Milliken Hatch Laboratory of Neuroendocrinology, The Rockefeller University, New York, NY, USA

1. INTRODUCTION

Estrogens induce spine synapses on pyramidal cell dendrites in the hippocampal CA1 region of the young adult female rat (Woolley et al., 1990). As seen in Figure 1, the remodeling of hippocampal pyramidal cells occurs over the female rats' 5 day estrous cycle: the synapses take several days to be induced under the influence of estrogens and then disappear within 12 h under the influence of the proestrous surge of progesterone (McEwen et al., 1995). Like rats, estradiol also produces changes in the mouse CA1 (Li et al., 2004). However, rather than increasing the density of dendritic spines, estradiol increases the number of mushroom-shaped spines which are thought to form mature, functional spine synapses (Li et al., 2004).

In both rats and mice, estrogen regulation of spine synapses involves the cooperation of between pre- and postsynaptic cellular compartments. In the presynaptic compartment, presynaptic proteins (e.g. synaptophysin and syntaxin), glutamatergic, gamma-aminobutyric acid (GABA)ergic, cholinergic, and growth factor receptors (e.g. TrkB) are modulated to various degrees by estrogen levels. Likewise, estrogens regulate proteins in the postsynaptic compartment various degrees. The postsynaptic compartment includes postsynaptic proteins (spinophilin and PSD95), glutamatergic receptors, GABAergic receptors (e.g. $GABA_A$), and glutamate receptors and peptide receptors. Studies in both rats and mice have shown that estrogen-sensitive receptors (e.g. estrogen receptors (ERs) α and β and progestin receptors (PRs)) are in both pre- and postsynaptic compartments (for review, see McEwen and Milner, 2007). Recent findings that ER α and β agonists differentially regulate spine formation (Phan et al., 2011) as well as pre- and postsynaptic proteins (Waters et al., 2009), emphasize the relative importance of each ER in this dynamic process.

Although estrogen-sensitive receptors are present in both females and males (Mitterling et al., 2010), the "privilege" of estrogen induction of spine synapses appears to be limited to females. Estrogens do not induce spine synapse formation in the hippocampal CA1 region of male castrated rats (Leranth et al., 2003); instead treatment with testosterone or dihydrotestosterone increases spine synapses in males

The Synapse
http://dx.doi.org/10.1016/B978-0-12-418675-0.00007-9

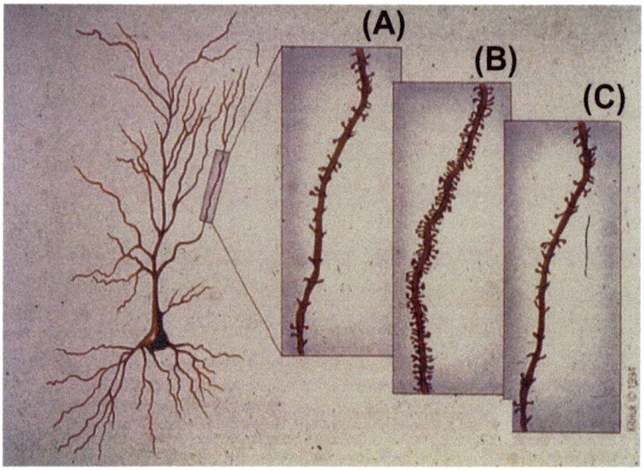

Figure 1 Representation of CA1 pyramidal neurons in the female rat hippocampus during the 4–5 days estrous cycle. (A) Diestrus, when estradiol levels are lowest. (B) Proestrus, when estradiol levels are highest. (C) Estrus, when estradiol levels are decreasing. *From McEwen and Schmeck. The Hostage Brain Rockefeller University Press, 1994. Drawing by Lidia Kibiuk.*

(Kovacs et al., 2003). However, estrogen can increase dendritic spines in the adult male hippocampus if sexual differentiation is blocked by administering aromatase inhibitors immediately after birth (Lewis et al., 1995). Along with evidence that spatial memory processes show sex differences programmed by the aromatization of estrogens early in neonatal life (Williams and Meck, 1991), these results indicate that the hippocampus undergoes sexual differentiation.

In females, the efficacy of estrogen to regulate spine formation varies at different life stages. This effect is manifested in young adulthood through reproductive senescence but changes after menopause. In aged rats (22 months), the ability of estrogens to induced spine synapse formation is lost (Adams et al., 2001). Although both ERs appear to be part of the mechanism for spine synapse formation (Yuen et al., 2011), differential expression of ERα and ERβ may contribute to the loss of estrogen-mediated synaptic plasticity with age.

This chapter will focus on estrogen regulation of synapses in the CA1 region of the hippocampus.

2.1. HIPPOCAMPAL SYNAPSES: ESTROGEN EFFECTS AND RECEPTORS

2.1.1. Presynaptic Compartment

2.1.1.1. Presynaptic Proteins

Synaptophysin is a presynaptic protein associated with small synaptic vesicles, whereas syntaxin is associated with the presynaptic membrane; both proteins are implicated in vesicular docking (Calakos and Scheller, 1994; Calakos et al., 1994). As seen in Figure 2,

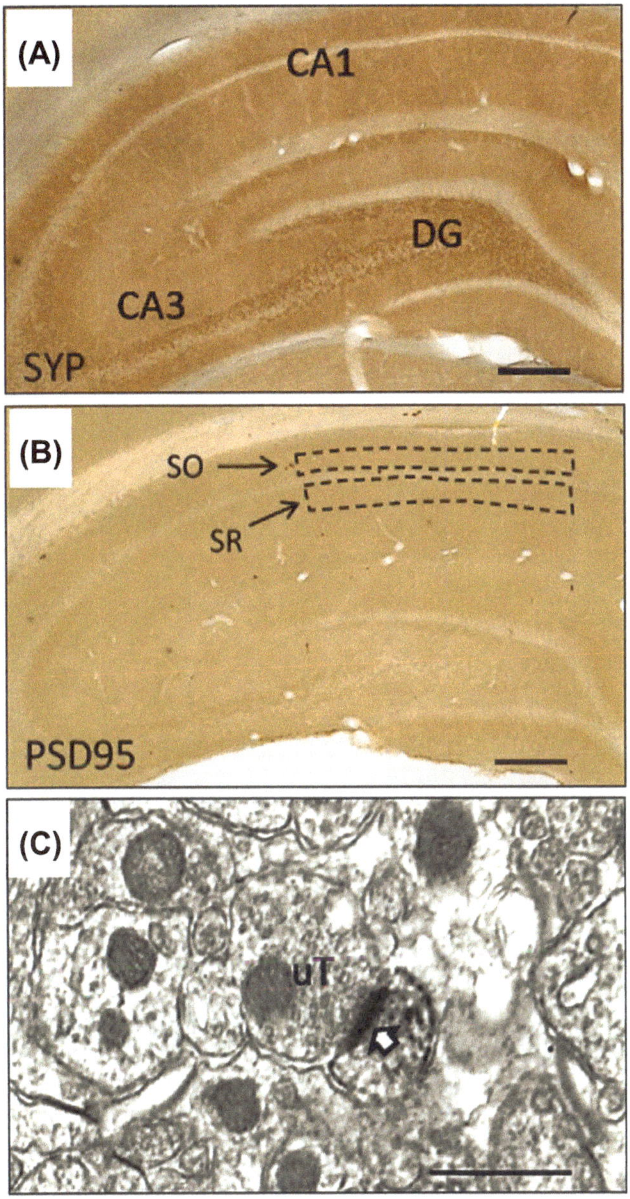

Figure 2 Localization of representative pre- and postsynaptic proteins in the rat hippocampus. (A) Light photomicrograph showing synaptophysin (SYP)-ir in the CA1 region as well as the dentate gyrus (DG) and CA3. (B) Light photomicrograph showing PSD-95-ir in the stratum orients (SO) and stratum radiatum (SR) of CA1. Dotted lines represent regions examined by densitometry in the CA1. (C) Electron photomicrograph showing an unlabeled terminal (uT) that forms an asymmetric synapse (arrow) on a PSD-95-containing dendritic spine in SR of CA1. Scale bars = 300 μm (A, B); 500 nm (C). All photomicrographs were taken from young females. *A and B from Williams and Milner, 2011.*

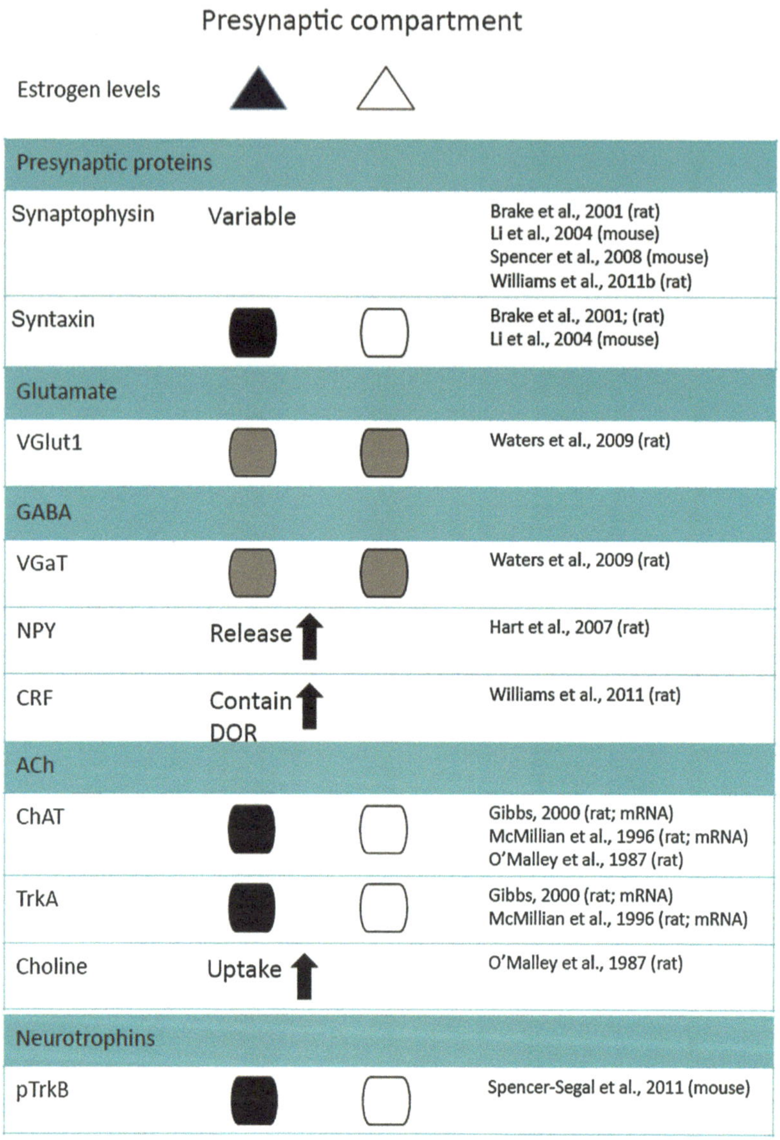

Figure 3 Summary of estrogen effects on proteins in the presynaptic compartment.

both synaptophysin-ir (Figure 2(A)) and syntaxin-ir are most intense in lamina lacking principal cell bodies in the hippocampus. Estrogens regulate the levels of synaptophysin and syntaxin in the hippocampal CA1 regions in both the female rats (Brake et al., 2001) and mice (Li et al., 2004; Spencer et al., 2008); however, the direction of the change in synaptophysin levels differs depending on the species and paradigm as seen in Figures 3 and 4.

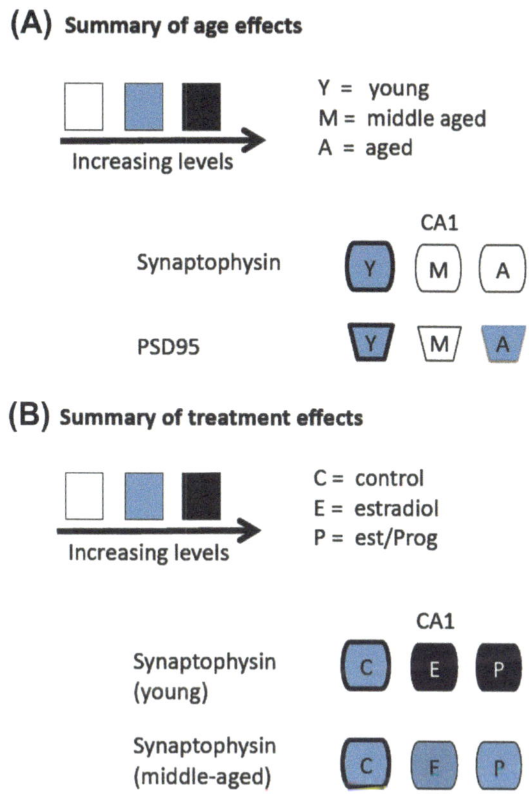

Figure 4 Summary of age and treatment effects on pre- and postsynaptic protein levels in the dorsal hippocampal formation. (A) With age, females showed regionally specific changes in levels of presynaptic protein synaptophysin and postsynaptic protein PSD-95. Significant changes with age, if evident, were found in comparison to the young (Y) group noted by the thick black outline. (B) Significant changes with hormone treatment (E or P), if evident, were found in comparison to the control (C) group noted by the thick black outline. *Data from Williams et al., 2011. Brain Res. 1379, 71–85.*

Synaptophysin levels in the CA1 region are elevated in ovariectomized (OVXed) rats after two consecutive days of estradiol benzoate (EB) injections (10 μg/kg, s.c.) (Brake et al., 2001) or 3 days after receiving silastic implants containing 17β-estradiol (Williams et al., 2011b), unchanged in OVXed rats 48 h after two consecutive EB injections (10 μg/kg, s.c.) or decreased in normal cycling mice examined at proestrus (high estrogen phase) (Spencer et al., 2008). In contrast, consistent elevations in syntaxin levels in the CA1 region have been reported in OVXed rats after 2 days of EB injections (Brake et al., 2001) and in OVXed mice after 5 days of EB injections (1 μg/0.1 ml, s.c.) (Li et al., 2004).

2.1.1.2. Glutamate

The major sources of glutamatergic afferents in the CA1 regions of the hippocampus are intrinsic pyramidal cells giving rise to commissural and associational projections and extrinsic septal and entorhinal cortical projections (for review, see Frotscher, 1991). Glutamatergic terminals form asymmetric (excitatory-type) synapses on the dendrites of pyramidal cells and interneurons. Glutamatergic terminals synapse on dendritic spines. Estrogens do not appear to directly affect the levels of glutamate in terminals. Neither administration of estradiol nor ER-selective agonists to ovariectomized rats alters the levels of vesicular glutamate transporter 1 (VGluT1) in the hippocampus (Figure 3) (Waters et al., 2009).

2.1.1.3. GABA

GABAergic afferents in the hippocampus emanate from two main sources: intrinsic GABAergic interneurons and the medial septum/diagonal band (for review, see Frotscher, 1991). GABAergic terminals form symmetric (inhibitory-type) synapses primarily on the soma and dendrites of principal cells. Unlike glutamatergic synapses, GABAergic terminals almost never synapse on dendritic spines. Immunoreactivity for the GABA transporter (VGaT) is restricted to the terminals and cell bodies of inhibitory neurons (Chaudhry et al., 1998). In the CA1 region, VGatT labeling is prominent in all lamina except the pyramidal cell layer (Waters et al., 2009).

Estrogens do not appear to directly affect the levels of GABA in terminals (Figure 3). Studies in cultured hippocampal neurons suggest that estrogen-induced disinhibition is related to blocking GABA synthesis or transport (Murphy et al., 1998). However, analysis of VGaT-ir in the hippocampi of young OVXed rats 48 h after receiving two injections of EB 24 h apart was identical to that of control rats (Waters et al., 2009). However, estradiol could indirectly regulate GABA release from terminals. Neuropeptide Y (NPY) is contained in GABAergic interneurons in the CA1 region (Milner and Veznedaroglu, 1992; Frotscher, 1991). Within terminals, ERα-ir is on NPY-containing large dense-core vesicles located away from synaptic specializations (Hart et al., 2007). However, estradiol administration releases NPY which in turn mobilizes a subset of vesicles toward the synaptic specialization (Hart et al., 2007). The estradiol-facilitated release of NPY is important for the attenuation of seizures (Ledoux et al., 2009).

Males and normal cycling females have a similar number of GABAergic corticotropin releasing factor (CRF)-labeled interneurons that colabel delta opioid receptor (DOR) (Williams and Milner, 2011). Although not studied in CA1, in the dentate hilus, proestrus females have a greater number of CRF-labeled terminals that lack DORs compared to males (Williams and Milner, 2011). These findings suggest that while DORs are positioned to modulate CRF-containing neuron activity and CRF peptide release, their ability to do so may be compromised in females when estrogen levels are high.

2.1.1.4. Acetylcholine

A major subcortical projection to the CA1 regions arises from the cholinergic septal complex (i.e. the medial septal nucleus and the nucleus of the diagonal band)

(Amaral and Kurz, 1985). These afferents can be identified by choline acetyltransferase (ChAT) and vesicular acetylcholine transporter (VAChT) as well as by the neurotropin receptors p75NTR and TrkA (Barker-Gibb et al., 2001; Dougherty and Milner, 1999). Our EM studies have demonstrated that cholinergic terminals form major synaptic contacts on GABAergic interneurons, especially those containing NPY, and pyramidal cells in the hippocampus (Dougherty and Milner, 1999; Towart et al., 2003). Importantly, deletion of cholinergic afferents to the hippocampus results in a loss of NPY-immunoreactive interneurons (Milner et al., 1997).

As reviewed by Gibbs (2009), septohippocampal cholinergic afferents play an important role in learning and attentional processes and their function declines with advanced age and in Alzheimer's disease. Several lines of evidence suggest that the effects of estrogens on neuronal plasticity and cognitive function are related to or rely upon their interactions with these cholinergic afferents (Figure 3). In particular, the messenger RNA (mRNA) levels of ChAT and TrkA in the septal complex are increased in OVXed rats following EB (McMillan et al., 1996; Gibbs, 2000) or in proestrus (the high estrogen phase) (Gibbs, 1996). Ten days of continuous EB treatment to OVXed rats increases high-affinity choline uptake and ChAT activity in the CA1 region of the hippocampus (O'Malley et al., 1987). ER activation of cholinergic afferents increases N-methyl-D-aspartate (NMDA) ER binding, activation of inhibitory GABAergic interneurons, and improvement in working memory (Daniel and Dohanich, 2001; Rudick et al., 2003). The observations that cholinergic septohippocampal afferents contain both ERα and GPR30 (Hammond et al., 2011; Towart et al., 2003) suggest that estrogen regulation presynaptic acetylcholine release or reuptake is involved in these effects. Moreover, it would appear that the ability of estrogen to increase NMDA receptor (NMDAR) expression, that is required for spine synapse formation in CA1, may be indirectly mediated by cholinergic activity (Section 3.2).

2.1.1.5. Neurotrophins

Neurotrophins, particularly those acting at the trkB neurotropin receptor, can modulate activity-dependent plasticity (reviewed by Chao, 2003). In particular, trkB activation can increase synaptic protein expression, induce long-term potentiation, modulate dendritic spine morphology, and increase synaptic strength (reviewed in Spencer-Segal et al., 2011b). The full-length form of the trkB receptor is prominent in axons and axon terminals in the CA1 region of the rat hippocampus (Drake et al., 1999). In mice, the number of axons and axon terminals containing the phosphorylated form of trkB (ptrkB) are elevated in proestrus females compared to males (Figure 3; Spencer-Segal et al., 2011b).

2.2. POSTSYNAPTIC COMPARTMENTS

2.2.1. Postsynaptic Proteins

The postsynaptic protein spinophilin is an actin-binding protein found predominantly in dendritic spines (Allen et al., 1997; Feng et al., 2000). PSD-95 is a scaffolding protein that is involved in the organization of glutamate receptors and other constituents of the postsynaptic density (Kim and Sheng, 2004). Both spinophilin-ir and PSD95-ir (Figure 2(B)) are concentrated in hippocampal lamina lacking cell bodies. At the electron microscopic level, PSD95-ir is prominent in dendritic spines (Figure 2(C)). Estradiol treatment increases both spinophilin-ir and PSD95-ir in OVXed mice (Li et al., 2004) (Spencer-Segal et al., 2011a) and rats (Brake et al., 2001; Waters et al., 2009). Consistent with these observations, PSD95-ir is increased at proestrus (the high-estrogen stage) in young C57/Bl6 wild-type mice (Spencer et al., 2008, 2010). In contrast, PSD95 is not increased in mice with the brain-derived neurotrophic factor variant Val66Met (Spencer et al., 2010). Interestingly, mice with the Val66Met variant had increased anxiety and decreased spatial memory compared to wild-type mice at proestrus (Spencer et al., 2010), suggesting a correlation between synaptic density and behavior.

In OVXed rats, ERα and ERβ agonists alone or in combination increase the levels of PSD-95 (Waters et al., 2009). In contrast, estradiol treatment did not change the levels of PSD95-ir in CA1 in either OVXed alpha or beta ER knock-out mice (AERKO and BERKO, respectively) (Spencer-Segal et al., 2011a).

2.2.2. Glutamate Receptors

In female rats, estrogen-induced spinogenesis on CA1 pyramidal cells critically involves activation of glutamatergic NMDA receptors (Woolley, 1999; McEwen et al., 1995). As seen in Figure 5, the effects of estrogen levels on glutamate receptors vary. Estradiol increases NMDAR binding in hippocampus of OVXed females (Weiland, 1992; Daniel and Dohanich, 2001) but does not alter NMDAR binding in male hippocampus (Romeo et al., 2005). The estrogen-induced increases in NMDA binding are blocked by antagonists of muscarinic M2 receptors (Daniel and Dohanich, 2001). Binding with an NMDAR 2B ligand was reduced after OVX in the CA1 stratum radiatum and both estradiol and an ERα agonist prevented decreased binding and loss of NMDAR 2B mRNA (Morissette et al., 2008). An earlier report indicated that estradiol could increase the levels of the NMDA NR1 subunit (Gazzaley et al., 1996) in the female rat hippocampal CA1 region; however, a recent study reported no changes in immunoreactivity for NMDAR1 or NMDAR2 A and B (Snyder et al., 2011). Only the ERβ agonist regulated the expression of AMPA receptor subunits GluR2 and 3 (Waters et al., 2009). Estrogen-induced spinogenesis is blocked

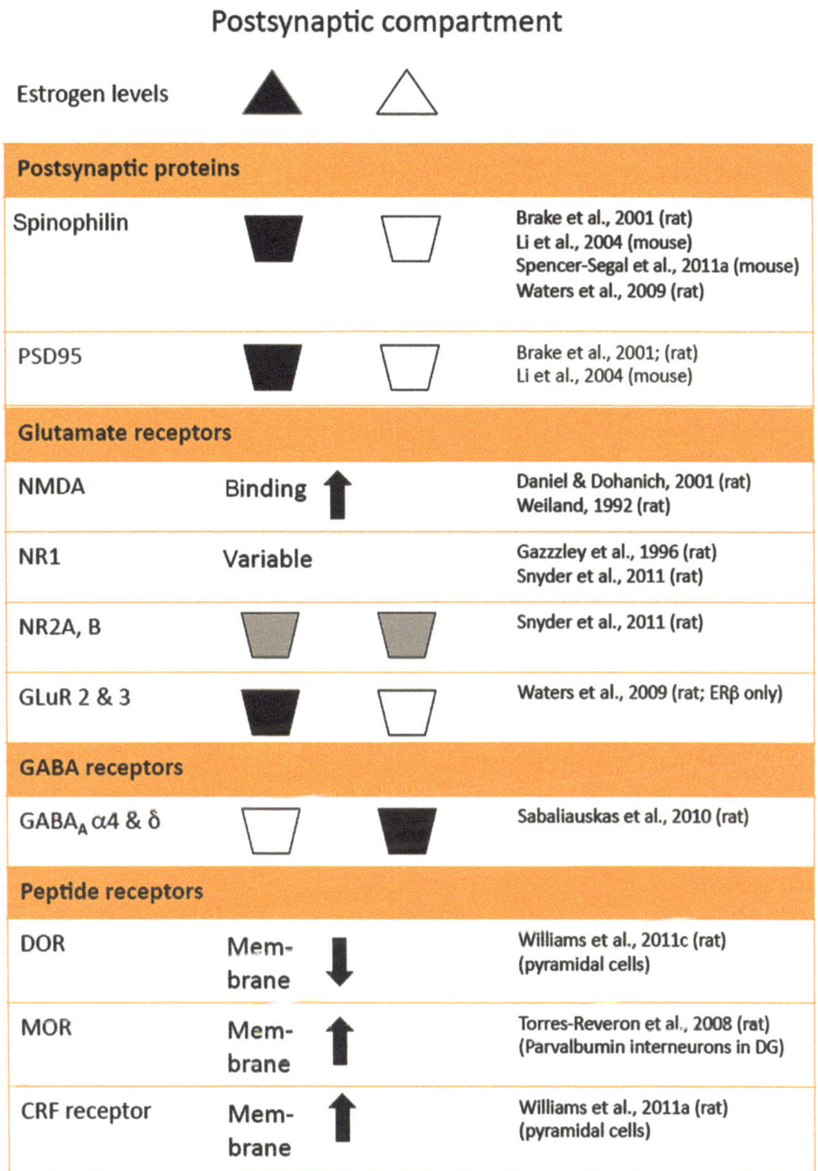

Figure 5 Summary of estrogen effects on proteins in the postsynaptic compartment.

by concurrent administration of NMDAR antagonists but not by AMPA-kainate receptor antagonists (Woolley and McEwen, 1994). Altogether, variations in the dose and duration of estrogen exposure used in these studies along with the modulatory actions of other receptors suggest that estrogen's effect on glutamate receptors is subject to regulation by multiple sources.

2.2.3. GABA Receptors

Estradiol indirectly influences disinhibition of pyramidal cells through activation of GABA interneurons (Rudick et al., 2003; Rudick and Woolley, 2003). Estradiol does not appear to alter GABA receptor kinetics globally since it does not affect GABA-evoked currents; however, estradiol may act to prolong currents at a subset of GABAergic synapses (Rudick et al., 2003). In particular, regulation of the α4 $GABA_A$ receptor subunit alters inhibitory postsynaptic current decay time (Smith et al., 1998). Recently, immunoreactivity for the α4 and δ $GABA_A$ receptor subunits in dendrites has been shown to decrease at the proestrus/estrus cuff (Figure 5) (Sabaliauskas et al., 2010). These changes correlate with increases in spatial learning (Sabaliauskas et al., 2010).

2.2.4. Peptide Receptors

Activation of DORs and mu opioid receptors (MORs) in the hippocampus regulates excitation (for review, see Drake et al., 2007). In the CA1 region, DORs are found in the somata and dendrites of interneurons and, to a lesser extent, principal cells (Commons and Milner, 1997). Compared to males, females have less DOR-ir in the distal apical dendrites of pyramidal cells; however, proestrus females exhibit greater internalization and trafficking of DORs toward the soma (Figure 5) (Williams et al., 2011c). Moreover, proestrus females had fewer NPY-labeled interneuron somata that colabeled with DOR (Williams et al., 2011c). However, although NPY-containing terminals frequently contained DORs found at or near the plasma membrane, no differences were noted by cycle phase or sex (Williams et al., 2011c).

In CA1 as well as other hippocampal subregions, MORs are primarily found in the dendrites and somata of GABAergic interneurons (Drake and Milner, 2002, 1999). Although we have not studied the effects of ovarian steroids on MOR trafficking in CA1, we have studied this in the dentate gyrus. Here, elevated levels of estradiol (either in proestrus rats or in OVXed rats 48 h after receiving two EB injections 24 h apart) increase the number of MORs on the plasmalemma of parvalbumin containing dendrites (Figure 5) (Torres-Reveron et al., 2009). Together with the above findings for DORs, these results suggest that opioid-mediated excitation could be increased when estrogen levels are elevated.

Activation of CRF receptors are also effect excitation and can induce synaptic plasticity in the hippocampus (Wang et al., 1998; Blank et al., 2002; Schierloh et al., 2007). CRF receptors are prominent in CA1 pyramidal cells, where they are concentrated at dendritic spine synapses (Chen et al., 2004). Our recent studies have provided evidence that the opioid–CRF systems intersect in the hippocampus. Proestrus females and males have comparable levels of CRF receptor-ir per DOR-labeled dendrites and similar cytoplasmic densities of CRF receptor-ir in DOR-labeled dendrites (Williams et al., 2011a). In contrast, proestrus females have an increased number of dendritic profiles dually labeled for DOR and CRF receptors and an increased membrane density of CRF receptor-ir in DOR dendrites compared to males (Figure 7.5) (Williams et al., 2011a). In functional assays using NG 108 cells, the DOR agonist SNC80 inhibited CRF-induced cyclic adenosine monophosphate accumulation (Williams et al., 2011a). These findings provide evidence for a sexual dimorphism to stress reactivity in the hippocampus.

2.3. ESTROGEN-SENSITIVE RECEPTORS AND ASSOCIATED SIGNALING MOLECULES

2.3.1. ERs and PRs

With the exception of nuclear ERαs in scattered interneurons (and nuclear PR in a few interneurons in mice), estrogen-sensitive receptors in the hippocampus are nonnuclear in both rats and mice (McEwen and Milner, 2007). Electron microscopic autoradiography studies with ^{125}I estradiol have demonstrated that nonnuclear ER's are capable of binding estradiol and thus may be functional receptors (Milner et al., 2008). Nonnuclear ERs, including the newly discovered GPR30 (Akama et al., 2013), and PRs are found in both pre- and postsynaptic compartments (see our review in McEwen and Milner, 2007). In particular, as seen in Figure 6, all four receptors are found in terminals and dendritic spines in both rats and mice (Mitterling et al., 2010; Milner et al., 2001, 2005; Waters et al., 2008). Moreover, ERβ-ir and GPR30-ir are prominent on the membranes of pyramidal cell somata and dendrites. In terminals, ERα-ir is often associated with synaptic vesicles (Hart et al., 2007; Tabatadze et al., 2013). Dual label electron microscopic studies have revealed that ERα-ir is found in cholinergic terminals (identified by VAChT-ir) (Towart et al., 2003) and in a subset of GABAergic (Hart et al., 2007) and NPY-containing (Ledoux et al., 2009). Moreover, ERβ-ir often is associated with mitochondria in both terminals and dendritic shafts (Hart et al., 2007).

As shown in Figures7 and 8, depending on the hippocampal lamina, the numbers and types of pre- and postsynaptic profiles containing ERα, ERβ, and PRs fluctuate across the estrous cycle in young mice and differ between females and males (Mitterling et al., 2010). Interestingly, in the CA1 stratum radiatum region (where dendritic spines are increased at proestrus (McEwen and Milner, 2007)), these changes are limited. ERα-containing dendritic spines are elevated at diestrus (low-estrogen phase) and lowest terminals and dendritic spines in males compared to females at all stages. PR-containing dendritic spines, and to a lesser extent dendritic spines, are elevated at proestrus and lowest in diestrus females compared to proestrus and estrus females and males. In addition to classic estrogen-sensitive receptors, the novel seven-transmembrane G-protein-coupled ER GPR30/GPER1 (Prossnitz et al., 2008) is expressed in the hippocampus (Brailoiu et al., 2007). Activation of GPR30 can regulate cholinergic synaptic efficacy and affect cognitive performance (Hammond and Gibbs, 2011). Like ERs and PRs, we have found that GPR30-ir is exclusively extranuclear in pyramidal cells, interneurons, and glia in the CA1 region. In neurons,

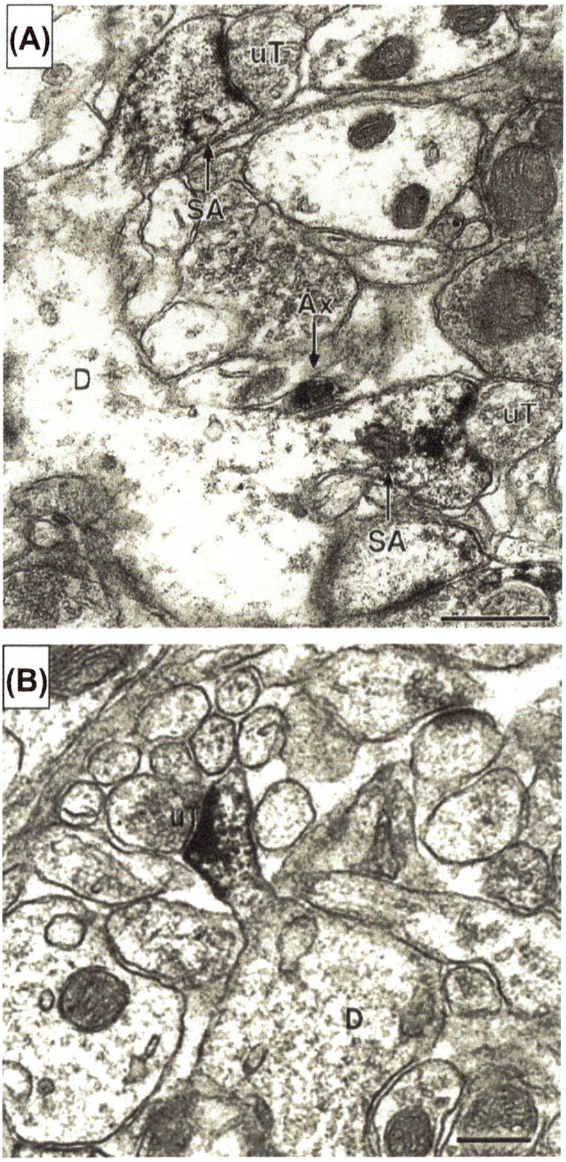

Figure 6 Electron micrographs show extranuclear ERs in postsynaptic profiles in the rat hippocampus. (A) ERα-ir is found in two dendritic spines emerging from a common dendritic process. Both spines contain a spine apparatus (SA) and are contacted by unlabeled terminals (uT). An ERα-labeled axon (Ax) abuts the dendritic profile. *From Milner et al., 2001. J. Comp. Neurol. 429, 355–371.* (B) An ERβ-labeled dendritic spine (arrowhead) is contacted by an unlabeled terminal (uT). Bars = 500 nm. *From Milner et al., 2005. J. Comp. Neurol. 491, 81–95.*

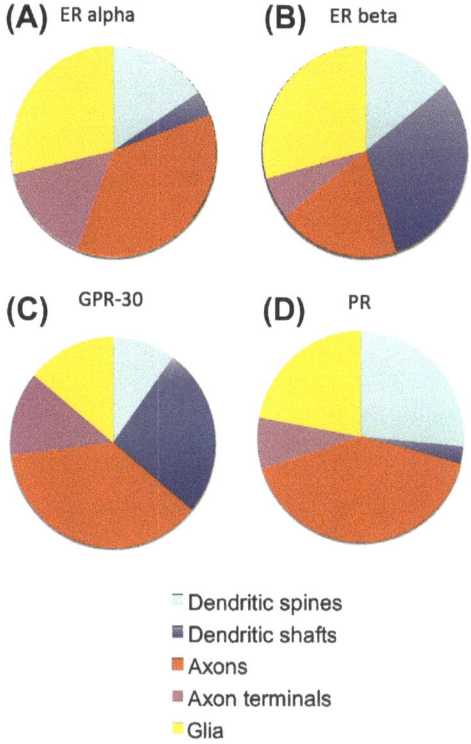

Figure 7 Schematic diagrams showing the distribution of the types of profiles containing estrogen-sensitive receptors in the mouse hippocampus. (A) ERα-ir. (B) ERβ-ir. (C) GPR30-ir. (D) PR-ir. The number of profiles are averaged from all hippocampal regions (CA1, CA3, and DG). A total of 3600 μm² per mouse (*N*=12 mice) was analyzed. Data was taken from stage with highest number of profiles (diestrus for A, B, and C; proestrus for D). *Data A, B, and D from Mitterling et al., 2010. J. Comp. Neurol. 518, 2729–2743; Data in C from Waters et al., unpublished.*

GPR30-ir is prominent on membranes. This observation agrees with the findings that GPER1 protein is localized at the plasma membrane (Funakoshi et al., 2006; Filardo et al., 2007) and in endoplasmic reticulum (Langer et al., 2010). However, unlike ERs and PRs, the numbers of GPR30-containing profiles do not fluctuate over the estrous cycle nor differ between females and males.

Estrogens also change the subcellular distribution of both ERα-ir and ERβ-ir in pre- and postsynaptic profiles. In particular, as shown in Figure 9, postembedding EM studies using an SynBin analysis have examined the cytoplasmic and synaptic distributions of both ERs in terminals and dendritic profiles in CA1. As shown in Figure 10 in aged OVXed rats, the percentage of synapses labeled for either ERα or ERβ is decreased compared to young OVXed rats. Moreover, estradiol treatment increases the percentage of ERβ, but not ERα, synapses in both young and aged rats (Figure 10). As shown in Figure 11, in young rats, estradiol decreases ERα-ir (Adams et al., 2002), but not ERβ-ir (Waters et al.,

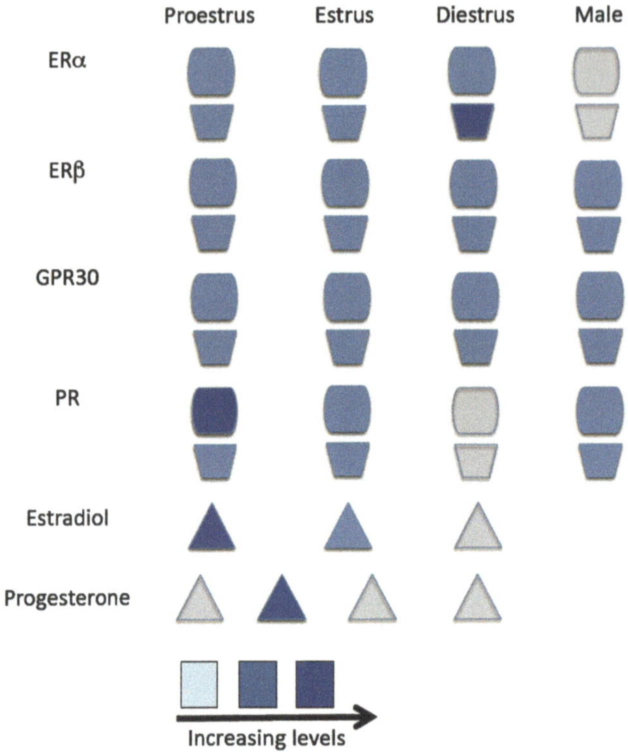

Figure 8 Summary diagram showing the levels of estrogen-sensitive receptors in the mouse CA1 region across the estrous cycle in female mice compared to male mice. Levels of know fluctuations in circulating estradiol and progesterone are indicated. *Data from Mitterling et al., 2010. J. Comp. Neurol. 518, 2729–2743.*

2011), in the distal synaptic compartments of both terminals and dendritic spines. However, as shown in Figure 12, estradiol increases ERβ-ir, but not ERα-ir, in the proximal pre- and postsynaptic compartments of young and aged OVXed rats (Adams et al., 2002; Waters et al., 2011). Estrogen regulation of the presynaptic cytoplasmic compartment, where synaptic vesicles reside, could affect neurotransmitter release (Ledoux et al., 2009; Becker, 1990) and reuptake (O'Malley et al., 1987). Estrogen regulation of the postsynaptic cytoplasmic compartment could promote synaptic potentiation and spine maturation.

Both the levels and subcellular distributions of estrogen-sensitive receptors likely contribute to the rapid effects of estrogens on synaptic transmission. ERα and ERβ have been shown to work in a complex, both complementary and sometimes antagonistic manner (Lindberg et al., 2003; Monroe et al., 2003). For example, treatment with ERα, but not ERβ, agonists rapidly increases dendritic spines in hippocampal CA1 stratum radiatum of OVXed females (Phan et al., 2011). Moreover, ERβ expression in hippocampal neurons attenuates estradiol-induced spine formation (Szymczak et al., 2006). As shown in

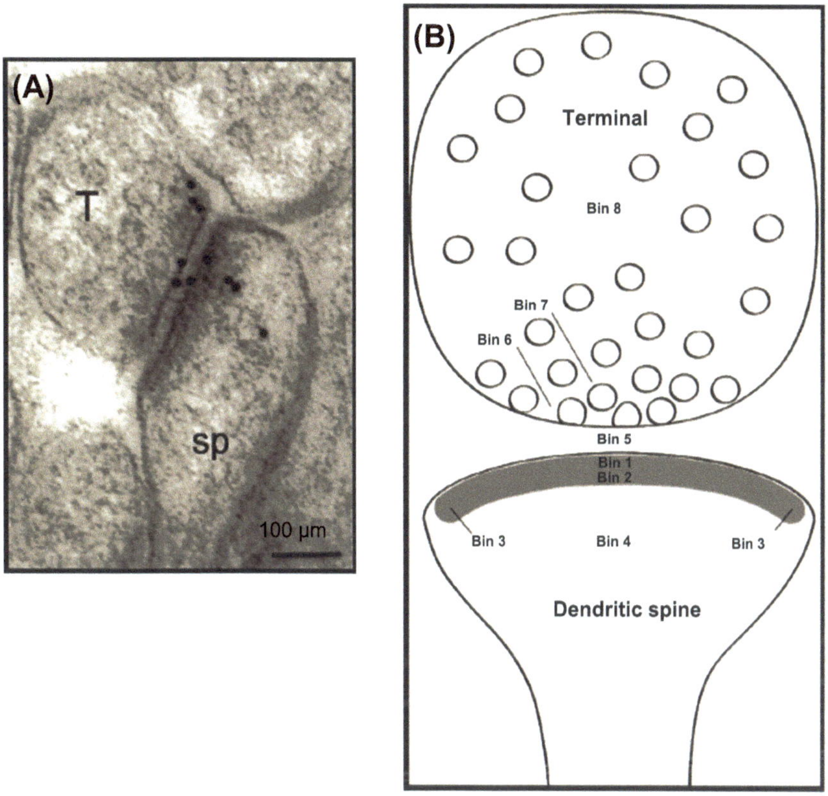

Figure 9 Representative electron micrograph of estrogen-sensitive receptors and schematic of Syn-bin analysis. (A) ERβ immunogold particles (black dots) are found in a terminal (T) and dendritic spine (sp). (B) Schematic drawing showing a pre- and postsynaptic profile with the eight Bin divisions used in the electron microscopic analysis. Eight zones were defined for each synapse: (1) two postsynaptic bins, the first bin 0–30 nm from the inner leaflet of the postsynaptic membrane [Bin 1] and the second bin 30–60 nm from the postsynaptic membrane [Bin 2]; (2) perisynaptic bins that were 15 nm lateral to both of the postsynaptic bins [Bin 3]; (3) a cytoplasmic bin that included immunogold particles >60 nm from the postsynaptic membrane [Bin 4]; (4) the synaptic cleft [Bin 5]; (5) two presynaptic zones, one extending 0–30 nm from the inner border of the presynaptic membrane [Bin 6] and the second extending 30–60 nm from the inner border of the presynaptic membrane [Bin 7]; and (6) a presynaptic bin that included immunogold particles >60 nm from the presynaptic membrane [Bin 8]. *From Waters et al., 2011. Brain Res. 1379, 86–97.*

Figure 13, these reports along with overlapping distribution of estrogen-sensitive receptor-containing profiles suggest a complex interplay between the various ERs to coordinate rapid and subchronic attributes of estrogen signaling and synaptic plasticity.

2.3.2. Signaling Molecules

In addition to direct and rapid effects on synaptic function and morphology, nonnuclear estrogen-sensitive receptors can initiate signaling pathways that regulate nuclear

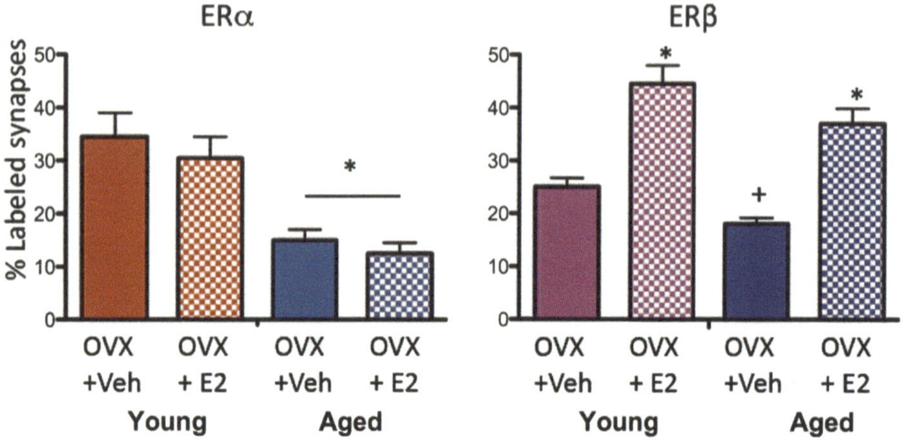

Figure 10 Comparison of the percentage of ERα- and ERβ-labeled synapses in young and aged Rats. Both the percentage of ERα- (left side) and ERβ (right side)-labeled synapses is decreased in aged OVXed rats compared to young OVXed rats. Estradiol treatment increases the percentage of ERβ-, but not ERα-, labeled synapses. *Data from Adams et al., 2002. J. Neurosci. 22, 3608–3614 (ERα) and Waters et al., 2011. Brain Res. 1379, 86–97 (ERβ).*

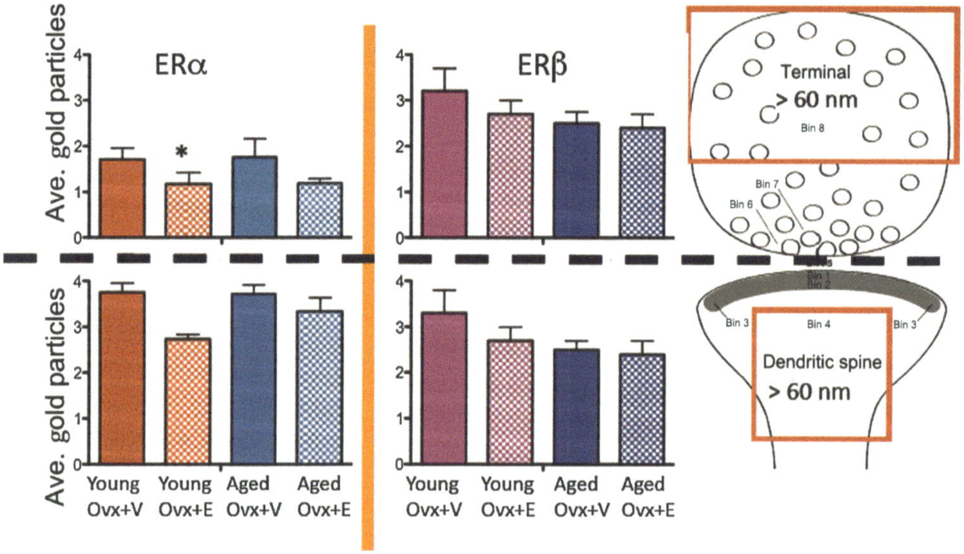

Figure 11 Estradiol regulates ERα but not ERβ in distal pre- and postsynaptic compartments. *Data from Adams et al., 2002. J. Neurosci. 22, 3608–3614 (ERα) and Waters et al., 2011. Brain Res. 1379, 86–97 (ERβ).*

transcriptional events (Razandi et al., 1999). Since a direct interaction between phosphoinositol-3 kinase (PI3K) and ERα was discovered (Simoncini et al., 2000), many kinase and phosphatases have been implicated in ERα and ERβ actions (Levin, 2001) likely mediated through an ER signaling complex. One estrogen-sensitive signaling pathway important for synapse function involves PI3K, and phosphorylated Akt (pAkt)

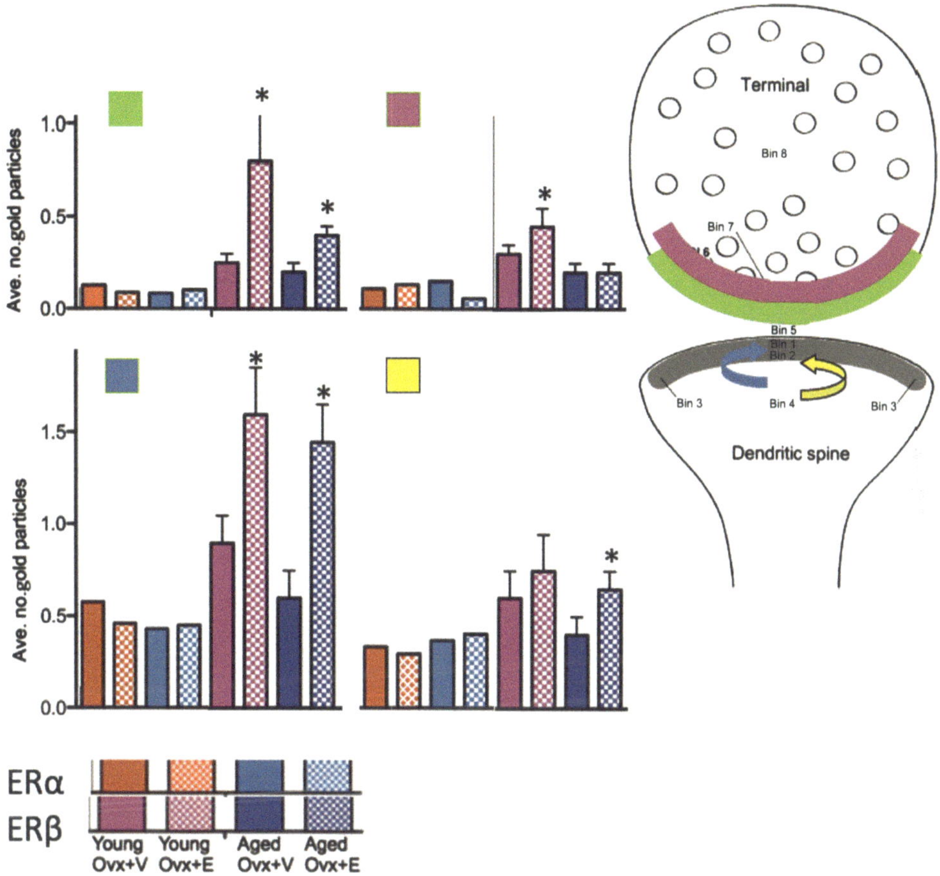

Figure 12 ERβ, not ERα, is upregulated by estradiol in proximal pre- and postsynaptic compartments. *Data from Adams et al., 2002. J. Neurosci. 22, 3608–3614 (ERα) and Waters et al., 2011. Brain Res. 1379, 86–97 (ERβ).*

(protein kinase B) (Akama and McEwen, 2003; Znamensky et al., 2003; Datta et al., 1999; Spencer et al., 2008; Yildirim et al., 2011). The pAkt pathway leads to PSD-95 de novo translation and spine maturation (Akama and McEwen, 2003). In the hippocampus, pAkt-ir is most prominent in the CA1 region where it is predominantly localized to the dendritic shafts and spines of pyramidal cells (Znamensky et al., 2003). The levels of pAkt-ir in CA1 stratum radiatum are highest in proestrus rats compared to estrus, diestrus, and male rats; these increases were particularly prominent on the plasma membranes of dendritic spines and shafts (Znamensky et al., 2003). Consistent with this observation, the levels of pAkt-ir are elevated in young OVXed wild-type mice 6 h, but not 48 h, after EB (Spencer-Segal et al., 2011a). In contrast, EB treatment of AERKO and BERKO mice did not affect the levels of pAKt-ir (Spencer-Segal et al., 2011a), suggesting that both ERs are critically involved in Akt signaling. In young OVXed rats, the overall percentage of pAkt immunoreactive axospinous synapses is increased following

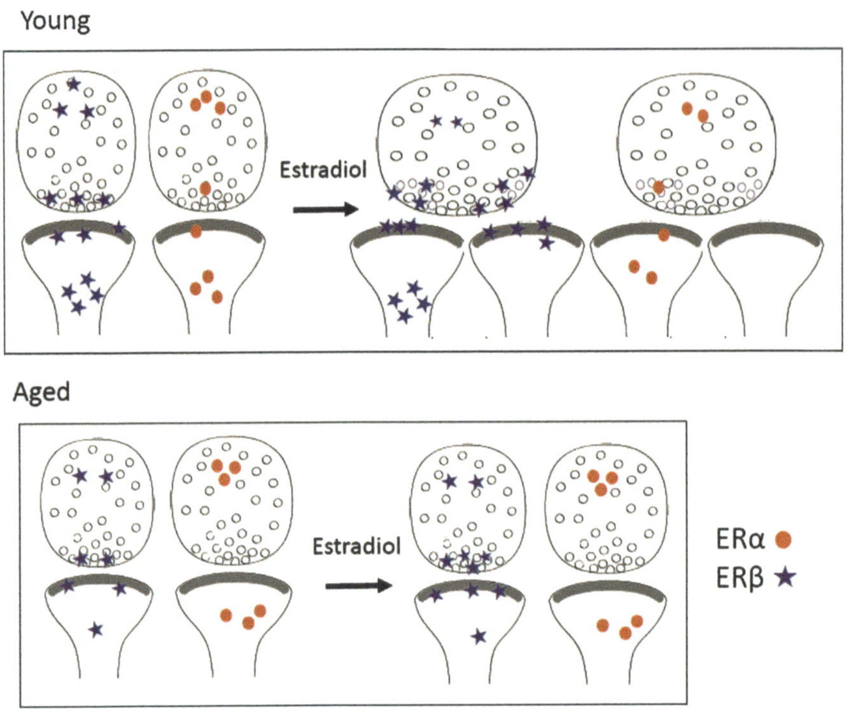

Figure 13 The changing ratio of ERα to ERβ may alter estradiol-induced plasticity in aged females. *Data from Adams et al., 2002. J. Neurosci. 22, 3608–3614 (ERα) and Waters et al., 2011. Brain Res. 1379, 86–97 (ERβ).*

estradiol treatment (Yildirim et al., 2011). Moreover, as shown in Figure 14, estradiol increases the levels of pAkt in both the synaptic and cytoplasmic postsynaptic compartments, synaptic cleft, and cytoplasmic presynaptic compartment in young OVXed rats (Yildirim et al., 2011).

Another signaling pathway important for synapse function involves phosphorylated LIM kinase (pLIMK), a serine/threonine kinase (Spencer et al., 2008; Yildirim et al., 2008; Yuen et al., 2011). The LIMK pathway is associated with actin polymerization and depolymerization through phosphorylation of cofilin (Arber et al., 1998; Yang et al., 1998; Yuen et al., 2011). Immunoreactivity for pLIMK is found predominantly in dendritic shafts and spines where it is often associated with the plasma membrane (Yildirim et al., 2008). Rarely, pLIMK is found in axon terminals (Yildirim et al., 2008). In naturally cycling mice, the levels of pLIMK in the hippocampal CA1 region are highest at proestrus females compared to estrus or diestrus females (Spencer et al., 2008). Consistent with this observation, estradiol increases the percentage of synaptic complexes containing pLIMK-ir in young OVXed rats (Figure 14; Yildirim et al., 2008). Moreover, estradiol increases the levels of pLIMK in the cytoplasmic compartment of dendritic spines in young OVXed rats (Yildirim et al., 2008).

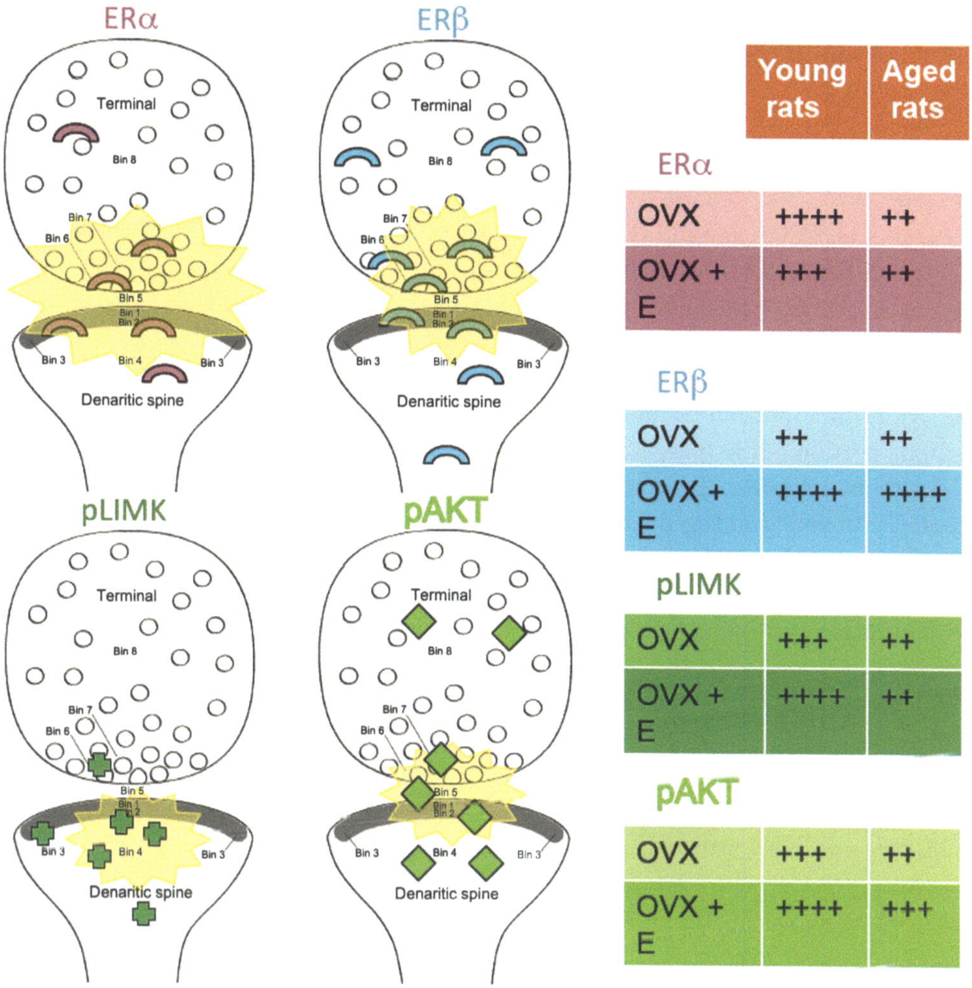

Figure 14 Schematic diagram showing that estrogen and aging differentially affect ERs and signaling molecules in pre- and postsynaptic compartments. Yellow indicated estrogen-sensitive regions. Plus symbols indicate relative differences. *Data from Adams et al., 2002. J. Neurosci. 22, 3608–3614 (ERα) and Waters et al., 2011. Brain Res. 1379, 86–97 (ERβ); Yildirim et al., 2008. Neuroscience 152, 360–370 (pLIMK); Yildirim et al., 2011. Brain Res. 1379, 98–108 (pAkt).*

2.4. CHANGES IN AGING

Estrogen-induced spine synapse formation is lost in aged (22 month-old) female rats (Adams et al., 2001). Moreover, synaptophysin-ir in the rat CA1 is decreased in middle-aged (11–12 months) and aged (22 months) old OVXed rats compared to young (4 month-old) rats (Figure 6; Williams et al., 2011b). However, in OVXed aged female rats (27–28 month-old), 5 days of 5 μg EB injections (5 μg/0.05 ml, s.c.) did increase synaptophysin in total hippocampal samples but a lower dose did not (1 μg/0.05 ml) (Frick et al., 2002). Age-related changes are accompanied by loss of synaptic complexes containing ERα-ir (Adams et al., 2002) and ERβ-ir (Waters et al., 2011) as well as the phosphorylated forms of the signaling molecules, Akt (Yildirim et al., 2011), and LIMK (Yildirim et al., 2008; Figure 14). Interestingly, the ability of estradiol to increase ERα-ir (Adams et al., 2002) and pLIMK (Yildirim et al., 2008), but not ERβ-ir (Waters et al., 2011) and pAkt (Yildirim et al., 2011), is lost in the aged hippocampus. However, in aged rats, the increases in ERβ-ir are limited to the synaptic cleft and postsynaptic compartments (Waters et al., 2011). Moreover, two differences in the effects of estradiol on the synaptic distributions of pAkt-ir are seen: estradiol decreases the levels of pAkt-ir in the cytoplasmic portion of the postsynaptic compartment and increases pAkt-ir in the synaptic portion of the presynaptic compartment (Yildirim et al., 2011).

In contrast to ERs, the overall synaptic levels of NR2A and NR2B in the rat CA1 are similar in young and aged female rats, regardless of treatment with estradiol. Moreover, estradiol restores NR2B levels back to young levels in the lateral portions of the active synaptic zone in the aged rats (Adams et al., 2004). Thus, estrogen may impact the mobility of NMDARs across the synapse and, in the process, restore a more youthful synaptic profile.

3. CONCLUSIONS

3.1. Summary

Estrogen regulation of spine synapses in the hippocampus involves distinct actions in both pre- and postsynaptic cellular compartments. The sum of these processes alters synaptic efficacy, synaptic structure, and network properties that could ultimately impact behavior. Moreover, as the hormonal milieu across the lifespan changes, the outcome of estrogen actions also is modified. The "estrogen-privilege" of females is established in early development and persists until menopause.

3.2. Unanswered Questions

The last two decades have yielded a wealth of information regarding estrogen's effects on hippocampal synapses. These discoveries have already begun to inform clinical practice. However, as more estrogen-sensitive receptors are characterized and more high-throughput techniques are created for characterizing synapses, more questions can now be addressed. These include the following:

- What is the interaction of classic ERs with estrogen-sensitive receptors (e.g. GPR30) and PRs in the regulation of synaptic function?
- Does the ratio of ERα:ERβ predict estrogen responses at synapses?
- What is the role of signaling pathways activated in the synapse in indirect genomic actions of estrogen?

- How do indirect genomic actions of estrogen alter direct local effects of estrogen on synaptic plasticity?
- How do hormonal changes in menopause and during the menopause transition affect synaptic function?
- Can hormone therapies be used in both males and females to facilitate successful aging and prevent neurodegeneration?

ACKNOWLEDGMENTS

GRANT SUPPORT: NIH grants HL098351, HL096571, DA08259 & AG016765 (TAM), AG039850 (EMW & TAM), NS07080, AG016765 (BSM)

REFERENCES

Adams, M.M., Fink, S.E., Janssen, W.G., Shah, R.A., Morrison, J.H., 2004. Estrogen modulates synaptic N-methyl-D-aspartate receptor subunit distribution in the aged hippocampus. J. Comp. Neurol. 474, 419–426.

Adams, M.M., Fink, S.E., Shah, R.A., Janssen, W.G., Hayashi, S., Milner, T.A., McEwen, B.S., Morrison, J.H., 2002. Estrogen and aging affect the subcellular distribution of estrogen receptor-alpha in the hippocampus of female rats. J. Neurosci. 22, 3608–3614.

Adams, M.M., Shah, R.A., Janssen, W.G., Morrison, J.H., 2001. Different modes of hippocampal plasticity in response to estrogen in young and aged female rats. Proc. Natl. Acad. Sci. USA 98, 8071–8076.

Akama, K.T., McEwen, B.S., 2003. Estrogen stimulates PSD-95 rapid protein synthesis via the Akt/PKB pathway. J. Neurosci. 23, 2333–2339.

Akama, K.T., Thompson, L.I., Milner, T.A., McEwen, B.S., 2013. Post-synaptic density-95 (PSD-95) binding capacity of G-protein-coupled receptor 30 (GPR30), an estrogen receptor that can be identified in hippocampal dendritic spines. J. Biol. Chem. 288, 6438–6450.

Allen, P.B., Ouimet, C.C., Greengard, P., 1997. Spinophilin, a novel protein phosphatase 1 binding protein localized to dendritic spines. Proc. Natl. Acad. Sci. USA 94, 9956–9961.

Amaral, D.G., Kurz, J., 1985. An analysis of the origins of the cholinergic and noncholinergic septal projections to the hippocampal formation of the rat. J. Comp. Neurol. 240, 37–59.

Arber, S., Barbayannis, F.A., Hanser, H., Schneider, C., Stanyon, C.A., Bernard, O., Caroni, P., 1998. Regulation of actin dynamics through phosphorylation of cofilin by LIM-kinase. Nature 393, 805–809.

Barker-Gibb, A.L., Dougherty, K.D., Einheber, S., Drake, C.T., Milner, T.A., 2001. Hippocampal tyrosine kinase A receptors are restricted primarily to presynaptic vesicle clusters. J. Comp. Neurol. 430, 182–199.

Becker, J.B., 1990. Direct effect of 17b-estradiol on striatum: sex differences in dopamine release. Synapse 5, 157–164.

Blank, T., Nijholt, I., Eckart, K., Spiess, J., 2002. Priming of long-term potentiation in mouse hippocampus by corticotropin-releasing factor and acute stress: implications for hippocampus-dependent learning. J. Neurosci. 22, 3788–3794.

Brailoiu, E., Dun, S.L., Brailoiu, G.C., Mizuo, K., Sklar, L.A., Oprea, T.I., Prossnitz, E.R., Dun, N.J., 2007. Distribution and characterization of estrogen receptor G protein-coupled receptor 30 in the rat central nervous system. J. Endocrinol. 193, 311–321.

Brake, W.G., Alves, S.E., Dunlop, J.C., Lee, S.J., Bulloch, K., Allen, P.B., Greengard, P., McEwen, B.S., 2001. Novel target sites for estrogen action in the dorsal hippocampus: an examination of synaptic proteins. Endocrinology 142, 1284–1289.

Calakos, N., Bennett, M.K., Peterson, K.E., Scheller, R.H., 1994. Protein-protein interactions contributing to the specificity of intracellular vesicular trafficking. Science 263, 1146–1149.

Calakos, N., Scheller, R.H., 1994. Vesicle-associated membrane protein and synaptophysin are associated on the synaptic vesicle. J. Biol. Chem. 269, 24534–24537.

Chao, M.V., 2003. Neurotrophins and their receptors: a convergence point for many signalling pathways. Nat. Rev. Neurosci. 4, 299–309.

Chaudhry, F.A., Reimer, R.J., Bellocchio, E.E., Danbolt, N.C., Osen, K.K., Edwards, R.H., Storm-Mathisen, J., 1998. The vesicular GABA transporter, VGAT, localizes to synaptic vesicles in sets of glycinergic as well as GABAergic neurons. J. Neurosci. 18, 9733–9750.

Chen, Y., Brunson, K.L., Adelmann, G., Bender, R.A., Frotscher, M., Baram, T.Z., 2004. Hippocampal corticotropin releasing hormone: pre- and postsynaptic location and release by stress. Neuroscience 126, 533–540.

Commons, K.G., Milner, T.A., 1997. Localization of delta opiate receptor immunoreactivity in interneurons and pyramidal cells in the rat hippocampus. J. Comp. Neurol. 381, 373–387.

Daniel, J.M., Dohanich, G.P., 2001. Acetylcholine mediates the estrogen-induced increase in NMDA receptor binding in CA1 of the hippocampus and the associated improvement in working memory. J. Neurosci. 21, 6949–6956.

Datta, S.R., Brunet, A., Greenberg, M.E., 1999. Cellular survival: a play in three Akts. Genes Dev. 13, 2905–2927.

Dougherty, K.D., Milner, T.A., 1999. Cholinergic septal afferent terminals preferentially contact neuropeptide Y-containing interneurons compared to parvalbumin-containing interneurons in the rat dentate gyrus. J. Neurosci. 19, 10140–10152.

Drake, C.T., Chavkin, C., Milner, T.A., 2007. Opioid systems in the dentate gyrus. Prog. Brain Res. 163, 245–814.

Drake, C.T., Milner, T.A., 1999. Mu opioid receptors are in somatodendritic and axonal compartments of GABAergic neurons in rat hippocampal formation. Brain Res. 849, 203–215.

Drake, C.T., Milner, T.A., 2002. Mu opioid receptors are in discrete hippocampal interneuron subpopulations. Hippocampus 12, 119–136.

Drake, C.T., Milner, T.A., Patterson, S.L., 1999. Ultrastructural localization of full-length trkB immunoreactivity in rat hippocampus suggests multiple roles in modulating activity-dependent synaptic plasticity. J. Neurosci. 19, 8009–8026.

Feng, J., Yan, Z., Ferreira, A., Tomizawa, K., Liauw, J.A., Zhuo, M., Allen, P.B., Ouimet, C.C., Greengard, P., 2000. Spinophilin regulates the formation and function of dendritic spines. Proc. Natl. Acad. Sci. USA 97, 9287–9292.

Filardo, E., Quinn, J., Pang, Y., Graeber, C., Shaw, S., Dong, J., Thomas, P., 2007. Activation of the novel estrogen receptor G protein-coupled receptor 30 (GPR30) at the plasma membrane. Endocrinology 148, 3236–3245.

Frick, K.M., Fernandez, S.M., Bulinski, S.C., 2002. Estrogen replacement improves spatial reference memory and increases hippocampal synaptophysin in aged female mice. Neuroscience 115, 547–558.

Frotscher, M., 1991. Target cell specificity of synaptic connections in the hippocampus. Hippocampus 1, 123–130.

Funakoshi, T., Yanai, A., Shinoda, K., Kawano, M.M., Mizukami, Y., 2006. G protein-coupled receptor 30 is an estrogen receptor in the plasma membrane. Biochem. Biophys. Res. Commun. 346, 904–910.

Gazzaley, A.H., Weiland, N.G., McEwen, B.S., Morrison, J.H., 1996. Differential regulation of NMDAR1 mRNA and protein by estradiol in the rat hippocampus. J. Neurosci. 16, 6830–6838.

Gibbs, R.B., 1996. Fluctuations in relative levels of choline acetyltransferase mRNA in different regions of the rat basal forebrain across the estrous cycle: effects of estrogen and progesterone. J. Neurosci. 16, 1049–1055.

Gibbs, R.B., 2000. Effects of gonadal hormone replacement on measures of basal forebrain cholinergic function. Neuroscience 101, 931–938.

Gibbs, R.B., 2009. Estrogen therapy and cognition: a review of the cholinergic hypothesis. Endocr. Rev. 31, 224–253.

Hammond, R., Gibbs, R.B., 2011. GPR30 is positioned to mediate estrogen effects on basal forebrain cholinergic neurons and cognitive performance. Brain Res. 1379, 53–60.

Hammond, R., Nelson, D., Gibbs, R.B., 2011. GPR30 co-localizes with cholinergic neurons in the basal forebrain and enhances potassium-stimulated acetylcholine release in the hippocampus. Psychoneuroendocrinology 36, 182–192.

Hart, S.A., Snyder, M.A., Smejkalova, T., Woolley, C.S., 2007. Estrogen mobilizes a subset of estrogen receptor-alpha-immunoreactive vesicles in inhibitory presynaptic boutons in hippocampal CA1. J. Neurosci. 27, 2102–2111.

Kim, E., Sheng, M., 2004. PDZ domain proteins of synapses. Nat. Rev. Neurosci. 5, 771–781.
Kovacs, E.G., MacLusky, N.J., Leranth, C., 2003. Effects of testosterone on hippocampal CA1 spine synaptic density in the male rat are inhibited by fimbria/fornix transection. Neuroscience 122, 807–810.
Langer, G., Bader, B., Meoli, L., Isensee, J., Delbeck, M., Noppinger, P.R., Otto, C., 2010. A critical review of fundamental controversies in the field of GPR30 research. Steroids 75, 603–610.
Ledoux, V.A., Smejkalova, T., May, R.M., Cooke, B.M., Woolley, C.S., 2009. Estradiol facilitates the release of neuropeptide Y to suppress hippocampus-dependent seizures. J. Neurosci. 29, 1457–1468.
Leranth, C., Petnehazy, O., MacLusky, N.J., 2003. Gonadal hormones affect spine synaptic density in the CA1 hippocampal subfield of male rats. J. Neurosci. 23, 1588–1592.
Levin, E.R., 2001. Cell localization, physiology, and nongenomic actions of estrogen receptors. J. Appl. Physiol. 91, 1860–1867.
Lewis, C., McEwen, B.S., Frankfurt, M., 1995. Estrogen-induction of dendritic spines in ventromedial hypothalamus and hippocampus: effects of neonatal aromatase blockade and adult GDX. Brain Res. Dev. Brain Res. 87, 91–95.
Li, C., Brake, W.G., Romeo, R.D., Dunlop, J.C., Gordon, M., Buzescu, R., Magarinos, A.M., Allen, P.B., Greengard, P., Luine, V., McEwen, B.S., 2004. Estrogen alters hippocampal dendritic spine shape and enhances synaptic protein immunoreactivity and spatial memory in female mice. Proc. Natl. Acad. Sci. USA 101, 2185–2190.
Lindberg, M.K., Moverare, S., Skrtic, S., Gao, H., Dahlman-Wright, K., Gustafsson, J.A., Ohlsson, C., 2003. Estrogen receptor (ER)-beta reduces ERalpha-regulated gene transcription, supporting a "ying yang" relationship between ERalpha and ERbeta in mice. Mol. Endocrinol. 17, 203–208.
McEwen, B.S., Gould, E., Orchinik, M., Weiland, N.G., Woolley, C.S., 1995. Oestrogens and the structural and functional plasticity of neurons: implications for memory, ageing and neurodegenerative processes. Ciba Found. Symp. 191, 52–66 discussion 66–73, 52–66.
McEwen, B.S., Milner, T.A., 2007. Hippocampal formation: shedding light on the influence of sex and stress on the brain. Brain Res. Rev. 55, 343–355.
McMillan, P.J., Singer, C.A., Dorsa, D.M., 1996. The effects of ovariectomy and estrogen replacement on trkA and choline acetyltransferase mRNA expression in the basal forebrain of the adult female Sprague-Dawley rat. J. Neurosci. 16, 1860–1865.
Milner, T.A., Ayoola, K., Drake, C.T., Herrick, S.P., Tabori, N.E., McEwen, B.S., Warrier, S., Alves, S.E., 2005. Ultrastructural localization of estrogen receptor beta immunoreactivity in the rat hippocampal formation. J. Comp. Neurol. 491, 81–95.
Milner, T.A., Lubbers, L.S., Alves, S.E., McEwen, B.S., 2008. Nuclear and extranuclear estrogen binding sites in the rat forebrain and autonomic medullary areas. Endocrinology 149, 3306–3312.
Milner, T.A., McEwen, B.S., Hayashi, S., Li, C.J., Reagan, L.P., Alves, S.E., 2001. Ultrastructural evidence that hippocampal alpha estrogen receptors are located at extranuclear sites. J. Comp. Neurol. 429, 355–371.
Milner, T.A., Veznedaroglu, E., 1992. Ultrastructural localization of neuropeptide Y-like immunoreactivity in the rat hippocampal formation. Hippocampus 2, 107–126.
Milner, T.A., Wiley, R.G., Kurucz, O.S., Prince, S.R., Pierce, J.P., 1997. Selective changes in hippocampal neuropeptide Y neurons following removal of the cholinergic septal inputs. J. Comp. Neurol. 386, 46–59.
Mitterling, K.L., Spencer, J.L., Dziedzic, N., Shenoy, S., McCarthy, K., Waters, E.M., McEwen, B.S., Milner, T.A., 2010. Cellular and subcellular localization of estrogen and progestin receptor immunoreactivities in the mouse hippocampus. J. Comp. Neurol. 518, 2729–2743.
Monroe, D.G., Johnsen, S.A., Subramaniam, M., Getz, B.J., Khosla, S., Riggs, B.L., Spelsberg, T.C., 2003. Mutual antagonism of estrogen receptors alpha and beta and their preferred interactions with steroid receptor coactivators in human osteoblastic cell lines. J. Endocrinol. 176, 349–357.
Morissette, M., Le, S.M., Di, P.T., 2008. Effect of oestrogen receptor alpha and beta agonists on brain N-methyl-D-aspartate receptors. J. Neuroendocrinol. 20, 1006–1014.
Murphy, D.D., Cole, N.B., Greenberger, V., Segal, M., 1998. Estradiol increases dendritic spine density by reducing GABA neurotransmission in hippocampal neurons. J. Neurosci. 18, 2550–2559.
O'Malley, C.A., Hautamaki, M., Kelley, M., Meyer, E.M., 1987. Effects of ovariectomy and estradiol benzoate on high affinity choline uptake, ACh synthesis, and release from rat cerebral cortical synaptosomes. Brain Res. 403, 389–392.

Phan, A., Lancaster, K.E., Armstrong, J.N., MacLusky, N.J., Choleris, E., 2011. Rapid effects of estrogen receptor alpha and beta selective agonists on learning and dendritic spines in female mice. Endocrinology 152, 1492–1502.
Prossnitz, E.R., Sklar, L.A., Oprea, T.I., Arterburn, J.B., 2008. GPR30: a novel therapeutic target in estrogen-related disease. Trends Pharmacol. Sci. 29, 116–123.
Razandi, M., Pedram, A., Greene, G.L., Levin, E.R., 1999. Cell membrane and nuclear estrogen receptors (ERs) originate from a single transcript: studies of ERa and ERb expressed in Chinese Hamster ovary cells. Mol. Endocrinol. 13, 307–319.
Romeo, R.D., McCarthy, J.B., Wang, A., Milner, T.A., McEwen, B.S., 2005. Sex differences in hippocampal estradiol-induced N-methyl-D-aspartic acid binding and ultrastructural localization of estrogen receptor-alpha. Neuroendocrinology 81, 391–399.
Rudick, C.N., Gibbs, R.B., Woolley, C.S., 2003. A role for the basal forebrain cholinergic system in estrogen-induced disinhibition of hippocampal pyramidal cells. J. Neurosci. 23, 4479–4490.
Rudick, C.N., Woolley, C.S., 2003. Selective estrogen receptor modulators regulate phasic activation of hippocampal CA1 pyramidal cells by estrogen. Endocrinology 144, 179–187.
Sabaliauskas, N., Molla, J., Aoki, C., Smith, S.S., 2010. Estrous cycle-related changes in CNS plasticity: role of $GABA_A$ receptors. Soc. Neurosci. Abstr. 443.5.
Schierloh, A., Deussing, J., Wurst, W., Zieglgansberger, W., Rammes, G., 2007. Corticotropin-releasing factor (CRF) receptor type 1-dependent modulation of synaptic plasticity. Neurosci. Lett. 416, 82–86.
Simoncini, T., Hafezi-Moghadam, A., Brazil, D.P., Ley, K., Chin, W.W., Liao, J.K., 2000. Interaction of oestrogen receptor with the regulatory subunit of phosphatidylinositol-3-OH kinase. Nature 407, 538–541.
Smith, S.S., Gong, Q.H., Hsu, F.C., Markowitz, R.S., Ffrench-Mullen, J.M.H., Li, X., 1998. GABA(A) receptor alpha4 subunit suppression prevents withdrawal properties of an endogenous steroid. Nature 392, 926–930.
Snyder, M.A., Cooke, B.M., Woolley, C.S., 2011. Estradiol potentiation of NR2B-dependent EPSCs is not due to changes in NR2B protein expression or phosphorylation. Hippocampus 21, 398–408.
Spencer, J.L., Waters, E.M., Milner, T.A., Lee, F.S., McEwen, B.S., 2010. BDNF variant Val66Met interacts with estrous cycle in the control of hippocampal function. Proc. Natl. Acad. Sci. USA 107, 4395–4400.
Spencer, J.L., Waters, E.M., Milner, T.A., McEwen, B.S., 2008. Estrous cycle regulates activation of hippocampal Akt, LIM kinase, and neurotrophin receptors in C57BL/6 mice. Neuroscience 155, 1106–1119.
Spencer-Segal, J.L., Tsuda, M.C., Mattei, L., Waters, E.M., Romeo, R.D., Milner, T.A., McEwen, B.S., Ogawa, S., 2011a. Estradiol acts via estrogen receptors alpha and beta on pathways important for synaptic plasticity in the mouse hippocampal formation. Neuroscience (e-pub).
Spencer-Segal, J.L., Waters, E.M., Bath, K.G., Chao, M.V., McEwen, B.S., Milner, T.A., 2011b. Distribution of phosphorylated TrkB receptor in the mouse hippocampal formation depends on sex and estrous cycle stage. J. Neurosci. 31, 6780–6790.
Szymczak, S., Kalita, K., Jaworski, J., Mioduszewska, B., Savonenko, A., Markowska, A., Merchenthaler, I., Kaczmarek, L., 2006. Increased estrogen receptor beta expression correlates with decreased spine formation in the rat hippocampus. Hippocampus 16, 453–463.
Tabatadze, N., Smejkalova, T., Woolley, C.S., 2013. Distribution and posttranslational modification of synaptic ERalpha in the adult female rat hippocampus. Endocrinology 154, 819–830.
Torres-Reveron, A., Williams, T.J., Chapleau, J.D., Waters, E.M., McEwen, B.S., Drake, C.T., Milner, T.A., 2009. Ovarian steroids alter mu opioid receptor trafficking in hippocampal parvalbumin GABAergic interneurons. Exp. Neurol. 219, 319–327.
Towart, L.A., Alves, S.E., Znamensky, V., Hayashi, S., McEwen, B.S., Milner, T.A., 2003. Cellular relationships between cholinergic terminals and estrogen receptor alpha in the dorsal hippocampus. J. Comp. Neurol. 463, 390–401.
Wang, H.L., Wayner, M.J., Chai, C.Y., Lee, E.H.Y., 1998. Corticotrophin-releasing factor produces a long-lasting enhancement of synaptic efficacy in the hippocampus. Euro. J. Neurosci. 10, 3428–3437.
Waters, E.M., Mitterling, K., Spencer, J.L., Mazid, S., McEwen, B.S., Milner, T.A., 2009. Estrogen receptor alpha and beta specific agonists regulate expression of synaptic proteins in rat hippocampus. Brain Res. 1290, 1–11.
Waters, E.M., Torres-Reveron, A., McEwen, B.S., Milner, T.A., 2008. Ultrastructural localization of extranuclear progestin receptors in the rat hippocampal formation. J. Comp. Neurol. 511, 34–46.

Waters, E.M., Yildirim, M., Janssen, W.G.M., Lou, W.Y., McEwen, B.S., Milner, T.A., 2011. Estrogen and aging affect the synaptic distribution of estrogen receptor beta-immunoreactivity in the CA1 region of the female rat hippocampus. Brain Res. 1379, 86–97.

Weiland, N.G., 1992. Estradiol selectively regulates agonist binding sites on the N-methyl-D-aspartate receptor complex in the CA1 region of the hippocampus. Endocrinology 131, 662–668.

Williams, C.L., Meck, W.H., 1991. The organizational effects of gonadal steroids on sexually dimorphic spatial ability. Psychoneuroendocrinology 16, 155–176.

Williams, T.J., Akama, K.T., Knudsen, M.G., McEwen, B.S., Milner, T.A., 2011a. Ovarian hormones influence corticotropin releasing factor receptor colocalization with delta opioid receptors in CA1 pyramidal cell dendrites. Exp. Neurol. 230, 186–196.

Williams, T.J., Milner, T.A., 2011. Delta opioid receptors colocalize with corticotropin releasing factor in hippocampal interneurons. Neuroscience 179, 9–22 Epub; 26 January 2011.

Williams, T.J., Mitterling, K.L., Thompson, L.I., Torres-Reveron, A., Waters, E.M., McEwen, B.S., Gore, A.C., Milner, T.A., 2011b. Age- and hormone-regulation of opioid peptides and synaptic proteins in the rat dorsal hippocampal formation. Brain Res. 1379, 71–85.

Williams, T.J., Torres-Reveron, A., Chapleau, J.D., Milner, T.A., 2011c. Hormonal regulation of delta opioid receptor immunoreactivity in interneurons and pyramidal cells in the rat hippocampus. Neurobiol. Learn. Mem. 95, 206–220.

Woolley, C.S., 1999. Effects of estrogen in the CNS. Curr. Opin. Neurobiol. 9, 349–354.

Woolley, C.S., Gould, E., Frankfurt, M., McEwen, B.S., 1990. Naturally occurring fluctuation in dendritic spine density on adult hippocampal pyramidal neurons. J. Neurosci. 10, 4035–4039.

Woolley, C.S., McEwen, B.S., 1994. Estradiol regulates hippocampal dendritic spine density via an N-methyl-D-aspartate receptor-dependent mechanism. J. Neurosci. 14, 7680–7687.

Yang, N., Higuchi, O., Ohashi, K., Nagata, K., Wada, A., Kangawa, K., Nishida, E., Mizuno, K., 1998. Cofilin phosphorylation by LIM-kinase 1 and its role in Rac-mediated actin reorganization. Nature 393, 809–812.

Yildirim, M., Janssen, W.G., Lou, W.Y., Akama, K.T., McEwen, B.S., Milner, T.A., Morrison, J.H., 2011. Effects of estrogen and aging on the synaptic distribution of phosphorylated Akt-immunoreactivity in the CA1 region of the female rat hippocampus. Brain Res. 1379, 98–108.

Yildirim, M., Janssen, W.G., Tabori, N.E., Adams, M.M., Yuen, G.S., Akama, K.T., McEwen, B.S., Milner, T.A., Morrison, J.H., 2008. Estrogen and aging affect synaptic distribution of phosphorylated LIM kinase (pLIMK) in CA1 region of female rat hippocampus. Neuroscience 152, 360–370.

Yuen, G.S., McEwen, B.S., Akama, K.T., 2011. LIM kinase mediates estrogen action on the actin depolymerization factor Cofilin. Brain Res. 1379, 44–52.

Znamensky, V., Akama, K.T., McEwen, B.S., Milner, T.A., 2003. Estrogen levels regulate the subcellular distribution of phosphorylated Akt in hippocampal CA1 dendrites. J. Neurosci. 23, 2340–2347.

CHAPTER EIGHT

Trafficking of Glutamate Receptors and Associated Proteins in Synaptic Plasticity

Martin Horak[1], Gail K. Seabold[2] and Ronald S. Petralia[2]
[1]Institute of Physiology AS CR, v.v.i., Videnska, Czech Republic
[2]NIDCD/NIH, Bethesda, MD

1. INTRODUCTION

Glutamate receptors (GluRs) are one of the key players in the processes mediating learning and memory, both during development and in adults. This is largely because GluRs are the main mediators of fast excitatory neurotransmission in the central nervous system, as well as being key regulators of changes in neuronal communications. Fast excitatory neurotransmission typically occurs at synapses between the axons of neurons and the dendrites of other neurons. The central mechanism involved in learning and memory begins with glutamate release from presynaptic terminals on the axon of one neuron, and consequent binding of glutamate to postsynaptic GluRs (there are presynaptic and adjacent glial GluRs, too). Postsynaptic GluRs are associated with and regulated by a large number of proteins that together constitute the postsynaptic density (PSD). Changes in the trafficking, localization and activation of GluRs and their associated proteins at synapses are a major mechanism of synaptic plasticity and underlie learning and memory. Not surprisingly, the central role of GluRs in normal brain functions also makes them key proteins affected in pathology of the nervous system.

In this review, we will examine the trafficking of GluRs and their associated proteins, especially as concerns the plasticity of synapses (see Gereau and Swanson, 2008, for a comprehensive review of GluRs). Broadly speaking, there are 3 kinds of GluRs in animals, including (1) metabotropic GluRs (mGluRs: G-protein coupled receptors that typically form dimers, and are made of subunits that have seven transmembrane domains; mGlu1-8 or mGluR1-8; Nicoletti et al., 2011), (2) glutamate-gated chloride channels found in some invertebrates and related to glycine, GABA, serotonin, and acetylcholine receptors (Cys-loop receptors; Hibbs and Gouaux, 2011), and (3) classical ionotropic GluRs (iGluRs; Gereau and Swanson, 2008); we will concentrate only on the latter group in this review. iGluRs are ion-passing channels that open in response to glutamate and related agonists, and pass sodium and potassium and sometimes calcium. iGluRs are found in both plants and animals (Janovjak et al.,

The Synapse
http://dx.doi.org/10.1016/B978-0-12-418675-0.00008-0

2011); among animals, they are found even in placozoans and basal eumetazoans such as hydras and sea anemones (Ryan and Grant, 2009; Alie and Manuel, 2010; Croset et al., 2010; Srivastava et al., 2010). They probably originated from simpler bacterial GluRs that pass potassium selectively (GluR0). A possible intermediate GluR called AvGluR1 is found only in bdelloid rotifers and has the complex form of an iGluR but is potassium selective (Janovjak et al., 2011). iGluRs are assembled as dimers of dimers (i.e., tetramers) that are assembled in the endoplasmic reticulum (ER). Each subunit begins with a large extracellular amino-terminal domain (NTD) followed by a ligand-binding domain (S1), a transmembrane domain (M1), and an ion pore-lining loop (M2) region. RNA editing of an amino acid residue (glutamine to arginine) in the M2 region renders the channel calcium impermeable. The M2 pore loop is followed by another transmembrane domain (M3), second extracellular ligand-binding domain (S2; glutamate binds between S1 and S2), final transmembrane domain (M4), and intracellular C-terminal domain. iGluRs also may have regulatory auxiliary subunit proteins such as TARPs, CNIH, and NETO-1/2 (Jackson and Nicoll 2011; described in a later section of this chapter).

1.1. iGluR classification

The iGluRs are divided into three major groups, based on their specific agonists-kainate, α-amino-3-hydroxy-5-methyl-4-isoxazolepropionic acid (AMPA) and *N*-methyl-D-aspartate (NMDA) (Figure 1; Table 1) (Traynelis et al., 2010). It is generally accepted that all functional iGluRs are composed of four subunits. There are five subtypes of kainate receptor (KAR) subunits, GluK1, GluK2, GluK3, GluK4 and GluK5, which are arranged in different ways to form a functional tetramer. The GluK1-3 subunits can form homomers and heteromers while GluK4 and GluK5 subunits form functional receptors only by combining with one of the GluK1-3 subunits. Four different genes, GluA1, GluA2, GluA3 and GluA4, encode AMPAR subunits. The AMPARs expressed in the mammalian brain form mainly GluA1/GluA2, GluA2/GluA3 and GluA1/GluA4 heterotetramers, although GluA1 homomeric receptors are likely present as well. Three major subunits of NMDARs have been identified, GluN1, GluN2 and GluN3. The GluN1 subunit is expressed from a single gene, but has eight different splice variants. There are four different genes for GluN2 subunits, GluN2A, GluN2B, GluN2C and GluN2D, and two genes for GluN3 subunits, GluN3A and GluN3B. The functional NMDARs are mainly made up of two GluN1 and two GluN2 subunits. The GluN3 subunit can replace one GluN2 subunit in the tetramer and thus helps to form a GluN1/GluN2/GluN3 receptor or it can form a functional glycine receptor composed of GluN1/GluN3 subunits. There are also the orphan glutamate delta receptors (GluD1 and GluD2) that belong to the family of iGluRs (Lomeli et al., 1993). GluD2 is expressed in the Purkinje cells of the cerebellum; GluD1 is expressed in neurons of the hippocampus.

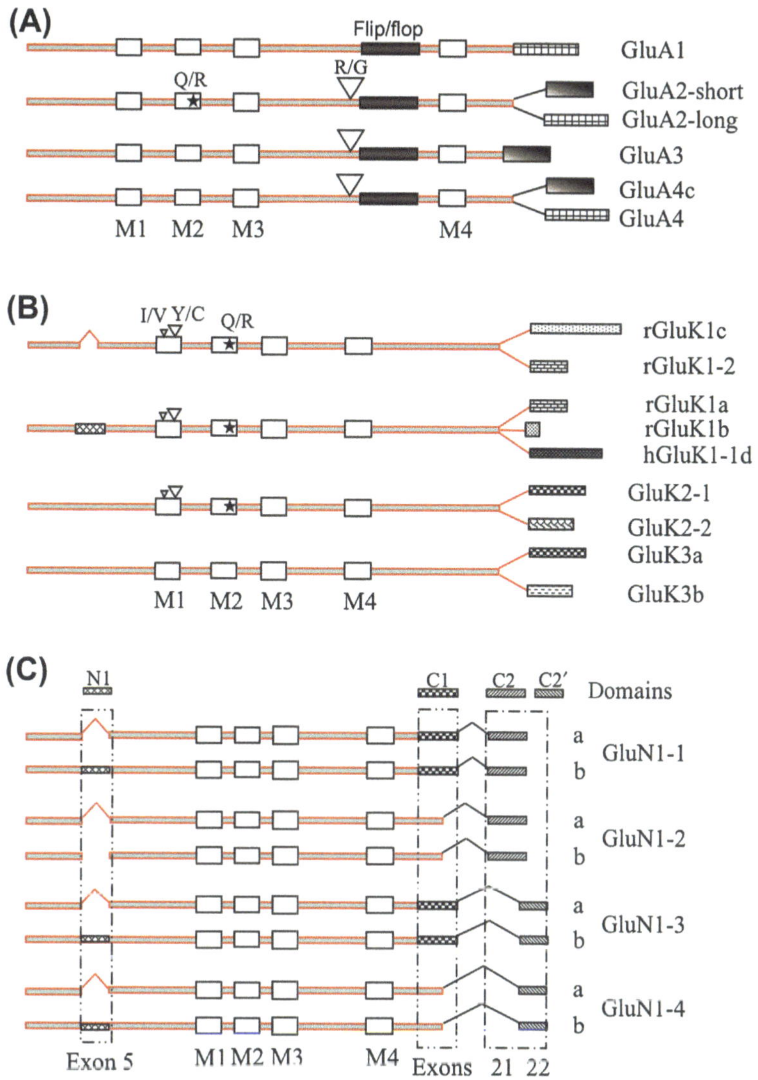

Figure 1 Alternative splicing and editing of glutamate receptor subunits. The subunits of glutamate receptors exhibiting alternative splicing and editing are shown. All subunits contain four membrane domains (M1–M4), an extracellular N-terminus, and an intracellular C-terminus (see text for more details). (A) All four AMPAR subunits are found in two alternatively spliced variants, flip and flop, encoded by different exons (e.g. exons 14 and 15 in GluA2) positioned close to the M4 domain. The C-termini of AMPAR subunits (GluA2 and GluA4) are alternatively spliced. The short versions of GluA4 and GluA2-short have homologous C-termini. The GluA2 subunit contains two editing sites that are indicated above the subunit diagram (Q/R, R/G). (B) Three of the KAR subunits (GluK1, GluK2, and GluK3) are alternatively spliced in their C-termini and furthermore the GluK1 contains a 15-amino acid-long exon in its N-terminus in some variants. The specific editing sites are indicated above each subunit diagram (I/V, Y/C, and Q/R). (C) The GluN1 subunit has three alternative exons: exon 5 in the N-terminus (N1 cassette) and exons 21 and 22 (C1 and C2 cassettes). Exon 22 contains an alternate acceptor splice site; when this site is used, an alternative cassette C2′ is expressed. *Figure based on Dingledine et al. (1999).*

Table 1 The Mammalian Subunits of Glutamate Ionotropic Receptors

IUPHAR* Names	Common Names
GluA1	GluR1, GluRA
GluA2	GluR2, GluRB
GluA3	GluR3, GluRC
GluA4	GluR4, GluRD
GluK1	GluR5
GluK2	GluR6
GluK3	GluR7
GluK4	KA1
GluK5	KA2
GluN1	NMDAR1, NR1, GluRξ1
GluN2A	NMDAR2A, NR2A, GluRε1
GluN2B	NMDAR2B, NR2B, GluRε2
GluN2C	NMDAR2C, NR2C, GluRε3
GluN2D	NMDAR2D, NR2D, GluRε4
GluN3A	NR3A
GluN3B	NR3B
GluD1	δ1, GluR delta-1
GluD2	δ2, GluR delta-2

*International Union of Basic and Clinical Pharmacology.
Based on Traynelis et al. (2010).

1.2. Synaptic organization

Postsynaptic iGluRs are only one component of a large complex of proteins that make up the PSD (Feng and Zhang, 2009; Ryan and Grant, 2009; Alie and Manuel, 2010; Zheng et al., 2011; Fig 2). In fact, most of the basic components of the PSD evolved in unicellular ancestors of multicellular animals, probably developed as a structure to sense the environment and to regulate calcium signaling and cytoskeleton organization (Ryan and Grant, 2009; Alie and Manuel, 2010; Srivastava et al., 2010). iGluRs became a component of these complexes with the earliest appearance of a central nervous system and neuronal synapses and thus of PSDs. Associated with the PSD are cytoplasmic proteins in the adjacent zone around the PSD and in the associated perisynaptic portion of the cell membrane (Perez-Otano et al., 2006; Newpher and Ehlers, 2008). Note that in this review, the perisynaptic zone is considered a subset of the extrasynaptic surface, which is all of the neuron surface outside of the postsynaptic membrane, as marked by the presence of the PSD (Petralia et al., 2010; Petralia, 2012). The side of a synaptic spine is a zone of great activity, especially where iGluRs and other synaptic membrane proteins can be trafficked via local exocytosis and endocytosis (Kennedy and Ehlers, 2011). Note also that there is a complex of presynaptic proteins (i.e., associated with release of glutamate), as well as a series of proteins that bridge and/or signal across the synaptic cleft, thus organizing pre- and postsynaptic components into a single functioning entity (Srivastava et al., 2010; Siddiqui and Craig, 2011).

The primary scaffolding proteins of the PSD and related protein complexes (reviewed in Petralia et al., 2009; Zheng et al,. 2011) are the membrane-associated guanylate kinases (MAGUKs) and proline-rich domain proteins such as Shank or Whirlin (Alie and Manuel, 2010; Zheng et al., 2011); indeed, they probably formed the prime scaffold of a protein complex of our unicellular ancestors. Both kinds of proteins contain PDZ domains (PSD-95, Dlg [discs large homolog], and ZO-1 [zona occludens or tight junction]) that can bind to many other proteins. The latter proteins have a PDZ-binding domain, typically on their C-terminus, and include many iGluRs. In addition, MAGUKs contain a GK (guanylate kinase) domain. The best-known MAGUKs are the PSD-95 group (SAP97, PSD-93 or Chapsyn 110, SAP102, and PSD-95 or SAP90; they also are designated DLG1-4); they contain 3 PDZ domains and single SH3 (Src homology 3) and GK domains. Other common ones are the MAGIs (including S-SCAM) and CASK, which is mainly presynaptic. MAGUKs and proline-rich domain PDZ proteins such as Shank can join together directly or through connecting proteins such as GKAP and IRSp53 (insulin receptor phosphotyrosine 53 kDa substrate). A common arrangement in PSDs is for GKAP to link the GK domain of PSD-95 to the PDZ domain of Shank. In addition, a number of signaling proteins such as SynGAP, CRIPT, NOS and Citron bind to MAGUKs. Common components linked in this system are various kinases that phosphorylate PSD components such as the iGluRs to affect their activity or density (i.e., synaptic plasticity); most prominent is calcium/calmodulin-dependent protein kinase II (CaMKII), which can contribute a large part of the total structure of the PSD. The scaffold is linked to the adjacent actin cytoskeleton via other proteins such as SPAR, which binds to the GK domain of MAGUKs, and cortactin, which binds to Shank. Shank can link to various receptors via dimeric Homer proteins, e.g., to mGluRs in the perisynaptic zone or IP3 receptors (a type of calcium channel) on adjacent smooth ER. MAGUKs also can be linked together, via adhesion proteins, across the synaptic cleft; e.g., neurexin/neuroligin pairs of adhesion proteins link a PDZ domain of postsynaptic PSD-95 or MAGI to that of presynaptic CASK. Another example is a link between postsynaptic MAGI and presynaptic actin via a cadherin/catenin pair of adhesion proteins. Thus, the PSD forms a highly-integrated complex linking iGluRs, MAGUKs, Shank, and various other receptors, adhesion proteins, and signaling proteins, maintaining a close arrangement for coordinating the initial signaling via ion entry to various secondary signaling pathways. All of this controls the plasticity of the synapse, making it an ideal mechanism for mediating learning and memory.

2.1. DEVELOPMENT

2.1.1. Synaptogenesis

During the initial formation of a synapse, one or more cell adhesion molecules (CAMs) make contact between the pre- and postsynaptic sides of the synapse. This contact then allows for the recruitment of the synaptic machinery necessary for signal transduction and synapse maturation. Synaptic vesicles are recruited on the presynaptic side and receptors on the postsynaptic membrane. The recruitment process occurs rapidly; a synapse with a postsynaptic density (PSD) is formed within 2 h (Friedman et al., 2000). Adhesion molecules are thought to regulate synaptogenesis in various ways (reviewed in more detail below and in Dalva et al. (2007); Giagtzoglou, (2009)). Ephrin and the ephrin-B receptor (EphB) have a synaptogenic effect resulting in the trans-synaptic induction of presynaptic specializations (Kayser et al., 2006). EphBs bind to syntenin-1 through a PDZ domain interaction leading to syntenin-1 accumulation at synaptic contacts (McClelland et al., 2009). Syntenin-1 may then be involved in recruitment of proteins to the presynaptic active zone (Jin and Garner, 2008) through interactions with proteins like ELKS/RAB6-interacting/-CAST family member2 (ERC2) (Ko et al., 2006b).

Similarly, synaptic CAM (SynCAM) can induce the formation of functional presynaptic specializations in the axons of cocultured neurons (Biederer, 2002). CAMs may also act in coordination with each other for synaptogenesis to occur. For example, N-cadherin and neuroligin-1 (NL-1) can promote the formation of glutamatergic synapses in concert (Stan et al., 2010; Aiga et al., 2011). They can also affect the lifetime of spines; disruption of N-cadherin interactions results in increased spine motility, decreased length, and then spine loss (Mysore et al., 2007). Many CAMs are directly or indirectly linked to glutamate receptors (GluRs) and may be involved in regulating their trafficking or surface expression, and thus spine maturation (Figure 1). In young cortical neurons, *N*-methyl-D-aspartate receptors (NMDARs) and α-amino-3-hydroxy-5-methyl-4-isoxazolepropionic acid receptors (AMPARs) are present in transport packets before and during synaptogenesis (Washbourne et al., 2002). NMDARs are recruited to points of axodendritic contacts faster than AMPARs. They are recruited within minutes, while AMPARs are recruited an hour later (Washbourne et al., 2002).

PSD-95 is usually not associated with these mobile NMDARs, but SAP102 is often present in NMDAR packets traveling along microtubules (Washbourne et al., 2002; Washbourne et al., 2004). NL-1 also associates with NMDARs in dendrites before synapses form, and both are cotransported to spines during synaptogenesis (Barrow et al., 2009). By postnatal day 2 (P2), hippocampal excitatory synapses contain most of the PSD components involved in synaptic function and/or development (Figures 2 and 3; Petralia et al., 2005), including GluN2B-containing NMDARs and SAP102, as well as GK-associated protein (GKAP), Shank, Homer, and SynGAP (Petralia et al., 2005; Carlisle et al., 2008; Foa and Gasperini, 2009).

2.1.2. Postnatal Development

The expression of PSD–membrane-associated guanylate kinases (MAGUKs) and GluRs is tightly regulated during postembryonic development (Sans et al., 2000). The PSD–MAGUK, SAP102, is highly expressed before postnatal day 10 (P10) in the rat brain; then its expression decreases. In contrast, the expression of PSD-93, PSD-95, and SAP97

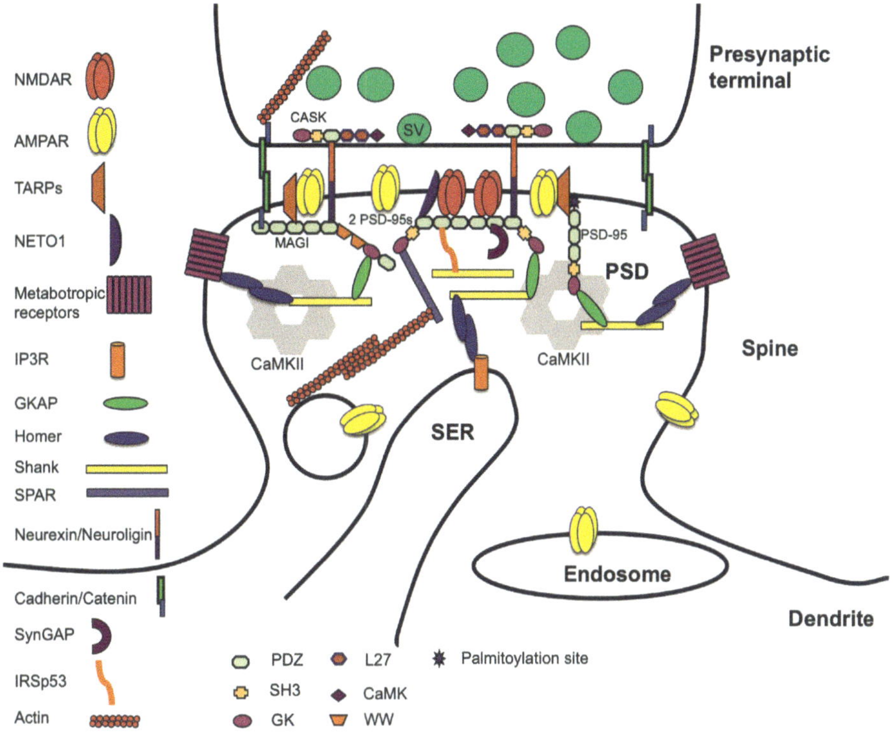

Figure 2 Diagram illustrates the position of MAGUKs in junctional complexes at the neuronal synapse, including both the postsynaptic density and presynaptic active zone. As described in the text, surface receptors and adhesion proteins and other membrane proteins in the postsynaptic membrane can bind via PDZ domains to MAGUKs that also interact with various signaling proteins, and bind through intermediate proteins such as GKAP to proline-rich domain-containing proteins such as Shank. Similar arrangements may be present in the presynaptic active zone, especially with the MAGUK, CASK. Adhesion proteins can link in the cleft and join presynaptic and postsynaptic complexes via their MAGUKs. Note that the exact orientation of MAGUKs in the PSD is not known (and not discussed in this review) and could be affected by their interactions with glutamate receptors, adhesion proteins, other MAGUKs, etc.; for example, there could be different orientations depending on binding partners and position in the scaffold. Putative orientations are shown in this figure and in Figures 5 and 8. SER, smooth endoplasmic reticulum.

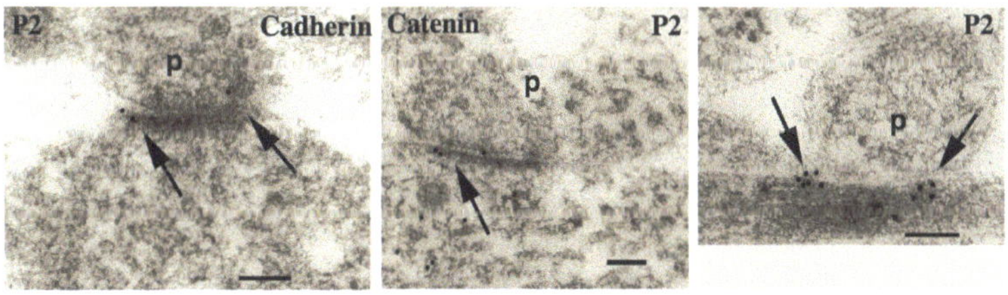

Figure 3 Immunogold localization (arrows) of cadherin (left micrograph) and catenin (right two micrographs) in newly formed synapses in the postnatal day 2 (P2) hippocampus (CA1 stratum radiatum). Gold particles typically are found near the edges of the synaptic active zone. Synapses in the two left micrographs show the thickening PSD characteristic of immature glutamatergic synapses. The right micrograph shows an even earlier contact, prior to formation of the PSD. p, presynaptic terminal. Scale bar is 100 nm. *Modified from part of Figure 3 in Petralia et al. (2005).*

is very low in newborn rat pups but their expression is enhanced during development. Finally, PSD-95 becomes a dominant PSD–MAGUK in the adult rat brain.

There is low expression of all AMPAR subunits at P2 in the hippocampus (Petralia et al., 2005, Figure 4). Indeed, the expression of AMPAR subunits increases and

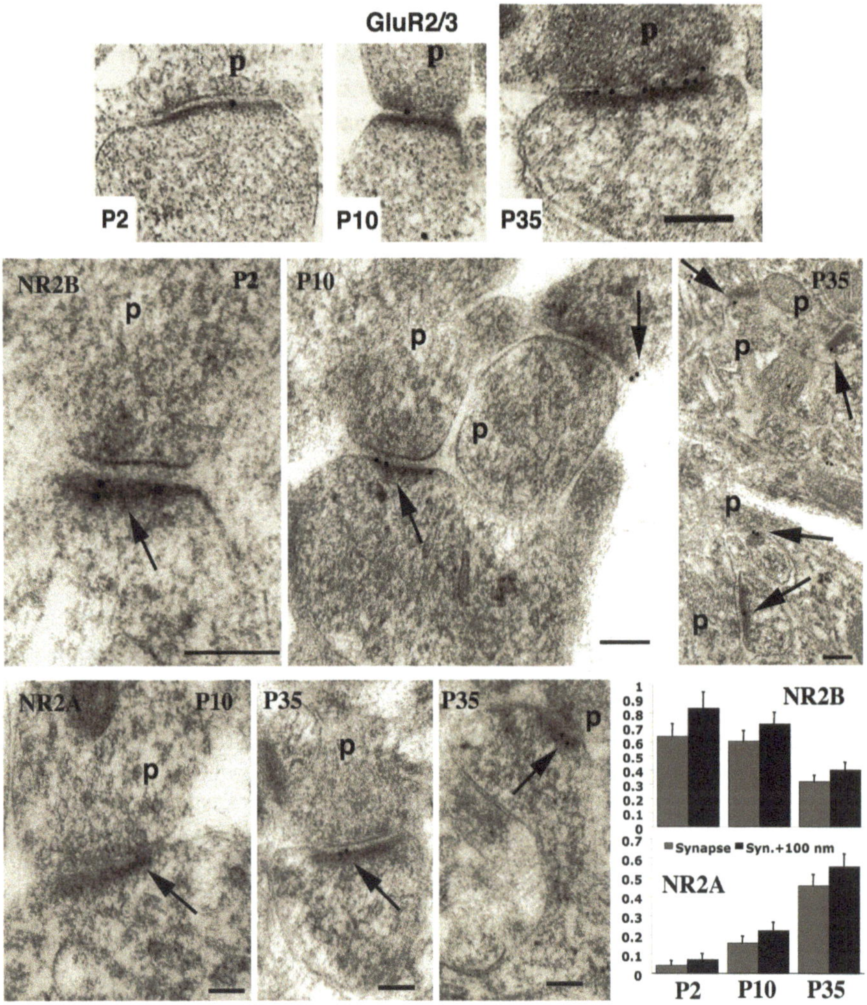

Figure 4 Immunogold localization of the AMPAR subunits GluA2/3 (GluR2/3), and GluN2A and GluN2B (NR2A and NR2B) NMDAR subunits in the hippocampus (CA1 stratum radiatum) during postnatal development (days 2, 10, 35 = P2, 10, 35). Note the great increase in postsynaptic labeling of GluA2/3 by P35. GluN2A also increases greatly during this time; in contrast, synaptic GluN2B decreases. Chart compares the change in gold density (gold per synapse) between GluN2A and GluN2B for the 3 ages (synapse area is the postsynaptic membrane and density; Syn.+100 nm also includes gold up to 100 nm below the postsynaptic membrane). p, presynaptic terminal. Scale bar is 100 nm. *GluA2/3 was modified from part of Figure 2 in Petralia et al. (1999), and GluN2A/GluN2B was published in Petralia et al. (2005).*

reaches the maximum between P14 and P28 and is then maintained into adulthood (Jensen et al., 2003). Hippocampal principal cells express mainly GluA1/2 subunits and less GluA3 subunits (Keinanen et al., 1990; Petralia and Wenthold, 1992). The GluA4 subunit seems to be expressed mainly during embryonic development and in hippocampal interneurons (Keinanen et al., 1990; Geiger et al., 1995). Interestingly, messenger RNAs (mRNAs) for the GluA2 subunit undergo editing at the Q/R site, resulting in the formation of low-conductance, Ca^{2+}-impermeable channels (CI-AMPARs). In contrast, the AMPARs lacking edited GluA2 form high-conductance and Ca^{2+}-permeable channels (CP-AMPARs) (Hollmann et al., 1991; Burnashev et al., 1992; Seeburg et al., 1998). In the mature central nervous system (CNS), CI-AMPARs are the dominant form expressed in synapses between principal neurons while CP-AMPARs participate in synapses early in development in several brain regions (Pickard et al., 2000; Aizenman et al., 2002; Kumar et al., 2002; Eybalin et al., 2004). However, CP-AMPARs can also be inserted into synapses in response to environmental factors, like exposure to drugs of abuse, and therefore can contribute to plasticity changes in adults (reviewed in Wolf and Ferrario (2010)). Thus, developmental expression of AMPAR subunits as well as their editing and trafficking regulates synaptic AMPAR functioning.

NMDARs also have subunit-specific regulation. After birth, the NMDARs are primarily complexes of GluN2B subunits on the neuronal cell surface. During postnatal development, there is a switch from a predominance of GluN2B-containing to GluN2A-containing receptors (Petralia et al., 2005, Figure 4; review by Paoletti et al. (2013)). Several studies proposed that the switch from GluN2B- to GluN2A-containing receptors occurs in conjunction with learning experience including, e.g. visual experience (Quinlan et al., 1999a,b). Interestingly, motoneurons in the hypoglossal, facial, and trigeminal motor nuclei express high levels of mRNAs for GluN2A early in development (Oshima et al., 2002). Probably, this phenomenon is related to their function in the suckling reflex immediately after birth. In hippocampal CA1 pyramidal neurons, the GluN2B to GluN2A switch is driven acutely by activation of NMDARs and mGlu1, and involves phospholipase C, Ca^{2+} release from IP(3)R-dependent stores, and protein kinase C (PKC) activity (Matta et al., 2011). The GluN2B to GluN2A switch is also regulated by casein kinase 2 (CK2) that phosphorylates GluN2B, but not GluN2A, resulting in GluN2B-endocytosis and replacement of the synaptic GluN2B with GluN2A (Sanz-Clemente et al., 2010). Interestingly, long-term potentiation (LTP)-inducing stimuli can cause a rapid switch from GluN2B- to GluN2A-containing receptors in young animals but not in older ones (Bellone and Nicoll, 2007). This process can be reversed by depotentiation.

GluN3A is prominently expressed during a narrow period that correlates with intense synaptogenesis (Wong et al., 2002; Perez-Otano et al., 2006; reviewed by Paoletti et al. (2013)). Although the mechanism is unknown, the sharp downregulation of GluN3A is required for synapse maturation and memory consolidation (Roberts et al., 2009). Thus, GluN3A likely acts as a molecular brake that prevents the stabilization of premature synapses. Moreover, presynaptic GluN3A-containing receptors are required to enhance spontaneous and evoked glutamate release in the juvenile mouse visual cortex and thus are crucial for spike timing-dependent long-term depression (LTD) (Larsen et al., 2011).

In cerebellar granule cells, GluN2C replaces GluN2B after the cells complete their migration and are innervated by the mossy fibers (Ozaki et al., 1997). Mossy fiber terminals release neuregulin that activates ErbB2 and ErbB4 receptors on cerebellar granule cells; they probably enhance the expression of GluN2C via a structural link with MAGUKs (Garcia et al., 2000; Hahn et al., 2006). Interestingly, IGF-1 upregulation of PI3K stimulates protein kinase B (PKB)/Akt, which phosphorylates GluN2C on S1096 (Chen and Roche, 2009). This phosphorylation promotes the association of 14-3-3 with GluN1/GluN2C heteromeric receptors resulting in their increased surface delivery. The GluN2A expression also likely is enhanced in cerebellar granule cells during development although the mechanism is unknown (Fu et al., 2005).

2.2. TRAFFICKING OF GLUTAMATE RECEPTORS IN SYNAPTIC PLASTICITY

The human brain contains over 100 billion (1×10^{11}) neurons, with each forming thousands of synapses. The glutamatergic excitatory synapses are the major type of synapses in the brain (Kessels and Malinow, 2009). Indeed, typically these synapses contain two major types of ionotropic GluRs—AMPARs and NMDARs (Table 1). The release of glutamate from presynaptic terminals leads to activation of AMPARs (fast component of postsynaptic current) resulting in a depolarization of the postsynaptic membrane that unblocks magnesium inhibition of NMDARs and activates NMDARs (slow component of postsynaptic current) (Elias and Nicoll, 2007; Kessels and Malinow, 2009). It is generally thought that short- and long-term changes of synaptic strength (synaptic plasticity) are responsible for our ability to learn and memorize. Synaptic plasticity can be regulated presynaptically or postsynaptically. Many studies show that changes in the number and/or the subunit composition of AMPARs, often associated with signaling mediated by NMDARs, is a prevalent mechanism in models of synaptic plasticity on the molecular level (Elias and Nicoll, 2007; Kessels and Malinow, 2009, Figure 5).

2.2.1. Synthesis, Assembly, and Trafficking

Before AMPARs and NMDARs reach the synapse, they first must be synthesized, assembled, and trafficked to the PSD. mRNAs for GluA1 and GluA2 are found in the neuronal cell body as well as in dendrites (Ju et al., 2004; Grooms et al., 2006). Activity blockade of rat cultured neurons may increase dendritic GluA1 levels, but not GluA2 levels (Ju et al., 2004). NMDAR activation depletes dendritic levels of AMPAR mRNAs via a rise in

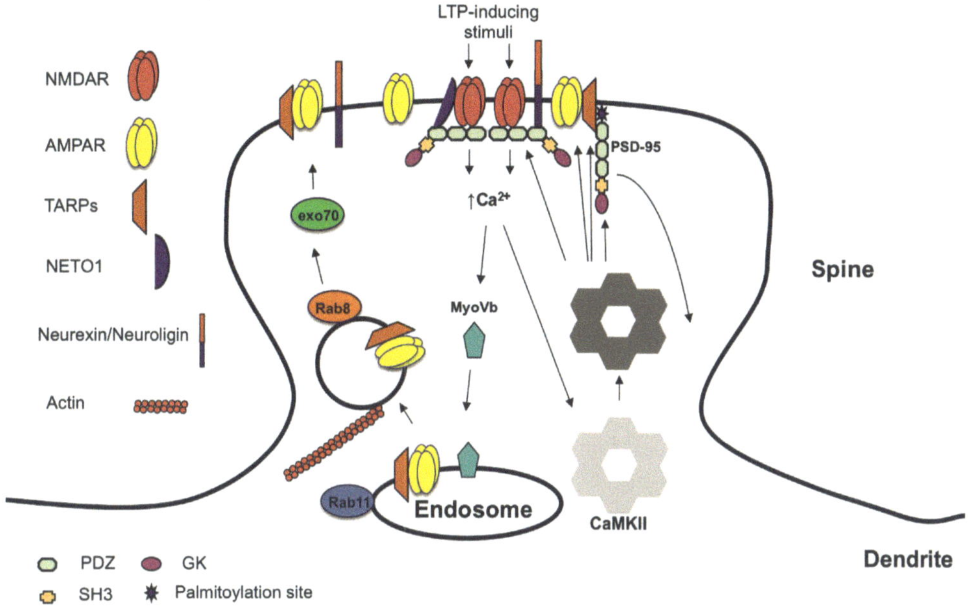

Figure 5 Synaptic plasticity in the dendritic spine. LTP-inducing stimuli activate synaptic NMDARs resulting in an increase in intracellular Ca^{2+} concentration. The high intracellular Ca^{2+} levels switch MyoVb from the folded to the extended conformation (Wang et al., 2008). Then, MyoVb forms a ternary complex with Rab11 and FIP2; this complex is believed to tether Rab11-recycling endosomes and moves them along actin microfilaments into the spine head where they supply AMPARs for the expression of LTP (Hales et al., 2002; Jagoe et al., 2006; Kelly et al., 2011). The raised Ca^{2+} level activates CaMKII that is then redistributed from the dendritic shaft to spines. Activated CaMKII fosters the phosphorylation of Ser831 of the GluA1 subunit, resulting in an increase in AMPAR activity. Activated CaMKII also associates with GluN2B-containing NMDARs; this is critical for expression of NMDAR-LTP (Barria and Malinow, 2005). Furthermore, CaMKII phosphorylates Stargazin, facilitating the binding of Stargazin to PSD-95 and the synaptic retention of this complex with AMPARs (Opazo et al., 2010). Also, phosphorylation of Ser73 of PSD-95 by CaMKII terminates spine growth by inducing the translocation of PSD-95 and Shank out of the active spine (Steiner et al., 2008).

intracellular Ca^{2+}, activation of extracellular signal-regulated kinase/mitogen-activated protein kinase signaling, and transcriptional arrest at the level of the nucleus (Grooms et al., 2006). Also mRNAs for GluN1 are found in dendrites (Steward and Schuman, 2003). Thus, both AMPARs and NMDARs are likely synthesized in the neuronal cell body as well as in dendrites.

AMPARs and NMDARs are synthesized as monomers in the endoplasmic reticulum (ER), where they assemble into tetramers (reviewed in Herguedas et al. (2013)). How is oligomerization regulated? In the case of AMPARs, it has been shown that channel function upon glutamate binding is monitored prior to release of the receptor from the ER, and determines ER-export kinetics (Penn et al., 2008). GluA1/GluA2 heteromers exit the ER more efficiently than GluA2/GluA3 heteromers (Greger et al., 2002). The majority of GluA2 subunits are retained in the ER; this retention is mediated by the Q/R-editing site. Interestingly, RNA editing can play a role in the regulation of assembly of AMPARs since it can alter channel functioning (Greger et al., 2002; Penn et al., 2008). The GluA2 C-terminus contains a PDZ-binding motif (-SVKI) that interacts with the protein, PICK1. This interaction is crucial for the release of GluA2 from the ER (Greger et al., 2002). Also, the interaction of AMPARs with transmembrane AMPAR regulating proteins (TARPs; discussed more below), likely in the ER, can enhance the surface delivery of AMPARs (Vandenberghe et al., 2005). It is not clear which domains are responsible for the TARP/AMPAR interaction; the C-terminal PDZ-binding motif of TARPs (-TTPV) does not seem to control the surface delivery of AMPARs, but controls their synaptic localization (Tomita, 2010). Since the nine amino acids in the extreme N-terminus of GluA1 are critical for anterograde trafficking of AMPARs (Xia et al., 2002), multiple independent mechanisms probably control proper assembly of functional tetramers and their trafficking from the ER.

The NMDARs also contain many different ER-retention/export signals that play roles in ensuring that only correctly folded heterotetramers reach the cell surface. First, the C-terminal region of GluN1 has been shown to contain the RRR and KKK ER-retention signals that are masked upon assembly of functional tetramers; these ER-retention signals are also negated by the presence of the PDZ-binding motif in the far C-terminus of the C2′ cassette (Standley et al., 2000; Horak and Wenthold, 2009). The C2′ cassette enhances the forward trafficking of NMDARs via binding of a divaline motif, within the C2′ cassette, to COPII coats; in contrast, the C2 cassette slows the trafficking of NMDARs (Mu et al., 2003; Horak and Wenthold, 2009). Interestingly, synaptic activity regulates GluN1 mRNA splicing at the C2/C2′ site and thus may regulate the surface targeting of NMDARs (Mu et al., 2003). The GluN2B subunit contains the ER-retention signal within its C-terminus, although the precise motif has not yet been identified (Hawkins et al., 2004).

Other regions within the NMDAR subunits have been identified recently as ER-retention signals, including the N-terminal domain within the GluN2A subunit (Qiu et al., 2009), the glycine-binding site within the GluN1 subunit (Kenny et al., 2009), and the third membrane domain (Horak et al., 2008). It is still not clear how these various regions collaborate so that only correctly assembled receptors leave the ER. Likely, these trafficking motifs interact with specific proteins that regulate the formation of heterotetramers or inhibit the forward trafficking of dimers and monomers. Similarly, previous work has also identified ER-retention motifs in the C-terminus and the intracellular loop region of GluK5 that prevent surface expression of GluK5 homomers (Nasu-Nishimura et al., 2006). Furthermore, two PKC phosphorylation sites on the GluK2 C-terminus regulate receptor trafficking through the biosynthetic pathway (Nasu-Nishimura et al., 2010). Interestingly, for KARs containing GluK2, there is an intracellular role for desensitization in controlling maturation and trafficking of GluRs, i.e. glutamate binds to these receptors while they are in the ER and subsequent desensitization is required for surface expression of these receptors (Priel et al., 2006).

After membrane receptors leave the quality-control machinery in the ER, they generally enter the Golgi apparatus and the trans-Golgi network (TGN) where they are processed into vesicular and tubulovesicular carriers that transport membrane proteins to the cell surface. These processes remain poorly understood. Both NMDARs and AMPARs reside in mobile transport packets in young dendrites. Interestingly, the NMDAR clusters mostly lack AMPARs, and are transported along microtubules on large tubulovesicular organelles where they colocalize with SAP102 (Washbourne et al., 2004). Moreover, prior to synapse formation, NMDARs cycle to/from the cell surface by exocytosis and endocytosis. However, such cycling of NMDARs has not been observed in mature neurons. In older neurons (>12 d.i.v.), AMPARs are trafficked via the conventional somatic Golgi network. However, NMDARs are diverted from the somatic ER into a specialized ER subcompartment that bypasses the somatic Golgi, trafficking instead through dendritic Golgi outposts (Jeyifous et al., 2009). This ER subcompartment is composed of highly mobile vesicles containing the NMDAR subunits—GluN1 and GluN2B, the microtubule-dependent motor protein—KIF17, and the postsynaptic adaptor proteins—CASK and SAP97.

The NMDARs are trafficked to the cell surface in specific protein complexes. They associate with different MAGUKs including SAP102 and PSD-95 along the secretory pathway (Standley et al., 2012). Through SAP102, which interacts with the PDZ-binding domain of Sec8, the NMDARs may interact with the exocyst complex (Sans et al., 2003). Indeed, the disruption of the exocyst complex by the dominant-negative form of Sec8 blocks NMDAR surface delivery. Interestingly, the deletion of the PDZ-binding domain on the GluN2B subunit allows the surface delivery of NMDARs. However, synaptic delivery is compromised. Furthermore, the SH3 and GK domains of SAP102 also bind mammalian homolog of *Drosophila melanogaster* partner of inscuteable (mPINs); this complex interacts with NMDARs in brain and likely is formed early in the secretory pathway (Sans et al., 2005). The mPINs–SAP102–NMDARs interaction enhances trafficking of NMDAR to the neuronal cell surface. As mPINs also interacts with Gαi, G-protein signaling may also regulate NMDAR surface delivery.

What molecular motors are involved in the movement of AMPARs and NMDARs in dendrites? Glutamate receptor interacting protein 1 (GRIP1; a multi-PDZ-domain containing protein) can directly interact with and steer kinesin heavy chains to dendrites; kinesins probably act as motors for AMPAR trafficking (Setou et al., 2002). Thus, the GluA2/3-containing AMPARs may associate with GRIP1 through their C-terminal PDZ-binding domains, and this complex likely transports GluA2/3-containing receptors on a microtubule track in dendrites (Dong et al., 1997; Srivastava et al., 1998). Liprin-alpha/SYD-2 is a multimodular scaffolding protein that interacts with the neuron-specific kinesin motor KIF1A (Shin et al., 2003). Interestingly, interaction between liprin-alpha/SYD-2 and GRIP1 is required for dendritic trafficking of GluA2-containing receptors (Wyszynski et al., 2002).

Finally, one member of the myosin family, myosin Vb, has been implicated as a motor that transports GluA1-containing receptors from the cell body to the dendritic cell surface (Lise et al., 2006). The presence of the small GTPase, Rab11, is also critical for dendritic transport of AMPARs. In the case of dendritic targeting of NMDARs, one possibility is that the GluN2B-containing receptor interacts with mLin-7 through its PDZ-binding motif; this complex interacts with mLin-2 and KIF17, a dendrite-specific microtubule-dependent molecular motor (Setou et al., 2000). Another possibility is that the kinesin superfamily motor protein KIF1B-alpha, which interacts with several MAGUK proteins, also transports NMDARs (Mok et al., 2002). The myosin motors are likely responsible for delivery of NMDARs to and from the surface membranes. Also, while the mechanism is not clear, NMDARs interact with myosin regulatory light chain (Amparan et al., 2005). Trafficking of AMPARs and NMDARs to spines will be discussed below.

Palmitoylation can dramatically regulate the trafficking of both AMPARs and NMDARs. All AMPAR subunits are palmitoylated on two cysteine residues in their M2 domain and in their C-terminal region. Palmitoylation on M2 leads to an accumulation of the receptor in the Golgi apparatus and a reduction of receptor surface expression. The C-terminal palmitoylation decreases interaction of the AMPARs with protein 4.1N and regulates AMPA- and NMDA-induced AMPAR internalization (Hayashi et al., 2005). Palmitoylation of GRIP1 (see above) causes the latter to be targeted to endosomes, resulting in increased AMPAR recycling (Thomas et al., 2012). Similar to AMPARs, GluN2A and GluN2B have two distinct clusters of palmitoylation sites in their C-terminal region (Hayashi et al., 2009). Palmitoylation within the first cluster on a membrane-proximal region enhances stable surface expression of NMDA receptors via increase in tyrosine phosphorylation of tyrosine-based internalization motifs by Src family protein tyrosine kinases. Palmitoylation of the second cluster in the middle of the C-terminus causes NMDARs to accumulate in the Golgi apparatus and reduces their surface expression.

2.2.2. Subunit Composition and Synaptic Targeting

Are there subunit-specific rules for synaptic delivery of GluRs? Previously, Wenthold et al. (1996) used a biochemical approach to show that there are two major populations of AMPAR complexes in hippocampal CA1 neurons, made up of GluA1/GluA2 and GluA2/GluA3. This study also suggested that a minor population of homomeric GluA1 receptors is present in these neurons. A recent electrophysiological study shows that in CA1 pyramidal neurons, GluA1/GluA2 heteromers prevail (~81%) (Lu et al., 2009). The rest of the synaptic receptors are composed of GluA2/GluA3 heteromers (~16%) and other AMPAR subtypes. Lu et al. (2009) also show that the majority of extrasynaptic receptors (~95%) are composed of GluA1/GluA2 heteromers, suggesting that there are different mechanisms for the regulation of synaptic and extrasynaptic AMPARs.

How is the synaptic targeting of GluRs regulated? An elegant study from Malinow's laboratory (Hayashi et al., 2000) showed, in rat hippocampal culture slices, that GluA1/GluA2 receptors are added to synapses during plasticity and that this requires interactions between GluA1 and group I PDZ-domain proteins. In contrast, GluA2/GluA3 receptors replace existing synaptic receptors continuously and this requires interactions of GluA2 with N-ethylmaleimide-sensitive factor (NSF) and group II PDZ-domain proteins (Shi et al., 2001). The synaptic stabilization of AMPARs by NSF involves the disruption of GluA2–PICK1 interactions by NSF ATPase activity, allowing AMPARs to cycle back to the synapse (Hanley et al., 2002). Passafaro et al. (2001) used a thrombin cleavage assay in cultured hippocampal neurons to demonstrate that surface insertion of GluA1 is slow under basal conditions and is accelerated by NMDAR and insulin stimulation, whereas GluA2 exocytosis is constitutive but rapid. However, a knockin mouse lacking the last seven residues of GluA1, comprising the PDZ ligand, does not exhibit any abnormalities in basal GluA1 synaptic localization, basal synaptic transmission, LTP, or LTD in CA1 hippocampal synaptic plasticity (Kim et al., 2005). This indicates that the C-terminus of GluA1 is not critical for synapse delivery at least in basal conditions. Recent evidence suggests that the synaptic targeting of AMPARs is also regulated by a CaMKII phosphorylation site in the first intracellular loop of GluA1 (Lu et al., 2010). Thus, other regions in addition to the C-termini may regulate synaptic targeting.

Several studies have demonstrated that protein kinases may regulate activity-dependent synapse delivery of AMPARs including CaMKII, PKA, and PKC. Interestingly, early in development, PKA-induced phosphorylation of GluA4 triggers the synaptic delivery of GluA4-containing receptors. Subsequently, these receptors are replaced with GluA2-containing receptors; this requires an interaction between GluA2 and NSF (Zhu et al., 2000). Around the second postnatal week, the long splice variant of GluA2, GluA2-long (see Figure 1), is the prevalent subunit that mediates LTP (GluA2-short is the main variant in most areas of the adult brain; Kolleker et al., 2003). Later, LTP is mediated by GluA1 addition into synapses upon phosphorylation by PKA and activation of CaMKII (Esteban et al., 2003). Thus, developmental changes in the subunit-specific synaptic insertion are consistent with the switch in synaptic cascades required for LTP. In experiments using native receptors, the LTP induced in CA1 hippocampal pyramidal neurons causes rapid incorporation of GluA2-lacking CP-AMPARs. These CP-AMPARs are replaced by GluA2-containing AMPARs approximately 25 min after LTP induction (Plant et al., 2006). However, a follow-up study reported that CP-AMPARs are not inserted into synapses during LTP (Adesnik and Nicoll, 2007). In mouse CA1 pyramidal neurons, there is no requirement of the GluA1 C-tail for LTP and, furthermore, its replacement with a GluA2 subunit or an artificially expressed kainate receptor not normally found at these synapses results in normal LTP (Granger et al., 2013). Nevertheless, it is clear that AMPARs are highly mobile receptors that can traffic into synapses in response to specific activity-dependent stimuli. In addition, some synapses may have a homeostatic mechanism that substitutes kainate receptors for lost AMPARs (or vice versa) to maintain neuronal activity (Yan et al., 2013).

The majority of synaptic receptors in the hippocampus are composed of GluN1/GluN2A and GluN1/GluN2B heterotetramers, although triheteromeric GluN1/GluN2A/GluN2B receptors are likely present as well (Al-Hallaq et al., 2007); in fact, triheteromeric NMDARs may be the most common type in the hippocampus (Tovar et al., 2013). Early in development, there is a majority of GluN2B-containing receptors that are subsequently replaced with GluN2A-containing receptors (see Section 2 and Figure 4). Although a preference of GluN2A-containing receptors over GluN2B-containing receptors for synaptic expression has been described, both are localized to synaptic as well as extrasynaptic membranes (Thomas et al., 2006; Petralia et al., 2010; Petralia, 2012). The mechanism underlying the regulation of number of synaptic NMDARs is unclear. In cerebellar granule cells, overexpression of GluN2A or GluN2B subunits increases the total number of surface NMDARs. However, the number of synaptic receptors is unchanged, indicating that different mechanisms control the extrasynaptic and synaptic delivery of NMDARs (Prybylowski et al., 2002).

Interestingly, the overexpression of GluN1 subunits does not increase total surface NMDARs suggesting that these cells express an excess of GluN1 subunits. In hippocampal slice cultures, synaptic incorporation of GluN2B-containing receptors is not limited by synaptic transmission nor enhanced by increased subunit expression. However, GluN2A-containing receptors may replace synaptic GluN2B-containing receptors upon enhanced expression of GluN2A, a transition that requires synaptic activity (Barria and Malinow, 2002). Both GluN2A and GluN2B subunits with deleted PDZ-binding motifs (lacking the last six amino acid residues of the C-terminus) show reduced delivery of recombinant receptors to synapses, suggesting a role for the PDZ-binding motif of the GluN2 subunits in the synaptic delivery or stabilization of the receptor at synapses. Mice that express GluN2A without the C-terminus exhibit reduced synaptic but not extrasynaptic delivery (Steigerwald et al., 2000). However, in cultured hippocampal neurons, GluN2A subunits without the C-terminus can be incorporated into synapses (Thomas et al., 2006).

Recently, it has been shown that a putative N-glycosylation site present in GluN2B, but not in GluN2A, is necessary and sufficient to drive NMDARs into synapses in an activity-independent manner (Storey et al., 2011). Also, the association between active CaMKII and GluN2B is required for different forms of synaptic LTP (Barria and Malinow, 2005). The LTD of synaptic NMDARs (NMDAR-LTD) using a specific stimulating protocol has been observed in rat hippocampus (Morishita et al., 2005). This form of LTD is blocked by inhibitors of protein phosphatase 1 but is not affected by endocytosis inhibitors, suggesting that phosphorylation but not endocytosis plays a role in NMDAR-LTD; in this case, NMDARs may be removed from the synapse via lateral diffusion or a dephosphorylation event, and/or their function may be reduced without their removal from the synapse.

2.2.3. Lateral Diffusion vs Intracellular Trafficking

Are synaptic GluRs supplied from extrasynaptic membranes by lateral diffusion or directly from intracellular compartments? An initial study showed that disrupting exocytosis causes a gradual reduction in the AMPAR-excitatory postsynaptic potential (EPSC)s, while inhibition of endocytosis causes a gradual increase in the AMPAR-EPSCs. Interestingly, these manipulations had no effect on the NMDAR-EPSCs but prevented the subsequent induction of LTD (Luscher et al., 1999), suggesting that AMPARs, but not NMDARs, cycle into and out of the synaptic membrane at a rapid rate. Fast cycling of surface AMPARs with receptors from internal stores occurs at extrasynaptic sites on the neuron cell body, whereas cycling of synaptic AMPARs is much slower (Adesnik et al., 2005). This indicates that the intracellular pool of AMPARs exchanges rapidly with extrasynaptic AMPARs, and that the extrasynaptic AMPARs travel laterally along dendrites to reside at synapses. Using an irreversible antagonist of AMPARs, ANQX, the most dynamic movement of AMPARs was shown to occur by lateral movement across the surface of neurons (Adesnik et al., 2005).

Using single-molecule tracking, it was demonstrated that local synaptic activity reduces diffusional exchange of GluA1 between synaptic and extrasynaptic domains, resulting in postsynaptic accumulation of GluA1 (Ehlers et al., 2007). This observation supports the idea that lateral movement of AMPARs is the important pathway for synaptic delivery of AMPARs. The importance of lateral movement for synapse delivery was confirmed by fluorescence recovery after photobleaching (FRAP) experiments in hippocampal neurons, which show that about half of synaptic GluA2 subunits are exchanged within ~15 min and that this exchange is slowed by extracellular application of anti-green fluorescent protein antibodies that inhibit movement of the subunit (Ashby et al., 2006). Thus, it is not likely, at least in this case, that the GluA2 subunit is exocytosed directly into synaptic regions from intracellular stores. Similar results were also observed with GluA1 subunits (Sharma et al., 2006). Single-particle tracking experiments confirmed that there are two populations of GluA2 subunits within synapses—one mobile and one immobile (Ehlers et al., 2007)—which is in agreement with the FRAP experiments. Also extrasynaptic GluA2 was seen to enter the synapse. Thus, extrasynaptic AMPARs are likely to be an important reservoir for synaptic AMPARs.

2.2.4. Synaptic Stabilization

NMDARs are likely more stable in synapses than AMPARs since ~15% of GluN1 subunits are exchanged within 5 min in FRAP experiments, in contrast to ~40% of GluA1 (Sharma et al., 2006). However, electrophysiological studies using the open-channel blocker MK-801 in dissociated cultures of hippocampal neurons showed that ~65% of synaptic NMDARs is exchanged with receptors from extrasynaptic pools within 7 min (Tovar and Westbrook, 2002). Using acute hippocampal slices, it was shown that extrasynaptic NMDARs comprise ~35% of the dendritic NMDAR pool (Harris and Pettit, 2007). However, in experiments with the MK-801, similar to ones described above, Harris and Pettit (2007) found that almost no exchange of NMDARs between extrasynaptic and synaptic pools occurred.

The reason for this discrepancy is not clear although possible explanations include the differences in neuronal preparation, technique, and age of neurons used for experiments. The surface mobility of NMDARs depends on the GluN2 subunit subtype, with GluN2A-containing NMDARs being more stable than GluN2B-containing receptors in dissociated cultures of hippocampal neurons (Groc et al., 2006). There is also higher surface stabilization of GluN2B-containing NMDARs in early synapses, when compared with mature synapses. Also, the residency time of GluN2A-containing NMDARs is higher than that of GluN2B-containing NMDARs, indicating a better stabilization of GluN2A-containing NMDARs within mature synapses. These data support the conclusion that GluN2A-containing receptors are preferentially targeted to synapses of mature neurons.

Is mobility of surface GluRs altered by synaptic activity? Many studies have shown that synaptic activity can alter surface diffusion of AMPARs (Tardin et al., 2003; Ehlers et al., 2007). As mentioned above, single-molecular imaging revealed that active synapses trap GluA1 subunits more efficiently than inactive synapses (Ehlers et al., 2007), suggesting that synapses compete for GluA1 from the extracellular pool. In contrast to AMPARs, NMDAR mobility in both synaptic and extrasynaptic compartments is not altered by application of TTX or KCl, demonstrating that NMDARs are more tightly attached to their surface positions (Groc et al., 2004).

In addition to lateral diffusion of AMPARs, several studies suggest that rapid exocytosis of intracellular AMPARs may occur during LTP. How is this exocytosis regulated? The Rab family of small GTPases is the key regulators of membrane trafficking between intracellular compartments. When active, Rab proteins mediate membrane trafficking events such as vesicle formation, motility, docking, and fusion (Kelly et al., 2011). The various stages in AMPAR endocytosis and recycling involve at least Rab4, 5, 8, and 11, along with other important trafficking proteins such as GRIP1, GRASP-1, PICK1, NSF, protein 4.1N, NEEP21, and KIBRA (reviewed in Anggono and Huganir (2012), and most are discussed elsewhere in this chapter). Initially, recycling endosomes, controlled by Rab11, were proposed to be critical for the activity-dependent delivery of GluA1 into synapses (Park et al., 2004). Later, it was shown that Rab8, which regulates TGN trafficking, is also crucial for synaptic insertion of GluA1 during LTP (Gerges et al., 2004).

Based on ultrastructural studies, it has been proposed that Rab11-containing recycling endosomes mediate entry of AMPARs into spines and Rab8-containing endosomes mediate the insertion of AMPARs into synaptic membranes (Gerges et al.,

2004). Interestingly, Rab8, but not Rab11, is critical for constitutive synaptic cycling of AMPARs. Thus, at least some trafficking events overlap between constitutive and activity-dependent delivery of synaptic AMPARs. Furthermore, another study showed that the exocyst is also critical for membrane insertion of AMPARs into synapses since deletion of the Exo70 subunit of the exocyst results in the accumulation of AMPARs in the PSD (Gerges et al., 2006). Exocytosis of AMPAR-containing recycling compartments during elevated synaptic activity occurs at microdomains enriched in the plasma membrane t-SNARE syntaxin 4 (Stx4), and disruption of Stx4 using short hairpin interfering RNA (shRNA) or a dominant-negative form impairs spine exocytosis, and the latter also impairs LTP at hippocampal synapses (Kennedy et al., 2010). Other studies show that AMPAR exocytosis during LTP requires complexin, a small soluble protein that binds to SNARE complexes (Ahmad et al., 2012), or syntaxin 3 and SNAP-47 (Jurado et al., 2013). In the case of NMDARs, it has been shown that SNAP-23, a ubiquitously expressed SNAP-25 homolog, regulates the surface expression of NMDARs, likely on the level of exocytosis (Suh et al., 2010).

2.2.5. Mechanisms of LTP and LTD

As discussed above, various mechanisms regulate the dendritic targeting of AMPARs. What is the mechanism of transportation of AMPARs that underlies the expression of LTP? LTP-inducing stimuli increase Ca^{2+} levels in dendritic spines via activation of NMDARs (Kessels and Malinow, 2009). A member of the myosin family of proteins, MyoVb, switches from folded to extended conformation in high Ca^{2+} levels (Wang et al., 2008). Then, MyoVb can form a ternary complex with Rab11 and FIP2; this complex is believed to tether Rab11-recycling endosomes and moves them along actin microfilaments into the spine head (Hales et al., 2002; Jagoe et al., 2006; Kelly et al., 2011). This way, AMPARs are supplied for the expression of LTP. Another member of the myosin family, MyoVa, also has been implicated as a critical protein for the expression of LTP-AMPARs (Correia et al., 2008). However, this conclusion is not consistent with that of a previous study and thus more experiments need to be performed to fully understand the mechanisms underlying the expression of LTP (Schnell and Nicoll, 2001).

Synaptic AMPARs can be internalized upon specific stimuli resulting in LTD. This process is dependent on the activation of Rab5 (upon activation of NMDARs) and requires clathrin-mediated endocytosis (Brown et al., 2005). Recent data showed that NMDAR activation leads to an initial decrease in surface GluA2 subunits followed by a delayed reduction in GluA2 from synaptic puncta (i.e. presumed synapses; Sanderson et al., 2011). In contrast, activation of mGluRs using dihydroxyphenylglycine results in a delayed decrease in GluA2 subunits from dendritic shafts, and interestingly, GluA2 fluorescence increases in some synaptic puncta and decreases in the others. Thus, there are at least two different forms of LTD mechanisms triggered by the activation of different kinds of GluRs in hippocampal neurons (Figure 6).

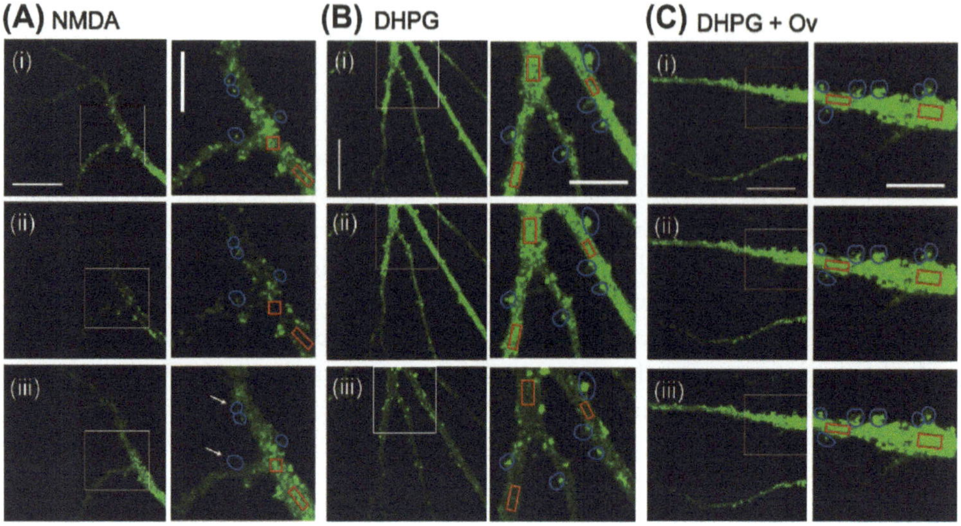

Figure 6 NMDA and DHPG (mGluR agonist) differentially regulate trafficking of super ecliptic pHluorin (SEP)-GluA2 (essentially fluoresces only when on the neuron surface) in cultured hippocampal neurons. (A) Representative images of cultured hippocampal neurons infected with SEP-GluA2 (i) prior to NMDA treatment, (ii) 5 min postaddition of NMDA, and (iii) 35 min postaddition of NMDA. (B) Representative images of cultured hippocampal neurons infected with SEP-GluA2 (i) prior to DHPG treatment, (ii) 5 min postaddition of DHPG, and (iii) 35 min postaddition of DHPG. (C) Representative images of cultured hippocampal neurons pretreated with 1 mM orthovanadate (i) prior to DHPG treatment, (ii) 5 min postaddition of DHPG, and (iii) 35 min postaddition of DHPG. Example regions of interest (ROI) of the kind that were used for analysis are highlighted in red (dendritic shaft) and blue (puncta). Calibration represents 20 μm on the main images and 10 μm on insets. *Figure and modified legend are reprinted from Sanderson et al. (2011).*

The mechanism of NMDAR-induced internalization involves the influx of calcium through the NMDAR; the calcium influx activates calcineurin and protein phosphatase 1, which dephosphorylate phosphatidylinositol 4-phosphate 5-kinaseγ661 (PIP5Kγ661). As a result, PIP5Kγ661 associates with the clathrin adaptor protein, AP-2, ultimately inducing AMPAR endocytosis (Unoki et al., 2012). In addition, S-nitrosylation of C875 of GluA1 (one consequence of calcium entry through NMDARs) enhances binding of the AMPAR to AP-2 (Selvakumar et al., 2013). mGluR-dependent LTD involves AMPAR endocytosis mediated by the guanine-exchange factor, BRAG2 (reviewed in Anggono and Huganir (2012)). Interestingly, internalization of synaptic AMPARs upon LTD does not seem to be dependent on their subunit composition (Esteban, 2008).

2.2.6. Recycling and Internalization of GluRs

Are internalized receptors recycled back to the synapse? The AMPARs are constitutively internalized from the postsynaptic membrane and recycled back via a Rab4-dependent pathway (Ehlers, 2000; Passafaro et al., 2001). The fate of internalized AMPARs depends on the mode of receptor stimulation. Upon stimulation of AMPARs, internalized receptors are directed from early/sorting endosomes into late endosomes and lysosomes for degradation (Ehlers, 2000). However, stimulation of NMDARs leads to trafficking of internalized AMPARs back to the postsynaptic membrane via recycling endosomes. The intracellular sorting of internalized AMPARs may depend on subunit composition since different protein–protein interactions during/after internalization may occur (Lee et al., 2004).

Probably, internalized AMPARs are recycled back to the synapse, to extrasynaptic membranes, or are degraded. For example, interaction of PICK1 with phosphorylated GluA2 is required for removal of AMPARs from hippocampal synapses (Kim et al., 2001). Subsequently, GluA2, PICK1, NSF, and a-/b-SNAPs form a complex in the presence of ATPgS that activates NSF ATPase activity, resulting in disruption of PICK1–GluA2 interactions (Hanley et al., 2002). In the nematode, *Caenorhabditis elegans*, internalized AMPARs (subunit GLR-1) may be sequestered into early/sorting endosomes by a complex of the Rab6 GTPase, RAB-6.2, and LIN-10, retromer complex, and RME-8; from the endosomes, RAB-6.2 then may traffic GLR-1 on cargo vesicles to the Golgi (both soma and dendrite outposts) for later recycling of the AMPARs back to the surface (Zhang et al., 2012).

NMDARs are internalized through a clathrin-dependent mechanism (Roche et al., 2001; Petralia et al., 2003; Lavezzari et al., 2004). Both GluN2A and GluN2B interact with the AP-2 adaptor complex, although, with different endocytic motifs. The GluN2A subunit is internalized via a C-terminal dileucine motif that interacts with AP-2 (Lavezzari et al., 2004). An additional AP-2 binding motif, YXXΦ, localized near the M4 domain of GluN2A, mediates the use-dependent rundown of NMDAR currents (Vissel et al., 2001). The GluN2B subunit has an AP-2 binding motif, YXXΦ (YEKL), in the far C-terminus close to the PDZ-binding domain (Roche et al., 2001). Phosphorylation of Tyr1472 within this motif, by Fyn kinase, prevents the internalization and thus increases synaptic NMDAR currents (Prybylowski et al., 2005). Interestingly, the Y1472F-knockin mice exhibit an increase in perisynaptic GluN2B-containing receptors (Nakazawa et al., 2006). Since Fyn interaction with PSD–MAGUKs enhances the phosphorylation of Tyr1472, PSD–MAGUKs may bring NMDARs in contact with Fyn kinase (Prybylowski et al., 2005). As mentioned above, the deletion of the GluN2B PDZ-binding domain increases the total NMDAR currents but not synaptic NMDAR currents (Prybylowski et al., 2002).

Together, these data suggest that removal of synaptic NMDARs requires their release from a synaptic anchor, likely PSD–MAGUKs, and then internalization via a clathrin-dependent process. In mature neurons, clathrin-mediated endocytosis has been observed mainly in the extrasynaptic area on the side of spines; in immature synapses, endocytosis has been observed commonly in the perisynaptic region (Blanpied et al., 2002; Petralia et al., 2003). It is not clear what mechanism(s) drive NMDARs from the PSDs to the site of endocytosis. NMDARs may be internalized via a clathrin-independent pathway, e.g. via a flotillin-dependent pathway (Swanwick et al., 2009). After internalization, both GluN2A- and GluN2B-containing receptors enter the early/sorting endosomes but then are differentially processed (Lavezzari et al., 2004). GluN2A-containing receptors preferentially traffic to late endosomes whereas GluN2B-containing receptors traffic to recycling endosomes. Thus, internalized GluN2B-containing receptors can return to the synapse in contrast to the GluN2A-containing receptors. Finally, GluN3A-containing receptors are specifically internalized via their interaction with PACSIN1 (PKC and CK substrate in neurons protein 1) in the endocytic zone on the side of spines (Perez-Otano et al., 2006).

The synaptic KARs containing GluK5 are internalized by a mechanism that includes a SNAP25/GluK5 interaction (Selak et al., 2009), and that is employed in basal conditions as well as during LTD and is regulated by PKC. This form of postsynaptic, KAR-mediated, SNAP25-dependent LTD occurs at hippocampal mossy fiber synapses on CA3 pyramidal neurons, and it also requires activation of mGluRs but not the subsequent release of calcium. Another specific KAR-mediated synaptic transmission (EPSC-KAR) LTD has been described in perirhinal cortex layer II/III neurons (Park et al., 2006). This form of LTD occurs via a mechanism involving mGlu1, PKC, and PICK1 PDZ-domain interactions and requires synaptic activation of inwardly rectifying KARs and the release of Ca^{2+} from stores. Furthermore, it has been found that SNAP25 is critical for the synaptic removal of KARs during LTD at MF-CA3 synapses (Selak et al., 2009). Another form of LTD, found at synapses between parallel fibers and Purkinje cells in the immature cerebellum, is mediated by D-serine activation of GluD2 receptors (Kakegawa et al., 2011).

2.2.7. Perisynaptic and Extrasynaptic Distributions

Many studies suggest that a specialized perisynaptic region is located within ~100–200 nm of the edge of the PSD (Newpher and Ehlers, 2008). This region contains several specific proteins such as homer, mGluRs, and GluN3A-containing receptors and may regulate the exchange of GluRs between extrasynaptic and synaptic regions (Tu et al., 1999; Perez-Otano et al., 2006; Newpher and Ehlers, 2008). For example, bath application of glutamate, that causes a dramatic reduction in the number of synaptic AMPARs, leads to an accumulation of GluA2-containing receptors in the perisynaptic region (Tardin et al., 2003). On the other hand, bicuculline/glycine treatment mimicking LTP stimuli results in a reduction in the number of perisynaptic AMPARs. The local lateral movement from the perisynaptic region can replace desensitized synaptic AMPARs with nondesensitized ones (Heine et al., 2008a). Concerning protein–protein interactions in the perisynaptic region, N-cadherin, which interacts with the extracellular region of AMPARs, is localized to the edges of the PSDs (Petralia et al., 2005; Nuriya and Huganir, 2006, Figures 3 and 7). Also, GluN3A-containing NMDARs are preferentially localized at the edges of the PSD and in perisynaptic regions (Perez-Otano et al., 2006).

The NMDARs are distributed widely in extrasynaptic areas of the dendritic shafts and spines (Hardingham and Bading, 2010; Petralia et al., 2010; Bard and Groc, 2011;

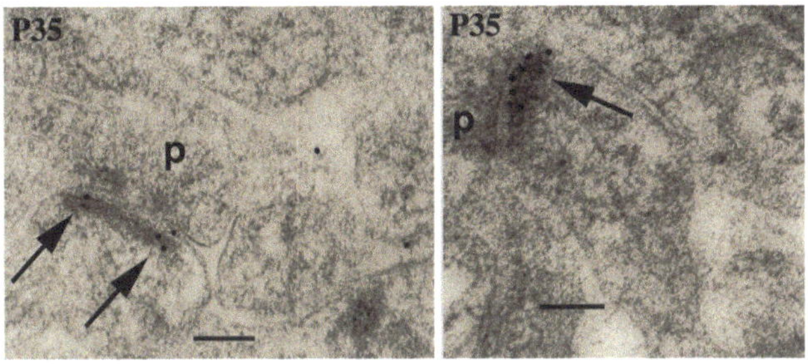

Figure 7 Immunogold labeling for cadherin (left) and catenin (right) at excitatory synapses in the P35 hippocampus (CA1 stratum radiatum); these show the features of mature glutamatergic synapses. Note the prevalence for labeling on the edge of the synapse in the left micrograph. p, presynaptic terminal. Scale bar in 100 nm. *Modified from part of Figure 3 in Petralia et al. (2005).*

Gladding and Raymond, 2011; Petralia, 2012). Extrasynaptic trafficking of GluN2-containing NMDARs may be mediated by their binding partner [G alpha interacting protein]-interacting protein C terminus (GAIP), a single PDZ-domain containing protein, that interacts preferentially with intracellular and surface extrasynaptic receptors (Yi et al., 2007). Synaptic and extrasynaptic NMDARs also are gated by different coagonists, D-serine and glycine, respectively (Papouin et al., 2012). Proper regulation of synaptic and extrasynaptic receptors is critical for normal functioning of the CNS. For example, in Huntington's disease, mutant huntingtin protein in GABAergic medium-sized spiny projection neurons (MSNs) of the striatum may cause the increased forward trafficking of NMDARs to the cytoplasmic membrane and the redistribution of synaptic NMDARs to extrasynaptic sites (Hardingham and Bading, 2010; Gladding and Raymond, 2011). extracellular signal-related kinase (ERK)-induced serine phosphorylation of the protein messenger, Jacob, results from synaptic but not extrasynaptic NMDAR stimulation; subsequent arrival of phosphorylated Jacob to the nucleus induces neuronal survival while the nonphosphorylated form induces cell death (Karpova et al., 2013).

2.2.8. Auxiliary Subunits of GluRs

TARPs are transmembrane proteins that associate with AMPARs, and act as auxiliary subunits, as noted above. The first TARP, stargazin (γ-2), was discovered through the characterization of stargazer, a spontaneous mutation in an inbred mouse line, in which AMPAR-mediated synaptic as well as extrasynaptic currents are largely absent (Letts et al., 1998; Chen et al., 2000). Later studies identified additional members of this family of proteins, γ-3, γ-4, γ-5, γ-7, and γ-8, that can interact with AMPARs (Jackson and Nicoll, 2011). In addition to their role in the early biosynthetic pathway of AMPARs, at least some TARPs are likely to be critical for AMPAR targeting to extrasynaptic membranes. TARPs also interact with PSD-MAGUKs through their PDZ-binding domains; this interaction retains AMPARs in synapses since its disruption leads to increased removal of AMPARs from synapses (Chen et al., 2000; Bats et al., 2007). The PSD-MAGUKs will be discussed in more detail in the next section.

Together, the data support the conclusion that extrasynaptic AMPARs are a reservoir for synaptic receptors. However, the deletion of γ-8, expressed mainly in the hippocampus, leads to a markedly decreased pool of extrasynaptic AMPARs with a minor effect on the synaptic pool of AMPARs (Rouach et al., 2005). Thus, different TARPs may be involved in multiple levels of regulation of AMPAR surface trafficking. Similarly, novel auxiliary subunits of AMPARs called cornichons (CNIH-2 and CNIH-3), have been identified using a proteomics approach (Schwenk et al., 2009). These proteins may act as ER chaperones rather than bona fide auxiliary subunits (Shi et al., 2010). Manipulating neuronal CNIH-2 levels modulates the electrophysiological properties of extrasynaptic and synaptic γ-8-containing AMPARs. This suggests that γ-8 and CNIH-2 probably interact with hippocampal AMPAR complexes (Kato et al., 2010). CNIH-2 modulates neuronal AMPAR auxiliary subunit assembly by regulating the number of TARPs within an AMPAR complex to modulate receptor gating and pharmacology (Gill et al., 2011). CNIH-2 and CNIH-3 bind selectively to GluA1 in hippocampal neurons, thus allowing GluA1/GluA2 receptors to reach the cell surface. Moreover, γ-8 may prevent the interaction of CNIH-2/3 with non-GluA1 subunits. Thus, trafficking and function of individual AMPARs are regulated by interplay of TARPs and cornichons with GluA subunits (Herring et al., 2013).

Of course, in addition to these transmembrane auxiliary subunits, other AMPAR-interacting proteins, including key intracellular interactors such as GRIP1, PICK1, and NSF, play important roles in AMPAR trafficking and functioning, and are discussed in other parts of this chapter (also reviewed in Anggono and Huganir (2012)). The KARs have been shown to directly interact with NETO2 and NETO1 proteins; these interactions alter their biophysical properties (Zhang et al., 2009; Copits et al., 2011; Tang et al., 2011). Furthermore, NETO2 targets GluK1-containing receptors to hippocampal excitatory synapses (Copits et al., 2011). Interestingly, the interaction of NETO1 with NMDARs also may be crucial for synaptic plasticity as well as for learning (Ng et al., 2009). Finally, there is a direct interaction between GluK2-containing receptors and KRIP6 (kainate receptor interacting protein for GluK2) (Laezza et al., 2007). Similar to other proteins, this interaction may alter the trafficking as well as the functional properties of KARs.

2.3. MAGUK PROTEIN COMPLEXES IN SYNAPTIC PLASTICITY

We have described the basic synaptic organization in an earlier section; as noted, postsynaptic proteins are integrated into a complex centered especially around the MAGUKs. Here we will concentrate on the control of the positioning and clustering of these proteins.

2.3.1. PSD Scaffolding Proteins and Receptor Distributions

The arrangement and composition of PSDs and protein–protein interactions with PSD-MAGUKs may regulate trafficking of GluRs within dendritic spines. The PSDs are composed of many proteins. Abundant proteins include CaMKII-α and -β that occupy 7.4 and 1.3% of the PSD mass, respectively (Peng et al., 2004; Cheng et al., 2006). In addition to their obvious structural role, both proteins are involved in signal transduction within the PSD. The PSD-MAGUKs and other PDZ-containing proteins are also abundant with ~50–400 molecules/PSD (Peng et al., 2004; Chen et al., 2005; Cheng et al., 2006). There is a laminar organization of scaffold proteins in the PSDs, with peak concentration of PSD-95 at a distance of ~12 nm from the postsynaptic membrane and of GKAP-Shank at ~25 nm (Valtschanoff and Weinberg, 2001; Petersen et al., 2003; Petralia et al., 2005). This indicates that Shank proteins act as a scaffold of scaffolds since they link together PSD-MAGUKs as well as other proteins such as homers and the actin cytoskeleton (Boeckers et al., 2002; Baron et al., 2006). Interestingly, Shank protein level and its clustering at the PSD depend on local Zn^{2+} concentration (Grabrucker et al., 2011). A decreased Zn^{2+} concentration probably contributes to Shank scaffold disruption and to synaptic instability. Since high concentrations of Zn^{2+} have been observed in amyloid deposits in Alzheimer's disease (AD) patients and Zn^{2+} promotes Aβ oligomerization, it has been suggested that trapping the Zn^{2+} in synaptic regions may be at least one of the pathomechanisms of AD.

GluRs are expressed in markedly smaller amounts in PSDs than the previously mentioned PSD proteins, with ~5–200 AMPARs (Peng et al., 2004; Tanaka et al., 2005; Cheng et al., 2006; Masugi-Tokita et al., 2007; Newpher and Ehlers, 2008). The number of NMDARs is likely even smaller than that of AMPARs (Nimchinsky et al., 2004). The excess of scaffold proteins over the GluRs indicates that there is an availability of slots for GluRs within PSDs that may be used in case of synaptic plasticity (Lisman and Raghavachari, 2006). AMPARs either are localized at the periphery or are homogeneous throughout the PSD, while NMDARs appear to be distributed at the center of the PSD (Racca et al., 2000; Masugi-Tokita et al., 2007). This conclusion is in agreement with FRAP experiments showing that the fluorescence recovery of GluA1–AMPARs is faster than that of GluN1–NMDARs (Sharma et al., 2006). MacGillavry et al. (2013) used a super-resolution light microscope technique to show that postsynaptic PSD-95 is concentrated in subsynaptic scaffolding domains within the PSD, and that these are preferentially enriched in AMPARs. However, other studies indicate that both AMPARs and NMDARs can have homogeneous distributions in the synapse (Petralia et al., 1999; Antal et al., 2008; Dani et al., 2010). Activation of GluRs may alter the composition of the PSDs. For example, it has been found that activation of NMDARs induces the synaptic translocation of endogenous IRSp53 in an actin cytoskeleton-dependent manner (Hori et al., 2005). IRSp53 is an effector of Cdc42, a small GTP-binding protein affecting actin dynamics. The translocation of IRSp53, which is regulated by PKC, enhances synaptic activity as detected by increased miniature EPSCs (mEPSCs).

PSD-95 was first identified as a component of the PSD (Cho et al., 1992). Later, three related proteins belonging to the PSD-MAGUK family of proteins—PSD-93, SAP102, SAP97—were characterized (Funke et al., 2005; Zheng et al., 2011). The PSD-MAGUKs share a domain structure organization with three PDZ domains, an Src-homology 3 domain (SH3), and an enzymatically silent guanylate kinase (GK) domain. The PDZ domains can bind to consensus binding motifs that are often localized in the extreme C-terminus of membrane proteins. For example, the first and the second PDZ domains of PSD-MAGUKs can interact with C-termini of the GluN2 subunits of the NMDARs (Kornau et al., 1995). Due to this interaction, the PSD-MAGUKs likely cluster surface NMDARs (Kim et al., 1996) and modulate the biophysical properties of NMDAR-mediated currents (Yamada et al., 1999; Lin et al., 2006).

Interestingly, the overexpression of PSD-95 in dissociated neuronal cultures as well as organotypic slice cultures produces a major increase in AMPAR-EPSCs, but only a minor or no increase in NMDAR-EPSCs (Elias and Nicoll, 2007). These observations are quite surprising considering that a biochemical approach reveals only the association of PSD-95 with NMDARs (Kornau et al., 1995). Similarly, overexpression of PSD-93 and SAP102 enhances AMPAR-EPSCs (Elias and Nicoll, 2007), while shRNAs against PSD-93 and PSD-95 reduce AMPAR-EPSCs. Interestingly, the analysis of mEPSCs of AMPARs in neurons expressing PSD-93 and PSD-95 shRNAs shows a reduction only in the frequency of events but not in their amplitude (Elias et al., 2006), suggesting that there is an increase in the number of silent synapses containing virtually no AMPARs. Coexpression of PSD-93 and PSD-95 shRNAs decreases AMPAR synaptic currents by ~75% indicating that PSD-93 and PSD-95 control synaptic delivery of AMPARs.

Expression of SAP102 shRNAs in hippocampal neurons has no effect on synaptic delivery of AMPARs in mature synapses of wild-type animals, but virtually depletes AMPARs as well as NMDARs from double knockout (KO) PSD-93 and PSD-95 animals. This suggests that (1) SAP102 is not critical for AMPAR synaptic trafficking in mature synapses; (2) SAP102 can functionally compensate for the loss of PSD-93 and PSD-95; and (3) NMDARs require a critical level of PSD-MAGUKs for their synaptic targeting (Elias and Nicoll, 2007). The expression of PSD-MAGUKs is developmentally regulated; SAP102 is dominantly expressed in immature neurons in contrast to PSD-93 and PSD-95 that are strongly expressed later in development (Sans et al., 2000). SAP102 mediates synaptic trafficking of AMPARs and NMDARs during synaptogenesis. Indeed, expression of SAP102 shRNAs in immature neurons reduces AMPAR-mediated currents (Elias et al., 2008). In contrast, PSD-95 assumes the functions of SAP102 later in development and is necessary for two aspects of synapse maturation: the developmental increase in AMPAR transmission and replacement of GluN2B-NMDARs with GluN2A-NMDARs. TARPs play a role in some of the PSD-MAGUK-mediated effects of AMPAR trafficking since TARPs form a physical link between AMPARs and PSD-MAGUKs, as described in an earlier section (reviewed by Jackson and Nicoll (2011)).

What is the role of PSD-MAGUKs in activity-dependent regulation of the number of synaptic AMPARs? Overexpression of PSD-95 occludes LTP and enhances LTD and, consistently, in PSD-95-KO animals, an enhancement of LTP and absence of LTD is observed (Migaud et al., 1998; Stein et al., 2003; Ehrlich and Malinow, 2004). The molecular mechanism underlying these phenomena remains to be determined.

SAP97 is the only PSD-MAGUK that directly interacts with the C-terminus of an AMPAR; it interacts only with GluA1 (Leonard et al., 1998). The GluA1–SAP97 interaction likely occurs early in the biosynthetic pathway although it is crucial for the delivery of homomeric GluA1 receptors into synapses (Sans et al., 2001). SAP97 also can interact with NMDARs through its first two PDZ domains. Thus, it is not surprising that the overexpression of SAP97 early in development increases the number of synaptic AMPARs as well as NMDARs in hippocampal slice cultures (Howard et al., 2010). SAP97 is likely a part of the machinery that traffics GluRs to synapses, although its function can be compensated by other PSD-MAGUKs when it is conditionally deleted from neurons.

The PSD-MAGUKs, especially PSD-95, form a key skeleton of the PSDs. Data show that ~95% of PSD-95 present in spines is located within the PSD in contrast to ~70% of SAP102 (Zheng et al., 2010). Furthermore, PSD-95 and SAP102 have different mobilities in spines as determined by FRAP experiments; the majority of PSD-95 is immobile in spines in contrast to SAP102, of which the majority is highly mobile in spines. This suggests that there are different roles for PSD-95 and SAP102 in spines. The stabilization of PSD-95 within the PSD requires the N-terminal palmitoylation and protein interactions mediated by the first and second PDZ domains, whereas formation of a stable lattice of PSD-95 molecules within the PSD additionally requires the C-terminal SH3 domain (Sturgill et al., 2009). It also has been shown that activity-dependent growth of dendritic spines of hippocampal neurons is accompanied by destabilization of the PSD, resulting in transient loss and rapid replacement of PSD-95 and Shank2 (Steiner et al., 2008).

Furthermore, an activity-dependent trafficking of PSD-95 is triggered by phosphorylation of a conserved CaMKII consensus phosphorylation site, described earlier in the text, which negatively regulates spine growth. The mobility of PSD-95 is not affected by the mobility of the actin cytoskeleton in basal conditions; on the other hand, the mobility of SAP102 is tied to the mobility of actin filaments. However, PSD-95 appears to regulate the mobility of actin filaments during synaptic plasticity (Zheng et al., 2010, 2011). Indeed, several PSD-95–actin complexes are known from the PSDs including PSD-95–IRSp53–Eps8–actin, PSD-95–spine-associated RapGAP (SPAR)–actin, PSD-95–GKAP–shank–cortactin–actin, and PSD-95–SPIN90/WISH–N-WASP–actin (Petralia et al., 2009; there also is the complex of SAP97–MyosinVI–actin). Obviously, PSD-95 plays a complex role in the regulation of spine morphology by regulation of the actin cytoskeleton within spines.

2.3.2. Protein Interactions of MAGUK Proteins

PSD-MAGUKs can cluster NMDARs within synapses (Zheng et al., 2010; Gladding and Raymond, 2011; Zheng et al., 2011), but what determines the clustering of PSD-MAGUKs there? In the case of PSD-95, at least, two regions are necessary for clustering: two N-terminal cysteines that can be palmitoylated and C-terminal tyrosine-based trafficking motifs (Craven and Bredt, 2000; El-Husseini et al., 2000a). On the other hand, SAP102 requires the SH3 and GK domains for synaptic clustering (Zheng et al., 2010). Thus, it is likely that different protein–protein interactions of PSD-MAGUKs control their synaptic clustering. PSD-MAGUKs also may help cluster extrasynaptic NMDARs, and thus may control whether GluRs are synaptic or extrasynaptic. For example, in Huntington's disease, enhanced binding of PSD-95 to the GluN2B subunit may lead to increased stability of extrasynaptic NMDARs (Gladding and Raymond, 2011).

However, PDZ domains are quite promiscuous in their protein–protein interactions and thus PSD-MAGUKs exhibit multiple protein–protein interactions within dendritic spines. It has been reported that the first and second PDZ domains of PSD-95 interact with the receptor tyrosine kinase ErbB4 (Garcia et al., 2000; Huang et al., 2000), semaphorin adhesion proteins (Burkhardt et al., 2005), and shaker-type voltage-gated potassium channels (Kim et al., 1995). Moreover, these domains interact with neuronal nitric oxide synthase, which is regulated by NMDAR-mediated Ca^{2+} influx (Brenman et al., 1996; Christopherson et al., 1999). Thus, PSD-95 is also a scaffold for Ca^{2+}-regulated enzymes. The third PDZ domain of PSD-95 interacts with NL (Irie et al., 1997) and at least three regulators of small GTPases, SynGAP (Kim et al., 1998), Kalirin7 (Penzes et al., 2000), and Citron-N (Zhang et al., 1999).

Indeed, NLs are key players in synapse functioning, and the NL–PSD-95 interaction regulates both excitatory and inhibitory synaptogenesis (Prange et al., 2004; see also below in the section on adhesion proteins). The SynGAP–PSD-95 interaction is crucial for the regulation of the cytoskeleton in dendritic spines (Vazquez et al., 2004). Interestingly, depolarization may cause migration of SynGAP out of the PSD to allow other proteins to bind to PSD-95 (Yang et al., 2011). The activation of Kalirin7 by the EphB2 receptor controls the actin cytoskeleton in spines; the PSD-95–Kalirin7 interaction likely regulates this process (Penzes et al., 2000, 2001, 2003). A recent report showed that Plk2 phosphorylates a quartet of Ras and Rap regulators—SynGAP, PDZ-GEF1, RasGRF1, and SPAR—resulting in bidirectional control over Rap and Ras activity (Lee et al., 2011). Thus, the GEFs and GAPs cooperate to regulate excitatory synapses, dendritic spines, and surface expression of AMPARs.

2.3.3. Cytoskeletal Associations and Splice Variants

The GK domain of PSD-95 interacts with a spine-associated RapGAP called SPAR. SPAR interacts with actin and regulates the actin cytoskeleton in spines by downregulation of

Rap activity (Pak et al., 2001). Furthermore, as mentioned in the Introduction, the GK domain associates with GKAPs, and the C-termini of GKAPs interact with the single PDZ domains of Shank family proteins (Naisbitt et al., 1999).

All members of the PSD-MAGUKs are extensively spliced on the mRNA level so that in addition to the shared domains, these proteins may contain multiple alternative domains. For example, PSD-95, PSD-93, and SAP97 contain on their N-termini either two cysteine residues that can be palmitoylated (α-isoform) or an L27 domain (β-isoform) (Funke et al., 2005; Schluter et al., 2006). Palmitoylation stabilizes PSD-95 and SAP97 in spines and alters synaptic transmission independently of synaptic activity; in contrast, the effect of β-isoforms on synaptic transmission is activity-dependent (Schluter et al., 2006; Waites et al., 2009). αSAP97 occludes LTP by enhancing the levels of postsynaptic AMPARs, whereas βSAP97 blocks LTP by reducing the synaptic localization of NMDARs (Li et al., 2011). This study also indicates that αSAP97 selectively regulates the synaptic pool of AMPARs, whereas βSAP97 regulates the extrasynaptic pools of both AMPARs and NMDARs.

Other N-terminal insertions in SAP97, I1A and I1B, facilitate multimerization of SAP97 and mediate interactions with tyrosine kinases (Nakagawa et al., 2004; Funke et al., 2005). Three splice variants of SAP102 have been described. In addition to the prototypic SAP102, one variant has an 18-aa deletion in the N-terminus close to the PDZ1 domain called I1 and the second variant has a 14-aa deletion between the SH3 and GK domains called I2 (Muller et al., 1996). Interestingly, the I1 region is recognized by the GluN2B subunit and this interaction promotes lengthening of dendritic spines and the formation of synapses (Chen et al., 2011). The Hook region (between GK and SH3 domains) also is alternatively spliced in PSD-93 and SAP97 (Muller et al., 1996; Rumbaugh et al., 2003). In the case of SAP97, the Hook region contains four alternative insertions—I2, I3, I4, and I5. The I3 insert interacts with protein 4.1N; this interaction is critical for delivery of SAP97 and AMPARs into synapses (Rumbaugh et al., 2003).

2.4. ADHESION PROTEINS IN SYNAPTIC PLASTICITY

CAMs are involved in regulating synaptogenesis and spine maturation, as well as any subsequent synaptic plasticity that occurs. They are also involved in developmental events such as neurite outgrowth and migration. However, in this section, we will focus on how they can regulate formation of excitatory synapses and the localization and density of GluRs. Direct or indirect interactions between GluRs and several CAMs have been reported, including neural cell adhesion molecule (NCAM), N-cadherin, SynCAMs, EphB receptors, NGL-2 (netrin-G ligand 2), synaptic adhesion-like molecules (SALMs), neurexins, NLs, and leucine-rich repeat transmembrane (LRRTM) proteins (Figure 8). However, synapses can still form when several CAMs

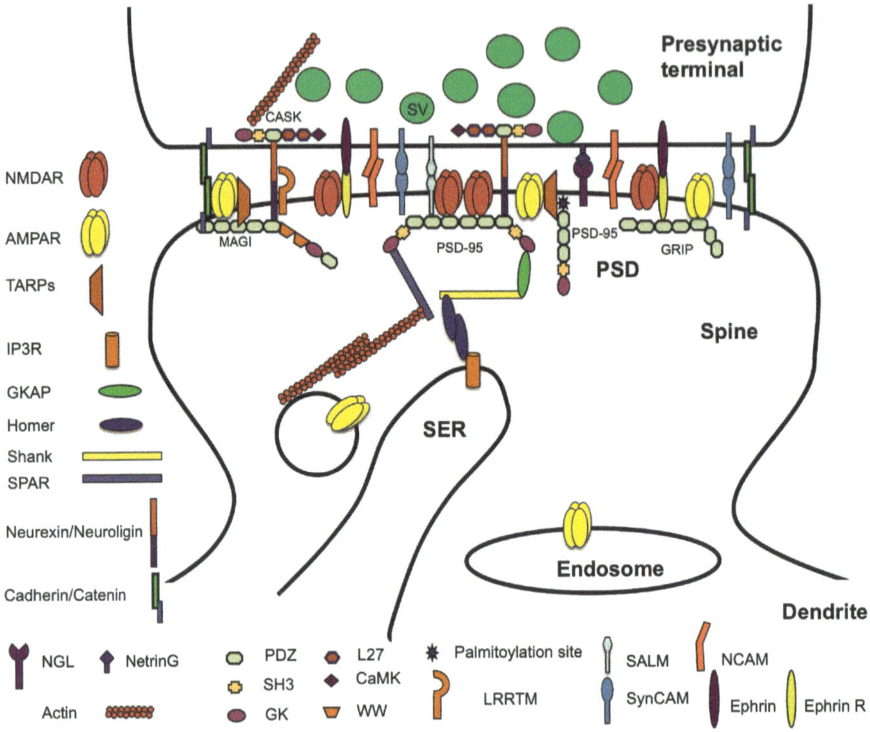

Figure 8 Interactions between cell adhesion molecules (CAMs) and glutamate receptors. several CAMs, including NCAM, N-cadherin, SynCAM, EphB receptors, NGL-2, SALMs, neuroligins, neurexins, and LRRTM proteins, interact with NMDAR and AMPARs through direct or indirect interactions. For example, the GluA2 subunit of AMPARs interacts with N-cadherin via a direct extracellular interaction (Saglietti et al., 2007; Nuriya and Huganir, 2006). Both EphB2 (Dalva et al., 2000) and SALM1 (Wang et al., 2006) interact with NMDARs through an extracellular interaction with the GluN1 subunit of the NMDAR. NL-1 interacts with the NMDAR through PSD-95 family members or MAGI-2 (Chih et al., 2005; Graf et al., 2004; Nam and Chen, 2005; Sudhof, 2008). Alternatively, there may be an uncharacterized extracellular interaction between NL-1 and the NMDARs. Interactions between CAMs, NMDARs, and AMPARs influence the activity and location of the GluR receptors, thereby modulating spine maturation, LTP, and LTD.

are removed in KOs, like EphBs (Kayser et al., 2008), neurexins (Missler et al., 2003), NLs (Varoqueaux et al., 2006), or SynCAM1 (Fujita et al., 2006), suggesting that different CAMs can compensate for one another. Nevertheless, loss of CAMs often causes changes in synaptic function and affects neurodevelopment. Mutation or deletion of neurexin genes (reviewed in Sudhof (2008)), NL genes (Jamain et al., 2003; Laumonnier et al., 2004), synapse-associated CAM gene (CADM1) (Zhiling et al., 2008), cadherin genes (Wang et al., 2009), netrin-G1 gene (Borg et al., 2005), and SALM genes (Wang et al., 2009; de Bruijn et al., 2010;Voineagu et al., 2011) all contribute to autism and/or mental retardation.

2.4.1. Synaptic NCAM

What roles do CAMs have in synapse development and GluR localization? NCAM is recruited to synaptic contacts very rapidly (Sytnyk et al., 2002) and can promote the stabilization of excitatory synapses (Dityatev et al., 2000). It is indirectly linked to NMDARs and CaMKII through spectrin (Sytnyk et al., 2006). As discussed earlier, NMDARs and CaMKII are important in establishing the initial phases of synaptic plasticity. NCAM-deficient neurons result in disruption of the spectrin scaffold, and CaMKII is no longer recruited to synapses after NMDAR activation (Sytnyk et al., 2006). Both NCAM and NMDARs can influence each other's function and localization at the synapse. NCAM-mediated synaptogenesis is blocked by the NMDAR antagonist AP5 (Diyatev et al., 2004) and NMDAR currents are decreased by polysialic acid (PSA) or PSA-NCAM when low concentrations of glutamate are applied, indicating that these may function as a competitive antagonist for NMDARs (Hammond et al., 2006). As a result, PSA-NCAM prevents excitotoxicity by inhibiting GluN2B (Hammond et al., 2006). In addition, LTP protocols can affect the location of NCAM and NMDARs at the synapse. Both are redistributed to the edges of the PSD after in vivo induction of LTP by tetanic stimulation of the perforant pathway of the dentate gyrus (Fux et al., 2003). Furthermore, NMDAR-mediated LTP is impaired in NCAM-KO mice (Muller et al., 1996; Bukalo et al., 2004). However, inhibition of GluN2B-containing receptors restores levels of LTP in slices from NCAM-KO mice or brain slices that are PSA-deficient due to enzymatic treatment with endosialidase of bacteriophage K1F (endoNF) (Kochlamazashvili et al., 2010). On the other hand, D-cyclo-serine, a partial coagonist of the NMDAR glycine-binding site, acts through GluN2A-containing receptors to also restore LTP in NCAM-KO mice (Kochlamazashvili et al., 2012), suggesting that NCAM might mediate a balance between the GluN2B and GluN2A subunits and regulate their involvement in synaptic plasticity. Meanwhile, mice overexpressing the NCAM extracellular proteolytic cleavage fragment have impaired LTP in the prefrontal cortex (Brennaman et al., 2011).

2.4.2. N-cadherin as an Activity Sensor

N-cadherins (Figures 8.3 and 8.7) mediate calcium-dependent adhesion through homophilic interactions, and may act as trans-synaptic activity sensors. Decreases in extracellular calcium disrupt cadherin interactions and allow for synaptic plasticity. When antibodies or peptides made to the extracellular region of cadherins are applied to hippocampal slices during LTP induction, LTP is inhibited (Tang et al., 1998). These results suggest that the stimulus for LTP induction is sufficient to deplete extracellular calcium by activation of L-type Ca^{2+} channels or NMDARs, allowing the peptides to bind to the cadherins and disrupt their function. If enough calcium is added to prevent the destabilization of cadherins, loss of LTP is prevented. As mentioned previously, the GluA2 subunit of AMPARs interacts with N-cadherin via a direct extracellular interaction (Saglietti et al.,

2007; Nuriya and Huganir, 2006). The interaction between AMPARs and N-cadherin causes a reciprocal recruitment of both proteins to synapses (Saglietti et al., 2007; Nuriya and Huganir, 2006). In addition, N-cadherins decrease the lateral diffusion of AMPARs (Saglietti et al., 2007) and AMPARs and N-cadherin interact intracellularly through δ-catenin/NPRAP and GRIP/ABP (Silverman et al., 2007). Afadin, a component of N-cadherin and β-catenin adhesion complexes, also interacts with the GluA2 subunit and helps maintain surface levels of AMPARs (Srivastava et al., 2012). Furthermore, the association of N-cadherin and β-catenin with AMPARs regulates AMPAR activity and trans-synaptic signaling. For example, loss of postsynaptic β-catenin decreases the amplitude of mEPSCs mediated by AMPARs, affecting postsynaptic activity (Okuda et al., 2007). On the other hand, postsynaptic expression of dominant-negative N-cadherin impairs presynaptic release probability, affecting presynaptic activity. However, this presynaptic effect can be blocked by knockdown of the GluA2 subunit with siRNA (Vitureira et al., 2011). Furthermore, N-cadherins can undergo activity-regulated endocytosis, which removes the CAM from the cell surface, and decreases adhesion (Tai et al., 2007; Yasuda et al., 2007). The rate of N-cadherin internalization is reduced by NMDAR activation (Tai et al., 2007). Another mechanism that may prevent N-cadherin internalization is the cleavage of its cytoplasmic tail by presenilin-1 (PS1)/γ-secretase, in response to NMDAR activation (Marambaud et al., 2003). In addition, other interactions with the C-terminal tail of N-cadherin are important in regulating LTD. Overexpression of a mutant β-catenin that binds N-cadherin with greater affinity prevents N-cadherin internalization and blocks LTD (Tai et al., 2007).

2.4.3. SynCAM Contribution to Vesicle Clustering and GluR Trafficking

SynCAM1 is involved in the differentiation of presynaptic terminals and can induce the clustering of presynaptic vesicles (Biederer et al., 2002). Overexpression of SynCAM1 in hippocampal neurons increases the function of excitatory synapses, but not their number (Sara et al., 2005). On the other hand, overexpression of SynCAM1 in mice increases the number of excitatory synapses and the frequency of synaptic events, while loss of SynCAM1 decreases them (Robbins et al., 2010). On the postsynaptic side, protein 4.1B localizes NMDARs to SynCAM aggregation sites, and enhances NMDAR-mediated synaptic events, while protein 4.1N can recruit AMPARs to SynCAM adhesions (Hoy et al., 2009). Interaction of SynCAM1 with the MAGUK CASK (Biederer et al., 2002) may help regulate NMDAR trafficking since the Veli/CASK/Mint1 complex links NMDAR to the motor protein KIF17 (Setou et al., 2000). SynCAM1 is also involved in the maintenance of new synapses and can regulate LTD. Knocking out its expression at different time points affects the ability of SynCAM to maintain excitatory synapse number. When SynCAM1 is overexpressed early, but then turned off, the increase of synaptic events normally seen with overexpression is lost. SynCAM overexpressors also fail to show LTD, possibly due to the stabilization of spines or GluRs, while SynCAM1-KO mice have increased LTD, consistent with spine shrinkage or loss of synaptic contacts (Robbins et al., 2010). Furthermore, KO mice learn spatial tasks better than wild type, suggesting that flexibility in synaptic contacts may help learning. When the Morris water maze is used (a hippocampus-dependent task of spatial reference learning), KO mice learn the correct quadrant faster than wild-type mice. On the other hand, overexpressors are unable to learn the task or remember the location of a hidden platform (Robbins et al., 2010). Therefore, SynCAM1 can regulate excitatory synapses by modulating their number and plasticity, contributing to changes in learning and memory.

2.4.4. Eph Receptors and Glutamatergic Synapses

In the case of Eph receptors, interactions between postsynaptic EphB and presynaptic Ephrin B1 or 2 also induce the formation of glutamatergic synapses (Kayser et al., 2006; McClelland et al., 2009). EphB2 binds to ephrinB2 and this association stabilizes NMDARs at the synapse (Dalva et al., 2000; Nolt et al., 2011), potentiates NMDAR-dependent calcium influx (Tasaku et al., 2002), and prevents calcium-dependent densitization of GluN2B-containing NMDARs at mature synapses (Nolt et al., 2011). EphBs directly interact with NMDARs through an interaction mediated by the extracellular domain of EphB2 binding to the extracellular N-terminus of GluN1 (Dalva et al., 2000). As a result, clustering of Eph receptors induces the aggregation of NMDARs. For example, overexpression of EphB2 in cortical neurons or incubation of neurons with soluble ephrinB2 increases the density of GluN1 clusters (Dalva et al., 2000). In addition, treatment of cells with ephrinB2 increases tyrosine phosphorylation of the NMDAR by Src family kinases (Tasaku et al., 2002; Antion et al., 2010). EphB receptors can also interact indirectly with AMPARs through GRIP (Torres et al., 1998) and regulate their trafficking (Irie et al., 2005; Kayser et al., 2006). During LTP, Ephrin–EphB interactions may synchronize changes in the pre- and postsynaptic sides of the synapse. The use of interfering peptides to disrupt signaling between the two blocks retrograde signaling through EphrinB and inhibits changes in release probability at the presynapse, thereby blocking LTP (Contractor et al., 2002). Postsynaptically, the interaction of the EphB receptors with GluRs has also been implicated in regulation of LTP. In tadpoles, these trans-synaptic interactions and adjustments in synaptic plasticity may regulate activity-dependent refinement of the retinotopic map (Lim et al., 2008). At CA3-CA1 hippocampal synapses, EphB2 mediates LTP in a manner that is independent of its kinase activity or C-terminal tail (Grunwald et al., 2001; Henderson et al., 2001), possibly through its extracellular interaction with the NMDAR (Henderson et al., 2001). However, regulation of LTP is also dependent on phosphotyrosine-dependent reverse signaling mediated by EphrinB2 (Bouzioukh et al., 2007). In addition, the PDZ-binding motif of EphrinB2 is critical for both LTP and LTD (Bouzioukh et al., 2007). During LTP at mossy fiber-CA3 synapses, postsynaptic EphB2 is involved in activating AMPARs, resulting in EphB2 clustering, possibly through their mutual interaction with GRIP. As a result, this clustering may mediate presynaptic ephrinB clustering, leading to increased glutamate release (Contractor et al., 2002). In addition, EphrinB2 signaling also has been reported to reduce AMPAR internalization postsynaptically (Essmann et al., 2008). EphrinB3 may also play a role in mediating LTP. Two lines of EphrinB3-KO mice showed defective LTP at CA1-CA3 synapses that is NMDAR-dependent (Grunwald et al., 2004; Rodenas-Ruano et al., 2006), while another study reported normal LTP (Armstrong et al., 2006). EphrinB3 also regulates AMPARs at synapses in vitro (Aoto et al., 2007; Essmann et al., 2008; McClelland et al., 2010) and EphrinB3-KO mice have decreased AMPAR activity (Rodenas-Ruano et al., 2006; Antion et al., 2010).

2.4.5. NGL Family Recruitment of Postsynaptic Proteins

The NGL family has an LRR, an immunoglobulin (Ig) domain, and an intracellular PDZ-binding motif. NGL-1, NGL-2, and NGL-3 are mainly postsynaptic and bind presynaptic Netrin-G1, Netrin-G2, and leukocyte common antigen-related (LAR), respectively (Lin et al., 2003; Kim et al., 2006; Nishimura-Akiyoshi et al., 2007; Woo et al., 2009). While the NGLs are expressed throughout the brain, their binding partners are limited to more discrete areas; this may help determine their targeting and function (reviewed in Kwon et al. (2010); DeNardo et al. (2012)). Bead-induced clustering of NGL-2 causes the recruitment of the postsynaptic proteins PSD-95, GKAP, Shank, and NMDARs, but not AMPARs (Kim et al., 2006), while clustering of NGL-3 mediates the clustering of all these postsynaptic proteins (Woo et al., 2009). In addition, NGL-2 coprecipitates with NMDARs, but not with AMPARs in vivo (Kim et al., 2006). NGL-3, but not NGL-1 or NGL-2, interacts with the receptor tyrosine phosphatase LAR (Woo et al., 2009). LAR also forms a complex with cadherin, β-catenin, and AMPARs (Dunah et al., 2005). LAR dephosphorylates β-catenin, enhancing its interaction with cadherin (Kypta et al., 1996; Dunah et al., 2005). Together, this complex is thought to be involved in trafficking cadherins with AMPARs to synaptic spines (Dunah et al., 2005). LAR and NGL-3 have bidirectional synaptogenic effects on both sides of the synapse (Woo et al., 2009; Kwon et al., 2010). On the other hand, netrin-G2, the ligand of NGL-2, is not sufficient to induce clustering of postsynaptic proteins (Kim et al, 2006). Knockdown of NGL-2 decreases the number of excitatory synapses, but not that of inhibitory synapses (Kim et al., 2006). Similarly, knockdown of NGL-3 decreases the number and function of excitatory synapses (Woo et al., 2009). In vivo, NGL-2 lentivirus shRNA electroporated into the hippocampus decreases both AMPA and NMDAR EPSCs in the stratum radium, but not in the stratum lacunosum moleculare, suggesting that NGL-2 activity is restricted to the Schaffer collateral synapses in the CA1, where netrin-G2 is localized (DeNardo et al., 2012). Furthermore, transgenic mice that only express the extracellular domain of LAR have impaired spatial learning and hyperactivity (Kolkman et al., 2004), while loss of the netrin-G1 gene was found in a patient that had atypical Rett syndrome (Borg et al., 2005). However, other patients with atypical Rett syndrome did not have symptomatic mutations in netrin-G1, so it may only contribute to a small number of cases (Archer et al., 2006; Nectoux et al., 2007).

2.4.6. SALM Structure and Role in Synaptic Trafficking

Like NGLs, SALMs have an extracellular LRR and Ig domain. They also have an additional FN3 adhesion domain. Similar to many CAMs, SALMs 1–3 have an intracellular PDZ-binding motif. SALM expression increases during the early postnatal period when neurite outgrowth and synapse formation occur. They also have been reported to influence the synaptic localization of GluRs (Wang et al., 2006; Ko et al., 2006a). Similar to Eph receptors, SALM1 interacts directly with the GluN1 subunit of the NMDAR through an extracellular association. The SALMs also interact indirectly with NMDARs through an intracellular association with PSD-95 MAGUK family members via their PDZ-binding motif (Wang et al., 2006; Ko et al., 2006a; Morimura et al., 2006). As a result of this interaction, SALM1 enhances the surface expression of NMDARs and recruits endogenous NMDARs to dendritic spines (Wang et al., 2006, Figure 9). Deletion of the SALM1 PDZ-binding motif decreases surface expression of SALM1 in dendrites (Seabold et al., 2012), as well as its colocalization with endogenous PSD-95 and NMDARs (Wang et al., 2006). SALM2, on the other hand, influences the postsynaptic localization of AMPARs. Bead-induced aggregation of SALM2 induces clustering of postsynaptic proteins such as the GluA1 subunit of AMPARs and the scaffolding proteins PSD-95 and GKAP (Ko et al., 2006a). The NMDAR is also affected, but to a lesser extent. Furthermore, overexpression of SALM2 increases the number of excitatory synapses in hippocampal neurons, while knockdown of SALM2 causes a decrease (Ko et al., 2006a). Bead-induced aggregation of SALM3 in dendrites also causes clustering of postsynaptic proteins (Mah et al., 2010). In contrast, SALM5, which does not have a PDZ-binding motif, does not cluster PSD-95 (Mah et al., 2010). However, both SALM3 and SALM5 can induce the formation of excitatory and inhibitory presynaptic terminals (Mah et al., 2010). While SALM4 and SALM5 can interact homophilically in *trans*, between cells, the trans-synaptic ligands of the other SALMs are unknown (Seabold et al., 2008). Therefore, it is not clear which presynaptic ligand enables the induction of presynaptic differentiation, at this point. Similar to SALM2, knockdown of SALM5 results in a decrease in excitatory synapse number and activity. However, SALM5 knockdown also decreases inhibitory synapses. Still, the mEPSCs are reduced to a greater degree than mIPSCs, suggesting that SALM5 expression may have a greater impact at excitatory synapses (Mah et al., 2010).

2.4.7. Neurexin, NL, and LRRTM Involvement in Synaptic Clustering

Extensive work has been done to examine how neurexin and its ligands, the NLs and LRRTMs, regulate glutamatergic synapses. Neurexins expressed in nonneuronal cells induce the clustering of PSD-95 and NMDARs, but not AMPARs. However,

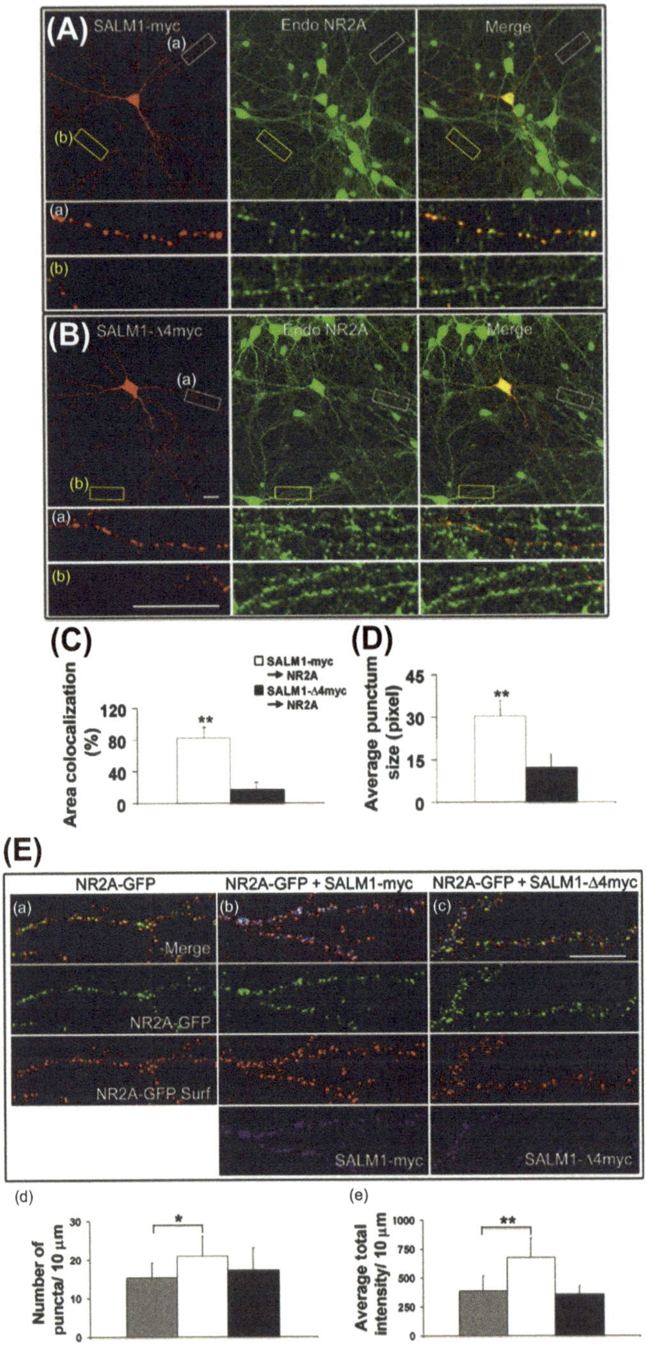

Figure 9 SALM1 regulates NMDAR localization through its interaction with PDZ proteins. Hippocampal neurons transfected at 14 DIV with (A) myc-tagged SALM1 or (B) SALM1-δ4, a construct that lacks the PDZ-binding motif, were permeabilized and immunostained with anti-GluN2A (NR2A) to

AMPARs are inserted into the membrane and associate with these PSD-95 clusters after application of glutamate or expression of the constitutively active form of CaMKII (Nam and Chen, 2005). NLs are involved in mediating the balance between excitatory and inhibitory synapses (Chih et al., 2005). NL-1 primarily localizes to excitatory synapses (Song et al., 1999), while NL-2 is concentrated at inhibitory synapses (Graf et al., 2004; Varoqueaux et al., 2004; Chih et al., 2005) and NL-3 is found at both (Budreck and Scheiffele, 2007). The distinct expression patterns of the NLs and their NRXN ligands in different cell types is thought to limit their trans-synaptic interactions and determines their separate functions at excitatory versus inhibitory synapses (Futai et al., 2013). A dominant-negative form of NL inhibits the activity of excitatory synapses and decreases the density and size of the PSD-95 and AMPAR clusters (Nam and Chen, 2005). Experiments using neurexin-coated beads, or cross-linked Fc fragments, coupled with electrophysiology recordings, further confirm that the formation of neurexin–NL-1 adhesions results in the recruitment of AMPARs to PSD-95 aggregations (Heine et al., 2008b; Mondin et al., 2011) and limits the diffusion of AMPARs (Mondin et al., 2011). This may be mediated by the AMPARs' interaction, via Stargazin/TARPs, with PSD-95 (El-Husseini et al., 2000b) or S-SCAM (Deng et al., 2006).

Trapping AMPARs at NL-1 adhesions is also dependent on the GluA2 subunit (Mondin et al., 2011), which is involved in the stabilization of presynaptic contacts that associate with AMPARs (Heine et al., 2008b; Ripley et al., 2011). These contacts are enhanced when NL-1 is coexpressed with the extracellular domain of GluA2 in heterologous cells (Ripley et al., 2011), suggesting an uncharacterized extracellular interaction between AMPARs, NLs, and/or an unknown protein. On the other hand, clustering of NL-1 results in the accumulation of both NMDARs and AMPARs (Nam and Chen, 2005; Heine et al., 2008b). NMDARs indirectly associate with NLs through PDZ domain-containing proteins like the MAGUKs (Chih et al., 2005; Graf et al., 2004; Nam and Chen, 2005; Sudhof, 2008). PDZ interactions can mediate the cotransport of NMDAR with NL-1 (Sans et al., 2003; Washbourne et al., 2004; Barrow et al., 2009). However,

detect endogenous NMDARs after 48 h. (C) The percentage of SALM1 or SALM1-δ4 colocalized with GluN2A staining is defined as the area of colocalization between the overexpressed SALM1 constructs and endogenous GluN2A. SALM1-δ4 has significantly decreased levels of colocalization with the NMDAR compared to SALM1. (D) Average punctum size of GluN2A is larger in neurons transfected with SALM1 than neurons with SALM1-δ4, indicating that NMDARs are recruited to SALM1 puncta. (E) Hippocampal neurons transfected with (a) GluN2A (with an extracellular GFP tag) and vector, (b) SALM1, or (c) SALM1-δ4 at 14 DIV were surface stained with anti-GFP (red), then permeabilized and stained with anti-myc (blue) for SALM1 or SALMδ4. The expression of GluN2A-GFP is shown in green. (d) The number of puncta for GluN2A-GFP surface staining and (e) the average total intensity of GluN2A-GFP surface staining show a significant increase with coexpression of SALM1. Scale bar, 10 μm. *Modified figure and legend are reprinted from Wang et al. (2006).*

NL-1 can still recruit NMDARs to synapses, even with the loss of the PDZ-binding motif, although not as effectively as in the wild type. In addition, PSD-95 is often recruited to synapses at a slower rate than NL-1 and NMDARs (Barrow et al., 2009). Therefore, NL-1 may interact with the NMDAR through another mechanism other than direct binding to PSD-95. For example, another PDZ protein that is related to PSD-95, S-SCAM (MAGI-2), can bind NL-1 through its WW domain (Iida et al., 2004) and NMDARs through a PDZ interaction (Iida et al., 2007). Alternatively, there may be an extracellular interaction between NL-1 and NMDARs. In fact, synaptic recruitment of the NMDAR is dependent on the extracellular domain of NL-1 (Chih et al., 2005; Barrow et al., 2009; Budreck et al., 2013) and the GluN1 subunit of the NMDAR (Budreck et al., 2013). Also, a previously unidentified region in the C-terminus of NL-1 and NL-3, that does not include the PDZ-binding motif or WW domain, has also been identified as critical for NL function at excitatory synapses (Shipman et al., 2011). Therefore, multiple domains may be important at different stages of synaptic development for NL-mediated transport and function of GluRs.

NLs are important for synapse maturation in vivo (Varoqueaux et al., 2006). Overexpression of NL-1 increases the density of dendritic spines with irregular spine heads that contain PSD-95 and GluN1, and associate with VGLUT1 puncta (Chih et al., 2005). It has been suggested that dimers of NL-1 bind monomeric axonal NRXN inducing the formation of functional presynaptic sites (Shipman and Nicoll, 2012a). Knockdown of NL-1, NL-2, or NL-3 inhibits synaptic maturation, shown by a decrease in the number of synaptic spines, and knockdown of all three reduces the number of excitatory synapses identified by the colocalization of VGLUT1 and GluA1 (Chih et al., 2005). Cleavage of the extracellular domain of NL-1 by proteases is dependent on NMDAR activation (Peixoto et al., 2012; Suzuki et al., 2012) and causes a decrease in synaptic transmission by reducing the presynaptic release probability (Peixoto et al., 2012), while inhibition of cleavage increases the number of dendritic spines (Suzuki et al., 2012). Loss of NL-1 decreases NMDAR-mediated currents and impairs LTP (Kim et al., 2008; Blundell et al., 2010; Budreck et al., 2013). These impairments can be rescued by expression of the extracellular domain of NL-1 (Shipman and Nicoll, 2012b; Budreck et al., 2013). Loss of NL-1 also decreases NMDAR-mediated calcium influx (Kwon et al., 2012). In addition, NL-1-KO mice display repetitive grooming, an autism-like phenotype, which can be rescued by D-cycloserine, an NMDAR partial coagonist (Blundell et al., 2010). Furthermore, overexpression of NL-1 in transgenic mice also causes decreased LTP (Dahlhaus et al., 2010). Interestingly, mice overexpressing NL-1 have impaired learning and behavioral flexibility, as well as more mature spines than wild-type mice (Hoy et al., 2013). On the other hand, NL-3 associates with Epac2 (exchange protein directly activated by cAMP, cAMP-GEFII, and RapGEF4), which initiates a signaling cascade that activates Rap, a Ras-like GTPase. As a result, AMPARs are removed from the synapse and spine motility increases (Woolfrey et al., 2009). Since a loss of AMPARs and spine shrinkage occurs with LTD, NL-3 may be important in regulating LTD. Further evidence supporting this theory includes knockdown of NL-3 in Purkinje cells of the cerebellum, which results in the loss of mGluR-dependent LTD and a constitutive increase in phospho-GluA2 (Baudouin et al., 2012).

Similar to many other CAMs discussed, neurexin's other ligand, the LRRTMs, also have synaptogenic activity and promote excitatory synapse formation (de Wit et al., 2009; Ko et al., 2009; Linhoff et al., 2009; Siddiqui et al., 2010). The LRR domain of LRRTM2 induces presynaptic differentiation, while bead-induced aggregation of LRRTM causes clustering of excitatory, but not inhibitory, postsynaptic proteins. In addition to its synaptogenic effect, the LRR domain is also required for increased synaptic activity at excitatory synapses (de Wit et al., 2009). This may be partially due to the fact that LRRTM2 interacts with AMPAR and NMDAR subunits (GluA1, GluA2, and GluN1) through its extracellular LRR domain. It does not interact with EphB2, indicating a specific interaction (de Wit et al., 2009). Knockdown of LRRTM2 decreases the surface expression of AMPARs in hippocampal neurons. Lentivirus injected into the dentate gyrus of P5–P6 rat pups also results in knockdown of LRRTM2 in granule cells and decreases AMPAR-mediated EPSCs and NMDAR-mediated EPSCs (de Wit et al., 2009). This may be due to a reduced number of synapses onto granule cells or reduced GluR density.

LRRTM and NLs may act redundantly in excitatory synapse maintenance. A triple knockdown (TKD) of LRRTM1, LRRTM2, and NL-3 in hippocampal neurons does not change synapse density. However, neurons cultured from NL-1 KO mice with knockdown of all three CAMs (NL-1 KO/TKD) have a significant loss of excitatory synapses, including both total number and functional synapses (Ko et al., 2011). Both NMDAR- and AMPAR-mediated currents are equally affected in NL-1 KO/TKD neurons. However, expression of NL-1 or LRRTM2 reverses this effect of decreased excitatory synaptic transmission (Ko et al., 2011). Synapse elimination by TKD is also dependent on synaptic activity. An AMPAR antagonist (NBQX) prevents synapse loss in TKD neurons, while AP5 only partially blocks the effect (Ko et al., 2011). Calcium influx through L-type Ca^{2+} channels and NMDARs is important, as well as calmodulin activity, suggesting that in the absence of these CAMs, activity-dependent synapse elimination may occur.

By using lentiviruses that control the expression of three shRNAs at the same time, NLs and LRRTMs in the CA1 hippocampal region can be knocked down to further examine their effect on excitatory synapse formation. Early knockdown of both LRRTM1 and LRRTM2 at P0, during periods of synaptogenesis, decreases AMPAR EPSCs, but not NMDAR EPSCs, while knockdown of NL-1 and NL-3 has no effect (Soler-Llavina et al., 2011). Therefore, LRRTMs regulate AMPAR activity during the first 2 weeks of development and may be involved in the recruitment or maintenance of synaptic receptors. Furthermore, NL-1-KO pups injected with lentivirus regulating the expression of LRRTM1, LRRTM2, and NL3 have larger deficits in evoked synaptic responses for AMPAR and NMDARs than knockdown of the LRRTMs alone. However, spine density does not significantly decrease (Soler-Llavina et al., 2011). At later stages (P21), when synapses are mature, there is a significant decrease in NMDAR/AMPAR ratios in NL-KO mice, consistent with other reports using these mice that suggest a decrease in the number of NMDARs at the synapse (Chubykin et al., 2007; Kim et al., 2008; Jung et al., 2010). Knockdown of NL-3 does not cause a further decrease, indicating that NL-1 is the major NL involved in maintaining NMDARs at mature excitatory synapses in the CA1 (Soler-Llavina et al., 2011). In addition, knockdown of LRRTM1 and LRRTM2 does not have a further effect, suggesting that LRRTMs do not influence transmission at mature excitatory synapses. Again, no change in spine density or morphology is observed in TKD/KO cultures. Therefore, NLs and LRRTM are probably not necessary for maintenance of spine structure at mature synapses (Soler-Llavina et al., 2011). Expression of the extracellular domain of LRRTM2 at P0 or of NL1 at P21 is sufficient to reverse the decrease in AMPAR-mediated currents or the NMDAR/AMPAR ratio, respectively (Soler-Llavina et al., 2011). Therefore, LRRTM and NL may have both redundant and distinct functions, depending on the age of the synapses where they are expressed.

3. CONCLUSIONS

Assessing the mechanism of GluR trafficking in synaptic plasticity is challenging, due to the explosion of new potential mechanisms involving more than 1500 proteins in the postsynaptic proteome alone (Ryan and Grant, 2009) and many other protein associations in other relevant parts of the neuron. Trafficking in each neuron is under the influence of signals, such as neurotransmitters and cell adhesion protein contacts, from the presynaptic terminals of other neurons. One common response to these signals is an influx of calcium from synaptic and/or extrasynaptic NMDARs. Calcium acts as a second messenger to initiate a variety of cascades of molecular responses in the neuron leading to plasticity, while the resultant depletion of calcium from the synaptic cleft can effect changes in cleft molecules such as adhesion proteins that help modulate the structural plasticity of the synapse and neuron. Other trafficking mechanisms may be instigated by activation of mGluRs. Among the many cytoplasmic molecules involved in GluR trafficking and plasticity are kinases (such as CaMKII), phosphatases, and PSD-MAGUKs such as PSD-95 and SAP102, as well as their associated proteins. Many subtleties of GluR trafficking and plasticity need to be worked out. For example, why do some protein interactions and trafficking mechanisms prevail in some circumstances and not in others? And no doubt, many more interactions and pathways are yet to be discovered.

ACKNOWLEDGMENTS

This work was supported by the NIDCD Intramural Research Program. This work also was supported by the Grant Agency of the Czech Republic (P303/11/0075, P304/12/G069, P303/12/1464), Marie Curie International Reintegration Grant (PIRG-GA-2010-276827) and Research Project of the AS CR AV0Z50110509 and RVO:67985823 (M.H.). We thank Katarina Lichnerova for help with figures, and Drs. Katherine Roche (NINDS/NIH) and Kenneth Pelkey (NICHD/NIH) for reviewing the manuscript.

REFERENCES

Adesnik, H., Nicoll, R.A., 2007. Conservation of glutamate receptor 2-containing AMPA receptors during long-term potentiation. J. Neurosci. 27, 4598–4602.

Adesnik, H., Nicoll, R.A., England, P.M., 2005. Photoinactivation of native AMPA receptors reveals their real-time trafficking. Neuron 48, 977–985.

Ahmad, M., Polepalli, J.S., Goswami, D., Yang, X., Kaeser-Woo, Y.J., Sudhof, T.C., Malenka, R.C., 2012. Post-synaptic complexin controls AMPA receptor exocytosis during LTP. Neuron 73, 260–267.

Aiga, M., Levinson, J.N., Bamji, S.X., 2011. N-cadherin and neuroligins cooperate to regulate synapse formation in hippocampal cultures. J. Biol. Chem. 286, 851–858.

Aizenman, C.D., Munoz-Elias, G., Cline, H.T., 2002. Visually driven modulation of glutamatergic synaptic transmission is mediated by the regulation of intracellular polyamines. Neuron 34, 623–634.

Al-Hallaq, R.A., Conrads, T.P., Veenstra, T.D., Wenthold, R.J., 2007. NMDA di-heteromeric receptor populations and associated proteins in rat hippocampus. J. Neurosci. 27, 8334–8343.

Alie, A., Manuel, M., 2010. The backbone of the post-synaptic density originated in a unicellular ancestor of choanoflagellates and metazoans. BMC Evol. Biol. 10, 34.

Amparan, D., Avram, D., Thomas, C.G., Lindahl, M.G., Yang, J., Bajaj, G., Ishmael, J.E., 2005. Direct interaction of myosin regulatory light chain with the NMDA receptor. J. Neurochem. 92, 349–361.

Anggono, V., Huganir, R.L., 2012. Regulation of AMPA receptor trafficking and synaptic plasticity. Curr. Opin. Neurobiol..

Antal, M., Fukazawa, Y., Eordogh, M., Muszil, D., Molnar, E., Itakura, M., Takahashi, M., Shigemoto, R., 2008. Numbers, densities, and colocalization of AMPA- and NMDA-type glutamate receptors at individual synapses in the superficial spinal dorsal horn of rats. J. Neurosci. 28, 9692–9701.

Antion, M.D., Christie, L.A., Bond, A.M., Dalva, M.B., Contractor, A., 2010. Ephrin-B3 regulates glutamate receptor signaling at hippocampal synapses. Mol. Cell Neurosci. 45, 378–388.

Aoto, J., Ting, P., Maghsoodi, B., Xu, N., Henkemeyer, M., Chen, L., 2007. Postsynaptic ephrinB3 promotes shaft glutamatergic synapse formation. J. Neurosci. 27, 7508–7519.

Araki, Y., Lin, D.T., Huganir, R.L., 2010. Plasma membrane insertion of the AMPA receptor GluA2 subunit is regulated by NSF binding and Q/R editing of the ion pore. Proc. Natl. Acad. Sci. U.S.A. 107, 11080–11085.

Archer, H.L., Evans, J.C., Millar, D.S., Thompson, P.W., Kerr, A.M., Leonard, H., Christodoulou, J., Ravine, D., Lazarou, L., Grove, L., Verity, C., Whatley, S.D., Pilz, D.T., Sampson, J.R., Clarke, A.J., 2006. NTNG1 mutations are a rare cause of Rett syndrome. Am. J. Med. Genet. A. 140, 691–694.

Armstrong, J.N., Saganich, M.J., Xu, N.J., Henkemeyer, M., Heinemann, S.F., Contractor, A., 2006. B-ephrin reverse signaling is required for NMDA-independent long-term potentiation of mossy fibers in the hippocampus. J. Neurosci. 26, 3474–3481.

Ashby, M.C., Maier, S.R., Nishimune, A., Henley, J.M., 2006. Lateral diffusion drives constitutive exchange of AMPA receptors at dendritic spines and is regulated by spine morphology. J. Neurosci. 26, 7046–7055.

Bard, L., Groc, L., 2011. Glutamate receptor dynamics and protein interaction: Lessons from the NMDA receptor. Mol. Cell. Neurosci. 48, 298–307.

Baron, M.K., Boeckers, T.M., Vaida, B., Faham, S., Gingery, M., Sawaya, M.R., Salyer, D., Gundelfinger, E.D., Bowie, J.U., 2006. An architectural framework that may lie at the core of the postsynaptic density. Science 311, 531–535.

Barria, A., Malinow, R., 2002. Subunit-specific NMDA receptor trafficking to synapses. Neuron 35, 345–353.

Barria, A., Malinow, R., 2005. NMDA receptor subunit composition controls synaptic plasticity by regulating binding to CaMKII. Neuron 48, 289–301.

Barrow, S.L., Constable, J.R., Clark, E., El-Sabeawy, F., McAllister, A.K., Washbourne, P., 2009. Neuroligin1: a cell adhesion molecule that recruits PSD-95 and NMDA receptors by distinct mechanisms during synaptogenesis. Neural Dev. 4, 17.

Bats, C., Groc, L., Choquet, D., 2007. The interaction between Stargazin and PSD-95 regulates AMPA receptor surface trafficking. Neuron 53, 719–734.

Baudouin, S.J., Gaudias, J., Gerharz, S., Hatstatt, L., Zhou, K., Punnakkal, P., Tanaka, K.F., Spooren, W., Hen, R., De Zeeuw, C.I., Vogt, K., Scheiffele, P., 2012. Shared synaptic pathophysiology in syndromic and nonsyndromic rodent models of autism. Science 338, 128–132.

Bellone, C., Nicoll, R.A., 2007. Rapid bidirectional switching of synaptic NMDA receptors. Neuron 55, 779–785.

Biederer, T., Sara, Y., Mozhayeva, M., Atasoy, D., Liu, X., Kavalali, E.T., Sudhof, T.C., 2002. SynCAM, a synaptic adhesion molecule that drives synapse assembly. Science 297, 1525–1531.

Blanpied, T.A., Kerr, J.M., Ehlers, M.D., 2008. Structural plasticity with preserved topology in the postsynaptic protein network. Proc. Natl. Acad. Sci. U.S.A. 105, 12587–12592.

Blanpied, T.A., Scott, D.B., Ehlers, M.D., 2002. Dynamics and regulation of clathrin coats at specialized endocytic zones of dendrites and spines. Neuron 36, 435–449.

Blundell, J., Blaiss, C.A., Etherton, M.R., Espinosa, F., Tabuchi, K., Walz, C., Bolliger, M.F., Sudhof, T.C., Powell, C.M., 2010. Neuroligin-1 deletion results in impaired spatial memory and increased repetitive behavior. J. Neurosci. 30, 2115–2129.

Boeckers, T.M., Bockmann, J., Kreutz, M.R., Gundelfinger, E.D., 2002. ProSAP/Shank proteins - a family of higher order organizing molecules of the postsynaptic density with an emerging role in human neurological disease. J. Neurochem. 81, 903–910.

Borg, I., Freude, K., Kubart, S., Hoffmann, K., Menzel, C., Laccone, F., Firth, H., Ferguson-Smith, M.A., Tommerup, N., Ropers, H.H., Sargan, D., Kalscheuer, V.M., 2005. Disruption of Netrin G1 by a balanced chromosome translocation in a girl with Rett syndrome. Eur. J. Hum. Genet. 13, 921–927.

Bouzioukh, F., Wilkinson, G.A., Adelmann, G., Frotscher, M., Stein, V., Klein, R., 2007. Tyrosine phosphorylation sites in ephrinB2 are required for hippocampal long-term potentiation but not long-term depression. J. Neurosci. 27, 11279–11288.

Brenman, J.E., Chao, D.S., Gee, S.H., McGee, A.W., Craven, S.E., Santillano, D.R., Wu, Z., Huang, F., Xia, H., Peters, M.F., Froehner, S.C., Bredt, D.S., 1996 March 8. Interaction of nitric oxide synthase with the postsynaptic density protein PSD-95 and alpha1-syntrophin mediated by PDZ domains. Cell 84 (5), 757–767.

Brenman, J.E., Chao, D.S., Gee, S.H., McGee, A.W., Craven, S.E., Santillano, D.R., Wu, Z., Huang, F., Xia, H., Peters, M.F., Froehner, S.C., Bredt, D.S., 1996. Interaction of nitric oxide synthase with the postsynaptic density protein PSD-95 and alpha1-syntrophin mediated by PDZ domains. Cell 84, 757–767.

Brennaman, L.H., Kochlamazashvili, G., Stoenica, L., Nonneman, R.J., Moy, S.S., Schachner, M., Dityatev, A., Maness, P.F., 2011. Transgenic mice overexpressing the extracellular domain of NCAM are impaired in working memory and cortical plasticity. Neurobiol. Dis. 43, 372–378.

Brose, N., 2009. Synaptogenic proteins and synaptic organizers: "many hands make light work. Neuron 61, 650–652.

Brown, T.C., Tran, I.C., Backos, D.S., Esteban, J.A., 2005. NMDA receptor-dependent activation of the small GTPase Rab5 drives the removal of synaptic AMPA receptors during hippocampal LTD. Neuron 45, 81–94.

Budreck, E.C., Kwon, O.B., Jung, J.H., Baudouin, S., Thommen, A., Kim, H.S., Fukazawa, Y., Harada, H., Tabuchi, K., Shigemoto, R., Scheiffele, P., Kim, J.H., 2013. Neuroligin-1 controls synaptic abundance of NMDA-type glutamate receptors through extracellular coupling. Proc. Natl. Acad. Sci. USA 110, 725–730.

Budreck, E.C., Scheiffele, P., 2007. Neuroligin-3 is a neuronal adhesion protein at GABAergic and glutamatergic synapses. Eur. J. Neurosci. 26, 1738–1748.

Bukalo, O., Fentrop, N., Lee, A.Y., Salmen, B., Law, J.W., Wotjak, C.T., Schweizer, M., Dityatev, A., Schachner, M., 2004. Conditional ablation of the neural cell adhesion molecule reduces precision of spatial learning, long-term potentiation, and depression in the CA1 subfield of mouse hippocampus. J. Neurosci. 24, 1565–1577.

Burkhardt, C., Muller, M., Badde, A., Garner, C.C., Gundelfinger, E.D., Puschel, A.W., 2005. Semaphorin 4B interacts with the post-synaptic density protein PSD-95/SAP90 and is recruited to synapses through a C-terminal PDZ-binding motif. FEBS Lett. 579, 3821–3828.

Burnashev, N., Monyer, H., Seeburg, P.H., Sakmann, B., 1992. Divalent ion permeability of AMPA receptor channels is dominated by the edited form of a single subunit. Neuron 8, 189–198.

Carlisle, H.J., Manzerra, P., Marcora, E., Kennedy, M.B., 2008. SynGAP regulates steady-state and activity-dependent phosphorylation of cofilin. J. Neurosci. 28, 13673–13683.

Chen, B.S., Braud, S., Badger 2nd, J.D., Isaac, J.T., Roche, K.W., 2006. Regulation of NR1/NR2C N-methyl-D-aspartate (NMDA) receptors by phosphorylation. J. Biol. Chem. 281, 16583–16590.

Chen, B.S., Roche, K.W., 2009. Growth factor-dependent trafficking of cerebellar NMDA receptors via protein kinase B/Akt phosphorylation of NR2C. Neuron 62, 471–478.

Chen, B.S., Thomas, E.V., Sanz-Clemente, A., Roche, K.W., 2011. NMDA receptor-dependent regulation of dendritic spine morphology by SAP102 splice variants. J. Neurosci. 31, 89–96.

Chen, L., Chetkovich, D.M., Petralia, R.S., Sweeney, N.T., Kawasaki, Y., Wenthold, R.J., Bredt, D.S., Nicoll, R.A., 2000. Stargazin regulates synaptic targeting of AMPA receptors by two distinct mechanisms. Nature 408, 936–943.

Chen, N., Luo, T., Raymond, L.A., 1999. Subtype-dependence of NMDA receptor channel open probability. J. Neurosci. 19, 6844–6854.

Chen, X., Nelson, C.D., Li, X., Winters, C.A., Azzam, R., Sousa, A.A., Leapman, R.D., Gainer, H., Sheng, M., Reese, T.S., 2011. PSD-95 is required to sustain the molecular organization of the postsynaptic density. J. Neurosci. 31, 6329–6338.

Chen, X., Vinade, L., Leapman, R.D., Petersen, J.D., Nakagawa, T., Phillips, T.M., Sheng, M., Reese, T.S., 2005. Mass of the postsynaptic density and enumeration of three key molecules. Proc. Natl. Acad. Sci. U.S.A. 102, 11551–11556.

Cheng, D., Hoogenraad, C.C., Rush, J., Ramm, E., Schlager, M.A., Duong, D.M., Xu, P., Wijayawardana, S.R., Hanfelt, J., Nakagawa, T., Sheng, M., Peng, J., 2006. Relative and absolute quantification of postsynaptic density proteome isolated from rat forebrain and cerebellum. Mol. Cell Proteomics 5, 1158–1170.

Chih, B., Engelman, H., Scheiffele, P., 2005. Control of excitatory and inhibitory synapse formation by neuroligins. Science 307, 1324–1328.

Cho, K.O., Hunt, C.A., Kennedy, M.B., 1992. The rat brain postsynaptic density fraction contains a homolog of the Drosophila discs-large tumor suppressor protein. Neuron 9, 929–942.

Christopherson, K.S., Hillier, B.J., Lim, W.A., Bredt, D.S., 1999. PSD-95 assembles a ternary complex with the N-methyl-D-aspartic acid receptor and a bivalent neuronal NO synthase PDZ domain. J. Biol. Chem. 274, 27467–27473.

Chubykin, A.A., Atasoy, D., Etherton, M.R., Brose, N., Kavalali, E.T., Gibson, J.R., Sudhof, T.C., 2007. Activity-dependent validation of excitatory versus inhibitory synapses by neuroligin-1 versus neuroligin-2. Neuron 54, 919–931.

Contractor, A., Rogers, C., Maron, C., Henkemeyer, M., Swanson, G.T., Heinemann, S.F., 2002. Trans-synaptic Eph receptor-ephrin signaling in hippocampal mossy fiber LTP. Science 296, 1864–1869.

Copits, B.A., Robbins, J.S., Frausto, S., Swanson, G.T., 2011. Synaptic targeting and functional modulation of GluK1 kainate receptors by the auxiliary neuropilin and tolloid-like (NETO) proteins. J. Neurosci. 31, 7334–7340.

Correia, S.S., Bassani, S., Brown, T.C., Lise, M.F., Backos, D.S., El-Husseini, A., Passafaro, M., Esteban, J.A., 2008. Motor protein-dependent transport of AMPA receptors into spines during long-term potentiation. Nat. Neurosci. 11, 457–466.

Craven, S.E., Bredt, D.S., 2000. Synaptic targeting of the postsynaptic density protein PSD-95 mediated by a tyrosine-based trafficking signal. J. Biol. Chem. 275, 20045–20051.

Croset, V., Rytz, R., Cummins, S.F., Budd, A., Brawand, D., Kaessmann, H., Gibson, T.J., Benton, R., 2010. Ancient protostome origin of chemosensory ionotropic glutamate receptors and the evolution of insect taste and olfaction. PLoS Genet. 6, e1001064.

Dahlhaus, R., Hines, R.M., Eadie, B.D., Kannangara, T.S., Hines, D.J., Brown, C.E., Christie, B.R., El-Husseini, A., 2010. Overexpression of the cell adhesion protein neuroligin-1 induces learning deficits and impairs synaptic plasticity by altering the ratio of excitation to inhibition in the hippocampus. Hippocampus 20, 305–322.

Dalva, M.B., McClelland, A.C., Kayser, M.S., 2007. Cell adhesion molecules: signalling functions at the synapse. Nat. Rev. Neurosci. 8, 206–220.

Dalva, M.B., Takasu, M.A., Lin, M.Z., Shamah, S.M., Hu, L., Gale, N.W., Greenberg, M.E., 2000. EphB receptors interact with NMDA receptors and regulate excitatory synapse formation. Cell 103, 945–956.

Dani, A., Huang, B., Bergan, J., Dulac, C., Zhuang, X., 2010. Superresolution imaging of chemical synapses in the brain. Neuron 68, 843–856.

de Bruijn, D.R., van Dijk, A.H., Pfundt, R., Hoischen, A., Merkx, G.F., Gradek, G.A., Lybaek, H., Stray-Pedersen, A., Brunner, H.G., Houge, G., 2010. Severe Progressive Autism Associated with Two de novo Changes: A 2.6-Mb 2q31.1 Deletion and a Balanced t(14;21)(q21.1;p11.2) Translocation with Long-Range Epigenetic Silencing of LRFN5 Expression. Mol Syndromol, 1, 46–57.

de Wit, J., Sylwestrak, E., O'Sullivan, M.L., Otto, S., Tiglio, K., Savas, J.N., Yates 3rd, J.R., Comoletti, D., Taylor, P., Ghosh, A., 2009. LRRTM2 interacts with Neurexin1 and regulates excitatory synapse formation. Neuron 64, 799–806.

DeNardo, L.A., de Wit, J., Otto-Hitt, S., Ghosh, A., 2012. NGL-2 regulates input-specific synapse development in CA1 pyramidal neurons. Neuron 76, 762–775.

Deng, F., Price, M.G., Davis, C.F., Mori, M., Burgess, D.L., 2006. Stargazin and other transmembrane AMPA receptor regulating proteins interact with synaptic scaffolding protein MAGI-2 in brain. J. Neurosci. 26, 7875–7884.

Dingledine, R., Borges, K., Bowie, D., Traynelis, S.F., 1999. The glutamate receptor ion channels. Pharmacol. Rev. 51, 7–61.

Dityatev, A., Dityateva, G., Schachner, M., 2000. Synaptic strength as a function of post- versus presynaptic expression of the neural cell adhesion molecule NCAM. Neuron 26, 207–217.

Dityatev, A., Dityateva, G., Sytnyk, V., Delling, M., Toni, N., Nikonenko, I., Muller, D., Schachner, M., 2004. Polysialylated neural cell adhesion molecule promotes remodeling and formation of hippocampal synapses. J. Neurosci. 24, 9372–9382.

Dong, H., O'Brien, R.J., Fung, E.T., Lanahan, A.A., Worley, P.F., Huganir, R.L., 1997. GRIP: a synaptic PDZ domain-containing protein that interacts with AMPA receptors. Nature 386, 279–284.

Dunah, A.W., Hueske, E., Wyszynski, M., Hoogenraad, C.C., Jaworski, J., Pak, D.T., Simonetta, A., Liu, G., Sheng, M., 2005. LAR receptor protein tyrosine phosphatases in the development and maintenance of excitatory synapses. Nat. Neurosci. 8, 458–467.

Eckhardt, M., Bukalo, O., Chazal, G., Wang, L., Goridis, C., Schachner, M., Gerardy-Schahn, R., Cremer, H., Dityatev, A., 2000. Mice deficient in the polysialyltransferase ST8SiaIV/PST-1 allow discrimination of the roles of neural cell adhesion molecule protein and polysialic acid in neural development and synaptic plasticity. J. Neurosci. 20, 5234–5244.

Ehlers, M.D., 2000. Reinsertion or degradation of AMPA receptors determined by activity-dependent endocytic sorting. Neuron 28, 511–525.

Ehlers, M.D., Heine, M., Groc, L., Lee, M.C., Choquet, D., 2007. Diffusional trapping of GluR1 AMPA receptors by input-specific synaptic activity. Neuron 54, 447–460.

Ehrlich, I., Malinow, R., 2004. Postsynaptic density 95 controls AMPA receptor incorporation during long-term potentiation and experience-driven synaptic plasticity. J. Neurosci. 24, 916–927.

El-Husseini, A.E., Craven, S.E., Chetkovich, D.M., Firestein, B.L., Schnell, E., Aoki, C., Bredt, D.S., 2000a. Dual palmitoylation of PSD-95 mediates its vesiculotubular sorting, postsynaptic targeting, and ion channel clustering. J. Cell Biol. 148, 159–172.

El-Husseini, A.E., Schnell, E., Chetkovich, D.M., Nicoll, R.A., Bredt, D.S., 2000b. PSD-95 involvement in maturation of excitatory synapses. Science 290, 1364–1368.

Elias, G.M., Elias, L.A., Apostolides, P.F., Kriegstein, A.R., Nicoll, R.A., 2008. Differential trafficking of AMPA and NMDA receptors by SAP102 and PSD-95 underlies synapse development. Proc. Natl. Acad. Sci. U.S.A. 105, 20953–20958.

Elias, G.M., Funke, L., Stein, V., Grant, S.G., Bredt, D.S., Nicoll, R.A., 2006. Synapse-specific and developmentally regulated targeting of AMPA receptors by a family of MAGUK scaffolding proteins. Neuron 52, 307–320.

Elias, G.M., Nicoll, R.A., 2007. Synaptic trafficking of glutamate receptors by MAGUK scaffolding proteins. Trends Cell Biol. 17, 343–352.

Essmann, C.L., Martinez, E., Geiger, J.C., Zimmer, M., Traut, M.H., Stein, V., Klein, R., Acker-Palmer, A., 2008. Serine phosphorylation of ephrinB2 regulates trafficking of synaptic AMPA receptors. Nat. Neurosci. 11, 1035–1043.

Esteban, J.A., 2008. Cellular Biology of AMPA Receptor Trafficking and Synaptic Plasticity. In: Hell, J.W., Ehlers, M.D. (Eds.), Structural and Functional Organization of the Synapse. Springer, New York, pp. 271–287.

Esteban, J.A., Shi, S.H., Wilson, C., Nuriya, M., Huganir, R.L., Malinow, R., 2003. PKA phosphorylation of AMPA receptor subunits controls synaptic trafficking underlying plasticity. Nat. Neurosci. 6, 136–143.

Eybalin, M., Caicedo, A., Renard, N., Ruel, J., Puel, J.L., 2004. Transient Ca2+-permeable AMPA receptors in postnatal rat primary auditory neurons. Eur. J. Neurosci. 20, 2981–2989.

Feng, W., Zhang, M., 2009. Organization and dynamics of PDZ-domain-related supramodules in the postsynaptic density. Nat. Rev. Neurosci. 10, 87–99.

Foa, L., Gasperini, R., 2009. Developmental roles for Homer: more than just a pretty scaffold. J. Neurochem. 108, 1–10.

Friedman, H.V., Bresler, T., Garner, C.C., Ziv, N.E., 2000. Assembly of new individual excitatory synapses: time course and temporal order of synaptic molecule recruitment. Neuron 27, 57–69.

Fu, Z., Logan, S.M., Vicini, S., 2005. Deletion of the NR2A subunit prevents developmental changes of NMDA-mEPSCs in cultured mouse cerebellar granule neurones. J. Physiol. 563, 867–881.

Fujita, E., Kouroku, Y., Ozeki, S., Tanabe, Y., Toyama, Y., Maekawa, M., Kojima, N., Senoo, H., Toshimori, K., Momoi, T., 2006. Oligo-astheno-teratozoospermia in mice lacking RA175/TSLC1/SynCAM/IGSF4A, a cell adhesion molecule in the immunoglobulin superfamily. Mol. Cell. Biol. 26, 718–726.

Funke, L., Dakoji, S., Bredt, D.S., 2005. Membrane-associated guanylate kinases regulate adhesion and plasticity at cell junctions. Annu. Rev. Biochem. 74, 219–245.

Futai, K., Doty, C.D., Baek, B., Ryu, J., Sheng, M., 2013. Specific trans-synaptic interaction with inhibitory interneuronal neurexin underlies differential ability of neuroligins to induce functional inhibitory synapses. J. Neurosci. 33, 3612–3623.

Fux, C.M., Krug, M., Dityatev, A., Schuster, T., Schachner, M., 2003. NCAM180 and glutamate receptor subtypes in potentiated spine synapses: an immunogold electron microscopic study. Mol. Cell Neurosci. 24, 939–950.

Garcia, R.A., Vasudevan, K., Buonanno, A., 2000. The neuregulin receptor ErbB-4 interacts with PDZ-containing proteins at neuronal synapses. Proc. Natl. Acad. Sci. U.S.A. 97, 3596–3601.

Geiger, J.R., Melcher, T., Koh, D.S., Sakmann, B., Seeburg, P.H., Jonas, P., Monyer, H., 1995. Relative abundance of subunit mRNAs determines gating and Ca2+ permeability of AMPA receptors in principal neurons and interneurons in rat CNS. Neuron 15, 193–204.

Gereau, I.V.R.W., Swanson, G.T., 2008. The Glutamate Receptors. Humana Press, Totowa, NJ 576.

Gerges, N.Z., Backos, D.S., Esteban, J.A., 2004. Local control of AMPA receptor trafficking at the postsynaptic terminal by a small GTPase of the Rab family. J. Biol. Chem. 279, 43870–43878.

Gerges, N.Z., Backos, D.S., Rupasinghe, C.N., Spaller, M.R., Esteban, J.A., 2006. Dual role of the exocyst in AMPA receptor targeting and insertion into the postsynaptic membrane. EMBO J. 25, 1623–1634.

Gerrow, K., Romorini, S., Nabi, S.M., Colicos, M.A., Sala, C., El-Husseini, A., 2006. A preformed complex of postsynaptic proteins is involved in excitatory synapse development. Neuron 49, 547–562.

Giagtzoglou, N., Ly, C.V., Bellen, H.J., 2009. Cell adhesion, the backbone of the synapse: "vertebrate" and "invertebrate" perspectives. Cold Spring Harb Perspect Biol 1, a003079.

Gill, M.B., Kato, A.S., Roberts, M.F., Yu, H., Wang, H., Tomita, S., Bredt, D.S., 2011. Cornichon-2 modulates AMPA receptor-transmembrane AMPA receptor regulatory protein assembly to dictate gating and pharmacology. J. Neurosci. 31, 6928–6938.

Gladding, C.M., Raymond, L.A., 2011. Mechanisms underlying NMDA receptor synaptic/extrasynaptic distribution and function. Mol. Cell Neurosci. 48, 308–320.

Grabrucker, A.M., Schmeisser, M.J., Schoen, M., Boeckers, T.M., 2011. Postsynaptic ProSAP/Shank scaffolds in the cross-hair of synaptopathies. Trends Cell Biol. 21, 594–603.

Graf, E.R., Zhang, X., Jin, S.X., Linhoff, M.W., Craig, A.M., 2004. Neurexins induce differentiation of GABA and glutamate postsynaptic specializations via neuroligins. Cell 119, 1013–1026.

Granger, A.J., Shi, Y., Lu, W., Cerpas, M., Nicoll, R.A., 2013. LTP requires a reserve pool of glutamate receptors independent of subunit type. Nature 493, 495–501.

Gray, N.W., Weimer, R.M., Bureau, I., Svoboda, K., 2006. Rapid redistribution of synaptic PSD-95 in the neocortex in vivo. PLoS Biol 4, e370.

Greger, I.H., Khatri, L., Ziff, E.B., 2002. RNA editing at arg607 controls AMPA receptor exit from the endoplasmic reticulum. Neuron 34, 759–772.

Groc, L., Heine, M., Cognet, L., Brickley, K., Stephenson, F.A., Lounis, B., Choquet, D., 2004. Differential activity-dependent regulation of the lateral mobilities of AMPA and NMDA receptors. Nat. Neurosci. 7, 695–696.

Groc, L., Heine, M., Cousins, S.L., Stephenson, F.A., Lounis, B., Cognet, L., Choquet, D., 2006. NMDA receptor surface mobility depends on NR2A-2B subunits. Proc. Natl. Acad. Sci. U.S.A. 103, 18769–18774.

Grooms, S.Y., Noh, K.M., Regis, R., Bassell, G.J., Bryan, M.K., Carroll, R.C., Zukin, R.S., 2006. Activity bidirectionally regulates AMPA receptor mRNA abundance in dendrites of hippocampal neurons. J. Neurosci. 26, 8339–8351.

Grunwald, I.C., Korte, M., Adelmann, G., Plueck, A., Kullander, K., Adams, R.H., Frotscher, M., Bonhoeffer, T., Klein, R., 2004. Hippocampal plasticity requires postsynaptic ephrinBs. Nat. Neurosci. 7, 33–40.

Grunwald, I.C., Korte, M., Wolfer, D., Wilkinson, G.A., Unsicker, K., Lipp, H.P., Bonhoeffer, T., Klein, R., 2001. Kinase-independent requirement of EphB2 receptors in hippocampal synaptic plasticity. Neuron 32, 1027–1040.

Hahn, C.G., Wang, H.Y., Cho, D.S., Talbot, K., Gur, R.E., Berrettini, W.H., Bakshi, K., Kamins, J., Borgmann-Winter, K.E., Siegel, S.J., Gallop, R.J., Arnold, S.E., 2006. Altered neuregulin 1-erbB4 signaling contributes to NMDA receptor hypofunction in schizophrenia. Nat. Med. 12, 824–828.

Hales, C.M., Vaerman, J.P., Goldenring, J.R., 2002. Rab11 family interacting protein 2 associates with Myosin Vb and regulates plasma membrane recycling. J. Biol. Chem. 277, 50415–50421.

Hammond, M.S., Sims, C., Parameshwaran, K., Suppiramaniam, V., Schachner, M., Dityatev, A., 2006. Neural cell adhesion molecule-associated polysialic acid inhibits NR2B-containing N-methyl-D-aspartate receptors and prevents glutamate-induced cell death. J. Biol. Chem. 281, 34859–34869.

Hanley, J.G., Khatri, L., Hanson, P.I., Ziff, E.B., 2002. NSF ATPase and alpha-/beta-SNAPs disassemble the AMPA receptor-PICK1 complex. Neuron 34, 53–67.

Hardingham, G.E., Bading, H., 2010. Synaptic versus extrasynaptic NMDA receptor signalling: implications for neurodegenerative disorders. Nat. Rev. Neurosci. 11, 682–696.

Harris, A.Z., Pettit, D.L., 2007. Extrasynaptic and synaptic NMDA receptors form stable and uniform pools in rat hippocampal slices. J. Physiol. 584, 509–519.

Hawkins, L.M., Prybylowski, K., Chang, K., Moussan, C., Stephenson, F.A., Wenthold, R.J., 2004. Export from the endoplasmic reticulum of assembled N-methyl-d-aspartic acid receptors is controlled by a motif in the c terminus of the NR2 subunit. J. Biol. Chem. 279, 28903–28910.

Hayashi, T., Rumbaugh, G., Huganir, R.L., 2005. Differential regulation of AMPA receptor subunit trafficking by palmitoylation of two distinct sites. Neuron 47, 709–723.

Hayashi, T., Thomas, G.M., Huganir, R.L., 2009. Dual palmitoylation of NR2 subunits regulates NMDA receptor trafficking. Neuron 64, 213–226.

Hayashi, Y., Shi, S.H., Esteban, J.A., Piccini, A., Poncer, J.C., Malinow, R., 2000. Driving AMPA receptors into synapses by LTP and CaMKII: requirement for GluR1 and PDZ domain interaction. Science 287, 2262–2267.

Heine, M., Groc, L., Frischknecht, R., Beique, J.C., Lounis, B., Rumbaugh, G., Huganir, R.L., Cognet, L., Choquet, D., 2008a. Surface mobility of postsynaptic AMPARs tunes synaptic transmission. Science 320, 201–205.

Heine, M., Thoumine, O., Mondin, M., Tessier, B., Giannone, G., Choquet, D., 2008b. Activity-independent and subunit-specific recruitment of functional AMPA receptors at neurexin/neuroligin contacts. Proc. Natl. Acad. Sci. U.S.A. 105, 20947–20952.

Henderson, J.T., Georgiou, J., Jia, Z., Robertson, J., Elowe, S., Roder, J.C., Pawson, T., 2001. The receptor tyrosine kinase EphB2 regulates NMDA-dependent synaptic function. Neuron 32, 1041–1056.

Herguedas, B., Krieger, J., Greger, I.H., 2013. Receptor heteromeric assembly—How it works and why it matters: the case of ionotropic glutamate receptors. Prog. Mol. Biol. Transl. Sci. 117, 361–386.

Herring, B.E., Shi, Y., Suh, Y.H., Zheng, C.-Y., Blankenship, S.M., Roche, K.W., Nicoll, R.A., 2013. Cornichon proteins determine the subunit composition of synaptic AMPA receptors. Neuron 77, 1083–1096.

Hibbs, R.E., Gouaux, E., 2011. Principles of activation and permeation in an anion-selective Cys-loop receptor. Nature 474, 54–60.

Hollmann, M., Hartley, M., Heinemann, S., 1991. Ca2+ permeability of KA-AMPA–gated glutamate receptor channels depends on subunit composition. Science 252, 851–853.

Horak, M., Chang, K., Wenthold, R.J., 2008. Masking of the endoplasmic reticulum retention signals during assembly of the NMDA receptor. J. Neurosci. 28, 3500–3509.

Horak, M., Wenthold, R.J., 2009. Different roles of C-terminal cassettes in the trafficking of full-length NR1 subunits to the cell surface. J. Biol. Chem. 284, 9683–9691.

Hori, K., Yasuda, H., Konno, D., Maruoka, H., Tsumoto, T., Sobue, K., 2005. NMDA receptor-dependent synaptic translocation of insulin receptor substrate p53 via protein kinase C signaling. J. Neurosci. 25, 2670–2681.

Howard, M.A., Elias, G.M., Elias, L.A., Swat, W., Nicoll, R.A., 2010. The role of SAP97 in synaptic glutamate receptor dynamics. Proc. Natl. Acad. Sci. U.S.A. 107, 3805–3810.

Hoy, J.L., Constable, J.R., Vicini, S., Fu, Z., Washbourne, P., 2009. SynCAM1 recruits NMDA receptors via protein 4.1B. Mol. Cell. Neurosci. 42, 466–483.

Hoy, J.L., Haeger, P.A., Constable, J.R., Arias, R.J., McCallum, R., Kyweriga, M., Davis, L., Schnell, E., Wehr, M., Castillo, P.E., Washbourne, P., 2013. Neuroligin1 drives synaptic and behavioral maturation through intracellular interactions. J. Neurosci. 33, 9364–9384.

Huang, Y.Z., Won, S., Ali, D.W., Wang, Q., Tanowitz, M., Du, Q.S., Pelkey, K.A., Yang, D.J., Xiong, W.C., Salter, M.W., Mei, L., 2000. Regulation of neuregulin signaling by PSD-95 interacting with ErbB4 at CNS synapses. Neuron 26, 443–455.

Iida, J., Hirabayashi, S., Sato, Y., Hata, Y., 2004. Synaptic scaffolding molecule is involved in the synaptic clustering of neuroligin. Mol. Cell. Neurosci. 27, 497–508.

Iida, J., Ishizaki, H., Okamoto-Tanaka, M., Kawata, A., Sumita, K., Ohgake, S., Sato, Y., Yorifuji, H., Nukina, N., Ohashi, K., Mizuno, K., Tsutsumi, T., Mizoguchi, A., Miyoshi, J., Takai, Y., Hata, Y., 2007. Synaptic scaffolding molecule alpha is a scaffold to mediate N-methyl-D-aspartate receptor-dependent RhoA activation in dendrites. Mol. Cell. Biol. 27, 4388–4405.

Irie, F., Okuno, M., Pasquale, E.B., Yamaguchi, Y., 2005. EphrinB-EphB signalling regulates clathrin-mediated endocytosis through tyrosine phosphorylation of synaptojanin 1. Nat. Cell Biol. 7, 501–509.

Irie, M., Hata, Y., Takeuchi, M., Ichtchenko, K., Toyoda, A., Hirao, K., Takai, Y., Rosahl, T.W., Sudhof, T.C., 1997. Binding of neuroligins to PSD-95. Science 277, 1511–1515.

Jackson, A.C., Nicoll, R.A., 2011. The expanding social network of ionotropic glutamate receptors: TARPs and other transmembrane auxiliary subunits. Neuron 70, 178–199.

Jagoe, W.N., Lindsay, A.J., Read, R.J., McCoy, A.J., McCaffrey, M.W., Khan, A.R., 2006. Crystal structure of rab11 in complex with rab11 family interacting protein 2. Structure 14, 1273–1283.

Jamain, S., Quach, H., Betancur, C., Rastam, M., Colineaux, C., Gillberg, I.C., Soderstrom, H., Giros, B., Leboyer, M., Gillberg, C., Bourgeron, T., 2003. Mutations of the X-linked genes encoding neuroligins NLGN3 and NLGN4 are associated with autism. Nat. Genet. 34, 27–29.

Janovjak, H., Sandoz, G., Isacoff, E.Y., 2011. A modern ionotropic glutamate receptor with a K(+) selectivity signature sequence. Nat. Commun. 2, 232.

Jensen, V., Kaiser, K.M., Borchardt, T., Adelmann, G., Rozov, A., Burnashev, N., Brix, C., Frotscher, M., Andersen, P., Hvalby, O., Sakmann, B., Seeburg, P.H., Sprengel, R., 2003. A juvenile form of postsynaptic hippocampal long-term potentiation in mice deficient for the AMPA receptor subunit GluR-A. J. Physiol. 553, 843–856.

Jeyifous, O., Waites, C.L., Specht, C.G., Fujisawa, S., Schubert, M., Lin, E.I., Marshall, J., Aoki, C., de Silva, T., Montgomery, J.M., Garner, C.C., Green, W.N., 2009. SAP97 and CASK mediate sorting of NMDA receptors through a previously unknown secretory pathway. Nat. Neurosci. 12, 1011–1019.

Jin, Y., Garner, C.C., 2008. Molecular mechanisms of presynaptic differentiation. Annu. Rev. Cell Dev. Biol. 24, 237–262.

Ju, W., Morishita, W., Tsui, J., Gaietta, G., Deerinck, T.J., Adams, S.R., Garner, C.C., Tsien, R.Y., Ellisman, M.H., Malenka, R.C., 2004. Activity-dependent regulation of dendritic synthesis and trafficking of AMPA receptors. Nat. Neurosci. 7, 244–253.

Jung, S.Y., Kim, J., Kwon, O.B., Jung, J.H., An, K., Jeong, A.Y., Lee, C.J., Choi, Y.B., Bailey, C.H., Kandel, E.R., Kim, J.H., 2010. Input-specific synaptic plasticity in the amygdala is regulated by neuroligin-1 via postsynaptic NMDA receptors. Proc. Natl. Acad. Sci. U.S.A. 107, 4710–4715.

Jurado, S., Goswami, D., Zhang, Y., Molina, A.J.M., Südhof, T.C., Malenka, R.C., 2013. LTP requires a unique postsynaptic SNARE fusion machinery. Neuron 77, 542–558.

Kakegawa, W., Miyoshi, Y., Hamase, K., Matsuda, S., Matsuda, K., Kohda, K., Emi, K., Motohashi, J., Konno, R., Zaitsu, K., Yuzaki, M., 2011. D-serine regulates cerebellar LTD and motor coordination through the delta2 glutamate receptor. Nat. Neurosci. 14, 603–611.

Karpova, A., Mikhaylova, M., Bera, S., Bär, J., Reddy, P.P., Behnisch, T., Rankovic, V., Spilker, C., Bethge, P., Sahin, J., Kaushik, R., Zuschratter, W., Kähne, T., Naumann, M., Gundelfinger, E.D., Kreutz, M.R., 2013. Encoding and transducing the synaptic or extrasynaptic origin of NMDA receptor signals to the nucleus. Cell 152, 1119–1133.

Kato, A.S., Gill, M.B., Ho, M.T., Yu, H., Tu, Y., Siuda, E.R., Wang, H., Qian, Y.W., Nisenbaum, E.S., Tomita, S., Bredt, D.S., 2010. Hippocampal AMPA receptor gating controlled by both TARP and cornichon proteins. Neuron 68, 1082–1096.

Kayser, M.S., McClelland, A.C., Hughes, E.G., Dalva, M.B., 2006. Intracellular and trans-synaptic regulation of glutamatergic synaptogenesis by EphB receptors. J. Neurosci. 26, 12152–12164.

Kayser, M.S., Nolt, M.J., Dalva, M.B., 2008. EphB receptors couple dendritic filopodia motility to synapse formation. Neuron 59, 56–69.

Keinanen, K., Wisden, W., Sommer, B., Werner, P., Herb, A., Verdoorn, T.A., Sakmann, B., Seeburg, P.H., 1990. A family of AMPA-selective glutamate receptors. Science 249, 556–560.

Kelly, E.E., Horgan, C.P., McCaffrey, M.W., Young, P., 2011. The role of endosomal-recycling in long-term potentiation. Cell Mol. Life Sci. 68, 185–194.

Kennedy, M.J., Davison, I.G., Robinson, C.G., Ehlers, M.D., 2010. Syntaxin-4 defines a domain for activity-dependent exocytosis in dendritic spines. Cell 141, 524–535.

Kennedy, M.J., Ehlers, M.D., 2011. Mechanisms and function of dendritic exocytosis. Neuron 69, 856–875.

Kenny, A.V., Cousins, S.L., Pinho, L., Stephenson, F.A., 2009. The integrity of the glycine co-agonist binding site of N-methyl-D-aspartate receptors is a functional quality control checkpoint for cell surface delivery. J. Biol. Chem. 284, 324–333.

Kessels, H.W., Malinow, R., 2009. Synaptic AMPA receptor plasticity and behavior. Neuron 61, 340–350.

Kim, C.H., Chung, H.J., Lee, H.K., Huganir, R.L., 2001. Interaction of the AMPA receptor subunit GluR2/3 with PDZ domains regulates hippocampal long-term depression. Proc. Natl. Acad. Sci. U.S.A. 98, 11725–11730.

Kim, C.H., Takamiya, K., Petralia, R.S., Sattler, R., Yu, S., Zhou, W., Kalb, R., Wenthold, R., Huganir, R., 2005. Persistent hippocampal CA1 LTP in mice lacking the C-terminal PDZ ligand of GluR1. Nat. Neurosci. 8, 985–987.

Kim, C.H., Takamiya, K., Petralia, R.S., Sattler, R., Yu, S., Zhou, W., Kalb, R., Wenthold, R., Huganir, R., 2005. Persistent hippocampal CA1 LTP in mice lacking the C-terminal PDZ ligand of GluR1. Nat. Neurosci. 8, 985–987.

Kim, E., Cho, K.O., Rothschild, A., Sheng, M., 1996. Heteromultimerization and NMDA receptor-clustering activity of Chapsyn-110, a member of the PSD-95 family of proteins. Neuron 17, 103–113.

Kim, E., Niethammer, M., Rothschild, A., Jan, Y.N., Sheng, M., 1995. Clustering of Shaker-type K+ channels by interaction with a family of membrane-associated guanylate kinases. Nature 378, 85–88.

Kim, J., Jung, S.Y., Lee, Y.K., Park, S., Choi, J.S., Lee, C.J., Kim, H.S., Choi, Y.B., Scheiffele, P., Bailey, C.H., Kandel, E.R., Kim, J.H., 2008. Neuroligin-1 is required for normal expression of LTP and associative fear memory in the amygdala of adult animals. Proc. Natl. Acad. Sci. U.S.A. 105, 9087–9092.

Kim, J.H., Liao, D., Lau, L.F., Huganir, R.L., 1998. SynGAP: a synaptic RasGAP that associates with the PSD-95/SAP90 protein family. Neuron 20, 683–691.

Kim, S., Burette, A., Chung, H.S., Kwon, S.K., Woo, J., Lee, H.W., Kim, K., Kim, H., Weinberg, R.J., Kim, E., 2006. NGL family PSD-95-interacting adhesion molecules regulate excitatory synapse formation. Nat. Neurosci. 9, 1294–1301.

Ko, J., Fuccillo, M.V., Malenka, R.C., Sudhof, T.C., 2009. LRRTM2 functions as a neurexin ligand in promoting excitatory synapse formation. Neuron 64, 791–798.

Ko, J., Kim, S., Chung, H.S., Kim, K., Han, K., Kim, H., Jun, H., Kaang, B.K., Kim, E., 2006a. SALM synaptic cell adhesion-like molecules regulate the differentiation of excitatory synapses. Neuron 50, 233–245.

Ko, J., Soler-Llavina, G.J., Fuccillo, M.V., Malenka, R.C., Sudhof, T.C., 2011. Neuroligins/LRRTMs prevent activity- and Ca2+/calmodulin-dependent synapse elimination in cultured neurons. J. Cell Biol. 194, 323–334.

Ko, J., Yoon, C., Piccoli, G., Chung, H.S., Kim, K., Lee, J.R., Lee, H.W., Kim, H., Sala, C., Kim, E., 2006b. Organization of the presynaptic active zone by ERC2/CAST1-dependent clustering of the tandem PDZ protein syntenin-1. J. Neurosci. 26, 963–970.

Kochlamazashvili, G., Bukalo, O., Senkov, O., Salmen, B., Gerardy-Schahn, R., Engel, A.K., Schachner, M., Dityatev, A., 2012. Restoration of synaptic plasticity and learning in young and aged NCAM-deficient mice by enhancing neurotransmission mediated by GluN2A-containing NMDA receptors. J. Neurosci. 32, 2263–2275.

Kochlamazashvili, G., Senkov, O., Grebenyuk, S., Robinson, C., Xiao, M.F., Stummeyer, K., Gerardy-Schahn, R., Engel, A.K., Feig, L., Semyanov, A., Suppiramaniam, V., Schachner, M., Dityatev, A., 2010. Neural cell adhesion molecule-associated polysialic acid regulates synaptic plasticity and learning by restraining the signaling through GluN2B-containing NMDA receptors. J. Neurosci. 30, 4171–4183.

Kolkman, M.J., Streijger, F., Linkels, M., Bloemen, M., Heeren, D.J., Hendriks, W.J., Van der Zee, C.E., 2004. Mice lacking leukocyte common antigen-related (LAR) protein tyrosine phosphatase domains demonstrate spatial learning impairment in the two-trial water maze and hyperactivity in multiple behavioural tests. Behav. Brain Res. 154, 171–182.

Kolleker, A., Zhu, J.J., Schupp, B.J., Qin, Y., Mack, V., Borchardt, T., Kohr, G., Malinow, R., Seeburg, P.H., Osten, P., 2003. Glutamatergic plasticity by synaptic delivery of GluR-B(long)-containing AMPA receptors. Neuron 40, 1199–1212.

Kornau, H.C., Schenker, L.T., Kennedy, M.B., Seeburg, P.H., 1995. Domain interaction between NMDA receptor subunits and the postsynaptic density protein PSD-95. Science 269, 1737–1740.

Kreienkamp, H.J., Dityatev, A., 2004. The magic of synaptogenesis. Meeting on spinogenesis and synaptic plasticity. EMBO Rep. 5, 1125–1129.

Kumar, S.S., Bacci, A., Kharazia, V., Huguenard, J.R., 2002. A developmental switch of AMPA receptor subunits in neocortical pyramidal neurons. J. Neurosci. 22, 3005–3015.

Kwon, H.B., Kozorovitskiy, Y., Oh, W.J., Peixoto, R.T., Akhtar, N., Saulnier, J.L., Gu, C., Sabatini, B.L., 2012. Neuroligin-1-dependent competition regulates cortical synaptogenesis and synapse number. Nat. Neurosci. 15, 1667–1674.

Kwon, S.K., Woo, J., Kim, S.Y., Kim, H., Kim, E., 2010. Trans-synaptic adhesions between netrin-G ligand-3 (NGL-3) and receptor tyrosine phosphatases LAR, protein-tyrosine phosphatase delta (PTPdelta), and PTPsigma via specific domains regulate excitatory synapse formation. J. Biol. Chem. 285, 13966–13978.

Kypta, R.M., Su, H., Reichardt, L.F., 1996. Association between a transmembrane protein tyrosine phosphatase and the cadherin-catenin complex. J. Cell Biol. 134, 1519–1529.

Laezza, F., Wilding, T.J., Sequeira, S., Coussen, F., Zhang, X.Z., Hill-Robinson, R., Mulle, C., Huettner, J.E., Craig, A.M., 2007. KRIP6: a novel BTB/kelch protein regulating function of kainate receptors. Mol. Cell. Neurosci. 34, 539–550.

Larsen, R.S., Corlew, R.J., Henson, M.A., Roberts, A.C., Mishina, M., Watanabe, M., Lipton, S.A., Nakanishi, N., Perez-Otano, I., Weinberg, R.J., Philpot, B.D., 2011. NR3A-containing NMDARs promote neurotransmitter release and spike timing-dependent plasticity. Nat. Neurosci. 14, 338–344.

Laumonnier, F., Bonnet-Brilhault, F., Gomot, M., Blanc, R., David, A., Moizard, M.P., Raynaud, M., Ronce, N., Lemonnier, E., Calvas, P., Laudier, B., Chelly, J., Fryns, J.P., Ropers, H.H., Hamel, B.C., Andres, C., Barthelemy, C., Moraine, C., Briault, S., 2004. X-linked mental retardation and autism are associated with a mutation in the NLGN4 gene, a member of the neuroligin family. Am. J. Hum. Genet. 74, 552–557.

Lavezzari, G., McCallum, J., Dewey, C.M., Roche, K.W., 2004. Subunit-specific regulation of NMDA receptor endocytosis. J. Neurosci. 24, 6383–6391.

Lee, K.J., Lee, Y., Rozeboom, A., Lee, J.Y., Udagawa, N., Hoe, H.S., Pak, D.T., 2011. Requirement for Plk2 in orchestrated ras and rap signaling, homeostatic structural plasticity, and memory. Neuron 69, 957–973.

Lee, S.H., Simonetta, A., Sheng, M., 2004. Subunit rules governing the sorting of internalized AMPA receptors in hippocampal neurons. Neuron 43, 221–236.

Leonard, A.S., Davare, M.A., Horne, M.C., Garner, C.C., Hell, J.W., 1998. SAP97 is associated with the alpha-amino-3-hydroxy-5-methylisoxazole-4-propionic acid receptor GluR1 subunit. J. Biol. Chem. 273, 19518–19524.

Letts, V.A., Felix, R., Biddlecome, G.H., Arikkath, J., Mahaffey, C.L., Valenzuela, A., Bartlett 2nd, F.S., Mori, Y., Campbell, K.P., Frankel, W.N., 1998. The mouse stargazer gene encodes a neuronal Ca2+-channel gamma subunit. Nat. Genet. 19, 340–347.

Li, D., Specht, C.G., Waites, C.L., Butler-Munro, C., Leal-Ortiz, S., Foote, J.W., Genoux, D., Garner, C.C., Montgomery, J.M., 2011. SAP97 directs NMDA receptor spine targeting and synaptic plasticity. J. Physiol. 589, 4491–4510.

Lim, B.K., Matsuda, N., Poo, M.M., 2008. Ephrin-B reverse signaling promotes structural and functional synaptic maturation in vivo. Nat. Neurosci. 11, 160–169.

Lin, J.C., Ho, W.H., Gurney, A., Rosenthal, A., 2003. The netrin-G1 ligand NGL-1 promotes the outgrowth of thalamocortical axons. Nat. Neurosci. 6, 1270–1276.

Lin, Y., Jover-Mengual, T., Wong, J., Bennett, M.V., Zukin, R.S., 2006. PSD-95 and PKC converge in regulating NMDA receptor trafficking and gating. Proc. Natl. Acad. Sci. U.S.A. 103, 19902–19907.

Linhoff, M.W., Lauren, J., Cassidy, R.M., Dobie, F.A., Takahashi, H., Nygaard, H.B., Airaksinen, M.S., Strittmatter, S.M., Craig, A.M., 2009. An unbiased expression screen for synaptogenic proteins identifies the LRRTM protein family as synaptic organizers. Neuron 61, 734–749.

Lise, M.F., Wong, T.P., Trinh, A., Hines, R.M., Liu, L., Kang, R., Hines, D.J., Lu, J., Goldenring, J.R., Wang, Y.T., El-Husseini, A., 2006. Involvement of myosin Vb in glutamate receptor trafficking. J. Biol. Chem. 281, 3669–3678.

Lisman, J., Raghavachari, S., 2006. A unified model of the presynaptic and postsynaptic changes during LTP at CA1 synapses. Sci. STKE 2006:re11.

Lomeli, H., Sprengel, R., Laurie, D.J., Kohr, G., Herb, A., Seeburg, P.H., Wisden, W., 1993. The rat delta-1 and delta-2 subunits extend the excitatory amino acid receptor family. FEBS Lett. 315, 318–322.

Lomeli, H., Sprengel, R., Laurie, D.J., Kohr, G., Herb, A., Seeburg, P.H., Wisden, W., 1993. The rat delta-1 and delta-2 subunits extend the excitatory amino acid receptor family. FEBS Lett. 315, 318–322.

Lu, W., Isozaki, K., Roche, K.W., Nicoll, R.A., 2010. Synaptic targeting of AMPA receptors is regulated by a CaMKII site in the first intracellular loop of GluA1. Proc. Natl. Acad. Sci. U.S.A. 107, 22266–22271.

Lu, W., Shi, Y., Jackson, A.C., Bjorgan, K., During, M.J., Sprengel, R., Seeburg, P.H., Nicoll, R.A., 2009. Subunit composition of synaptic AMPA receptors revealed by a single-cell genetic approach. Neuron 62, 254–268.

Luscher, C., Xia, H., Beattie, E.C., Carroll, R.C., von Zastrow, M., Malenka, R.C., Nicoll, R.A., 1999. Role of AMPA receptor cycling in synaptic transmission and plasticity. Neuron 24, 649–658.

MacGillavry, H.D., Song, Y., Raghavachari, S., Blanpied, T.A., 2013. Nanoscale scaffolding domains within the postsynaptic density concentrate synaptic AMPA receptors. Neuron 78, 615–622.

Mah, W., Ko, J., Nam, J., Han, K., Chung, W.S., Kim, E., 2010. Selected SALM (synaptic adhesion-like molecule) family proteins regulate synapse formation. J. Neurosci. 30, 5559–5568.

Marambaud, P., Wen, P.H., Dutt, A., Shioi, J., Takashima, A., Siman, R., Robakis, N.K., 2003. A CBP binding transcriptional repressor produced by the PS1/epsilon-cleavage of N-cadherin is inhibited by PS1 FAD mutations. Cell 114, 635–645.

Masugi-Tokita, M., Tarusawa, E., Watanabe, M., Molnar, E., Fujimoto, K., Shigemoto, R., 2007. Number and density of AMPA receptors in individual synapses in the rat cerebellum as revealed by SDS-digested freeze-fracture replica labeling. J. Neurosci. 27, 2135–2144.

Matta, J.A., Ashby, M.C., Sanz-Clemente, A., Roche, K.W., Isaac, J.T., 2011. mGluR5 and NMDA receptors drive the experience- and activity-dependent NMDA receptor NR2B to NR2A subunit switch. Neuron 70, 339–351.

McClelland, A.C., Hruska, M., Coenen, A.J., Henkemeyer, M., Dalva, M.B., 2010. Trans-synaptic EphB2-ephrin-B3 interaction regulates excitatory synapse density by inhibition of postsynaptic MAPK signaling. Proc. Natl. Acad. Sci. U.S.A. 107, 8830–8835.

McClelland, A.C., Sheffler-Collins, S.I., Kayser, M.S., Dalva, M.B., 2009. Ephrin-B1 and ephrin-B2 mediate EphB-dependent presynaptic development via syntenin-1. Proc. Natl. Acad. Sci. U.S.A. 106, 20487–20492.

Migaud, M., Charlesworth, P., Dempster, M., Webster, L.C., Watabe, A.M., Makhinson, M., He, Y., Ramsay, M.F., Morris, R.G., Morrison, J.H., O'Dell, T.J., Grant, S.G., 1998. Enhanced long-term potentiation and impaired learning in mice with mutant postsynaptic density-95 protein. Nature 396, 433–439.

Missler, M., Zhang, W., Rohlmann, A., Kattenstroth, G., Hammer, R.E., Gottmann, K., Sudhof, T.C., 2003. Alpha-neurexins couple Ca2+ channels to synaptic vesicle exocytosis. Nature 423, 939–948.

Mok, H., Shin, H., Kim, S., Lee, J.R., Yoon, J., Kim, E., 2002. Association of the kinesin superfamily motor protein KIF1Balpha with postsynaptic density-95 (PSD-95), synapse-associated protein-97, and synaptic scaffolding molecule PSD-95/discs large/zona occludens-1 proteins. J. Neurosci. 22, 5253–5258.

Mondin, M., Labrousse, V., Hosy, E., Heine, M., Tessier, B., Levet, F., Poujol, C., Blanchet, C., Choquet, D., Thoumine, O., 2011. Neurexin-neuroligin adhesions capture surface-diffusing AMPA receptors through PSD-95 scaffolds. J. Neurosci. 31, 13500–13515.

Morimura, N., Inoue, T., Katayama, K., Aruga, J., 2006. Comparative analysis of structure, expression and PSD95-binding capacity of Lrfn, a novel family of neuronal transmembrane proteins. Gene 380, 72–83.

Morishita, W., Marie, H., Malenka, R.C., 2005. Distinct triggering and expression mechanisms underlie LTD of AMPA and NMDA synaptic responses. Nat. Neurosci. 8, 1043–1050.

Mu, Y., Otsuka, T., Horton, A.C., Scott, D.B., Ehlers, M.D., 2003. Activity-dependent mRNA splicing controls ER export and synaptic delivery of NMDA receptors. Neuron 40, 581–594.

Muller, B.M., Kistner, U., Kindler, S., Chung, W.J., Kuhlendahl, S., Fenster, S.D., Lau, L.F., Veh, R.W., Huganir, R.L., Gundelfinger, E.D., Garner, C.C., 1996. SAP102, a novel postsynaptic protein that interacts with NMDA receptor complexes in vivo. Neuron 17, 255–265.

Muller, D., Wang, C., Skibo, G., Toni, N., Cremer, H., Calaora, V., Rougon, G., Kiss, J.Z., 1996. PSA-NCAM is required for activity-induced synaptic plasticity. Neuron 17, 413–422.

Mysore, S.P., Tai, C.Y., Schuman, E.M., 2007. Effects of N-cadherin disruption on spine morphological dynamics. Front. Cell. Neurosci. 1, 1.

Naisbitt, S., Kim, E., Tu, J.C., Xiao, B., Sala, C., Valtschanoff, J., Weinberg, R.J., Worley, P.F., Sheng, M., 1999. Shank, a novel family of postsynaptic density proteins that binds to the NMDA receptor/PSD-95/GKAP complex and cortactin. Neuron 23, 569–582.

Nakagawa, T., Futai, K., Lashuel, H.A., Lo, I., Okamoto, K., Walz, T., Hayashi, Y., Sheng, M., 2004. Quaternary structure, protein dynamics, and synaptic function of SAP97 controlled by L27 domain interactions. Neuron 44, 453–467.

Nakazawa, T., Komai, S., Watabe, A.M., Kiyama, Y., Fukaya, M., Arima-Yoshida, F., Horai, R., Sudo, K., Ebine, K., Delawary, M., Goto, J., Umemori, H., Tezuka, T., Iwakura, Y., Watanabe, M., Yamamoto, T., Manabe, T., 2006. NR2B tyrosine phosphorylation modulates fear learning as well as amygdaloid synaptic plasticity. EMBO J. 25, 2867–2877.

Nam, C.I., Chen, L., 2005. Postsynaptic assembly induced by neurexin-neuroligin interaction and neurotransmitter. Proc. Natl. Acad. Sci. U.S.A. 102, 6137–6142.

Nasu-Nishimura, Y., Hurtado, D., Braud, S., Tang, T.T., Isaac, J.T., Roche, K.W., 2006. Identification of an endoplasmic reticulum-retention motif in an intracellular loop of the kainate receptor subunit KA2. J. Neurosci. 26, 7014–7021.

Nasu-Nishimura, Y., Jaffe, H., Isaac, J.T., Roche, K.W., 2010. Differential regulation of kainate receptor trafficking by phosphorylation of distinct sites on GluR6. J. Biol. Chem. 285, 2847–2856.
Nectoux, J., Girard, B., Bahi-Buisson, N., Prieur, F., Afenjar, A., Rosas-Vargas, H., Chelly, J., Bienvenu, T., 2007. Netrin G1 mutations are an uncommon cause of atypical Rett syndrome with or without epilepsy. Pediatr. Neurol. 37, 270–274.
Newpher, T.M., Ehlers, M.D., 2008. Glutamate receptor dynamics in dendritic microdomains. Neuron 58, 472–497.
Ng, D., Pitcher, G.M., Szilard, R.K., Sertie, A., Kanisek, M., Clapcote, S.J., Lipina, T., Kalia, L.V., Joo, D., McKerlie, C., Cortez, M., Roder, J.C., Salter, M.W., McInnes, R.R., 2009. Neto1 is a novel CUB-domain NMDA receptor-interacting protein required for synaptic plasticity and learning. PLoS Biol. 7, e41.
Nicoletti, F., Bockaert, J., Collingridge, G.L., Conn, P.J., Ferraguti, F., Schoepp, D.D., Wroblewski, J.T., Pin, J.P., 2011. Metabotropic glutamate receptors: from the workbench to the bedside. Neuropharmacol. 60, 1017–1041.
Niethammer, P., Delling, M., Sytnyk, V., Dityatev, A., Fukami, K., Schachner, M., 2002. Cosignaling of NCAM via lipid rafts and the FGF receptor is required for neuritogenesis. J. Cell Biol. 157, 521–532.
Nimchinsky, E.A., Yasuda, R., Oertner, T.G., Svoboda, K., 2004. The number of glutamate receptors opened by synaptic stimulation in single hippocampal spines. J. Neurosci. 24, 2054–2064.
Nishimura-Akiyoshi, S., Niimi, K., Nakashiba, T., Itohara, S., 2007. Axonal netrin-Gs transneuronally determine lamina-specific subdendritic segments. Proc. Natl. Acad. Sci. U.S.A. 104, 14801–14806.
Nolt, M.J., Lin, Y., Hruska, M., Murphy, J., Sheffler-Colins, S.I., Kayser, M.S., Passer, J., Bennett, M.V., Zukin, R.S., Dalva, M.B., 2011. EphB controls NMDA receptor function and synaptic targeting in a subunit-specific manner. J. Neurosci. 31, 5353–5364.
Nuriya, M., Huganir, R.L., 2006. Regulation of AMPA receptor trafficking by N-cadherin. J. Neurochem. 97, 652–661.
Okuda, T., Yu, L.M., Cingolani, L.A., Kemler, R., Goda, Y., 2007. beta-Catenin regulates excitatory postsynaptic strength at hippocampal synapses. Proc. Natl. Acad. Sci. U. S. A. 104, 13479–13484.
Opazo, P., Labrecque, S., Tigaret, C.M., Frouin, A., Wiseman, P.W., De Koninck, P., Choquet, D., 2010. CaMKII triggers the diffusional trapping of surface AMPARs through phosphorylation of stargazin. Neuron 67, 239–252.
Oshima, S., Fukaya, M., Masabumi, N., Shirakawa, T., Oguchi, H., Watanabe, M., 2002. Early onset of NMDA receptor GluR epsilon 1 (NR2A) expression and its abundant postsynaptic localization in developing motoneurons of the mouse hypoglossal nucleus. Neurosci. Res. 43, 239–250.
Ozaki, M., Sasner, M., Yano, R., Lu, H.S., Buonanno, A., 1997. Neuregulin-beta induces expression of an NMDA-receptor subunit. Nature 390, 691–694.
Pak, D.T., Yang, S., Rudolph-Correia, S., Kim, E., Sheng, M., 2001. Regulation of dendritic spine morphology by SPAR, a PSD-95-associated RapGAP. Neuron 31, 289–303.
Paoletti, P., Bellone, C., Zhou, Q., 2013. NMDA receptor subunit diversity: impact on receptor properties, synaptic plasticity and disease. Nat. Rev. Neurosci. 14, 383–400.
Papouin, T., Ladepeche, L., Ruel, J., Sacchi, S., Labasque, M., Hanini, M., Groc, L., Pollegioni, L., Mother, J.-P., Oliet, S.H.R., 2012. Synaptic and extrasynaptic NMDA receptors are gated by different endogenous coagonists. Cell 150, 633–646.
Park, M., Penick, E.C., Edwards, J.G., Kauer, J.A., Ehlers, M.D., 2004. Recycling endosomes supply AMPA receptors for LTP. Science 305, 1972–1975.
Park, Y., Jo, J., Isaac, J.T., Cho, K., 2006. Long-term depression of kainate receptor-mediated synaptic transmission. Neuron 49, 95–106.
Passafaro, M., Piech, V., Sheng, M., 2001. Subunit-specific temporal and spatial patterns of AMPA receptor exocytosis in hippocampal neurons. Nat. Neurosci. 4, 917–926.
Peixoto, R.T., Kunz, P.A., Kwon, H., Mabb, A.M., Sabatini, B.L., Philpot, B.D., Ehlers, M.D., 2012. Trans-synaptic signaling by activity-dependent cleavage of neuroligin-1. Neuron 76, 396–409.
Peng, J., Kim, M.J., Cheng, D., Duong, D.M., Gygi, S.P., Sheng, M., 2004. Semiquantitative proteomic analysis of rat forebrain postsynaptic density fractions by mass spectrometry. J. Biol. Chem. 279, 21003–21011.

Penn, A.C., Williams, S.R., Greger, I.H., 2008. Gating motions underlie AMPA receptor secretion from the endoplasmic reticulum. EMBO J. 27, 3056–3068.
Penzes, P., Beeser, A., Chernoff, J., Schiller, M.R., Eipper, B.A., Mains, R.E., Huganir, R.L., 2003. Rapid induction of dendritic spine morphogenesis by trans-synaptic ephrinB-EphB receptor activation of the Rho-GEF kalirin. Neuron 37, 263–274.
Penzes, P., Johnson, R.C., Alam, M.R., Kambampati, V., Mains, R.E., Eipper, B.A., 2000. An isoform of kalirin, a brain-specific GDP/GTP exchange factor, is enriched in the postsynaptic density fraction. J. Biol. Chem. 275, 6395–6403.
Penzes, P., Johnson, R.C., Sattler, R., Zhang, X., Huganir, R.L., Kambampati, V., Mains, R.E., Eipper, B.A., 2001. The neuronal Rho-GEF Kalirin-7 interacts with PDZ domain-containing proteins and regulates dendritic morphogenesis. Neuron 29, 229–242.
Perez-Otano, I., Lujan, R., Tavalin, S.J., Plomann, M., Modregger, J., Liu, X.B., Jones, E.G., Heinemann, S.F., Lo, D.C., Ehlers, M.D., 2006. Endocytosis and synaptic removal of NR3A-containing NMDA receptors by PACSIN1/syndapin1. Nat. Neurosci. 9, 611–621.
Petersen, J.D., Chen, X., Vinade, L., Dosemeci, A., Lisman, J.E., Reese, T.S., 2003. Distribution of postsynaptic density (PSD)-95 and Ca2+/calmodulin-dependent protein kinase II at the PSD. J. Neurosci. 23, 11270–11278.
Petralia, R.S., 2012. Distribution of extrasynaptic NMDA receptors on neurons. TSW J. Neurosci. (in press).
Petralia, R.S., Al-Hallaq, R.A., Wenthold, R.J., 2009. Trafficking and Targeting of NMDA Receptors. In: VanDongen, A.M. (Ed.), Biology of the NMDA Receptor. Taylor and Francis, New York, pp. 149–200.
Petralia, R.S., Esteban, J.A., Wang, Y.X., Partridge, J.G., Zhao, H.M., Wenthold, R.J., Malinow, R., 1999. Selective acquisition of AMPA receptors over postnatal development suggests a molecular basis for silent synapses. Nat. Neurosci. 2, 31–36.
Petralia, R.S., Sans, N., Wang, Y.X., Wenthold, R.J., 2005. Ontogeny of postsynaptic density proteins at glutamatergic synapses. Mol. Cell. Neurosci. 29, 436–452.
Petralia, R.S., Wang, Y.X., Hua, F., Yi, Z., Zhou, A., Ge, L., Stephenson, F.A., Wenthold, R.J., 2010. Organization of NMDA receptors at extrasynaptic locations. Neuroscience 167, 68–87.
Petralia, R.S., Wang, Y.X., Wenthold, R.J., 2003. Internalization at glutamatergic synapses during development. Eur. J. Neurosci. 18, 3207–3217.
Petralia, R.S., Wenthold, R.J., 1992. Light and electron immunocytochemical localization of AMPA-selective glutamate receptors in the rat brain. J. Comp. Neurol. 318, 329–354.
Petralia, R.S., Wenthold, R.J., 1999. Immunocytochemistry of NMDA receptors. Methods Mol. Biol. 128, 73–92.
Pickard, L., Noel, J., Henley, J.M., Collingridge, G.L., Molnar, E., 2000. Developmental changes in synaptic AMPA and NMDA receptor distribution and AMPA receptor subunit composition in living hippocampal neurons. J. Neurosci. 20, 7922–7931.
Plant, K., Pelkey, K.A., Bortolotto, Z.A., Morita, D., Terashima, A., McBain, C.J., Collingridge, G.L., Isaac, J.T., 2006. Transient incorporation of native GluR2-lacking AMPA receptors during hippocampal long-term potentiation. Nat. Neurosci. 9, 602–604.
Prange, O., Wong, T.P., Gerrow, K., Wang, Y.T., El-Husseini, A., 2004. A balance between excitatory and inhibitory synapses is controlled by PSD-95 and neuroligin. Proc. Natl. Acad. Sci. U.S.A. 101, 13915–13920.
Priel, A., Selak, S., Lerma, J., Stern-Bach, Y., 2006. Block of kainate receptor desensitization uncovers a key trafficking checkpoint. Neuron 52, 1037–1046.
Prybylowski, K., Chang, K., Sans, N., Kan, L., Vicini, S., Wenthold, R.J., 2005. The synaptic localization of NR2B-containing NMDA receptors is controlled by interactions with PDZ proteins and AP-2. Neuron 47, 845–857.
Prybylowski, K., Fu, Z., Losi, G., Hawkins, L.M., Luo, J., Chang, K., Wenthold, R.J., Vicini, S., 2002. Relationship between availability of NMDA receptor subunits and their expression at the synapse. J. Neurosci. 22, 8902–8910.
Qiu, S., Zhang, X.M., Cao, J.Y., Yang, W., Yan, Y.G., Shan, L., Zheng, J., Luo, J.H., 2009. An endoplasmic reticulum retention signal located in the extracellular amino-terminal domain of the NR2A subunit of N-Methyl-D-aspartate receptors. J. Biol. Chem. 284, 20285–20298.

Quinlan, E.M., Olstein, D.H., Bear, M.F., 1999a. Bidirectional, experience-dependent regulation of N-methyl-D-aspartate receptor subunit composition in the rat visual cortex during postnatal development. Proc. Natl. Acad. Sci. U.S.A. 96, 12876–12880.

Quinlan, E.M., Philpot, B.D., Huganir, R.L., Bear, M.F., 1999b. Rapid, experience-dependent expression of synaptic NMDA receptors in visual cortex in vivo. Nat. Neurosci. 2, 352–357.

Racca, C., Stephenson, F.A., Streit, P., Roberts, J.D., Somogyi, P., 2000. NMDA receptor content of synapses in stratum radiatum of the hippocampal CA1 area. J. Neurosci. 20, 2512–2522.

Ripley, B., Otto, S., Tiglio, K., Williams, M.E., Ghosh, A., 2011. Regulation of synaptic stability by AMPA receptor reverse signaling. Proc. Natl. Acad. Sci. U.S.A. 108, 367–372.

Robbins, E.M., Krupp, A.J., Perez de Arce, K., Ghosh, A.K., Fogel, A.I., Boucard, A., Sudhof, T.C., Stein, V., Biederer, T., 2010. SynCAM 1 adhesion dynamically regulates synapse number and impacts plasticity and learning. Neuron 68, 894–906.

Roberts, A.C., Diez-Garcia, J., Rodriguiz, R.M., Lopez, I.P., Lujan, R., Martinez-Turrillas, R., Pico, E., Henson, M.A., Bernardo, D.R., Jarrett, T.M., Clendeninn, D.J., Lopez-Mascaraque, L., Feng, G., Lo, D.C., Wesseling, J.F., Wetsel, W.C., Philpot, B.D., Perez-Otano, I., 2009. Downregulation of NR3A-containing NMDARs is required for synapse maturation and memory consolidation. Neuron 63, 342–356.

Roche, K.W., Ly, C.D., Petralia, R.S., Wang, Y.X., McGee, A.W., Bredt, D.S., Wenthold, R.J., 1999. Postsynaptic density-93 interacts with the delta2 glutamate receptor subunit at parallel fiber synapses. J. Neurosci. 19, 3926–3934.

Roche, K.W., Standley, S., McCallum, J., Dune Ly, C., Ehlers, M.D., Wenthold, R.J., 2001. Molecular determinants of NMDA receptor internalization. Nat. Neurosci. 4, 794–802.

Roche, K.W., Tu, J.C., Petralia, R.S., Xiao, B., Wenthold, R.J., Worley, P.F., 1999. Homer 1b regulates the trafficking of group I metabotropic glutamate receptors. J. Biol. Chem. 274, 25953–25957.

Rodenas-Ruano, A., Perez-Pinzon, M.A., Green, E.J., Henkemeyer, M., Liebl, D.J., 2006. Distinct roles for ephrinB3 in the formation and function of hippocampal synapses. Dev. Biol. 292, 34–45.

Rouach, N., Byrd, K., Petralia, R.S., Elias, G.M., Adesnik, H., Tomita, S., Karimzadegan, S., Kealey, C., Bredt, D.S., Nicoll, R.A., 2005. TARP gamma-8 controls hippocampal AMPA receptor number, distribution and synaptic plasticity. Nat. Neurosci. 8, 1525–1533.

Rumbaugh, G., Sia, G.M., Garner, C.C., Huganir, R.L., 2003. Synapse-associated protein-97 isoform-specific regulation of surface AMPA receptors and synaptic function in cultured neurons. J. Neurosci. 23, 4567–4576.

Ryan, T.J., Grant, S.G., 2009. The origin and evolution of synapses. Nat. Rev. Neurosci. 10, 701–712.

Saglietti, L., Dequidt, C., Kamieniarz, K., Rousset, M.C., Valnegri, P., Thoumine, O., Beretta, F., Fagni, L., Choquet, D., Sala, C., Sheng, M., Passafaro, M., 2007. Extracellular interactions between GluR2 and N-cadherin in spine regulation. Neuron 54, 461–477.

Sanderson, T.M., Collingridge, G.L., Fitzjohn, S.M., 2011. Differential trafficking of AMPA receptors following activation of NMDA receptors and mGluRs. Mol. Brain 4, 30.

Sans, N., Petralia, R.S., Wang, Y.X., Blahos 2nd, J., Hell, J.W., Wenthold, R.J., 2000. A developmental change in NMDA receptor-associated proteins at hippocampal synapses. J. Neurosci. 20, 1260–1271.

Sans, N., Prybylowski, K., Petralia, R.S., Chang, K., Wang, Y.X., Racca, C., Vicini, S., Wenthold, R.J., 2003. NMDA receptor trafficking through an interaction between PDZ proteins and the exocyst complex. Nat. Cell Biol. 5, 520–530.

Sans, N., Racca, C., Petralia, R.S., Wang, Y.X., McCallum, J., Wenthold, R.J., 2001. Synapse-associated protein 97 selectively associates with a subset of AMPA receptors early in their biosynthetic pathway. J. Neurosci. 21, 7506–7516.

Sans, N., Wang, P.Y., Du, Q., Petralia, R.S., Wang, Y.X., Nakka, S., Blumer, J.B., Macara, I.G., Wenthold, R.J., 2005. mPins modulates PSD-95 and SAP102 trafficking and influences NMDA receptor surface expression. Nat. Cell Biol. 7, 1179–1190.

Sanz-Clemente, A., Matta, J.A., Isaac, J.T., Roche, K.W., 2010. Casein kinase 2 regulates the NR2 subunit composition of synaptic NMDA receptors. Neuron 67, 984–996.

Sara, Y., Biederer, T., Atasoy, D., Chubykin, A., Mozhayeva, M.G., Sudhof, T.C., Kavalali, E.T., 2005. Selective capability of SynCAM and neuroligin for functional synapse assembly. J. Neurosci. 25, 260–270.

Schluter, O.M., Xu, W., Malenka, R.C., 2006. Alternative N-terminal domains of PSD-95 and SAP97 govern activity-dependent regulation of synaptic AMPA receptor function. Neuron 51, 99–111.

Schnell, E., Nicoll, R.A., 2001. Hippocampal synaptic transmission and plasticity are preserved in myosin Va mutant mice. J. Neurophysiol. 85, 1498–1501.

Schwenk, J., Harmel, N., Zolles, G., Bildl, W., Kulik, A., Heimrich, B., Chisaka, O., Jonas, P., Schulte, U., Fakler, B., Klocker, N., 2009. Functional proteomics identify cornichon proteins as auxiliary subunits of AMPA receptors. Science 323, 1313–1319.

Seabold, G.K., Wang, P.Y., Chang, K., Wang, C.Y., Wang, Y.X., Petralia, R.S., Wenthold, R.J., 2008. The SALM family of adhesion-like molecules forms heteromeric and homomeric complexes. J. Biol. Chem. 283, 8395–8405.

Seabold, G.K., Wang, P.Y., Petralia, R.S., Chang, K., Zhou, A., McDermott, M.I., Wang, Y.-X., Milgram, S.L., Wenthold, R.J., 2012. Dileucine and PDZ-binding motifs mediate synaptic adhesion-like molecule 1 (SALM1) trafficking in hippocampal neurons. J. Biol. Chem. 287, 4470–4484.

Seeburg, P.H., Higuchi, M., Sprengel, R., 1998. RNA editing of brain glutamate receptor channels: mechanism and physiology. Brain Res. Rev. 26, 217–229.

Selak, S., Paternain, A.V., Aller, M.I., Pico, E., Rivera, R., Lerma, J., 2009. A role for SNAP25 in internalization of kainate receptors and synaptic plasticity. Neuron 63, 357–371.

Selvakumar, B., Jenkins, M.A., Hussain, N.K., Huganir, R.L., Traynelis, S.F., Snyder, S.H., 2013. S-nitrosylation of AMPA receptor GluA1 regulates phosphorylation, single-channel conductance, and endocytosis. Proc. Natl. Acad. Sci. U.S.A. 110, 1077–1082.

Setou, M., Nakagawa, T., Seog, D.H., Hirokawa, N., 2000. Kinesin superfamily motor protein KIF17 and mLin-10 in NMDA receptor-containing vesicle transport. Science 288, 1796–1802.

Setou, M., Seog, D.H., Tanaka, Y., Kanai, Y., Takei, Y., Kawagishi, M., Hirokawa, N., 2002. Glutamate-receptor-interacting protein GRIP1 directly steers kinesin to dendrites. Nature 417, 83–87.

Sharma, K., Fong, D.K., Craig, A.M., 2006. Postsynaptic protein mobility in dendritic spines: long-term regulation by synaptic NMDA receptor activation. Mol. Cell. Neurosci. 31, 702–712.

Shi, S., Hayashi, Y., Esteban, J.A., Malinow, R., 2001. Subunit-specific rules governing AMPA receptor trafficking to synapses in hippocampal pyramidal neurons. Cell 105, 331–343.

Shi, S.H., Hayashi, Y., Petralia, R.S., Zaman, S.H., Wenthold, R.J., Svoboda, K., Malinow, R., 1999. Rapid spine delivery and redistribution of AMPA receptors after synaptic NMDA receptor activation. Science 284, 1811–1816.

Shi, Y., Suh, Y.H., Milstein, A.D., Isozaki, K., Schmid, S.M., Roche, K.W., Nicoll, R.A., 2010. Functional comparison of the effects of TARPs and cornichons on AMPA receptor trafficking and gating. Proc. Natl. Acad. Sci. U.S.A. 107, 16315–16319.

Shin, H., Wyszynski, M., Huh, K.H., Valtschanoff, J.G., Lee, J.R., Ko, J., Streuli, M., Weinberg, R.J., Sheng, M., Kim, E., 2003. Association of the kinesin motor KIF1A with the multimodular protein liprin-alpha. J. Biol. Chem. 278, 11393–11401.

Shipman, S.L., Nicoll, R.A., 2012a. Dimerization of postsynaptic neuroligin drives synaptic assembly via transsynaptic clustering of neurexin. Proc. Natl. Acad. Sci. U. S. A. 109, 19432–19437.

Shipman, S.L., Nicoll, R.A., 2012b. A subtype-specific function for the extracellular domain of neuroligin 1 in hippocampal LTP. Neuron 76, 309–316.

Shipman, S.L., Schnell, E., Hirai, T., Chen, B.S., Roche, K.W., Nicoll, R.A., 2011. Functional dependence of neuroligin on a new non-PDZ intracellular domain. Nat. Neurosci. 14, 718–726.

Siddiqui, T.J., Craig, A.M., 2011. Synaptic organizing complexes. Curr. Opin. Neurobiol. 21, 132–143.

Siddiqui, T.J., Pancaroglu, R., Kang, Y., Rooyakkers, A., Craig, A.M., 2010. LRRTMs and neuroligins bind neurexins with a differential code to cooperate in glutamate synapse development. J. Neurosci. 30, 7495–7506.

Silverman, J.B., Restituito, S., Lu, W., Lee-Edwards, L., Khatri, L., Ziff, E.B., 2007. Synaptic anchorage of AMPA receptors by cadherins through neural plakophilin-related arm protein AMPA receptor-binding protein complexes. J. Neurosci. 27, 8505–8516.

Soler-Llavina, G.J., Fuccillo, M.V., Ko, J., Sudhof, T.C., Malenka, R.C., 2011. The neurexin ligands, neuroligins and leucine-rich repeat transmembrane proteins, perform convergent and divergent synaptic functions in vivo. Proc. Natl. Acad. Sci. U.S.A. 108, 16502–16509.

Song, J.Y., Ichtchenko, K., Sudhof, T.C., Brose, N., 1999. Neuroligin 1 is a postsynaptic cell-adhesion molecule of excitatory synapses. Proc. Natl. Acad. Sci. U.S.A. 96, 1100–1105.
Srivastava, D.P., Copits, B.A., Xie, Z., Huda, R., Jones, K.A., Mukherji, S., Cahill, M.E., VanLeeuwen, J.E., Woolfrey, K.M., Rafalovich, I., Swanson, G.T., Penzes, P., 2012. Afadin is required for maintenance of dendritic structure and excitatory tone. J. Biol. Chem. 287, 35964–35974.
Srivastava, M., Simakov, O., Chapman, J., Fahey, B., Gauthier, M.E., Mitros, T., Richards, G.S., Conaco, C., Dacre, M., Hellsten, U., Larroux, C., Putnam, N.H., Stanke, M., Adamska, M., Darling, A., Degnan, S.M., Oakley, T.H., Plachetzki, D.C., Zhai, Y., Adamski, M., Calcino, A., Cummins, S.F., Goodstein, D.M., Harris, C., Jackson, D.J., Leys, S.P., Shu, S., Woodcroft, B.J., Vervoort, M., Kosik, K.S., Manning, G., Degnan, B.M., Rokhsar, D.S., 2010. The Amphimedon queenslandica genome and the evolution of animal complexity. Nature 466, 720–726.
Srivastava, S., Osten, P., Vilim, F.S., Khatri, L., Inman, G., States, B., Daly, C., DeSouza, S., Abagyan, R., Valtschanoff, J.G., Weinberg, R.J., Ziff, E.B., 1998. Novel anchorage of GluR2/3 to the postsynaptic density by the AMPA receptor-binding protein ABP. Neuron 21, 581–591.
Stan, A., Pielarski, K.N., Brigadski, T., Wittenmayer, N., Fedorchenko, O., Gohla, A., Lessmann, V., Dresbach, T., Gottmann, K., 2010. Essential cooperation of N-cadherin and neuroligin-1 in the transsynaptic control of vesicle accumulation. Proc. Natl. Acad. Sci. U.S.A. 107, 11116–11121.
Standley, S., Petralia, R.S., Gravell, M., Hamilton, R., Wang, Y.-X., Schubert, M., Wenthold, R.J., 2012. Trafficking of the NMDAR2B receptor subunit distal cytoplasmic tail from endoplasmic reticulum to the synapse. PLoS One 7 (e39585), 1–21.
Standley, S., Roche, K.W., McCallum, J., Sans, N., Wenthold, R.J., 2000. PDZ domain suppression of an ER retention signal in NMDA receptor NR1 splice variants. Neuron 28, 887–898.
Steigerwald, F., Schulz, T.W., Schenker, L.T., Kennedy, M.B., Seeburg, P.H., Kohr, G., 2000. C-Terminal truncation of NR2A subunits impairs synaptic but not extrasynaptic localization of NMDA receptors. J. Neurosci. 20, 4573–4581.
Stein, V., House, D.R., Bredt, D.S., Nicoll, R.A., 2003. Postsynaptic density-95 mimics and occludes hippocampal long-term potentiation and enhances long-term depression. J. Neurosci. 23, 5503–5506.
Steiner, P., Higley, M.J., Xu, W., Czervionke, B.L., Malenka, R.C., Sabatini, B.L., 2008. Destabilization of the postsynaptic density by PSD-95 serine 73 phosphorylation inhibits spine growth and synaptic plasticity. Neuron 60, 788–802.
Steward, O., Schuman, E.M., 2003. Compartmentalized synthesis and degradation of proteins in neurons. Neuron 40, 347–359.
Storey, G.P., Opitz-Araya, X., Barria, A., 2011. Molecular determinants controlling NMDA receptor synaptic incorporation. J. Neurosci. 31, 6311–6316.
Sturgill, J.F., Steiner, P., Czervionke, B.L., Sabatini, B.L., 2009. Distinct domains within PSD-95 mediate synaptic incorporation, stabilization, and activity-dependent trafficking. J. Neurosci. 29, 12845–12854.
Sudhof, T.C., 2008. Neuroligins and neurexins link synaptic function to cognitive disease. Nature 455, 903–911.
Suh, Y.H., Terashima, A., Petralia, R.S., Wenthold, R.J., Isaac, J.T., Roche, K.W., Roche, P.A., 2010. A neuronal role for SNAP-23 in postsynaptic glutamate receptor trafficking. Nat. Neurosci. 13, 338–343.
Suzuki, K., Hayashi, Y., Nakahara, S., Kumazaki, H., Prox, J., Horiuchi, K., Zeng, M., Tanimura, S., Nishiyama, Y., Osawa, S., Sehara-Fujisawa, A., Saftig, P., Yokoshima, S., Fukuyama, T., Matsuki, N., Koyama, R., Tomita, T., Iwatsubo, T., 2012. Activity-dependent proteolytic cleavage of neuroligin-1. Neuron 76, 410–422.
Swanwick, C.C., Shapiro, M.E., Yi, Z., Chang, K., Wenthold, R.J., 2009. NMDA receptors interact with flotillin-1 and -2, lipid raft-associated proteins. FEBS Lett. 583, 1226–1230.
Sytnyk, V., Leshchyns'ka, I., Delling, M., Dityateva, G., Dityatev, A., Schachner, M., 2002. Neural cell adhesion molecule promotes accumulation of TGN organelles at sites of neuron-to-neuron contacts. J. Cell Biol. 159, 649–661.
Sytnyk, V., Leshchyns'ka, I., Nikonenko, A.G., Schachner, M., 2006. NCAM promotes assembly and activity-dependent remodeling of the postsynaptic signaling complex. J. Cell Biol. 174, 1071–1085.
Tai, C.Y., Mysore, S.P., Chiu, C., Schuman, E.M., 2007. Activity-regulated N-cadherin endocytosis. Neuron 54, 771–785.

Takasu, M.A., Dalva, M.B., Zigmond, R.E., Greenberg, M.E., 2002. Modulation of NMDA receptor-dependent calcium influx and gene expression through EphB receptors. Science 295, 491–495.

Tanaka, J., Matsuzaki, M., Tarusawa, E., Momiyama, A., Molnar, E., Kasai, H., Shigemoto, R., 2005. Number and density of AMPA receptors in single synapses in immature cerebellum. J. Neurosci. 25, 799–807.

Tang, L., Hung, C.P., Schuman, E.M., 1998. A role for the cadherin family of cell adhesion molecules in hippocampal long-term potentiation. Neuron 20, 1165–1175.

Tang, M., Pelkey, K.A., Ng, D., Ivakine, E., McBain, C.J., Salter, M.W., McInnes, R.R., 2011. Neto1 is an auxiliary subunit of native synaptic kainate receptors. J. Neurosci. 31, 10009–10018.

Tardin, C., Cognet, L., Bats, C., Lounis, B., Choquet, D., 2003. Direct imaging of lateral movements of AMPA receptors inside synapses. EMBO J. 22, 4656–4665.

Thomas, C.G., Miller, A.J., Westbrook, G.L., 2006. Synaptic and extrasynaptic NMDA receptor NR2 subunits in cultured hippocampal neurons. J. Neurophysiol. 95, 1727–1734.

Thomas, G.M., Hayashi, T., Chiu, S.L., Chen, C.M., Huganir, R.L., 2012. Palmitoylation by DHHC5/8 Targets GRIP1 to Dendritic Endosomes to Regulate AMPA-R Trafficking. Neuron 73, 482–496.

Tomita, S., 2010. Regulation of ionotropic glutamate receptors by their auxiliary subunits. Physiol. (Bethesda) 25, 41–49.

Torres, R., Firestein, B.L., Dong, H., Staudinger, J., Olson, E.N., Huganir, R.L., Bredt, D.S., Gale, N.W., Yancopoulos, G.D., 1998. PDZ proteins bind, cluster, and synaptically colocalize with Eph receptors and their ephrin ligands. Neuron 21, 1453–1463.

Tovar, K.R., Westbrook, G.L., 2002. Mobile NMDA receptors at hippocampal synapses. Neuron 34, 255–264.

Traynelis, S.F., Wollmuth, L.P., McBain, C.J., Menniti, F.S., Vance, K.M., Ogden, K.K., Hansen, K.B., Yuan, H., Myers, S.J., Dingledine, R., 2010. Glutamate receptor ion channels: structure, regulation, and function. Pharmacol. Rev. 62, 405–496.

Traynelis, S.F., Wollmuth, L.P., McBain, C.J., Menniti, F.S., Vance, K.M., Ogden, K.K., Hansen, K.B., Yuan, H., Myers, S.J., Dingledine, R., 2010. Glutamate receptor ion channels: structure, regulation, and function. Pharmacol Rev 62, 405–496.

Tu, J.C., Xiao, B., Naisbitt, S., Yuan, J.P., Petralia, R.S., Brakeman, P., Doan, A., Aakalu, V.K., Lanahan, A.A., Sheng, M., Worley, P.F., 1999. Coupling of mGluR/Homer and PSD-95 complexes by the Shank family of postsynaptic density proteins. Neuron 23, 583–592.

Unoki, T., Matsuda, S., Kakegawa, W., Van, N.T., Kohda, K., Suzuki, A., Funakoshi, Y., Hasegawa, H., Yuzaki, M., Kanaho, Y., 2012. NMDA Receptor-Mediated PIP5K Activation to Produce PI(4,5)P(2) Is Essential for AMPA Receptor Endocytosis during LTD. Neuron 73, 135–148.

Valtschanoff, J.G., Weinberg, R.J., 2001. Laminar organization of the NMDA receptor complex within the postsynaptic density. J. Neurosci. 21, 1211–1217.

Vandenberghe, W., Nicoll, R.A., Bredt, D.S., 2005. Interaction with the unfolded protein response reveals a role for stargazin in biosynthetic AMPA receptor transport. J. Neurosci. 25, 1095–1102.

Varoqueaux, F., Aramuni, G., Rawson, R.L., Mohrmann, R., Missler, M., Gottmann, K., Zhang, W., Sudhof, T.C., Brose, N., 2006. Neuroligins determine synapse maturation and function. Neuron 51, 741–754.

Varoqueaux, F., Jamain, S., Brose, N., 2004. Neuroligin 2 is exclusively localized to inhibitory synapses. Eur. J. Cell Biol. 83, 449–456.

Vazquez, L.E., Chen, H.J., Sokolova, I., Knuesel, I., Kennedy, M.B., 2004. SynGAP regulates spine formation. J. Neurosci. 24, 8862–8872.

Vissel, B., Krupp, J.J., Heinemann, S.F., Westbrook, G.L., 2001. A use-dependent tyrosine dephosphorylation of NMDA receptors is independent of ion flux. Nat. Neurosci. 4, 587–596.

Vitureira, N., Letellier, M., White, I.J., Goda, Y., 2011. Differential control of presynaptic efficacy by postsynaptic N-cadherin and βs-catenin. Nat. Neurosci. 15, 81–89.

Voineagu, I., Wang, X., Johnston, P., Lowe, J.K., Tian, Y., Horvath, S., Mill, J., Cantor, R.M., Blencowe, B.J., Geschwind, D.H., 2011. Transcriptomic analysis of autistic brain reveals convergent molecular pathology. Nature 474, 380–384.

Waites, C.L., Specht, C.G., Hartel, K., Leal-Ortiz, S., Genoux, D., Li, D., Drisdel, R.C., Jeyifous, O., Cheyne, J.E., Green, W.N., Montgomery, J.M., Garner, C.C., 2009. Synaptic SAP97 isoforms regulate AMPA receptor dynamics and access to presynaptic glutamate. J. Neurosci. 29, 4332–4345.

Wang, C.Y., Chang, K., Petralia, R.S., Wang, Y.X., Seabold, G.K., Wenthold, R.J., 2006. A novel family of adhesion-like molecules that interacts with the NMDA receptor. J. Neurosci. 26, 2174–2183.
Wang, K., Zhang, H., Ma, D., Bucan, M., Glessner, J.T., Abrahams, B.S., Salyakina, D., Imielinski, M., Bradfield, J.P., Sleiman, P.M., Kim, C.E., Hou, C., Frackelton, E., Chiavacci, R., Takahashi, N., Sakurai, T., Rappaport, E., Lajonchere, C.M., Munson, J., Estes, A., Korvatska, O., Piven, J., Sonnenblick, L.I., Alvarez Retuerto, A.I., Herman, E.I., Dong, H., Hutman, T., Sigman, M., Ozonoff, S., Klin, A., Owley, T., Sweeney, J.A., Brune, C.W., Cantor, R.M., Bernier, R., Gilbert, J.R., Cuccaro, M.L., McMahon, W.M., Miller, J., State, M.W., Wassink, T.H., Coon, H., Levy, S.E., Schultz, R.T., Nurnberger, J.I., Haines, J.L., Sutcliffe, J.S., Cook, E.H., Minshew, N.J., Buxbaum, J.D., Dawson, G., Grant, S.F., Geschwind, D.H., Pericak-Vance, M.A., Schellenberg, G.D., Hakonarson, H., 2009. Common genetic variants on 5p14.1 associate with autism spectrum disorders. Nature 459, 528–533.
Wang, P.Y., Seabold, G.K., Wenthold, R.J., 2008. Synaptic adhesion-like molecules (SALMs) promote neurite outgrowth. Mol. Cell Neurosci. 39, 83–94.
Wang, Z., Edwards, J.G., Riley, N., Provance Jr., D.W., Karcher, R., Li, X.D., Davison, I.G., Ikebe, M., Mercer, J.A., Kauer, J.A., Ehlers, M.D., 2008. Myosin Vb mobilizes recycling endosomes and AMPA receptors for postsynaptic plasticity. Cell 135, 535–548.
Washbourne, P., Bennett, J.E., McAllister, A.K., 2002. Rapid recruitment of NMDA receptor transport packets to nascent synapses. Nat. Neurosci. 5, 751–759.
Washbourne, P., Dityatev, A., Scheiffele, P., Biederer, T., Weiner, J.A., Christopherson, K.S., El-Husseini, A., 2004. Cell adhesion molecules in synapse formation. J. Neurosci. 24, 9244–9249.
Washbourne, P., Liu, X.B., Jones, E.G., McAllister, A.K., 2004. Cycling of NMDA receptors during trafficking in neurons before synapse formation. J. Neurosci. 24, 8253–8264.
Wenthold, R.J., Blahos 2nd, J., Huh, K.H., Petralia, R.S., 1999. Detergent solubilization and immunoprecipitation of native NMDA receptors. Methods Mol. Biol. 128, 113–119.
Wenthold, R.J., Petralia, R.S., Blahos II, J., Niedzielski, A.S., 1996. Evidence for multiple AMPA receptor complexes in hippocampal CA1/CA2 neurons. J. Neurosci. 16, 1982–1989.
Wolf, M.E., Ferrario, C.R., 2010. AMPA receptor plasticity in the nucleus accumbens after repeated exposure to cocaine. Neurosci. Biobehav. Rev. 35, 185–211.
Wong, H.K., Liu, X.B., Matos, M.F., Chan, S.F., Perez-Otano, I., Boysen, M., Cui, J., Nakanishi, N., Trimmer, J.S., Jones, E.G., Lipton, S.A., Sucher, N.J., 2002. Temporal and regional expression of NMDA receptor subunit NR3A in the mammalian brain. J. Comp. Neurol. 450, 303–317.
Woo, J., Kwon, S.K., Choi, S., Kim, S., Lee, J.R., Dunah, A.W., Sheng, M., Kim, E., 2009. Trans-synaptic adhesion between NGL-3 and LAR regulates the formation of excitatory synapses. Nat. Neurosci. 12, 428–437.
Woolfrey, K.M., Srivastava, D.P., Photowala, H., Yamashita, M., Barbolina, M.V., Cahill, M.E., Xie, Z., Jones, K.A., Quilliam, L.A., Prakriya, M., Penzes, P., 2009. Epac2 induces synapse remodeling and depression and its disease-associated forms alter spines. Nat. Neurosci. 12, 1275–1284.
Wyszynski, M., Kim, E., Dunah, A.W., Passafaro, M., Valtschanoff, J.G., Serra-Pages, C., Streuli, M., Weinberg, R.J., Sheng, M., 2002. Interaction between GRIP and liprin-alpha/SYD2 is required for AMPA receptor targeting. Neuron 34, 39–52.
Xia, H., von Zastrow, M., Malenka, R.C., 2002. A novel anterograde trafficking signal present in the N-terminal extracellular domain of ionotropic glutamate receptors. J. Biol. Chem. 277, 47765–47769.
Yamada, Y., Chochi, Y., Takamiya, K., Sobue, K., Inui, M., 1999. Modulation of the channel activity of the epsilon2/zeta1-subtype N-methyl D-aspartate receptor by PSD-95. J. Biol. Chem. 274, 6647–6652.
Yan, D., Yamasaki, M., Straub, C., Watanabe, M., Tomita, S., 2013. Homeostatic control of synaptic transmission by distinct glutamate receptors. Neuron 78, 687–699.
Yang, Y., Tao-Cheng, J.H., Reese, T.S., Dosemeci, A., 2011. SynGAP moves out of the core of the postsynaptic density upon depolarization. Neuroscience 192, 132–139.
Yasuda, S., Tanaka, H., Sugiura, H., Okamura, K., Sakaguchi, T., Tran, U., Takemiya, T., Mizoguchi, A., Yagita, Y., Sakurai, T., De Robertis, E.M., Yamagata, K., 2007. Activity-induced protocadherin arcadlin regulates dendritic spine number by triggering N-cadherin endocytosis via TAO2beta and p38 MAP kinases. Neuron 56, 456–471.

Yi, Z., Petralia, R.S., Fu, Z., Swanwick, C.C., Wang, Y.X., Prybylowski, K., Sans, N., Vicini, S., Wenthold, R.J., 2007. The role of the PDZ protein GIPC in regulating NMDA receptor trafficking. J. Neurosci. 27, 11663–11675.

Zhang, D., Isack, N.R., Glodowski, D.R., Liu, J., Chen, C.C., Xu, X.Z., Grant, B.D., Rongo, C., 2012. RAB-6.2 and the retromer regulate glutamate receptor recycling through a retrograde pathway. J. Cell Biol. 196, 85–101.

Zhang, W., St-Gelais, F., Grabner, C.P., Trinidad, J.C., Sumioka, A., Morimoto-Tomita, M., Kim, K.S., Straub, C., Burlingame, A.L., Howe, J.R., Tomita, S., 2009. A transmembrane accessory subunit that modulates kainate-type glutamate receptors. Neuron 61, 385–396.

Zhang, W., Vazquez, L., Apperson, M., Kennedy, M.B., 1999. Citron binds to PSD-95 at glutamatergic synapses on inhibitory neurons in the hippocampus. J. Neurosci. 19, 96–108.

Zheng, C.Y., Petralia, R.S., Wang, Y.X., Kachar, B., Wenthold, R.J., 2010. SAP102 is a highly mobile MAGUK in spines. J. Neurosci. 30, 4757–4766.

Zheng, C.Y., Seabold, G.K., Horak, M., Petralia, R.S., 2011. MAGUKs, synaptic development, and synaptic plasticity. Neuroscientist 17, 493–512.

Zhiling, Y., Fujita, E., Tanabe, Y., Yamagata, T., Momoi, T., Momoi, M.Y., 2008. Mutations in the gene encoding CADM1 are associated with autism spectrum disorder. Biochem. Biophys. Res. Commun. 377, 926–929.

Zhu, J.J., Esteban, J.A., Hayashi, Y., Malinow, R., 2000. Postnatal synaptic potentiation: delivery of GluR4-containing AMPA receptors by spontaneous activity. Nat. Neurosci. 3, 1098–1106.

CHAPTER NINE

Structural Alterations of Synapses in Psychiatric and Neurodegenerative Disorders

Peter Penzes[1,2], Christine Remmers[1], Michael E. Cahill[1], Kelly A. Jones[1], Jon-Eric VanLeeuwen[1] and Kevin M. Woolfrey[1]

[1]Department of Physiology, Northwestern University Feinberg School of Medicine, Chicago, IL, USA

[2]Department of Psychiatry and Behavioral Sciences, Northwestern University Feinberg School of Medicine, Chicago, IL, USA

1. INTRODUCTION

In the mammalian forebrain, most glutamatergic excitatory synapses occur on protrusions from dendrites called dendritic spines. Spines are highly plastic: changes in spine morphology occur in synaptogenesis, maintenance, plasticity, and elimination. During development, dynamic spines allow dendrites to actively participate in synaptogenesis, thereby contributing to the establishment of connectivity within neuronal circuits (Holtmaat and Svoboda, 2009). Activity-dependent spine maintenance or elimination contributes to the remodeling of neuronal circuits during postnatal development and adolescence (Alvarez and Sabatini, 2007; Zuo et al., 2005). Spines undergo experience-dependent morphological changes in live animals (Alvarez and Sabatini, 2007; Holtmaat and Svoboda, 2009), and spine morphology is intimately linked to human cognitive development and function (Lewis and Gonzalez-Burgos, 2008).

Many neurological disorders are associated with disruptions in spine numbers and morphology. These include neurodevelopmental disorders such as fragile X (Irwin et al., 2000), Down's syndrome (Fiala et al., 2002), mental retardation, and autism spectrum disorders (ASDs) (Hutsler and Zhang, 2010); psychiatric and neurological disorders such as schizophrenia (SZ) (Glantz and Lewis, 2000), addiction, and depression; and neurodegenerative disorders, including Alzheimer's disease (AD) (Tackenberg et al., 2009). Deficits in synaptic function and connectivity are emerging as important contributors to the pathogenesis of these disorders, as recent studies support altered synaptic connectivity and plasticity in the brains of patients (Geschwind and Levitt, 2007; He et al., 2009). Because spine dysmorphogenesis reflects abnormal synaptic connectivity and function, spine pathology may be a key cellular determinant of these disorders, and may underlie cognitive deficits.

Spine dynamics are regulated by molecular pathways that control numerous processes, such as cytoskeletal remodeling, trans-synaptic adhesion, receptor trafficking, protein translation, ubiquitination, and gene expression (Penzes et al., 2008; Penzes and

The Synapse
http://dx.doi.org/10.1016/B978-0-12-418675-0.00009-2

Jones, 2008; Tada and Sheng, 2006). Recent genetic studies have implicated dendritic spine regulators in the etiology of psychiatric and neurological disorders. Concurrently, proteins initially identified in a disease context have been shown to affect spine morphology. Thus, synapse pathology research is at an exciting cross-roads: uncovering the molecular mechanisms of spine plasticity will facilitate the understanding and treatment of a range of disorders and will allow newly identified disease genes to be placed into a molecular context at the spine.

Here we discuss the role of spine structural abnormalities in three neuropathologies—ASDs, SZ, and AD—whose symptoms manifest at different stages of life (e.g. early childhood, adolescence, and late adulthood, respectively). We review the postmortem neuropathological studies that reveal the pathophysiological outcome of the disease process, genetic studies that address etiology, and animal/cellular studies that investigate the underlying neurobiological mechanisms. We explore the idea that although these disorders have largely distinct symptom profiles and characteristic neurobiological correlates, they share the common structural substrate of abnormal dendritic spine morphogenesis. Spine loss or overgrowth may result from genetic, epigenetic, or environmental insults to any of the components of particular molecular pathways that regulate spine dynamics at a certain stage in development, resulting in a disease-specific spine phenotype that may underlie the cognitive deficits characteristic of these disorders.

2.1. AUTISM SPECTRUM DISORDERS

ASDs, characterized by deficits in social interactions, verbal communication, and the presence of repetitive behavior, affect 0.9% of children. Accumulating evidence from human neuropathological, genetic, and model system studies suggests that autism may most accurately be conceptualized as a disease of the synapse (Geschwind and Levitt, 2007; Walsh et al., 2008a).

2.1.1. Neuropathological Studies

Recent evidence from Golgi-impregnated postmortem ASD human brain tissue revealed an increase in spine density on apical dendrites of pyramidal neurons from cortical layers II and V (Hutsler and Zhang, 2010). Density was inversely correlated with cognitive function. Such findings are consistent with the emerging hypothesis that the autistic brain is characterized by hyperconnectivity in local circuits and hypoconnectivity between brain regions (Geschwind and Levitt, 2007).

Further evidence of spine pathology is observed in tissue from patients with diseases comorbid with autism. Similar to "pure" autism, fragile X is characterized by elevated spine density, thought to result from pruning deficits (Irwin et al., 2001), along with elongated, tortuous spines (Wisniewski et al., 1991), indicative of altered function. In contrast, brain tissue from patients with Rett syndrome and Angelman's syndrome show reduced numbers of spines and dendrites (Belichenko et al., 1994; Jay et al., 1991). Collectively, these findings underscore the profound spine pathology exhibited by ASDs and comorbid disorders. It is likely that spine dysmorphology contributes to circuit-level abnormalities, which in turn underlie the sociocognitive impairments characteristic of these disorders.

2.1.2. Genetics and Molecular Mechanisms

The high degree of heritability of ASDs has sparked a great deal of research over the last decade. Multiple reports have identified rare mutations in synaptic proteins in autistic individuals. Though these mutations may not account for a large percentage of cases, consistent with a "common disease-many rare alleles" model (McClellan and King, 2010), they provide valuable insight into the molecular pathways underlying pathogenesis (Figure 1).

Rare variants in the synaptic cell adhesion proteins neuroligin-3 and -4 (NLGN3 and NLGN4) (Daoud et al., 2009; Jamain et al., 2003; Laumonnier et al., 2004), and their presynaptic ligand neurexin1 (NRXN1) (Kim et al., 2008) have been genetically linked

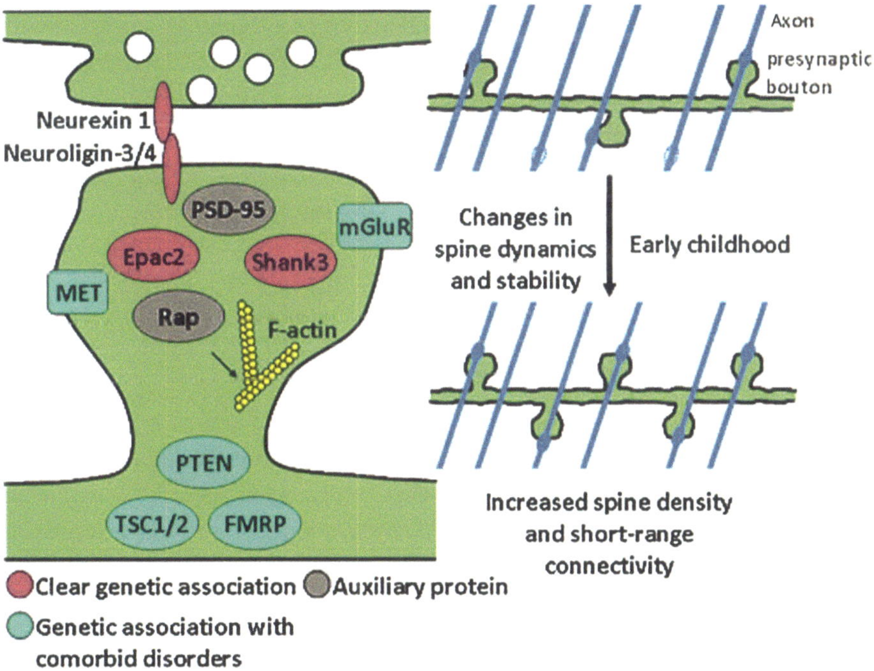

Figure 1 Model of molecular mechanisms of spine pathology in ASD. Proteins with genetic associations with ASD and comorbid disorders participate in pathways that regulate spine morphogenesis. Their disruption may alter spine dynamics and stability, leading to an increase in spine density, and increased connectivity with nearby axons (blue lines), during early childhood.

with autism. Point mutants in NLGN3 and NLGN4, a truncation of NLGN4 and a promoter mutation in NLGN4 have been identified. Wild-type NLGN3 or NLGN4 increases synapse number in hippocampal neurons and the NLGN3-R451C variant fails to induce novel synapses (Chih et al., 2004). α-Neurexin loss reduces dendrite length and total spine number in cortex. Autism-associated mutations in the postsynaptic scaffolding molecule shank3 (Durand et al., 2007) include a variety of frameshift, truncation, and missense mutations. Shank3 controls spine maintenance in the forebrain (Roussignol et al., 2005) and may complex with NLGNs. NLGN3 also complexes with the synaptically localized Rap-GEF Epac2 and enhances its signaling activity (Woolfrey et al., 2009). Rare structural variants of Epac2 have been identified in autistic individuals. EPAC2 missense mutations affect dendritic spine number (Epac2-T809S) and morphology (Epac2-V646F) (Woolfrey et al., 2009). Thus, NLGN3, neurexin1, shank3, and Epac2 constitute members of a synaptic regulatory pathway, disruption of which could confer autism susceptibility.

Genetic studies investigating monogenic disorders comorbid with ASDs may also shed light on ASD etiology. Patients with mutations of tuberous sclerosis proteins 1 and 2 (TSC1 and 2) and the phosphatase and tensin homolog (*PTEN*) gene are frequently diagnosed with autism (Butler et al., 2005). These proteins are involved in synaptic signaling pathways, disruptions of which have serious consequences for neuronal morphology and connectivity (Pardo and Eberhart, 2007). PTEN, TSC1, and TSC2 inhibit the Akt/mTOR/FMRP signaling pathway. Mutation or loss of each of these signaling molecules yields deficits in neuronal morphology. PTEN deficiency results in dendritic hypertrophy and elevated spine density (Kwon et al., 2006). Loss of TSC1 or TSC2 is characterized by an enlarged spine phenotype (Tavazoie et al., 2005).

A promoter variant of the receptor tyrosine kinase Met has been found in ASD (Campbell et al., 2006). Met is enriched at forebrain excitatory synapses, ideally positioning it to regulate dendrite outgrowth and synaptic protein clustering, but the effect of Met signaling on spine morphology remains unknown.

2.1.3. Animal Models

Surprisingly, spine morphology has not yet been examined in ASD animal models; however, NL3 R451C transgenic mice have abnormal excitatory/inhibitory synaptic balance (Tabuchi et al., 2007). There are well established mouse models of two syndromes comorbid with ASDs: Rett and fragile X. Rett syndrome can be effectively modeled using three MECP2-deficient mouse lines (Armstrong, 2005). Using animals both deficient in and overexpressing MeCP2, it has been demonstrated that MeCP2 controls the number of functional, excitatory synapses (Chao et al., 2007). Modeling fragile X syndrome using FMR1 knockout mice has recapitulated the immature spine morphologies observed in human brain (Irwin et al., 2002). Thus synaptic pathology is a prominent phenotype in mouse models based on disorders comorbid with ASDs. Remarkably, pharmacological and genetic manipulations have successfully rescued both structural and behavioral phenotypes in PTEN (Zhou et al., 2009), FMRP1 (Dolen et al., 2007), and MECP2 (Guy et al., 2007) knockout mice. That behavioral recovery in each of these rescue strategies was paralleled by reversal of spine pathologies underscores the importance of these structural perturbations in the pathogenesis of ASDs and comorbid disorders.

2.2. SCHIZOPHRENIA

SZ is a heterogenous disorder affecting thought, perceptions of reality, effect, and cognition, which affects approximately 0.5–1% of the population. Symptoms typically emerge in late adolescence or early adulthood.

2.2.1. Neuropathological Studies

Functional imaging studies have consistently shown gray matter loss in SZ, which are accelerated during periadolescence (Gogtay, 2008). Among the ultrastructural changes thought to directly contribute to gray matter loss are reductions in spine density (Selemon and Goldman-Rakic, 1999). Several postmortem studies have examined spine density in brain regions showing the greatest indices of gray matter loss. The dorsolateral prefrontal cortex (DLPFC) shows severe dysfunction in SZ as patients show reduced activity of this region during cognitive tasks (Tan et al., 2007). Indeed, spine loss in the DLPFC has been reported, particularly in layer 3 neurons (Glantz and Lewis, 2000). A reduction in superior temporal gyrus (STG) gray matter volume is one of the most consistently reported alterations in the brains of SZ patients (Yoshida et al., 2009), and is associated with auditory hallucinations (Barta et al., 1990). SZ patients show a profound reduction in spine density on pyramidal neurons in the STG, particularly in the primary auditory cortex (Sweet et al., 2009).

Several studies have shown reductions in hippocampal volume in SZ (Steen et al., 2006) and among relatives of SZ patients. SZ patients show hypoactivity of the hippocampus during memory recall (Weiss et al., 2004). Within the hippocampus, reduced spine density on subicular dendrites has been reported. Reduced CA3 spine density (Kolomeets et al., 2005; Law et al., 2004), as well as reduced spine size (Kolomeets et al., 2005), has been a consistent finding. This is also associated with a lowered density of synaptic contacts formed by individual mossy fiber tracts on CA3 pyramidal neurons (Kolomeets et al., 2007).

2.2.2. Genetics

A handful of genes have shown consistent associations with SZ, including but not limited to, NRG1, ERBB4, DISC1, and KALRN (Figure 2). In addition, rare but highly penetrant copy number variants (CNVs) have been found in many genes encoding synaptic proteins (Walsh et al., 2008b).

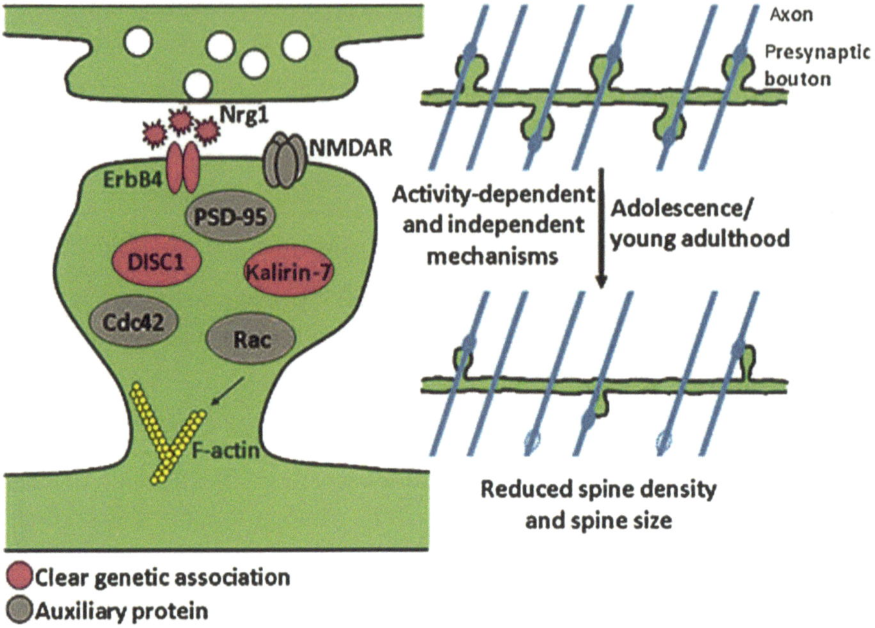

Figure 2 Model of mechanisms contributing to spine dysfunction in SZ. Molecules genetically or neuropathologically implicated in SZ interact with regulators of spine plasticity and maintenance. Their disruption may lead to exaggerated spine loss and loss of connectivity with axons (blue lines) in late adolescence or early adulthood.

Polymorphisms in neuregulin-1 (NRG1) are associated with SZ (Mei and Xiong, 2008). NRGs are trophic factors that exist in both membrane-bound and soluble form. ErbB receptors are postsynaptic receptor tyrosine kinases and are activated upon NRG binding (Mei and Xiong, 2008). ErbB4 is thought to be the predominant receptor for NRG-1. Common variants in ERBB4 are also associated with SZ (Shiota et al., 2008; Silberberg et al., 2006). Interestingly, a rare CNV for ERBB4 has been identified in SZ (Walsh et al., 2008b). This mutant is a 399 kb deletion that would result in a protein lacking most of its intracellular kinase domain, akin to a dominant-negative protein. NRG1 and erbB4 regulate synaptic structure and function. ErbB4 is expressed in interneurons, and less abundantly, in cortical pyramidal cells and in spines. Long-term NRG1 treatment increases pyramidal neuronal spine density and the preponderance of spines with mature phenotypes (Barros et al., 2009). ErbB4 overexpression increases spine density, area, and excitatory synaptic transmission (Li et al., 2007). Conversely, erbB4 knockdown reduces spine density and size (Li et al., 2007).

The initial link of the disrupted in SZ 1 (*DISC1*) gene to SZ was identified in a Scottish pedigree with a disruption of the DISC1 open reading frame. Polymorphisms and frameshift mutations of DISC1 have been linked to SZ in other lineages (Schumacher et al., 2009). As DISC1 has been associated with several psychiatric disorders, including bipolar disorder, depression, and autism, it seems likely that it constitutes a general psychiatric vulnerability gene. DISC1 is highly abundant in spines (Kirkpatrick et al., 2006). In cortical neurons, long-term knockdown reduces spine area (Hayashi-Takagi et al.), whereas its short-term knockdown increases spine density and size.

A recent genome-wide association study detected association signals with SZ at the region of the *KALRN* gene (St Jean, 2008). This association was also supported in an independent study in a Japanese population (Ikeda et al., 2011). Although single locus analysis did not reach genome-wide significance, the study confirmed a shared polygenic risk of SZ between the Japanese and the Caucasian samples. Following up on these findings, Kushima et al. (2010) resequenced all exons of the KALRN gene and identified a number of rare missense mutations enriched in patients with SZ as compared to nonpsychiatric controls. Several of these sequence alterations are predicted to have functional and damaging consequences. The authors detected a significant association of the P2255T mutation (OR = 2.09) as well as combined association of all mutations (OR = 2.07) with SZ. Interestingly, several of these mutations are in exons encoding protein domains present in kalirin-9 or -12 but not kalirin-7. Thus, rare missense mutations in the KALRN gene may contribute to genetic risk in SZ.

The 22q11.2 microdeletion syndrome is the most common CNV associated with SZ, accounting for 1–2% of cases (Stark et al., 2008). Primary hippocampal neurons from mice engineered to carry the 1.3 Mb orthologous chromosomal region (Df(16)A+/−) showed reduced spine density and sizes (Mukai et al., 2008). Interestingly, loss of either of two genes within this region (ZDHHC8 and DGCR8) was sufficient to impair spine and dendrite morphology (Mukai et al., 2008; Stark et al., 2008). ZDHHC8 is a palmitoyl transferase which palmitoylates PSD-95; its loss results in reduced spine density and simpler dendrites, and its replacement into Df(16)A+/− neurons rescued spine and dendrite deficiency (Mukai et al., 2008). Dgcr8 is involved in miRNA processing, and its loss results in smaller spines and simpler dendrites (Stark et al., 2008).

2.2.3. Molecular Mechanisms

Postmortem expression studies revealed changes in molecules that regulate spine morphology in SZ subjects. In parallel, neurobiological studies have uncovered interactions of SZ-associated molecules with spine regulators (Figure 2).

While DISC1 mRNA levels seem unaffected in SZ patients (Dean et al., 2007; Lipska et al., 2006), the expression of DISC1-interacting proteins was reduced in patients carrying high-risk DISC1 SNPs (Lipska et al., 2006), suggesting that DISC1 function might be affected in SZ. Disruption of DISC1's ability to scaffold proteins in spines would be expected to have deleterious consequences on spine morphogenesis. Indeed, DISC1 is known to interact with several well-established regulators of spine morphogenesis, most prominently the RacGEF kalirin-7 (Millar et al., 2003). Recently kalirin-7 was found to directly regulate the effects of DISC1 on spine morphology via activation of its downstream effector Rac1 (Hayashi-Takagi et al.). Interestingly, the expression of kalirin mRNA was reduced in the DLPFC of patients with SZ, irrespective of antipsychotic treatment (Hill et al., 2006). This suggests a potential role for small GTPase pathways in spine pathology in SZ. Indeed, the expression of Cdc42 mRNA was also reduced in postmortem schizophrenic DLPFC (Hill et al., 2006; Ide and Lewis, 2010). Loss of kalirin and Cdc42 strongly correlates with spine loss in layer 3 PFC neurons (Hill et al., 2006).

Components of the glutamatergic receptor system, such as NMDA receptor subunits, known to modulate basal spine morphology and activity-dependent spine plasticity, are reduced in postmortem schizophrenic brains (Hill et al., 2006). Scaffolding proteins function as organizing molecules in spines and provide a structural link between surface receptors and intracellular signaling networks. Interestingly, erbB4 and DISC1 interact with the PDZ-containing scaffolding molecule PSD-95 in spines (Garcia et al., 2000; Hayashi-Takagi et al.). Notably, PSD-95 protein levels are reduced in the SZ cortex (Kristiansen et al., 2006).

2.2.4. Animal Models

Support for the contribution of spine loss to SZ comes from genetic animal models, which reproduce SZ-related behavioral phenotypes along with the spine alterations in the disease.

Mice deficient in NRG1 type III show reductions in spine density in hippocampal neurons (Chen et al., 2008). Mice lacking erbB2 and erbB4 in the CNS show reduced spine density in both the hippocampus and cortex (Barros et al., 2009). In both these mice, spine morphological deficits accompany SZ-related behavioral phenotypes.

The effects of DISC1 mutations on spine density in mice reflect brain region and developmentally influenced effects. Namely, spine numbers in dentate gyrus (DG) granule cells are reduced in a mouse model of disease-associated chromosomal translocation (Kvajo et al., 2008), and dendrite complexity is reduced by early postnatal expression of DISC1 C-terminus in mice (Li et al., 2007). Spine density in cortical pyramidal neurons was increased by prenatal expression of mutant DISC1, while combined prenatal and postnatal expression increased spine density in hippocampal granule cells (Ayhan et al., 2011).

Because of kalirin's important synaptic functions, interactions with DISC1, and reduced expression in SZ, recent studies have examined how kalirin loss impacts spines and behavior. Interestingly, kalirin knockout mice show severe reductions in spine density and dendrite complexity in the frontal cortex, as well as SZ-related impairments in working memory, sociability, and prepulse inhibition (Cahill et al., 2009; Xie et al., 2010). Remarkably, both spine loss and behavioral dysfunction emerged during adolescence and were absent in juvenile knockout mice (Cahill et al., 2009). This is interesting given the onset of SZ symptoms in adolescence in humans, and points to a tight association between the onset of spine loss and the onset of behavioral impairments in these animals.

Mice modeling the 22q11.2 microdeletion syndrome (Df(16)A+/−) showed reduced hippocampal spine density and sizes (Mukai et al., 2008). Mice deficient in individual genes within this region (ZDHHC8 and DGCR8) showed simplified dendritic trees and reduced spine density (Mukai et al., 2008), or smaller spines (Stark et al., 2008), respectively.

2.3. ALZHEIMER'S DISEASE

AD represents the most common form of dementia that affects 13% of Americans over 65. Although amyloid deposits, neurofibrillary tangles (NFTs), and cell death remain defining characteristics of AD, findings from genetic and molecular studies have provided strong support for a central role for synapse degeneration in AD (Selkoe, 2002).

2.3.1. Neuropathological Studies

Synapse loss is a prominent and consistent finding in postmortem tissue samples from patients diagnosed with AD (DeKosky and Scheff, 1990; Scheff et al., 1990). Analysis of cortical biopsies in AD patients demonstrated a 25–35% reduction in synapse density and synapse number per neuron (Davies et al., 1987). This synapse loss was greater than what would be expected from neuronal death. Interestingly, synapse and dendrite loss demonstrate a stronger correlation to cognitive decline than do NFTs or neuronal loss (DeKosky and Scheff, 1990; Terry et al., 1991). Stereologic sampling of autopsy tissue demonstrated synapse loss in hippocampal CA1 (Scheff et al., 2007). The authors found an 18% reduction of synapse numbers in patients with mild cognitive impairment (MCI) and a 55% loss in AD patients, indicating that synapse loss occurs early in the progression of AD and worsens as the disease advances.

2.3.2. Genetics

The genetic contributions underlying AD are complex and incompletely understood. Familial AD (FAD), which has an autosomal dominant form of inheritance, has been associated with mutations in three genes that are involved with Aβ production (Bertram and Tanzi, 2008). Amyloid precursor protein (APP), when cleaved sequentially by β- and γ-secretase, produces the Aβ peptide. All FAD-associated mutations are found in APP or genes that make up the γ-secretase complex: presenilin 1 (PSEN1) and presenilin 2 (PSEN2). Mutations found in FAD are well known to increase Aβ production, and cellular studies provide compelling evidence that oligomeric Aβ disrupts synaptic signaling (reviewed in Klein (2006); Selkoe (2008)). Furthermore, Aβ has been found to target spines, induce spine dysgenesis, and reduce spine density (Lacor et al., 2007; Shankar et al., 2007).

The vast majority of AD cases, however, are considered late-onset. Although hundreds of genes have been proposed as late-onset AD risk factors, the apolipoprotein E-ε4 (APOE-ε4) allele is widely accepted as the most important risk factor. Studies with transgenic mice recently revealed that ApoE isoforms differentially influence dendrite and spine morphology. Mice expressing human ApoE-ε4 display reduced spine density in the DG when compared to WT mice and mice expressing human ApoE-ε3 (Ji et al., 2003). The authors also found an inverse correlation between ApoE-ε4 dose and DG spine density in human brain. In another study, human ApoE-ε4 was found to reduce dendritic length and branching in mouse cortex and hippocampus (Dumanis et al., 2009). However, this study did not find that ApoE isoforms affect spine morphology in the hippocampus. Previous studies, though, reported that expression of the ApoE-ε2 isoform in a mouse model of AD can restore spine density to control levels (Lanz et al., 2003). Despite some conflicting results, it is interesting that the major genetic risk factor for late-onset AD affects dendrite and spine morphology.

2.3.3. Animal Models

Based on genetic findings, numerous transgenic mouse models have been generated to recapitulate specific aspects of AD pathology. The widely used Tg2576 mouse model (Hsiao et al., 1996) expresses mutant human APP and displays decreased spine density in the CA1 and DG, well before the development of amyloid plaques (Jacobsen et al., 2006; Lanz et al., 2003). Animal models expressing mutations in both APP and PS1 yield neurons with diminished frequency of large spines and dendritic abnormalities (Knafo et al., 2009). Dendrites near amyloid deposits are reported to experience shaft atrophy, neurite breakage, and greater reductions in spine density (Grutzendler et al., 2007; Tsai et al., 2004). Similarly, amyloid plaques in Tg2576 mice alter neurites and reduce spine density on dendrites nearby (Spires et al., 2005). Such AD animal models support the concept that synaptic degeneration is central to the disease and may serve as a driving force, rather than a byproduct, of AD pathology that leads to memory impairment. Such structural alterations are reversible pharmacologically, opening new therapeutic directions in AD (Smith et al., 2009).

2.3.4. Molecular Mechanisms

AD patients demonstrate altered expression of many synaptic proteins (for a review, see Arendt (2009)). Presynaptic proteins such as synaptophysin are reduced in patients with MCI or AD (Masliah et al., 2001; Selkoe, 2002) and several postsynaptic signaling molecules have also been shown to be affected in AD. Although the precise mechanisms that cause spine degeneration in AD remain unknown, recent findings suggest that signaling pathways regulating actin dynamics may play a crucial role (Figure 3).

Cofilin and drebrin are actin-binding proteins that exert opposite effects on actin dynamics, but both are affected in AD. Active cofilin causes actin destabilization, and much evidence supports a role for cofilin in neurodegeneration, including AD (Maloney and Bamburg, 2007). Drebrin, a postsynaptic protein that binds and stabilizes actin in spines, has been observed to be reduced in the hippocampus (Harigaya et al., 1996) and

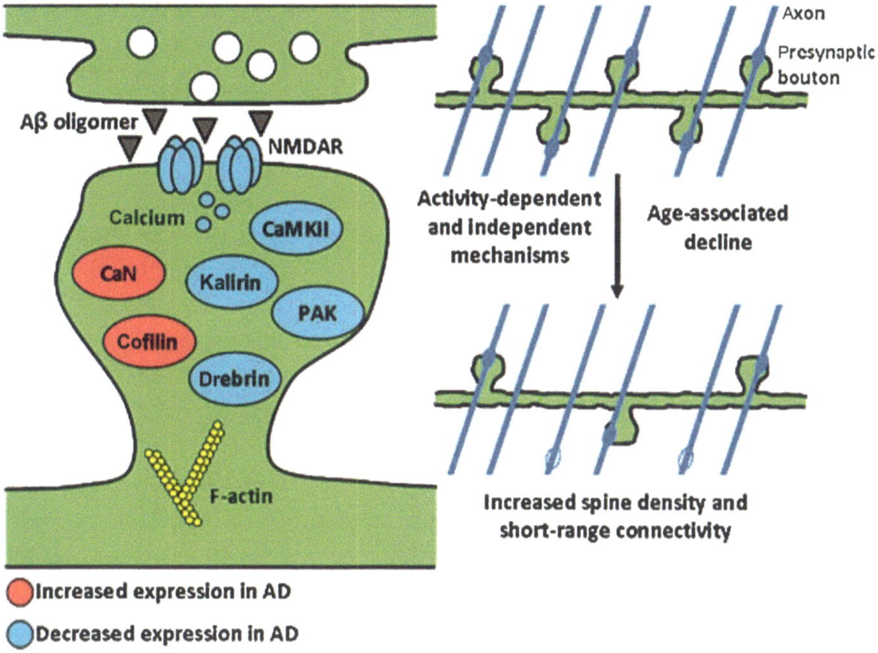

Figure 3 Model of molecular mechanisms involved in spine pathology in AD. Aβ oligomers inhibit LTP and induce spine dysgenesis, likely by interfering with NMDAR-dependent regulation of the spine cytoskeleton, causing synapse loss and decreased connectivity with nearby axons (blue lines) later in life.

cortex (Counts et al., 2006) of patients with AD. Surprisingly, drebrin levels in the frontal cortex of MCI patients were increased by ~30%, a potential indication of compensation that may occur early in the disease (Counts et al., 2006).

A critical regulator of actin assembly in spines is p21-activated kinase (PAK), a downstream effector molecule of Rac (Penzes et al., 2003; Zhao et al., 2006). In the hippocampus of AD patients, and in animal models of AD, PAK activation is markedly reduced and mislocalized (Ma et al., 2008; Zhao et al., 2006). Furthermore, pharmacological inhibition of PAK in mice was sufficient to cause memory impairment, cofilin pathology, and drebrin loss. Interestingly, kalirin-7, a key regulator of spine morphogenesis and an upstream activator of PAK in spines, was found to be the most consistently underexpressed mRNA in the hippocampus of AD patients (Xie et al., 2007; Youn et al., 2007). That both kalirin-7 mRNA and protein levels are significantly diminished in AD patients suggests a role for the kalirin-7/Rac1/PAK signaling pathway in AD-associated spine pathology.

Calcineurin (or PP2B) is a calcium-sensitive phosphatase involved in synaptic plasticity, and its activation leads to synaptic weakening (Xia and Storm, 2005). Interestingly, calcineurin activation, rather than inactivation, has been reported in AD patients and animal models (Liu et al., 2005; Norris et al., 2005). Aβ oligomer-induced spine loss and dendritic dystrophies can be prevented by calcineurin inhibition (Wu et al., 2010). Furthermore, long-term depression (LTD) induced by Aβ oligomers in hippocampal CA1 requires calcineurin activity (Li et al., 2009).

Many of the signaling molecules identified as having a potential role in AD pathogenesis lend support to the emerging hypothesis that, in AD, Aβ oligomers promote synaptic degeneration by acting on spines to cause a shift toward LTD, creating an imbalance of synaptic plasticity mechanisms (Hsieh et al., 2006; Selkoe, 2008) (Figure 3). Dampened NMDAR activation can induce LTD and, indeed, most of the signaling molecules discussed above are known to be downstream of NMDARs. Moreover, Aβ reduces surface expression of NMDARs (Snyder et al., 2005). This hypothesis is bolstered by studies that demonstrated synaptotoxic effects of Aβ could be prevented by NMDAR antagonists (Shankar et al., 2007).

Although much progress has been made, precise signaling cascades underlying synapse loss and cognitive decline need to be further elucidated. Determining the molecular processes underlying synapse degeneration in AD is critical for understanding the disease and for developing effective therapeutics.

3. CONCLUSIONS

Across the lifespan, spine proliferation occurring perinatally and postnatally is followed by a period of synapse elimination which lasts throughout childhood and into adolescence in some brain regions. This is followed by spine maintenance in the adult brain aimed at preserving circuitry established earlier in life. Disruption of spine morphogenesis during a particular time in development, or in a particular brain region, may dictate the neuropathological symptoms that arise. For example, ASD may involve delayed or reduced pruning during early childhood, while SZ symptoms may arise from accelerated pruning during adolescence. Similarly, SZ and AD are both associated with reduced spine density, but the acceleration of synapse loss occurs during two distinct stages of life: late childhood or adolescence in SZ, and late adulthood in AD. Thus, the timing of spine pathology, and whether spine number is upregulated or downregulated, can lead to disease-specific synaptic and cellular dysfunction, neuronal circuit alterations, and ultimately cognitive and behavioral symptoms.

The cellular and molecular characteristics of spine pathology seem partly distinct for ASD, SZ, and AD, and may underlie disease-specific profiles of cognitive disruptions (Figures 3). ASD is characterized by increased spine density, perhaps due to abnormal spine formation and stability. Molecularly, adhesion proteins like neuroligins and molecules that control spine dynamics, exemplified by Epac2, are strong candidates for regulators of ASD-associated spine pathologies. SZ is associated with reduced spine density in the prefrontal and auditory cortices and hippocampus, potentially mediated by exaggerated activity-dependent spine elimination. Spine pathologies in SZ may also be affected by synaptotrophic signaling, for which NRG1/erbB4 signaling is a promising candidate. Spine pathology in AD involves a reduction in spine density in late adulthood, suggesting impaired activity-dependent and -independent spine maintenance. Disruptions in NMDAR signaling and spine maintenance pathways such as kalirin-Rac-PAK may play important roles in mediating AD-related spine changes.

Disorders such as ASD, SZ, and AD have complex etiologies with heterogeneous symptomatology. The majority of candidate genes for each of these diseases only account for a fraction of cases, and many of these mutations and CNVs are rare, yet highly penetrant (Guilmatre et al., 2009; Sebat et al., 2007; Walsh et al., 2008a). This inherent heterogeneity highlights the importance of determining common pathways of disease-associated genes, as no single gene will account for the majority of cases. Recent breakthroughs in the characterization of regulators of spine morphogenesis have provided opportunities to identify molecular changes that could directly contribute to spine pathology. The molecular networks that control spines provide a framework for understanding how a large number of rare genetic perturbations can interact to disrupt synaptic function, neuronal circuit organization, and behavioral output in a disease-specific manner.

The development of effective treatments for these disorders will be facilitated by a better understanding of the causal links between synaptic pathology and disease pathophenotypes, the availability of better animal models specifically linking synaptic pathology with pathophenotypes, and by the identification of novel drug targets based on these links. It will be important to determine whether abnormal spine morphology is an etiological cause, pathophysiological effect, or confound in a particular disease. Continued work to determine the relationship between proteins that regulate spine morphology and disease is required to understand molecular disruptions that cause devastating neuropathologies. Concurrently, it would be instructive to elucidate the functions of newly identified disease-associated genes within synaptic molecular networks. Additionally, we should strive to understand how these molecules and spine plasticity contribute to disease endophenotypes; cognitive phenotypes, such as working memory (SZ), social cognition (ASD), and reference memory (AD), are of particular interest. How does altered structural plasticity at the synaptic or cellular level manifest at the circuit or behavioral level, and where and when are the candidate molecules relevant to a disease-associated cognitive deficit? Finally, some newly characterized synaptic regulators may be favorable drug targets, due to inherent enzymatic activity that could be easily modulated by small molecules. Novel drugs may prevent the emergence of symptoms in genetically susceptible patients, delay the progression of symptoms in the early stages of the disease, or mitigate symptoms or promote functional recovery after the disease is fully manifested. Specifically, drugs that target dendritic spine regulation might aim to promote spine maturation and restore spine stability in ASD, to fortify existing synapses and restore spine plasticity in SZ, or prevent synapse loss in AD.

ABBREVIATIONS

Epac2 exchange protein directly activated by cAMP
Shank3 SH3 and multiple ankyrin repeat domains 3
TSC1 tuberous sclerosis protein 1
TSC2 tuberous sclerosis protein 2
FMRP fragile X mental retardation protein
MeCP2 methyl CpG-binding protein 2
PTEN phosphatase and tensin homolog
MET met proto-oncogene
DISC1 disrupted-in-schizophrenia 1
Cdc42 cell division control protein 42 homolog
PSD-95 postsynaptic density-95
NRG1 neuregulin 1
ErbB4 v-erb-a erythroblastic leukemia viral oncogene homolog 4
PAK1 p21-activated kinase 1
PAK3 p21-activated kinase 3
ApoE-ε4 ε4 allele of apolipoprotein E

ACKNOWLEDGMENTS

This work was supported by grants from NIH-NIMH (MH071316, MH071533, and MH097216), and Alzheimer's Association (to P.P.), NIH 1F31AG031621 (M.E.C.), NIH 1F31MH085362 (K.A.J.), NIH 1F31MH087043 (J.V.), and a predoctoral American Heart Association (AHA) fellowship (K.M.W).

REFERENCES

Alvarez, V.A., Sabatini, B.L., 2007. Anatomical and physiological plasticity of dendritic spines. Annu. Rev. Neurosci. 30, 79–97.
Arendt, T., 2009. Synaptic degeneration in Alzheimer's disease. Acta Neuropathol. 118, 167–179.
Armstrong, D.D., 2005. Neuropathology of Rett syndrome. J. Child Neurol. 20, 747–753.
Ayhan, Y., Abazyan, B., Nomura, J., Kim, R., Ladenheim, B., Krasnova, I.N., Sawa, A., Margolis, R.L., Cadet, J.L., Mori, S., Vogel, M.W., Ross, C.A., Pletnikov, M.V., 2011. Differential effects of prenatal and postnatal expressions of mutant human DISC1 on neurobehavioral phenotypes in transgenic mice: evidence for neurodevelopmental origin of major psychiatric disorders. Mol. Psychiatry 16 (3), 293–306.
Barros, C.S., Calabrese, B., Chamero, P., Roberts, A.J., Korzus, E., Lloyd, K., Stowers, L., Mayford, M., Halpain, S., Muller, U., 2009. Impaired maturation of dendritic spines without disorganization of cortical cell layers in mice lacking NRG1/ErbB signaling in the central nervous system. Proc. Natl. Acad. Sci. USA.
Barta, P.E., Pearlson, G.D., Powers, R.E., Richards, S.S., Tune, L.E., 1990. Auditory hallucinations and smaller superior temporal gyral volume in schizophrenia. Am. J. Psychiatry 147, 1457–1462.
Belichenko, P.V., Oldfors, A., Hagberg, B., Dahlstrom, A., 1994. Rett syndrome: 3-D confocal microscopy of cortical pyramidal dendrites and afferents. Neuroreport 5, 1509–1513.
Bertram, L., Tanzi, R.E., 2008. Thirty years of Alzheimer's disease genetics: the implications of systematic meta-analyses. Nat. Rev. Neurosci. 9, 768–778.

Butler, M.G., Dasouki, M.J., Zhou, X.P., Talebizadeh, Z., Brown, M., Takahashi, T.N., Miles, J.H., Wang, C.H., Stratton, R., Pilarski, R., Eng, C., 2005. Subset of individuals with autism spectrum disorders and extreme macrocephaly associated with germline PTEN tumour suppressor gene mutations. J. Med. Genet. 42, 318–321.

Cahill, M.E., Xie, Z., Day, M., Barbolina, M.V., Miller, C.A., Weiss, C., Radulovic, J., Sweatt, J.D., Disterhoft, J.F., Surmeier, D.J., Penzes, P., 2009. Kalirin regulates cortical spine morphogenesis and disease-related behavioral phenotypes. Proc. Natl. Acad. Sci. USA 106, 13058–13063.

Campbell, D.B., Sutcliffe, J.S., Ebert, P.J., Militerni, R., Bravaccio, C., Trillo, S., Elia, M., Schneider, C., Melmed, R., Sacco, R., Persico, A.M., Levitt, P., 2006. A genetic variant that disrupts MET transcription is associated with autism. Proc. Natl. Acad. Sci. USA 103, 16834–16839.

Chao, H.T., Zoghbi, H.Y., Rosenmund, C., 2007. MeCP2 controls excitatory synaptic strength by regulating glutamatergic synapse number. Neuron 56, 58–65.

Chen, Y.J., Johnson, M.A., Lieberman, M.D., Goodchild, R.E., Schobel, S., Lewandowski, N., Rosoklija, G., Liu, R.C., Gingrich, J.A., Small, S., Moore, H., Dwork, A.J., Talmage, D.A., Role, L.W., 2008. Type III neuregulin-1 is required for normal sensorimotor gating, memory-related behaviors, and corticostriatal circuit components. J. Neurosci. 28, 6872–6883.

Chih, B., Afridi, S.K., Clark, L., Scheiffele, P., 2004. Disorder-associated mutations lead to functional inactivation of neuroligins. Hum. Mol. Genet. 13, 1471–1477.

Counts, S.E., Nadeem, M., Lad, S.P., Wuu, J., Mufson, E.J., 2006. Differential expression of synaptic proteins in the frontal and temporal cortex of elderly subjects with mild cognitive impairment. J. Neuropathol. Exp. Neurol. 65, 592–601.

Daoud, H., Bonnet-Brilhault, F., Vedrine, S., Demattei, M.V., Vourc'h, P., Bayou, N., Andres, C.R., Barthelemy, C., Laumonnier, F., Briault, S., 2009. Autism and nonsyndromic mental retardation associated with a de novo mutation in the NLGN4X gene promoter causing an increased expression level. Biol. Psychiatry 66, 906–910.

Davies, C.A., Mann, D.M., Sumpter, P.Q., Yates, P.O., 1987. A quantitative morphometric analysis of the neuronal and synaptic content of the frontal and temporal cortex in patients with Alzheimer's disease. J. Neurol. Sci. 78, 151–164.

Dean, B., Keriakous, D., Scarr, E., Thomas, E.A., 2007. Gene expression profiling in Brodmann's area 46 from subjects with schizophrenia. Aust. N. Z. J. Psychiatry 41, 308–320.

DeKosky, S.T., Scheff, S.W., 1990. Synapse loss in frontal cortex biopsies in Alzheimer's disease: correlation with cognitive severity. Ann. Neurol. 27, 457–464.

Dolen, G., Osterweil, E., Rao, B.S., Smith, G.B., Auerbach, B.D., Chattarji, S., Bear, M.F., 2007. Correction of fragile X syndrome in mice. Neuron 56, 955–962.

Dumanis, S.B., Tesoriero, J.A., Babus, L.W., Nguyen, M.T., Trotter, J.H., Ladu, M.J., Weeber, E.J., Turner, R.S., Xu, B., Rebeck, G.W., Hoe, H.S., 2009. ApoE4 decreases spine density and dendritic complexity in cortical neurons in vivo. J. Neurosci. 29, 15317–15322.

Durand, C.M., Betancur, C., Boeckers, T.M., Bockmann, J., Chaste, P., Fauchereau, F., Nygren, G., Rastam, M., Gillberg, I.C., Anckarsater, H., Sponheim, E., Goubran-Botros, H., Delorme, R., Chabane, N., Mouren-Simeoni, M.C., de Mas, P., Bieth, E., Roge, B., Heron, D., Burglen, L., Gillberg, C., Leboyer, M., Bourgeron, T., 2007. Mutations in the gene encoding the synaptic scaffolding protein SHANK3 are associated with autism spectrum disorders. Nat. Genet. 39, 25–27.

Fiala, J.C., Spacek, J., Harris, K.M., 2002. Dendritic spine pathology: cause or consequence of neurological disorders? Brain Res. Rev. 39, 29–54.

Garcia, R.A., Vasudevan, K., Buonanno, A., 2000. The neuregulin receptor ErbB-4 interacts with PDZ-containing proteins at neuronal synapses. Proc. Natl. Acad. Sci. USA 97, 3596–3601.

Geschwind, D.H., Levitt, P., 2007. Autism spectrum disorders: developmental disconnection syndromes. Curr. Opin. Neurobiol. 17, 103–111.

Glantz, L.A., Lewis, D.A., 2000. Decreased dendritic spine density on prefrontal cortical pyramidal neurons in schizophrenia. Arch. Gen. Psychiatry 57, 65–73.

Gogtay, N., 2008. Cortical brain development in schizophrenia: insights from neuroimaging studies in childhood-onset schizophrenia. Schizophr. Bull. 34, 30–36.

Grutzendler, J., Helmin, K., Tsai, J., Gan, W.B., 2007. Various dendritic abnormalities are associated with fibrillar amyloid deposits in Alzheimer's disease. Ann. N.Y. Acad. Sci. 1097, 30–39.

Guilmatre, A., Dubourg, C., Mosca, A.L., Legallic, S., Goldenberg, A., Drouin-Garraud, V., Layet,V., Rosier,A., Briault, S., Bonnet-Brilhault, F., Laumonnier, F., Odent, S., LeVacon, G., Joly-Helas, G., David,V., Bendavid, C., Pinoit, J.M., Henry, C., Impallomeni, C., Germano, E., Tortorella, G., Di Rosa, G., Barthelemy, C., Andres, C., Faivre, L., Frebourg, T., SaugierVeber, P., Campion, D., 2009. Recurrent rearrangements in synaptic and neurodevelopmental genes and shared biologic pathways in schizophrenia, autism, and mental retardation. Arch. Gen. Psychiatry 66, 947–956.
Guy, J., Gan, J., Selfridge, J., Cobb, S., Bird, A., 2007. Reversal of neurological defects in a mouse model of Rett syndrome. Science 315, 1143–1147.
Harigaya, Y., Shoji, M., Shirao, T., Hirai, S., 1996. Disappearance of actin-binding protein, drebrin, from hippocampal synapses in Alzheimer's disease. J. Neurosci. Res. 43, 87–92.
Hayashi-Takagi, A., Takaki, M., Graziane, N., Seshadri, S., Murdoch, H., Dunlop, A.J., Makino, Y., Seshadri, A.J., Ishizuka, K., Srivastava, D.P., Xie, Z., Baraban, J.M., Houslay, M.D., Tomoda, T., Brandon, N.J., Kamiya, A., Yan, Z., Penzes, P., Sawa, A. Disrupted-in-Schizophrenia 1 (DISC1) regulates spines of the glutamate synapse via Rac1. Nat. Neurosci. 13, 327–332.
He, Y., Chen, Z., Gong, G., Evans, A., 2009. Neuronal networks in Alzheimer's disease. Neuroscientist 15, 333–350.
Hill, J.J., Hashimoto, T., Lewis, D.A., 2006. Molecular mechanisms contributing to dendritic spine alterations in the prefrontal cortex of subjects with schizophrenia. Mol. Psychiatry 11, 557–566.
Holtmaat, A., Svoboda, K., 2009. Experience-dependent structural synaptic plasticity in the mammalian brain. Nat. Rev. Neurosci. 10, 647–658.
Hsiao, K., Chapman, P., Nilsen, S., Eckman, C., Harigaya, Y., Younkin, S., Yang, F., Cole, G., 1996. Correlative memory deficits, Abeta elevation, and amyloid plaques in transgenic mice. Science 274, 99–102.
Hsieh, H., Boehm, J., Sato, C., Iwatsubo, T., Tomita, T., Sisodia, S., Malinow, R., 2006. AMPAR removal underlies Abeta-induced synaptic depression and dendritic spine loss. Neuron 52, 831–843.
Hutsler, J.J., Zhang, H., 2010. Increased dendritic spine densities on cortical projection neurons in autism spectrum disorders. Brain Res. 1309, 83–94.
Ide, M., Lewis, D.A., 2010. Altered cortical CDC42 signaling pathways in schizophrenia: implications for dendritic spine deficits. Biol. Psychiatry 68 (1), 25–32.
Ikeda, M., Aleksic, B., Kinoshita, Y., Okochi, T., Kawashima, K., Kushima, I., Ito, Y., Nakamura, Y., Kishi, T., Okumura, T., Fukuo, Y., Williams, H.J., Hamshere, M.L., Ivanov, D., Inada, T., Suzuki, M., Hashimoto, R., Ujike, H., Takeda, M., Craddock, N., Kaibuchi, K., Owen, M.J., Ozaki, N., O'Donovan, M.C., Iwata, N., 2011. Genome-wide association study of schizophrenia in a Japanese population. Biol. Psychiatry 69, 472–478.
Irwin, S.A., Galvez, R., Greenough, W.T., 2000. Dendritic spine structural anomalies in fragile-X mental retardation syndrome. Cereb. Cortex 10, 1038–1044.
Irwin, S.A., Idupulapati, M., Gilbert, M.E., Harris, J.B., Chakravarti, A.B., Rogers, E.J., Crisostomo, R.A., Larsen, B.P., Mehta, A., Alcantara, C.J., Patel, B., Swain, R.A., Weiler, I.J., Oostra, B.A., Greenough, W.T., 2002. Dendritic spine and dendritic field characteristics of layer V pyramidal neurons in the visual cortex of fragile-X knockout mice. Am. J. Med. Genet. 111, 140–146.
Irwin, S.A., Patel, B., Idupulapati, M., Harris, J.B., Crisostomo, R.A., Larsen, B.P., Kooy, F., Willems, P.J., Cras, P., Kozlowski, P.B., Swain, R.A., Weiler, I.J., Greenough, W.T., 2001. Abnormal dendritic spine characteristics in the temporal and visual cortices of patients with fragile-X syndrome: a quantitative examination. Am. J. Med. Genet. 98, 161–167.
Jacobsen, J.S., Wu, C.C., Redwine, J.M., Comery, T.A., Arias, R., Bowlby, M., Martone, R., Morrison, J.H., Pangalos, M.N., Reinhart, P.H., Bloom, F.E., 2006. Early-onset behavioral and synaptic deficits in a mouse model of Alzheimer's disease. Proc. Natl. Acad. Sci. USA 103, 5161–5166.
Jamain, S., Quach, H., Betancur, C., Rastam, M., Colineaux, C., Gillberg, I.C., Soderstrom, H., Giros, B., Leboyer, M., Gillberg, C., Bourgeron, T., 2003. Mutations of the X-linked genes encoding neuroligins NLGN3 and NLGN4 are associated with autism. Nat. Genet. 34, 27–29.
Jay, V., Becker, L.E., Chan, F.W., Perry Sr., T.L., 1991. Puppet-like syndrome of Angelman: a pathologic and neurochemical study. Neurology 41, 416–422.
Ji, Y., Gong, Y., Gan, W., Beach, T., Holtzman, D.M., Wisniewski, T., 2003. Apolipoprotein E isoform-specific regulation of dendritic spine morphology in apolipoprotein E transgenic mice and Alzheimer's disease patients. Neuroscience 122, 305–315.

Kim, H.G., Kishikawa, S., Higgins, A.W., Seong, I.S., Donovan, D.J., Shen, Y., Lally, E., Weiss, L.A., Najm, J., Kutsche, K., Descartes, M., Holt, L., Braddock, S., Troxell, R., Kaplan, L., Volkmar, F., Klin, A., Tsatsanis, K., Harris, D.J., Noens, I., Pauls, D.L., Daly, M.J., MacDonald, M.E., Morton, C.C., Quade, B.J., Gusella, J.F., 2008. Disruption of neurexin 1 associated with autism spectrum disorder. Am. J. Hum. Genet. 82, 199–207.

Kirkpatrick, B., Xu, L., Cascella, N., Ozeki, Y., Sawa, A., Roberts, R.C., 2006. DISC1 immunoreactivity at the light and ultrastructural level in the human neocortex. J. Comp. Neurol. 497, 436–450.

Klein, W.L., 2006. Synaptic targeting by A beta oligomers (ADDLS) as a basis for memory loss in early Alzheimer's disease. Alzheimers Dement. 2, 43–55.

Knafo, S., Alonso-Nanclares, L., Gonzalez-Soriano, J., Merino-Serrais, P., Fernaud-Espinosa, I., Ferrer, I., DeFelipe, J., 2009. Widespread changes in dendritic spines in a model of Alzheimer's disease. Cereb. Cortex 19, 586–592.

Kolomeets, N.S., Orlovskaya, D.D., Rachmanova, V.I., Uranova, N.A., 2005. Ultrastructural alterations in hippocampal mossy fiber synapses in schizophrenia: a postmortem morphometric study. Synapse 57, 47–55.

Kolomeets, N.S., Orlovskaya, D.D., Uranova, N.A., 2007. Decreased numerical density of CA3 hippocampal mossy fiber synapses in schizophrenia. Synapse 61, 615–621.

Kristiansen, L.V., Beneyto, M., Haroutunian, V., Meador-Woodruff, J.H., 2006. Changes in NMDA receptor subunits and interacting PSD proteins in dorsolateral prefrontal and anterior cingulate cortex indicate abnormal regional expression in schizophrenia. Mol. Psychiatry 11, 737–747 705.

Kushima, I., Nakamura, Y., Aleksic, B., Ikeda, M., Ito, Y., Shiino, T., Okochi, T., Fukuo, Y., Ujike, H., Suzuki, M., Inada, T., Hashimoto, R., Takeda, M., Kaibuchi, K., Iwata, N., Ozaki, N., 2010. Resequencing and association analysis of the KALRN and EPHB1 genes and their contribution to schizophrenia susceptibility. Schizophr Bull.

Kvajo, M., McKellar, H., Arguello, P.A., Drew, L.J., Moore, H., MacDermott, A.B., Karayiorgou, M., Gogos, J.A., 2008. A mutation in mouse Disc1 that models a schizophrenia risk allele leads to specific alterations in neuronal architecture and cognition. Proc. Natl. Acad. Sci. USA 105, 7076–7081.

Kwon, C.H., Luikart, B.W., Powell, C.M., Zhou, J., Matheny, S.A., Zhang, W., Li, Y., Baker, S.J., Parada, L.F., 2006. Pten regulates neuronal arborization and social interaction in mice. Neuron 50, 377–388.

Lacor, P.N., Buniel, M.C., Furlow, P.W., Clemente, A.S., Velasco, P.T., Wood, M., Viola, K.L., Klein, W.L., 2007. Abeta oligomer-induced aberrations in synapse composition, shape, and density provide a molecular basis for loss of connectivity in Alzheimer's disease. J. Neurosci. 27, 796–807.

Lanz, T.A., Carter, D.B., Merchant, K.M., 2003. Dendritic spine loss in the hippocampus of young PDAPP and Tg2576 mice and its prevention by the ApoE2 genotype. Neurobiol. Dis. 13, 246–253.

Laumonnier, F., Bonnet-Brilhault, F., Gomot, M., Blanc, R., David, A., Moizard, M.P., Raynaud, M., Ronce, N., Lemonnier, E., Calvas, P., Laudier, B., Chelly, J., Fryns, J.P., Ropers, H.H., Hamel, B.C., Andres, C., Barthelemy, C., Moraine, C., Briault, S., 2004. X-linked mental retardation and autism are associated with a mutation in the NLGN4 gene, a member of the neuroligin family. Am. J. Hum. Genet. 74, 552 557.

Law, A.J., Weickert, C.S., Hyde, T.M., Kleinman, J.E., Harrison, P.J., 2004. Reduced spinophilin but not microtubule-associated protein 2 expression in the hippocampal formation in schizophrenia and mood disorders: molecular evidence for a pathology of dendritic spines. Am. J. Psychiatry 161, 1848–1855.

Lewis, D.A., Gonzalez-Burgos, G., 2008. Neuroplasticity of neocortical circuits in schizophrenia. Neuropsychopharmacology 33, 141–165.

Li, B., Woo, R.S., Mei, L., Malinow, R., 2007. The neuregulin-1 receptor erbB4 controls glutamatergic synapse maturation and plasticity. Neuron 54, 583–597.

Li, S., Hong, S., Shepardson, N.E., Walsh, D.M., Shankar, G.M., Selkoe, D., 2009. Soluble oligomers of amyloid Beta protein facilitate hippocampal long-term depression by disrupting neuronal glutamate uptake. Neuron 62, 788–801.

Lipska, B.K., Peters, T., Hyde, T.M., Halim, N., Horowitz, C., Mitkus, S., Weickert, C.S., Matsumoto, M., Sawa, A., Straub, R.E., Vakkalanka, R., Herman, M.M., Weinberger, D.R., Kleinman, J.E., 2006. Expression of DISC1 binding partners is reduced in schizophrenia and associated with DISC1 SNPs. Hum. Mol. Genet. 15, 1245–1258.

Liu, F., Grundke-Iqbal, I., Iqbal, K., Oda, Y., Tomizawa, K., Gong, C.X., 2005. Truncation and activation of calcineurin A by calpain I in Alzheimer disease brain. J. Biol. Chem. 280, 37755–37762.

Ma, Q.L., Yang, F., Calon, F., Ubeda, O.J., Hansen, J.E., Weisbart, R.H., Beech, W., Frautschy, S.A., Cole, G.M., 2008. p21-activated kinase-aberrant activation and translocation in Alzheimer disease pathogenesis. J. Biol. Chem. 283, 14132–14143.

Maloney, M.T., Bamburg, J.R., 2007. Cofilin-mediated neurodegeneration in Alzheimer's disease and other amyloidopathies. Mol. Neurobiol. 35, 21–44.

Masliah, E., Mallory, M., Alford, M., DeTeresa, R., Hansen, L.A., McKeel Jr., D.W., Morris, J.C., 2001. Altered expression of synaptic proteins occurs early during progression of Alzheimer's disease. Neurology 56, 127–129.

McClellan, J., King, M.C., 2010. Genetic heterogeneity in human disease. Cell 141, 210–217.

Mei, L., Xiong, W.C., 2008. Neuregulin 1 in neural development, synaptic plasticity and schizophrenia. Nat. Rev. Neurosci. 9, 437–452.

Millar, J.K., Christie, S., Porteous, D.J., 2003. Yeast two-hybrid screens implicate DISC1 in brain development and function. Biochem. Biophys. Res. Commun. 311, 1019–1025.

Mukai, J., Dhilla, A., Drew, L.J., Stark, K.L., Cao, L., MacDermott, A.B., Karayiorgou, M., Gogos, J.A., 2008. Palmitoylation-dependent neurodevelopmental deficits in a mouse model of 22q11 microdeletion. Nat. Neurosci. 11, 1302–1310.

Norris, C.M., Kadish, I., Blalock, E.M., Chen, K.C., Thibault, V., Porter, N.M., Landfield, P.W., Kraner, S.D., 2005. Calcineurin triggers reactive/inflammatory processes in astrocytes and is upregulated in aging and Alzheimer's models. J. Neurosci. 25, 4649–4658.

Pardo, C.A., Eberhart, C.G., 2007. The neurobiology of autism. Brain Pathol. 17, 434–447.

Penzes, P., Beeser, A., Chernoff, J., Schiller, M.R., Eipper, B.A., Mains, R.E., Huganir, R.L., 2003. Rapid induction of dendritic spine morphogenesis by trans-synaptic ephrinB-EphB receptor activation of the Rho-GEF kalirin. Neuron 37, 263–274.

Penzes, P., Cahill, M.E., Jones, K.A., Srivastava, D.P., 2008. Convergent CaMK and RacGEF signals control dendritic structure and function. Trends Cell. Biol. 18, 405–413.

Penzes, P., Jones, K.A., 2008. Dendritic spine dynamics – a key role for kalirin-7. Trends Neurosci. 31, 419–427.

Roussignol, G., Ango, F., Romorini, S., Tu, J.C., Sala, C., Worley, P.F., Bockaert, J., Fagni, L., 2005. Shank expression is sufficient to induce functional dendritic spine synapses in aspiny neurons. J. Neurosci. 25, 3560–3570.

Scheff, S.W., DeKosky, S.T., Price, D.A., 1990. Quantitative assessment of cortical synaptic density in Alzheimer's disease. Neurobiol. Aging 11, 29–37.

Scheff, S.W., Price, D.A., Schmitt, F.A., DeKosky, S.T., Mufson, E.J., 2007. Synaptic alterations in CA1 in mild Alzheimer disease and mild cognitive impairment. Neurology 68, 1501–1508.

Schumacher, J., Laje, G., Abou Jamra, R., Becker, T., Muhleisen, T.W., Vasilescu, C., Mattheisen, M., Herms, S., Hoffmann, P., Hillmer, A.M., Georgi, A., Herold, C., Schulze, T.G., Propping, P., Rietschel, M., McMahon, F.J., Nothen, M.M., Cichon, S., 2009. The DISC locus and schizophrenia: evidence from an association study in a central European sample and from a meta-analysis across different European populations. Hum. Mol. Genet. 18, 2719–2727.

Sebat, J., Lakshmi, B., Malhotra, D., Troge, J., Lese-Martin, C., Walsh, T., Yamrom, B., Yoon, S., Krasnitz, A., Kendall, J., Leotta, A., Pai, D., Zhang, R., Lee, Y.H., Hicks, J., Spence, S.J., Lee, A.T., Puura, K., Lehtimaki, T., Ledbetter, D., Gregersen, P.K., Bregman, J., Sutcliffe, J.S., Jobanputra, V., Chung, W., Warburton, D., King, M.C., Skuse, D., Geschwind, D.H., Gilliam, T.C., Ye, K., Wigler, M., 2007. Strong association of de novo copy number mutations with autism. Science 316, 445–449.

Selemon, L.D., Goldman-Rakic, P.S., 1999. The reduced neuropil hypothesis: a circuit based model of schizophrenia. Biol. Psychiatry 45, 17–25.

Selkoe, D.J., 2002. Alzheimer's disease is a synaptic failure. Science 298, 789–791.

Selkoe, D.J., 2008. Soluble oligomers of the amyloid beta-protein impair synaptic plasticity and behavior. Behav. Brain Res. 192, 106–113.

Shankar, G.M., Bloodgood, B.L., Townsend, M., Walsh, D.M., Selkoe, D.J., Sabatini, B.L., 2007. Natural oligomers of the Alzheimer amyloid-beta protein induce reversible synapse loss by modulating an NMDA-type glutamate receptor-dependent signaling pathway. J. Neurosci. 27, 2866–2875.

Shiota, S., Tochigi, M., Shimada, H., Ohashi, J., Kasai, K., Kato, N., Tokunaga, K., Sasaki, T., 2008. Association and interaction analyses of NRG1 and ERBB4 genes with schizophrenia in a Japanese population. J. Hum. Genet. 53, 929–935.

Silberberg, G., Darvasi, A., Pinkas-Kramarski, R., Navon, R., 2006. The involvement of ErbB4 with schizophrenia: association and expression studies. Am. J. Med. Genet. B Neuropsychiatr. Genet. 141B, 142–148.
Smith, D.L., Pozueta, J., Gong, B., Arancio, O., Shelanski, M., 2009. Reversal of long-term dendritic spine alterations in Alzheimer disease models. Proc. Natl. Acad. Sci. USA 106, 16877–16882.
Snyder, E.M., Nong, Y., Almeida, C.G., Paul, S., Moran, T., Choi, E.Y., Nairn, A.C., Salter, M.W., Lombroso, P.J., Gouras, G.K., Greengard, P., 2005. Regulation of NMDA receptor trafficking by amyloid-beta. Nat. Neurosci. 8, 1051–1058.
Spires, T.L., Meyer-Luehmann, M., Stern, E.A., McLean, P.J., Skoch, J., Nguyen, P.T., Bacskai, B.J., Hyman, B.T., 2005. Dendritic spine abnormalities in amyloid precursor protein transgenic mice demonstrated by gene transfer and intravital multiphoton microscopy. J. Neurosci. 25, 7278–7287.
St Jean, P. 2008. Genes associated with schizophrenia identified using a whole genome scan.
Stark, K.L., Xu, B., Bagchi, A., Lai, W.S., Liu, H., Hsu, R., Wan, X., Pavlidis, P., Mills, A.A., Karayiorgou, M., Gogos, J.A., 2008. Altered brain microRNA biogenesis contributes to phenotypic deficits in a 22q11-deletion mouse model. Nat. Genet. 40, 751–760.
Steen, R.G., Mull, C., McClure, R., Hamer, R.M., Lieberman, J.A., 2006. Brain volume in first-episode schizophrenia: systematic review and meta-analysis of magnetic resonance imaging studies. Br. J. Psychiatry 188, 510–518.
Sweet, R.A., Henteleff, R.A., Zhang, W., Sampson, A.R., Lewis, D.A., 2009. Reduced dendritic spine density in auditory cortex of subjects with schizophrenia. Neuropsychopharmacology 34, 374–389.
Tabuchi, K., Blundell, J., Etherton, M.R., Hammer, R.E., Liu, X., Powell, C.M., Sudhof, T.C., 2007. A neuroligin-3 mutation implicated in autism increases inhibitory synaptic transmission in mice. Science 318, 71–76.
Tackenberg, C., Ghori, A., Brandt, R., 2009. Thin, stubby or mushroom: spine pathology in Alzheimer's disease. Curr. Alzheimer Res. 6, 261–268.
Tada, T., Sheng, M., 2006. Molecular mechanisms of dendritic spine morphogenesis. Curr. Opin. Neurobiol. 16, 95–101.
Tan, H.Y., Callicott, J.H., Weinberger, D.R., 2007. Dysfunctional and compensatory prefrontal cortical systems, genes and the pathogenesis of schizophrenia. Cereb. Cortex 17 (Suppl. 1), i171–i181.
Tavazoie, S.F., Alvarez, V.A., Ridenour, D.A., Kwiatkowski, D.J., Sabatini, B.L., 2005. Regulation of neuronal morphology and function by the tumor suppressors Tsc1 and Tsc2. Nat. Neurosci. 8, 1727–1734.
Terry, R.D., Masliah, E., Salmon, D.P., Butters, N., DeTeresa, R., Hill, R., Hansen, L.A., Katzman, R., 1991. Physical basis of cognitive alterations in Alzheimer's disease: synapse loss is the major correlate of cognitive impairment. Ann. Neurol. 30, 572–580.
Tsai, J., Grutzendler, J., Duff, K., Gan, W.B., 2004. Fibrillar amyloid deposition leads to local synaptic abnormalities and breakage of neuronal branches. Nat. Neurosci. 7, 1181–1183.
Walsh, C.A., Morrow, E.M., Rubenstein, J.L., 2008a. Autism and brain development. Cell 135, 396–400.
Walsh, T., McClellan, J.M., McCarthy, S.E., Addington, A.M., Pierce, S.B., Cooper, G.M., Nord, A.S., Kusenda, M., Malhotra, D., Bhandari, A., Stray, S.M., Rippey, C.F., Roccanova, P., Makarov, V., Lakshmi, B., Findling, R.L., Sikich, L., Stromberg, T., Merriman, B., Gogtay, N., Butler, P., Eckstrand, K., Noory, L., Gochman, P., Long, R., Chen, Z., Davis, S., Baker, C., Eichler, E.E., Meltzer, P.S., Nelson, S.F., Singleton, A.B., Lee, M.K., Rapoport, J.L., King, M.C., Sebat, J., 2008b. Rare structural variants disrupt multiple genes in neurodevelopmental pathways in schizophrenia. Science 320, 539–543.
Weiss, A.P., Zalesak, M., DeWitt, I., Goff, D., Kunkel, L., Heckers, S., 2004. Impaired hippocampal function during the detection of novel words in schizophrenia. Biol. Psychiatry 55, 668–675.
Wisniewski, K.E., Segan, S.M., Miezejeski, C.M., Sersen, E.A., Rudelli, R.D., 1991. The Fra(X) syndrome: neurological, electrophysiological, and neuropathological abnormalities. Am. J. Med. Genet. 38, 476–480.
Woolfrey, K.M., Srivastava, D.P., Photowala, H., Yamashita, M., Barbolina, M.V., Cahill, M.E., Xie, Z., Jones, K.A., Quilliam, L.A., Prakriya, M., Penzes, P., 2009. Epac2 induces synapse remodeling and depression and its disease-associated forms alter spines. Nat. Neurosci. 12, 1275–1284.
Wu, H.Y., Hudry, E., Hashimoto, T., Kuchibhotla, K., Rozkalne, A., Fan, Z., Spires-Jones, T., Xie, H., Arbel-Ornath, M., Grosskreutz, C.L., Bacskai, B.J., Hyman, B.T., 2010. Amyloid beta induces the morphological neurodegenerative triad of spine loss, dendritic simplification, and neuritic dystrophies through calcineurin activation. J. Neurosci. 30, 2636–2649.

Xia, Z., Storm, D.R., 2005. The role of calmodulin as a signal integrator for synaptic plasticity. Nat. Rev. Neurosci. 6, 267–276.

Xie, Z., Cahill, M.E., Penzes, P., 2010. Kalirin loss results in cortical morphological alterations. Mol. Cell. Neurosci. 43, 81–89.

Xie, Z., Srivastava, D.P., Photowala, H., Kai, L., Cahill, M.E., Woolfrey, K.M., Shum, C.Y., Surmeier, D.J., Penzes, P., 2007. Kalirin-7 controls activity-dependent structural and functional plasticity of dendritic spines. Neuron 56, 640–656.

Yoshida, T., McCarley, R.W., Nakamura, M., Lee, K., Koo, M.S., Bouix, S., Salisbury, D.F., Morra, L., Shenton, M.E., Niznikiewicz, M.A., 2009. A prospective longitudinal volumetric MRI study of superior temporal gyrus gray matter and amygdala–hippocampal complex in chronic schizophrenia. Schizophr. Res. 113, 84–94.

Youn, H., Jeoung, M., Koo, Y., Ji, H., Markesbery, W.R., Ji, I., Ji, T.H., 2007. Kalirin is under-expressed in Alzheimer's disease hippocampus. J. Alzheimers Dis. 11, 385–397.

Zhao, L., Ma, Q.L., Calon, F., Harris-White, M.E., Yang, F., Lim, G.P., Morihara, T., Ubeda, O.J., Ambegaokar, S., Hansen, J.E., Weisbart, R.H., Teter, B., Frautschy, S.A., Cole, G.M., 2006. Role of p21-activated kinase pathway defects in the cognitive deficits of Alzheimer disease. Nat. Neurosci. 9, 234–242.

Zhou, J., Blundell, J., Ogawa, S., Kwon, C.H., Zhang, W., Sinton, C., Powell, C.M., Parada, L.F., 2009. Pharmacological inhibition of mTORC1 suppresses anatomical, cellular, and behavioral abnormalities in neural-specific Pten knock-out mice. J. Neurosci. 29, 1773–1783.

Zuo, Y., Lin, A., Chang, P., Gan, W.B., 2005. Development of long-term dendritic spine stability in diverse regions of cerebral cortex. Neuron 46, 181–189.

CHAPTER TEN

Synaptic Correlates of Aging and Cognitive Decline

Yuko Hara[1,2] and John H. Morrison[1,2,3,4]
[1]Friedman Brain Institute, Mount Sinai School of Medicine, New York, NY, USA
[2]Fishberg Department of Neuroscience and Kastor Neurobiology of Aging Laboratories, Mount Sinai School of Medicine, New York, NY, USA
[3]Department of Geriatrics and Palliative Medicine, Mount Sinai School of Medicine, New York, NY, USA
[4]Computational Neurobiology and Imaging Center, Mount Sinai School of Medicine, New York, NY, USA

1. INTRODUCTION

1.1. Aging and Cognitive Decline

Normal aging is associated with a significant decline in cognitive function in mammals. Executive function, which includes processes such as cognitive flexibility, cognitive tracking, set maintenance, divided attention, and working memory, is a cognitive domain impaired in aged humans and monkeys and is thought to be one of the first functions to decline with aging (Moore et al., 2006; Rapp and Amaral, 1989). Recognition memory, a function that is generally attributed to the hippocampus and related cortical regions, is also vulnerable to the effects of aging (Moss et al., 1988; Presty et al., 1987; Rapp and Amaral, 1989). These declines in cognitive function can occur in normal aging in the absence of neurodegenerative disorders such as Alzheimer's disease. As in humans, cognitive decline is variable in the aging population of monkeys and rodents and some aged individuals function as well as the young (Bachevalier et al., 1991; Gallagher et al., 1993; Peters et al., 1996; Presty et al., 1987; Rapp and Amaral, 1991). Thus identification of neurobiological features that underlie age-related cognitive decline has been the subject of intense investigation.

One of the major focuses of this chapter will be on the prefrontal cortex (PFC), a brain area thought to play an important role in executive functions. Area 46 of the primate dorsolateral PFC has complex connections with somatosensory, visual, visuomotor, motor, and limbic systems and plays a critical role in the mediation of working memory tasks (Funahashi et al., 1993; Goldman-Rakic, 1995; Luebke et al., 2010). Working memory is a type of memory that temporarily stores, manages, and updates information in mind in order to guide goal-directed behavior. In primates, working memory can be assessed using the spatial delayed response test (Figure 1), where subjects have to remember the spatial location of a previously cued object after a delay period (Bachevalier et al., 1991; Bartus et al., 1978; Goldman-Rakic, 1995; Rapp and Amaral, 1989; Voytko and Tinkler, 2004). Aged monkeys require significantly more trials to

The Synapse
http://dx.doi.org/10.1016/B978-0-12-418675-0.00010-9

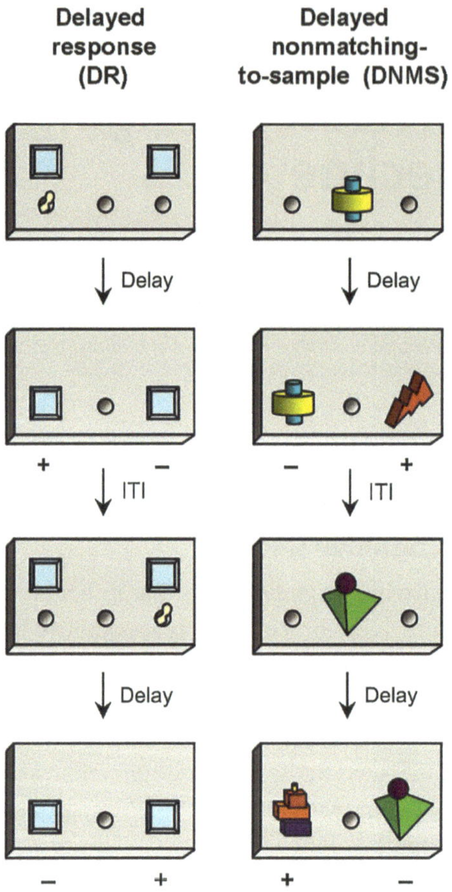

Figure 1 Schematic diagrams illustrating the delayed response (DR) test and the delayed nonmatching-to-Sample (DNMS) test. (Left) DR is a spatial working memory test in which the subject is rewarded for remembering the spatial location of a cued object, under which a small food item is hidden (e.g. peanut). An opaque screen hides the stimulus tray from view during the entire duration of the delay interval. (Right) In the DNMS test of visual recognition memory, the subject is presented with an object, and after a delay interval, the subject is rewarded for choosing a novel object when presented together with the originally displayed object. +, rewarded stimulus; –, unrewarded stimulus; ITI, intertrial interval.

acquire this task, and they perform disproportionately worse at longer delay intervals compared to young monkeys (Bachevalier et al., 1991; Bartus et al., 1978; Rapp and Amaral, 1989). Lesion studies have shown that area 46 is critical for performance on delayed response (Funahashi et al., 1993), and many neurons within this region show persistent spatially tuned firing during the delay period (Funahashi et al., 1989; Goldman-Rakic, 1995). Thus, the "memory field" generated by these area 46 neurons is thought to be the cellular basis for working memory (Funahashi et al., 1989; Goldman-Rakic, 1995).

In the rat, infralimbic and prelimbic PFC subregions are most likely the functional homolog of the primate dorsolateral PFC and are also critical for working memory (Brown and Bowman, 2002; Fritts et al., 1998; Kesner et al., 1996).

The medial temporal lobe system including the hippocampus and related cortical regions will be the other main focus of this chapter. These structures play a critical role in learning and memory and their functions decline with aging (Barnes and McNaughton, 1985; Corkin, 1984; Rapp and Amaral, 1991; Zola-Morgan et al., 1986). Humans with medial temporal lobe lesions display severe memory impairment (Corkin, 1984; Zola-Morgan et al., 1986), and the pattern of memory deficits observed in aged monkeys resembles that of monkeys with medial temporal lobe damage (Rapp and Amaral, 1991). Delayed nonmatching-to-sample (DNMS) test is a well-characterized visual recognition memory test that is sensitive to the effects of aging and is commonly used in nonhuman primates (Moss et al., 1988; Presty et al., 1987; Rapp and Amaral, 1991). In this test, a sample object is placed over a food reward and when the subject retrieves the reward, an opaque screen is lowered to impose the retention interval (Figure 1). After the delay interval, the subject is rewarded for picking a novel object when presented together with the originally displayed object. Aged monkeys require more trials to acquire the DNMS task and score lower on the delay portion of the test (Moss et al., 1988; Presty et al., 1987; Shamy et al., 2006). In rodents, spatial memory can be tested using the Morris water maze, which is very sensitive in detecting age-related effects and can also identify aged individuals that are unimpaired and perform equally to young adults (Gallagher et al., 1993; Morris, 1984).

This chapter will review many studies that have sought to identify the neurobiological measures within the PFC and medial temporal lobe structures that may underlie age-related cognitive decline.

2.1. METHODS USED FOR QUANTITATIVE ANALYSES OF STRUCTURAL AND IMMUNOCYTOCHEMICAL DATA

There is a long history of attempts at determining the morphological signatures of normal aging. Before the advent of stereological methods, measures such as neuronal number and synapse number were often biased. The disector method, commonly used now to accurately assess synapse density in an unbiased manner, is described in Section 2.1.1.

While some of the first attempts at examining the molecular signatures of aging and cognitive decline assessed gross changes in total protein levels or receptor binding sites, more recent work has focused on changes within subfields and laminae of cortical and hippocampal regions. Furthermore, electron microscopic immunocytochemistry enables examination of protein distributions within individual synapses. Our lab has developed a custom-designed software named SynBin that assigns immunogold particles into synaptic compartments and allows quantification of the number and distribution of immunogold particles for comparison across groups. This software and its applications are described in detail in Section 2.1.2.

2.1.1. The Disector Method

Unbiased stereological methods are critical in accurately assessing densities of objects, such as neurons and synapses, which are variable in size, shape, and orientation. Larger objects and those with complex shapes have a higher likelihood of being transected during sectioning, and thus are relatively more likely to appear in 2D micrographs than smaller objects. A long object can be transected many more times if sectioned perpendicular to the long axis than parallel to it. All these variables result in biases when trying to determine the number or density of objects in 3D space using 2D images.

The development of the disector method has made it possible to circumvent these issues (de Groot and Bierman, 1986; Geinisman et al., 1995; Sterio, 1984). A disector is a three-dimensional stereological probe that consists of two parallel sections that can be either two serial sections (physical disector) or two focal planes within a thicker tissue slice (optical disector). The disector height is chosen based on the object of interest and its size. For example, serial ultrathin sections ranging from 40 to 90 nm in thickness are typically used for assessing synapse density, as shown in Figure 2. The postsynaptic density (PSD) is a protein-rich specialization lying adjacent to the postsynaptic membrane

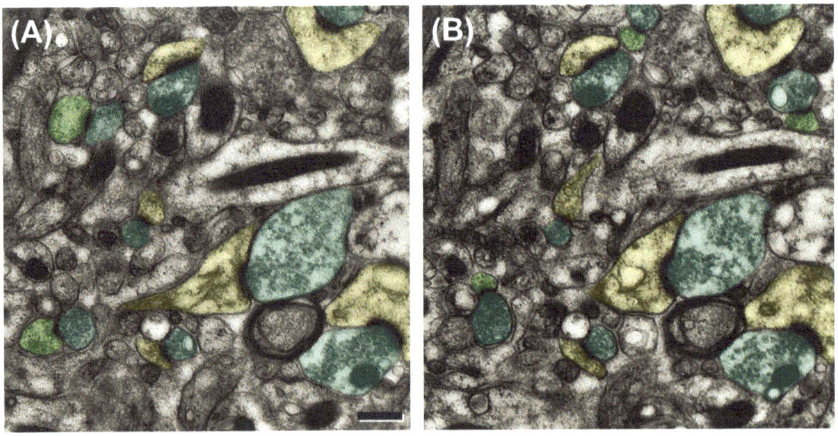

Figure 2 Serial electron micrographs illustrating the disector method for obtaining axospinous synapse density. (A, B) One section is used as the "reference" and the adjacent section is used as the "look-up". The postsynaptic density (PSD) is used as the counting unit. Only PSDs that are present in the "reference" but not the "look-up" are counted and divided by the disector volume. This procedure is repeated by switching the "reference" and "look-up" sections. Yellow, spines with PSDs that appear in both A and B; green, spines with PSDs that appear in A, but not B (or in B, but not A); light blue, axonal boutons. Scale bar, 500 nm.

and can be identified readily by electron microscopy due to its dark gray appearance. The PSD, therefore, can be used as the counting unit for determining the density of excitatory-type synapses. In a pair of serial sections, one section is used as the "reference" and the adjacent section is used as the "look-up". Only PSDs appearing in the "reference" section and not the "look-up" are counted and used to obtain the density measure. The logic behind this is that when tissue is sectioned serially, each object can only be counted once, at its leading edge. If a PSD is present in one section and not in the adjacent section, then the PSD is being transected at its edge. By this method, objects of any size, shape, or orientation will be counted in proportion to its density. These counts are then divided by the volume (counting frame area × disector height) in order to obtain the numerical density per unit volume.

The same logic can be applied for determining neuron density or density of other structures. For neuron density measurements, thicker serial sections or optical slices can be used and the neuronal nucleus is often used as the counting unit. By performing disector analyses for synapses and neurons (the double-disector approach), one can compute the number of synapses per neuron (Geinisman et al., 1992a; Newton et al., 2008). In addition, the disector method for determining synapse density can be used in conjunction with the stereological Cavalieri method. The Cavalieri principle allows estimation of a brain region's volume by using a series of optical or physical slices. After selecting a random starting point before the appearance of the brain structure, every *n*th section is used to measure the structure's cross-sectional area and from these measures, the total volume can be estimated. By obtaining the total volume and synapse density, the total synapse number within the brain region can be calculated (Geinisman et al., 1996). This method becomes relevant when the volume of a brain region is expected to be different across groups. For example, if an experimental manipulation shrinks a brain region, an increase in synapse density may be observed without a change in the total synapse number.

2.1.2. Binning of Immunogold Particles into Postsynaptic Bins (SynBin)

The SynBin software was originally designed for examining synaptic and subcellular distributions of neurotransmitter receptors and associated proteins. For example, not all ionotropic glutamate receptors (discussed in Section 2.4) within a dendritic spine are functional as only those localized on or near the synaptic membrane can bind to glutamate released from presynaptic terminals. Therefore, receptors located at the synaptic membrane are likely to be directly involved in neurotransmission, while those in the cytoplasm may reflect a newly synthesized reserve pool or those that have been internalized and awaiting to be recycled. Assembled receptors may be laterally transported along the dendritic membrane to the spine synapse or exocytosed from endosomes to perisynaptic sites (Carroll and Zukin, 2002; Henley et al., 2011). Thus, receptors located on perisynaptic membranes may indicate a pool that is deliverable to the synapse and may be in transit to and from the synapse.

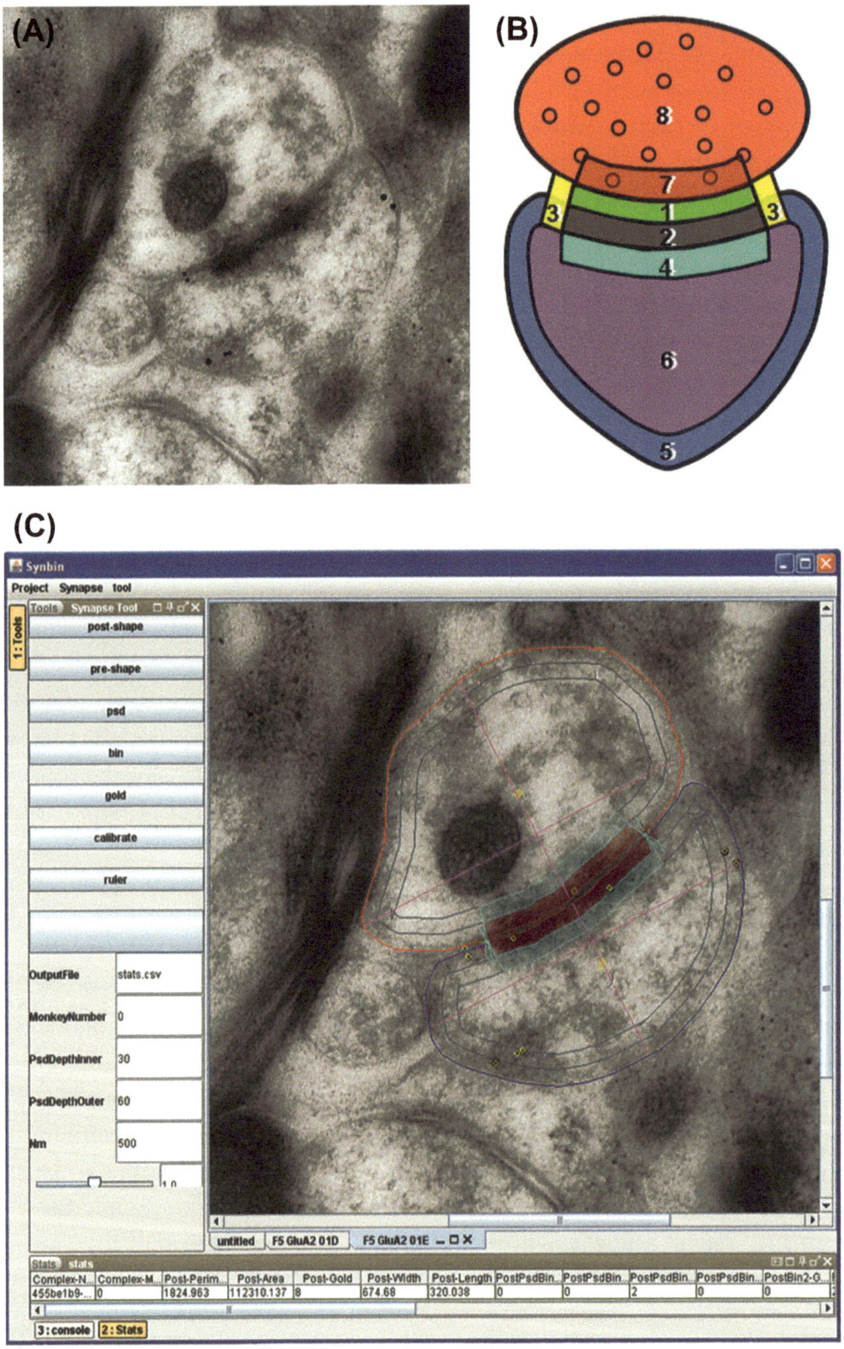

Figure 3 Illustration of the SynBin Software used for immunogold analysis. (A) An electron micrograph of a hippocampal axospinous synapse immunolabeled with the GluA2 subunit of the AMPA receptor. (B) A schematic diagram demonstrating eight bins to which immunogold particles are assigned.

While most bins in SynBin are customizable, pre- and postsynaptic bins are generally designated based on the proximity of the gold particle to the plasma membrane and the PSD, as shown in Figure 3. Immunogold localization has a theoretical resolution of 15–25 nm in radius when accounting for the size of the gold particle (typically 10 nm) and the length of primary and secondary antibodies, combined. Thus 30 nm is used as a conservative bin width. The synaptic cleft bin includes gold particles located between the presynaptic and postsynaptic membrane and directly apposing the PSD. The PSD bin includes an area inside the PSD and within 30 nm deep from the postsynaptic membrane. The subsynaptic bin is the region directly underneath the PSD and 30–60 nm from the postsynaptic membrane. This bin may represent an area in which receptors lie in reserve to be exocytosed, or those that have been internalized from the synapse. Additionally, this region may be where receptors interact with downstream signaling molecules and/or scaffolding proteins. The perisynaptic bins are located 0–30 nm lateral to the PSD and labeling in this region may represent either synaptic proteins or those that are extrasynaptic but readily deliverable to the synapse. The plasmalemmal bin includes gold particles within 30 nm from the plasma membrane, but outside the synaptic and perisynaptic bins. Proteins found in this bin are likely being trafficked to and from the synapse, laterally along the plasma membrane. The cytoplasmic bin includes all gold particles in the spine head core at least 30 nm away from the plasma membrane and outside the subsynaptic bin. Cytoplasmic labeling most likely reflects proteins that have been newly synthesized, internalized, or those waiting to be recycled or degraded.

The most current version of SynBin (shown above in Figure 3) also allows measurement of PSD length and sizes of the dendritic spine and axon terminal (cross-sectional area, perimeter, and diameter). These measures are important as spines come in different shapes and sizes, as seen in Figure 4. Thin spines have small PSDs, are highly motile and plastic, and have a high turnover rate (Holtmaat et al., 2005; Kasai et al., 2003). In contrast, large mushroom spines mediate strong synaptic currents and have a high degree of structural stability (Geinisman et al., 1992b, 1993; Kasai et al., 2003; Nusser et al., 1998). When examining molecular profiles across groups, it is critical to take the spine and synapse size into account in order to compare similar classes of spines across

Bin 1 (green) is the synaptic cleft bin located between the presynaptic and postsynaptic membrane and directly apposing the PSD. Bin 2 (gray) is the PSD bin, which includes an area inside the PSD and within 30 nm deep from the postsynaptic membrane. Bin 3 (yellow) is the perisynaptic bin, located 0–30 nm lateral to the PSD. Bin 4 (light blue) represents the subsynaptic PSD bin, directly underneath the PSD and 30–60 nm from the postsynaptic membrane. Bin 5 (blue) is the plasmalemmal bin, which includes gold particles within 30 nm from the plasma membrane, but outside of the synaptic and perisynaptic bins. Bin 6 (purple) is the cytoplasmic bin located in the spine head core at least 30 nm away from the plasma membrane and outside of the subsynaptic bin. Bin 7 (red) represents the presynaptic active zone, 30 nm from the presynaptic membrane. Bin 8 (light red) represents presynaptic labeling outside of the active zone. (C) A demonstration of the SynBin analysis (version 1.6.7).

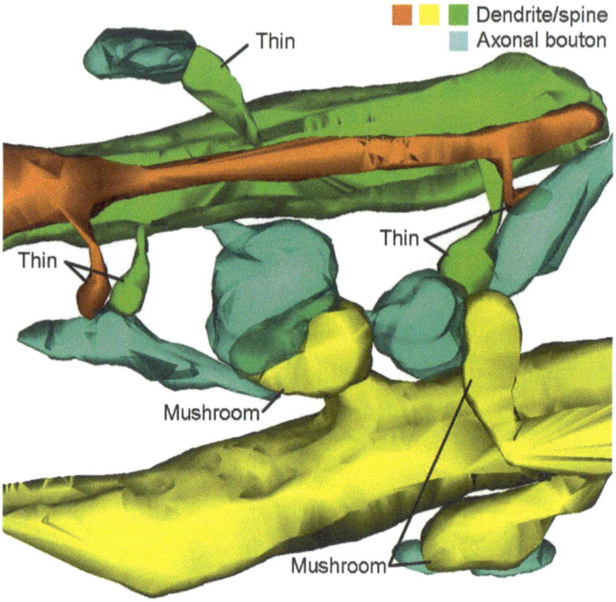

Figure 4 Three-dimensional reconstructions of dendrites and axonal boutons illustrating the wide range of sizes and shapes of dendritic spines. Sixteen serial sections from the monkey dentate gyrus were reconstructed using the free software developed by Dr Kristen Harris and her laboratory. *Reconstruct, version 1.1.0.0;* http://synapses.clm.utexas.edu.

groups. The most accurate way of assessing synapse size involves using serial sections so that individual spines can be followed through the series to obtain estimates of PSD area (sum of PSD lengths across all sections × section thickness). The size indices will allow conversion of immunogold particle labeling into density measures. Without these size measures, effects of age (or other experimental conditions) may be obscured, particularly if molecular changes are only occurring in a specific class of synapses.

2.2. THE EFFECTS OF AGING ON THE PREFRONTAL CORTEX

For many years, it was assumed that the decline in PFC-dependent cognition with normal aging was a consequence of neuronal loss, but more recent studies using unbiased stereological techniques have shown that there is no global neuronal loss in monkey neocortex with aging (Hof et al., 2000; Peters et al., 1994, 1998a; Peters and Sethares, 2002). One exception is area 8a of the PFC, where a 32% loss of neurons has been shown in aged compared with young monkeys (Smith et al., 2004). Area 46 of the monkey dorsolateral PFC (Figure 5) is a brain area associated with working memory (Funahashi et al., 1993). The volume of area 46 decreases with aging (Shamy et al., 2010); however, the total number of neurons does not change with age and fails to correlate with cognitive performance (Peters et al., 1994; Smith et al., 2004).

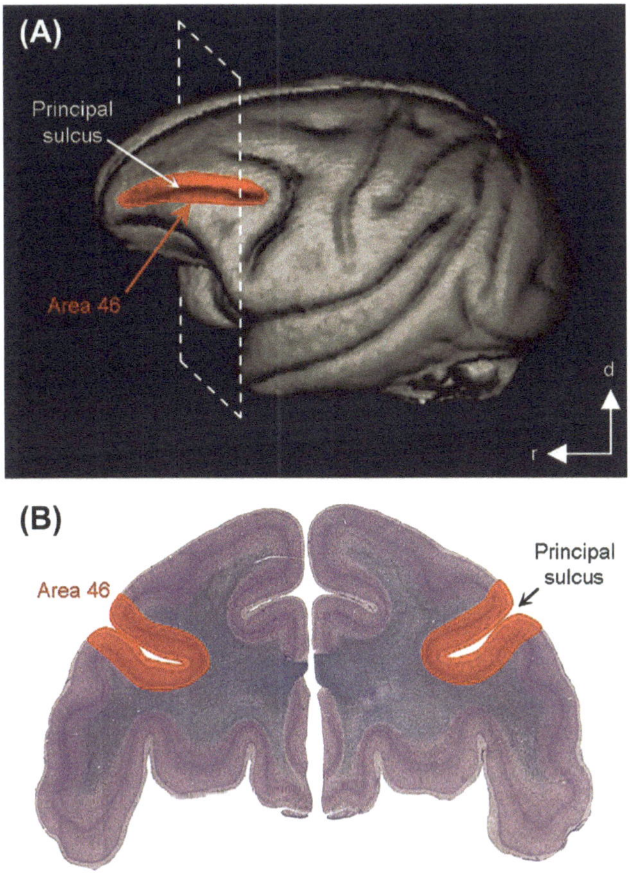

Figure 5 Area 46 of the monkey dorsolateral PFC. (A) An MRI image showing the lateral view of a monkey brain. Area 46 (shaded in red) lies in the dorsolateral PFC and extends over the upper and lower banks of the principal sulcus. White dotted lines indicate the approximate level where the coronal section in B corresponds to d, dorsal; r, rostral. (B) A Nissl-stained coronal section of the monkey PFC. Area 46 is highlighted in red. The original MRI image showing 3D surface renderings of the rhesus monkey brain was generously provided by Dr Jul Lea Shamy. The original coronal monkey section was processed and imaged by Mr William G. M. Janssen.

Recent effort has focused on investigating the density and structure of synapses. Dendritic spines, small protrusions from neuronal dendrites, are the major recipients of excitatory input to cortical neurons and are key sites for synaptic plasticity. One of the earliest studies in primates examining the effects of normal aging on PFC synapses explored area 9 of the dorsolateral PFC using Golgi staining (Cupp and Uemura, 1980). The Golgi method stains a small fraction of neurons at random and the mechanism by which this occurs is largely unknown. Although the Golgi method stains the entire neuronal soma and dendritic extent, this method is not reliable for quantitative assessments. Using this method, Cupp and Uemura reported that dendrites of PFC area 9 neurons continue to grow between young adult and middle-aged, but very old monkeys exhibit significantly shorter dendritic length, fewer numbers of dendritic branches, and lower spine densities (Cupp and Uemura, 1980). Aged monkeys also showed lower synapse densities in area 9 compared to young monkeys (Uemura, 1980), although this study was performed before the advent of unbiased methods (Sterio, 1984). In contrast, in human area 9, an unbiased stereological study showed maintenance of synaptic density and size in lamina III and V across a wide age range (20–89 years old) (Scheff et al., 2001). Unbiased stereological methods are critical in accurately assessing densities of objects, such as synapses, that are variable in size, shape, and orientation. These methods are described in detail in Section 2.1.1.

More recently, area 46 of the dorsolateral PFC has become the primary focus with respect to synaptic changes in the context of aging and cognitive decline. A study employing the unbiased disector method revealed that in layer I of monkey area 46, aging results in a 30–60% decrease in synapse density, and this decrease is accompanied by a reduction in the number of apical dendritic branches (Peters et al., 1998b). Despite these alterations in dendrites and dendritic spines, unmyelinated axons and axon terminals in layer I show no age-related changes (Peters et al., 1998b). Similar decreases in synapse density are also observed in layers II/III of monkey area 46 (Dumitriu et al., 2010; Peters et al., 2008), and these changes are discussed in more detail, in relation to cognitive function, in Section 2.2.1.

The rat ventral medial PFC, including the infralimbic and prelimbic subregions, is most likely the functional homolog of the primate dorsolateral PFC (Brown and Bowman, 2002). Rats with PFC lesions that include the prelimbic and/or infralimbic area have significant working memory deficits (Fritts et al., 1998; Kesner et al., 1996). Aging does not result in significant changes in the total volume or neuronal number in the rat ventral medial PFC (Stranahan et al., 2011; Yates et al., 2008). Furthermore, the apical dendritic length and branch numbers of rat prelimbic PFC pyramidal neurons remain stable with aging (Bloss et al., 2010). In the absence of gross anatomical or neuronal changes, aged rats compared to young adults have significantly lower density of dendritic spines on prelimbic PFC pyramidal neurons, and in particular, the small thin spines are the most vulnerable to aging (Bloss et al., 2011).

2.2.1. Synaptic Correlates of Prefrontal Cortical Aging and Cognitive Decline

Rhesus monkeys provide a valuable model for studying the neurobiological bases of cognitive aging because they are vulnerable to age-related cognitive decline in a manner similar to humans. In the last few decades, quantitative neuroanatomical studies in behaviorally characterized young and aged monkeys have emerged in search of synaptic parameters that are coupled to age-related cognitive decline. One of the first such studies revealed that the loss of synapses in PFC area 46 layer I correlates with measures of cognitive decline, including the acquisition rate and memory scores for the DNMS test of recognition memory (Peters et al., 1998b).

In addition to layer I, there is a marked 30% decrease in synapses from age 5 to 30 years (young to old age) in layers II/III of PFC area 46, and both asymmetric (excitatory) and symmetric (inhibitory) synapses are lost at the same rate (Peters et al., 2008). Strong correlations have been found between layers II/III asymmetric synapse density and cognitive scores including task acquisition rate for a memory test and a cognitive impairment index (Peters et al., 2008). This age-related decrease in synapse density was confirmed using unbiased methods; an age-related loss of dendritic spines averaging 33% was observed in monkey area 46 layer III pyramidal neurons (Dumitriu et al., 2010). Furthermore, aging was associated with a 46% loss of thin spines and an overall shift in spine size distribution toward larger spines (Dumitriu et al., 2010), as shown in Figure 6. The shift in spine size is of significance because spine morphology can predict spine stability and synaptic strength. Small spines are relatively more motile and plastic and form weaker synapses (Holtmaat et al., 2005; Kasai et al., 2003), whereas larger spines contain more glutamate amino-3-hydroxy-5-methyl-4-isoxazolepropionate (AMPA) receptors, form stronger synapses, and persist for weeks to months, if not longer (Matsuzaki et al., 2001; Nusser et al., 1998). In monkeys, faster memory task acquisition was strongly correlated with higher axospinous synapse density, increased thin spine density, and smaller volume of thin spines on area 46 layer III neurons (Dumitriu et al., 2010). Indeed, the average size of these thin spines showed the tightest correlation with task acquisition rate (Dumitriu et al., 2010), as shown in Figure 6. Thus in monkey area 46, the class of spines selectively vulnerable to aging and essential for learning a new task appears to be the smallest of the thin spines (Dumitriu et al., 2010).

Presynaptic changes with aging have also been observed in monkey area 46 layers II/III. Axon terminals forming axosomatic and axodendritic synapses become larger with age (Soghomonian et al., 2010). In axosomatic axon terminals, the number of synaptic vesicles increases with age and correlates with cognitive impairment (Soghomonian et al., 2010). Therefore, enhanced axosomatic GABAergic input to pyramidal neurons in the aged monkey area 46 may be another neuronal mechanism contributing to decreased cognitive performance. In contrast to area 46 layer II/III pyramidal neurons that make corticocortical connections, layer V neurons contribute to the cortico-striatal-thalamic-cortical circuitry (Barbas and Rempel-Clower, 1997). In area 46 layer V, there is an age-related decrease in synapses by 20% and this reduction is due entirely to a loss of asymmetric synapses (Peters et al., 2008). In contrast to findings in layers II/III, synapse density in layer V does not correlate with cognitive measures in monkeys (Peters et al., 2008). Thus, within PFC area 46, it is the layer II/III corticocortical pyramidal neurons that are thought to be crucial for cognitive capacities vulnerable to the effects of aging (Luebke et al., 2010; Luebke and Chang, 2007).

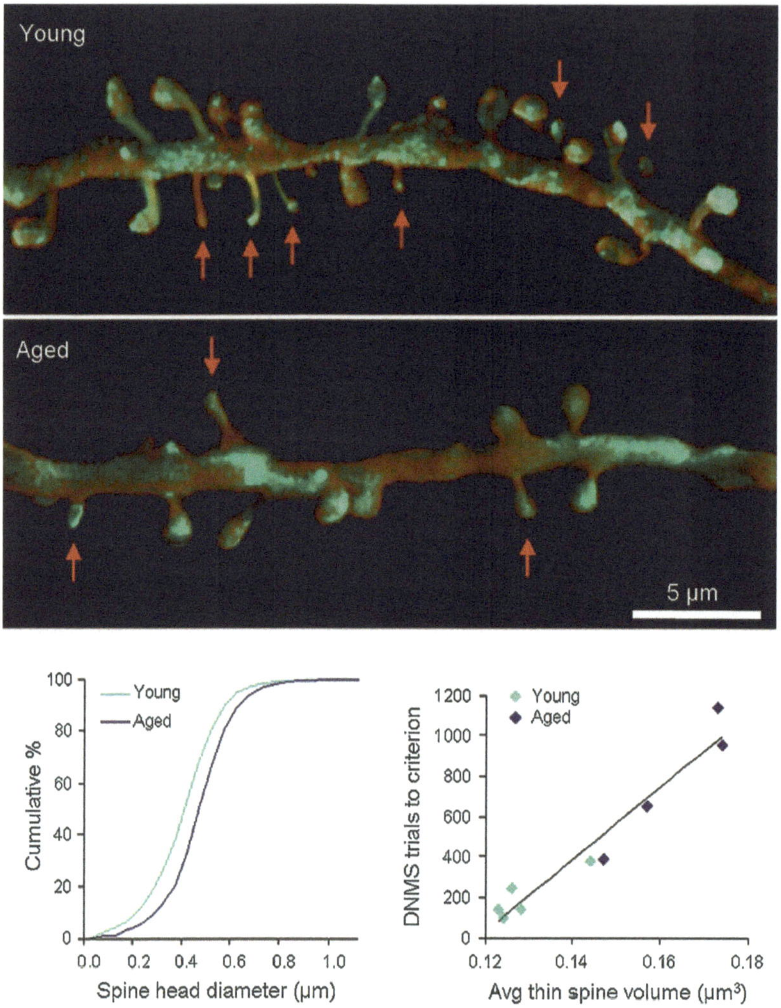

Figure 6 Age-related loss of the smallest class of spines in monkey dorsolateral PFC (area 46) correlates with task acquisition rate on the DNMS test. (Top) Apical dendritic segments of area 46 layer III pyramidal neurons in young and aged monkeys. Young monkeys display a higher density of thin spines (red arrows) than aged monkeys, while the densities of large mushroom spines are comparable across age. The thin spines that remain in aged monkeys are larger in volume than those in the young. (Bottom left) A cumulative frequency plot of dendritic spine head diameter by age groups. Aging results in a shift in the spine size distribution toward larger spines. (Bottom right) A scatter plot showing a positive correlation between the average thin spine volume and DNMS acquisition (number of trials to reach criterion). Individuals with the smallest thin spines learned the DNMS task the fastest. Pearson correlation: $n=9$, $r=0.97$, $p=0.1\times10^{-6}$. The original confocal microscopic images of dendritic segments were obtained by Mr William G. M. Janssen and were subsequently reconstructed and visualized by theVisualMD. *Bottom graphs are replotted from data presented in Dumitriu et al. (2010).*

Age-related synaptic changes in monkeys also occur outside the PFC region. Studies examining the effects of aging on dendritic trees and spines of layer III pyramidal neurons forming long corticocortical projections from the superior temporal cortex to area 46 have reported a 31% decrease in apical dendritic length, a 43% loss of spines from apical dendrites, and a 27% loss of spines from the basal dendrites (Duan et al., 2003; Kabaso et al., 2009). In the primary visual cortex (layer I of area 17), there is also an age-related decrease in synapse density as determined using unbiased methods, but the loss of synapses does not correlate with cognitive impairment (Peters et al., 2001). Together, multiple lines of evidence point to the PFC synapses mediating corticocortical connections as important for cognitive function.

2.3. THE EFFECTS OF AGING ON THE HIPPOCAMPUS

The medial temporal lobe contains anatomically and functionally related structures that include the hippocampus (CA fields, dentate gyrus (DG), and the subiculum) and adjacent perirhinal cortex, entorhinal cortex (EC), and parahippocampal cortex. Medial temporal lobe structures are important for declarative memory (or explicit memory), which refers to memory for facts, events, and knowledge. Humans with medial temporal lobe lesions exhibit severe memory impairment (Corkin, 1984; Zola-Morgan et al., 1986), and the pattern of memory deficits observed in aged monkeys resembles that of monkeys with medial temporal lobe damage (Rapp and Amaral, 1991). These observations have led to many studies in search of structures or circuits that are selectively vulnerable to age-related decline.

One of the most vulnerable circuits in normal aging is thought to be the perforant path, which originates in layer II of the EC and terminates in the DG and CA3 of the hippocampus (Morrison and Hof, 1997; Witter et al., 1989; Yassa et al., 2010). The EC receives converging inputs from the polysensory associational regions of the cortex, and funnels the highly processed neocortical information to the hippocampus via this path (Squire and Zola-Morgan, 1988). This path constitutes the first portion of the hippocampal trisynaptic circuit. The second part consists of the DG granule cell projection to CA3 via mossy fibers; and the third portion is comprised by pyramidal cells of CA3 which send their axons to CA1 via the Schaffer collaterals (Rosene and Van Hoesen, 1987; Spruston and McBain, 2007). CA1 neurons then project to the subiculum, which in turn projects out of the hippocampus, back to the EC (Rosene and Van Hoesen, 1987; Spruston and McBain, 2007). The major inputs and outputs within these hippocampal regions are illustrated in Figure 7.

In studies using unbiased stereological methods, no age-dependent differences have been observed in neuronal number or density in the subiculum, CA1, CA2, CA3, DG, and EC in rhesus monkeys (Amaral, 1993; Gazzaley et al., 1997; Keuker et al., 2003; Merrill et al., 2000; Peters et al., 1996; Rapp, 1995; West et al., 1993). Although

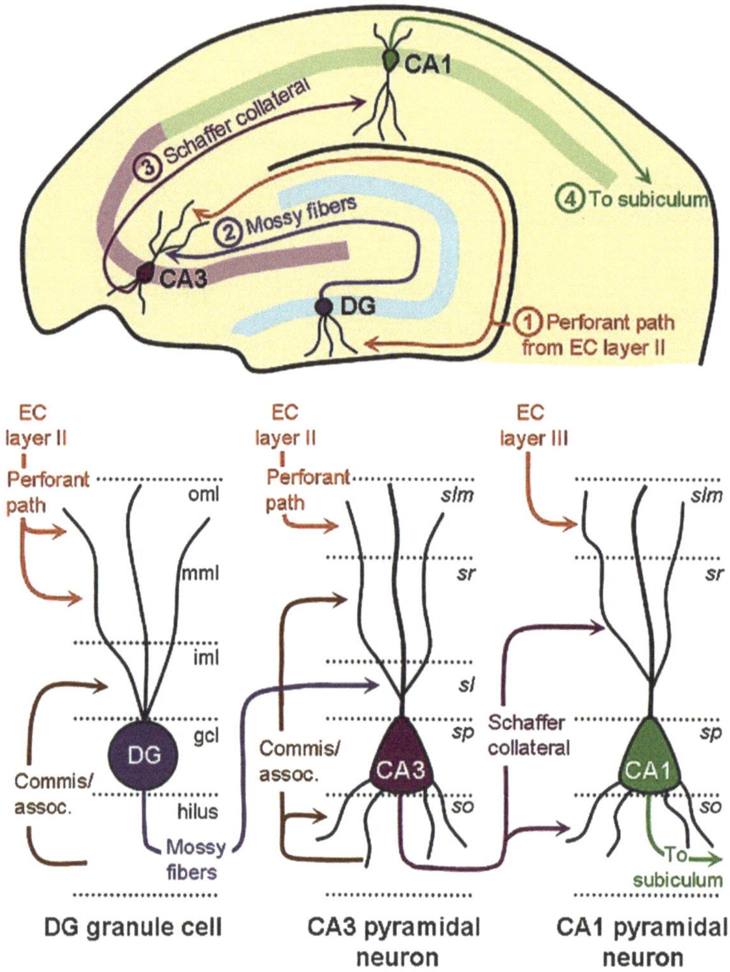

Figure 7 Schematic diagrams illustrating the hippocampal circuit and the major inputs to and outputs from the DG, CA3, and CA1. The perforant path from the EC layer II projects to the DG middle/outer molecular layers and the CA3 stratum lacunosum/moleculare. In rodents, the medial and lateral perforant paths project to the middle and outer molecular layer, respectively. In the monkey brain, there is no clear distinction between the middle and outer portions, and therefore, the outer two-thirds of the molecular layer are collectively called the outer molecular layer. The DG inner molecular layer receives commissural/associational input from the mossy cells in the DG hilus. The DG granule cells project to the CA3 stratum lucidum via the mossy fiber tract. The CA3 stratum radiatum and oriens receive axon collaterals from CA3 pyramidal neurons. The CA3 pyramidal neurons project to the CA1 stratum radiatum and oriens via the Schaffer collaterals. The CA1 stratum lacunosum/moleculare receives input from the EC layer III. Abbreviations: commis/assoc., commissural/associational input; DG, dentate gyrus; EC, entorhinal cortex; gcl, granule cell layer; iml, inner molecular layer; mml, middle molecular layer; oml, outer molecular layer; sl, stratum lucidum; slm, stratum lacunosum/moleculare; so, stratum oriens; sp, stratum pyramidale; sr, stratum radiatum.

based on a small number of monkeys, there is no evidence for a relationship between neuronal numbers in these regions and recognition memory scores (Rapp, 1995; West et al., 1993).

In humans, however, some studies have reported changes in neuronal numbers with aging. For example, West used stereological techniques to examine the hippocampus of 32 male human subjects spanning 13–85 years old and reported an age-related loss of neurons in the DG hilus and the subiculum (West, 1993). Simic et al. (1997) performed a similar quantitative analysis in 18 human subjects ranging from 16 to 99 years old and found an age-associated decrease in neuronal numbers in the subiculum and CA1. One caveat to aging studies in humans, however, is that there is potential inclusion of undiagnosed Alzheimer's disease patients. Even in mild stages of Alzheimer's disease, neuronal loss is observed in the EC (35% overall, 50% in layer II) and CA1 (46%), while no significant decrease in neuronal number is observed in healthy nondemented, aged individuals (Price et al., 2001). Thus in the human hippocampus, there is no convincing evidence for a significant loss of neurons during the normal, nonpathological, aging process (Peters et al., 1996).

In rhesus monkeys, the DG molecular layer consists of the outer and inner molecular layers, which receive perforant path input from the EC (Witter et al., 1989), and intrinsic projections from the DG hilus (Kondo et al., 2008), respectively. In contrast to the age-related changes observed in PFC synapses (Dumitriu et al., 2010; Peters et al., 2008), axospinous synapse density and PSD length in the monkey DG molecular layer (both outer and inner) are unchanged with aging and fail to correlate with DNMS task acquisition or recognition memory scores (Hara et al., 2012a; Tigges et al., 1996). Aged monkeys, however, display lower axodendritic synapse density in the DG outer molecular layer, but because axospinous synapses predominate, this age-related decrease amounts to a mere 3% reduction in total synapses (Tigges et al., 1995).

The subiculum is another hippocampal brain region where synaptic changes in the context of normal aging have been examined in monkeys. Apical dendrites continue to grow between young adulthood (7 years) to middle age (20 years old), but very old monkeys (27–28 years old) display markedly shorter dendritic length and branches in both apical and basal dendrites (Uemura, 1985a). Although no significant differences have been observed from young adulthood to middle age, subiculum pyramidal neurons in old subjects exhibit significantly lower spine and synapse density compared to young (by 19%) and middle-aged (by 13%) rhesus monkeys (Uemura, 1985b).

In the rat hippocampus, however, synapse density is preserved in some regions while decreased in others. The unbiased physical disector method revealed significant age-related losses of axospinous synapses in the DG middle (by 24%) and inner molecular layers (by 24%) (Geinisman et al., 1992a), which receive medial perforant path input and intrinsic projections from the DG hilus, respectively. In the CA3, synapses are lost with age in the *stratum radiatum* (de Groot and Bierman, 1987) and *stratum lacunosum moleculare* (Adams et al., 2010). In the rat CA1, synapse density is preserved with aging (Geinisman et al., 2004); however, changes in the morphological characteristics of synapses have been observed (Nicholson et al., 2004), as discussed in Section 2.3.1.

2.3.1. Perforated Synapses and Memory Performance

Perforated synapses are morphologically characterized by a discontinuity in the PSD. There are three different subtypes of perforated synapses based on the morphology of the PSD: a fenestrated synapse has a hole in the PSD, a horseshoe-shaped synapse has a U-shaped PSD, and a segmented synapse has two or more completely separate PSD plates (Geinisman et al., 1992b). An example of each subtype is shown in Figure 8.

Perforated synapses are typically on large mushroom spines, are implicated in memory-related plasticity, and are proposed to be a structural correlate of synaptic efficacy (Geinisman et al., 1986a, 1991; Greenough et al., 1978; Peters and Kaiserman-Abramof, 1969; Toni et al., 2001). Conditions that increase perforated synapses include complex environments, long-term potentiation (LTP), and hippocampal kindling (Geinisman et al., 1991, 1992c; Greenough et al., 1978). The segmented subtype has received particular attention due to its selective increase after LTP induction (Geinisman et al., 1992b, 1993), its large PSD size compared to other perforated synapse subtypes, and its exceptionally high level of AMPA receptors (Ganeshina et al., 2004b).

In the rat, both perforated and nonperforated axospinous synapse numbers are decreased with age in the DG molecular layer (Geinisman et al., 1992a, 1995), but only the loss of perforated synapses in the subregion receiving perforant path input correlates with the degree of hippocampus-dependent spatial memory impairment (Geinisman et al., 1986a,b). While synapse density remains stable with aging in the CA1, aged rats with spatial learning impairment have significantly smaller PSD area of all three subtypes of perforated synapses compared to both young and aged unimpaired rats (Nicholson et al., 2004). With regards to the segmented subtype, aged unimpaired rats had smaller PSD size than the young, but larger PSD size compared to the aged impaired rats (Nicholson et al., 2004). In the monkey DG outer molecular layer, total perforated synapse density is unaltered with normal aging, but it is decreased after menopause and correlates with the decline in recognition memory accuracy (Hara et al., 2012a).

In contrast to the PFC, where the highly plastic thin spines are most tightly linked to variability in cognitive function, large perforated synapses appear to be the key morphological correlate of memory capacity in the hippocampus.

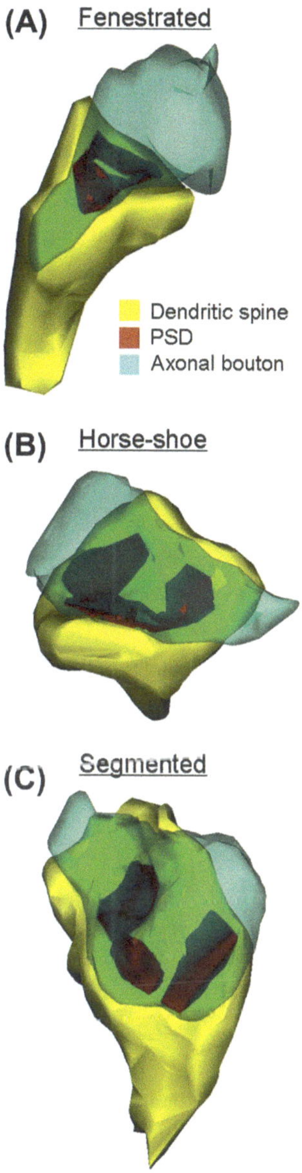

Figure 8 Three-dimensional reconstructions of three subtypes of perforated synapses. (A) A fenestrated perforated synapse has a hole in the postsynaptic density (PSD). (B) A horseshoe-shaped perforated synapse has a U-shaped PSD. (C) A segmented perforated synapse has two or more completely separate PSD plates. Serial sections from the monkey dentate gyrus were reconstructed using the free software developed by Dr Kristen Harris and her laboratory. *Reconstruct, version 1.1.0.0;* http://synapses.clm.utexas.edu.

2.3.2. Nonsynaptic Boutons, Multiple-Synaptic Boutons, and Learning and Memory

In the absence of gross morphological changes in monkey DG dendritic spines, we tested whether synaptic characteristics of DG axonal boutons are coupled to age (Hara et al., 2011a). Using serial section electron microscopy, frequencies of single-synaptic boutons (SSBs), multiple-synaptic boutons (MSBs), and boutons with no apparent synaptic contacts (nonsynaptic boutons (NSBs)) were determined in the subregion of DG where perforant path inputs terminate (DG outer molecular layer). MSBs are axonal boutons synaptically connected to more than one dendrite or dendritic spine (shown in Figure 9) and potentially act as a morphological substrate to increase coupling between presynaptic and postsynaptic neurons (Geinisman et al., 2001; Harris, 1995; Toni et al., 1999; Yankova et al., 2001). MSBs have been implicated in structural plasticity and hippocampus-dependent associative learning (Geinisman et al., 2001; Jones et al., 1999; Toni et al., 1999).

While there are no age effects on bouton density or average size in the DG outer molecular layer, the proportion of NSBs is doubled in aged compared to young monkeys (Hara et al., 2011a). Thus, aging may result in a lower turnover or higher retraction of spines, leading to a greater proportion of boutons that fail to make or maintain a synaptic contact. Aged monkeys also display a lower frequency of MSBs and fewer synaptic contacts per MSB compared to young adults (Hara et al., 2011a). While an MSB of a young monkey can make as many as six synaptic contacts, this number of connections is not observed in aged individuals. While menopause is not associated with changes in the proportions of NSBs, SSBs, or MSBs, postmenopausal monkeys have a lower proportion of MSBs that contact segmented perforated synapses compared to both young and aged premenopausal monkeys (Hara et al., 2011a). A summary of age- and menopause-related changes in DG synaptic connections is shown in Figure 10.

Most notably, the proportions of NSBs correlate with both task acquisition and recognition accuracy on the DNMS test such that individuals with higher NSB frequencies are slower to learn the task initially and exhibit worse memory (Hara et al., 2011a). In contrast, subjects with a higher number of synaptic contacts per MSB acquired the DNMS task faster (Hara et al., 2011a). Thus, increased NSB density in the DG appears to interfere with both task acquisition and recognition memory, while boutons with many synaptic contacts facilitate learning (Figure 11). In addition, higher proportions of MSBs contacting large segmented perforated synapses correlate with better DNMS recognition memory scores (Hara et al., 2011a). MSBs contacting large segmented perforated synapses may possess the greatest synaptic strength (Geinisman et al., 1992b, 1993) and the ability to simultaneously depolarize multiple postsynaptic neurons (Toni et al., 1999; Yankova et al., 2001). Thus, despite the increase in NSBs with

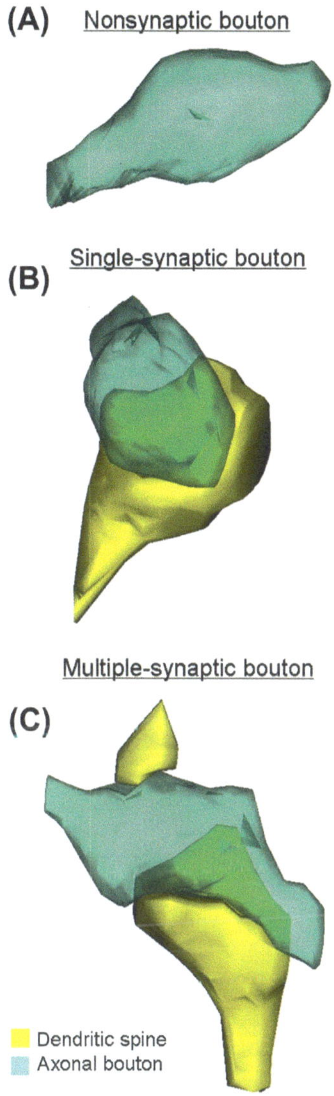

Figure 9 Three-dimensional reconstructions of three major types of axonal boutons. (A) A nonsynaptic bouton is an axonal bouton that makes no synaptic contacts. (B) A single-synaptic bouton is an axonal bouton that makes one synaptic contact, typically with a dendritic spine or shaft. (C) A multiple-synaptic bouton makes synaptic contacts with two or more dendritic spines or shafts. Serial sections from the monkey dentate gyrus were reconstructed using the free software developed by Dr Kristen Harris and her laboratory. *Reconstruct, version 1.1.0.0;* http://synapses.clm.utexas.edu.

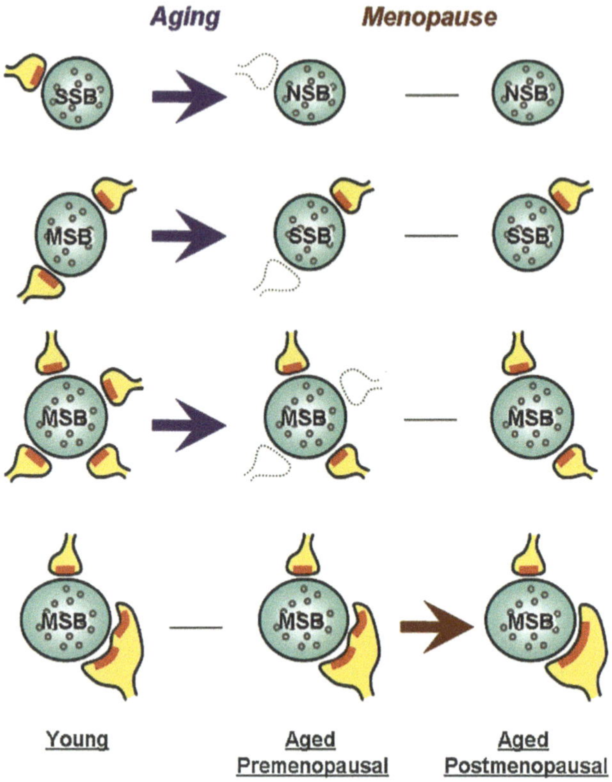

Figure 10 Schematic diagrams illustrating the changes in the types of synaptic connections in the monkey dentate gyrus with aging and menopause. Aging results in an increase in nonsynaptic boutons (NSB), a decrease in multiple-synaptic boutons (MSB), and a decrease in the number of synaptic contacts per MSB. While menopause does not result in changes in the proportions of NSBs or MSBs, it is associated with a selective decrease in MSBs contacting one or more segmented perforated synapses. Proportions of single-synapse boutons (SSBs) are comparable across age and menses groups. Arrows indicate possible shifts in bouton subtypes. Blue, axonal bouton; red, postsynaptic density; yellow, dendritic spine. *Diagrams adapted and modified from Hara et al. (2011a).*

aging, continued menstrual cycles or hormone replacement therapies may preserve this bouton subtype and maintain memory functions (Hara et al., 2011a). The types of DG synaptic connections that are correlated with cognitive functions are summarized in Figure 11.

2.4. LOCATION AND FUNCTION OF GLUTAMATE RECEPTORS IN THE AGING BRAIN

Glutamate is the major excitatory neurotransmitter released at cortical and hippocampal synapses and glutamate receptor subtype, number, and localization potently regulate synaptic strength and plasticity (Heynen et al., 2000; Kessels and Malinow, 2009;

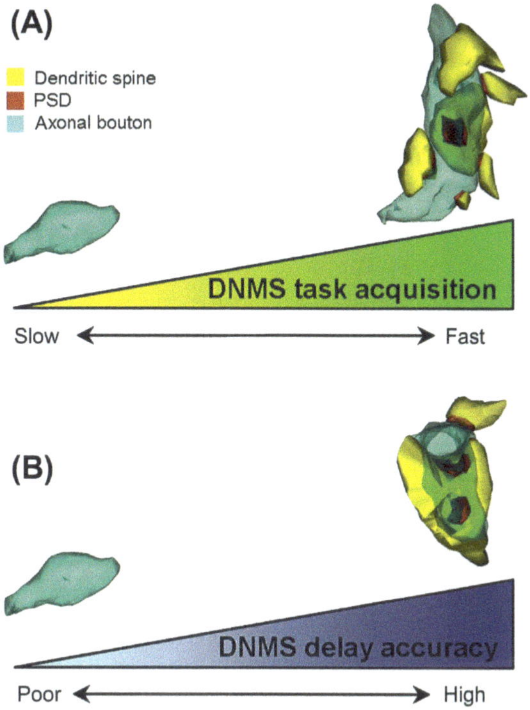

Figure 11 Schematic diagrams illustrating the relationships between types of synaptic connections in the monkey dentate gyrus and delayed nonmatching-to-sample (DNMS) measures. (A) A higher frequency of nonsynaptic boutons (NSBs) is associated with slower acquisition of the DNMS task, while a higher number of synaptic contacts per MSB is correlated with faster learning. (B) Higher NSB frequency is also associated with lower recognition memory scores, while MSBs contacting one or more segmented perforated synapses are associated with better memory scores. Serial sections from the monkey dentate gyrus were reconstructed using the free software developed by Dr Kristen Harris and her laboratory. *Reconstruct, version 1.1.0.0;* http://synapses.clm.utexas.edu. *Diagrams were made based on findings from Hara et al. (2011a).*

Nusser et al., 1998). The ionotropic glutamate receptors consist of three families based on preferential affinities for ligands: N-methyl-D-aspartate (NMDA), AMPA, and kainate.

NMDA receptors play an important role in synaptic plasticity and synaptogenesis (Bliss and Collingridge, 1993; Woolley and McEwen, 1994). In the DG and CA1, NMDA receptors are critical for induction of LTP, a functional correlate of learning and memory (Collingridge, 1992), and blocking these receptors prevent hippocampus-dependent spatial learning (Davis et al., 1992; Morris et al., 1986). AMPA receptors are key mediators of fast excitatory synaptic transmission (Kessels and Malinow, 2009), and synapses that lack these receptors are functionally silent at resting potentials (Liao et al., 1995). While NMDA receptors are involved in the initiation of LTP, AMPA receptors are rapidly inserted into the synapse following this induction and contribute to expression of LTP (Davies et al., 1989; Muller et al., 1988; Yang et al., 2008). Thus, AMPA receptors are highly mobile proteins and undergo dynamic trafficking into and out of the synapse in a tightly regulated and activity-dependent manner (Carroll and Zukin, 2002; Groc and Choquet, 2006; Kessels and Malinow, 2009).

Kainate receptors are less abundant than NMDA or AMPA receptors in the adult brain and the distribution pattern is distinct from those of NMDA and AMPA receptors (Johnson et al., 1996). Kainate receptors are best known for mediating seizure and epilepsy activities in the hippocampus but also play a role in synaptic plasticity (Huettner, 2003). Together, the integrity of glutamate receptors is likely to be important for memory and other cognitive functions that deteriorate with aging. Thus, there is considerable interest in determining the specific roles of each glutamate receptor subtype in normal aging and cognitive decline.

2.4.1. Age-Related Changes in NMDA Receptors

The NMDA receptor has been the most studied glutamate receptor and appears to be more vulnerable to the aging process than other glutamate receptors (Magnusson et al., 2010). NMDA receptors are ligand-gated ion channels that are typically composed of the essential GluN1 (NR1) subunit and one or more of the GluN2A-D (NR2A-D) subunits (Hollmann and Heinemann, 1994; Malenka and Nicoll, 1993; Monyer et al., 1992; Moriyoshi et al., 1991). GluN2B is highly expressed throughout the brain in the early stages of development, but a developmental switch downregulates GluN2B as GluN2A expression increases (Liu et al., 2004; Monyer et al., 1994). Each GluN2A-D subunit confers different properties to the NMDA receptor, such as changes in affinities to agonists/antagonists, gating properties, and electrophysiological responses (Kutsuwada et al., 1992; Monyer et al., 1992). Interestingly, mice lacking the GluN2A subunit have impaired spatial working memory (Bannerman et al., 2008), and overexpression of GluN2B enhances LTP and learning abilities (Tang et al., 1999).

Many of the earliest NMDA receptor-binding assays examined age effects across broad regions of the brain. In both the cerebral cortex and hippocampus, NMDA receptor-binding sites, as measured by NMDA-displaceable [^{3}H]glutamate binding, are decreased with aging in rats (Tamaru et al., 1991). In aged monkeys, similar widespread declines in NMDA binding are seen in the hippocampus and cortex, particularly in the frontal and temporal lobes (Wenk et al., 1991). Subsequent studies have assessed individual subregions within the hippocampus and cortex to determine their relative contributions and vulnerabilities to aging. In humans, an autoradiography study has shown that NMDA receptor binding in the CA3 (*stratum oriens*, *pyramidale*, and *radiatum*) decreases with age, but binding is not altered from adulthood to old age in any other hippocampal or adjacent cortical regions (Johnson et al., 1996).

In rats, aging results in decreased NMDA binding in the CA1 (*stratum oriens* and *radiatum*), CA3 (*stratum oriens*, *lucidum*, and *radiatum*), and DG (granule cell layer, inner, and outer molecular layers) (Clark et al., 1992). In mice, NMDA-binding sites are reduced with age in many parts of the cortex (frontal, cingulate, parietal, occipital/temporal, and EC), CA1 (*stratum oriens* and *lacunosum moleculare*), CA3 (*stratum oriens*), and DG molecular layer (Magnusson, 1998; Magnusson and Cotman, 1993), although the magnitude of change depends on the strain of mice. While binding assays do not provide subcellular resolution, perhaps the most consistent finding across these studies may be that the extent of age-related changes in NMDA receptors is greater in the cerebral cortex compared to the hippocampus (Magnusson et al., 2010). However, it is currently unknown whether this is due to the greater magnitude of synapse loss in the PFC as compared to the hippocampus.

An in situ hybridization study has shown that mRNA levels for the GluN2B subunit decreased with age in the monkey PFC, but not in the hippocampal regions such as the subiculum, CA1, CA3, CA4, and DG, or the parahippocampal gyrus (Bai et al., 2004). However, a trend for an age-related decrease in GluN2B expression is reported in the DG granule cell layer and the CA3 pyramidal cell layer (Bai et al., 2004). In rats, GluN2B mRNA levels are reduced with age in the prelimbic PFC and CA1 (Dyall et al., 2007). In aged mice, GluN1 mRNA is decreased in the frontal cortex and DG granule cells, and GluN2B mRNA is decreased in the DG granule cells and throughout the cortex, including frontal and ECs (Magnusson, 2000). In contrast, GluN2A mRNA levels appear to stay constant throughout the cortex and hippocampus with advancing age (Magnusson, 2000).

The production of antibodies specific to individual NMDA receptor subunits has enabled Western blot and immunocytochemical experiments that probe age effects on subunit protein levels. In the mouse hippocampus, GluN1 levels are decreased in aged mice, GluN2B levels are higher in middle-aged compared to young and aged mice, and GluN2A levels remain stable across all ages (Magnusson et al., 2002). In the cerebral cortex of aged mice, GluN1, GluN2A, and GluN2B protein levels are decreased compared to young and middle-aged mice (Magnusson et al., 2002). In aged rats, both GluN2A and GluN2B protein levels are reduced in the hippocampus (Clayton and Browning, 2001; Sonntag et al., 2000).

More recent studies have examined NMDA receptor subunits in individual hippocampal and cortical subregions. In the rat CA3, age-related declines in GluN1, GluN2A, and GluN2B protein levels are observed (Adams et al., 2008). In the rat CA1, GluN2B protein levels decline with age (Dyall et al., 2007). A reduction in GluN1 expression is also observed in the aged rat EC and postrhinal cortex (Liu et al., 2008). GluN2A expression is decreased in the aged rat EC, perirhinal, and postrhinal cortices, but is unchanged in the hippocampus or PFC (Liu et al., 2008). In the rat DG middle molecular layer, where the perforant path input terminates, decreased GluN1 protein levels are observed with aging, but GluN2A and GluN2B levels remain stable (Newton et al., 2008).

A similar observation has been made in monkeys, where aged monkeys exhibit a significantly lower expression of GluN1 protein (by 30%), selectively in the DG region that receives perforant path input (Gazzaley et al., 1996; Witter et al., 1989). Because the number of EC layer II neurons remains stable in aging monkeys (Gazzaley et al., 1997), the selective decrease in GluN1 expression in the DG is unlikely to be a consequence of perforant path degeneration (Gazzaley et al., 1996). Also in monkeys, aging results in a significant reduction of GluN1 expression in corticocortical neurons that project to PFC area 46 (Hof et al., 2002). In summary, while the effects of aging on NMDA receptor subunit expression depends on the species and the brain region, the greatest changes are observed with GluN1 and GluN2B expression, while GluN2A expression appears to be relatively stable.

Some of these studies on aging and NMDA receptors have been examined in the context of memory decline. In a study that combined Morris water maze testing and autoradiography, NMDA-binding sites were decreased in the CA3 (*stratum radiatum*) and in the outer laminae of frontal, parietal, and temporal cortices, in cognitively unimpaired, but not in impaired, aged rats (Le Jeune et al., 1996). Aside from changes in receptor binding, aging appears to increase basal glutamate release in the hippocampus (Freeman and Gibson, 1987), decrease the rate and capacity of glutamate uptake via transporters in the cortex (Wheeler and Ondo, 1986), and decrease protein levels of glutamate uptake transporters in the hippocampus (Potier et al., 2010). Together, these alterations may lead to increased extracellular glutamate levels and higher vulnerability of neurons to NMDA receptor-induced excitotoxicity. Thus, the decrease in NMDA receptor-binding sites in the unimpaired rats may be part of a compensatory mechanism to preserve cognitive function.

Other studies have suggested the opposite, where higher levels of NMDA receptor were associated with better memory. In the rat CA3, no global effect of age on GluN1 protein levels is observed; however, better spatial learning ability as measured by Morris water maze learning index correlated with higher GluN1 immunofluorescence levels in CA3 dendrites (*stratum lucidum*, *radiatum*, and *lacunosum moleculare*) (Adams et al., 2001). Furthermore, rats with lower NMDA receptor antagonist binding density required more trials to reach criterion on the water maze task (Pelleymounter et al., 1990). In mice, an age-related decrease in NMDA-displaceable glutamate binding in the frontal cortex and CA1 (*stratum oriens* and *lacunosum moleculare*) was associated with poorer memory performance indicated by farther distance to platform locations (Magnusson, 1998). These studies suggest that the optimal level of NMDA receptors for cognitive functioning may shift during the aging process and levels that are lower or higher may be associated with poorer memory performance. Consistent with this view, a recent study has shown that in middle-aged mice, higher protein expression of GluN1 and GluN2B in hippocampal synaptic membrane fractions was associated with better spatial memory on the Morris water maze; conversely, in aged mice, higher synaptic expression of these two proteins was associated with poorer memory (Zhao et al., 2009).

2.4.2. Age-Related Changes in AMPA Receptors

AMPA receptors are tetrameric and are composed of four subunits, GluA1–4 (GluR1–4). The most abundant subunit of AMPA receptors in the adult brain is GluA2 and forms heteromeric receptors with GluA1, GluA3, or less commonly with GluA4 (Mead and Stephens, 2003; Nakagawa, 2010). GluA2 is a critical subunit in determining AMPA receptor properties, including receptor kinetics, channel conductance, and calcium permeability (Geiger et al., 1995; Jonas et al., 1994). In addition, GluA2 expression is associated with spine morphology as overexpression of GluA2 increases hippocampal spine and synapse size (Passafaro et al., 2003), while GluA2 knock-out mice have decreased proportions of large mushroom spines and spines with segmented perforated synapses (Medvedev et al., 2008). As AMPA receptors play crucial roles in synaptic plasticity and multiple forms of memory, the expression, assembly, and trafficking of these receptors are tightly regulated (Henley et al., 2011; Isaac et al., 2007; Migues et al., 2010).

Generally, while changes in AMPA receptor binding are observed with aging, the magnitude of change appears to be less pronounced than what is observed with NMDA receptor binding (Magnusson et al., 2010). In humans, an autoradiography study showed that AMPA receptor binding decreases with age in most layers of CA1 and CA2, in the DG, subiculum, and parahippocampal gyrus (Johnson et al., 1996). In the rat DG/hilus region, an age-related decrease in AMPA receptor binding is observed (Wenk and Barnes, 2000). In another rat autoradiography study, a decrease in AMPA receptor binding is shown in the CA3 (*stratum lucidum* and *radiatum*) and DG (granule cell layer, inner, and outer molecular layer) with aging, but these measures are not correlated with Morris water maze escape latency or days to criterion (Clark et al., 1992).

In another study that combined Morris water maze testing and autoradiography, AMPA binding sites are increased in CA1 (*stratum oriens*, *radiatum*, and *lacunosum moleculare*), CA3 (*stratum pyramidale* and *radiatum*), and DG in cognitively impaired, but not unimpaired, aged rats (Le Jeune et al., 1996). These inconsistencies in the literature may be due to differences in strains and ages of the rats, the materials used (brain sections versus membrane fractions), and assay conditions. In aged mice, AMPA binding sites are decreased in the CA1 (*stratum lacunosum moleculare*) (Magnusson and Cotman, 1993). Furthermore, higher AMPA binding densities in the frontal cortex, CA1 (*stratum lacunosum moleculare*), and CA3 (*stratum oriens* and *radiatum*) are correlated with better spatial memory, as measured by proximity to the water maze platform (Magnusson, 1998).

In situ hybridization studies have shown that mRNA levels for GluA1 and GluA2 are decreased in the rat DG, CA1, and CA3 (Dyall et al., 2007; Pagliusi et al., 1994). GluA2 mRNA is also reduced with age in the rat prelimbic PFC (Dyall et al., 2007). In monkeys, aging results in a significant reduction of GluA2 expression in corticocortical neurons that project to area 46, including those located within area 46 (Hof et al., 2002). No age effects in the AMPA receptor GluA2/3 subunits are observed in the monkey DG outer molecular layer (Gazzaley et al., 1996). In the rat DG middle molecular layer, where the perforant path input terminates, decreased GluA1 protein levels are observed with aging, but GluA2 levels remain stable (Newton et al., 2008).

In contrast, GluA2 protein levels are decreased with aging in the rat CA1, postrhinal cortex (Liu et al., 2008), and prelimbic PFC (Dyall et al., 2007). In the rat CA3, age-related declines in GluA1 and GluA2 protein levels are observed, as measured by Western blots (Adams et al., 2008). An immunofluorescence study, on the other hand, has shown an absence of age effect on GluA2 protein levels in rat CA3 dendrites (*stratum lucidum*, radiatum, and *lacunosum moleculare*), and GluA2 immunoreactivity fails to correlate with spatial learning ability as measured by Morris water maze (Adams et al., 2001).

While age-related changes in mRNA and protein levels of glutamate receptor subunits offer important clues, these studies do not provide the subcellular resolution to pinpoint which subtype of synapses have altered molecular profiles and in which postsynaptic compartments do the molecular changes take place. Although most excitatory synapses in cortical regions occur at dendritic spines, not all spines are equal. Thin spines have small synapses, are highly motile and plastic, and are associated with learning (Dumitriu et al., 2010; Holtmaat et al., 2005; Kasai et al., 2003). In contrast, large mushroom spines contain strong, stable synapses, have abundant AMPA receptors, and are hypothesized to be critical for memory (Ganeshina et al., 2004a,b; Geinisman et al., 1986a; Hara et al., 2012a; Kasai et al., 2003; Nusser et al., 1998). Furthermore, not all glutamate receptors within a dendritic spine are functional as only those localized on or near the synaptic membrane can bind to glutamate released from presynaptic terminals. Glutamate receptors on extrasynaptic plasma membrane may be in transit to and from the synapse, while cytoplasmic receptors may reflect the reserve pool (newly synthesized or internalized and awaiting to be recycled) and those destined for degradation.

To address these questions, electron microscopic immunocytochemistry was performed on hippocampal serial sections from young and aged monkeys that have undergone cognitive assessment, including the DNMS test of recognition memory (Hara et al., 2012b). Two proteins were examined at the ultrastructural level in the monkey DG outer molecular layer: the GluA2 subunit and the atypical ζ isoform of protein kinase C (PKMζ), as shown in Figure 12.

PKMζ is a constitutively active kinase that is highly expressed in the hippocampus and neocortex (Crary et al., 2006; Hernandez et al., 2003; Oster et al., 2004; Sacktor et al., 1993). PKMζ plays a critical role in storage of memory by blocking the removal of GluA2-containing AMPA receptors from the synapse, and consequently, promoting persistent GluA2 expression at the synapse (Hardt et al., 2010; Migues et al., 2010; Pastalkova et al., 2006; Sacktor, 2011; Serrano et al., 2008; Shema et al., 2009). While many short-acting molecules have been implicated in memory formation (e.g. CaMKII, MAPK, CREB, etc.), PKMζ is unique in that its persistent activity is critical for storage of memory, and its effects can last for months, if not longer (Sacktor, 2011; Shema et al., 2009).

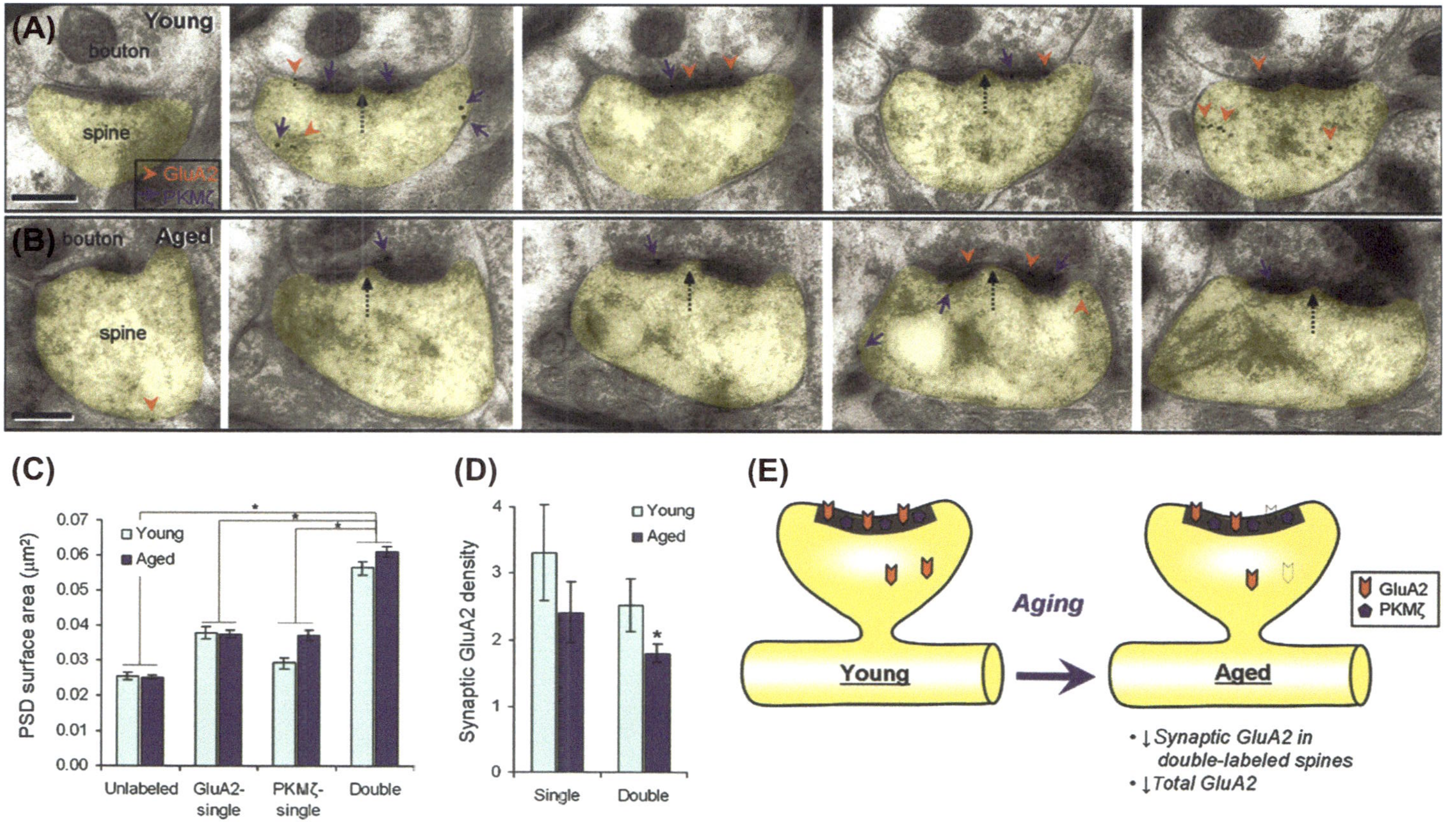

Figure 12 The effects of aging on GluA2 and PKMζ distributions in monkey DG dendritic spines. (A, B) Electron micrographs of five serial sections containing GluA2 and PKMζ double-labeled perforated synapse spines (shaded in yellow) from young (A) and aged (B) monkeys. The 10 and 15 nm immunogold particles represent GluA2 (red arrowheads) and PKMζ (blue arrows) labeling, respectively. Dashed black arrows point to perforations in the PSD. (C) Independent of age, dendritic spines coexpressing GluA2 and PKMζ have significantly larger synapses, as measured by PSD area, than unlabeled and single-labeled spines. (D) The density of synaptic GluA2 immunogold particles (per 0.1 μm² of PSD area) is lower in aged compared with young adults in double-labeled, but not single-labeled spines. *$p < 0.05$. (E) Schematic diagrams illustrating GluA2 and PKMζ distributions in relation to aging. Compared to young adults, aged monkeys have decreased densities of synaptic GluA2 in spines coexpressing PKMζ. Aged monkeys also have a lower density of total GluA2 in both single- (not shown) and double-labeled spines. *Micrographs, histograms, and schematic diagrams are adapted and modified from Hara et al. (2012b).*

Several key findings emerged from this study (Hara et al., 2012b). First, dendritic spines coexpressing GluA2 and PKMζ have significantly larger synapses (PSD area) than those that contain just one or neither (Figure 12). Indeed, a great majority of perforated synapse spines (84%) were colabeled with GluA2 and PKMζ. These double-labeled spines appear to be a class of spines important for cognitive functions as monkeys with faster task acquisition and more accurate recognition memory on the DNMS test exhibit higher frequencies of these spines (Figure 13). Within this population of double-labeled spines, aged monkeys compared to young adults express lower densities of synaptic and total GluA2, which correlate with lower recognition memory accuracy and slower task acquisition, respectively (Figures 12 and 13). Additionally, in the absence of a significant age effect, the density of synaptic PKMζ in double-labeled spines correlates with both task acquisition and memory accuracy such that individuals with higher synaptic PKMζ density learn the task faster and score higher on the memory test (Figure 13). Together, these findings suggest that age-related impairment in maintenance of GluA2 at large DG synapses correlates with recognition memory deficits. Furthermore, PKMζ may promote continual synaptic delivery of GluA2-containing AMPA receptors and protect aged individuals from memory decline. The key molecular correlates of learning and recognition memory presented in this study are summarized in Figure 13.

2.4.3. Age-Related Changes in Kainate Receptors

Each kainate receptor consists of one or more of the following subunits: GluK1 (GluR5), GluK2 (GluR6), GluK3 (GluR7), GluK4 (KA1), and GluK5 (KA2) (Bloss and Hunter, 2010; Hollmann and Heinemann, 1994). Kainate receptors can be homomeric or heteromeric and their subunit composition and posttranslational modifications determine their affinity for glutamate, gating properties, and current characteristics (Bloss and Hunter, 2010; Hollmann and Heinemann, 1994). Presynaptic kainate receptors modulate glutamate release while postsynaptic kainate receptors regulate excitatory transmission (Chittajallu et al., 1996; Huettner, 2003; Represa et al., 1987). Receptor binding studies have shown that kainate receptors are most abundant in the CA3 *stratum lucidum* (Clark et al., 1992; Le Jeune et al., 1996; Monaghan and Cotman, 1982; Nicolle et al., 1996), where mossy fiber inputs from the DG terminate.

In humans, kainate binding peaks at fetal and neonatal stages but no changes in binding are observed following adulthood in any hippocampal or adjacent cortical regions (Johnson et al., 1996). No age effects in the protein levels of kainate receptor GluK1/2/3 subunits are observed in the monkey DG (Gazzaley et al., 1996). In mice, kainate binding sites decline with aging in the frontal cortex, CA1 (*stratum lacunosum moleculare*), and DG (middle and outer molecular layer) (Magnusson and Cotman, 1993). In the rat, kainate binding decreases with age in the CA3 (*stratum oriens* and *lucidum*), DG (molecular layer and hilus), and in the inner layers of temporal, parietal,

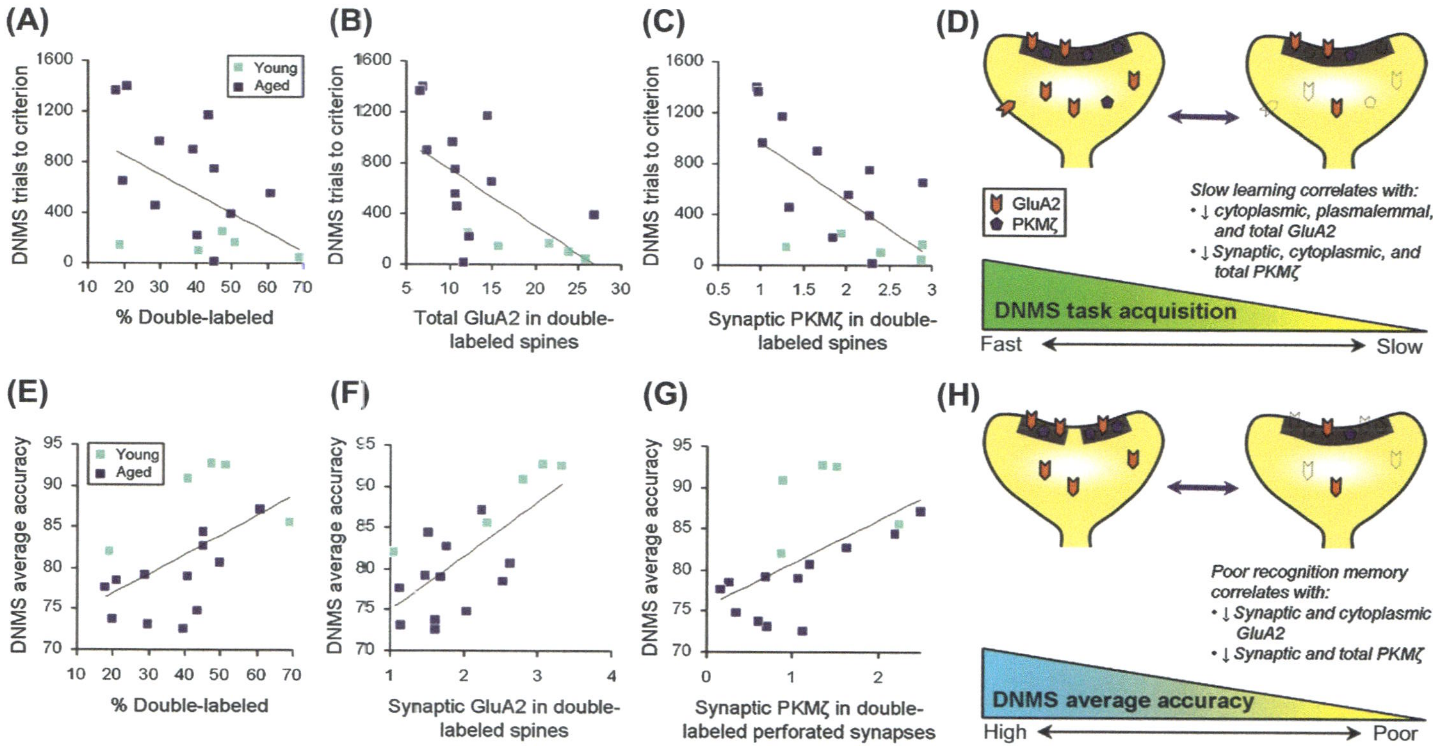

Figure 13 GluA2 and PKMζ distributions in monkey DG spines are correlated with (A–D) task acquisition and (E–H) recognition memory on the DNMS test. (A–C) The number of trials to criterion on the DNMS inversely correlates with (A) the percentage of dendritic spines double-labeled with GluA2 and PKMζ, (B) the density of total GluA2 in double-labeled spines, and (C) the density of synaptic PKMζ in double-labeled spines. (D) A schematic representation of the relationship between DNMS task acquisition and GluA2 and PKMζ distributions. (E–G) DNMS average accuracy (across all delay intervals) positively correlates with (E) the percentage of dendritic spines double-labeled with GluA2 and PKMζ, (F) the density of synaptic GluA2 in double-labeled spines, and (G) the density of synaptic PKMζ in double-labeled perforated synapse spines. (H) A schematic representation of the relationship between DNMS recognition memory and GluA2 and PKMζ distributions. *Scatter plots and schematic diagrams are adapted and modified from Hara et al. (2012b).*

and occipital cortices (Clark et al., 1992; Le Jeune et al., 1996). Interestingly, kainate binding in the aged rat CA3 *stratum lucidum* correlates with spatial learning such that subjects with higher kainate binding sites exhibit poorer learning in the Morris water maze (Nicolle et al., 1996).

2.5. INTERACTIVE EFFECTS OF AGING AND ESTROGEN ON CORTICAL SYNAPSES

Menopause is characterized by the permanent cessation of menstrual cycles and the loss of ovarian hormone production, signaling the end of a woman's reproductive phase. During the menopausal transition, which can last many years, many women experience symptoms that include hot flush, mood changes, and memory problems. It is becoming increasingly evident that a strong interaction exists between age-related cognitive decline and endocrine changes (Hao et al., 2007; Henderson, 2008). Female macaque monkeys are valuable models for studying the effects of menopause on cognition because they share many reproductive and endocrine features with women and also undergo menopause in a manner similar to women (Gilardi et al., 1997; Walker and Herndon, 2008). Furthermore, menopause-related cognitive impairment seen in women is also present in female monkeys. For example, peri/postmenopausal female monkeys show significant impairments in working memory (delayed response) and recognition memory (DNMS) compared to premenopausal monkeys after accounting for the effect of age (Hara et al., 2012a; Roberts et al., 1997).

While cognitive symptoms associated with menopause in women are likely to be caused by multiple factors, estrogen deficiency is thought to be one of the culprits. In fact, some of these symptoms are attenuated by hormone replacement therapy in nonhuman primate models of menopause. For example, in aged ovariectomized female monkeys, cyclic estradiol replacement consisting of a single injection every 3 weeks (partially mimicking the natural fluctuations of estrogen in premenopausal monkeys) reverses the age-related impairment in working memory and modestly improves recognition memory scores (Rapp et al., 2003).

Improvements in cognitive function with cyclic estradiol administration are accompanied by alterations in neuronal morphology in monkey area 46 (Hao et al., 2007; Rapp et al., 2003), a PFC region strongly implicated in working memory (Funahashi et al., 1993; Luebke et al., 2010). The effects of estrogen on area 46 pyramidal neurons have been experimentally tested in a cohort of young and aged ovariectomized female monkeys receiving cyclic estradiol or vehicle treatments (Hao et al., 2007). In the delayed response test of working memory, young monkeys receiving estradiol or vehicle, and aged monkeys receiving estradiol perform equally well (Hao et al., 2007). Reconstructions of pyramidal neurons in PFC area 46 revealed that estradiol increases the density of the highly plastic and motile small spines in both young and aged monkeys (Hao et al.,

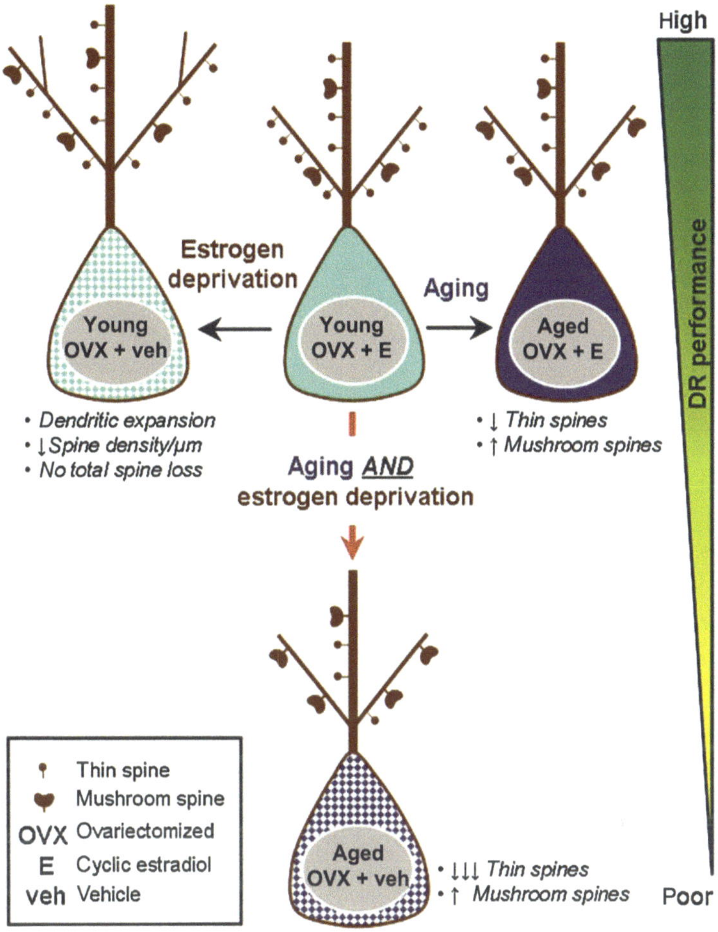

Figure 14 Schematic diagrams illustrating the neuronal and spine morphology of layer III pyramidal neurons in monkey area 46 and their relationships to aging, estrogen deprivation, and working memory performance. Young ovariectomized monkeys treated with cyclic estradiol or vehicle (young OVX + E or OVX + veh, respectively) and aged ovariectomized monkeys treated with cyclic estradiol (aged OVX + E) perform equally well on the delayed response (DR) test of working memory, while aged OVX + veh monkeys perform poorly. Compared to young OVX + E group, the young OVX + veh group has lower spine density, but longer dendritic length; consequently, these two young groups have comparable numbers of spines per neuron. Aging results in decreased spine density, a selective loss of small thin spines, and a relative increase in mushroom spines. Estrogen deprivation results in a further loss of thin spines in the aged OVX + veh group such that this group only has one-third the density of thin spines as compared to the young OVX + E group. Only apical dendrites are drawn for simplicity, but similar changes occur in basal dendrites and throughout distal higher order dendrites. *Diagrams were drawn based on findings from Hao et al. (2007).*

2007). Although young vehicle-treated monkeys display lower spine density, their dendritic length is increased such that the total number of spines per neuron is equivalent to the young estradiol-treated group (Hao et al., 2007). This dendritic expansion may be a compensatory mechanism that protects cognitive function in young monkeys deprived of estrogen. In contrast, only vehicle-treated aged monkeys exhibit significant working memory impairment, and this deficit occurs concomitantly with a dramatic decrease in area 46 layer III spine density in the absence of a compensatory extension of dendrites (Hao et al., 2007). Because both aging and estrogen deprivation result in decreased spine density and a shift toward larger spines, the aged vehicle-treated monkeys have a dramatic loss of small thin spines to a level that is only one-third of that observed in the young estradiol-treated monkeys (Hao et al., 2007). These dendritic and spine changes with aging and estradiol treatment are summarized in Figure 14. (Note: Text that goes with Figures 12 and 13 has been embargoed until the journal article is published. Once it is published, we will update this chapter to include those data.)

In addition, we have begun to investigate the molecular mechanisms underlying the beneficial effects of estrogen on cognitive function in aged monkeys. The localization of estrogen receptor α (ERα) in area 46 was examined with electron microscopy and SynBin (see Section 2.1.2 for details) in the same behaviorally characterized young and aged ovariectomized monkeys (Wang et al., 2010). Although ERα levels remain stable across age and hormone treatment groups, the abundance of ERα in the subsynaptic region (30–60 nm from the synaptic membrane) correlates with working memory scores exclusively in aged monkeys receiving cyclic estradiol (Wang et al., 2010). Therefore, the cognitive benefits observed in aged monkeys with estradiol treatment may be mediated by the activation of ERα in a subsynaptic domain coupled to secondary messengers and signaling cascades involved in spine/synapse formation and stabilization (Spencer et al., 2008).

3. CONCLUSIONS

3.1. Summary of Synaptic Correlates of Aging and Cognitive Decline

Several important studies examining the morphological and molecular correlates of aging have been carried out in behaviorally characterized young and aged animals. Multiple lines of evidence now suggest that the critical synaptic subtypes associated with cognitive decline are different between the PFC and hippocampus and are dependent on the types of cognitive demands (Dumitriu et al., 2010; Hao et al., 2007; Hara et al., 2011a, 2012a). In the monkey dorsolateral PFC (area 46), a significant decline in spine/synapse density occurs with age, and the dendritic spine subtype most vulnerable to the effects of aging and coupled to task learning is the smallest class of thin spines (Dumitriu et al., 2010).

In contrast, the density and size of axospinous synapses remain stable with aging in the monkey hippocampal DG, but the density of large mushroom spines with perforated synapses correlates with recognition memory scores (Hara et al., 2011a, 2012a). In the absence of overt synapse loss, age-related rearrangements in synaptic connections are observed in the DG, including a significant increase in the proportion of NSBs, a decrease in MSBs, and a decrease in the number of contacts per MSB (Hara et al., 2011a). Higher frequency of NSBs in the DG, which may reflect an impairment in making and/or maintaining synaptic contacts, is correlated with slower task acquisition rates and lower memory scores on the DNMS (Hara et al., 2011a).

These striking differences between prefrontal and hippocampal synapses and how they differentially relate to aging and cognitive decline are in line with observations that subjects impaired on a PFC-reliant cognitive test are not necessarily impaired on a hippocampus-associated test (Dumitriu et al., 2010; Rapp and Amaral, 1989). Thus, interventions to slow or reverse cognitive decline need to be explored with the recognition that the PFC and hippocampus may age independently. Different spine subtypes are associated with different cognitive domains such that interventions that promote youthful synaptic morphology in the PFC may have detrimental effects on hippocampal synapses and functions.

3.2. Future Directions

Molecular targets of aging explored to date have been studied primarily by light microscopy and autoradiography. In light of recent evidence demonstrating that the sizes of dendritic spines are strong indicators of their plasticity, motility, and synaptic strength (Ganeshina et al., 2004b; Holtmaat et al., 2005; Kasai et al., 2003), future immunocytochemical studies will be most informative if serial section electron microscopy is combined with the use of a software like SynBin. By these methods, molecular targets can be examined at the subcellular level while accounting for synapse size and morphology. These studies will likely offer important clues as to which subcellular compartment the age-related changes occur in and may also identify a specific subtype of spines in which the changes take place.

While glutamate receptors have received considerable attention with regards to aging, other neurotransmitter systems including the monoamine and cholinergic systems are also highly implicated in cognitive functions (Arnsten, 2006; Arnsten and Li, 2005; Brozoski et al., 1979; Hara et al., 2011b; Voytko et al., 1995). Thus, receptors and transporters for dopamine, norepinephrine, and acetylcholine are of significant interest. Furthermore, age-related changes in receptor trafficking may be a consequence of dysregulations in protein kinases, scaffolding proteins, and cell adhesion molecules (Benson and Huntley, 2010; Henley et al., 2011; Kessels and Malinow, 2009; Nakagawa, 2010), many of which are present in or near the PSD. These proteins also merit closer examination.

Finally, the most important area for future research may be to determine how these morphological and molecular changes with normal aging relate to neurodegenerative disorders such as Alzheimer's disease. Alzheimer's disease is characterized by dementia, formation of plaques and tangles, and extensive neuronal loss. Furthermore, a stereological electron microscopic study using postmortem human tissue has demonstrated a significant loss of CA1 synapses in Alzheimer's patients (by 55%) compared to individuals without the disease (Scheff et al., 2007). This loss of synapses significantly correlates with several cognitive tests and appears to precede the extensive neuronal loss (Scheff et al., 2007). While the neurodegenerative process of Alzheimer's disease is distinctive, it remains possible that synaptic and morphological alterations that occur with normal aging may render neurons more susceptible to the disease. Thus, identification of synaptic signatures associated with neuronal vulnerability may lead to development of treatments that maintain synaptic health. Early clinical intervention may, in turn, reduce the risk or delay the onset of Alzheimer's disease.

ACKNOWLEDGMENTS

Original research and chapter preparation was supported by National Institute on Aging grants R37 AG06647, R01 AG010606, and P01 AG16765 to J.H.M. We thank Dr Jul Lea Shamy for providing us with a monkey brain MRI image. The imaging and reconstruction of monkey cortical neurons were performed by Mr William G. M. Janssen and theVisualMD, respectively.

REFERENCES

Adams, M.M., Donohue, H.S., Linville, M.C., Iversen, E.A., Newton, I.G., Brunso-Bechtold, J.K., 2010. Age-related synapse loss in hippocampal CA3 is not reversed by caloric restriction. Neuroscience 171, 373–382.

Adams, M.M., Shi, L., Linville, M.C., Forbes, M.E., Long, A.B., Bennett, C., Newton, I.G., Carter, C.S., Sonntag, W.E., Riddle, D.R., Brunso-Bechtold, J.K., 2008. Caloric restriction and age affect synaptic proteins in hippocampal CA3 and spatial learning ability. Exp. Neurol. 211, 141–149.

Adams, M.M., Smith, T.D., Moga, D., Gallagher, M., Wang, Y., Wolfe, B.B., Rapp, P.R., Morrison, J.H., 2001. Hippocampal dependent learning ability correlates with N-methyl-D-aspartate (NMDA) receptor levels in CA3 neurons of young and aged rats. J. Comp. Neurol. 432, 230–243.

Amaral, D.G., 1993. Morphological analyses of the brains of behaviorally characterized aged nonhuman primates. Neurobiol. Aging 14, 671–672.

Arnsten, A.F., 2006. Fundamentals of attention-deficit/hyperactivity disorder: circuits and pathways. J. Clin. Psychiatry 67 (Suppl. 8), 7–12.

Arnsten, A.F., Li, B.M., 2005. Neurobiology of executive functions: catecholamine influences on prefrontal cortical functions. Biol. Psychiatry 57, 1377–1384.

Bachevalier, J., Landis, L.S., Walker, L.C., Brickson, M., Mishkin, M., Price, D.L., Cork, L.C., 1991. Aged monkeys exhibit behavioral deficits indicative of widespread cerebral dysfunction. Neurobiol. Aging 12, 99–111.

Bai, L., Hof, P.R., Standaert, D.G., Xing, Y., Nelson, S.E., Young, A.B., Magnusson, K.R., 2004. Changes in the expression of the NR2B subunit during aging in macaque monkeys. Neurobiol. Aging 25, 201–208.

Bannerman, D.M., Niewoehner, B., Lyon, L., Romberg, C., Schmitt, W.B., Taylor, A., Sanderson, D.J., Cottam, J., Sprengel, R., Seeburg, P.H., Kohr, G., Rawlins, J.N., 2008. NMDA receptor subunit NR2A is required for rapidly acquired spatial working memory but not incremental spatial reference memory. J. Neurosci. 28, 3623–3630.

Barbas, H., Rempel-Clower, N., 1997. Cortical structure predicts the pattern of corticocortical connections. Cereb. Cortex 7, 635–646.

Barnes, C.A., McNaughton, B.L., 1985. An age comparison of the rates of acquisition and forgetting of spatial information in relation to long-term enhancement of hippocampal synapses. Behav. Neurosci. 99, 1040–1048.

Bartus, R.T., Fleming, D., Johnson, H.R., 1978. Aging in the rhesus monkey: debilitating effects on short-term memory. J. Gerontol. 33, 858–871.

Benson, D.L., Huntley, G.W., 2010. Building and remodeling synapses. Hippocampus.

Bliss, T.V., Collingridge, G.L., 1993. A synaptic model of memory: long-term potentiation in the hippocampus. Nature 361, 31–39.

Bloss, E.B., Hunter, R.G., 2010. Hippocampal kainate receptors. Vitam. Horm. 82, 167–184.

Bloss, E.B., Janssen, W.G., McEwen, B.S., Morrison, J.H., 2010. Interactive effects of stress and aging on structural plasticity in the prefrontal cortex. J. Neurosci. 30, 6726–6731.

Bloss, E.B., Janssen, W.G., Ohm, D.T., Yuk, F.J., Wadsworth, S., Saardi, K.M., McEwen, B.S., Morrison, J.H., 2011. Evidence for reduced experience-dependent dendritic spine plasticity in the aging prefrontal cortex. J. Neurosci. 31, 7831–7839.

Brown, V.J., Bowman, E.M., 2002. Rodent models of prefrontal cortical function. Trends Neurosci. 25, 340–343.

Brozoski, T.J., Brown, R.M., Rosvold, H.E., Goldman, P.S., 1979. Cognitive deficit caused by regional depletion of dopamine in prefrontal cortex of rhesus monkey. Science 205, 929–932.

Carroll, R.C., Zukin, R.S., 2002. NMDA-receptor trafficking and targeting: implications for synaptic transmission and plasticity. Trends Neurosci. 25, 571–577.

Chittajallu, R., Vignes, M., Dev, K.K., Barnes, J.M., Collingridge, G.L., Henley, J.M., 1996. Regulation of glutamate release by presynaptic kainate receptors in the hippocampus. Nature 379, 78–81.

Clark, A.S., Magnusson, K.R., Cotman, C.W., 1992. In vitro autoradiography of hippocampal excitatory amino acid binding in aged Fischer 344 rats: relationship to performance on the Morris water maze. Behav. Neurosci. 106, 324–335.

Clayton, D.A., Browning, M.D., 2001. Deficits in the expression of the NR2B subunit in the hippocampus of aged Fisher 344 rats. Neurobiol. Aging 22, 165–168.

Collingridge, G.L., 1992. The Sharpey-Schafer Prize Lecture. The mechanism of induction of NMDA receptor-dependent long-term potentiation in the hippocampus. Exp. Physiol. 77, 771–797.

Corkin, S., 1984. Lasting consequences of bilateral medial temporal lobectomy: clinical course and experimental findings in H.M. Semin. Neurol. 4, 249–259.

Crary, J.F., Shao, C.Y., Mirra, S.S., Hernandez, A.I., Sacktor, T.C., 2006. Atypical protein kinase C in neurodegenerative disease I: PKMzeta aggregates with limbic neurofibrillary tangles and AMPA receptors in Alzheimer disease. J. Neuropathol. Exp. Neurol. 65, 319–326.

Cupp, C.J., Uemura, E., 1980. Age-related changes in prefrontal cortex of *Macaca mulatta*: quantitative analysis of dendritic branching patterns. Exp. Neurol. 69, 143–163.

Davies, S.N., Lester, R.A., Reymann, K.G., Collingridge, G.L., 1989. Temporally distinct pre- and post-synaptic mechanisms maintain long-term potentiation. Nature 338, 500–503.

Davis, S., Butcher, S.P., Morris, R.G., 1992. The NMDA receptor antagonist D-2-amino-5-phosphonopentanoate (D-AP5) impairs spatial learning and LTP in vivo at intracerebral concentrations comparable to those that block LTP in vitro. J. Neurosci. 12, 21–34.

de Groot, D.M., Bierman, E.P., 1986. A critical evaluation of methods for estimating the numerical density of synapses. J. Neurosci. Methods 18, 79–101.

de Groot, D.M., Bierman, E.P., 1987. Numerical changes in rat hippocampal synapses. An effect of "ageing"? Acta Stereol. 6, 53–58.

Duan, H., Wearne, S.L., Rocher, A.B., Macedo, A., Morrison, J.H., Hof, P.R., 2003. Age-related dendritic and spine changes in corticocortically projecting neurons in macaque monkeys. Cereb. Cortex 13, 950–961.

Dumitriu, D., Hao, J., Hara, Y., Kaufmann, J., Janssen, W.G., Lou, W., Rapp, P.R., Morrison, J.H., 2010. Selective changes in thin spine density and morphology in monkey prefrontal cortex correlate with aging-related cognitive impairment. J. Neurosci. 30, 7507–7515.

Dyall, S.C., Michael, G.J., Whelpton, R., Scott, A.G., Michael-Titus, A.T., 2007. Dietary enrichment with omega-3 polyunsaturated fatty acids reverses age-related decreases in the GluR2 and NR2B glutamate receptor subunits in rat forebrain. Neurobiol. Aging 28, 424–439.

Freeman, G.B., Gibson, G.E., 1987. Selective alteration of mouse brain neurotransmitter release with age. Neurobiol. Aging 8, 147–152.

Fritts, M.E., Asbury, E.T., Horton, J.E., Isaac, W.L., 1998. Medial prefrontal lesion deficits involving or sparing the prelimbic area in the rat. Physiol. Behav. 64, 373–380.

Funahashi, S., Bruce, C.J., Goldman-Rakic, P.S., 1989. Mnemonic coding of visual space in the monkey's dorsolateral prefrontal cortex. J. Neurophysiol. 61, 331–349.

Funahashi, S., Bruce, C.J., Goldman-Rakic, P.S., 1993. Dorsolateral prefrontal lesions and oculomotor delayed-response performance: evidence for mnemonic "scotomas". J. Neurosci. 13, 1479–1497.

Gallagher, M., Burwell, R., Burchinal, M., 1993. Severity of spatial learning impairment in aging: development of a learning index for performance in the Morris water maze. Behav. Neurosci. 107, 618–626.

Ganeshina, O., Berry, R.W., Petralia, R.S., Nicholson, D.A., Geinisman, Y., 2004a. Differences in the expression of AMPA and NMDA receptors between axospinous perforated and nonperforated synapses are related to the configuration and size of postsynaptic densities. J. Comp. Neurol. 468, 86–95.

Ganeshina, O., Berry, R.W., Petralia, R.S., Nicholson, D.A., Geinisman, Y., 2004b. Synapses with a segmented, completely partitioned postsynaptic density express more AMPA receptors than other axospinous synaptic junctions. Neuroscience 125, 615–623.

Gazzaley, A.H., Siegel, S.J., Kordower, J.H., Mufson, E.J., Morrison, J.H., 1996. Circuit-specific alterations of N-methyl-D-aspartate receptor subunit 1 in the dentate gyrus of aged monkeys. Proc. Natl. Acad. Sci. USA 93, 3121–3125.

Gazzaley, A.H., Thakker, M.M., Hof, P.R., Morrison, J.H., 1997. Preserved number of entorhinal cortex layer II neurons in aged macaque monkeys. Neurobiol. Aging 18, 549–553.

Geiger, J.R., Melcher, T., Koh, D.S., Sakmann, B., Seeburg, P.H., Jonas, P., Monyer, H., 1995. Relative abundance of subunit mRNAs determines gating and Ca2+ permeability of AMPA receptors in principal neurons and interneurons in rat CNS. Neuron 15, 193–204.

Geinisman, Y., Berry, R.W., Disterhoft, J.F., Power, J.M., Van der Zee, E.A., 2001. Associative learning elicits the formation of multiple-synapse boutons. J. Neurosci. 21, 5568–5573.

Geinisman, Y., deToledo-Morrell, L., Morrell, F., 1986a. Aged rats need a preserved complement of perforated axospinous synapses per hippocampal neuron to maintain good spatial memory. Brain Res. 398, 266–275.

Geinisman, Y., deToledo-Morrell, L., Morrell, F., 1986b. Loss of perforated synapses in the dentate gyrus: morphological substrate of memory deficit in aged rats. Proc. Natl. Acad. Sci. USA 83, 3027–3031.

Geinisman, Y., deToledo-Morrell, L., Morrell, F., 1991. Induction of long-term potentiation is associated with an increase in the number of axospinous synapses with segmented postsynaptic densities. Brain Res. 566, 77–88.

Geinisman, Y., deToledo-Morrell, L., Morrell, F., Heller, R.E., 1995. Hippocampal markers of age-related memory dysfunction: behavioral, electrophysiological and morphological perspectives. Prog. Neurobiol. 45, 223–252.

Geinisman, Y., deToledo-Morrell, L., Morrell, F., Heller, R.E., Rossi, M., Parshall, R.F., 1993. Structural synaptic correlate of long-term potentiation: formation of axospinous synapses with multiple, completely partitioned transmission zones. Hippocampus 3, 435–445.

Geinisman, Y., deToledo-Morrell, L., Morrell, F., Persina, I.S., Rossi, M., 1992a. Age-related loss of axospinous synapses formed by two afferent systems in the rat dentate gyrus as revealed by the unbiased stereological dissector technique. Hippocampus 2, 437–444.

Geinisman, Y., deToledo-Morrell, L., Morrell, F., Persina, I.S., Rossi, M., 1992b. Structural synaptic plasticity associated with the induction of long-term potentiation is preserved in the dentate gyrus of aged rats. Hippocampus 2, 445–456.

Geinisman, Y., Morrell, F., deToledo-Morrell, L., 1992c. Increase in the number of axospinous synapses with segmented postsynaptic densities following hippocampal kindling. Brain Res. 569, 341–347.

Geinisman, Y., Ganeshina, O., Yoshida, R., Berry, R.W., Disterhoft, J.F., Gallagher, M., 2004. Aging, spatial learning, and total synapse number in the rat CA1 stratum radiatum. Neurobiol. Aging 25, 407–416.

Geinisman, Y., Gundersen, H.J., van der Zee, E., West, M.J., 1996. Unbiased stereological estimation of the total number of synapses in a brain region. J. Neurocytol. 25, 805–819.

Gilardi, K.V., Shideler, S.E., Valverde, C.R., Roberts, J.A., Lasley, B.L., 1997. Characterization of the onset of menopause in the rhesus macaque. Biol. Reprod. 57, 335–340.
Goldman-Rakic, P.S., 1995. Cellular basis of working memory. Neuron 14, 477–485.
Greenough, W.T., West, R.W., DeVoogd, T.J., 1978. Subsynaptic plate perforations: changes with age and experience in the rat. Science 202, 1096–1098.
Groc, L., Choquet, D., 2006. AMPA and NMDA glutamate receptor trafficking: multiple roads for reaching and leaving the synapse. Cell Tissue Res. 326, 423–438.
Hao, J., Rapp, P.R., Janssen, W.G., Lou, W., Lasley, B.L., Hof, P.R., Morrison, J.H., 2007. Interactive effects of age and estrogen on cognition and pyramidal neurons in monkey prefrontal cortex. Proc. Natl. Acad. Sci. USA 104, 11465–11470.
Hara, Y., Park, C.S., Janssen, W.G., Punsoni, M., Rapp, P.R., Morrison, J.H., 2011a. Synaptic characteristics of dentate gyrus axonal boutons and their relationships with aging, menopause, and memory in female rhesus monkeys. J. Neurosci. 31, 7737–7744.
Hara, Y., Rapp, P.R., Morrison, J.H., 2011b. Neuronal and morphological bases of cognitive decline in aged rhesus monkeys. Age (Dordrecht).
Hara, Y., Park, C.S., Janssen, W.G., Roberts, M.T., Morrison, J.H., Rapp, P.R., 2012a. Synaptic correlates of memory and menopause in the hippocampal dentate gyrus in rhesus monkeys. Neurobiol. Aging 33 (421), e17–28.
Hara, Y., Punsoni, M., Yuk, F., Park, C.S., Janssen, W.G., Rapp, P.R., Morrison, J.H., 2012b. Synaptic distributions of GluA2 and PKMζ in the monkey dentate gyrus and their relationships with aging and memory. J. Neurosci. 32, 7336–7344.
Hardt, O., Migues, P.V., Hastings, M., Wong, J., Nader, K., 2010. PKMzeta maintains 1-day- and 6-day-old long-term object location but not object identity memory in dorsal hippocampus. Hippocampus 20, 691–695.
Harris, K.M., 1995. How multiple-synapse boutons could preserve input specificity during an interneuronal spread of LTP. Trends Neurosci. 18, 365–369.
Henderson, V.W., 2008. Cognitive changes after menopause: influence of estrogen. Clin. Obstet. Gynecol. 51, 618–626.
Henley, J.M., Barker, E.A., Glebov, O.O., 2011. Routes, destinations and delays: recent advances in AMPA receptor trafficking. Trends Neurosci. 34, 258–268.
Hernandez, A.I., Blace, N., Crary, J.F., Serrano, P.A., Leitges, M., Libien, J.M., Weinstein, G., Tcherapanov, A., Sacktor, T.C., 2003. Protein kinase M zeta synthesis from a brain mRNA encoding an independent protein kinase C zeta catalytic domain. Implications for the molecular mechanism of memory. J. Biol. Chem. 278, 40305–40316.
Heynen, A.J., Quinlan, E.M., Bae, D.C., Bear, M.F., 2000. Bidirectional, activity-dependent regulation of glutamate receptors in the adult hippocampus in vivo. Neuron 28, 527–536.
Hof, P.R., Duan, H., Page, T.L., Einstein, M., Wicinski, B., He, Y., Erwin, J.M., Morrison, J.H., 2002. Age-related changes in GluR2 and NMDAR1 glutamate receptor subunit protein immunoreactivity in corticocortically projecting neurons in macaque and patas monkeys. Brain Res. 928, 175–186.
Hof, P.R., Nimchinsky, E.A., Young, W.G., Morrison, J.H., 2000. Numbers of meynert and layer IVB cells in area V1: a stereologic analysis in young and aged macaque monkeys. J. Comp. Neurol. 420, 113–126.
Hollmann, M., Heinemann, S., 1994. Cloned glutamate receptors. Annu. Rev. Neurosci. 17, 31–108.
Holtmaat, A.J., Trachtenberg, J.T., Wilbrecht, L., Shepherd, G.M., Zhang, X., Knott, G.W., Svoboda, K., 2005. Transient and persistent dendritic spines in the neocortex in vivo. Neuron 45, 279–291.
Huettner, J.E., 2003. Kainate receptors and synaptic transmission. Prog. Neurobiol. 70, 387–407.
Isaac, J.T., Ashby, M.C., McBain, C.J., 2007. The role of the GluR2 subunit in AMPA receptor function and synaptic plasticity. Neuron 54, 859–871.
Johnson, M., Perry, R.H., Piggott, M.A., Court, J.A., Spurden, D., Lloyd, S., Ince, P.G., Perry, E.K., 1996. Glutamate receptor binding in the human hippocampus and adjacent cortex during development and aging. Neurobiol. Aging 17, 639–651.
Jonas, P., Racca, C., Sakmann, B., Seeburg, P.H., Monyer, H., 1994. Differences in Ca2+ permeability of AMPA-type glutamate receptor channels in neocortical neurons caused by differential GluR-B subunit expression. Neuron 12, 1281–1289.

Jones, T.A., Chu, C.J., Grande, L.A., Gregory, A.D., 1999. Motor skills training enhances lesion-induced structural plasticity in the motor cortex of adult rats. J. Neurosci. 19, 10153–10163.
Kabaso, D., Coskren, P.J., Henry, B.I., Hof, P.R., Wearne, S.L., 2009. The electrotonic structure of pyramidal neurons contributing to prefrontal cortical circuits in macaque monkeys is significantly altered in aging. Cereb. Cortex 19, 2248–2268.
Kasai, H., Matsuzaki, M., Noguchi, J., Yasumatsu, N., Nakahara, H., 2003. Structure–stability–function relationships of dendritic spines. Trends Neurosci. 26, 360–368.
Kesner, R.P., Hunt, M.E., Williams, J.M., Long, J.M., 1996. Prefrontal cortex and working memory for spatial response, spatial location, and visual object information in the rat. Cereb. Cortex 6, 311–318.
Kessels, H.W., Malinow, R., 2009. Synaptic AMPA receptor plasticity and behavior. Neuron 61, 340–350.
Keuker, J.I., Luiten, P.G., Fuchs, E., 2003. Preservation of hippocampal neuron numbers in aged rhesus monkeys. Neurobiol. Aging 24, 157–165.
Kondo, H., Lavenex, P., Amaral, D.G., 2008. Intrinsic connections of the macaque monkey hippocampal formation: I. Dentate gyrus. J. Comp. Neurol. 511, 497–520.
Kutsuwada, T., Kashiwabuchi, N., Mori, H., Sakimura, K., Kushiya, E., Araki, K., Meguro, H., Masaki, H., Kumanishi, T., Arakawa, M., et al., 1992. Molecular diversity of the NMDA receptor channel. Nature 358, 36–41.
Le Jeune, H., Cecyre, D., Rowe, W., Meaney, M.J., Quirion, R., 1996. Ionotropic glutamate receptor subtypes in the aged memory-impaired and unimpaired Long-Evans rat. Neuroscience 74, 349–363.
Liao, D., Hessler, N.A., Malinow, R., 1995. Activation of postsynaptically silent synapses during pairing-induced LTP in CA1 region of hippocampal slice. Nature 375, 400–404.
Liu, P., Smith, P.F., Darlington, C.L., 2008. Glutamate receptor subunits expression in memory-associated brain structures: regional variations and effects of aging. Synapse 62, 834–841.
Liu, X.B., Murray, K.D., Jones, E.G., 2004. Switching of NMDA receptor 2A and 2B subunits at thalamic and cortical synapses during early postnatal development. J. Neurosci. 24, 8885–8895.
Luebke, J., Barbas, H., Peters, A., 2010. Effects of normal aging on prefrontal area 46 in the rhesus monkey. Brain Res. Rev. 62, 212–232.
Luebke, J.I., Chang, Y.M., 2007. Effects of aging on the electrophysiological properties of layer 5 pyramidal cells in the monkey prefrontal cortex. Neuroscience 150, 556–562.
Magnusson, K.R., 1998. Aging of glutamate receptors: correlations between binding and spatial memory performance in mice. Mech. Ageing Dev. 104, 227–248.
Magnusson, K.R., 2000. Declines in mRNA expression of different subunits may account for differential effects of aging on agonist and antagonist binding to the NMDA receptor. J. Neurosci. 20, 1666–1674.
Magnusson, K.R., Brim, B.L., Das, S.R., 2010. Selective vulnerabilities of N-methyl-D-aspartate (NMDA) receptors during brain aging. Front. Aging Neurosci. 2, 11.
Magnusson, K.R., Cotman, C.W., 1993. Age-related changes in excitatory amino acid receptors in two mouse strains. Neurobiol. Aging 14, 197–206.
Magnusson, K.R., Nelson, S.E., Young, A.B., 2002. Age-related changes in the protein expression of subunits of the NMDA receptor. Brain Res. Mol. Brain Res. 99, 40–45.
Malenka, R.C., Nicoll, R.A., 1993. NMDA-receptor-dependent synaptic plasticity: multiple forms and mechanisms. Trends Neurosci. 16, 521–527.
Matsuzaki, M., Ellis-Davies, G.C., Nemoto, T., Miyashita, Y., Iino, M., Kasai, H., 2001. Dendritic spine geometry is critical for AMPA receptor expression in hippocampal CA1 pyramidal neurons. Nat. Neurosci. 4, 1086–1092.
Mead, A.N., Stephens, D.N., 2003. Involvement of AMPA receptor GluR2 subunits in stimulus-reward learning: evidence from glutamate receptor gria2 knock-out mice. J. Neurosci. 23, 9500–9507.
Medvedev, N.I., Rodriguez-Arellano, J.J., Popov, V.I., Davies, H.A., Tigaret, C.M., Schoepfer, R., Stewart, M.G., 2008. The glutamate receptor 2 subunit controls post-synaptic density complexity and spine shape in the dentate gyrus. Eur. J. Neurosci. 27, 315–325.
Merrill, D.A., Roberts, J.A., Tuszynski, M.H., 2000. Conservation of neuron number and size in entorhinal cortex layers II, III, and V/VI of aged primates. J. Comp. Neurol. 422, 396–401.
Migues, P.V., Hardt, O., Wu, D.C., Gamache, K., Sacktor, T.C., Wang, Y.T., Nader, K., 2010. PKMzeta maintains memories by regulating GluR2-dependent AMPA receptor trafficking. Nat. Neurosci. 13, 630–634.

Monaghan, D.T., Cotman, C.W., 1982. The distribution of [3H]kainic acid binding sites in rat CNS as determined by autoradiography. Brain Res. 252, 91–100.
Monyer, H., Burnashev, N., Laurie, D.J., Sakmann, B., Seeburg, P.H., 1994. Developmental and regional expression in the rat brain and functional properties of four NMDA receptors. Neuron 12, 529–540.
Monyer, H., Sprengel, R., Schoepfer, R., Herb, A., Higuchi, M., Lomeli, H., Burnashev, N., Sakmann, B., Seeburg, P.H., 1992. Heteromeric NMDA receptors: molecular and functional distinction of subtypes. Science 256, 1217–1221.
Moore, T.L., Killiany, R.J., Herndon, J.G., Rosene, D.L., Moss, M.B., 2006. Executive system dysfunction occurs as early as middle-age in the rhesus monkey. Neurobiol. Aging 27, 1484–1493.
Moriyoshi, K., Masu, M., Ishii, T., Shigemoto, R., Mizuno, N., Nakanishi, S., 1991. Molecular cloning and characterization of the rat NMDA receptor. Nature 354, 31–37.
Morris, R., 1984. Developments of a water-maze procedure for studying spatial learning in the rat. J. Neurosci. Methods 11, 47–60.
Morris, R.G., Anderson, E., Lynch, G.S., Baudry, M., 1986. Selective impairment of learning and blockade of long-term potentiation by an N-methyl-D-aspartate receptor antagonist, AP5. Nature 319, 774–776.
Morrison, J.H., Hof, P.R., 1997. Life and death of neurons in the aging brain. Science 278, 412–419.
Moss, M.B., Rosene, D.L., Peters, A., 1988. Effects of aging on visual recognition memory in the rhesus monkey. Neurobiol. Aging 9, 495–502.
Muller, D., Joly, M., Lynch, G., 1988. Contributions of quisqualate and NMDA receptors to the induction and expression of LTP. Science 242, 1694–1697.
Nakagawa, T., 2010. The biochemistry, ultrastructure, and subunit assembly mechanism of AMPA receptors. Mol. Neurobiol. 42, 161–184.
Newton, I.G., Forbes, M.E., Linville, M.C., Pang, H., Tucker, E.W., Riddle, D.R., Brunso-Bechtold, J.K., 2008. Effects of aging and caloric restriction on dentate gyrus synapses and glutamate receptor subunits. Neurobiol. Aging 29, 1308–1318.
Nicholson, D.A., Yoshida, R., Berry, R.W., Gallagher, M., Geinisman, Y., 2004. Reduction in size of perforated postsynaptic densities in hippocampal axospinous synapses and age-related spatial learning impairments. J. Neurosci. 24, 7648–7653.
Nicolle, M.M., Bizon, J.L., Gallagher, M., 1996. In vitro autoradiography of ionotropic glutamate receptors in hippocampus and striatum of aged Long-Evans rats: relationship to spatial learning. Neuroscience 74, 741–756.
Nusser, Z., Lujan, R., Laube, G., Roberts, J.D., Molnar, E., Somogyi, P., 1998. Cell type and pathway dependence of synaptic AMPA receptor number and variability in the hippocampus. Neuron 21, 545–559.
Oster, H., Eichele, G., Leitges, M., 2004. Differential expression of atypical PKCs in the adult mouse brain. Brain Res. Mol. Brain Res. 127, 79–88.
Pagliusi, S.R., Gerrard, P., Abdallah, M., Talabot, D., Catsicas, S., 1994. Age-related changes in expression of AMPA-selective glutamate receptor subunits: is calcium-permeability altered in hippocampal neurons? Neuroscience 61, 429–433.
Passafaro, M., Nakagawa, T., Sala, C., Sheng, M., 2003. Induction of dendritic spines by an extracellular domain of AMPA receptor subunit GluR2. Nature 424, 677–681.
Pastalkova, E., Serrano, P., Pinkhasova, D., Wallace, E., Fenton, A.A., Sacktor, T.C., 2006. Storage of spatial information by the maintenance mechanism of LTP. Science 313, 1141–1144.
Pelleymounter, M.A., Beatty, G., Gallagher, M., 1990. Hippocampal 3H-CPP binding and spatial learning deficits in aged rats. Psychobiology 18, 298–304.
Peters, A., Kaiserman-Abramof, I.R., 1969. The small pyramidal neuron of the rat cerebral cortex. The synapses upon dendritic spines. Z. Zellforsch. Mikrosk. Anat. 100, 487–506.
Peters, A., Leahu, D., Moss, M.B., McNally, K.J., 1994. The effects of aging on area 46 of the frontal cortex of the rhesus monkey. Cereb. Cortex 4, 621–635.
Peters, A., Morrison, J.H., Rosene, D.L., Hyman, B.T., 1998a. Feature article: are neurons lost from the primate cerebral cortex during normal aging? Cereb. Cortex 8, 295–300.
Peters, A., Sethares, C., Moss, M.B., 1998b. The effects of aging on layer 1 in area 46 of prefrontal cortex in the rhesus monkey. Cereb. Cortex 8, 671–684.
Peters, A., Moss, M.B., Sethares, C., 2001. The effects of aging on layer 1 of primary visual cortex in the rhesus monkey. Cereb. Cortex 11, 93–103.

Peters, A., Rosene, D.L., Moss, M.B., Kemper, T.L., Abraham, C.R., Tigges, J., Albert, M.S., 1996. Neurobiological bases of age-related cognitive decline in the rhesus monkey. J. Neuropathol. Exp. Neurol. 55, 861–874.

Peters, A., Sethares, C., 2002. The effects of age on the cells in layer 1 of primate cerebral cortex. Cereb. Cortex 12, 27–36.

Peters, A., Sethares, C., Luebke, J.I., 2008. Synapses are lost during aging in the primate prefrontal cortex. Neuroscience 152, 970–981.

Potier, B., Billard, J.M., Riviere, S., Sinet, P.M., Denis, I., Champeil-Potokar, G., Grintal, B., Jouvenceau, A., Kollen, M., Dutar, P., 2010. Reduction in glutamate uptake is associated with extrasynaptic NMDA and metabotropic glutamate receptor activation at the hippocampal CA1 synapse of aged rats. Aging Cell 9, 722–735.

Presty, S.K., Bachevalier, J., Walker, L.C., Struble, R.G., Price, D.L., Mishkin, M., Cork, L.C., 1987. Age differences in recognition memory of the rhesus monkey (*Macaca mulatta*). Neurobiol. Aging 8, 435–440.

Price, J.L., Ko, A.I., Wade, M.J., Tsou, S.K., McKeel, D.W., Morris, J.C., 2001. Neuron number in the entorhinal cortex and CA1 in preclinical Alzheimer disease. Arch. Neurol. 58, 1395–1402.

Rapp, P.R., 1995. Cognitive neuroscience perspectives on aging in nonhuman primates. In: Nakajima, T., Ono, T. (Eds.), Emotion, Memory and Behavior: Studies on Human and Nonhuman Primates. Japan Scientific Societies Press, Tokyo, pp. 197–211.

Rapp, P.R., Amaral, D.G., 1989. Evidence for task-dependent memory dysfunction in the aged monkey. J. Neurosci. 9, 3568–3576.

Rapp, P.R., Amaral, D.G., 1991. Recognition memory deficits in a subpopulation of aged monkeys resemble the effects of medial temporal lobe damage. Neurobiol. Aging 12, 481–486.

Rapp, P.R., Morrison, J.H., Roberts, J.A., 2003. Cyclic estrogen replacement improves cognitive function in aged ovariectomized rhesus monkeys. J. Neurosci. 23, 5708–5714.

Represa, A., Tremblay, E., Ben-Ari, Y., 1987. Kainate binding sites in the hippocampal mossy fibers: localization and plasticity. Neuroscience 20, 739–748.

Roberts, J.A., Gilardi, K.V., Lasley, B., Rapp, P.R., 1997. Reproductive senescence predicts cognitive decline in aged female monkeys. Neuroreport 8, 2047–2051.

Rosene, D.L., Van Hoesen, G.W., 1987. The hippocampal formation of the primate brain: a review of some comparative aspects of cytoarchitecture and connections. In: Jones, E.G., Peters, A. (Eds.), Cerebral Cortex: Further Aspects of Cortical Function, Including Hippocampus. Plenum Press, New York, pp. 345–456.

Sacktor, T.C., 2011. How does PKMzeta maintain long-term memory? Nat. Rev. Neurosci. 12, 9–15.

Sacktor, T.C., Osten, P., Valsamis, H., Jiang, X., Naik, M.U., Sublette, E., 1993. Persistent activation of the zeta isoform of protein kinase C in the maintenance of long-term potentiation. Proc. Natl. Acad. Sci. USA 90, 8342–8346.

Scheff, S.W., Price, D.A., Schmitt, F.A., DeKosky, S.T., Mufson, E.J., 2007. Synaptic alterations in CA1 in mild Alzheimer disease and mild cognitive impairment. Neurology 68, 1501–1508.

Scheff, S.W., Price, D.A., Sparks, D.L., 2001. Quantitative assessment of possible age-related change in synaptic numbers in the human frontal cortex. Neurobiol. Aging 22, 355–365.

Serrano, P., Friedman, E.L., Kenney, J., Taubenfeld, S.M., Zimmerman, J.M., Hanna, J., Alberini, C., Kelley, A.E., Maren, S., Rudy, J.W., Yin, J.C., Sacktor, T.C., Fenton, A.A., 2008. PKMzeta maintains spatial, instrumental, and classically conditioned long-term memories. PLoS Biol. 6, 2698–2706.

Shamy, J.L., Buonocore, M.H., Makaron, L.M., Amaral, D.G., Barnes, C.A., Rapp, P.R., 2006. Hippocampal volume is preserved and fails to predict recognition memory impairment in aged rhesus monkeys (*Macaca mulatta*). Neurobiol. Aging 27, 1405–1415.

Shamy, J.L., Habeck, C., Hof, P.R., Amaral, D.G., Fong, S.G., Buonocore, M.H., Stern, Y., Barnes, C.A., Rapp, P.R., 2010. Volumetric correlates of spatiotemporal working and recognition memory impairment in aged rhesus monkeys. Cereb. Cortex 21 (7), 1559–1573.

Shema, R., Hazvi, S., Sacktor, T.C., Dudai, Y., 2009. Boundary conditions for the maintenance of memory by PKMzeta in neocortex. Learn. Mem. 16, 122–128.

Simic, G., Kostovic, I., Winblad, B., Bogdanovic, N., 1997. Volume and number of neurons of the human hippocampal formation in normal aging and Alzheimer's disease. J. Comp. Neurol. 379, 482–494.

Smith, D.E., Rapp, P.R., McKay, H.M., Roberts, J.A., Tuszynski, M.H., 2004. Memory impairment in aged primates is associated with focal death of cortical neurons and atrophy of subcortical neurons. J. Neurosci. 24, 4373–4381.

Soghomonian, J.J., Sethares, C., Peters, A., 2010. Effects of age on axon terminals forming axosomatic and axodendritic inhibitory synapses in prefrontal cortex. Neuroscience 168, 74–81.

Sonntag, W.E., Bennett, S.A., Khan, A.S., Thornton, P.L., Xu, X., Ingram, R.L., Brunso-Bechtold, J.K., 2000. Age and insulin-like growth factor-1 modulate N-methyl-D-aspartate receptor subtype expression in rats. Brain Res. Bull. 51, 331–338.

Spencer, J.L., Waters, E.M., Romeo, R.D., Wood, G.E., Milner, T.A., McEwen, B.S., 2008. Uncovering the mechanisms of estrogen effects on hippocampal function. Front. Neuroendocrinol. 29, 219–237.

Spruston, N., McBain, C., 2007. Structural and functional properties of hippocampal neurons. In: Anderson, P. (Ed.), The Hippocampus Book. Oxford University Press, Inc., New York, pp. 133–242.

Squire, L.R., Zola-Morgan, S., 1988. Memory: brain systems and behavior. Trends Neurosci. 11, 170–175.

Sterio, D.C., 1984. The unbiased estimation of number and sizes of arbitrary particles using the disector. J. Microsc. 134, 127–136.

Stranahan, A.M., Jiam, N.T., Stocker, A.M., Gallagher, M., 2011. Aging reduces total neuron number in the dorsal component of the rodent prefrontal cortex. J. Comp. Neurol.

Tamaru, M., Yoneda, Y., Ogita, K., Shimizu, J., Nagata, Y., 1991. Age-related decreases of the N-methyl-D-aspartate receptor complex in the rat cerebral cortex and hippocampus. Brain Res. 542, 83–90.

Tang, Y.P., Shimizu, E., Dube, G.R., Rampon, C., Kerchner, G.A., Zhuo, M., Liu, G., Tsien, J.Z., 1999. Genetic enhancement of learning and memory in mice. Nature 401, 63–69.

Tigges, J., Herndon, J.G., Rosene, D.L., 1995. Mild age-related changes in the dentate gyrus of adult rhesus monkeys. Acta Anat. (Basel) 153, 39–48.

Tigges, J., Herndon, J.G., Rosene, D.L., 1996. Preservation into old age of synaptic number and size in the supragranular layer of the dentate gyrus in rhesus monkeys. Acta Anat. (Basel) 157, 63–72.

Toni, N., Buchs, P.A., Nikonenko, I., Bron, C.R., Muller, D., 1999. LTP promotes formation of multiple spine synapses between a single axon terminal and a dendrite. Nature 402, 421–425.

Toni, N., Buchs, P.A., Nikonenko, I., Povilaitite, P., Parisi, L., Muller, D., 2001. Remodeling of synaptic membranes after induction of long-term potentiation. J. Neurosci. 21, 6245–6251.

Uemura, E., 1980. Age-related changes in prefrontal cortex of *Macaca mulatta*: synaptic density. Exp. Neurol. 69, 164–172.

Uemura, E., 1985a. Age-related changes in the subiculum of *Macaca mulatta*: dendritic branching pattern. Exp. Neurol. 87, 412–427.

Uemura, E., 1985b. Age-related changes in the subiculum of *Macaca mulatta*: synaptic density. Exp. Neurol. 87, 403–411.

Voytko, M.L., Sukhov, R.R., Walker, L.C., Breckler, S.J., Price, D.L., Koliatsos, V.E., 1995. Neuronal number and size are preserved in the nucleus basalis of aged rhesus monkeys. Dementia 6, 131–141.

Voytko, M.L., Tinkler, G.P., 2004. Cognitive function and its neural mechanisms in nonhuman primate models of aging, Alzheimer disease, and menopause. Front. Biosci. 9, 1899–1914.

Walker, M.L., Herndon, J.G., 2008. Menopause in nonhuman primates? Biol. Reprod. 79, 398–406.

Wang, A.C., Hara, Y., Janssen, W.G., Rapp, P.R., Morrison, J.H., 2010. Synaptic estrogen receptor-alpha levels in prefrontal cortex in female rhesus monkeys and their correlation with cognitive performance. J. Neurosci. 30, 12770–12776.

Wenk, G.L., Barnes, C.A., 2000. Regional changes in the hippocampal density of AMPA and NMDA receptors across the lifespan of the rat. Brain Res. 885, 1–5.

Wenk, G.L., Walker, L.C., Price, D.L., Cork, L.C., 1991. Loss of NMDA, but not GABA-A, binding in the brains of aged rats and monkeys. Neurobiol. Aging 12, 93–98.

West, M.J., 1993. Regionally specific loss of neurons in the aging human hippocampus. Neurobiol. Aging 14, 287–293.

West, M.J., Amaral, D.G., Rapp, P.R., 1993. Preserved hippocampal cell number in aged monkeys with recognition memory deficits. Soc. Neurosci. Abstr. 19, 599.

Wheeler, D.D., Ondo, J.G., 1986. Time course of the aging of the high affinity L-glutamate transporter in rat cortical synaptosomes. Exp. Gerontol. 21, 159–168.

Witter, M.P., Van Hoesen, G.W., Amaral, D.G., 1989. Topographical organization of the entorhinal projection to the dentate gyrus of the monkey. J. Neurosci. 9, 216–228.
Woolley, C.S., McEwen, B.S., 1994. Estradiol regulates hippocampal dendritic spine density via an N-methyl-D-aspartate receptor-dependent mechanism. J. Neurosci. 14, 7680–7687.
Yang, Y., Wang, X.B., Frerking, M., Zhou, Q., 2008. Delivery of AMPA receptors to perisynaptic sites precedes the full expression of long-term potentiation. Proc. Natl. Acad. Sci. USA 105, 11388–11393.
Yankova, M., Hart, S.A., Woolley, C.S., 2001. Estrogen increases synaptic connectivity between single presynaptic inputs and multiple postsynaptic CA1 pyramidal cells: a serial electron-microscopic study. Proc. Natl. Acad. Sci. USA 98, 3525–3530.
Yassa, M.A., Muftuler, L.T., Stark, C.E., 2010. Ultrahigh-resolution microstructural diffusion tensor imaging reveals perforant path degradation in aged humans in vivo. Proc. Natl. Acad. Sci. USA 107, 12687–12691.
Yates, M.A., Markham, J.A., Anderson, S.E., Morris, J.R., Juraska, J.M., 2008. Regional variability in age-related loss of neurons from the primary visual cortex and medial prefrontal cortex of male and female rats. Brain Res. 1218, 1–12.
Zhao, X., Rosenke, R., Kronemann, D., Brim, B., Das, S.R., Dunah, A.W., Magnusson, K.R., 2009. The effects of aging on N-methyl-D-aspartate receptor subunits in the synaptic membrane and relationships to long-term spatial memory. Neuroscience 162, 933–945.
Zola-Morgan, S., Squire, L.R., Amaral, D.G., 1986. Human amnesia and the medial temporal region: enduring memory impairment following a bilateral lesion limited to field CA1 of the hippocampus. J. Neurosci. 6, 2950–2967.

CHAPTER ELEVEN

Activity-Induced Fine Structural Changes of Synapses in Mammalian Central Nervous System

Jung-Hwa Tao-Cheng
National Institute of Neurological Disorders and Stroke, NIH, Bethesda, MD, USA

1. INTRODUCTION

The fine structure of synapse revealed by electron microscopy (EM) (Figure 1(A); reviewed in Harris and Weinberg, 2012) is one of the best illustrations of association between structure and function. The presynaptic terminal is filled with many synaptic vesicles (SVs) of a uniform size (~35–50 nm in diameter), providing the structural basis for quantum release of transmitters (Fatt and Katz, 1952). The SVs are released at specialized locations called active zones (AZ, reviewed in Gundelfinger and Fejtova, 2011) facing the postsynaptic density (PSD). The PSD is the dense cytoplasmic material beneath the postsynaptic membrane, and contains high concentrations of receptors and scaffold proteins as well as signaling enzymes (reviewed in Sheng and Kim, 2012). A third structural feature of the synapse is the synaptic cleft (~20 nm) between the pre- and postsynaptic membranes, and the cleft contains the extracellular portions of receptors and adhesion molecules (reviewed in Craig and Kang, 2007).

The great majority of excitatory synapses in mammalian central nervous system (CNS) use glutamate as the transmitter, while the majority of inhibitory synapses use gamma-aminobutyric acid (GABA). Glutamatergic excitatory synapses are called "asymmetric" due to their dense PSDs (arrows in Figure 1(B), (C)), and GABAergic inhibitory synapses are "symmetric" due to similar density of their pre- and postsynaptic membranes (double arrows in Figure 1(C); Gray, 1969; Peters and Palay, 1996). The majority of excitatory synapses in brain are located on spines, which are extended projections from dendritic shafts (Figure 1(B), (C)).

Studies on the classic preparation of the frog neuromuscular junctions provided valuable information on SV recycling of the peripheral nervous system by EM examination (Heuser and Reese, 1973; Ceccarelli et al., 1973). Such studies also demonstrated the critical importance of monitoring activity levels of any excitable tissues in order to achieve meaningful structural analysis. The study of fine structure of synapses in intact mammalian brain is made possible after the success of transcardial perfusion fixation

The Synapse
http://dx.doi.org/10.1016/B978-0-12-418675-0.00011-0

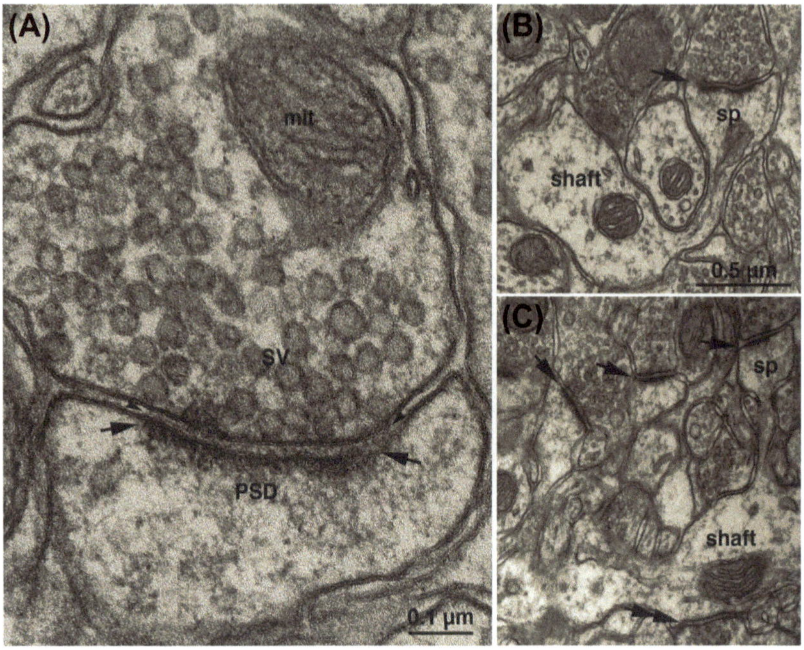

Figure 1 (A) A typical asymmetric excitatory synapse from a perfusion-fixed mouse brain in the CA1 region of the hippocampus. The presynaptic terminal contains numerous synaptic vesicles (SV) and mitochondria (mit) are often present. The postsynaptic density (PSD, area between two arrows) is composed of dense material underneath the postsynaptic membrane and makes the synapse "asymmetric" in appearance. The synaptic cleft is the extracellular gap (marked between two arrowheads) between the pre- and postsynaptic membranes, which are typically rigidly apposed to each other. In hippocampus, excitatory synapses (arrows in B and C) are typically located on spines (sp) arising from dendritic shafts, while inhibitory synapses (double arrows in C) are typically located on dendritic shafts. (B) and (C) share the same scale bar.

(Palay et al., 1962). With the advancement of tissue culture techniques, dissociated neuronal cultures provide an additional experimental model to study CNS neurons in vitro. Immunogold labeling allows localization of various synaptic proteins at the resolution of EM, as well as quantitative analyses on the redistribution of these proteins upon synaptic activity. This chapter reviews the structural changes as well as the redistribution of synaptic proteins of the mammalian CNS synapses.

1.1. Excitatory Synapses in Dissociated Neuronal Cultures Closely Resemble Those in Brain

Dissociated neuronal cultures are grown from various regions of the embryonic brain, and the resulting monolayers indeed form synapses that resemble the parent brain regions in structure and function. This chapter focuses on the dissociated cultures from

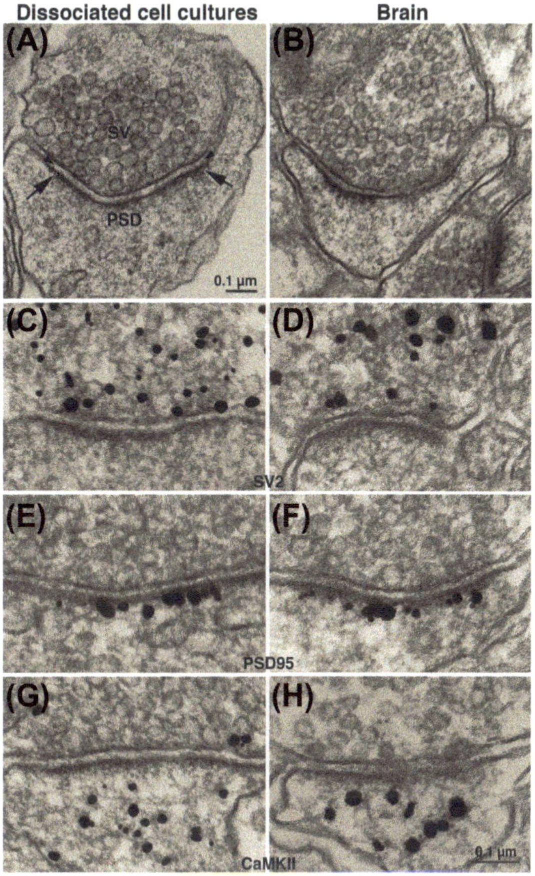

Figure 2 The fine structure of excitatory synapses is very similar in (A) hippocampal dissociated neuronal cultures and (B) hippocampus. All three synapse-defining features are similarly present: (1) the presynaptic terminal is filled with numerous synaptic vesicles (SV); (2) the postsynaptic density (PSD, area between two arrows in A) is well-defined with dense material; and (3) the synaptic cleft (small arrowheads in A) spans a uniform gap between the pre- and postsynaptic membranes. (C–H) Patterns of immunogold labeling of three different marker proteins at synapses in hippocampal dissociated cultures (left column) and in intact hippocampus (right column) are identical. Black dots are silver-enhanced gold particles conjugated to secondary antibodies that bind to primary antibodies to localize the specific proteins. (C, D) SV2, an SV integral membrane protein, is a presynaptic marker and is localized to SV. (E, F) PSD95, a PSD scaffold protein, is localized to the PSD close to the postsynaptic membrane. (G, H) CaMKII, calcium-calmodulin dependent kinase II, is localized in the spine cytoplasm. All synapses in this figure are under basal conditions. (A) and (B) share the same scale bar while (C–H) share the same scale bar.

rat hippocampus. Asymmetric excitatory synapses in dissociated hippocampal cultures (Figure 2(A)) have very similar fine structure as those in perfusion-fixed hippocampus (Figure 2(B)). Furthermore, the distribution patterns of synaptic proteins are also identical in excitatory synapses from dissociated cultures (Figure 2(C), (E), (G)) and from brain (Figure 2(D), (F), (H)). These similarities validate the use of dissociated cultures as an experimental model system to study dynamic changes of the synapses under different stimulation conditions.

2.1. ACTIVITY-DEPENDENT CHANGES AT THE PSD OF EXCITATORY SYNAPSES IN DISSOCIATED NEURONAL CULTURES

2.1.1. Activity Induces Increases in PSD Thickness and Curvature

In dissociated hippocampal neuronal cultures, pharmacological treatments can be applied to elicit uniform excitation or inhibition of synaptic activity. Upon high K^+ (90 mM, 2–3 min) depolarization, the most conspicuous structural changes are the increases of thickness and curvature of the PSD. Under control conditions (Figure 3, left column), the curvature of the PSD is typically flat or concave. After high K^+ depolarization (Figure 3, right column), the PSD increases in thickness (Dosemeci et al., 2001), and its curvature becomes convex and arches into the presynaptic terminal (Tao-Cheng et al., 2010). Similar PSD thickness and curvature changes occur after NMDA treatment (30–50 μM for 2–3 min, Dosemeci et al., 2002), demonstrating that these activity-induced structural changes are mediated by NMDA receptors. The increases in thickness and density of the dense material at the PSD complex suggest that more proteins have moved into this particular region during activity.

Upon depolarization, the thickened PSD complex can be subdivided into two layers (Figure 3(B); Reese et al., 2011): a PSD core that extends to ~30–40 nm from the postsynaptic membrane, and a PSD pallium that is contiguous with the PSD core but located further away and extends to approximately ~100–120 nm from the postsynaptic membrane. The PSD pallium is typically undetectable under resting conditions (Figure 3(A)), but becomes electron dense after stimulation (Figure 3(B)), giving the appearance that PSD has thickened. Interestingly, the density of the PSD pallium is most conspicuous when samples are treated with special chemicals (Voltschanoff and Weinberg, 2001) or osmium tetroxide alone (e.g. Figure 3(B)) for postfixation, and less detectable with reduced osmium-containing potassium ferrocyanide (unpublished observation).

2.1.2. Activity Induces Redistribution of CaMKII, Shanks and SynGAP at the PSD

Using pre-embedding immunogold labeling methods (Tanner et al., 1996), various PSD-associated proteins are localized under basal and depolarized conditions. Labeling for CaMKII, an abundant cytoplasmic kinase, is typically dispersed in the spine under control conditions (Figure 3(C)), but becomes concentrated at the PSD pallium after depolarization (Figure 3(D); Dosemeci et al., 2001) or NMDA receptor activation (Dosemeci et al., 2002). This NMDA-induced CaMKII redistribution is blocked by a specific CaMKII inhibitor, tatCN21 (Tao-Cheng et al., 2013). Interestingly, CYLD, a

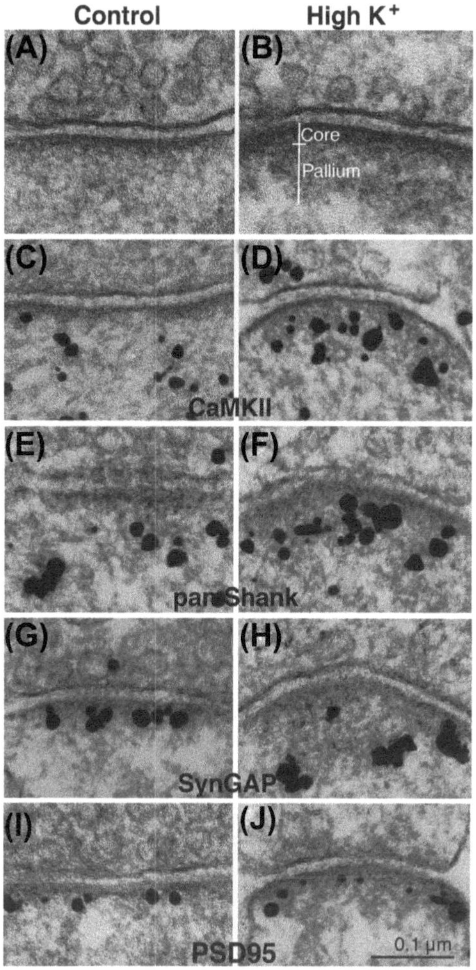

Figure 3 Structural changes and redistribution of postsynaptic proteins upon depolarization. Dissociated hippocampal neuronal cultures under basal conditions (control) and after depolarization (high K^+) are shown in the left and right columns, respectively. After depolarization, the dense material underneath the postsynaptic membrane becomes thicker (right column), revealing the PSD pallium (B). Labeling intensity increases in the PSD pallium for (D) CaMKII and (F) Shank. In contrast, SynGAP moves away from the postsynaptic membrane (H), and the location of PSD95 labeling remains unchanged (J) after depolarization.

deubiquitinase specific for lysine 63-linked polyubiquitins, has a similar activity-dependent redistribution pattern as CaMKII (Dosemeci et al., 2013). Labeling for Shank, a PSD scaffold protein, is typically concentrated at the PSD pallium as well as present in spine cytoplasm under control conditions (Figure 3(E)), but further increases in concentration at the PSD pallium after depolarization (Figure 3(F); Tao-Cheng et al., 2010). Labeling for

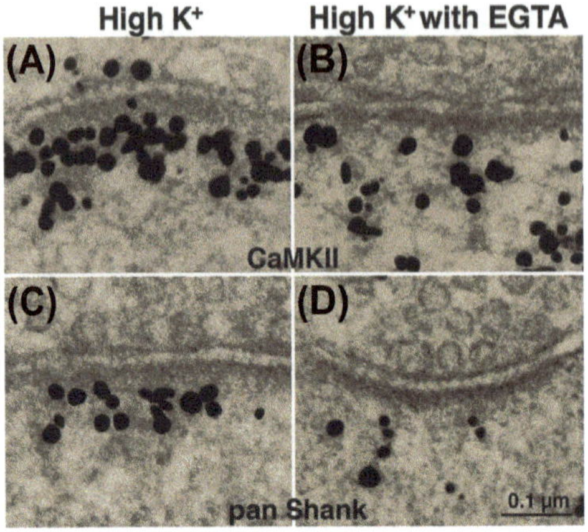

Figure 4 Upon depolarization with high K^+, labeling for (A) CaMKII and (C) Shank become tightly aggregated at the PSD. In the presence of EGTA, high K^+ treatment does not induce PSD thickening or curvature change. Instead, the bulk of (B) CaMKII and many of the (D) Shank labels remain dispersed in the cytoplasm.

SynGAP, a Ras GTPase-activating protein that interacts with PSD95, is typically at the PSD core close to the postsynaptic membrane under control conditions (Figure 3(G)), but moves away from the postsynaptic membrane after depolarization (Figure 3(H); Yang et al., 2011). In contrast, labeling for PSD95, a major component of the PSD scaffold thought to anchor SynGAP, did not change its location close to the postsynaptic membrane upon depolarization (Figure 3(I), (J); Yang et al., 2011). Thus, various components of the PSD move in and out of the PSD complex dynamically during activity, each with potential functional implications.

2.1.3. Activity-Induced Changes at the PSD are Calcium-Dependent and Reversible

When extracellular calcium is chelated by EGTA, high K^+ treatment fails to induce PSD thickness or curvature increase (Figure 4). Furthermore, CaMKII (Figure 4(B)) or Shank (Figure 4(D), Tao-Cheng et al., 2010) also do not translocate to the PSD (Tao-Cheng et al., 2010). These results indicate that calcium is required for the dynamic structural changes as well as the redistribution of CaMKII and Shank at the PSD upon depolarization.

Depolarization-induced structural changes at the PSD and the concurrent redistribution of PSD-associated proteins are reversible after a wash-out of the high K^+ medium and a 30 min incubation in Ringer or tissue culture medium. CaMKII labeling concentration at the PSD significantly increases after depolarization, but returns to control levels after recovery (Figure 5(A); Dosemeci et al., 2001). Concentration of pan Shank labeling at the PSD also increases from basal levels after depolarization, and returns to control levels after recovery (Figure 5(B); Tao-Cheng et al., 2010). Redistribution

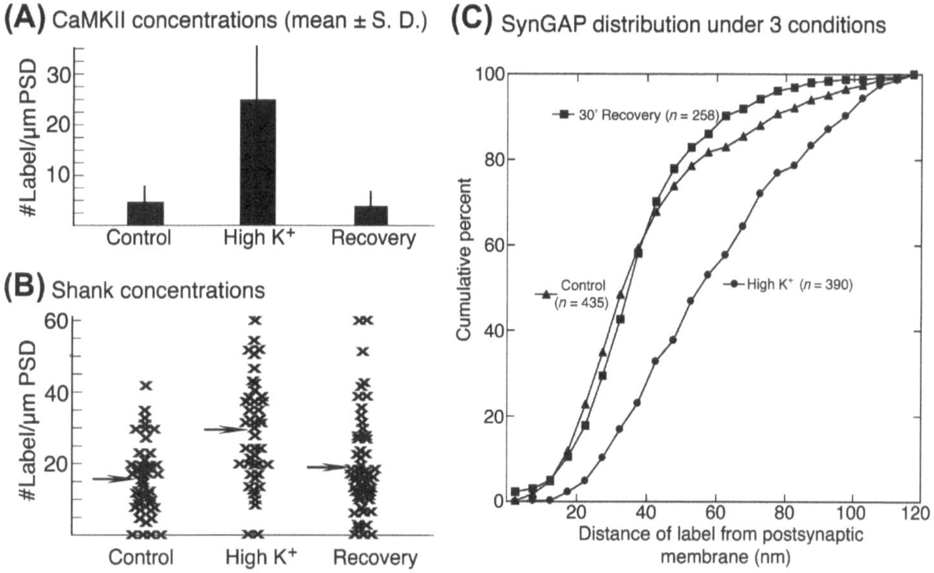

Figure 5 Depolarization-induced redistributions of CaMKII, Shank and SynGAP are reversible. (A) Bar graphs of mean CaMKII concentration at the PSD under three conditions. (B) Scattergram of Shank concentration at the PSD under three conditions, with each cross representing one data point from one synapse. Arrows point to mean values. (C) Cumulative frequency graph showing SynGAP distribution under three conditions (*n* = number of labels measured).

of SynGAP labels is measured in their distance from the postsynaptic membrane. The cumulative frequency curves show that SynGAP moves away from the postsynaptic membrane after depolarization and returns to control positions after recovery (Figure 5(C); Yang et al., 2011). Thus, these activity-induced changes do not represent structural damage but a reversible event (Figure 5).

2.2. ACTIVITY-INDUCED CHANGES OF AMPA RECEPTORS AT EXCITATORY SYNAPSES

2.2.1. Synaptic AMPA Receptors Increase in Concentration upon Depolarization

Trafficking of one subtype of glutamate receptor, the AMPA receptor, largely determines synaptic efficacy. Figure 6 shows immunogold labeling of an AMPA receptor subunit, GluR2, at excitatory synapses in dissociated cultures under control (Figure 6(A)) and depolarization (Figure 6(B)) conditions. GluR2 labels are located in the synaptic cleft due to the extracellular epitope of this particular antibody. Labeling intensity at the PSD increases significantly after high K^+ depolarization (Tao-Cheng et al., 2011), indicating rapid incorporation of AMPA receptors into the synaptic locations after neuronal activity.

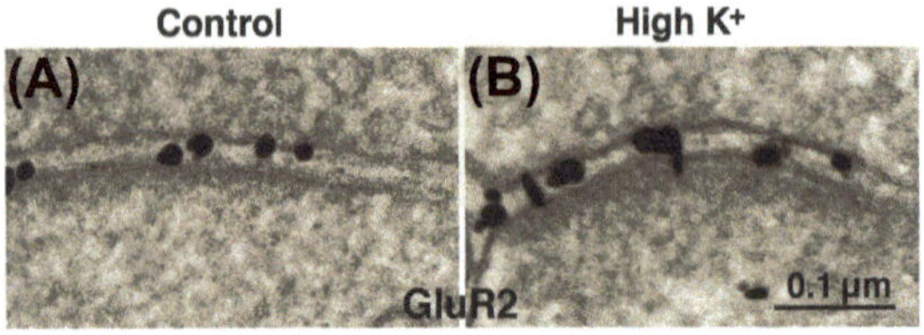

Figure 6 AMPA receptor labeling at excitatory synapses in dissociated hippocampal neurons under control (A) and high K^+ (B) depolarization conditions. GluR2 labeling concentration significantly increases after depolarization.

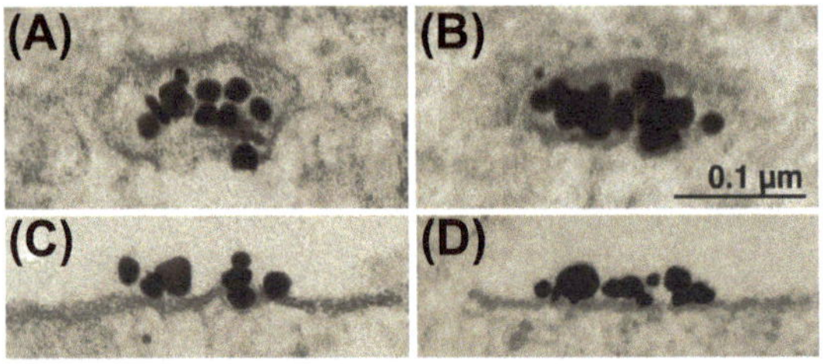

Figure 7 GluR2 labeling in dissociated hippocampal neurons showing possible candidates as precursor of AMPA receptor Exocytosis (A and B). Vesicles containing concentrated AMPA receptors near the surface of (A) soma and (B) dendrite could fuse with the plasma membrane and leave densely labeled patches on the surface (C and D). These patches of concentrate AMPA receptors have a fairly uniform size under different activity states (C from control, and D from depolarized samples). The occurrence frequency of these patches increases with stimulation intensity (Tao-Cheng et al., 2011).

2.2.2. EM Illustrations of AMPA Receptor Exocytosis and Endocytosis

The number of AMPA receptors at synapses could depend on receptor cycling, lateral diffusion, and synaptic trapping (reviewed in Opazo and Choquet, 2011). Live observations via light microscopy elegantly illustrated AMPA receptor exocytosis as vesicles packed with high concentration of receptors fuse with plasma membrane. Due to the instantaneous nature of exocytosis, EM micrographs only managed to capture the potential precursor (Figure 7(A) & (B)) and aftermath (Figure 7(C) & (D)) of, but not AMPA receptor exocytosis itself (Tao-Cheng et al., 2011). To date, no credible EM evidence exists to demonstrate direct exocytosis of AMPA receptors at the PSD.

The vast majority of AMPA receptor exocytosis occurs throughout the somal and dendritic surfaces and is not restricted to spines. Depolarization also induces rapid increases of AMPA receptor concentration in the extrasynaptic plasma membranes (Tao-Cheng et al., 2011). Activity-induced increase of synaptic AMPA receptors most likely occurs through lateral diffusion of more extrasynaptic receptors being trapped at the synaptic sites (Opazo and Choquet, 2011).

Endocytosis of AMPA receptors also occurs throughout the somal and dendritic surfaces, and is clathrin-mediated. Interestingly, coated pits sometimes occur

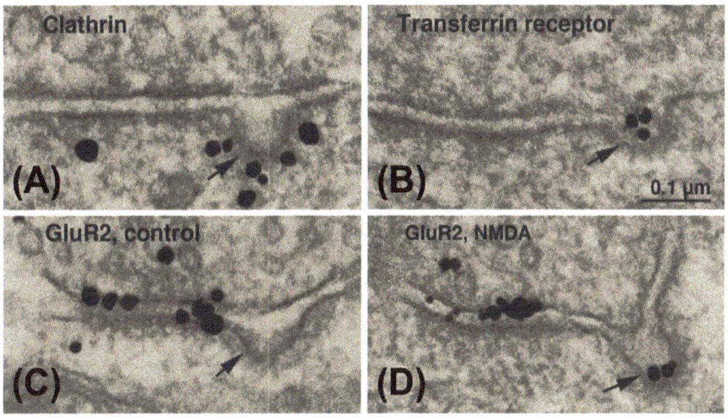

Figure 8 Coated vesicles immediately adjacent to PSDs are consistently labeled for (A) clathrin and (B) transferrin receptor, but typically lack GluR2 labeling under control conditions (C). After NMDA treatment, more AMPA receptors are endocytosed (D).

at peri-PSD locations (Rácz et al., 2004) signifying endocytosis immediately adjacent to the PSD. Virtually all peri-PSD pits label for clathrin (Figure 8(A)) and transferrin receptor (Figure 8(B)), two reliable markers for endocytosis. However, the great majority (94%) of such pits lack GluR2 labeling under basal conditions (Figure 8(C)). After NMDA treatment, a condition known to increase endocytosis of AMPA receptors, the occurrence frequency of peri-PSD pits almost doubled from that in control samples (24 vs 46 pits per 1000 synaptic profiles, respectively), and the percentage of GluR2-labeled peri-PSD pits increased form 6–42% (Figure 8(D)). Thus, peri-PSD endocytosis of AMPA receptors is highly regulated by activity, just as the vast majority of endocytosis of AMPA receptors that occur in somal and dendritic surfaces (Tao-Cheng et al., 2011).

2.3. ACTIVITY-INDUCED CHANGES AT PRESYNAPTIC TERMINALS OF EXCITATORY SYNAPSES IN DISSOCIATED NEURONAL CULTURES

2.3.1. Depolarization-Induced Redistribution of Synaptic Vesicles and Associated Proteins near the Active Zone

The presynaptic AZ is a well-defined area where exocytosis of SVs takes place. In this section, all morphometry on SVs and their associated proteins is measured in defined areas related to the AZ as shown in Figure 9A. Zones I and II contain the first two rows of vesicles; and zone III contains SVs further away from AZ. Upon depolarization, vesicle numbers changed little in zones I and II as vesicles are quickly replenished in this region, but SVs are depleted from zone III (Figure 9(C)). As expected, distribution pattern of SV integral membrane proteins, such as SV2 and synaptophysin, follows the distribution of SVs (Tao-Cheng, 2006).

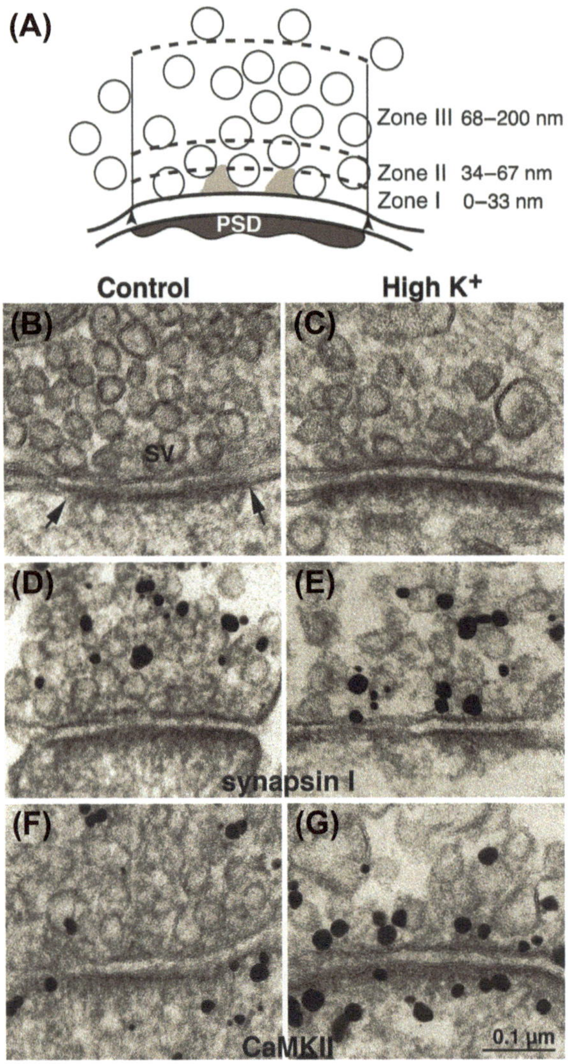

Figure 9 (A) A schematic diagram traced from an EM micrograph illustrates the morphometry criteria (Tao-Cheng, 2006; diagram by Evelyn Ralston). The measurement area near AZ is boxed in by (1) tracing the presynaptic membrane, (2) drawing two parallel lines (arrowheads) perpendicular to the arc of the PSD, and (3) sketching a dashed line at 200 nm from the presynaptic membrane. Zone I contains the AZ cytomatrix (light gray triangles) and the first row of SVs. Zone II contains the second row of SVs from the presynaptic membrane, and zone III contain SVs further away. Under control conditions (left column), synaptic vesicles (SV) are clustered facing the PSD (area between two arrows in B). Upon depolarization, there are little changes in the number of SVs in zones I and II, but depletion of SVs in zone III (B). Labeling for synapsin I is absent in zone I under control conditions (D), but present after depolarization (E). Labeling for CaMKII, a kinase often present in presynaptic terminals as well as in spines, is at the periphery of SV clusters and absent from the AZ under control conditions (F), but present in zone I after depolarization (G).

Interestingly, labeling for synapsin I, an SV-associated protein, is absent in zone I under control conditions (Figure 9(D)), but move into it after depolarization (Figure 9(E)). Synapsin I has been shown to cluster SVs in the reserve pool under resting conditions, and becomes dissociated from the SVs upon stimulation (reviewed in Cesca et al., 2010). These SVs without synapsin I are more mobile and can be recruited for exocytosis. Lack of synapsin I in zone I under basal conditions is consistent with the hypotheses that the vesicles without synapsin I near AZ may be more readily releasable. The presence of synapsin I in zone I upon depolarization suggests that the dissociation from SVs may be incomplete, and some synapsin I may be carried by the replenishing vesicles into zone I, or that the dissociated synapsin I may diffuse into zone I. The possible roles that synapsin I can play at the AZ are extensively reviewed in Cesca et al. (2010).

CaMKII is present in many presynaptic terminals. Labeling of CaMKII is typically absent from tightly clustered SV at the AZ under control conditions (Figure 9(F)), and moves into the AZ upon depolarization (Figure 9(G)). This activity-induced redistribution may leave CaMKII in a position to facilitate phosphorylation of additional proteins at the AZ (Tao-Cheng et al., 2006).

2.3.2. Active Zone Cytomatrix Proteins—Bassoon and Piccolo—Remain Unchanged upon Depolarization

Bassoon and Piccolo/Aczonin, two AZ matrix proteins (reviewed in Gundelfinger and Fejtova, 2011), appear to be stable components of the AZ cytomatrix complex in synapses under different activity states. Both proteins are specifically localized within narrow bands (between 35 and 80 nm; Limbach et al., 2011) from the presynaptic membrane (Figure 10), and their location and concentration hardly changed upon depolarization (Tao-Cheng, 2006). Piccolo is a large protein (550 kDa), and has been elegantly proven to have a compact molecular topology instead of extending from AZ as a long filament (Limbach et al., 2011). Both Bassoon and Piccolo are present early in development, and may play a role in stabilizing incipient synapses (Tao-Cheng, 2007; reviewed in Owald and Sigrist, 2009).

2.3.3. Depolarization-Induced Redistribution of Synaptic Vesicles, CaMKII, and Synapsin I in the Distal Reserve Pool

Under control conditions, SVs are also frequently clustered in locations distal to the AZ (Figure 11(A)), and these vesicles may represent the reserve pool of SVs. Under basal conditions, labeling for CaMKII typically surrounds the vesicle cluster and is absent in the center (Figure 11(C)). In contrast, labeling for synapsin I is present throughout the vesicle cluster (Figure 11(E)). Upon depolarization, SVs become unclustered (Figure 11(B)), and CaMKII labels are now among the dispersed vesicles (Figure 11(D)), occupying the same general domain as synapsin I (Figure 11(F)). CaMKII-mediated phosphorylation of synapsin I has been reported to mobilize SVs

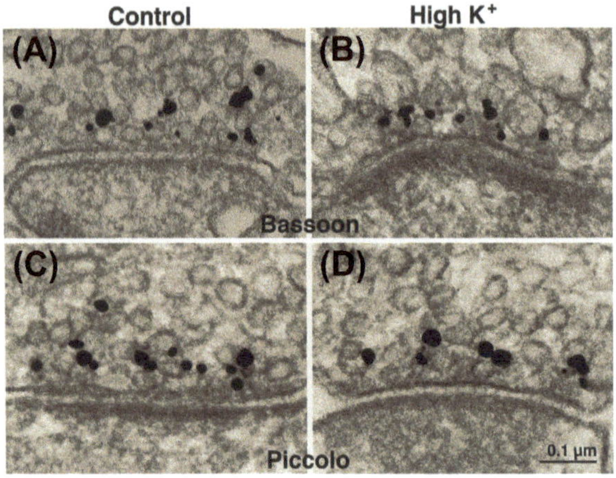

Figure 10 Distribution of bassoon (A, B) and Piccolo (C, D) Remain the same after high K^+ depolarization. *Piccolo antibody is a gift from Dr. Echart Gundelfinger.*

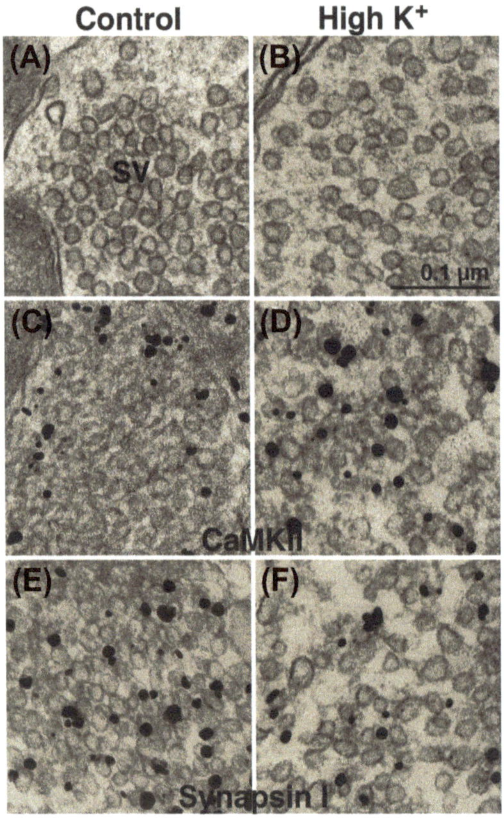

Figure 11 Under control conditions (left column), SVs are often tightly clustered (A) with CaMKII labeling surrounding the cluster (C) and synapsin I labeling inside the cluster (E). Upon depolarization (right column), SVs dispersed into looser arrangement (B) with both CaMKII (D) and synapsin I (F) labels intermingled in the same compartment.

from the reserve pool. These results indicate that CaMKII is typically separated from its substrate, synapsin I, under basal conditions, and moves into the same compartment as synapsin I upon depolarization (Tao-Cheng et al., 2006). Thus, it is likely that as CaMKII molecules become activated and progress toward the center of the SV clusters, synapsin I molecules become phosphorylated and detached from SV, and SVs become unclustered and mobile.

2.4. RAPID TURNOVER OF SYNAPTIC SPINULES IN ORGANOTYPIC HIPPOCAMPAL SLICE CULTURES

Synaptic spinule is a double-membrane invagination inside synaptic terminals (Tarrant and Routtenberg, 1977). Synaptic spinules are extremely rare in dissociated neuronal cultures, but are sometimes seen in acute brain slices and perfusion-fixed brains. In order to investigate the effect of activity on spinule formation, the activity states of all synapses in the sample has to be reliably and uniformly controlled. Organotypic slice cultures prove to be the most suitable experimental system for this purpose.

In hippocampal slice cultures, synaptic spinules are absent in synapses under basal conditions (Figure 12(A)), and form rapidly after high K^+ depolarization (Figure 12(B)–(G)). The most typical synaptic spinule consists of an outer membrane belonging to the axon terminal and an inner membrane belonging to the dendritic element (Figure 12(C)). Spinules exist in various sizes and shapes (Figure 12(C)–(G)), typically have a clathrin coating on their leading edge (arrows in Figure 12(C)–(G)), and can form in both excitatory (Figure 12B) and inhibitory (Figure 12G) synapses.

Synaptic spinules form and disappear quickly within minutes after depolarization (Tao-Cheng et al., 2009). This time course is within the range of that of the formation and resolution of clathrin-coated vesicles, but different from the time course of PSD thickening and recovery. It is possible that synaptic spinules might quickly remove excess axonal membranes and simultaneously reduce spine volumes in the face of intense synaptic activity.

2.5. INHIBITORY SYNAPSES UNDER DIFFERENT ACTIVITY CONDITIONS

2.5.1. Structural Characteristics of Inhibitory Synapses in the Hippocampus

Inhibitory synapses are typically located on dendritic shafts rather than on spines. Unlike excitatory synapses, inhibitory synapses lack a prominent PSD. Instead, the postsynaptic complex (area between two double arrows in Figure 13(A)) is defined by a uniform synaptic cleft (small arrowheads in Figure 13(A)). Two other features are

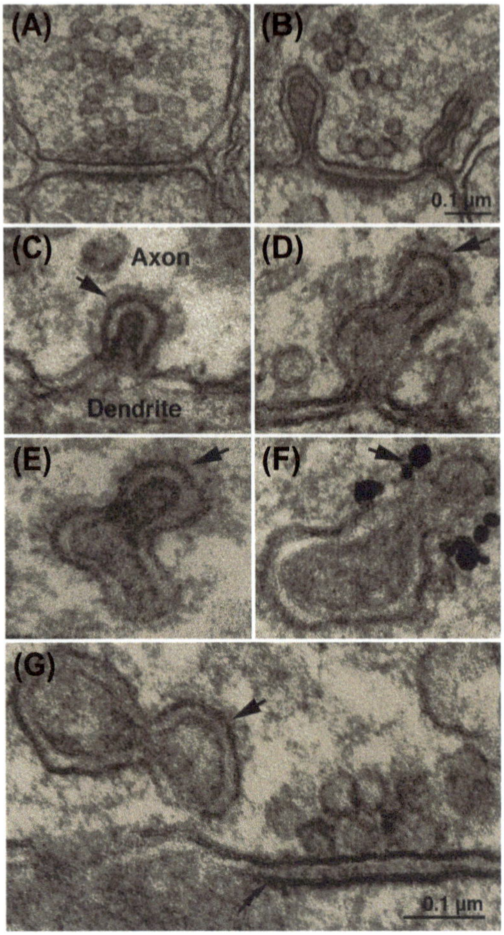

Figure 12 Depolarization-induced formation of synaptic spinules in hippocampal slice cultures. Spinules are absent under basal conditions (A), but quickly appear after high K^+ depolarization (B), often at peri-PSD locations. (C–G) Sampling of spinules from high K^+ depolarized samples. Spinules can also bud from other locations in the presynaptic terminals (C, D) or pinch off from plasma membrane and become free-standing in the cytoplasm (E, F). The leading edges of spinules typically contain a distinctive clathrin coat (arrows in C–G; labeled with a clathrin antibody in F). Spinules form in both excitatory (B) and inhibitory synapses (G, double arrow points to the symmetric postsynaptic membrane). (A) and (B) share the same scale bar; (C–G) share the same scale bar.

not unique to, but more frequent in inhibitory synapses than in excitatory synapses in the hippocampus: (1) peri-PSD pit (large arrowheads in Figure 13(A) & (F)), a coated vesicle in the process of clathrin-mediated endocytosis adjacent to the postsynaptic specialization, and (2) adherence junction (small arrows to the right in Figure 13(A)), a junction with symmetric patches of dark material, often near the synaptic AZ (Peters et al., 1991).

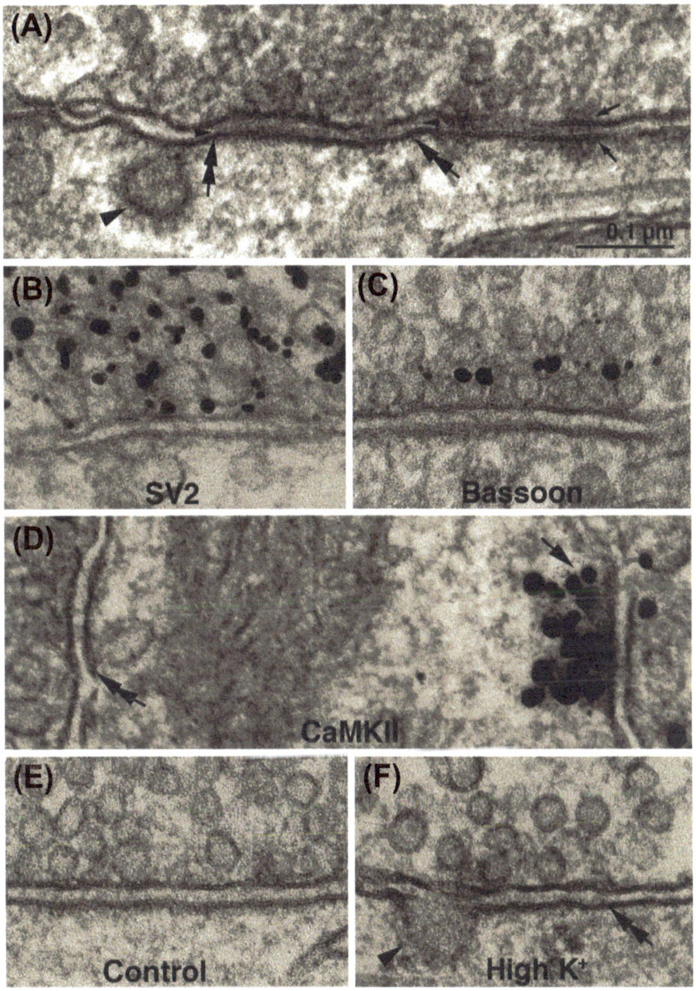

Figure 13 Structural specialization and protein distribution of inhibitory synapses. (A) An inhibitory synapse on a dendritic shaft from a perfusion-fixed hippocampus. The postsynaptic complex is marked between two double arrows; there is little readily visible dense material underneath the postsynaptic membrane. Clathrin-coated pit (arrowhead), indicative of endocytosis, is often seen immediately adjacent to the postsynaptic complex. Adherence junction (small arrows on right) is also often present near inhibitory synapses. (B–F) Inhibitory synapses from cultures of dissociated hippocampal cells. As in excitatory synapses, SV2 labels the synaptic vesicles (B), and Bassoon labels the AZ cytomatrix (C) in inhibitory synapses. Upon high K^+ depolarization, CaMKII only translocates to excitatory synapses (arrow on right in D) but not to inhibitory synapses (double arrows on left in D) located on the same dendritic shaft. High K^+ depolarization does not induce any visible structural changes (F), and the thickness and curvature of the postsynaptic complex remain the same as the control samples (E).

2.5.2. Inhibitory and Excitatory Synapses Share Similar Presynaptic as well as Cleft Proteins

Labeling for SV2, an SV membrane protein, is similarly localized on SVs in inhibitory synapses (Figure 13(B)) as in excitatory synapses (Figure 2(C)). Labeling pattern for Bassoon, an AZ cytomatrix protein, is also similar in inhibitory synapses (Figure 13(C)) as in excitatory synapses (Figure 10(A)). Both types of synapses also share similar adhesion molecules such as the pairing of presynaptic neurexin and postsynaptic neuroligin in their synaptic clefts (Craig and Kang, 2007). However, certain presynaptic proteins, such as synapsin I, although present in both types of synapses, may play different roles (reviewed in Cesca et al., 2010).

2.5.3. Inhibitory and Excitatory Synapses Have Different Transmitter Receptors, Scaffold Proteins, and Signaling Proteins

As expected, the two types of synapses have different transmitter receptors located in the postsynaptic membrane facing their respective AZ. Furthermore, inhibitory synapses have specific scaffold proteins at the postsynaptic complex such as gephrin and collybistin, and lack the PSD scaffold proteins of excitatory synapses, such as PSD95, GKAP, Shank, Homer, and SynGAP (Sheng and Kim, 2012). Interestingly, CaMKII, an abundant kinase linked to memory and learning, only translocates to excitatory synapses but not to inhibitory synapses (Figure 12(D)) after depolarization.

2.5.4. No Activity-Induced Structural Changes are Detected at the Postsynaptic Complex of Inhibitory Synapses

Upon depolarization, although SV depletion occurred in inhibitory synapses (Figure 13(F)) as in excitatory synapses (Figure 9(B)), no conspicuous thickening or curvature change was seen in the postsynaptic complex of inhibitory synapses (Figure 13 E vs F; double arrows in Figure 12(G) & 13(F)). Thus, either no protein was added to or deleted from the inhibitory postsynaptic complex, or too little was altered to make a visible difference that can be detected with the standard EM technique.

2.6. STUDYING SYNAPSES WITH DIFFERENT EXPERIMENTAL MODEL SYSTEMS

The advantages and caveats of studying synapses by EM are evaluated here for four different experimental model systems: (1) perfusion-fixed brain, (2) acute brain slices, (3) organotypic slice cultures, and (4) dissociated neuronal cultures.

2.6.1. Perfusion-Fixed Brains

A well-fixed brain offers the ultimate model system to study synapses in its precise anatomical location and with intact connections. However, perfusion fixation itself can induce structural changes (Tao-Cheng et al., 2007) that can be misidentified as resulting from experimental designs including genetic manipulations, drug treatments, disease models, electrical stimulations, or behavioral training. Thus, perfusion fixation must be carried out with utmost care and critical scrutiny. Dynamic changes at synapses that occur within minutes, such as increases in PSD thickness and curvature, depletion of SVs, and spinule formation are indications of short-term synaptic activity. Other structural features that are more stable, such as spine apparatus or perforations in PSD, are nevertheless location-specific and require critical monitoring of sampling for valid morphometric comparison under different experimental designs.

2.6.1.1. Structural Differences among Excitatory Synapses from Different Regions of the Brain

In mammalian CNS, not all excitatory synapses have the same structural details or size and shape. Thus, when describing synaptic structure, it is critical to specify the exact locations the synapses are sampled from. For example, excitatory synapses in stratum radiatum of the CA1 region of the hippocampus (Figure 14(A)) and those in layer III of the cortex (Figure 14(B)) are both from glutamatergic pyramidal neurons, and have synapses that are similar in shape and structure. In these forebrain synapses, PSDs can be "macula/nonperforated" (a disc-shaped density with no discontinuity, small arrows in Figure 14(A)), typically associated with thin spines; or "perforated" (a disc-like density with holes or discontinuity, large arrows in Figure 14(B)), and associated with larger mushroom spines (Harris et al., 1992). There are typically more mushroom spines in layer III of cerebral cortex (Figure 14(B)) than in the proximal stratum radiatum of the CA1 region of the hippocampus (Figure 14(A)). The cerebellar Purkinje spines (Figure 14(C)) have a distinctive rounded shape and contain PSDs that are more curved than excitatory synapses in the forebrain (Figure 14(A), (B)).

Forebrain pyramidal neurons and cerebellar Purkinje cells also have different types of endoplasmic reticulum (ER) in their spines. Forebrain spines do not often contain ER (small arrowheads in Figure 14(A); Harris and Weinberg, 2012), and when they do, the ER sometimes form a particular structural complex called the "spine apparatus" (double arrowheads in Figure 14(B)), consisting of layers of membranous cisternae interspersed with a dense lamella (Gray, 1959; reviewed in Segal et al., 2010). The spine apparatus is often seen in large mushroom type synapses in the forebrain (Figure 14(B)) but never in cerebellum (Figure 14(C)). The spine apparatus contains synaptopodin, an actin-binding protein, and is also coupled with ryanodine receptor (Vlachoe et al., 2009), a receptor that regulates calcium store. The occurrence frequency of spine apparatus is higher in layer III of the cerebral cortex (~35% of all spine profiles sampled; Figure 14(B)) than in the proximal stratum radiatum of the hippocampal CA1 region (~8%; Figure 14(A); Tao-Cheng et al., 2007). On the other hand, all

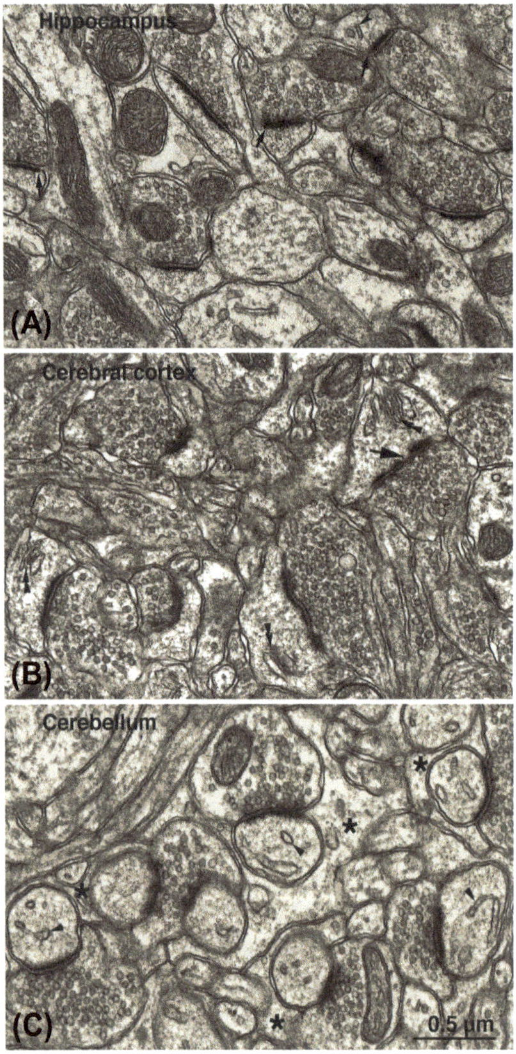

Figure 14 Electron micrographs from different regions of perfusion-fixed mouse brain: (A) proximal stratum radiatum of the CA1 region of the hippocampus, (B) layer III of the cerebral cortex, and (C) the molecular layer of the cerebellum. The spines from forebrain pyramidal neuron (A, B) contain synapses that are similar in shape and structure, but differ in proportion of macula (small arrows in A) and perforated (large arrows in B) PSDs. The cerebellar Purkinje spines (C) are more uniform in size and shape than the forebrain spines, and are distinctive in their more rounded shape and a more extensive ensheathment by glia processes (*). ER (small arrowhead in A) are infrequent in forebrain spines, but present in all cerebellar Purkinje spines (small arrowheads in C). A special ER complex, the spine apparatus (double arrowheads in B) are only seen in forebrain, and at a higher frequency in cortex (B) than in hippocampus (A).

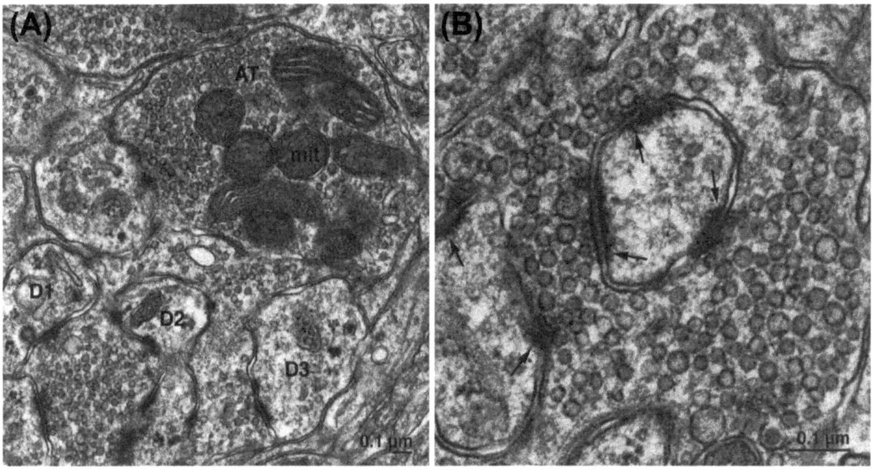

Figure 15 Mossy fiber synapses in CA3 region of the hippocampus. The axon terminal (AT) is very large and contains many mitochondria (mit), and forms synaptic contact with multiple dendritic profiles (D1, D2, and D3 in A). The axon terminal may also form multiple synaptic contacts (arrows in B) with the same dendritic element.

cerebellar Purkinje cell spines contain ER (small arrowheads in Figure 14(C); Harris and Weinberg, 2012), which contains concentrated IP3 receptors, another type of calcium store regulator (Takei et al., 1994).

A particular type of structurally distinctive synapses is formed by mossy fiber axons from granular cells in the dentate gyrus of the hippocampus with dendrites of the pyramidal neurons in the CA3 region. These presynaptic terminals are very large and form synapses with multiple dendritic elements (Figure 15(A)), or form multiple synapses with one dendritic element (Figure 15(B)).

2.6.1.2. Distance-Dependent Difference in Number of Perforated PSDs within the Stratum Radiatum of the Hippocampal CA1 Region

The stratum radiatum of the hippocampal CA1 region contains the synapses that are potentiated in the well-studied LTP model induced by particular stimulation patterns of the CA3 neurons. The name radiatum stems from the appearance of the radiating apical dendrites of the CA1 neurons (Figure 16(A)), and in adult mice, this region is ~200–250 μm deep. In the stratum radiatum, mushroom spines (Matsuzaki et al., 2001) with perforated PSDs contain more AMPA receptors than macula PSDs (Desmond and Weinberg, 1998; Ganeshina et al., 2004), and have long been associated with synaptic plasticity, such as LTP (reviewed in Bosch and Hayashi, 2011; Harris and Weinberg, 2012). The size of perforated PSDs also changes with age-related spatial learning impairments (Nicholson et al., 2004). Thus, perforated PSDs in mushroom spines are often singled out for measurement due to their functional implications.

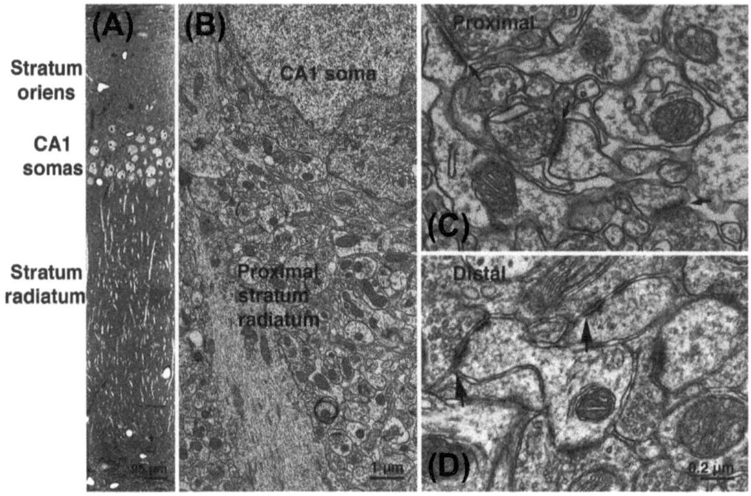

Figure 16 Perfusion-fixed CA1 region of mouse hippocampus. (A) Light micrograph showing clustered CA1 neuronal somas and the two stratified layers of dendrites and synapses above and below: stratum oriens and stratum radiatum (micrograph by Paul Gallant). (B) Electron micrograph showing one neuronal soma and the proximal region of the stratum radiatum. (C) The proximal stratum radiatum (<10 μm from the edge of the soma cluster) contain a high percentage of spines with macular PSDs (small arrows). (D) The distal stratum radiatum (>100 μm from soma) contains more spines with perforated PSD (large arrows in D). (C) and (D) share the same scale bar.

However, the number of perforated PSDs in the proximal region is only ~50% of that in the distal regions of the stratum radiatum (Nicholson et al., 2006). Thus, it is of extreme importance to document the exact sampling location (e.g. Fiala et al., 2003) when counting the prevalence of perforated PSDs in this particular part of the hippocampus.

2.6.1.3. Cytoskeleton and Polyribosomes in Hippocampal Spines under Different Conditions

Spines are actin-rich (reviewed in Bosch and Hayashi, 2011) and microtubules are typically excluded from spines regardless of the spine type (e.g. mushroom spines in Figure 17(A), or thin spines in Figure 17(B)). However, dynamic microtubules can protrude into spines transiently, and this process is activity-induced (reviewed in Hoogenraad and Akhmanova, 2010; Dent et al., 2011).

In hippocampus, polyribosomes are sometimes located at the base of the spine (double arrowheads in Figure 17(B); Steward and Levy, 1982), and can move into enlarged spines during LTP (Ostroff et al., 2002). The presence of polyribosomes suggests local protein synthesis, and their increased presence in enlarged spines suggests that these are the spines expressing LTP.

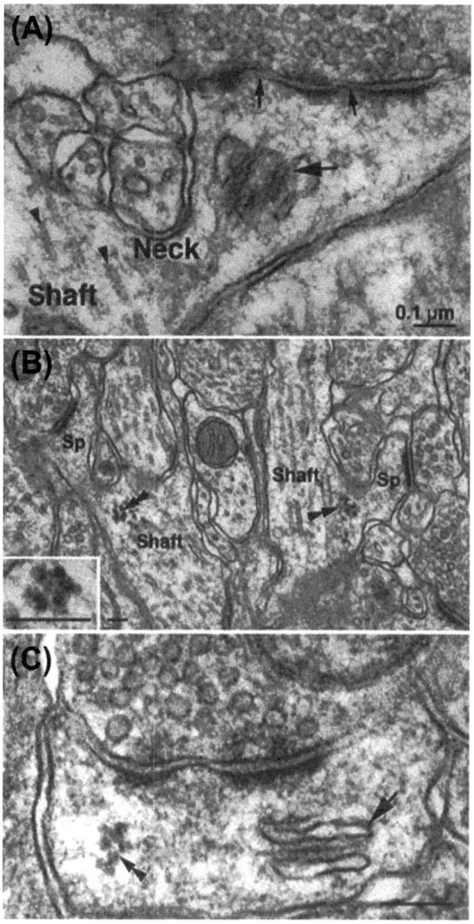

Figure 17 Structural details of synapses from perfusion-fixed hippocampus in the middle portion of stratum radiatum of CA1 region. (A) A spine with perforated PSD (small arrows pointing to perforations) arises from the dendritic shaft, with a spine apparatus (large arrow) extending through the spine neck. Arrowheads point to microtubules, which are abundant in shaft but typically absent in spines. (B) Polyribosomes (double arrowheads, enlarged in inset) are typically located in dendritic shafts near the base of the spines (sp). (C) A spine with perforated PSD containing a spine apparatus (arrow) and polyribosome (double arrowheads). Scale bar = 0.1 μm.

2.6.1.4. PSD Thickness and Curvature are Significantly Affected by the Quality of Perfusion Fixation

There are established benchmarks to assess the activity states of the synapses from perfusion-fixed brains (Tao-Cheng et al., 2007). Similar to activity-induced structural changes at synapses in dissociated neuronal cultures described in Section 2.1, the most consistent

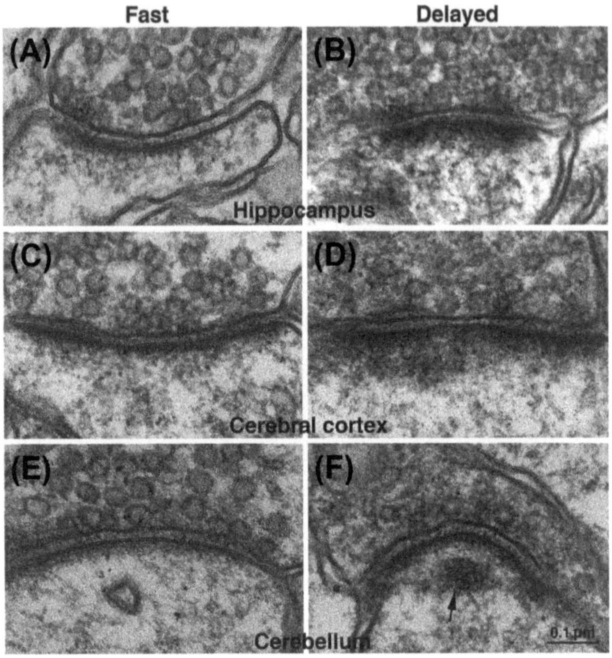

Figure 18 Synapses from perfusion-fixed stratum radiatum from the CA1 region of the hippocampus (A, B), layer III of the cerebral cortex (C, D), and the molecular layer of the cerebellum (E, F). After delayed perfusion fixation (right column), both the thickness and the curvature of the PSDs increase in all three regions of the brain. The thickness increase is higher in the forebrain (B, D) than in the cerebellum (F), where dense materials sometimes form a sub-PSD sphere (arrow in F).

indicators for synaptic activity are the increases in thickness and curvature of the PSDs. With "fast perfusion" fixation (Figure 18, left column), when the time between opening of the animal's chest cavity and flow of fixative through the heart is kept within 100 s, PSDs are typically flat or concave in curvature in the hippocampus (Figure 18(A)) and cerebral cortex (Figure 18(C)), and slightly convex in Purkinje spines in cerebellum (Figure 18(E)). With "delayed perfusion" fixation (Figure 18, right column), when the flow of fixative was delayed for 5–8 min, PSDs are consistently thickened, and their curvature becomes more convex and arches into the presynaptic terminal. In Purkinje spines, sub-PSD spheres (arrow in Figure 18(F)) sometimes form after delayed perfusion fixation (Tao-Cheng et al., 2007). These structural changes at synapses after delayed perfusion fixation are most likely results of heightened synaptic activity due to ischemia and glutamate release during the period of delay in perfusion fixation.

2.6.1.5. Activity-Dependent Structural Changes in Cerebellar Purkinje ER

Another structural difference brought about by delayed perfusion fixation is the formation of ER stacks in cerebellar Purkinje cells (Takei et al., 1994), including the ER in the

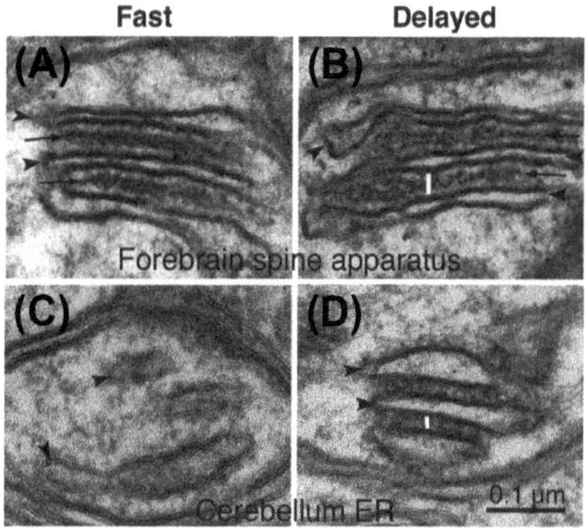

Figure 19 There is no detectible structural difference in spine apparatus from cerebral cortex after (A) fast and (B) delayed perfusion fixation. The white bar in (B) indicates the 35 nm cleft between membrane cisternae (arrowheads in A and B), and arrows point to a dense lamina in the cleft. In contrast, ER cisternae (arrowheads in C and D) in cerebellar Purkinje spines are randomly dispersed after fast perfusion (C) and stacked with a uniform distance of 20 nm (white bar) after delayed perfusion fixation (D).

spines (Figure 19(D)). With fast perfusion fixation, ER is present in Purkinje spines as randomly dispersed membranous cisternae (arrowheads in Figure 19(C)). After delayed perfusion fixation, these ER cisternae become stacked in orderly arrays with a uniform distance of 20 nm (white bar in Figure 19(D)). The cerebellar ER contains concentrated IP3 receptors that regulate calcium stores (Takei et al., 1994), and the rapid change in ER configuration may indicate different modes of calcium regulation under different levels of short-term synaptic activity. On the other hand, the spine apparatus in forebrain spines is coupled with another type of calcium store regulator, the ryanodine receptor (Vlachos et al., 2009). The spine apparatus does not change frequency or structural arrangement either after fast (Figure 19(A)) or delayed perfusion fixation (Figure 19(B); Tao-Cheng et al., 2007), suggesting that this structure is not involved in short-term structural changes, but may be involved in long-term plasticity (Vlachos et al., 2009).

2.6.2. Acute Brain Slices

Acute brain slices from various parts of the young or adult brains are used for electrophysiological or pharmacological experiments, and parallel structural changes can be documented. However, neuronal tissues show signs of extreme stimulation immediately after the slices are cut due to mechanical damages as well as lack of oxygen from the dissection process. In order for the brain slices to return to a basal state, the slices need to be carefully recovered in oxygenated medium over a period of ~1–2 h. Structural changes have been documented during this recovery period (Fiala et al., 2003).

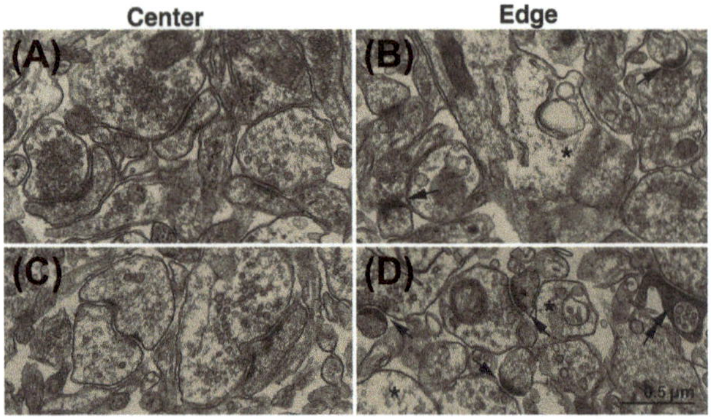

Figure 20 Regional difference in sampling from the same acute brain slices after recovery. Even within the same slices (A and B are from one slice, C and D are from the second slice), the center region (A and C) contains synapses that appear normal in structure and at basal conditions, while the area near the cut edge (B & D) contains dark (double arrow) or swollen (*) processes, and synapses with thick and curved PSD (arrows).

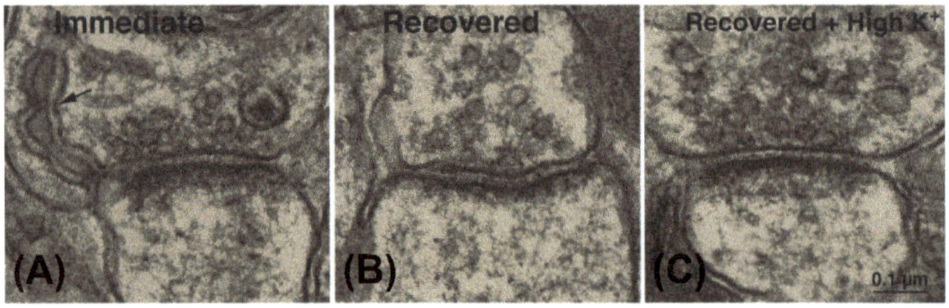

Figure 21 Synapses sampled from the central region of acute brain slices that are fixed immediately after dissection (A), after 1 h of recovery (B), and after 1 h of recovery and then depolarized with high K^+ (C). Synapses show signs of intense stimulation with spinule formation (arrow in A) and thickened PSD immediately after dissection (A). After recovery, synapses resemble those under basal conditions with no spinules and a relatively thin PSD (B). Upon depolarization, the recovered slices again show synapses with thickened PSD (C).

2.6.2.1. Location of Sampling from Acute Brain Slices

The cut surfaces of acute brain slices contain many cut and damaged neurons and their processes. During the recovery period, some cut processes can reseal and recover, and retain their physiological competency. However, some cut processes are damaged beyond repair, and will become dark and shrunken (double arrows in Figure 20(D)) or watery and swollen (asterisks in Figure 20(B) & (D)). There is a gradient of decreasing frequency of such damaged processes from the edge (Figure 20(B), (D)) toward the center of the slice (Figure 20(A), (C)) over a distance of ~100 µm. Furthermore, synapses in these edge areas of the slice also contain thick and curved PSDs (arrows in Figure 20(B), (D)), indicating that these synapses did not return to a basal state. Thus, when sampling synapses from recovered acute brain slices, the area within 100 µm of the cut surfaces should be excluded.

2.6.2.2. Assessment of Quality of Recovered Slices

Even within the central region of the acute brain slices with no or few mechanically damaged processes, the health of the synapses needs to be carefully assessed. The synapses should resemble those in a fast-perfusion fixed brain (cf. Figure 1, 14(A), 16, 17, 18(A)), with relatively thin PSDs that are slightly concave in curvature (Figure 20(A) & (C)). Compared to synapses from slices that are fixed immediately after dissection (Figure 21(A)), a recovered synapse has a thinner PSD (Figure 21(B)), and such recovered synapses are competent to be stimulated again, and show PSD thickening after depolarization (Figure 21(C); Dosemeci et al., 2001).

2.6.3. Organotypic Slice Cultures

Hippocampal slice cultures are typically prepared from week-old rat brain slices and then grown for 10–15 days in vitro on a support film. At best, they may be equivalent to 3-week-old brain in maturity, and are structurally more similar to a young than an adult brain as described in Harris et al. (1992). These slice cultures retain their neuronal circuitry and can be induced to express LTP in the stratum radiatum of the CA1 region (Otmakhov et al., 2004). The synapses in slice cultures resemble their counterparts in brain, both in structure and in function.

2.6.3.1. Slice Cultures Serve as a Reliable Model System to Study Synapses under Different Activity Levels

One advantage of using the organotypic slice cultures over acute brain slices or perfusion-fixed brains is that synapses can be reliably maintained at a basal state, or become activated by electrophysiological and pharmacological means. Interestingly, slice cultures that are brushed off from the support film and then fixed for EM examination (Figure 22(C)) contain synapses that are structurally identical to those seen in control slices that are fixed while still attached to the support film (Figure 22(A)). Thus, excitatory synapses in slice cultures remain at the basal state even after this mechanical dislodging from the tissue culture substrate. These dislodged slice cultures can serve as a unique and useful source of synapses at basal states for biochemical studies. To date, such a source of isolated synapses at basal states is still lacking because preparations of conventional synaptosomes and PSDs are isolated from brains, which always contain activated synapses due to extensive stimulations brought on by decapitation, dissections, and ischemia delays (Hu and Wieloch, 1995; Dosemeci et al., 2000). Thus, PSDs isolated from control slice cultures will be extremely interesting, and such a preparation is attainable since a protocol exists for isolating PSDs from acute brain slices (Dosemeci et al., 2006).

Pharmacological manipulations can be applied to induce precise and uniform activity for all synapses in the slice cultures. For example, slice cultures respond to high K^+ depolarization or NMDA treatment rapidly and, thus, offer a unique model system to study the time course of synaptic structural changes from initial response to recovery after cessation of stimulation in specified populations of synapses. Indeed, morphometry on formation and disappearance of synaptic spinules obtained in slice cultures (Tao-Cheng et al., 2009) demonstrates the rapid turnover of spinules,

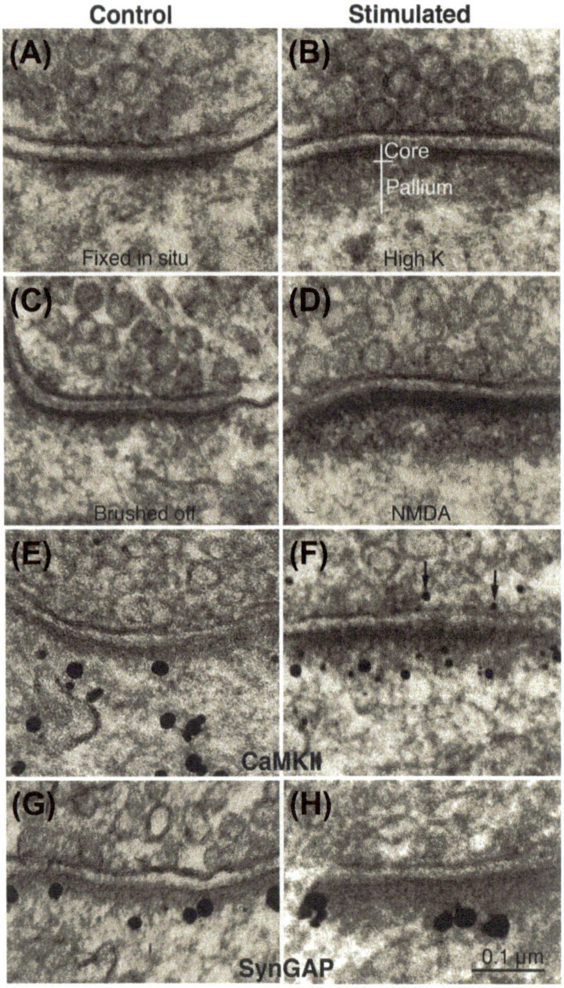

Figure 22 Excitatory synapses in hippocampal slice cultures undergo dynamic structural changes. PSDs are flat or concave under control conditions (left column), but increase in thickness and curvature after stimulation (right column). The two layers of the PSD complex, PSD core and PSD pallium, become more evident after stimulation (B). Control samples are treated with Ringer (A, E, and G), or brushed off from the supporting film before fixation (C). Stimulated samples are treated with high K^+ (B, F, and H) or NMDA (D). Labeling for CaMKII is dispersed in the spine and absent from synaptic vesicle clusters near the AZ under control conditions (E), but moves toward the PSD and becomes concentrated in the PSD pallium, as well as into the AZ (arrows in F) upon depolarization (F). Labeling for SynGAP is at the PSD core under control conditions (G), but moves away upon depolarization (H).

and any potential functions attributed to these curious looking entities will have to conform to this time course.

2.6.3.2. Synapses in Slice Cultures Have Similar Activity-Induced Changes as Those in Dissociated Cultures

Similar to excitatory synapses in dissociated neuronal cultures (Figure 3), excitatory synapses in stratum radiatum of the CA1 region of hippocampal slice cultures undergo dynamic structural changes after stimulation. PSDs are typically flat or concave in curvature under control conditions (Figure 22(A) & (C)), but increase in thickness and in curvature after high K^+ depolarization (Figure 22(B)) or NMDA treatment (Figure 22(D)). Synapses in slice cultures also have identical activity-dependent redistribution of synaptic proteins to those seen in dissociated cultures or brains. For example, CaMKII labels are dispersed in the spine (Figure 22(E)), but become concentrated in the PSD pallium upon depolarization (Figure 22(F)). CaMKII in presynaptic terminals also moved into the AZ upon depolarization (arrows in Figure 22(F); Tao-Cheng et al., 2006). SynGAP labels are in the PSD core under control conditions (Figure 22(G)), but move away from the postsynaptic membrane upon depolarization (Figure 22(H); Yang et al., 2011). Thus, these activity-dependent changes first observed in dissociated cells have been confirmed in synapses that are sampled form the stratum radiatum of the CA1 region in slice cultures.

2.6.3.3. Caveats in Sampling Synapses from Hippocampal Slice Cultures

The surface of slice cultures is covered by layers of glia (Figure 23(B)), and only the central portion resembles the brain in structural integrity and anatomical organization (Figure 23(A)). Thus, for studies that are related to the circuitry in hippocampus, such as LTP, sampling of synapses should be restricted to the central portion of the slice cultures.

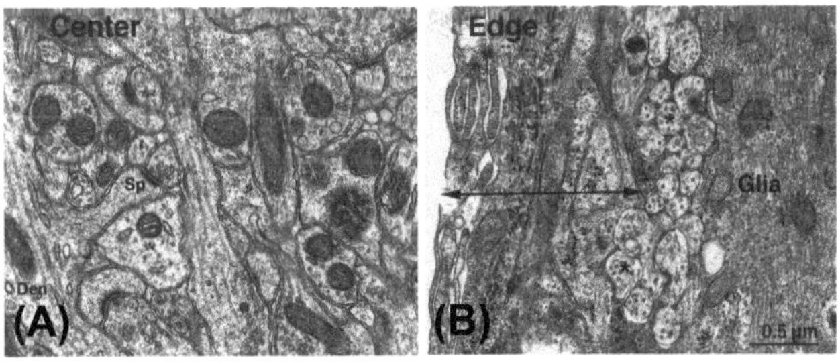

Figure 23 Hippocampal slice cultures sampled from (A) the center and (B) the edge of the slice that contain the stratum radiatum of the CA1 region. Neuronal structure in the central portion resembles the brain, with radiating dendrites (den) and spines (sp) sprouting from it. Arrows in (B) indicate the glia layers on the surface of the slice culture. Many neuronal processes (*) are wedged between the surface layers and a large glia process at right in (B).

2.6.4. Dissociated Hippocampal Neuronal Cell Cultures

Dissociated hippocampal cultures are typically prepared from embryos 1–2 days before birth, and then grown in vitro for ~2–3 weeks. These cultures are composed of mixed neuronal cell types from all parts of the hippocampus, and cannot form a three-dimensional organization resembling the brain. On the other hand, monolayers allow easy penetration of pharmacological and immunolabeling reagents. Furthermore, excitatory and inhibitory synapses have structural and compositional characteristics that resemble their counterparts in brain (Figure 2, 13). Thus, dissociated cultures are valid experimental models for studying synaptic structure, and with careful manipulations, for studying aspects of long-term synaptic modifications (reviewed in Molnár, 2011).

2.6.4.1. Dissociated Cultures Serve as the Easiest Model System to Study Activity-Dependent Structural Changes

Along with organotypic slice cultures, the dissociated cultures serve as a reliable model system where pharmacological manipulation can generate defined and uniform states of activity in all synapses. In addition, because dissociated cells can be easily and rapidly fixed by immersion, this model system can also reliably provide synapses at the basal state in control samples. In fact, it is in this system that immunolabeling of CaMKII was first shown, at the EM level, to be largely dispersed in the spine cytoplasm rather than attached to PSDs in synapses that are at the basal state (cf. Figure 2(G), 3(C); Dosemeci et al., 2001). Until then, most biochemical studies concluded that CaMKII is the major component of the PSD. However, EM immunolabeling demonstrates that CaMKII is

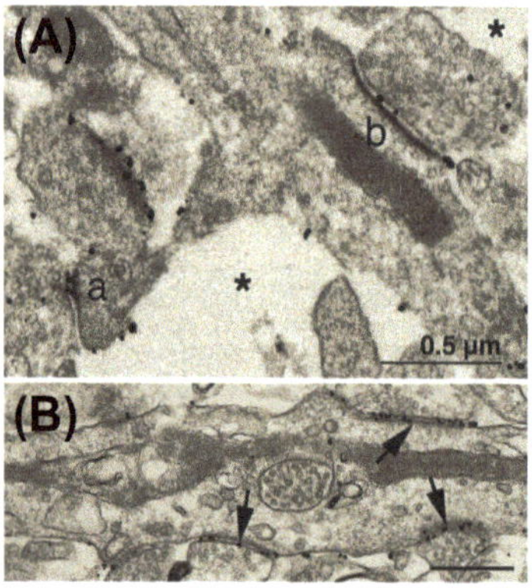

Figure 24 Dissociated hippocampal cultures labeled for GluR2. (A) (a and b) denote a spine and a shaft synapse, respectively. Neurites in dissociated cultures are not compactly organized, and there is a lot of extracellular space (*) among processes. (B) A longitudinal section through a dendritic shaft that receives three excitatory synapses (arrows), which are common in dissociated cultures but rare in adult brain. Scale bar = 0.5 μm.

only heavily attached to the PSD in activated synapses (Figure 3(D), 4(A), 9(G), 13(D)). With this understanding, similar activity-dependent redistribution of CaMKII labeling was later confirmed in slice cultures (Figure 22 (E) vs (F)), and in perfusion-fixed brains (Tao-Cheng et al., 2007).

2.6.4.2. Excitatory Synapses are Often Located on Dendritic Shafts in Dissociated Cultures

Unlike their counterparts in brain, excitatory synapses in dissociated cultures are more frequently located on dendritic shafts (Figure 24; Tao-Cheng et al., 2011) than at spines. Shaft synapses are also prevalent in young brains (Harris et al., 1992), consistent with the idea that dissociated cultures at 3 weeks in vitro resemble immature brains. Shaft synapses are on average larger than spine synapses and contain more AMPA receptors (Tao-Cheng et al., 2011).

2.6.4.3. Synaptic Cleft is More Permeable in Dissociated Cultures

Dissociated cells are easily accessible by reagents, and thus, especially suited for pre-embedding immunolabeling studies that rely on thorough penetration of reagents. For example, there is no noticeable difference in the appearance of the synaptic cleft of excitatory synapses in dissociated cultures (Figure 25(A)) or in slice cultures (Figure 25(B)). However, the cleft material is much more accessible for immunolabeling in dissociated cultures. Antibodies against the external portion of the AMPA receptor GluR2 label the receptors in the cleft successfully in dissociated cultures (Figure 25(C)), but fail to penetrate the synaptic cleft in synapses from slice cultures (Figure 25(D)) or in adult brain (unpublished observations). The different degrees of accessibility may be due to

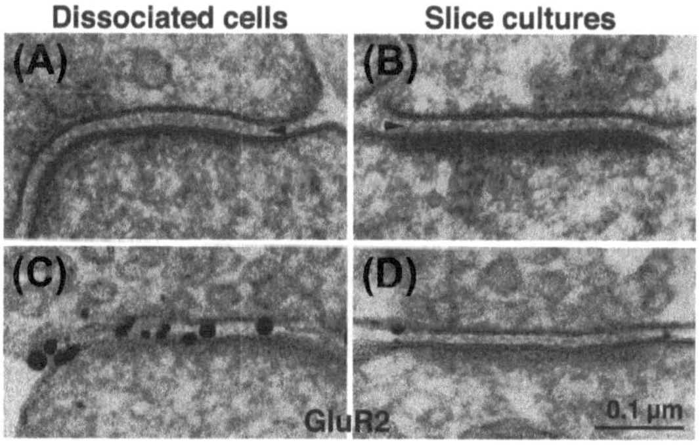

Figure 25 Synaptic cleft from dissociated cells (left column) and from slice cultures (right column) are similar in appearance, and contain filamentous materials (arrowheads in A and B) between the rigidly apposed pre- and postsynaptic membranes. Panels (A) and (B) are from samples fixed with glutaraldehyde, and treated with or without tannic acid (A and B, respectively) after the initial fixation. Tannic acid appears to slightly enhance the cleft materials (A). Panels (C) and (D) are fixed with paraformaldehyde, and labeled with an antibody against an extracellular portion of an AMPA receptor subunit (GluR2). The cleft is accessible to immunolabeling in dissociated cells (C) but not in slice cultures (D).

neuritis being loosely arranged in dissociated cultures (cf. Figure 24(A)) as opposed to their compact organization in the brain (Figure 14(A)) and slice cultures (Figure 23(A)). Alternatively, these results could suggest that configuration of the synaptic cleft may be less mature in dissociated cultures.

2.6.4.4. Structural Benchmarks for Spine Maturity in Dissociated Cultures

Mushroom spines with perforated PSDs are infrequent in dissociated cells (Figure 26(A)), and contain fewer organelles than their counterparts in brain (cf. Figure 17(A) & (C)). Spine apparatus are extremely rare in dissociated cultures (Segal et al., 2010), consistent

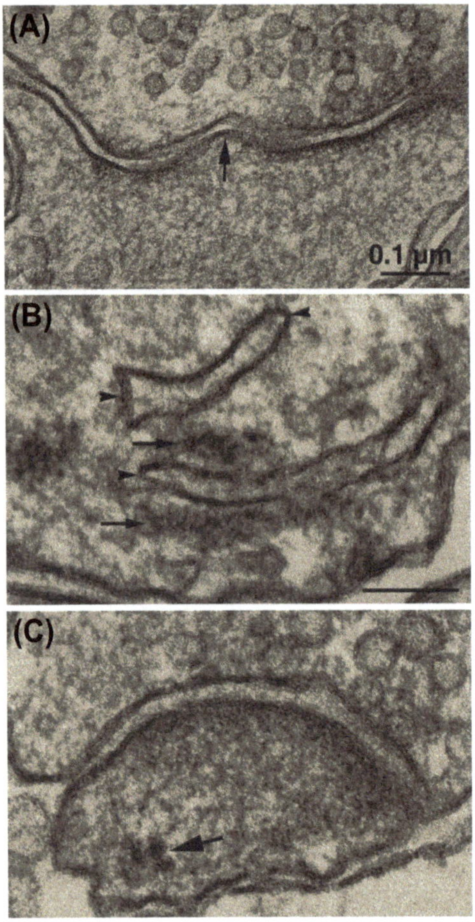

Figure 26 Atypical synaptic structures in dissociated cultures. (A) An infrequent example of mushroom spine with perforated PSD. Arrow points to the perforation. (B) A rare example of spine apparatus. Arrowheads point to ER cisternae and arrows point to dense lamellae. (C) However, polyribosomes (arrow) are frequently seen in spines. (B) and (C) share the same scale bar. Scale bar = 0.1 μm, (B) and (C) share the same scale bar.

with the observation that spine apparatus are few in young brains (Harris et al., 1992). Figure 26(B) shows one rare spine apparatus in a large spine from a 3-week-old dissociated culture. The spacing of ER cisternae and dense lamellae is wider and less orderly than that in mature brains (cf. Figure 19(A) & (B)). Interestingly, polyribosomes are more frequent and more abundant in dendritic spines (Figure 26(C)) and shafts in dissociated cultures than in mature brains or in organotypic slice cultures of similar age in vitro (unpublished observation). This unexpected finding suggests that local protein synthesis at dendritic locations near synapses may be more robust in dissociated cultures than in brains or slice cultures.

3. CONCLUSIONS

The elaborate, fine structure of synapses continues to fascinate since its first unveiling by EM, and keeps on providing clues to synaptic functions. Information on distribution of synaptic proteins, made available by immunolabeling, further advances our understanding of the structure and function of synapses.

The CNS glutamatergic excitatory synapses undergo dynamic structural changes upon stimulation. These changes include increases in PSD thickness and curvature, and redistribution of synaptic proteins, including CaMKII, Shanks, SynGAP, and AMPA receptors at the PSD, and SV2, synapsin I, and CaMKII in the presynaptic terminals. Other proteins such as PSD95 at the PSD, and Bassoon and Piccolo at the AZ, are more stable and remain unchanged under different activity states.

Dissociated neuronal cultures prove to be the easiest model system to study these activity-dependent changes because the activity levels of all synapses can be uniformly controlled and the monolayers are readily accessible to immunolabeling reagents. Identical synaptic structural changes have been confirmed in organotypic slice cultures, where synapses retain their circuitry and anatomical organization as in brain. Many of these findings are also verified in perfusion-fixed brains; however, the activity states of the synapses are more difficult to ascertain.

With the fresh insight that synaptic structure and protein composition are changing dynamically with activity, interpretation of synaptic structural studies may benefit from integrating these benchmarks of activity-induced changes at the synapses.

ACKNOWLEDGMENTS

The author would like to dedicate this chapter to Dr Yasuko Nakajima, who trained him as an electron microscopist, and together with Dr Shigehiro Nakajima, set examples of doing excellent science with honor. He wants to thank Dr Jack Rosenbluth who instilled in him the importance of morphometry and Dr Milton W. Brightman for showing him the gold standard of a well-fixed brain. His special gratitude goes to Dr Ayse Dosemeci who nudged him into studying the subject of her intense interests, the PSD, and for numerous lively discussions along our journey of ongoing collaboration; and to Dr Paul Gallant who provided superbly perfusion-fixed brains, as well as advice on neurobiology and biostatistics. He would like to thank Virginia Tanner Crocker and Rita Azzam for their excellent EM technical support, in processing and sectioning thousands of samples; and Christine A. Winters for countless gorgeous hippocampal dissociated as well as slice cultures. Finally, He thanks Dr Tom S. Reese for his generous support and encouragement, as well as helpful discussions. Funds are provided by NINDS intramural research.

REFERENCES

Bosch, M., Hayashi, Y., 2011. Structural plasticity of dendritic spines. Curr. Opin. Neurobiol. 22, 1–6.

Ceccarelli, B., Hurlbut, W.P., Mauro, A., 1973. Turnover of transmitter and synaptic vesicles at the frog neuromuscular junction. J. Cell Biol. 57, 499–524.

Cesca, F., Baldelli, P., Valtorta, F., Benfenati, F., 2010. The synapsins: key actors of synapse function and plasticity. Prog. Neurobiol. 91, 313–348.

Craig, A.M., Kang, Y., 2007. Neurexin-neuroligin signaling in synapse development. Curr. Opin. Neurobiol. 17, 43–52.

Dent, E.W., Merriam, E.B., Hu, X., 2011. The dynamic cytoskeleton: backbone of dendritic spine plasticity. Curr. Opin. Neurobiol. 21, 175–181.

Desmond, N.L., Weinberg, R.J., 1998. Enhanced expression of AMPA receptor protein at perforated axospinous synapses. Neuroreport 9, 857–860.

Dosemeci, A., Reese, T.S., Peterson, J., Tao-Cheng, J.-H., 2000. CaMKII clusters: a novel particulate form of ca^{2+}/calmodulin-dependent protein kinase II in neurons. J. Neurosci. 20, 3076–3084.

Dosemeci, A., Tao-Cheng, J.-H., Vinade, L., Winters, C.A., Pozzo-Miller, L., Reese, T.S., 2001. Glutamate-induced transient modification of the postsynaptic density. Proc. Natl. Acad. Sci. USA, 98, 10428–10432.

Dosemeci, A., Vinade, L., Winters, C., Reese, T.S., Tao-Cheng, J.-H., 2002. Inhibition of phosphatase activity prolongs NMDA-induced modification of the postsynaptic density. J. Neurocytol. 31, 605–612.

Dosemeci, A., Tao-Cheng, J.-H., Vinade, L., Jaffe, H., 2006. Preparation of postsynaptic density fraction from hippocampal slices and proteomic analysis. Biochem. Biophys. Res. Commun. 339, 687–694.

Dosemeci, A., Thein, S., Yang, Y., Reese, T.S., Tao-Cheng, J.-H., 2013. CYLD, a deubiquitinase specific for lysine 63-linked polyubiquitins, accumulates at the postsynaptic density in an activity-dependent manner. Biochem. Biophys. Res. Commun. 430, 245–249.

Fatt, P., Katz, B., 1952. Spontaneous subthreshold activity at motor nerve endings. J. Physiol. 117, 109–128.

Fiala, J.C., Kirov, S.A., Feinberg, M.D., Petrak, L.J., George, P., Goddard, C.A., Harris, K.M., 2003. Timing of neuronal and glial ultrastructure disruption during brain slice preparation and recovery in vitro. J. Comp. Neurol. 465, 90–103.

Ganeshina, O., Berry, R.W., Petralia, R.S., Nicholson, D.A., Geinisman, Y., 2004. Differences in the expression of AMPA and NMDA receptors between axospinous perforated and nonperforated synapses are related to the configuration and size of postsynaptic densities. J. Comp. Neurol. 468, 86–95.

Gray, E.G., 1959. Electron microscopy of synaptic contacts on dendrite spines of the cerebral cortex. Nature 183, 1592–1593.

Gray, E.G., 1969. Electron microscopy of excitatory and inhibitory synapses: a brief review. Prog. Brain Res. 31, 141–155.

Gundelfinger, E.D., Fejtova, A., 2011. Molecular organization and plasticity of the cytomatrix at the active zone. Curr. Opin. Neurobiol. 22, 1–8.

Harris, K.M., Jensen, F.E., Tsao, B., 1992. Three-dimensional structure of dendritic spines and synapses in rat hippocampus (CA1) at postnatal day 15 and adult ages: implications for the maturation of synaptic physiology and long-term potentiation. J. Neurosci. 12, 2685–2705.

Harris, K.M., Weinberg, R.J., 2012. Ultrastructure of synapses in the mammalian brain. Cold Spring Harb. Perspect. Biol. 3 (E pub ahead of print).

Heuser, J.E., Reese, T.S., 1973. Evidence for recycling of synaptic vesicle membrane during transmitter release at the frog neuromuscular junction. J. Cell Biol. 57, 315–344.

Hoogenraad, C.C., Akhmanova, A., 2010. Dendritic spine plasticity: new regulatory roles of dynamic microtubules. Neuroscientist 16, 650–661.

Hu, B.R., Wieloch, T., 1995. Persistent translocation of Ca2+/calmodulin-dependent protein kinase II to synaptic junctions in the vulnerable hippocampal CA1 region following transient ischemia. J. Neurochem. 64, 277–284.

Limbach, C., Laue, M.M., Wang, X., Hu, B., Thiede, N., Hultqvist, G., Kilimann, M.W., 2011. Molecular in situ topology of Aczonin/Piccolo and associated proteins at the mammalian neurotransmitter release site. Proc. Natl. Acad. Sci. USA 108, E392–E401.

Matsuzaki, M., Ellis-Davies, G.C., Nemoto, T., Miyashita, Y., Iino, M., Kasai, H., 2001. Dendritic spine geometry is critical for AMPA receptor expression in hippocampal CA1 pyramidal neurons. Nat. Neurosci. 4, 1086–1092.

Molnár, E., 2011. Long-term potentiation in cultured hippocampal neurons. Semin. Cell Dev. Biol. 22, 506–513.

Nicholson, D.A., Yoshida, R., Berry, R.W., Gallagher, M., Geinisman, Y., 2004. Reduction in size of perforated postsynaptic densities in hippocampal axospinous synapses and age-related spatial learning impairments. J. Neurosci. 24, 7648–7653.

Nicholson, D.A., Trana, R., Katz, Y., Kath, W.L., Spruston, N., Geinisman, Y., 2006. Distance-dependent differences in synapse number and AMPA receptor expression in hippocampal CA1 pyramidal neurons. Neuron 50, 431–442.

Opazo, P., Choquet, D., 2011. A three-step model for the synaptic recruitment of AMPA receptors. Mol. Cell Neurosci. 46, 1–8.

Ostroff, L.E., Fiala, J.C., Allwardt, B., Harris, K.M., 2002. Polyribosomes redistribute from dendritic shafts into spines with enlarged synapses during LTP in developing rat hippocampal slices. Neuron 35, 535–545.

Otmakhov, N., Tao-Cheng, J.-H., Carpenter, S., Asrican, B., Dosemeci, A., Reese, T.S., Lisman, J., 2004. Persistent accumulation of CaMKII Kinase II in dendritic spines after induction of chemical LTP. J. Neurosci. 24, 9324–9331.

Owald, D., Sigrist, S.J., 2009. Assembling the presynaptic active zone. Curr. Opin. Neurobiol. 19, 311–318.

Palay, S.L., McGee-Russell, S.M., Gordon Jr., S., Grillo, M.A., 1962. Fixation of neural tissues for electron microscopy by perfusion with solutions of osmium tetroxide. J. Cell Biol. 12, 385–410.

Peters, A., Palay, S.L., Webster, H.D., 1991. The Fine Structure of the Nervous System. Oxford University Press, New York.

Peters, A., Palay, S.L., 1996. The morphology of synapses. J. Neurocytol. 25, 687–700.

Rácz, B., Blanpied, T.A., Ehlers, M.D., Weinberg, R.J., 2004. Lateral organization of endocytic machinery in dendritic spines. Nat. Neurosci. 7, 917–918.

Reese, T.S., Tao-Cheng, J.-H., Dosemeci, A., 2011. Two Layers with Different Dynamic Properties in Postsynaptic Densities at Glutamatergic Synapses. Program number 658.17, 2011 Neuroscience Meeting Planner. Society for Neuroscience, Washington, DC Online.

Segal, M., Vlachos, A., Korkotian, E., 2010. The spine apparatus, synaptopodin, and dendritic spine plasticity. Neuroscientist 16, 125–131.

Sheng, M., Kim, E., 2012. The postsynaptic organization of synapses. Cold Spring Harb. Perspect. Biol. 3 (E pub ahead of print).

Steward, O., Levy, W.B., 1982. Preferential localization of polyribosomes under the base of dendritic spines in granule cells of the dentate gyrus. J. Neurosci. 2, 284–291.

Takei, K., Mignery, G.A., Mugnaini, E., Südhof, T.C., De Camilli, P., 1994. Inositol 1,4,5-trisphosphate receptor causes formation of ER cisternal stacks in transfected fibroblasts and in cerebellar Purkinje cells. Neuron 12, 327–342.

Tanner, V.A., Ploug, T., Tao-Cheng, J.-H., 1996. Subcellular localization of SV2 and other secretory vesicle components in PC12 cells by an efficient method of preembedding EM immunocytochemistry for cell cultures. J. Histochem. Cytochem. 44, 1481–1488.

Tao-Cheng, J.-H., 2006. Activity-related redistribution of presynaptic proteins at the active zone. Neuroscience 141, 1217–1224.

Tao-Cheng, J.-H., 2007. Ultrastructural localization of active zone and synaptic vesicle proteins in a preassembled multi-vesicle transport aggregate. Neuroscience 150, 575–584.

Tao-Cheng, J.-H., Dosemeci, A., Winters, C.A., Reese, T.S., 2006. Changes in the distribution of calcium calmodulin-dependant protein kinase II at the presynaptic bouton after depolarization. Brain Cell Biol. 35, 117–124.

Tao-Cheng, J.-H., Gallant, P.E., Brightman, M.W., Dosemeci, A., Reese, T.S., 2007. Effects of delayed perfusion fixation on postsynaptic density and CaMKII clustering in different regions of the mouse brain. J. Comp. Neurol. 501, 731–740.

Tao-Cheng, J.-H., Dosemeci, A., Gallant, P.E., Miller, S., Galbraith, J.A., Winters, C.A., Azzam, R., Reese, T.S., 2009. Rapid turnover of spinules at synaptic terminals. Neuroscience 160, 42–50.

Tao-Cheng, J.-H., Dosemeci, A., Gallant, P.E., Smith, C., Reese, T.S., 2010. Activity induced changes in the distribution of Shanks at hippocampal synapses. Neuroscience 168, 11–17.

Tao-Cheng, J.-H., Crocker, V.T., Winters, C.A., Azzam, R., Chludzinski, J., Reese, T.S., 2011. Trafficking of AMPA receptors at plasma membranes of hippocampal neurons. J. Neurosci. 31, 4834–4843.

Tao-Cheng, J.-H., Yang, Y., Bayer, K.U., Reese, T.S., Dosemeci, A., 2013. Effects of CaMKII inhibitor tatCN21 on activity-dependent redistribution of CaMKII in hippocampal neurons. Neuroscience 244, 188–196.

Tarrant, S.B., Routtenberg, A., 1977. The synaptic spinule in the dendritic spine: electron microscopic study of the hippocampal dentate gyrus. Tissue Cell 9, 461–473.

Valtschanoff, J.G., Weinberg, R.J., 2001. Laminar organization of the NMDA receptor complex within the postsynaptic density. J. Neurosci. 21, 1211–1217.

Vlachos, A., Korkotian, E., Schonfeld, E., Copanaki, E., Deller, T., Segal, M., 2009. Synaptopodin regulates plasticity of dendritic spines in hippocampal neurons. J. Neurosci. 29, 1017–1033.

Yang, Y., Tao-Cheng, J.-H., Reese, T.S., Dosemeci, A., 2011. SynGAP moves out of the core of the postsynaptic density upon depolarization. Neuroscience 192, 132–139.

CHAPTER TWELVE

Activity-Mediated Structural Plasticity of Dendritic Spines

Dominique Muller, Mathias De Roo and Irina Nikonenko
Department and Center of Neuroscience, University of Geneva, Geneva, Switzerland

1. INTRODUCTION

Learning and memory mechanisms and information processing by brain networks are usually interpreted within the framework of Hebbian plasticity and the notion that changes in synaptic strength play a key role in regulating the activity of neuronal populations. While this concept is strongly supported by in vitro and in vivo data, accumulating evidence indicates that induction of plasticity is also associated with an important structural reorganization of synaptic connectivity. In vivo work has shown that excitatory synapses on dendritic spines are highly plastic; they can be formed and eliminated throughout life through activity-dependent mechanisms and a major consequence of learning-related activity is therefore to promote a rewiring of brain circuits. This structural synaptic reorganization has strong potential implications for our understanding of brain development and learning and memory mechanisms and suggests notably that structural modifications of brain circuits could represent the substrate of acquired new skills and of long-lasting memory traces. In this chapter, some of the recent evidence pointing to the importance of these mechanisms are reviewed.

2.1. GENERAL MORPHOLOGICAL CHARACTERISTICS OF DENDRITIC SPINES

Dendritic spines are specialized protrusions receiving in most cases excitatory synaptic contacts and present mainly on excitatory neurons, but also on different interneurons throughout the brain. Their morphology shows a high level of variability affecting the size, shape, and organization of the spine. In the rat hippocampus, a structure where many studies of spine plasticity have been carried out, dendritic spines show variations in head diameters between 0.2 and 1.3 μm, which roughly corresponds to volumes between 0.004 and 1.2 μm^3, thus variations by a factor of around 300 times. Electron microscopic (EM) studies have shown that these variations in volume are usually also correlated with variations in the size of the postsynaptic density (PSD), size of the presynaptic terminal, and presence of various organelles in the spine (Bourne and Harris, 2008). For example, large spines often carry large PSDs that may have complex shapes

The Synapse
http://dx.doi.org/10.1016/B978-0-12-418675-0.00012-2

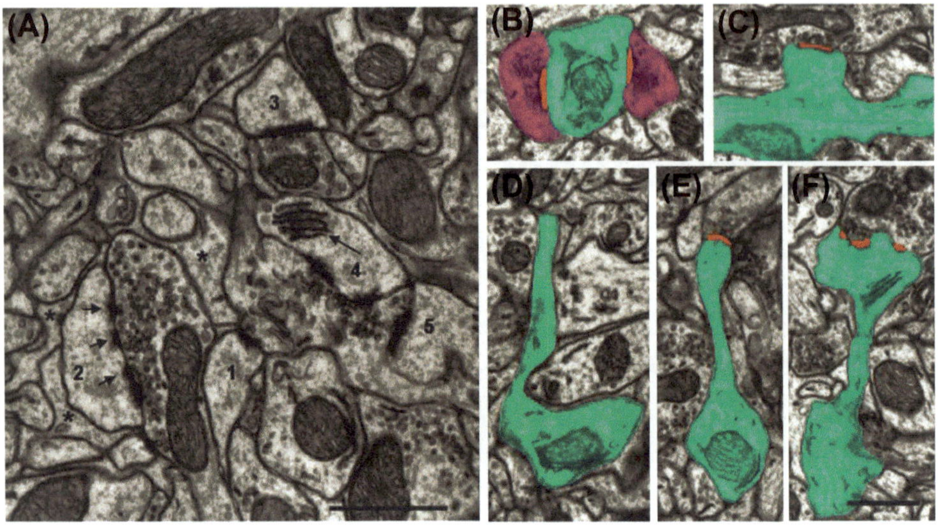

Figure 1 Electron microscopic (EM) illustrations of the variability of spine and PSD size and shape. (A) Spine heads of different sizes and shapes (No. 1–5). Spine no. 1 is of a thin type; spines no. 2–5 are of mushroom type. PSDs in spines 2 and 4 are of a complex shape. Arrow points at spine apparatus in spine no. 4. Spines no. 1, 4, and 5 contact the same presynaptic bouton (multisynaptic bouton). Arrows in spine 2 point at synaptic cleft in the complex PSD. Astrocytic processes surrounding this synapse are marked with asterisks. Scale bar: 0.5 μm. (B–F) EM illustrations of different types of excitatory synapses. (B) Two presynaptic terminals (magenta) making shaft synapses (red) on the dendritic shaft (blue), cut transversally, with a mitochondria in the center surrounded with endoplasmic reticulum. (C) Stubby spine (blue) characterized by a short thick neck and a macular PSD (red). (D) Filopodium. (E) Thin spine with a macular PSD. (E) Mushroom spine with a complex PSD and spine apparatus (arrow). Scale bar: 0.5 μm.

with sometimes multiple or segmented receptor zones (Figure 1). These complex shapes have been particularly reported under conditions of plasticity and increased synaptic remodeling (Geinisman et al., 1993; Toni et al., 1999, 2001; Ganeshina et al., 2004). Large spines also often contain a spine apparatus and/or ribosomes. The spine apparatus corresponds to smooth endoplasmic reticulum (Spacek and Harris, 1997) which forms a widespread network in dendrites and spines (Cooney et al., 2002) and might play a role in the regulation of calcium movements and/or local protein synthesis. Ribosomes are also implicated in local protein synthesis and recent work has shown a redistribution of free polyribosomes from dendritic shafts into enlarged spines after induction of long-term potentiation (LTP; Ostroff et al., 2002; Bourne and Harris, 2008).

All these results have led to the concept that large mushroom spines could be characterized by different properties than smaller, more elongated spines. In particular, mushroom spines have often been referred to as "memory spines", or spines that would have undergone a process of plasticity such as LTP that would change their functional and structural properties. In contrast, spines with smaller heads also sometimes referred to as "learning spines" would be more labile, they could be younger or recently generated and their maintenance in the network could depend upon activity factors and possibly induction of plasticity. This concept has led to the notion that dendritic spines are evolving structures that may acquire or loose specific characteristics over time depending upon their involvement in supporting network activity. This concept may explain the very large variability existing between spine synapses since the structure/function characteristics of each synapse would reflect its functional history.

Finally, an additional and important parameter to take into consideration for synaptic function is the glial component. Spine synapses are often surrounded by fine astrocytic processes (Figure 12.1(A)) that may contribute to synaptic network functions by controlling the local synaptic environment and coordinating synaptic activity and plasticity (Haydon, 2001; Volterra and Meldolesi, 2005). Although debated, recent work has provided evidence for a possible existence of gliotransmission and a role of astrocytes in regulating synaptic function and plasticity (Dityatev and Rusakov, 2011). Astrocytic processes show structural plasticity properties and in particular their coverage of excitatory synapses is modified upon induction of plasticity, which might contribute to modify the properties of the spine and of the synapse (Genoud et al., 2006; Witcher et al., 2007; Lushnikova et al., 2009).

2.2. SYNAPSE FORMATION THROUGH FILOPODIA

Time lapse confocal imaging of synapses over 2001–2010 has provided evidence that spines continuously change in shape over time but can also be formed and eliminated throughout life. Synapses must therefore be considered as dynamic structures susceptible to undergo various forms of transformation and evolution. The first step is certainly the formation of the dendritic spine itself. There appear to exist at least two different processes through which new excitatory spine synapses can be formed. One includes as an initial step the growth of a very motile structure, the filopodium, which extends from the dendrite on several microns (Figure 12.1(D)). Filopodia are mainly seen during early phases of synaptic development when the distance to potential partners is greater (Ziv and Smith, 1996). They usually remain highly motile and repeatedly establish contacts with nearby axons, probably though a mechanism that involves the generation of calcium transients (Lohmann and Bonhoeffer, 2008).

This phenomenon does not seem to implicate signaling through the neurotransmitter glutamate since glutamate receptor antagonists do not affect contact formation by filopodia. These contacts, however, probably implicate specific recognition mechanisms since filopodia are able to discriminate between partners and never make stabilized contacts with inhibitory axons (Lohmann and Bonhoeffer, 2008). In most cases, filopodia retract and disappear, but sometimes they can be transformed into spines (Maletic-Savatic et al., 1999; Marrs et al., 2001; Trachtenberg et al., 2002; Zuo et al., 2005; De Roo et al., 2008a). Several time lapse imaging studies suggest that this transformation is not very frequent. In living, young mice or in hippocampal slice cultures, only 10–20% of filopodia seem to be transformed into spines and very often, these spines will also disappear within a few days (Zuo et al., 2005; De Roo et al., 2008a). During early development, the role of filopodia might however be more important as suggested by evidence showing that decreasing filopodia motility by interfering with EphrinB signaling significantly affects mechanisms of synaptogenesis (Kayser et al., 2008).

At the molecular level, the motility and behavior of filopodia are regulated by a number of different molecules, some promoting filopodia formation such as calcium/calmodulin-dependent protein kinase II (CaMKII) (Jourdain et al., 2003), syndecan-2 (Ethell and Yamaguchi, 1999; Lin et al., 2007), or paralemmin-1 (Arstikaitis et al., 2008), while others, conversely, maintain dendritic filopodia, slowing spine maturation (Pak and Sheng, 2003; Vazquez et al., 2004; Kumar et al., 2005; Matsuno et al., 2006; Furutani et al., 2007; Yoshihara et al., 2009).

2.3. SYNAPSE FORMATION THROUGH THE GROWTH OF NEW SPINES

In more mature tissue, new spines usually appear directly as such (Engert and Bonhoeffer, 1999; Trachtenberg et al., 2002). This process occurs within minutes and accounts for about half of all protrusions formed in young hippocampal tissue (Engert and Bonhoeffer, 1999; Jourdain et al., 2003; De Roo et al., 2008a) and even a greater fraction in mature cortex. An interesting question is to understand where these new spines grow and why they grow at specific locations. One suggestion has been that they could extend from initially formed shaft synapses (Ethell and Pasquale, 2005). While such a mechanism has indeed been reported, it probably concerns only a minority of cases. Another possibility is that spines are formed in a nonspecific manner that cannot really be predicted, and mainly depends upon postsynaptic signaling events. This seems to be the case in the hippocampus during the period of intense synapse development, where spine growth seems to occur randomly and can be significantly affected by postsynaptic signaling mechanisms (De Roo et al., 2008a; Dubos et al., 2012). A third possibility is that spines growth occurs in regions where glutamate is released at high concentrations from active presynaptic structures.

There is indeed evidence that presynaptic activity and glutamate can promote spine formation (Maletic-Savatic et al., 1999; Richards et al., 2005). Recent work in which glutamate was focally uncaged on a dendritic segment revealed that this could trigger the growth of a new protrusion (Kwon and Sabatini, 2011). Also new protrusions seem to preferentially grow toward axonal boutons with active synapses, thus forming multi-synaptic boutons (Knott et al., 2006; Toni et al., 2007). Conversely, activated presynaptic boutons may grow filopodia-like protrusions that establish synaptic contacts (Nikonenko et al., 2003). These observations therefore suggest that spine formation may be regulated by multiple mechanisms, including postsynaptic factors that may be important to control the number of synapses on a neuron, and presynaptic glutamate release, which may act as a trophic influence guiding new protrusions to active boutons.

Another interesting observation provided by time lapse imaging experiments is that not all spines have the same life span. Many new spines appear to be very transient, while others may persist for prolonged periods of time, up to years under in vivo conditions (Grutzendler et al., 2002; Holtmaat et al., 2005; Cruz-Martin et al., 2010). The reason why some spines are only transient is still unknown, but could be linked to a failure to establish a synaptic contact with a proper partner. In many cases, these newly formed spines do not express immediately a PSD (Knott et al., 2006; Nagerl et al., 2007; De Roo et al., 2008a). Based on morphological criteria, a 3D EM reconstruction (is this a 3D EM reconstruction in live mice??) study in the cortex of living mice suggests that new spines do not become morphologically mature before 1–4 days. Under in vitro conditions in slice cultures, detection of PSD-95-EGFP expression in newly formed spines was only possible after about 5 h.

Consistent with this, EM analyses in the cortex or hippocampus have shown the existence of a small population of spine-like protrusions devoid of PSD or even without presynaptic partner, suggesting that spine growth could precede synapse formation by several hours (Knott et al., 2006; Arellano et al., 2007; De Roo et al., 2008a). These observations therefore support the idea that at least a fraction of spines grow in a random fashion, without initially forming a synapse, which could account for their high probability to disappear.

Synapse formation could also depend upon activity. A recent study in which new spines were stimulated through glutamate uncaging protocols showed that a new protrusion could respond to released transmitter within tens of minutes (Zito et al., 2009). Furthermore, the new spines formed following uncaging of glutamate have been shown to be rapidly functional (Kwon and Sabatini, 2011). Thus the process of contact formation could be promoted by activity and thereby affects the fate of newly formed spines.

2.4. FAST MOTILITY OF DENDRITIC SPINES

Analyses of spine behavior over periods of minutes under in vitro and in vivo conditions have revealed various forms of motility and morphological remodeling of dendritic spines. The first type of motility initially reported in cultures, but also described in vivo, is a fast (seconds to minutes) twitching of spine heads. This phenomenon has been linked to a continuous rearrangement of the actin cytoskeleton (Fischer et al., 1998; Majewska et al., 2006) and was proposed to be associated with activity and experience-dependent plasticity (Oray et al., 2004).

In the visual cortex, a brief monocular deprivation resulted in rapid structural changes of spines that could be mimicked by degradation of the extracellular matrix, suggesting that it could be important to permit structural remodeling. However, its significance in terms of synaptic function and properties remains unclear. In addition to rearrangements of the cytoskeleton, this spine twitching could also be linked to mechanisms of protein trafficking, which is particularly high in dendritic spines. Proteins such as PSD-95, Ras, or Shank3 have been shown to redistribute within minutes to hours between synaptic and dendritic pools (Gray et al., 2006; Tsuriel et al., 2006). Furthermore, properties of plasticity also depend upon trafficking mechanisms, which may rapidly modify receptor numbers through exo- and endocytotic processes directly occurring at synapses (Newpher and Ehlers, 2009). Together this fast twitching of spines could reflect the highly dynamic nature of the molecular organization of the synapse.

2.5. ACTIVITY-MEDIATED SPINE ENLARGEMENT

Early EM studies that analyzed the morphological correlate of synaptic plasticity were the first to suggest that the size and morphology of dendritic spines could be modified by the patterns of activity that induce LTP (Fifkova and Anderson, 1981; Buchs and Muller, 1996). However, this possibility was only confirmed when the development of confocal imaging approaches allowed to repetitively visualize the same synaptic structures before and after inducing plasticity. Through glutamate uncaging experiments, it was shown that local spine stimulation is accompanied by an increased sensitivity to glutamate and an enlargement of the dendritic spine head (Matsuzaki et al., 2004). This effect is rapid, it can last for several hours, and it closely correlates with the accumulation of α-amino-3-hydroxy-5-methyl-4-isoxazolepropionic acid receptor (AMPAR) at the synapse (Zito et al., 2009) and a reorganization of the actin cytoskeleton (Honkura et al., 2008). It has been reported mainly in in vitro studies, but there is also evidence that learning paradigms in vivo can be associated with an enlargement of spines (Roberts et al., 2010).

The results of these experiments therefore strongly support the idea that spine size and synaptic efficacy are tightly linked and that the changes in size correlate with the increased expression of postsynaptic receptors. This could be accounted for by what is referred to as the "slot hypothesis" which proposes that an expansion of the PSD allowing the accommodation of a greater number of AMPAR underlies the enlargement and increase in strength of synapses following LTP induction (Lisman et al., 2002; Bats et al., 2007).

Another parameter that has been associated with spine enlargement is the structural stability of the spine. Repetitive confocal imaging in live mouse cortex showed that large spines may be highly stable and persist for months while spines with small heads may be more transient (Grutzendler et al., 2002; Trachtenberg et al., 2002). Kasai et al. (2003) therefore proposed that the size of the spine could correlate with its structural stability. Consistent with this, work by Holtmaat et al. (2005, 2006) found a clear correlation between spine size and spine stability in vivo.

Several studies have since reproduced this finding, suggesting that the spine enlargement observed following LTP induction could reflect a mechanism of spine stabilization. Indeed, when analyzing the stability of identified spines after LTP induction, De Roo et al. (2008b) showed that activity promoted spine enlargement and switched these spines to a highly stable state (Figure 2). The effect was very specific to stimulated spines and did not affect nonstimulated spines present on the same dendrite. The stabilization of new spines is also a major feature reported in in vivo experiments following a behavioral learning paradigm. In motor skill learning experiments, new spines grow on selective populations of neurons; they are preferentially stabilized during subsequent training and they persist long after training has stopped (Xu et al., 2009; Yang et al., 2009).

Important also, the number of new spines stabilized following training shows a strong correlation with the retention of the learned behavior and thus memory. Similarly, whisker trimming or exposure to enriched environment is also associated with a stabilization of spines (Holtmaat et al., 2006; Bednarek and Caroni, 2011). In birds, similar mechanisms have been uncovered during song learning. They include an enlargement and a

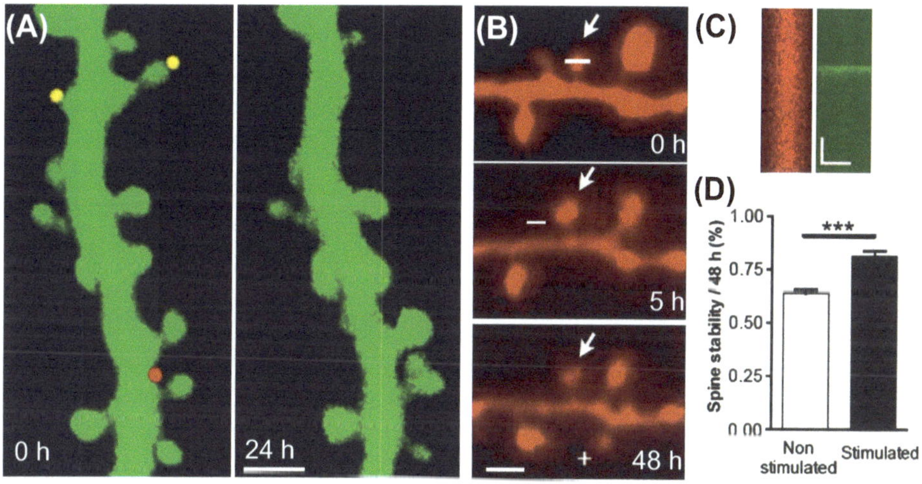

Figure 2 Structural plasticity of excitatory spine synapses in hippocampal slice cultures. (A) Dendritic segment of a CA1 pyramidal neuron expressing eGFP and imaged at 24 h interval. Red and yellow dots point newly formed and lost spines, respectively. Scale bar: 1 μm. (B) Enlargement and stabilization of a stimulated spine (arrow) in a cell expressing mRFP and loaded with the calcium indicator Fluo4-AM. Line scan analysis of this spine at the beginning of the experiment (0 h) showed an increased calcium transient upon electrical stimulation of a group of CA3 cells (shown in C). Theta burst activation of this spine resulted in an enlargement of the spine head 5 h later and promoted the persistence of the spine over the next 48 h. Note the formation and elimination of other spines in the vicinity (+ and − signs). Scale bar: 1 μm. (C) Line scan analyses through the spine illustrated in (B) and showing the red (mRFP) and green (Fluo4-AM) signals. Scale bars: 0.5 μm; 0.5 s. (D) Stimulated spines show a significantly increased stability over the next 48 h than nonstimulated spines ($n = 37$ and 44 spines analyzed).

rapid stabilization of dendritic spines, which in addition show an enhancement of synaptic activity (Roberts et al., 2010).

Together these different studies provide strong support to the idea that the stabilization of selective populations of spines could represent a structural basis for memory retention and that this phenomenon could be linked to the enlargement and strengthening of synapses. A specific demonstration that this stabilization process is the result of LTP induction during behavioral training is however still missing.

Another open question that remains regarding the link between spine enlargement and spine stability concerns the duration of the effect. Repetitive imaging experiments have shown that the size of spine heads shows continuous fluctuations over time (Holtmaat et al., 2005; De Roo et al., 2008b; Yasumatsu et al., 2008). Small spines can enlarge and/or retract over periods of days. Does this indicate that spine stability, but also PSD size and strength continuously fluctuate according to the activity history of the synapse? In the analyses made after LTP induction at identified synapses in vitro, it was observed that the spines that enlarged after LTP decreased in size over the next days and could no longer be distinguished by their size from nonstimulated spines (De Roo et al., 2008b). However unlike nonstimulated spines, they remained stable even though their size had decreased.

This experiment suggests that while enlargement could be a critical step to increase spine stability, the size increase does not necessarily need to be maintained to ensure the persistence of the synapse. Indeed morphological analyses have not reported global increases of spine size during development, while spines become more and more stable.

2.6. MOLECULAR MECHANISMS OF SPINE ENLARGEMENT AND STABILIZATION

Multiple mechanisms contribute to the regulation of spine enlargement and stabilization and to a large extent they show a significant overlap with mechanism implicated in LTP. Among major contributors downstream of *N*-methyl-D-aspartate (NMDA) receptor activation are protein kinases and particularly CaMKII. Strong evidence supports a role for CaMKII in LTP induction and recent results also indicate a contribution of a lasting activation of CaMKII through autophosphorylation in LTP maintenance (Sanhueza et al., 2007; Lisman et al., 2012).

Furthermore, CaMKII activity is required for activity-mediated spine enlargement; CaMKII regulates the insertion of NMDAR and AMPAR subunits and the trapping of surface AMPARs at the synapse and it is essential for learning (Correia et al., 2008; Steiner et al., 2008; Opazo et al., 2010; Murakoshi et al., 2011). Similarly protein kinase C, and particularly protein kinase M zeta, has been directly implicated in LTP maintenance; it regulates AMPAR incorporation during LTP (Tomita et al., 2005; Boehm et al., 2006), induces morphological changes of dendritic spines (Ron et al., 2011), and contributes to in vivo spine stabilization (Bednarek and Caroni, 2011; Sacktor, 2011).

Another mechanism contributing to the regulation of spine morphology involves the actin cytoskeleton. Much evidence indicates that interfering with actin polymerization blocks LTP, that actin-based mechanisms are required for activity-mediated spine enlargement and that the stabilization of spines under in vivo conditions requires the phosphorylation of the cytoskeleton-stabilizing protein beta-adducin (Lisman, 2003; Honkura et al., 2008; Hotulainen and Hoogenraad, 2010; Bosch and Hayashi, 2011).

Consistent with this important role played by the actin cytoskeleton, several studies have provided direct evidence for the implication of a number of molecules regulating actin polymerization and notably the Rho GTPases Rac1 and Cdc42 and their upstream or downstream modulators, including kalirin-7, DISC-1, PAKs, LIMK, and actin regulatory proteins such as cofilin. Interference with this signaling affects LTP mechanisms and the capacity of spines to enlarge (Xie et al., 2007; Hayashi-Takagi et al., 2010; Murakoshi et al., 2011; Dubos et al., 2012). Many of these molecules are also implicated in cognitive disability or even schizophrenia and part of their implication in these diseases could be linked to alterations in spine morphology (Kao et al., 2010; Penzes et al., 2011).

Another important system contributing to synaptic plasticity involves the regulation of protein synthesis at synapses. Translation of various mRNAs, including Arc and CaMKII mRNAs, has been reported at synapses following LTP induction and protein synthesis has been shown by many studies to be required for the maintenance of LTP. Interference with protein synthesis or Arc translation prevents spine enlargement and in vivo blockade of protein synthesis results in synapse destabilization (Bramham, 2008; Tanaka et al., 2008; Bednarek and Caroni, 2011; Shepherd and Bear, 2011).

Finally, several proteins expressed in the PSD play an important role in regulating synapse size and stability and could contribute to the slot hypothesis of LTP. This is typically the case of PSD-95 and AMPARs which are required not only for changes in spine size but also for activity-driven stabilization of the synapse (Ehrlich and Malinow, 2004; Kopec et al., 2006). Furthermore, several adhesion molecule systems including neuroligins, ephrins, and N-cadherin have been linked to changes in spine morphology and stability. They could mediate trans-synaptic signals regulating the actin cytoskeleton, but could also directly participate to increase trans-synaptic contacts and thus the stability of the synapse (Zuber et al., 2005).

For example, activity-mediated expression of N-cadherin correlates with changes in spine size, but is also required for the long-term stabilization of spines (Mendez et al., 2010a). Another group of molecules that could contribute to these changes include metalloproteases such as matrix metalloproteinase 9 (MMP9), which degrade the extracellular matrix and markedly affect synaptic structure and function. MMP9 is required for LTP at hippocampal synapses and mediates spine enlargement and synaptic potentiation through integrin signaling (Wang et al., 2008). The regulation of spine size and stability involves therefore complex interactions between various signaling systems affecting the actin cytoskeleton, protein synthesis, and mechanisms of trafficking at the synapse.

2.7. ACTIVITY-MEDIATED NETWORK REWIRING THROUGH SPINE TURNOVER

On a time scale of days and weeks, spines have also been shown to undergo a continuous turnover through a process of formation and elimination (Figures 2, 3). The rate of spine growth and disappearance is particularly high during critical periods of development when the main organization of cortical synaptic networks is to be established (Lendvai et al., 2000; Zuo et al., 2005; Cruz-Martin et al., 2010). Spine turnover then decreases with brain maturation but remains effective at a low level even in adult mice (Grutzendler et al., 2002; Trachtenberg et al., 2002).

Quantification of spine turnover in various brain regions has been difficult and shows significant variability. In mice, during the first weeks after birth, turnover rate can affect as many as 5–20% of total spines per hour, decreasing to about 5–10% per week at 1 month of age and 1–2% per month in adult tissue (Holtmaat and Svoboda, 2009; Cruz-Martin et al., 2010). These values however vary quite substantially depending on cortical region or cell type analyzed, criteria used for analysis and experimental approach, including possible effects of anesthesia (De Roo et al., 2009). Despite this variability, the important message provided by these studies is that remodeling of synaptic connections is very substantial during development and remains a significant phenomenon throughout life.

Spine turnover is not only developmentally regulated, but it is also controlled by various factors including neuronal activity, hormones, and growth factors. The experiments that analyzed by repetitive imaging approaches the effects of stimulation protocols that induce LTP first showed that activity could trigger the growth of new protrusions (Engert and Bonhoeffer, 1999; Maletic-Savatic et al., 1999; Toni et al., 1999; Nagerl et al., 2007). Later experiments then provided evidence that this growth process could also be associated with an increased elimination of spines and thus involved a more subtle control of spine dynamics (Nagerl et al., 2004; De Roo et al., 2008b; Minerbi et al., 2009).

More recent in vivo experiments have further confirmed the notion that neuronal and synaptic activity can regulate spine turnover. This was observed following sensory stimulation or motor learning paradigms (Holtmaat et al., 2006; Xu et al., 2009; Yang et al., 2009; Wilbrecht et al., 2010), this has also been reported in the visual cortex, following monocular deprivation and restoration of binocular vision (Hofer et al., 2009). All these results therefore provide a strong link between functional plasticity and specific synaptic rearrangements (Figure 3).

The question however is to understand the specificity of these synaptic rearrangements and how they may affect brain networks at a functional level. The evidence from in vivo studies following environment enrichment or learning paradigms suggests that these structural reorganizations affect large ensembles of synapses that they occur specifically in the brain regions implicated in the new learning and that their magnitude

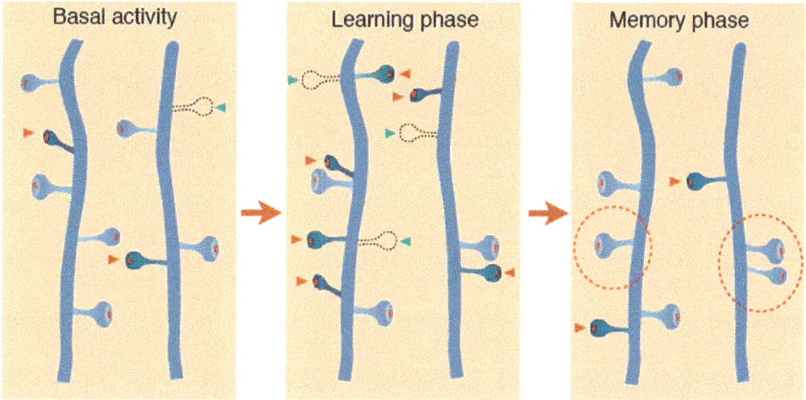

Figure 3 Schematic model of how learning related paradigms may affect synaptic network rewiring. Under basal conditions, most spine synapses remain stable, while a small fraction of them shows a continuous turnover (red and blue arrow heads for newly formed and lost spines). During a learning activity, spine dynamics is increased promoting network rewiring through an increased formation and elimination of spines, notably favoring a clustered growth of spines in active dendritic regions. Those spines implicated in synaptic activity during the learning phase become persistent, ensuring the maintenance of a long-lasting memory trace.

correlates with the learning performance (Xu et al., 2009; Yang et al., 2009; Roberts et al., 2010; Bednarek and Caroni, 2011; Ruediger et al., 2011; Lai et al., 2012). This structural rewiring could therefore play an important role in the acquisition of new skills or behaviors by promoting the adaptation of brain circuits to new forms of activity, while at the same time allowing the persistence in the network of the most functional contacts (Figure 3).

Interestingly, this activity-mediated synapse reorganization seems to be at least in part regulated by local mechanisms. Indeed, LTP induction at identified spines promotes the growth of new spines close to these activated spines (De Roo et al., 2008b). This phenomenon of clustered growth is further supported by two recent studies. Acquisition of a motor task through repetitive training was found to result in the formation and stabilization of new spines emerging in clusters as neighboring spine pairs (Fu et al., 2012). Furthermore, analyses of synaptic activity through calcium imaging indicate that neighboring synapses are more likely to be coactive than synapses farther from each other (Kleindienst et al., 2011). The clustering of spines has also been shown to represent an important aspect for information processing on small and distant dendrites (Nevian et al., 2007). This clustering mechanism could therefore depend upon a local regulation of spine turnover.

Regulation of spine turnover may also lead under some circumstances to modifications of spine density. This can be produced when the changes in spine growth and elimination are differentially affected. In animals submitted to an enriched environment protocol, spine growth was markedly enhanced and resulted in an increase in the absolute number of spines (Hofer et al., 2009; Bednarek and Caroni, 2011). Also application of estrogens to hippocampal or cortical tissue results in a more pronounced effect on spine growth than elimination and therefore also leads to an increase in spine density (Srivastava et al., 2008; Mendez et al., 2010b). Regulation of spine dynamics may thus not only promote rewiring but also control the level of connectivity of the network and affect its function.

2.8. MOLECULAR MECHANISMS REGULATING SPINE DYNAMICS

Little is known yet on the mechanisms that regulate spine turnover during development and activity. One important factor however could be the balance between excitation and inhibition. Alterations of this balance during critical periods of development significantly affect mechanisms of structural plasticity and the specificity with which synaptic networks are formed (Hensch, 2005). Interestingly this control seems to remain operative even later on in adulthood, since reduction of inhibition through various manipulations may restore plasticity in the adult visual cortex to levels comparable to those observed during development (Bavelier et al., 2010; Morishita et al., 2010). The mechanisms underlying these regulations remain unclear, but they probably involve changes in spine dynamics. During development, it has been shown that a short anesthesia or drugs that enhance GABAergic inhibition trigger in rapid and marked increase in spine growth and synapse gain (De Roo et al., 2009).

Several growth promoting factors or hormones also appear to affect the level of spine formation and turnover. One of the most studied examples is probably estrogens which can rapidly shift the balance of spine turnover toward increased growth and stabilization resulting in increases in spine density (Kramar et al., 2009). Interestingly the effect is reversible and the newly formed spines are also those that are preferentially eliminated when the hormonal level decreases (Mendez et al., 2010b). Brain-derived neurotrophic factor (BDNF) is another important candidate for regulating spine formation although its effects on spine dynamics and activity-mediated spine formation remain poorly characterized under in vivo conditions (Horch et al., 1999; Horch and Katz, 2002). BDNF could probably control spinogenesis through its regulation of protein synthesis and its activation of the PI3K and mammalian target of rapamycin (mTOR) signaling pathways (Cuesto et al., 2011). In the mouse model of fragile X syndrome, protein synthesis, mTOR signaling, and spine turnover are indeed affected and have been proposed to contribute to the physiopathology of the disease (Sharma et al., 2010).

Finally, mechanisms regulating the cytoskeleton including Rho GTPases and their regulatory proteins are likely to play an important role in the control of spine turnover. Consistent with this, one of the main alterations observed following inhibition of PAK3, a downstream effector of Rac1 and Cdc42, is the promotion of spine formation that results in an excessive increase in synaptic networks dynamics (Dubos et al., 2012). Additionally, interference with WAVE-1 or the Arp2/3 complex also leads to changes in spine density and interferes with synaptic plasticity (Soderling et al., 2007; Nakamura et al., 2011). The cytoskeleton therefore plays a critical role in structural plasticity both at the level of stimulated spines, by contributing to their enlargement and stabilization, but also in nearby dendrites, by promoting spine growth and changes in spine dynamics in response to cell activity. More work will however be needed in order to better identify the molecular mechanisms contributing to activity-mediated rewiring of synaptic networks.

2.9. SPINE ALTERATIONS AND BRAIN DISEASE

Alterations of dendritic spine morphology or function have been frequently associated with brain diseases, notably cognitive disorders and neurodegenerative diseases. Studies from 1990 to 2010 have identified a number of genetic alterations associated with intellectual disability and autism spectrum disorders and in many instances these defects affect genes coding for synaptic proteins or proteins involved in the regulation of synaptic properties (Ropers, 2006; Laumonnier et al., 2007; Bourgeron, 2009). Analyses of the underlying mechanisms using gain and loss of function approaches have revealed different types of alterations affecting spine morphology, spine density, or synaptic plasticity.

A typical example is provided by the fragile X syndrome linked to the silencing of the fmr1 gene (Bagni and Greenough, 2005) that codes for a protein regulating RNA granule transport in dendrite and at synapses and believed to play a role in the control of local translation at individual synapses. Various synaptic defects have indeed been reported in FMR1 knockout mice, including an increased spine density with more long, thin, tortuous spines, defects in synaptic LTP together with an enhanced metabotropic glutamate receptor (mGluR)-mediated long-term depression (LTD). A current hypothesis, therefore, is that this protein negatively regulates mGluR mediated LTD and when mutated or absent could lead to an enhanced endocytosis of AMPARs (Huber et al., 2002; Dolen et al., 2007).

Mutations of several other synaptic proteins associated with cognitive disability, autism spectrum disorders, or even schizophrenia also lead to dendritic spine abnormalities. Examples include oligophrenin, PAK3, ARGHEF6, SynGAP, Shank, EPAC2, and DISC1 (Sala et al., 2001; Boda et al., 2004; Govek et al., 2004; Vazquez et al., 2004; Node-Langlois et al., 2006; Woolfrey et al., 2009). Interestingly, the synaptic defects that have been observed often include an increase in the fraction of long, thin, and immature spines. Furthermore, these different proteins are implicated in regulation mechanisms that control either the cytoskeleton or the trafficking of receptors, suggesting that alterations in spine dynamics, enlargement, or plasticity could underlie the pathogenesis of intellectual disability. One interpretation of these results is that even subtle defects in spine turnover or activity-mediated spine stabilization could interfere with a correct wiring and specificity of synaptic networks (Boda et al., 2010).

Spine alterations have also been reported in association with degenerative diseases such as Alzheimer's disease (Selkoe, 2002; Spires et al., 2005), Parkinson's disease (Day et al., 2006), or prion disease (Fuhrmann et al., 2007). In mouse models of Alzheimer's disease, the presence of amyloid plaques results in the formation of dysmorphic neurites and spine loss and one hypothesis is that amyloid oligomers could interfere with LTP mechanisms and promote LTD accounting in this way for the loss of spines (Wei et al., 2010). A similar situation of spine loss seems to also occur at striatal medium spiny neurons in models of Parkinson's disease (Day et al., 2006). Thus alterations of spine formation and elimination are considered as important possible pathophysiological mechanisms underlying these degenerative diseases.

3. CONCLUSIONS

Dendritic spines are highly specialized structures that exhibit a high degree of structural and functional plasticity. They are tightly controlled by activity and signaling mechanisms that determine their strength, their size, the rate at which they are formed and replaced and their capacity to remain durably in brain networks. The complexity of these regulations and their impact on the construction and function of brain circuits only starts to be understood and clearly highlights the key role played by dendritic spines in the processing of information by neuronal networks.

REFERENCES

Arellano, J.I., Espinosa, A., Fairen, A., Yuste, R., DeFelipe, J., 2007. Non-synaptic dendritic spines in neocortex. Neuroscience 145, 464–469.

Arstikaitis, P., Gauthier-Campbell, C., Carolina Gutierrez Herrera, R., Huang, K., Levinson, J.N., Murphy, T.H., Kilimann, M.W., Sala, C., Colicos, M.A., El-Husseini, A., 2008. Paralemmin-1, a modulator of filopodia induction is required for spine maturation. Mol. Biol. Cell 19, 2026–2038.

Bagni, C., Greenough, W.T., 2005. From mRNP trafficking to spine dysmorphogenesis: the roots of fragile X syndrome. Nat. Rev. Neurosci. 6, 376–387.

Bats, C., Groc, L., Choquet, D., 2007. The interaction between Stargazin and PSD-95 regulates AMPA receptor surface trafficking. Neuron 53, 719–734.

Bavelier, D., Levi, D.M., Li, R.W., Dan, Y., Hensch, T.K., 2010. Removing brakes on adult brain plasticity: from molecular to behavioral interventions. J. Neurosci. 30, 14964–14971.

Bednarek, E., Caroni, P., 2011. Beta-adducin is required for stable assembly of new synapses and improved memory upon environmental enrichment. Neuron 69, 1132–1146.

Boda, B., Alberi, S., Nikonenko, I., Node-Langlois, R., Jourdain, P., Moosmayer, M., Parisi-Jourdain, L., Muller, D., 2004. The mental retardation protein PAK3 contributes to synapse formation and plasticity in hippocampus. J. Neurosci. 24, 10816–10825.

Boda, B., Dubos, A., Muller, D., 2010. Signaling mechanisms regulating synapse formation and function in mental retardation. Curr. Opin. Neurobiol. 20, 519–527.

Boehm, J., Kang, M.G., Johnson, R.C., Esteban, J., Huganir, R.L., Malinow, R., 2006. Synaptic incorporation of AMPA receptors during LTP is controlled by a PKC phosphorylation site on GluR1. Neuron 51, 213–225.

Bosch, M., Hayashi, Y., 2011. Structural plasticity of dendritic spines. Curr. Opin. Neurobiol.
Bourgeron, T., 2009. A synaptic trek to autism. Curr. Opin. Neurobiol. 19, 231–234.
Bourne, J.N., Harris, K.M., 2008. Balancing structure and function at hippocampal dendritic spines. Annu. Rev. Neurosci. 31, 47–67.
Bramham, C.R., 2008. Local protein synthesis, actin dynamics, and LTP consolidation. Curr. Opin. Neurobiol. 18, 524–531.
Buchs, P.A., Muller, D., 1996. Induction of long-term potentiation is associated with major ultrastructural changes of activated synapses. Proc. Natl. Acad. Sci. USA 93, 8040–8045.
Cooney, J.R., Hurlburt, J.L., Selig, D.K., Harris, K.M., Fiala, J.C., 2002. Endosomal compartments serve multiple hippocampal dendritic spines from a widespread rather than a local store of recycling membrane. J. Neurosci. 22, 2215–2224.
Correia, S.S., Bassani, S., Brown, T.C., Lise, M.F., Backos, D.S., El-Husseini, A., Passafaro, M., Esteban, J.A., 2008. Motor protein-dependent transport of AMPA receptors into spines during long-term potentiation. Nat. Neurosci. 11, 457–466.
Cruz-Martin, A., Crespo, M., Portera-Cailliau, C., 2010. Delayed stabilization of dendritic spines in fragile X mice. J. Neurosci. 30, 7793–7803.
Cuesto, G., Enriquez-Barreto, L., Carames, C., Cantarero, M., Gasull, X., Sandi, C., Ferrus, A., Acebes, A., Morales, M., 2011. Phosphoinositide-3-kinase activation controls synaptogenesis and spinogenesis in hippocampal neurons. J. Neurosci. 31, 2721–2733.
Day, M., Wang, Z., Ding, J., An, X., Ingham, C.A., Shering, A.F., Wokosin, D., Ilijic, E., Sun, Z., Sampson, A.R., Mugnaini, E., Deutch, A.Y., Sesack, S.R., Arbuthnott, G.W., Surmeier, D.J., 2006. Selective elimination of glutamatergic synapses on striatopallidal neurons in Parkinson disease models. Nat. Neurosci. 9, 251–259.
De Roo, M., Klauser, P., Briner, A., Nikonenko, I., Mendez, P., Dayer, A., Kiss, J.Z., Muller, D., Vutskits, L., 2009. Anesthetics rapidly promote synaptogenesis during a critical period of brain development. PLoS One 4, e7043.
De Roo, M., Klauser, P., Mendez, P., Poglia, L., Muller, D., 2008a. Activity-dependent PSD formation and stabilization of newly formed spines in hippocampal slice cultures. Cereb. Cortex 18, 151–161.
De Roo, M., Klauser, P., Muller, D., 2008b. LTP promotes a selective long-term stabilization and clustering of dendritic spines. PLoS Biol. 6, e219.
Dityatev, A., Rusakov, D.A., 2011. Molecular signals of plasticity at the tetrapartite synapse. Curr. Opin. Neurobiol. 21, 353–359.
Dolen, G., Osterweil, E., Rao, B.S., Smith, G.B., Auerbach, B.D., Chattarji, S., Bear, M.F., 2007. Correction of fragile X syndrome in mice. Neuron 56, 955–962.
Dubos, A., Combeau, G., Bernardinelli, Y., Barnier, J.V., Hartley, O., Gaertner, H., Boda, B., Muller, D., 2012. Alteration of synaptic network dynamics by the intellectual disability protein PAK3. J. Neurosci. 32, 519–527.
Ehrlich, I., Malinow, R., 2004. Postsynaptic density 95 controls AMPA receptor incorporation during long-term potentiation and experience-driven synaptic plasticity. J. Neurosci. 24, 916–927.
Engert, F., Bonhoeffer, T., 1999. Dendritic spine changes associated with hippocampal long-term synaptic plasticity. Nature 399, 66–70.
Ethell, I.M., Pasquale, E.B., 2005. Molecular mechanisms of dendritic spine development and remodeling. Prog. Neurobiol. 75, 161–205.
Ethell, I.M., Yamaguchi, Y., 1999. Cell surface heparan sulfate proteoglycan syndecan-2 induces the maturation of dendritic spines in rat hippocampal neurons. J. Cell Biol. 144, 575–586.
Fifkova, E., Anderson, C.L., 1981. Stimulation-induced changes in dimensions of stalks of dendritic spines in the dentate molecular layer. Exp. Neurol. 74, 621–627.
Fischer, M., Kaech, S., Knutti, D., Matus, A., 1998. Rapid actin-based plasticity in dendritic spines. Neuron 20, 847–854.
Fu, M., Yu, X., Lu, J., Zuo, Y., 2012. Repetitive motor learning induces coordinated formation of clustered dendritic spines in vivo. Nature 483, 92–95.
Fuhrmann, M., Mitteregger, G., Kretzschmar, H., Herms, J., 2007. Dendritic pathology in prion disease starts at the synaptic spine. J. Neurosci. 27, 6224–6233.
Furutani, Y., Matsuno, H., Kawasaki, M., Sasaki, T., Mori, K., Yoshihara, Y., 2007. Interaction between telencephalin and ERM family proteins mediates dendritic filopodia formation. J. Neurosci. 27, 8866–8876.

Ganeshina, O., Berry, R.W., Petralia, R.S., Nicholson, D.A., Geinisman, Y., 2004. Differences in the expression of AMPA and NMDA receptors between axospinous perforated and nonperforated synapses are related to the configuration and size of postsynaptic densities. J. Comp. Neurol. 468, 86–95.
Geinisman, Y., de Toledo-Morrell, L., Morrell, F., Heller, R.E., Rossi, M., Parshall, R.F., 1993. Structural synaptic correlate of long-term potentiation: formation of axospinous synapses with multiple, completely partitioned transmission zones. Hippocampus 3, 435–445.
Genoud, C., Quairiaux, C., Steiner, P., Hirling, H., Welker, E., Knott, G.W., 2006. Plasticity of astrocytic coverage and glutamate transporter expression in adult mouse cortex. PLoS Biol. 4, e343.
Govek, E.E., Newey, S.E., Akerman, C.J., Cross, J.R., Van der Veken, L., Van Aelst, L., 2004. The X-linked mental retardation protein oligophrenin-1 is required for dendritic spine morphogenesis. Nat. Neurosci. 7, 364–372.
Gray, N.W., Weimer, R.M., Bureau, I., Svoboda, K., 2006. Rapid redistribution of synaptic PSD-95 in the neocortex in vivo. PLoS Biol. 4, e370.
Grutzendler, J., Kasthuri, N., Gan, W.B., 2002. Long-term dendritic spine stability in the adult cortex. Nature 420, 812–816.
Hayashi-Takagi, A., Takaki, M., Graziane, N., Seshadri, S., Murdoch, H., Dunlop, A.J., Makino, Y., Seshadri, A.J., Ishizuka, K., Srivastava, D.P., Xie, Z., Baraban, J.M., Houslay, M.D., Tomoda, T., Brandon, N.J., Kamiya, A., Yan, Z., Penzes, P., Sawa, A., 2010. Disrupted-in-Schizophrenia 1 (DISC1) regulates spines of the glutamate synapse via Rac1. Nat. Neurosci.
Haydon, P.G., 2001. GLIA: listening and talking to the synapse. Nat. Rev. Neurosci. 2, 185–193.
Hensch, T.K., 2005. Critical period plasticity in local cortical circuits. Nat. Rev. Neurosci. 6, 877–888.
Hofer, S.B., Mrsic-Flogel, T.D., Bonhoeffer, T., Hubener, M., 2009. Experience leaves a lasting structural trace in cortical circuits. Nature 457, 313–317.
Holtmaat, A., Svoboda, K., 2009. Experience-dependent structural synaptic plasticity in the mammalian brain. Nat. Rev. Neurosci. 10, 647–658.
Holtmaat, A., Wilbrecht, L., Knott, G.W., Welker, E., Svoboda, K., 2006. Experience-dependent and cell-type-specific spine growth in the neocortex. Nature 441, 979–983.
Holtmaat, A.J., Trachtenberg, J.T., Wilbrecht, L., Shepherd, G.M., Zhang, X., Knott, G.W., Svoboda, K., 2005. Transient and persistent dendritic spines in the neocortex in vivo. Neuron 45, 279–291.
Honkura, N., Matsuzaki, M., Noguchi, J., Ellis-Davies, G.C., Kasai, H., 2008. The subspine organization of actin fibers regulates the structure and plasticity of dendritic spines. Neuron 57, 719–729.
Horch, H.W., Katz, L.C., 2002. BDNF release from single cells elicits local dendritic growth in nearby neurons. Nat. Neurosci. 5, 1177–1184.
Horch, H.W., Kruttgen, A., Portbury, S.D., Katz, L.C., 1999. Destabilization of cortical dendrites and spines by BDNF. Neuron 23, 353–364.
Hotulainen, P., Hoogenraad, C.C., 2010. Actin in dendritic spines: connecting dynamics to function. J. Cell Biol. 189, 619–629.
Huber, K.M., Gallagher, S.M., Warren, S.T., Bear, M.F., 2002. Altered synaptic plasticity in a mouse model of fragile X mental retardation. Proc. Natl. Acad. Sci. USA 99, 7746–7750.
Jourdain, P., Fukunaga, K., Muller, D., 2003. Calcium/calmodulin-dependent protein kinase II contributes to activity-dependent filopodia growth and spine formation. J. Neurosci. 23, 10645–10649.
Kao, D.I., Aldridge, G.M., Weiler, I.J., Greenough, W.T., 2010. Altered mRNA transport, docking, and protein translation in neurons lacking fragile X mental retardation protein. Proc. Natl. Acad. Sci. USA 107, 15601–15606.
Kasai, H., Matsuzaki, M., Noguchi, J., Yasumatsu, N., Nakahara, H., 2003. Structure–stability–function relationships of dendritic spines. Trends Neurosci. 26, 360–368.
Kayser, M.S., Nolt, M.J., Dalva, M.B., 2008. EphB receptors couple dendritic filopodia motility to synapse formation. Neuron 59, 56–69.
Kleindienst, T., Winnubst, J., Roth-Alpermann, C., Bonhoeffer, T., Lohmann, C., 2011. Activity-dependent clustering of functional synaptic inputs on developing hippocampal dendrites. Neuron 72, 1012–1024.
Knott, G.W., Holtmaat, A., Wilbrecht, L., Welker, E., Svoboda, K., 2006. Spine growth precedes synapse formation in the adult neocortex in vivo. Nat. Neurosci. 9, 1117–1124.
Kopec, C.D., Li, B., Wei, W., Boehm, J., Malinow, R., 2006. Glutamate receptor exocytosis and spine enlargement during chemically induced long-term potentiation. J. Neurosci. 26, 2000–2009.

Kramar, E.A., Chen, L.Y., Brandon, N.J., Rex, C.S., Liu, F., Gall, C.M., Lynch, G., 2009. Cytoskeletal changes underlie estrogen's acute effects on synaptic transmission and plasticity. J. Neurosci. 29, 12982–12993.
Kumar, V., Zhang, M.X., Swank, M.W., Kunz, J., Wu, G.Y., 2005. Regulation of dendritic morphogenesis by Ras-PI3K-Akt-mTOR and Ras-MAPK signaling pathways. J. Neurosci. 25, 11288–11299.
Kwon, H.B., Sabatini, B.L., 2011. Glutamate induces de novo growth of functional spines in developing cortex. Nature 474, 100–104.
Lai, C.S., Franke, T.F., Gan, W.B., 2012. Opposite effects of fear conditioning and extinction on dendritic spine remodelling. Nature 483, 87–91.
Laumonnier, F., Cuthbert, P.C., Grant, S.G., 2007. The role of neuronal complexes in human x-linked brain diseases. Am. J. Hum. Genet. 80, 205–220.
Lendvai, B., Stern, E.A., Chen, B., Svoboda, K., 2000. Experience-dependent plasticity of dendritic spines in the developing rat barrel cortex in vivo. Nature 404, 876–881.
Lin, Y.L., Lei, Y.T., Hong, C.J., Hsueh, Y.P., 2007. Syndecan-2 induces filopodia and dendritic spine formation via the neurofibromin-PKA-Ena/VASP pathway. J. Cell Biol. 177, 829–841.
Lisman, J., 2003. Actin's actions in LTP-induced synapse growth. Neuron 38, 361–362.
Lisman, J., Schulman, H., Cline, H., 2002. The molecular basis of CaMKII function in synaptic and behavioural memory. Nat. Rev. Neurosci. 3, 175–190.
Lisman, J., Yasuda, R., Raghavachari, S., 2012. Mechanisms of CaMKII action in long-term potentiation. Nat. Rev. Neurosci. 13, 169–182.
Lohmann, C., Bonhoeffer, T., 2008. A role for local calcium signaling in rapid synaptic partner selection by dendritic filopodia. Neuron 59, 253–260.
Lushnikova, I., Skibo, G., Muller, D., Nikonenko, I., 2009. Synaptic potentiation induces increased glial coverage of excitatory synapses in CA1 hippocampus. Hippocampus 19, 753–762.
Majewska, A.K., Newton, J.R., Sur, M., 2006. Remodeling of synaptic structure in sensory cortical areas in vivo. J. Neurosci. 26, 3021–3029.
Maletic-Savatic, M., Malinow, R., Svoboda, K., 1999. Rapid dendritic morphogenesis in CA1 hippocampal dendrites induced by synaptic activity. Science 283, 1923–1927.
Marrs, G.S., Green, S.H., Dailey, M.E., 2001. Rapid formation and remodeling of postsynaptic densities in developing dendrites. Nat. Neurosci. 4, 1006–1013.
Matsuno, H., Okabe, S., Mishina, M., Yanagida, T., Mori, K., Yoshihara, Y., 2006. Telencephalin slows spine maturation. J. Neurosci. 26, 1776–1786.
Matsuzaki, M., Honkura, N., Ellis-Davies, G.C., Kasai, H., 2004. Structural basis of long-term potentiation in single dendritic spines. Nature 429, 761–766.
Mendez, P., De Roo, M., Poglia, L., Klauser, P., Muller, D., 2010a. N-cadherin mediates plasticity-induced long-term spine stabilization. J. Cell Biol. 189, 589–600.
Mendez, P., Garcia-Segura, L.M., Muller, D., 2010b. Estradiol promotes spine growth and synapse formation without affecting pre-established networks. Hippocampus.
Minerbi, A., Kahana, R., Goldfeld, L., Kaufman, M., Marom, S., Ziv, N.E., 2009. Long-term relationships between synaptic tenacity, synaptic remodeling, and network activity. PLoS Biol. 7, e1000136.
Morishita, H., Miwa, J.M., Heintz, N., Hensch, T.K., 2010. Lynx1, a cholinergic brake, limits plasticity in adult visual cortex. Science.
Murakoshi, H., Wang, H., Yasuda, R., 2011. Local, persistent activation of Rho GTPases during plasticity of single dendritic spines. Nature 472, 100–104.
Nagerl, U.V., Eberhorn, N., Cambridge, S.B., Bonhoeffer, T., 2004. Bidirectional activity-dependent morphological plasticity in hippocampal neurons. Neuron 44, 759–767.
Nagerl, U.V., Kostinger, G., Anderson, J.C., Martin, K.A., Bonhoeffer, T., 2007. Protracted synaptogenesis after activity-dependent spinogenesis in hippocampal neurons. J. Neurosci. 27, 8149–8156.
Nakamura, Y., Wood, C.L., Patton, A.P., Jaafari, N., Henley, J.M., Mellor, J.R., Hanley, J.G., 2011. PICK1 inhibition of the Arp2/3 complex controls dendritic spine size and synaptic plasticity. EMBO J. 30, 719–730.
Nevian, T., Larkum, M.E., Polsky, A., Schiller, J., 2007. Properties of basal dendrites of layer 5 pyramidal neurons: a direct patch-clamp recording study. Nat. Neurosci. 10, 206–214.
Newpher, T.M., Ehlers, M.D., 2009. Spine microdomains for postsynaptic signaling and plasticity. Trends Cell Biol. 19, 218–227.

Nikonenko, I., Jourdain, P., Muller, D., 2003. Presynaptic remodeling contributes to activity-dependent synaptogenesis. J. Neurosci. 23, 8498–8505.

Node-Langlois, R., Muller, D., Boda, B., 2006. Sequential implication of the mental retardation proteins ARHGEF6 and PAK3 in spine morphogenesis. J. Cell Sci. 119, 4986–4993.

Opazo, P., Labrecque, S., Tigaret, C.M., Frouin, A., Wiseman, P.W., De Koninck, P., Choquet, D., 2010. CaMKII triggers the diffusional trapping of surface AMPARs through phosphorylation of stargazin. Neuron 67, 239–252.

Oray, S., Majewska, A., Sur, M., 2004. Dendritic spine dynamics are regulated by monocular deprivation and extracellular matrix degradation. Neuron 44, 1021–1030.

Ostroff, L.E., Fiala, J.C., Allwardt, B., Harris, K.M., 2002. Polyribosomes redistribute from dendritic shafts into spines with enlarged synapses during LTP in developing rat hippocampal slices. Neuron 35, 535–545.

Pak, D.T., Sheng, M., 2003. Targeted protein degradation and synapse remodeling by an inducible protein kinase. Science 302, 1368–1373.

Penzes, P., Cahill, M.E., Jones, K.A., VanLeeuwen, J.E., Woolfrey, K.M., 2011. Dendritic spine pathology in neuropsychiatric disorders. Nat. Neurosci. 14, 285–293.

Richards, D.A., Mateos, J.M., Hugel, S., de Paola, V., Caroni, P., Gahwiler, B.H., McKinney, R.A., 2005. Glutamate induces the rapid formation of spine head protrusions in hippocampal slice cultures. Proc. Natl. Acad. Sci. USA 102, 6166–6171.

Roberts, T.F., Tschida, K.A., Klein, M.E., Mooney, R., 2010. Rapid spine stabilization and synaptic enhancement at the onset of behavioural learning. Nature 463, 948–952.

Ron, S., Dudai, Y., Segal, M., 2011. Overexpression of PKMzeta alters morphology and function of dendritic spines in cultured cortical neurons. Cereb. Cortex.

Ropers, H.H., 2006. X-linked mental retardation: many genes for a complex disorder. Curr. Opin. Genet. Dev. 16, 260–269.

Ruediger, S., Vittori, C., Bednarek, E., Genoud, C., Strata, P., Sacchetti, B., Caroni, P., 2011. Learning-related feedforward inhibitory connectivity growth required for memory precision. Nature 473, 514–518.

Sacktor, T.C., 2011. How does PKMzeta maintain long-term memory? Nat. Rev. Neurosci. 12, 9–15.

Sala, C., Piech, V., Wilson, N.R., Passafaro, M., Liu, G., Sheng, M., 2001. Regulation of dendritic spine morphology and synaptic function by Shank and Homer. Neuron 31, 115–130.

Sanhueza, M., McIntyre, C.C., Lisman, J.E., 2007. Reversal of synaptic memory by Ca2+/calmodulin-dependent protein kinase II inhibitor. J. Neurosci. 27, 5190–5199.

Selkoe, D.J., 2002. Alzheimer's disease is a synaptic failure. Science 298, 789–791.

Sharma, A., Hoeffer, C.A., Takayasu, Y., Miyawaki, T., McBride, S.M., Klann, E., Zukin, R.S., 2010. Dysregulation of mTOR signaling in fragile X syndrome. J. Neurosci. 30, 694–702.

Shepherd, J.D., Bear, M.F., 2011. New views of Arc, a master regulator of synaptic plasticity. Nat. Neurosci. 14, 279–284.

Soderling, S.H., Guire, E.S., Kaech, S., White, J., Zhang, F., Schutz, K., Langeberg, L.K., Banker, G., Raber, J., Scott, J.D., 2007. A WAVE-1 and WRP signaling complex regulates spine density, synaptic plasticity, and memory. J. Neurosci. 27, 355–365.

Spacek, J., Harris, K.M., 1997. Three-dimensional organization of smooth endoplasmic reticulum in hippocampal CA1 dendrites and dendritic spines of the immature and mature rat. J. Neurosci. 17, 190–203.

Spires, T.L., Meyer-Luehmann, M., Stern, E.A., McLean, P.J., Skoch, J., Nguyen, P.T., Bacskai, B.J., Hyman, B.T., 2005. Dendritic spine abnormalities in amyloid precursor protein transgenic mice demonstrated by gene transfer and intravital multiphoton microscopy. J. Neurosci. 25, 7278–7287.

Srivastava, D.P., Woolfrey, K.M., Jones, K.A., Shum, C.Y., Lash, L.L., Swanson, G.T., Penzes, P., 2008. Rapid enhancement of two-step wiring plasticity by estrogen and NMDA receptor activity. Proc. Natl. Acad. Sci. USA 105, 14650–14655.

Steiner, P., Higley, M.J., Xu, W., Czervionke, B.L., Malenka, R.C., Sabatini, B.L., 2008. Destabilization of the postsynaptic density by PSD-95 serine 73 phosphorylation inhibits spine growth and synaptic plasticity. Neuron 60, 788–802.

Tanaka, J., Horiike, Y., Matsuzaki, M., Miyazaki, T., Ellis-Davies, G.C., Kasai, H., 2008. Protein synthesis and neurotrophin-dependent structural plasticity of single dendritic spines. Science 319, 1683–1687.

Tomita, S., Stein, V., Stocker, T.J., Nicoll, R.A., Bredt, D.S., 2005. Bidirectional synaptic plasticity regulated by phosphorylation of stargazin-like TARPs. Neuron 45, 269–277.

Toni, N., Buchs, P.A., Nikonenko, I., Bron, C.R., Muller, D., 1999. LTP promotes formation of multiple spine synapses between a single axon terminal and a dendrite. Nature 402, 421–425.

Toni, N., Buchs, P.A., Nikonenko, I., Povilaitite, P., Parisi, L., Muller, D., 2001. Remodeling of synaptic membranes after induction of long-term potentiation. J. Neurosci. 21, 6245–6251.

Toni, N., Teng, E.M., Bushong, E.A., Aimone, J.B., Zhao, C., Consiglio, A., van Praag, H., Martone, M.E., Ellisman, M.H., Gage, F.H., 2007. Synapse formation on neurons born in the adult hippocampus. Nat. Neurosci. 10, 727–734.

Trachtenberg, J.T., Chen, B.E., Knott, G.W., Feng, G., Sanes, J.R., Welker, E., Svoboda, K., 2002. Long-term in vivo imaging of experience-dependent synaptic plasticity in adult cortex. Nature 420, 788–794.

Tsuriel, S., Geva, R., Zamorano, P., Dresbach, T., Boeckers, T., Gundelfinger, E.D., Garner, C.C., Ziv, N.E., 2006. Local sharing as a predominant determinant of synaptic matrix molecular dynamics. PLoS Biol. 4, e271.

Vazquez, L.E., Chen, H.J., Sokolova, I., Knuesel, I., Kennedy, M.B., 2004. SynGAP regulates spine formation. J. Neurosci. 24, 8862–8872.

Volterra, A., Meldolesi, J., 2005. Astrocytes, from brain glue to communication elements: the revolution continues. Nat. Rev. Neurosci. 6, 626–640.

Wang, X.B., Bozdagi, O., Nikitczuk, J.S., Zhai, Z.W., Zhou, Q., Huntley, G.W., 2008. Extracellular proteolysis by matrix metalloproteinase-9 drives dendritic spine enlargement and long-term potentiation coordinately. Proc. Natl. Acad. Sci. USA 105, 19520–19525.

Wei, W., Nguyen, L.N., Kessels, H.W., Hagiwara, H., Sisodia, S., Malinow, R., 2010. Amyloid beta from axons and dendrites reduces local spine number and plasticity. Nat. Neurosci. 13, 190–196.

Wilbrecht, L., Holtmaat, A., Wright, N., Fox, K., Svoboda, K., 2010. Structural plasticity underlies experience-dependent functional plasticity of cortical circuits. J. Neurosci. 30, 4927–4932.

Witcher, M.R., Kirov, S.A., Harris, K.M., 2007. Plasticity of perisynaptic astroglia during synaptogenesis in the mature rat hippocampus. Glia 55, 13–23.

Woolfrey, K.M., Srivastava, D.P., Photowala, H., Yamashita, M., Barbolina, M.V., Cahill, M.E., Xie, Z., Jones, K.A., Quilliam, L.A., Prakriya, M., Penzes, P., 2009. Epac2 induces synapse remodeling and depression and its disease-associated forms alter spines. Nat. Neurosci. 12, 1275–1284.

Xie, Z., Srivastava, D.P., Photowala, H., Kai, L., Cahill, M.E., Woolfrey, K.M., Shum, C.Y., Surmeier, D.J., Penzes, P., 2007. Kalirin-7 controls activity-dependent structural and functional plasticity of dendritic spines. Neuron 56, 640–656.

Xu, T., Yu, X., Perlik, A.J., Tobin, W.F., Zweig, J.A., Tennant, K., Jones, T., Zuo, Y., 2009. Rapid formation and selective stabilization of synapses for enduring motor memories. Nature 462, 915–919.

Yang, G., Pan, F., Gan, W.B., 2009. Stably maintained dendritic spines are associated with lifelong memories. Nature 462, 920–924.

Yasumatsu, N., Matsuzaki, M., Miyazaki, T., Noguchi, J., Kasai, H., 2008. Principles of long-term dynamics of dendritic spines. J. Neurosci. 28, 13592–13608.

Yoshihara, Y., De Roo, M., Muller, D., 2009. Dendritic spine formation and stabilization. Curr. Opin. Neurobiol. 19, 146–153.

Zito, K., Scheuss, V., Knott, G., Hill, T., Svoboda, K., 2009. Rapid functional maturation of nascent dendritic spines. Neuron 61, 247–258.

Ziv, N.E., Smith, S.J., 1996. Evidence for a role of dendritic filopodia in synaptogenesis and spine formation. Neuron 17, 91–102.

Zuber, B., Nikonenko, I., Klauser, P., Muller, D., Dubochet, J., 2005. The mammalian central nervous synaptic cleft contains a high density of periodically organized complexes. Proc. Natl. Acad. Sci. USA 102, 19192–19197.

Zuo, Y., Lin, A., Chang, P., Gan, W.B., 2005. Development of long-term dendritic spine stability in diverse regions of cerebral cortex. Neuron 46, 181–189.

CHAPTER THIRTEEN

Experience-Dependent Synaptic Plasticity in the Developing Cerebral Cortex

Chiye Aoki[1] and Alev Erisir[2]
[1]Center for Neural Science, New York University, New York, NY, USA
[2]Department of Psychology, University of Virginia, Charlottesville, VA, USA

1. INTRODUCTION

Plasticity is probably the most lavishly used term in neuroscience, ascribed to a plethora of phenomena from large-scale axonal reorganizations to changes in receptor turnover, from axon regeneration to neurogenesis, and from spine motility to neurotransmitter upregulation. This is not surprising given that several decades of research have made it clear to us that the brain is endowed with multiple capacities that allow it to adapt to changes in the environment, and compensate for the perils of peripheral input loss. Brain plasticity can be desirable when a new pattern of function is to be acquired, such as learning a language or the path to our new house. Or it can be a means to cope or compensate when our sensory input is suddenly altered (amputation, ear infection, and cataract).

However, plasticity can also be vulnerability, a peril, which aligns the neural function with the demands of circulating hormones or drugs (stress and addiction). While the ultimate outcome in each of these phenomena is the same, that is, the neural circuits respond to a different stimulus than before, or respond to the same stimulus much better than before, the biological culprits enabling these changes can be diverse. Excellent reviews, summarizing the multiplicity of plastic phenomena that can be observed at any moment at different circuitries of the developing brain, as well as the diversity of underlying mechanisms are available (Fields, 2005; Hensch, 2005; Lee and Kirkwood, 2011; Smith et al., 2009; Taha and Stryker, 2005).

In this chapter, we will concentrate on mechanisms of synaptic plasticity in sensory pathways, as revealed by neuroanatomical studies of the cerebral cortex. The model system that we describe in greatest detail is the ocular dominance columns of the cat and ferret visual cortex during the critical period (CP). Where possible, we also describe parallel processes in the visual cortex of rats, mice, monkeys, and humans and of the auditory cortex of gerbils and somatosensory cortex of rodents, as well as the hippocampal CA1 region of rodents during adolescence. From such comparative neuroanatomical

The Synapse
http://dx.doi.org/10.1016/B978-0-12-418675-0.00013-4

analyses, we have sought to look for commonalities and thus features salient to the experience-dependent synaptic plasticity of the cerebral cortex. Electron microscopic data capture synaptogenesis and synapse pruning in detail.

We will also describe the maturation and distinct roles played by four neurotransmitter systems—glutamate, gamma-aminobutyric acid (GABA), norepinephrine, and acetylcholine—during early postnatal ages, including the developmental or CP plasticity, and the hippocampal CA1 region in adolescence, when the brain is highly sensitive for adaptation. These are by no means the only neurotransmitter systems important for cortical and hippocampal development during early life. Not only neurotransmitters but hormones also play pivotal roles in postnatal development. Adolescence is marked by the developmental surge of the frontal cortex and other cortical regions, as well as the regions beyond the CA1 of the hippocampus. This is just a partial list of topics that this chapter will not cover. Our selection of the topics included in this chapter can best be described as those to which the authors have had the most direct research contribution.

1.1. Ocular Dominance Column Plasticity

Ocular dominance column plasticity is the most widely used paradigm of developmental plasticity since it was introduced by Hubel and Wiesel in 1960s (Wiesel and Hubel, 1963). When balance of sensory input from two eyes is altered, most notably during a critical CP in early postnatal ages, long-lasting anatomical and physiological changes occur in visual cortex (Sherman and Spear, 1982). Physiological changes are mainly at the eye specificity of individual cortical neurons in the granular and supragranular layers, while large-scale anatomical changes occur at the first synapses that bring the primary input on layer 4 cortical cells.

Anatomical studies using transneuronal radioactive tracers injected into one eye of kittens which are monocularly deprived until end of their CP have demonstrated that cortical regions devoted to the deprived eye undergo a substantial shrinking while those of the nondeprived eye expand (Hubel and Wiesel, 1979; Hubel et al., 1977; Shatz and Stryker, 1978; Swindale et al., 1981). The shrinkage and expansion of eye-specific domains are solely due to reorganization of the thalamocortical (TC) pathway: while axons serving the deprived eye's pathway withdraw their branches and lose their synapses, axons serving the nondeprived eye invade the vacated space and synapse on neurons originally contacted by the deprived eye axons.

Postsynaptic layer 4 neurons, through their existing connections in supragranular layers, are situated to transfer the shift in ocularity onto the rest of the cortical circuitry. Anatomical changes therefore, once established, can sufficiently explain the physiological changes in downstream neurons. However, TC axon reorganization is not immediate, and it occurs in two stages: retraction of deprived axons and expansion of nondeprived axons. Using light microscope reconstruction of intracellularly filled TC axons, Antonini and Stryker (1996) demonstrated that the geniculate relay cells from the deprived eye's pathway lose their axon branches after a brief monocular deprivation; a longer deprivation is required for axons of the nondeprived eye's pathway to expand into the vacated space, and to synapse on neurons originally contacted by the deprived eye axons.

Using electron microscopy and immunolabeling to identify TC terminals, Coleman et al. (2010) also demonstrated that the ratio of synapse-bearing TC terminals to all terminals in layer 4 was reduced after 3 days of monocular deprivation, but this ratio was restored after 7 days of deprivation, indicating that retraction and expansion of TC axons occur in two consecutive stages. Acute changes (within hours of monocular lid suture) are also observed in both cortical physiology and anatomy of glutamatergic synapses, indicating the complexity of the cascade of events that we observe as experience-dependent cortical plasticity in cortical circuits. Several good reviews of the extensive literature on ocular dominance column plasticity are present (Hensch, 2005; Hooks and Chen, 2007; Movshon and Van Sluyters, 1981).

2.1. DE NOVO SYNAPTOGENESIS

2.1.1. The Role of Developmental Synaptogenesis upon Cortical Plasticity during the CP and Puberty

2.1.1.1. General Remarks about the Anatomy of the Cortical Synaptogenesis and Developmental Synapse Elimination

Synaptogenesis is by definition one of the hallmarks of brain development (neurogen esis, migration, and path finding are others), and a major surge for synapse formation occurs during postnatal ages during which sensory pathways are sensitive for modification based on neural activity. Furthermore, evidence exists for elimination of synapses before adulthood when the capacity for plasticity is greatly reduced. Because the CP plasticity, by definition, also involves modification of sensory pathways with elimination of unused synapses and formation of new ones, the capacity for synaptogenesis and synapse pruning has been thought to be involved intricately in mechanisms of postnatal cortical plasticity. However, questions remain regarding the nature of this involvement: is the surge of synaptogenesis also the onset of the postnatal plasticity? To what extent can developmental synapse elimination be equated with the loss of plasticity of cortical maps? If synaptogenesis and synapse elimination are critical for postnatal plasticity, then does the structural plasticity observed after developmental pruning, in adolescence, or in the adulthood rely on different mechanisms?

In the most radical form of the relationship between synaptogenesis and postnatal plasticity, the developmental mechanism of circuit formation itself provides the mechanism for morphological changes in neural circuits. Patterned activity is required for newly generated synapses to engage in circuitries that are compatible with the available sensory environment. Elimination of previously exuberant synapses can then be taken as an indication that environmentally correct connections are strengthened, cortical maps are hardwired, and environmentally compatible skills are learned. In this scenario, once pruning is completed, what we observe as plasticity should also diminish. There is considerable evidence that the segregation of retinogeniculate axons in the lateral geniculate laminae are attained through synaptogenesis and pruning. In the plasticity of postnatal cortical map, this relationship is not as evident.

A second way synaptogenesis can be involved in postnatal plasticity is that the CP plasticity highjacks a property of newly formed synapses. That property is not crucial for establishing the template of the cortical maps, yet it renders synapses very vulnerable in completing their program. Sensory activity that is strongly deviated from its normative pattern reguides axons to a new set of targets. In this scenario, pruning is not necessary for ending plasticity, but the elimination of the vulnerability factor from existing synapses is. If this is the case, then the interpretations of anatomical studies that infer the presence of plasticity from changing number of synapses are complicated by the cooccurrence of two independent processes; developmental synaptogenesis and pruning, and plastic reorganization, each of which changes the number of synapses in cortex.

This section summarizes the findings that relate to the relationship between the rate and the timetable of postnatal synaptogenesis, pruning, and the sensory map plasticity. There are several caveats in evaluating anatomical data that quantify synapse numbers. Electron microscopy, which confirms the presence of a synapse and allows studying terminal bouton structure, is advantageous over light microscopy, yet individually filled axon and dendrite morphology is most efficiently evaluated using light microscopy. Light microscopic visualization of dendritic spines and filopodia falls short of determining the type, number, or even the presence of any synapse on individual spines. Synapse density measurements are unavoidably influenced by other developmental events that change the volume of the brain, including myelination, dendritic tree elaboration, gliogenesis, and tightening of extracellular space.

2.1.1.2. Exuberant Synaptogenesis and Pruning as the Mechanisms for Target Finding and Map Formation during Development

A mechanism involving exuberant synapse formation and subsequent activity-dependent pruning has been demonstrated in several systems that maintain spatial or sensory maps, including selective innervation of alpha motor neurons on muscle fibers, segregation of retinogeniculate axons in eye-specific geniculate laminae, and formation of retinotopic maps in midbrain (Lichtman, 1995; Luo and O'Leary, 2005; McLaughlin et al., 2003; Shatz, 1983). In these systems, initial imprecision in target finding is rectified by correlated intrinsic activity of neurons that lead to strengthening of map-appropriate synapses, and subsequent degeneration of axon branches that initially synapsed on map-inappropriate neurons. During this stage, prevention of intrinsic activity in presynaptic neurons, as well as the blockade of postsynaptic receptors or blockade of postsynaptic activity, prevents pruning of axon branches in appropriate map locations, although axons continue to branch within both correct and incorrect locations (Balice-Gordon and Lichtman, 1994; Shatz and Stryker, 1988; Sretavan et al., 1988).

As studied most intensely in retinogeniculate system, the plasticity for refinement of axon fields in eye-specific geniculate laminae is contributed by competitive influences arising from presynaptic activity, as well as molecular axon guidance cues (Huberman, 2007), underlining cooccurrence of multiple mechanisms that can be observed as plastic structural changes in developing structures (Hong and Chen, 2011).

As the axons go through exuberant synapse formation and pruning stages, terminal morphology also rapidly changes from being small profiles with few vesicles to large boutons with large synaptic zones and more vesicles (Campbell and Shatz, 1992; Hall and Sanes, 1993). Retinogeniculate axons that branch in lateral geniculate laminae before pruning show clear synaptic zones. However, these do not yet display the characteristic morphology of retinogeniculate terminals, such as round, large, pale (RLP) morphology or engagement in glomerular arrangements (Campbell and Shatz, 1992; Colonnier and Guillery, 1964; Hall and Sanes, 1993; Mason, 1982a,b). For several more weeks following the stage of pruning, axon branches continue to elaborate in complexity and bouton frequency and retinogeniculate terminal morphology maturation continues (Mason, 1982a,b).

2.1.1.3. How Does the Surge of Synapse Formation and Elimination Correlate with Sensitive Periods in Virtual Cortex?

In sensory cortical areas that also maintain a map, a stage of exuberant synapse formation and pruning is also noted. In mammals, functional synapses are present at birth in all layers of the sensory cortex, yet a major epoch for synaptogenesis occurs perinatally (late prenatal and early postnatal, depending on the species). The developmental timetable for synaptogenesis was described in detail in the cat visual cortex over three decades ago (Anker and Cragg, 1974; Cragg, 1975; O'Kusky, 1985; Winfield, 1981). In this species, while synapses are present in the visual cortex a few weeks before birth, synapse density precipitously increases between postnatal day 8 (P8; eye opening) and P37, and the average number of synapses per neuron peaks at around postnatal week 7 (PW7). This initial wave of postnatal synapse formation also corresponds to ages during which long-range horizontal connections of layer 2/3 cells and axons of layer 4 spiny stellate cells are elaborated (Burkhalter et al., 1993; Callaway and Katz, 1990, 1992; Durack and Katz, 1996; Kleinschmidt et al., 1987; Luhmann et al., 1990; White et al., 1994), and orientation and direction selectivity maps are established (Chapman et al., 1996; Li et al., 2006; White et al., 2001).

The CP for ocular dominance plasticity in cat falls between postnatal weeks 4 and 6 (Hubel and Wiesel, 1970); that is, it is toward the end of the fast synaptogenesis period that the CP for ocular dominance plasticity opens (P28). By the time for the peak synapse density, CP sensitivity had already diminished (PW6). Similarly in monkeys, in which visual cortex is most vulnerable to monocular deprivation between PW1 and postnatal month 2 (PM2) (Horton and Hocking, 1997; LeVay et al., 1980), the number of synapses increases exponentially until PM3 to reach a plateau that is maintained for about 2 years (Bourgeois and Rakic, 1993; O'Kusky and Colonnier, 1982b).

Following the first phase of synaptogenesis, the second distinct phase for synaptic development was described in both cats and monkeys (Bourgeois and Rakic, 1993; O'Kusky and Colonnier, 1982b; O'Kusky, 1985; Winfield, 1981). In the cat, about 20–40% of synapses in general are gradually eliminated between PM3 and PM8 (i.e. during preadult ages). In the monkey, cortical synapse density is reduced by 40% after age 2, around the time of puberty. The pattern of an overshoot followed by a decline in synapse numbers is also evident in rats (Aghajanian and Bloom, 1967) and humans (Huttenlocher and Dabholkar, 1997; Huttenlocher et al., 1982). Similar to the rate of new synapse formation and the time course of the surge, the timing and the rate of synapse elimination differ in species. In the monkey, the surge starts in E144 (20 days before birth and the first patterned light exposure), and the decline starts at puberty, bringing synapse density in the adult to the levels at birth (Bourgeois and Rakic, 1993; O'Kusky and Colonnier, 1982b).

In the cat, the synapse density surges at the eye opening, and it increases about 35 times to reach adult levels (Cragg, 1975; Winfield, 1981). Then, several months before the onset of puberty, it declines rapidly to about 30% of its peak at P70. Because there was no change in synapse–neuron ratios, Cragg attributed this reduction to continuing development of glial cells rather than synapse or cell loss. Similarly, the reduction in synapse density is not correlated with puberty in the humans. A notable surge starts at 2 months of age, peaks at 8 months of age, and declines by about 60% by 3 years of age (Huttenlocher et al., 1982).

2.1.1.4. Morphological Properties of Developing and Plastic Synapses

Lamina or circuit specific synaptogenesis: Inconsistencies in the timeline for synaptogenesis and synapse elimination may be due to the fact that the sensory cortex is composed of multiple circuitries that are dedicated for specific functions, each with a unique developmental program, yet mutually interdependent for their individual maturation. When a massive synapse pruning occurs, it is rather selective for certain circuits of the visual cortex. For example, while supragranular layers are most vulnerable, TC-recipient layers show minimal decline or mature early. Further, asymmetric and symmetric (excitatory and inhibitory) synapses develop and mature at different rates such that when a synapse pruning is detected, it is primarily the excitatory synapses that are eliminated. Based on results of synapse density in individual lamina or identified inputs (Bourgeois and Rakic, 1993; O'Kusky and Colonnier, 1982a; O'Kusky and Colonnier, 1982b), the time course of developmental synapse elimination in cortex is not consistent with the notion that exuberant synapse formation underlies plastic capacity of the cortical circuits.

The rate of cortical synaptogenesis is studied in ferrets, a species that is born with a relatively immature visual system. In ferrets, eyes open at PD32, a few days before the onset of the CP (Issa et al., 1999). By comparison, the CP starts at birth in primates. In the ferret, the rate of synaptogenesis in neuropil regions of visual cortex (that is, excluding areas of cell bodies, blood vessels, and myelinated axons) was quantified individually for layer 4 and layers 2/3 (Erisir and Harris, 2003). Immediately upon eyelid opening, which precedes the onset of CP by a few days, the synapse density almost doubles, followed by a persistent increase throughout the weeks when ocular dominance column plasticity is evident (P35–P56), reaching a plateau after P60. Furthermore, two layers display significantly different synaptic densities at any given age, and the changes in synapse densities occur at different rates (Erisir and Harris, 2003) (Figure 1). In this

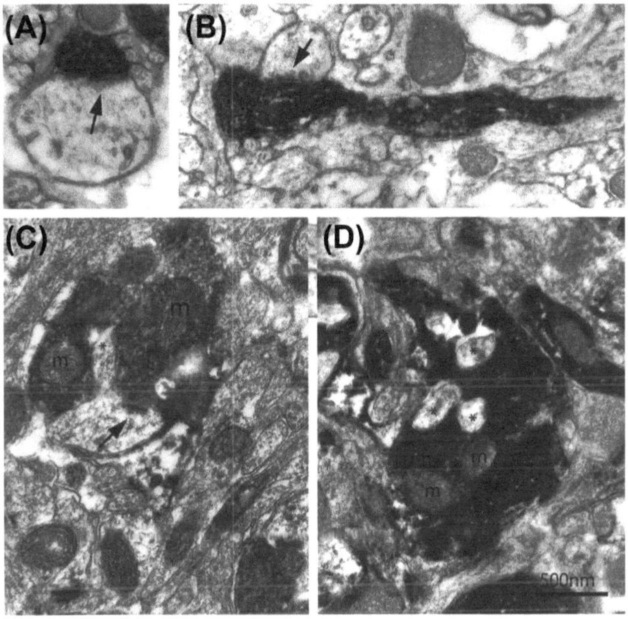

Figure 1 Thalamocortical morphology. The morphology of thalamocortical (TC) terminal boutons in ferret visual cortex, identified by VGluT2 immunostaining (Nahmani and Erisir, 2005; Erisir and Dreusicke, 2005). (A and B) At postnatal day 35, at the onset of the critical period for plasticity in the ferret (Issa et al., 1999), TC synapses are formed from en passant axon branches. Terminal cross-section area is small, postsynaptic protrusion are few. (C) By the end of the period when cortical circuitry is most vulnerable to monocular deprivation (P60), terminals attain the mature morphology that will be maintained through the adult ages. Mature TC boutons are large, and they may form multiple synapses. Postsynaptic protrusions into the TC terminals are frequent and these appear as unlabeled patches within cross-sections. (D) Dark rearing does not influence the rate of TC bouton maturation. The VGluT2-labeled terminal here was obtained from a ferret, which was dark-reared until P60. m, mitochondria; *, postsynaptic protrusions into TC terminal; arrows point to synapses from postsynaptic side.

material, where synapse density was measured irrespective of the changes that may happen in neuron size, no noticeable synaptic pruning was found in layers 2–4. Thus, while developmental programs for de novo synapse formation and elimination overlap with postnatal processes for the critical or sensitive periods for cortical circuits, a tight correlation is not obvious when cortical layers that are most the plastic are concerned.

What if structural plasticity is not a feature of neurons as a whole, but of only a select subset of synapses onto those neurons? Then, examination of all synapses, even in a given layer, could overlook subtle structural changes, and proliferation or elimination of specific inputs. Among the multiple circuits in the visual cortex, TC axons are the objects of most drastic axonal reorganization found after patterned visual deprivation. These bring the driver input to cortical layer 4, synapsing primarily on spiny stellate cells, and less densely on aspinous dendrites of cortical interneurons. In order to reveal the elements of synaptic morphology that may contribute to the capacity for plasticity in these inputs, TC axon development was studied (Erisir and Dreusicke, 2005) using extracellular tract tracing and immunostaining for vesicular glutamate transporter type 2, which is specific to TC axons in ferret, mice, and primate cortex (Coleman et al., 2010; Kaneko and Fujiyama, 2002; Kaneko et al., 2002; Nahmani and Erisir, 2005). Between the peak of the CP and adult ages in ferret, TC terminals become notably larger, and they make more elaborated contacts with their postsynaptic partners (Figures 1 and 2) (Erisir and Dreusicke, 2005).

A notable morphological feature of TC terminals is the postsynaptic protrusions that invaginate into the bouton and, in many cases, extend into TC axon's stalk. These protrusions appear as unlabeled inclusions within tracer or immunolabeled terminals, and emerge either from the synapse protrusions or from nonsynaptic portions of postsynaptic dendrites and spines. Interestingly, the structural maturation of postsynaptic protrusions closely correlates with the end of the CP for TC axons. The number of the protrusions and the area of apposing membranes from the presynaptic terminal and the postsynaptic dendrite increase after the peak of the CP, and reach a plateau at the end of the CP. These postsynaptic protrusions can be a result of continuing spine motility after the TC terminals have lost their capacity to retract from the synapses. Alternatively, by forming extensive surface adhesions between the TC terminal and its postsynaptic targets, postsynaptic protrusions can also be the basis of synaptic stability TC axons profess after the waning of ocular dominance column plasticity in cortex.

Examining the frequency at which TC axons make synapses in ferrets also reveals no evidence that TC axons prune their synapses after the peak of the period in which they display the most plasticity. On the contrary, they continue to make more synapses. Compared to the peak of the CP, adult TC axons form twice as many synapses along their lengths. Furthermore, these newly added synapses are selective for excitatory cells of cortex. In the adult cortex, about 90% of all TC synapses are formed on non-GABAergic, excitatory neurons, whereas at the peak of the CP, only 75% of TC synapses are formed on excitatory cells (Erisir and Dreusicke, 2005), suggesting that at the peak of plasticity, the inhibitory circuitry is driven more strongly, thus shifting the excitatory–inhibitory balance in favor of inhibition.

In summary, patterned visual activity induces massive synaptogenesis in visual cortex that is initiated by eyelid opening, or by birth in species born with open eyelids. While this surge is necessary for the ensuing CP plasticity, it is not sufficient. Several other factors need to be in place for sensory plasticity, and these factors may also need patterned sensory activity. A notable phase of synapse elimination also occurs in sensory circuits at times coinciding with loss or reduction in the capacity for plastic changes in cortex in general. However, a direct correlation between synapse pruning and circuits that shows most plasticity is not evident (Figure 2).

Figure 2, below, is a 3D animated video. Click on the arrow at the bottom-left corner of the video window to play it. (Depending on your operating system, you may need to download the free Apple Quicktime player.)

Supplementary video related to this article can be found at http://dx.doi.org/10.1016/B978-0-12-418675-0.00013-4.

The following is the supplementary video related to this article:

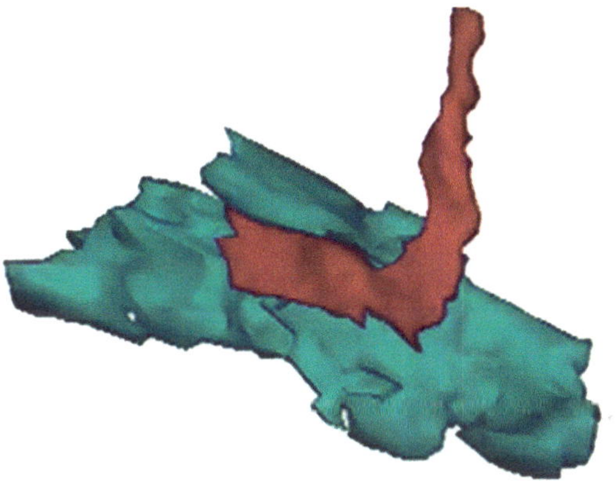

Figure 2 Three-dimensional reconstruction of thalamocortical (TC) terminals, their postsynaptic partner and postsynaptic protrusion extending into the TC axon. Reconstruction of over 60 adjacent ultrathin sections through VGluT2-labeled TC terminals and their postsynaptic partner from a P60 ferret. Gray, TC axon and terminal; blue, dendrite of a layer 4 cell that receives a synaptic contact from the gray axon; burgundy, another TC terminal that forms a synapse on the same blue dendrite. Synaptic zones are not shown. Two filamentous, finger-like protrusions emerge from the postsynaptic dendrite, and inset themselves into the presynaptic bouton, dragging membranes from pre- and postsynaptic neurons into the axons. These protrusions can extend long segments into TC axons in the adult brains, probably increasing the surface area of apposing membranes between the pre- and postsynaptic partners.

2.1.1.5. General Measures of Synaptic Density in Cortex are Unaffected by Activity—But Synapses of Specific Circuits Can be Affected

Activity dependence is the hallmark property of developmental cortical plasticity, and this is best revealed by a dark rearing (DR) paradigm. In the DR paradigm, animals are placed into complete darkness from birth, and thereby preventing patterned as well as nonpatterned visual activities. This contrasts with eyelid suture, which prevents patterned visual activity but does not prevent the unpatterned visual activity. DR delays the onset of sensitive periods; return to normal day–night cycle restores plasticity. Enucleation, premature birth, and monocular deprivation are also conditions that alter the timing of the first visual experience and the quality of the sensory activity. Examining the effect of sensory activity on the rate and the schedule of synaptogenesis revealed that synapse numbers in visual cortex are unaffected by the environmental changes, even though those same environmental changes lead to long-lasting perceptual deficits and physiological abnormalities: monocular deprivation (O'Kusky and Colonnier, 1982a; O'Kusky and Colonnier, 1982b), enucleation (Bourgeois and Rakic, 1993), and DR (Gabbott et al., 1986; O'Kusky, 1985) do not change the volumetric density of synapses in the visual cortex.

The timetable and the rate of synaptogenesis are closely tied to the time of conception, rather than the time of birth or the activity imposed on the neural pathways. Bourgeois et al. (1989) studied synapses in normal and preterm monkeys; the latter were exposed to normal light–day cycle of animals born at term. While the synapse accumulation in preterm visual cortex did not change, the morphology of terminals and their laminar distribution was affected, suggesting that the effect of activity on cortex maturation is via strengthening or modification of existing synapses, rather than regulating the rate of synapse production.

While the rate of synaptogenesis is independent of visual activity, Bourgeois and Rakic (1996) demonstrated that the pattern of connectivity on layer 4 cells (such as localization of synapses on spines vs shafts) could be dependent on activity. Lamina-specific analysis in mice also revealed that DR reduced the number of spines on apical dendrites of layer 5 cells coursing through layer 4 (Valverde, 1971), the spines of basal dendrites of layer 3 cells (Wallace and Bear, 2004), and the total synapse density in layer 4 (Freire, 1978). Thus, the activity-dependent changes in synapses can be limited to certain circuits or selective inputs, thereby eluding general synaptic density measurements.

2.2. CHEMICAL MATURATION OF GLUTAMATERGIC SYNAPSES

2.2.1. Role of NMDA Type Glutamate Receptors in Ocular Dominance Column Plasticity

Implicated by their capacity to work as a coincidence detector, and their ability to induce synaptic changes in hippocampal slices, NMDA type glutamate receptors (NMDARs) were studied intensely to uncover their role in ocular dominance column plasticity. Initial evidence for the involvement of NMDAR in ocular dominance column plasticity came from physiological studies. The blockade of NMDARs prevented monocular lid suture-induced physiological changes (Bear et al., 1990; Kleinschmidt et al., 1987), particularly by preventing the disconnection of the deprived pathway (Gu et al., 1989; Krahe and Medina, 2010). The role of NMDARs in ocular dominance column plasticity was subsequently confirmed in experiments infusing antisense oligonucleotides, instead of 5-amino phosphovaleric acid (APV), which has wider effects in suppressing the synaptic activity in cortical neurons (Roberts et al., 1998).

The developmental pattern of NMDAR function in cortical circuits is also in agreement with the timeframe for waning of plasticity in older animals: the NMDAR component of responses in layers 4–6 responses are stronger in young animals than in the adults (Fox et al., 1989, 1990; Tsumoto et al., 1987), suggesting that NMDARs play a crucial role in rendering granular layer synapses plastic. Similar to the ocular dominance column plasticity, the development of NMDAR action in the cortex itself is also activity dependent: DR prevents the developmental decline of NMDAR function, and several days of light after long-term DR restores the decline (Fox et al., 1992).

NMDARs are, of course, not exclusive to ocular dominance column circuitry. In fact, NMDARs are present in most synapses in central nervous system, and with the aid of subtle variations in their subunit compositions and localization, they are involved in diverse functions including neural development, plasticity, and pain perception (Cull-Candy and Leszkiewicz, 2004). As well as being key players in pre- and postnatal development, NMDARs are critically implicated in synaptic plasticity of sensory circuits (Constantine-Paton et al., 1990; Lau and Zukin, 2007). Being a notable means of Ca^{2+} entry to postsynaptic sites, NMDAR is a crucial variable in activity-dependent postsynaptic changes that are dependent on Ca^{2+}. Their role in trafficking α-amino-3-hydroxy-5-methyl-4-isoxazolepropionic acid receptors (AMPARs) in and out of the postsynaptic density (PSD) (Malinow and Malenka, 2002), and in inducing structural modification in the architecture of spines (Alvarez and Sabatini, 2007; Carlisle and Kennedy, 2005) are well studied. Furthermore, NMDARs themselves need to be modified to avoid saturation of synaptic strength, in a process defined as homeostatic plasticity (Abraham and Bear, 1996; Turrigiano and Nelson, 2004) via differential synthesis and trafficking of NMDAR subunits in an activity-dependent manner (Perez-Otano and Ehlers, 2005).

The amount of NMDARs at synapses is also autoregulated by NMDAR binding in adult brains: using electron microscopy and immunogold visualization, Aoki et al. (2003) demonstrated that the blockade of NMDAR with local injection of AP5 increased the number of 2A type NMDARs at layer 1 synaptic spines, suggesting that a compensatory upregulation of synaptic receptors can occur in response to reduced activity. From a sensory feature-processing point of view, in addition to ocular dominance plasticity, NMDARs also have a role in the acquisition of orientation selectivity (Ramoa et al., 2001; Tanaka et al., 2009) and directional tuning (White and Fitzpatrick, 2007), as well as in the normal sensory responses of cortical cells (Daw et al., 1993; Fox et al., 1989, 1990; Kasamatsu et al., 1998; Miller et al., 1989). The multiplicity of NMDAR function in developing cortex indicates that during postnatal development, NMDAR are present in multiple circuitries that occupy cortical layers, and they may have diverse functions as the mediators of experience-dependent synaptic changes and as the mediators of glutamatergic activity, selectively in different circuits.

2.2.2. Anatomical Studies of NMDAR Localization

Excellent reviews on the molecular properties of NMDAR are available (Kohr, 2006; Wenthold et al., 2003). Functional NMDARs are heterotetramers of an obligatory, channel-forming NR1 subunit and one or more types of the NR2A-D subunit family. The NR2 subunits, in particular, by setting the voltage dependence of the pore Mg^{2+} block, and the time course of NMDAR currents, define many receptor properties essential for synaptic plasticity. NR2B-containing receptors differ from NR2A-containing receptors in that the former affords longer current duration (Li et al., 1998) and larger Ca^{2+} currents through the channel (Sobczyk et al., 2005).

As is summarized in the previous section, NMDARs enable monocular deprivation-induced physiological and anatomical changes and this function is most prevalent during the CP of plasticity. These findings are supported by the finding that the localization of these receptors changes by the end of the CP (Aoki, 1997; Aoki et al., 1994b; Fox and Daw, 1993). Besides the relative prevalence of NR1 subunits in axon terminals and of their enrichment in the infragranular layers during early life, the obligatory subunit NR1 expression was found to be fairly stable over development (Aoki, 1997; Aoki et al., 1994b). In contrast, NR2 subunit expression during development drew much attention: long-duration, NR2B-mediated currents dominate in early development, while the NR2A-mediated short-duration currents and NR2A protein expression become more effective as the animals age (Monyer et al., 1994; Portera-Cailliau et al., 1996; Sheng et al., 1994; Williams et al., 1993).

Biochemical and light microscopic protein quantification approaches provide evidence that NR2A subunit expression supersedes the initially more prevalent subunit NR2B, and that this switch occurs sometimes around or after the ages when CP plasticity is most robust (Cao et al., 2000a,b; Chen et al., 2000; Quinlan et al., 1999a,b; Roberts and Ramoa, 1999). Because the increase in the NR2A–NR2B ratio is closely correlated with the shortening of cortical NMDA currents, the NR2A–NR2B switch model reasoned that insertion of NR2A subunits decreased the effectiveness of synaptic NMDAR in regulating activity-dependent plasticity (Carmignoto and Vicini, 1992; Flint et al., 1997; Roberts and Ramoa, 1999).

While the anatomical evidence for NR2A/B switch hypothesis remains elusive, spatial and circuit selective localization of NR2B- or NR2A-containing receptors at different glutamatergic inputs to individual neurons should also be considered (Kohr, 2006). Several lines of evidence indicate that the involvement of NR2 subunits in rendering cortical synapses plastic is more complicated than a general NR2A subunit dominance in mature circuits, and that the localization and developmental pattern of NR2 subunits is not uniform across cortical layers and thus cortical circuits. The surge of NR2A subunits occurs at the onset, rather than the end of the CP (Erisir and Harris, 2003; Quinlan et al., 1999a,b; Roberts and Ramoa, 1999). The reduction of plasticity is better correlated with the timing for the loss of NR2B-subunit containing NMDARs at layer 4 synapses, while such a loss does not occur at layer 2/3 synapses (Erisir and Harris, 2003).

A similar correlation between synaptic NR2B reduction and the end of the CP was also demonstrated in mouse somatosensory cortex (Liu et al., 2004). Furthermore, activity-dependent NR2 subunit modulation follows different principles for 2A and 2B: studying the prevalence of NR2 subunits in cortical layers upon DR during development, light deprivation in adulthood, and restoration of light after both conditions, we have shown that NR2A subunits rapidly downregulate with reduction of sensory activity and upregulate when sensory activity is initiated, both during development and in the adult (Corson et al., 2009), suggesting that activity-dependent NR2 modulation is a life-long property. The regulation of NR2B subunits on the other hand, is a developmental property: DR or light deprivation does not affect NR2B prevalence in cortex, but DR delays the developmental decline that happens several weeks after initiation of patterned sensory activity in normal animals (Corson et al., 2009). These anatomical findings are consistent with the idea that NR2A and NR2B subunit containing NMDARs are located at synapses serving different glutamatergic inputs onto the cortical cells, and their individual involvements in diverse plastic functions of cortex are not uniform.

2.3. MATURATION OF THE GABAERGIC SYSTEM AND ITS ROLE IN DEVELOPMENTAL PLASTICITY OF THE CEREBRAL CORTEX

2.3.1. General Remarks about the Anatomy of the Cortical GABAergic System

While the great majority of neurons in the cortex are excitatory, utilizing glutamate as their neurotransmitter, the second most prevalent population of neurons are GABAergic, comprising 20–25% of all cortical neurons (Gabbott and Somogyi, 1986; Hendry and Jones, 1988; Keller and White, 1987). Neurogenesis of cortical GABAergic neurons is strikingly different from that of pyramidal neurons: most of the GABA progenitor neurons originate from the medial ganglionic eminences in the basal ganglia, enter the cortex at the intermediate zone, and migrate tangentially, traversing multiple cortical columns (Anderson et al., 2001). In contrast, the pyramidal neurons and a smaller subset of GABAergic neurons containing parvalbumin (PV) enter the cortex from the subventricular zone (surface of the cerebral ventricle) and migrate radially, toward the pia mater, in parallel with single cortical columns (Marin et al., 2010).

In spite of these contrasting modes of origin, the pathways formed by the two neurotransmitter systems must be coordinated in time and place, so as to form functional cortical circuitry with the proper balance of excitatory and inhibitory synapses. This balance generates receptive field properties mirroring the sensory experience imposed upon the organism during the CP. Perhaps because of the distinct modes of entry of the neuronal populations, pyramidal neurons are relatively more "column-centric" in their connectivity, while GABAergic neurons diversify by becoming predominantly intercolumnar or intracolumnar in their synaptic circuitry, and with identifiable electrophysiological properties and peptide content (Bacci et al., 2005). For example, the intercolumnar excitatory flow within layer 5 is regulated by the fast-spiking (FS) GABAergic interneurons that contain PV, while the intracolumnar excitatory flow is regulated by GABAergic neurons are typically the non-FS, low-threshold spike (LTS) interneurons.

Evidence based on the expression of GABARs and the GABA-synthesizing enzyme, glutamic acid decarboxylase (GAD) suggests that GABAergic neurons become functional earlier than the glutamatergic neurons. Indeed, Purpura et al. (1965) reported that the most prominent synaptic activity evoked in immature (newborn to 1 month) sensorimotor cortex neurons by thalamic stimulation were the extraordinarily prolonged (>600 ms) inhibitory postsynaptic potentials (IPSPs) which summated readily following repetitive stimulation. In some of the cells, the earliest signals provided by GABA may also be excitatory due to the equilibrium potential for Cl^- being more positive than the resting membrane potential (Ben-Ari et al., 2004). Although these earliest depolarizing signals precede the onset of the CP, the idea that GABA may influence glutamatergic synaptic circuitry is a theme that will be considered throughout this section.

In addition to the excitatory role that GABA may play during the earliest phase of postnatal development, GABA can mediate inhibition synaptically and nonsynaptically, through activation of GABARs that are clustered immediately opposite to GABAergic axon terminals or through activation of GABARs that are located more distantly from the GABA release site. All these modes of inhibition will be considered in this section. The use of the term, "GABAergic system", instead of "GABAergic synapses" in the title of this section reflects the authors' intention to emphasize the multifaceted mode of action by GABA, rather than to focus only on GABAergic synapses. This chapter will not be covering the subject of the metabotropic $GABA_B$ receptors. Therefore, the abbreviation, GABAR, will refer to $GABA_A$ receptors, only.

2.3.2. Evidence that the GABAergic System Contributes to the Sculpting of Receptive Field Properties

Perhaps the first indication that the GABAergic system is involved in the CP for ocular dominance plasticity was provided by Mower et al. (1984) and Sillito et al. (1980), who showed that blockade of GABAergic receptors by the GABAR antagonist, bicuculline, unmasked the binocularity of neurons that were rendered monocular following monocular deprivation. Up to that point, most models explaining the ocular dominance shift were depicted as a result of competition between two sets of excitatory pathways—one serving the open eye's pathway and another serving the deprived eye's pathway.

Findings such as these supported the idea that ocular dominance shift might be generated by suppression of the inactive pathway via an endogenous inhibitory agent rather than, or in addition to deafferentation of the excitatory pathway serving the inactive eye (Movshon, 1981). However, an alternative interpretation that could not be dismissed by the bicuculline treatment was that all neurons became highly excitable and even epileptic, thereby masking the ocular dominance shift contributed by the excitatory pathways.

The question regarding the role played by the GABAergic system in the CP continued to be pursued and was significantly revitalized by findings from the Hensch laboratory two decades later, after it became established that the mouse visual system can be used effectively to examine the molecular mechanisms underlying the CP (Gordon and Stryker, 1996). Hensch's approach was to take advantage of the mouse model to knock out one of the two GABA-synthesizing enzymes (GAD65 KO). This manipulation was found to decrease, but importantly not cause a complete loss, of GABA release. The visual cortical neurons of GAD65 KO retained much of the normalcy of visual cortical neurons' receptive field properties, such as the level of spontaneous activity, retinotopic organization, and orientation and directional selectivity (Fagiolini and Hensch, 2000; Hensch, 2005; Hensch and Fagiolini, 2005; Hensch et al., 1998).

Importantly, ocular dominance was also normal but could not be shifted by monocular deprivation during the chronological age for the CP (c. fourth postnatal week for mice), unless the GABAergic transmission was boosted by exogenously administered benzodiazepines, the coagonists of GABA. Hensch et al. (1998) reported that benzodiazepines not only reinstate ocular dominance plasticity within the GAD65 KO cortex but can also enhance ocular dominance plasticity, even within the cortex of WT mice, by boosting GABAergic transmission.

2.3.3. Plasticity of the Receptive Field Properties of GABA Cells

Recent work indicates that GABAergic neurons are more binocular than are the excitatory neurons and are strongly responsive to monocular deprivation during the CP (Kameyama et al., 2010), although how they compare in their sensitivity to monocular deprivation, relative to the glutamatergic neurons, may be a matter that requires further studies (Gandhi et al., 2008). By measuring the excitability of individual GABAergic neurons, Kameyama et al. (2010) were able to demonstrate that the receptive field properties of GABAergic neurons exhibit ocular dominance shifts through depression of excitatory input from the deprived eye, and not due to the enhancement of responses from the nondeprived eye.

This differs from the pattern observed for the excitatory neurons, which show ocular dominance shifts that result from the enhancement of responses to the nondeprived eye, without any decrease in the responsiveness to the deprived eye. These descriptions of the ocular dominance shift suggest that the structural units exhibiting CP plasticity are still the excitatory cells, and more specifically the axon terminals of excitatory neurons impinging upon GABAergic neurons.

However, there are equally compelling findings that also point to plasticity of the GABAergic neurons, themselves. Findings supporting this notion are summarized below. The GABAergic interneurons continue to exhibit modifiability beyond the end of the CP, due primarily to the diminished excitability of these neurons to afferents subserving the deprived eye. The excitatory neurons of post-CP brains again contrast the GABAergic interneurons, in that they exhibit slight enhancement in their responsiveness to afferents driven by the nondeprived eye, while the responsiveness of the GABAergic neurons to the nondeprived eye remains unchanged following monocular deprivation (MD) in adulthood.

2.3.4. Activity-Dependence of GABAergic Synaptogenesis

Does the rearrangement of GABAergic synapses contribute to the ocular dominance shift? The GABAergic neurons are remarkably heterogeneous in terms of the mode of transmitter release, neuropeptides and gases that they corelease, spike adaptation, dendritic integration, and axonal innervation, and prevalence across the different layers of the cortex. Does sensory deprivation during the CP evoke changes in the GABAergic synapses formed perisomatically by PV cells, at the axon initial segment formed by the PV Chandelier cells, and/or at distal dendrites formed by somatostatin (SS)-containing GABA cells or even at dendritic spines? Are there differences in the sensitivity to sensory deprivation upon the formation of GABAergic synapses onto pyramidal cells, layer 4 stellate neurons or other GABAergic neurons? The following section will consider some of these questions to which answers are available.

As might be expected, GABAergic synapses are like the glutamatergic synapses in their reliance upon afferent activity for maturation (Kotak et al., 2008; Morales et al., 2002). Electrophysiology reveals that maturation of GABAergic synapses is characterized by an increase of the maximally evoked inhibitory postsynaptic current (IPSC) amplitude, shortened duration, and incorporation of the alpha 1-subunit-containing adult form of GABARs at the plasma membrane. All three of these changes reflect maturational events on the postsynaptic side. These postsynaptic changes are accompanied by a decrease in the number of spontaneous IPSC, reflecting maturational events on the presynaptic side. Following complete sensory deprivation, as with cochlear ablation at a prehearing stage of postnatal day (P10) for gerbils or by DR from birth, all these pre- and postsynaptic phenotypes become arrested at the pre-CP stage. Results obtained from electron microscopic immunocytochemistry parallels many of these points shown by electrophysiology.

Following cochlear ablation at P10 (sensorineural hearing loss), the gerbil auditory cortex has decreased areal density (frequency of encounter per unit area) of GABAergic axon terminals but increased levels of GAD within each axon terminal, quantified by measuring GAD immunoreactivity within individual terminals forming axosomatic synapses. Sensory deprivation also leads to a decrease in the proportion of beta2/3 subunits of GABARs incorporated at the plasma membrane (Sarro et al., 2008) (Figures 3). The reduction in the levels of beta2/3 subunits at axosomatic synapses was specific for the plasma membrane of pyramidal cell somata and not for the GABAergic interneuron somata (Sarro et al., 2008) (Figures 4). This failure in the exocytosis of key subunits to the plasma membrane would explain, in part, the decreased amplitudes of spontaneous and evoked IPSCs and an increase in the overall excitation of the cortical circuit following sensory deprivation (Kotak et al., 2005, 2008). Once matured, deafferentation does not cause the synaptic phenotypes to revert, indicating that there is a CP for the activity-dependent maturation of GABAergic synapses (Takesian et al., 2012) (see Section 2.4.4 for further details about GABAR changes during the development).

2.3.4.1. GABAergic Synaptogenesis at Dendrites

GABAergic inhibition occurs at multiple functionally distinct subcellular sites: somata, dendritic shafts, dendritic spines, and the axon initial segment. Not only are the sites of GABA inhibition diverse, but the isoform of the GAD, the synthetic enzyme for GABA, is also expressed differentially across these subcellular sites. Of the two isoforms of GAD, GAD65 occurs predominantly at dendritic shafts, while GAD67 occurs more predominantly at somata (Fukuda et al., 1998). Since GAD65 KO leads to the loss of ocular dominance plasticity (Fagiolini and Hensch, 2000; Hensch et al., 1998), axodendritic inhibition may be one of the loci that contributes significantly to sensory experience-dependent plasticity. In support of this idea, the SS-positive, regular spiking nonpyramidal neurons (RSNP) that target distal dendrites of pyramidal neurons in cortex become much more excitable during the third postnatal week, the age corresponding to the beginning of the CP for the mouse visual cortex (Lazarus and Huang, 2011).

On the other hand, Jiao et al. (2011), who studied GABAergic inhibition in layer 4 of the mouse barrel cortex report that the axosomatic and axodendritic synapses formed by RSNP neurons do not contribute significantly to the perturbation evoked by sensory deprivation, involving whisker trimming. RSNP neurons may still participate in sculpting receptive field properties during the CP. For example, sensory experience may modify synaptic inputs to these neurons. There is as yet little structural data and no ultrastructural data available on the activity-dependent maturational events that occur specifically at axodendritic synapses formed by these RSNP neurons or upon these neurons. Yet another dimension that cannot be neglected is the developmental changes that may occur upon the short-term plasticity properties of the various inhibitory interneurons which, in turn, depend on the cell types with which they form synapses (Ma et al., 2012) and the history of the animal's sensory experience during the CP (Liu et al., 2012).

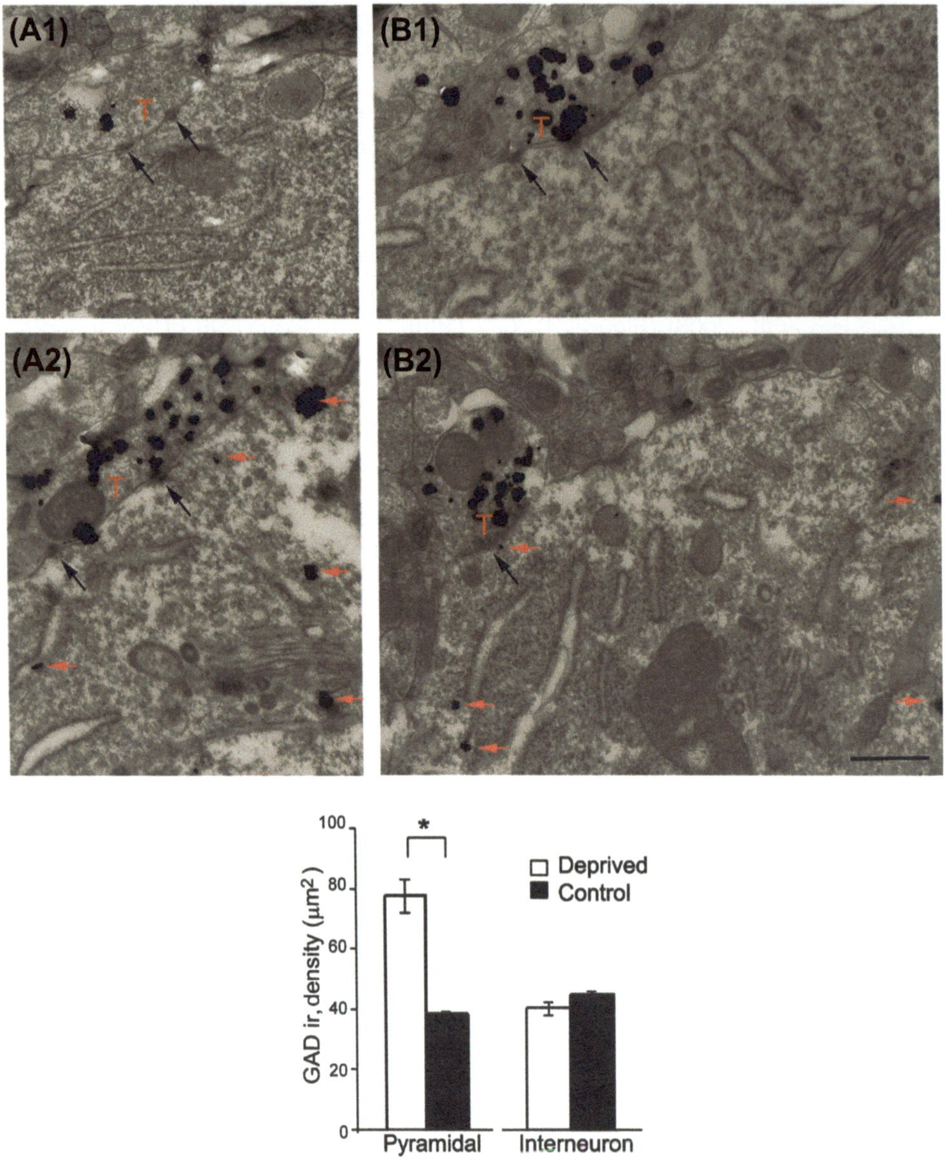

Figure 3 GAD content within axon terminals targeting cortical pyramidal neurons is higher within the cortex of animals that have undergone sensory deprivation during early life. Gerbils were deprived of hearing by bilateral cochlear ablation at an age preceding hearing onset (postnatal day 10, P10). GABAergic innervation of somata in the auditory cortex was analyzed at P17 of normally reared gerbils (A1 and A2) and age-matched gerbils deprived of hearing (B1 and B2). GABAergic axon terminals (T) are identifiable by the presence of silver-intensified colloidal gold (SIG), reflecting immunoreactivity (ir) to GAD, the GABA-synthesizing enzyme. Black arrows point to the postsynaptic plasma membrane of axosomatic synaptic junctions. The postsynaptic cells in A2 and B2 exhibit GAD-ir in the cytoplasm (red arrowheads), indicating that the postsynaptic cells are GABAergic interneurons. Pyramidal neurons are identified by the absence of GAD-ir in the postsynaptic cell's cytoplasm (A1 and B1). GAD-ir is visibly less within the axon terminal innervating the pyramidal cell from a normally reared animal (A1) than of the deprived animal (B1). In contrast, GAD-ir within axon terminals innervating interneurons are similar. This visual impression was verified by determining the density of SIG particles within the area of axon terminal profiles forming axosomatic synapses and is shown in the graph. Calibration bar = 400 nm. $P < 0.05$, determined by Kolmogorov–Smirnov test of independent groups. Further details of the methods and results can be found in Sarro et al. (2008). *The graph is modified, with permission from Oxford University Press.*

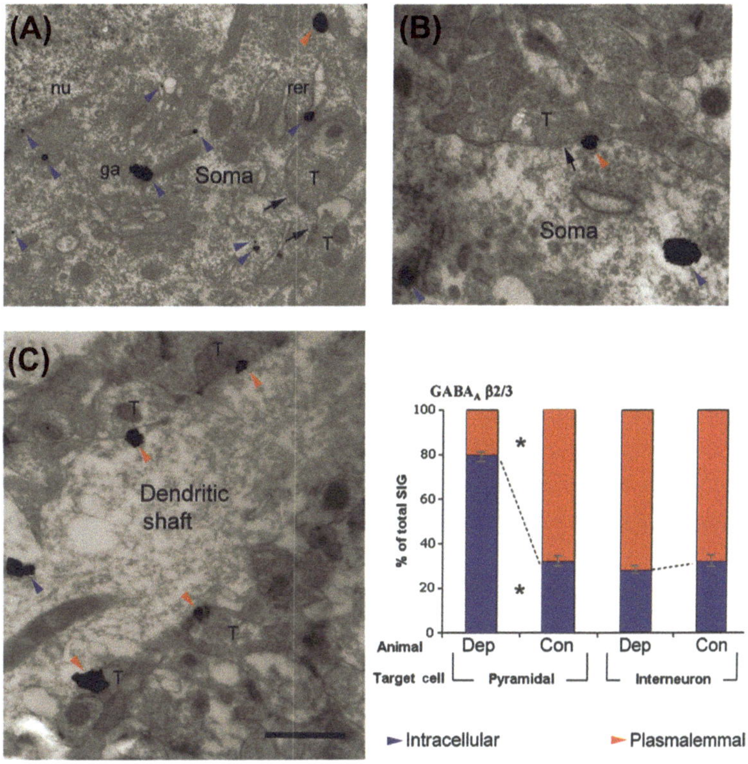

Figure 4 GABAergic inhibition upon cortical pyramidal cells is limited by the level of GABAR subunits at the plasma membrane. The three panels show examples of the localization of beta2/3 subunits of GABARs in the auditory cortex of P17 gerbils that were deprived of hearing starting from P10. The beta2/3 subunits were visualized by electron microscopic immunocytochemistry, using silver-intensified gold (SIG) as the label. SIG, reflecting beta2/3 subunit-ir, occurs along the cytoplasmic membranes of the Golgi apparatus (GA) and the endoplasmic reticulum (RER) (blue arrowheads in panels A and B). The red arrowheads point to beta2/3 subunit-ir occurring along the plasma membrane of a soma (panel B) and a dendritic shaft (panel C), most of which are associated with axon terminals (T). The black arrows point to morphologically identifiable symmetric axosomatic synaptic junctions. The graph depicts the result of a quantitative analysis, indicating that sensory deprivation reduced the proportion of plasmalemmal SIG particles of pyramidal neurons but not of the inhibitory interneurons. Asterisks mark significance of $p < 0.002$, determined by two-tailed Student's t-test. Further details of the methods and results can be found in Sarro et al. (2008). *The graph is modified, with permission from Oxford University Press.*

2.3.4.2. Activity-Dependence of GABAergic Synaptogenesis at Axosomatic Synapses Formed by PV Cells

Relatively more is known about the activity-dependent maturation involving innervations by PV-containing as compared to other GABAergic interneurons. This is due, at least in part, to their distinct biophysical signature (FS), the presence of molecular markers (PV and P–Q calcium channels in the axon terminals), their prevalence, and large cell body that makes these neurons more amenable for intracellular recording (Kawaguchi and Kubota, 1997). In contrast to the neuropeptide Y-, SS-, and nitric oxide-expressing GABAergic interneurons which enter the cortex through tangential migration, PV cells are more like the pyramidal neurons. They are born in the ganglionic eminence, like other GABAergic neurons, but enter the cortex from the subventricular zone in an inside-out sequence. They migrate radially and outwardly, and then land in their destination layer isochronously with the pyramidal neurons (Rymar and Sadikot, 2007).

In the mouse cortex, axosomatic inhibitory synapses which are primarily formed by PV-positive interneurons increase steadily in an inside-out sequence, from as early as P3, and gradually plateau at around P30 (Miller, 1986). Similarly, in the cat visual cortex, Mower's group reported that GAD65 is clearly evident within neurons at birth and that their processes follow the inside-out pattern (Guo et al., 1997). Perhaps the inside-out pattern in the appearance of GAD65-immunoreactive puncta reflects the sequential maturation of synapses formed by PV cells onto pyramidal neurons, soon after the PV cells and pyramidal neurons have completed their inside-out pattern of migration.

The time window of the inside-out sequence of GABAergic puncta supports the idea that GABAergic inputs are present before the onset of the CP and increase in number during the CP. As an added note to this description of the GABAergic synaptogenesis, a subset of basket cells (PV positive but also cholecystokinin-positive) actually forms synapses first on apical dendrites of pyramidal cells and shift a few days later to the soma. Although this is the temporal pattern described for hippocampal pyramidal neurons, it may apply to neocortical pyramidal neurons as well (Ben-Ari et al., 2004). What event coincides with and potentially triggers the shift from the apical dendrite to the soma remains unknown.

In the barrel cortex of the mouse, it is the PV cells forming perisomatic inhibition (identified based on their sensitivity to the P–Q channel blocker and direct visualization of the perisomatic PV-puncta) that are strongly dependent on sensory experience for maturation, while the RSNP neurons (identified by their expression of N-type calcium channels) are not activity-dependent for their maturation (Jiao et al., 2011). Takesian et al. (2012) also report the activity dependence in the maturation of IPSPs in the auditory cortex of gerbils for the FS PV cells and not for the LTS cells.

For the barrel cortex, brain-derived neurotrophic factor (BDNF) has been demonstrated to mediate a strong link between sensory experience and maturation of GABAergic axon terminals (Jiao et al., 2011). In fact, the expression of BDNF and the release of BDNF are both strongly linked to neuronal activity, including sensory experience-evoked neuronal activity (Tan et al., 2008). Not only the maturation but also the capacity for long-lasting, BDNF-dependent potentiation of inhibitory synapses is also dependent on experience during the CP (Xu et al., 2010). This later finding suggests that sensory deprivation during the CP can have long-lasting impact on not only sensory perception but also perceptual learning.

Since BDNF is not released from the GABAergic terminals themselves or from the TC axons (Gorba and Wahle, 1999), the source of BDNF is likely to be the postsynaptic excitatory neurons of the cortex. Recall that the maturation of axosomatic synapses on excitatory neurons is arrested by sensory deprivation, while the GABAergic synapses on GABAergic somata are not perturbed (Sarro et al., 2008). This differential dependence on activity across the two postsynaptic cell types may also be due to BDNF, which is released in an activity-dependent manner from excitatory neurons but not from GABAergic neurons.

2.3.4.3. Evidence that Ocular Dominance Shift by Parvalbumin Cells Mediates the Earliest Ocular Dominance Shift of Pyramidal Neurons

In studies that preceded Kameyama's, it was shown that FS (PV) interneurons are driven strongly by thalamic axons (Cruikshank et al., 2007). Electron microscopic analysis provided the cellular substrate for this strong thalamic drive by demonstrating that GABAergic interneurons, in general, are contacted by thalamic axons more frequently during the CP than after, and that the target sites are proximal (rather than distal) dendrites (Erisir and Dreusicke, 2005). The PV neurons are binocular (Yazaki-Sugiyama et al., 2009) and respond to monocular deprivation with rapid ocular dominance shifts (Yazaki-Sugiyama et al., 2009). Surprisingly, the direction of their shift is the opposite of that reported in earlier studies of ocular dominance shift (which could not identify the recorded neurons as glutamatergic or GABAergic and were probably mostly glutamatergic).

Although the mechanism bringing about change in this direction is counterintuitive, the functional significance of this ocular dominance shift among the PV neurons is simple to follow: inhibition driven by the nondeprived eye is weakened, resulting in an enhanced *E*-to-*I* ratio of the pathway driven by the nondeprived eye. In the same study, they demonstrated that the ocular dominance shift of pyramidal neurons is, indeed, dictated by alterations in the GABAergic inputs, because almost all pyramidal neurons responded to intracellular blockade of GABARs with a reversal of ocular dominance, as was seen many years ago in the cat visual cortex (Duffy et al., 1976). Application of a GABAR agonist, muscimol, also induces a robust reverse shift in favor of the deprived eye for the TC axons and for the cortical neurons (Hata and Stryker, 1994; Reiter and Stryker, 1988). Both these findings indicate that the ocular dominance shift is strongly dependent on the "*I*" portion of the *E*-to-*I* ratio.

Finally, GABA input-dependent inversion in the ocular dominance of pyramidal neurons in the mouse visual cortex does not occur following monocular deprivation at ages older than P60, indicating that this GABA-dependent mechanism of ocular dominance shift has a CP. When monocular deprivation during the CP is prolonged (from 2 days to 2 weeks), these PV neurons slowly switch their ocular dominance to becoming progressively less driven by the deprived eye and more by the active eye. Although Yazaki-Sugiyama et al. do not explain the functional significance of this slower change in the opposite direction, it presumably participates in sharpening the receptive field properties driven by the active eye.

2.3.4.4. Activity-Dependence of GABAergic Synaptogenesis at Spines during the Critical Period

Micheva and Beaulieu (1995) performed a rigorous serial EM analysis of the GABAergic axon terminals at the end of a 2-month-long somatosensory deafferentation that began from P1. Their quantitative EM analysis of all layers of the somatosensory cortex demonstrated a statistically significant decrement only of the axospinous synaptic junctions formed by GABAergic axon terminals in layer 4. The GABAergic axon terminals that course next to dendritic shafts but near the axospinous junctions would allow for shunting inhibition of a select subset of excitatory inputs that share the same parent dendritic shaft. GABAergic axons rarely form axospinous synaptic junctions but when they do, they are on spines that are targeted simultaneously by the excitatory axons arising from the thalamus (Kubota et al., 2007). The role of GABAergic input onto spines might be to provide selective inhibition of cortical excitability generated by the external world (via the thalamus), relative to the internal world (generated via corticocortical pathways).

2.3.4.5. Lack of Evidence for the Activity-Dependence of GABAergic Synaptogenesis at the Initial Segment

Chandelier cells express PV, are FS, and exert powerful inhibition upon multiple excitatory neurons via inhibition at the axon initial segment. The synapses formed at the initial segment can be evaluated based on the expression of the alpha2 subunits of GABARs there. The emergence of alpha2-mRNA is not activity dependent (Huntsman et al., 1994) and perturbation of alpha2-gene does not alter the onset of CP or sensitivity to monocular deprivation (Fagiolini et al., 2004). These findings suggest that the mode of inhibition via the initial segment may not be as important as the axosomatic synapses for regulating the sensory experience-dependent plasticity during the CP.

2.3.5. Activity-Dependence of GABAR Expression

GABA–benzodiazepine binding proteins appear early in the visual cortex and can precede the formation of morphologically identifiable GABAergic synapses to mediate GABA's paracrine, modulatory role in the formation of synaptic circuitry (Ben-Ari et al., 2004). The distribution of GABARs becomes highly stratified by the beginning of the CP (P30 for cat visual cortex) and remains stably highest in layer 4 through all later stages (Shaw et al., 1985, 1987). Complete blockade of retinal ganglion cell activity by the injection of tetrodotoxin into one eye causes reduction in the expression of GABAR subunits, even in adulthood (Hendry et al., 1990). In contrast, DR from birth, which deprives both form and nonform visions but leaves spontaneous activity intact, postpones the onset of the CP for ocular dominance shifts, yet has no overt effect on this macroscopic, laminar distribution. Closer examination of the receptor-binding properties of the GABARs revealed that DR increases the maximal binding concentration (Bmax) and GABA modulation of the dissociation constant, Kd, for the benzodiazepine binding site on GABARs (Shaw et al., 1987). Thus, if the change in *E*-to-*I* ratio is mediated through weakening of inhibition driven by the deprived eye (Section 2.4.2), this is not because of an alteration in the intrinsic property or concentration of the GABARs.

It has since become clear that GABARs are quite heterogeneous, through the 19 or more GABAR subunits that combine to form functional GABAR pentamers: the exact composition of the GABAR subunits can profoundly affect the kinetics of IPSPs as well as their synaptic versus nonsynaptic localization and clustering along the plasma membrane. Therefore, a possibility remained that the composition of GABARs could be activity-dependent. This possibility was examined by probing for the GABAR subunit composition by using zolpidem to test for the presence of a1 subunits (Kralic et al., 2002; Sanna et al., 2002) and benzodiazepines to test for the presence of alpha1/2/3/5 and gamma2 subunits. In the gerbil auditory cortex, GABARs contain alpha2 or alpha3 subunits at the prehearing stage and these can begin to form GABAR clusters via the help of gephyrin (Huang and Scheiffele, 2008; Tretter et al., 2008).

In the presence of sensory experience, these GABARs switch to become alpha1-containing at around P12—an age around sensory experience onset but before the onset of the CP: this subunit switch dramatically increases the IPSC kinetics and requires sensory experience during the CP, since cochlear ablation at the prehearing age of P10 for gerbils prevents the emergence of zolpidem sensitivity (Kotak et al., 2008). Once established, hearing is no longer required to maintain alpha1 subunits, since sensory deprivation in adulthood does not evoke any measurable change in the physiological properties of the IPSPs (Takesian et al., 2012), although it is possible that changes in GABARs do occur following prolonged hearing loss, as with aging (Caspary et al., 2008).

The failure to downregulate alpha3 subunits and replace these with alpha1 subunits would, in itself lead to slower, immature kinetics of IPSCs (Bosman et al., 2005; Kotak et al., 2008). Immaturity extends also to the degree of receptor clustering. If the only step in the GABAR maturation affected by sensory deprivation was the failure to switch from alpha2/3 to alpha1 subunits, the total amount of GABARs might not be expected to be altered. However, results based on quantitative electron microscopic immunocytochemistry of beta2/3 subunits indicate otherwise (Sarro et al., 2008). Beta2/3 subunits can form heteromers of GABARs with alpha1 as well as alpha2/3 subunits.

Electron microscopic immunocytochemical studies reveal that sensory deafferentation during the CP leads to a reduction in the proportion of beta2/3 subunits occurring at the plasma membrane, but not in the total amount, because a larger proportion of beta2/3 subunit immunoreactivity remains in the cytosol of dendritic shafts and somata. This indicates that activity regulates the trafficking of GABARs to the plasma membrane, more than the overall level of the receptor subunit proteins. Further analysis of the somatic immunolabeling indicates that the activity-dependence in the plasmalemmal labeling is specific for the somata of glutamatergic neurons (Sarro et al., 2008). This lack of effect of sensory deprivation upon GABAergic somata may be because GABAergic interneurons do not express the signature of activity, such as the releasable form of BDNF.

In order to assay for the receptor subtype responsible for ocular dominance plasticity, Fagiolini et al. developed a paradigm in which ocular dominance shift could be induced prematurely by boosting the endogenous GABA with benzodiazepines. This approach was combined with knock-in of three separate genes, each rendered insensitive to benzodiazepine via mutation of alpha1, alpha2, or alpha3 subunits. They found that the premature ocular dominance shift by benzodiazepines failed only for the mouse rendered insensitive to benzodiazepine at the alpha1 subunit. Based on this finding, they conclude that insertion of the alpha1 subunit is the important step for initiating the CP (Fagiolini et al., 2004).

2.3.6. Intracellular Mechanisms for the Stabilization and Maturation of GABAergic Synapses

GABAergic synapses are stabilized through the trans-synaptic adhesion of axon terminals and the postsynaptic membrane. Two classes of proteins have been identified as playing central roles in GABAergic synaptogenesis. They are neurexin on the presynaptic side and neuroligin on the postsynaptic side that, together, form the trans-synaptic adhesion of the pre- and postsynaptic membranes. In comparison to the excitatory synapses, which exhibit over 1500 synaptic proteins on the postsynaptic side (Ryan and Grant, 2009), the number of proteins occurring specifically at postsynaptic sides of GABAergic synapses is relatively scarce (Heller, 2009) but includes gephyrin, collybistin, dystrophin, and GABAR associated protein (GABARAP), among many others (Luscher et al., 2011). How these proteins, together with GABAR and GABA, contribute to GABAergic synaptogenesis during the CP will be considered.

GABA itself plays a critical role in GABAergic synaptogenesis. This finding was demonstrated through knock-down of GAD67, one of the two rate-limiting enzymes synthesizing GABA which is more tightly regulated by neuronal activity levels. Because GAD67 knock-out (KO) is lethal, the role of GAD67 in synaptogenesis has been explored through knock-down of GAD67 within organotypic slice cultures. GAD67 knock-down perturbs synapse formation and axonal outgrowth. These perturbations are rescued by enhancing GABAAR function, either by blocking GABA reuptake or by boosting the residual GABA with benzodiazepines (Chattopadhyaya et al., 2007).

Among the 19 GABAR subunits so far identified, certain subunits occur reliably at synapses, indicating that they may be required for synapse stabilization. Selective knock-down experiments have revealed that gamma2 subunits of GABARs are essential for the accumulation of GABARs to the synapse as well for the stable innervation by GABAergic axon terminals (Alldred et al., 2005). This suggests that a retrograde signal must influence the stabilization of the GABAergic axon terminal to the plasma membrane surface where GABARs are expressed. One retrograde messenger identified to be associated with GABAergic synaptogenesis is BDNF, described in Section 2.4.4.2, although the link between GABAR activation and BDNF release from the postsynaptic-to-be neuron remains unknown, as is the molecular mechanism through which the precise alignment of GABAergic axon terminals with GABAR clusters occurs.

Of the known molecules, the neurexin–neuroligin pair of membranous proteins rises as the best candidate bringing about the pre- to postalignment. Among the neuroligin class of proteins, genetic deletion and overexpression studies indicate that neuroligin 2 (NL2) is especially important for GABAergic synapse formation. Moreover, coexpression of NL2 with GABAR subunits alpha2, beta3, and gamma2 in nonneuronal cells indicate that NL2 can coaggregate with gamma2 subunits of GABARs and that spontaneous IPSC amplitude and frequency as well as the presynaptic GABAergic differentiation are strongly dependent on the coexpression of NL2 with gamma2, alpha, and beta3 subunits (Dong et al., 2007).

Alpha2 subunits occur at the highest concentrations at the axon initial segment. KO of alpha2 subunits strongly reduces the expression of gephyrin, together with NL2. Closer examination of the subcellular distribution pattern of alpha2 KO reveals that the expression of both NL2 and gephyrin as well as the alpha1 subunit are completely eliminated at the axon initial segment. In contrast, the expression of NL2, together with alpha1 subunits and dystrophin at the cell bodies are preserved. Based on these findings, authors conclude that alpha2-containing GABARs are obligatory for clustering of GABARs and gephyrin at the initial segment, while dystrophin is the protein more involved in axosomatic synaptogenesis (Panzanelli et al., 2011). How these proteins participate in the translation between experience-dependent afferent activity and GABAR clustering is a topic that remains to be studied.

2.3.7. Extracellular Mechanisms for Stabilizing GABAergic Synapses

Sensitivity to monocular deprivation can be reinstated in the adult cortex following removal of the extracellular matrix, called the perineuronal nets (PNs), a term coined by the foundational work of Emilio Golgi more than 120 years ago (Spreafico et al., 1999; Pizzorusso et al., 2002). This indicated that the formation of PNs plays an important role in ending the CP. This study was predated by an earlier work demonstrating that PNs occur enriched around the PV interneurons (Celio, 1993; Celio and Chiquet-Ehrismann, 1993) and of these PV interneurons, specifically of the basket cells that form axosomatic inhibitory synapses and not the chandelier cells forming axoaxonic synapses at the axon initial segment (Naegele and Katz, 1990). These studies are predated by an even earlier series of studies demonstrating that the emergence of PNs is dependent on sensory experience during the CP (Guimaraes et al., 1990; Lander et al., 1997; Sur et al., 1988).

Composition of the PNs has also been elucidated. Tenascin was the first extracellular matrix glycoprotein reported in the PNs (Celio, 1993; Celio and Chiquet-Ehrismann, 1993). Aggrecan, recognized by the antibody, Cat-315, in rodent brains and by the antibody Cat-301 in cat brains, is enriched in layers 4 and 6 of developing somatosensory cortex and is more selectively expressed around PV neurons, especially in layer 4 (McRae et al., 2007). The density of neurons expressing CAT-315-immunoreactivity is dependent on sensory experience during the CP. In contrast, the density of cells expressing chondroitin sulfate chains, also prevalent in the PNs, is not altered by sensory deprivation.

A possibility remains, however, that the level of chondroitin sulfate chains is activity-dependent. The hippocampus also exhibits PNs selectively around the cell bodies, proximal dendrites, and the axon initial segment of PV cells (Dityatev et al., 2007) and for these and presumably all PV cells of the neocortex, it is the calcium influx through GluR2-lacking AMPARs and L-channels that is critical for the formation of PNs. Equally important to note is that NMDAR activity and extracellular BDNF are less important for the formation of PNs. Although chemical removal of the extracellular matrix proteins constituting the PNs does not alter the number of GABAergic axosomatic inputs to the cells, excitability is increased (Dityatev et al., 2007). Thus, decrease in excitability of PV cells, as would happen as the PN emerges around these neurons, would appear to be related most directly to the end of the CP.

2.3.8. The Role of GABA in the Hippocampal CA1 during Adolescence

In contrast to the vast amount of research conducted for the CP of ocular dominance plasticity, relatively little work has addressed the developmental events follow afterward, during adolescence (Pinto et al., 2010). For female rodents, puberty is readily identifiable as the age past the opening of vagina, occurring at around P30–34, followed some later by the first systemic release of estrogen and progestin steroid hormones. However, it is not until P50–60 or so that the estrous cycle becomes regular, at which time rodents can be categorized as fully mature adults. Using these endocrinological hallmarks to define "adolescence", this last phase of development corresponds to ages well beyond the most sensitive period for sensory-experience-dependent developmental plasticity, such as ocular dominance shifts.

The CA1 pyramidal neurons exhibit extensive growth during adolescence—as much as 100% increase in 7 days, within stratum radiatum (Figure 5). This phase is marked by a transient increase of a rare type of GABAR—those composed of alpha4 and delta subunits, instead of the alpha1 and gamma2 subunits. Alpha4betadelta GABARs are expressed at constitutively high levels in the dentate gyrus and thalamus but are of low abundance in the CA1 of hippocampus and neocortex. The rise, then the rapid fall of progestin at puberty onset triggers a parallel rise and fall of allopregnanolone

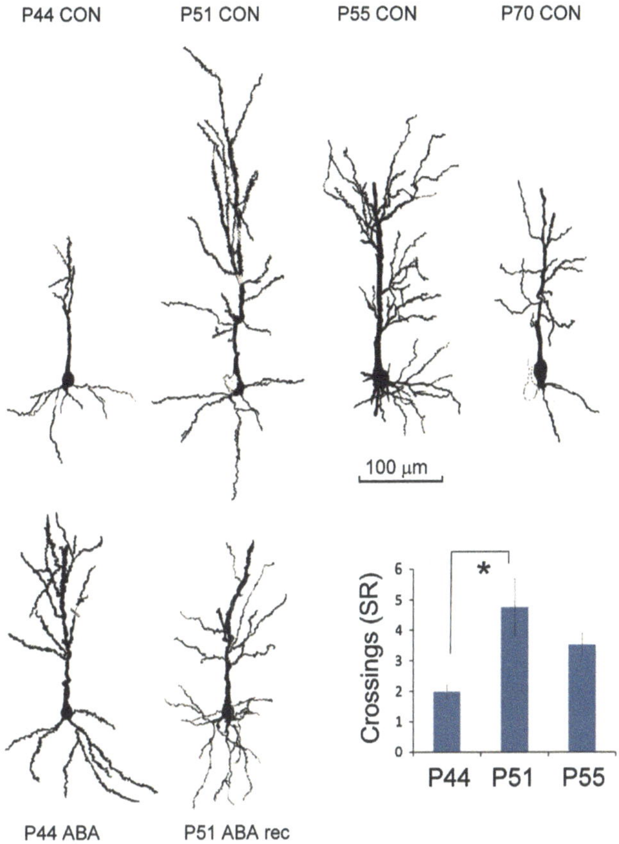

Figure 5 Dendritic growth of hippocampal pyramidal cells during adolescence. Adolescence is a CP for experience-dependent plasticity of hippocampal neurons. These CA1 pyramidal neurons were stained by the Golgi method, then traced by camera lucida and the neurolucida system. The graph summarizes complexity of the apical dendrites measured at P44, P51, P55, and P70 from the caudal CA1 region of the hippocampus of normally reared control female rats (upper row). Sholl analysis of dendritic branches in stratum radiatum was conducted for CA1 pyramidal neurons of three to five animals of each age and is summarized as a graph. The results indicate doubling of the dendritic complexity from P44 to P51, measured as the number of crossings of the apical dendrites' branches with concentric circles centered over the soma at radii ranging from 80 to 200 μm. The asterisk indicates significance of difference by a one way ANOVA, followed by post-hoc Fisher's test ($p < 0.01$). Animals reared from P41 to P44 under food restriction and with a running wheel become hyperactive and voluntarily seek running rather than food, even during the hour of food availability. This paradigm, called activity-based anorexia (ABA) is being used as an animal model to study the neurobiology of anorexia nervosa (Aoki et al., 2012). The P44 ABA is an example of a pyramidal neuron from the caudal CA1 of such an animal. Note the increased branching of the dendritic tree, compared to that from an age-matched control (P44 CON). Sholl analysis indicates significance of the difference between CON and ABA in the stratum radiatum zone 80–200 μm from the somata ($p < 0.05$). One week of recovery from the ABA-inducing environment allows for weight restoration and normalization of the dendritic arbor (P51 ABA rec), relative to the P51 CON group. (Brain tissue of ABA rats generated and provided by Nicole C. Barbarich-Marsteller; Golgi staining of neurons and morphometric analysis conducted by Tara Chowdhury, Mariel Rios, Daniela Cassataro, and Aoki et al., 2012.)

(3alpha-OH-5alpha[beta]-pregnan-20-one, THP) and with this, an increased expression of alpha4betadelta GABARs in the CA1, generating multiple profound physiological effects (Shen et al., 2007, 2010; Smith et al., 2009). In brief, the emergence of alpha4betadelta GABARs at puberty onset increases tonic inhibition by GABA, thereby reducing excitability of pyramidal neurons. Besides inducing the expression of the alpha4 and delta subunits, THP also has a more acute effect of becoming elevated during stress and reducing tonic inhibition mediated by these GABARs. The net effect is that the acute rise of THP during stress can rapidly increase excitability of the CA1 pyramidal neurons. As a key brain region of the limbic circuit, excitability of CA1 pyramidal neurons is closely linked to the behavioral state of anxiety. This paradoxical rise in expression but of the inhibition of the same alpha4betadelta GABARs during stress could be a component contributing to mood swings and anxiety. Indeed, pubertal female mice exhibit increased anxiety traits following restraint stress, while pubertal female littermates with genetic deletion of the delta subunits do not.

Due to the absence of the gamma2 subunit, the alpha4betadelta GABARs do not become localized to GABAergic synapses (Alldred et al., 2005). Instead, these receptors are activated by ambient GABA and mediate tonic inhibition. What electron microscopic immunocytochemistry revealed was that these receptors can occur at the plasma membrane of dendritic spines, adjacent to excitatory inputs (Figure 6). Based on this location, alpha4betadelta GABARs would be expected to cause shunting inhibition and reduce NMDAR activation. Indeed, electrophysiological measurements confirm these predictions since the threshold for evoking NMDAR currents increased by 100 nA under the condition that blocked synaptic (but not the nonsynaptic) GABARs.

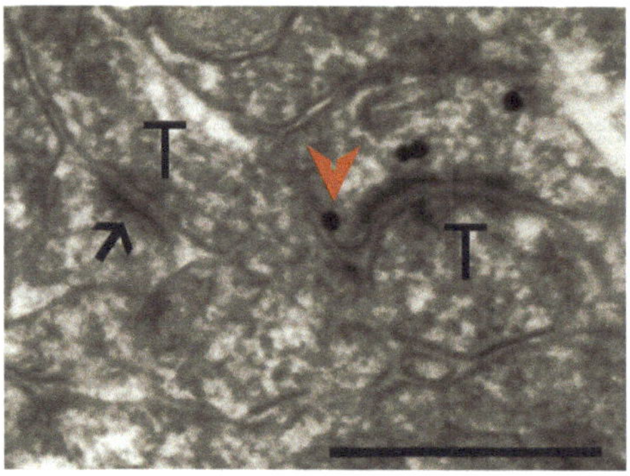

Figure 6 GABARs containing alpha4 and delta subunits occur near axospinous excitatory synapses in the CA1 region of the hippocampus. Electron microscopic immunocytochemistry was used to localize the alpha4 subunits of alpha4betadelta GABARs in the CA1-stratum radiatum of P44 female rat hippocampus. SIG particles, reflecting alpha4 immunoreactivity, occur in the dendritic spine's cytosol and on the plasma membrane (red arrowhead), immediately adjacent to the thick postsynaptic density of the excitatory axospinous synaptic junction. T, axon terminal. These GABAR subunits can be activated by ambient GABA to exert shunting inhibition. The black arrow points to the postsynaptic density formed by an unlabeled spine, for comparison. Bar = 500 nm. Further details of the methods and results can be found in Aoki et al. (2012).

The behavioral consequence of these biophysical effects is that the animal exhibits impairment in spatial memory formation. However, in the presence of THP, these receptors in the CA1 hippocampal pyramidal neurons become inactive, and this inactivity brings back excitability to pyramidal neurons and improvements in spatial memory formation. Electron microscopic immunocytochemistry reveals that the rise of alpha4betadelta GABARs is mediated through trafficking of the cytoplasmic pool of GABARs to the plasma membrane, not via the up- and downregulation of receptor subunit mRNAs. These subcellular parameters allow for a rapid modification of tonic inhibition mediated by ambient levels of GABA that are tightly linked to the animal's stress level. Besides the role the alpha4betadelta GABARs play in acute stress-evoked anxiety and learning improvements, the emergence of these receptors in the CA1 following chronic stress during adolescence is hypothesized to contribute to anxiety-related disorders, such as anorexia nervosa (Aoki et al., 2012).

2.4. ROLE OF NOREPINEPHRINE IN OCULAR DOMINANCE PLASTICITY

2.4.1. Resurrection of the Norepinephrine Theory of Critical Period Leads to the Recognition of the Importance of Acetylcholine

One of the earliest clues to the neurochemical basis of the CP for ocular dominance shift plasticity came from the findings by Kasamatsu and Pettigrew (Itakura et al., 1981; Kasamatsu, 1991; Kasamatsu and Pettigrew, 1976, 1979; Pettigrew and Kasamatsu, 1978). Their findings preceded the vast literature that now points to the involvement of the glutamatergic receptors and GABA. The major finding by Kasamatsu and Pettigrew was that the intracortical infusion of 6-hydroxydopamine (6-OHDA) directly into the visual cortex of kittens rendered the neurons no longer sensitive to monocular deprivation, in that the ocular dominance shift was prevented. Since this drug was well known to cause lesion of the catecholaminergic afferents (reviewed by Kasamatsu, 1991; Breese et al., 2005), their findings indicated that catecholamines are required for the cellular and molecular mechanisms that convert sensory experience-dependent neuronal activity into synaptic changes underlying ocular dominance shifts during the CP.

This idea was supported by findings using other manipulations, such as the blockade of beta-adrenergic receptors by propranolol that also rendered the visual cortex no longer responsive to monocular deprivation. Also, infusion of drugs activating the protein kinase A (PKA) cascade restored sensitivity of the 6-OHDA-treated visual cortex of kittens or of intact adult visual cortex to monocular deprivation (Imamura et al., 1999, 2007).

However, these findings were questioned by others, who demonstrated that the systemic depletion of norepinephrine or KO of one of the four regulatory subunits of PKA (PKA R1beta) did not abolish the ocular dominance plasticity during the CP (Daw et al., 1984, 1985; Hensch et al., 1998). These same investigators also could not identify any qualitative or quantitative difference between the CP and adulthood with regard to the noradrenergic modulation of visual cortical neurons' spontaneous or evoked activities (Daw et al., 1984). If local depletion of norepinephrine abolished plasticity during the CP, why would not a more global depletion of norepinephrine be more effective? And why would not the norepinephrine (NE) modulation of cortical neurons exhibit some special property during the CP?

This puzzle was solved a few years later by the demonstration that the administration of 6-OHDA (the toxin used by Kasamatsu and Pettigrew to deplete norepinephrine) directly into the visual cortex had a secondary effect, which was to block the muscarinic acetylcholine receptors. The same report demonstrated that ocular dominance plasticity during the CP required the presence of both the noradrenergic and cholinergic afferents: neither the depletion of norepinephrine from noradrenergic fibers nor the basal forebrain lesion of cholinergic afferents to the cortex, alone, altered plasticity. Rather, depletion of norepinephrine from the cortex by using a method that avoided direct infusion of 6-OHDA into the visual cortex, combined with lesions of cholinergic afferents did abolish the ocular dominance plasticity (Bear and Singer, 1986). This finding has been corroborated by the original group, with the caveat that the cholinergic participation in ocular dominance plasticity requires beta-adrenergic receptor activity (Kasamatsu, 1991).

That both the cholinergic and noradrenergic afferents to the neocortex would play necessary, permissive roles in developmental plasticity contributes toward the general notion that both systems play critical roles in attention, vigilance, and synaptic plasticity throughout life (Aston-Jones et al., 1991; Hasselmo and McGaughy, 2004; Klinkenberg et al., 2011; Turchi and Sarter, 1997). In the sections that follow, the developmental changes in the noradrenergic and cholinergic synapses will be reviewed, highlighting their relevance to the developmental plasticity of the neocortex.

2.4.2. Developmental Changes in the Ultrastructure of the Noradrenergic System Suggest a Shift in Its Role from Supporting Synapse Stabilization to Supporting Synaptic Transmission

All noradrenergic afferents to the neocortex arise from a single brainstem region, the locus coeruleus. Noradrenergic fibers can be identified reliably by the presence of two molecular markers—dopamine beta-hydroxylase (DBH), the rate-limiting synthetic enzyme, and the norepinephrine transporter (NET). Using either DBH or NET as the marker, noradrenergic fibers have been shown to arrive to the cortex even before the thalamic afferents, in time to "prepare" the cortex for activity-dependent synaptogenesis and synapse rearrangements. NET expression at the noradrenergic terminals in the rat cortex rises more than fivefold between P10 and P15 (Sanders et al., 2005), perfectly timed for modulating synaptogenesis.

The results of Pettigrew and Kasamatsu pointed to the importance of the cyclic adenosine monophosphate generation through the noradrenergic system. Of the four major receptor classes activated by norepinephrine, the beta1- and beta2-adrenergic receptors are the ones linked positively to adenylyl cyclase. Analysis of the cat visual cortex by receptor autoradiography shows that the distribution pattern of the beta1- and beta2-adrenergic receptors are laminarly distinct by the age of onset for the CP for ocular dominance plasticity, being highest in layers 1 through 3, lowest in layer 4 and intermediate in layers 5 and 6. This enrichment of beta-adrenergic receptors within the supragranular layers has been observed for the developing and adult rodent cortex as well as for the developing primate cortex (Lidow and Rakic, 1992; Vos et al., 1990).

Quantitative receptor autoradiography reveals that the concentration of beta-adrenergic receptors increases during the CP and reaches a plateau value by the end of the CP (Herdon et al., 1983; Jonsson and Kasamatsu, 1983). Concurrent with this gentle, monotonic rise in the concentration of beta-adrenergic receptors, the cortex is also undergoing a dramatic shift in the cellular components that are modulated via activation of beta-adrenergic receptors. Specifically, by using electron microscopic immunocytochemistry to detect beta-adrenergic receptors, it has been shown that the immunoreactivity to the beta-adrenergic receptors is sparse in neonatal cortex but where present, they are mostly on postsynaptic sites of asymmetric (i.e. excitatory) synapses. From the first to the second week, excitatory synaptic junctions are very rare, but approximately 50% of the excitatory synapses encountered are immunoreactive for the beta-adrenergic receptors. During week 2, the number of synapses per unit volume has increased 30-fold, but the number of synapses with immunoreactivity for beta-adrenergic receptors has barely doubled. The net result is that the portion synapses with beta-adrenergic immunoreactivity drops dramatically from the first to the third postnatal weeks. The same neuropil also begins to exhibit beta-adrenergic receptor immunoreactivity along the astrocytic plasma membrane starting the second postnatal week. By week 3, there is a 200-fold increase in the areal density of astrocytes immunoreactive for beta-adrenergic receptors, but relatively little change in the expression of beta-adrenergic receptors at synapses (bar graph of Figure 7).

These cellular and subcellular changes in the expression pattern of beta-adrenergic receptors would be expected to alter the nature of noradrenergic modulation of the cortical neuropil dramatically across the days spanning the CP for ocular dominance plasticity. Activation of beta-adrenergic receptors has been reported to increase the signal-to-noise ratio in the firing of some neurons and to have little influence on others. One simplistic prediction from the ultrastructural analysis of beta-adrenergic receptor distribution is that noradrenergic modulation targets the newly forming synapses during the first postnatal week. The localization specifically along the thick postsynaptic densities suggests that the clustering of glutamatergic receptors may be facilitated through a PKA-dependent step. This is followed a week later by the increase of presynaptic beta-adrenergic receptors. These receptors may be enhancing the release of glutamate through PKA-dependent phosphorylation of the voltage-dependent calcium channel. As the end of the CP approaches, noradrenergic modulation may shift from events involving synaptogenesis to the support

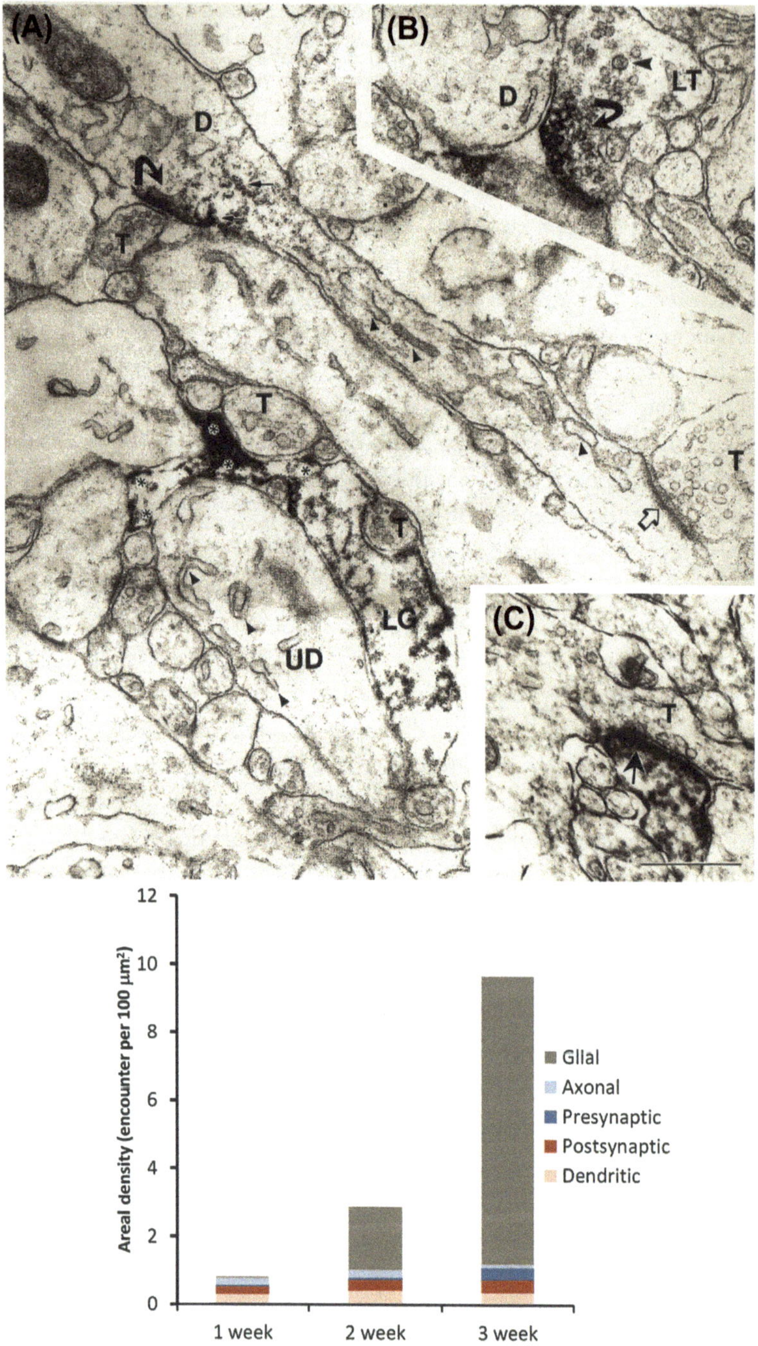

Figure 7 Beta-adrenergic receptors occur postsynaptically, presynaptically and in astrocytes of P10 rat visual cortex. Beta-adrenergic receptors were localized using HRP-DAB as the electron dense label. DAB labeling occurs at the postsynaptic membrane of a dendritic shaft (D, curved arrow, upper left in

of synaptic transmission, through mobilization of glycogen stores within the astrocytic processes (Stone and Ariano, 1989) or through the polymerization of actin which, in turn, leads to the retraction of astrocytic processes (Goldman and Abramson, 1990), thereby allowing for greater cross-talk among axons and dendrites.

Activation of beta-adrenergic receptors on astrocytes also induces the release of nerve growth factor (Bache et al., 1988), which could be important for supporting the expression of the nicotinic receptors that are activated by another neuromodulator important for synaptic plasticity during development and throughout life—acetylcholine. How the neuronal and astrocytic beta-adrenergic receptors work in tandem to modulate cortical plasticity is an intriguing question that deserves to be pursued in the future.

2.4.3. Sensory Deprivation Evokes Disturbance to the Tangential, but not Laminar Distribution Pattern of Beta-Adrenergic Receptors in the Cerebral Cortex

While the CP for ocular dominance shift is stereotypical for each species, DR from birth delays the onset of the CP (Mower et al., 1981). This indicates that initiation of the cellular machinery supporting the activity-dependent rearrangement and stabilization of the synapses in the cortex requires a visual experience-evoked event or process. Can the postponement of the onset of the CP be due to disturbance of the noradrenergic system? To date, there is little known about the effect of DR upon the noradrenergic afferents in the visual cortex or about the alpha-adrenergic receptors. However, there are number of studies pointing to the involvement of the beta-adrenergic receptors in activity-dependent modification of cortical circuitry.

One study characterized the laminar distribution pattern of beta-adrenergic receptors within the visual cortex of cat brains throughout and toward the end of the CP following normal visual rearing and DR. Results of this study showed that the distinct laminar distribution pattern within the visual cortex was well established by the onset of the CP (1 month) and maintained throughout the months of the CP, regardless of the rearing condition (Aoki et al., 1986) (Figure 8).

panel A) and associated with the smooth endoplasmic reticulum in the vicinity (small arrows). Another asymmetric synapse formed by the same dendrite (D) is unlabeled for the receptor (open straight arrow, lower right). An astrocytic process in the vicinity exhibits intense immunoreactivity (asterisks in the cytoplasm; LG = labeled glia). Arrowheads point to the unlabeled smooth endoplasmic reticulum of another unlabeled dendrite (UD) that is immediately adjacent to the labeled glia. Panel B shows an example of presynaptic labeling in an axon terminal (LT, labeled terminal) containing an electron dense vesicle (arrowhead). T points to unlabeled terminals, for comparison. Panel C shows an example of an immunolabeled dendritic spine forming an axospinous asymmetric synapse. Calibration bar = 500 nm for Panels A and B and 400 nm for Panel C. The bar graph summarizes the ontogenetic change in beta-adrenergic receptor expression across different subcellular domains within the rat visual cortex over the first three postnatal weeks. Further details of the methods and results can be found in Aoki (1997). Panels A and B are modified, with permission.

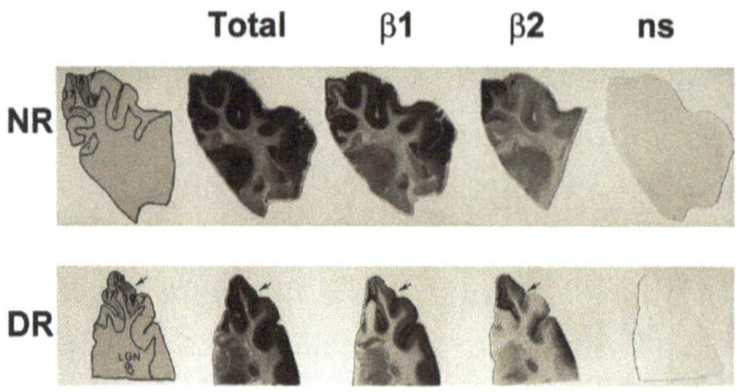

Figure 8 The laminar distribution pattern of beta-adrenergic receptors is not disturbed by dark rearing. The distribution patterns of beta1 and beta2 adrenergic receptors in the visual cortex of cats spanning the ages of the CP for ocular dominance plasticity were examined autoradiographically, using [^{125}I] iodo-pindolol as a ligand in combination with displacers specific for the beta1 and beta2 subtypes of adrenergic receptors. The upper row shows the total beta, beta1, beta2, and nonspecific (NS) labeling pattern at 3 months postnatal, which is toward the end of the CP. Beta1 and beta2 adrenergic receptors are enriched in layers 1 through 3, low in layer 4, and of moderate levels in layers 5 and 6. Beta2 adrenergic receptors differentiate the primary visual areas 17 and 18 from the associational visual area 19, just lateral to the arrow. Comparisons of the visual areas of kittens normally reared (NR, top row) and dark reared for 3 months from birth (DR, lower row) indicate that these laminar and regional patterns of expression are not disturbed DR. Further details of the methods and results can be found in Aoki et al. (1986).

On the other hand, sensory deprivation can affect the tangential distribution pattern of beta-adrenergic receptors within the supragranular layers of the cerebral cortex. In the "barrel" region of rat somatosensory cortex (S1 barrel cortex), each cluster of neurons in layer 4 represents one mystical vibrissa, easily visualized by the histochemical reaction of cytochrome oxidase or of succinate dehydrogenase. In the absence of sensory deprivation, beta2-adrenergic receptors occur clustered in the supragranular layers immediately above the cluster of neurons forming the barrels in layer 4 (Vos et al., 1990). Removal of the vibrissa and the underlying follicles at P2 causes irreversible peripheral deafferentation and long-lasting alterations of the S1 barrel cortex, as is indicated by the succinate dehydrogenase and cytochrome oxidase reactivity patterns (Catalano et al., 1995) and direct visualization of the TC axonal arbors (Catalano et al., 1995). The disruption of the S1 barrel cortex is accompanied by a disruption in the patterned distribution of beta2-adrenergic receptors, observed 8–10 weeks later by receptor autoradiography (Vos et al., 1990).

While this change in the beta-adrenergic receptors is certainly overlapping with the CP for S1 barrel formation in layer 4 (Catalano et al., 1995), what remains to be tested is whether perturbation of sensory afferents at later stages of S1 barrel development would also perturb the tangential distribution of beta2-adrenergic receptors. This remains a distinct possibility since monocular enucleation of adult monkeys, a manipulation analogous to vibrissal removal, reduces cytochrome oxidase reactivity within the ocular dominance columns representing the deprived eye and with it, an upregulation beta-adrenergic receptor immunoreactivity in the deprived ocular dominance columns (Aoki et al., 1994a) (Figure 9).

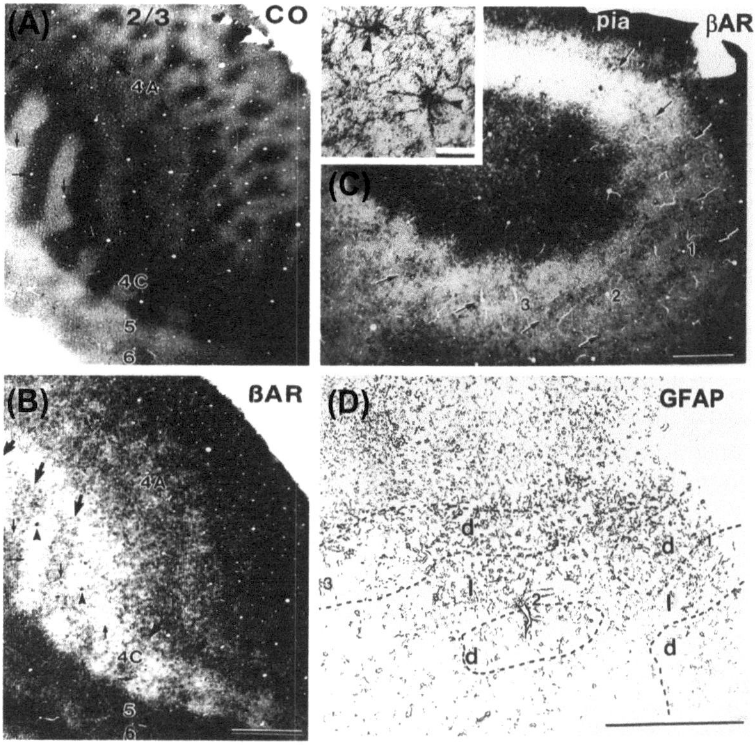

Figure 9 Monocular deprivation in adulthood causes upregulation of beta-adrenergic receptors within astrocytic processes of the deprived ocular dominance. Beta-adrenergic receptors (BARs) were localized by immunocytochemistry to be predominantly astrocytic (insert of Panel C). Macroscopic inspection of tangentially oriented section of the visual cortical area V1 of monkeys allows visualization of all layers of the cortex as well as the ocular dominance columns, identified as serving the deprived eye wherever the cytochrome oxidase (CO) reactivity is light and as serving the nondeprived eye wherever the CO reactivity is dark (Panel A). Panels B and C highlight multiple darken bands of beta-adrenergic receptor immunoreactivity (BAR, large arrows in B, small arrows in C). The numbers 1, 2, and 3 point to blood vessel profiles used as fiduciary marks to align with adjacent sections that were immunolabeled for the astrocytic marker, glial fibrillary acidic protein (GFAP, panel D), or for CO. Panel D shows a detailed map of GFAP labeling surrounding blood vessels 2 and 3 in panel C. The dashed lines depict boundaries between the zones with light (l) and dark (d) beta-adrenergic receptor immunoreactivity. The comparison of GFAP (panel D) and beta-adrenergic receptor (panel C) highlights the homogeneity of GFAP labeling that contrasts the banded pattern of beta-adrenergic receptor labeling. Comparisons with the immediately adjacent section processed for CO histochemistry reveals that the darker bands of beta-adrenergic receptor immunoreactivity corresponds to the deprived (lighter CO) ocular dominance columns. Together, these results indicate that sensory deprivation prompts increased trafficking of beta-adrenergic receptors within astrocytic processes residing within the inactive neuropil. Calibration bar = 1 mm for all panels and 20 μm for the detail in panel C. Further details of the methods and results can be found in Aoki (1994a). Modified with permission.

This finding was based on immunocytochemical labeling for beta-adrenergic receptors, which revealed that immunoreactive elements in the deprived ocular dominance columns were almost all astrocytic processes. This might be expected, based on the results described above, indicating that beta-adrenergic receptor immunoreactivity is much more prevalent in astrocytes than in neurons within the mature cortex. In contrast, immunocytochemistry of the adjacent sections with glial fibrillary acidic protein, the molecular marker of the major processes of astrocytes revealed no difference in immunoreactivity across the ocular dominance columns. This indicates that the expression level of beta-adrenergic receptor is independent of the growth or retraction of major astrocytic processes. More importantly, comparison of cytochrome oxidase histochemistry (which is used to locate the ocular dominance columns) with beta-adrenergic receptor immunocytochemistry revealed that the columns exhibiting higher concentrations of beta-adrenergic receptors corresponded to the deprived columns. This particularly surprising finding indicated that the intracellular trafficking and/or turnover of beta-adrenergic receptors might be inversely correlated with neuronal activity of the surrounding neuropil. What might be the physiological significance of this distribution pattern? One possibility is that the upregulation of beta-adrenergic receptors reflects a homeostatic response to prolonged neuronal inactivity: Activation of astrocytic beta-adrenergic receptors is known to increase GABA uptake but decrease glutamate uptake (Hansson and Ronnback, 1992), both of which would increase excitability of the surrounding neurons. The impact of this may be to maximize excitability of the cortical neurons that have been rendered inactive due to enucleation.

2.4.4. The Role of Acetylcholine in Ocular Dominance Plasticity during the Critical Period

The unexpected finding from the original work by Kasamatsu and Pettigrew then followed up by Bear and Singer demonstrated the importance of the convergent norepinephrine and acetylcholine inputs upon ocular dominance plasticity during the CP (reviewed in Section 2.4.1 above). To what extent are the cholinergic afferents and receptors present in the developing neocortex?

Work preceding the findings by Bear and Singer indicated that the cholinergic axons, like the noradrenergic system, are already present in the cortex by the time the thalamic axons arrive. Postnatally, these axons proliferate rapidly to exceed adult levels by the beginning of the CP for ocular dominance plasticity (1 month postnatal for kittens) (Bear et al., 1985). Blockade of muscarinic M1 receptors through continuous intracortical infusion of drugs into the visual cortex of kittens suppresses the ocular dominance shift following monocular deprivation. However, blockade of nicotinic receptors or of M2 muscarinic receptors is without effect on ocular dominance shifts (Goto et al., 1993). None of the drug infusions have an appreciable effect on the visually evoked or spontaneous activity of visual cortical neurons during the CP (Goto et al., 1993), even though adult cortex relies strongly on acetylcholine for visually evoked neuronal activity (Sato et al., 1987). Nevertheless, results from these pharmacological assays indicate that the cholinergic receptors are in place in time to modulate visual cortical neurons and experience-dependent synaptic plasticity during the CP.

2.4.4.1. Muscarinic Receptors in the Cortex during the Critical Period

Receptor autoradiography reveals that muscarinic receptors, labeled using a radioligand [^{3}H]QNB, are present by postnatal day 3 in the kitten visual cortex and preferentially in layer 4, the TC recipient zone. During the CP for ocular dominance plasticity (first through the third postnatal month for kittens), the distribution pattern of muscarinic receptors becomes reversed, so that by the end of the CP, layer 4 exhibits the least density and the supragranular layers exhibit the highest density (Shaw et al., 1985). Cellular analysis of the muscarinic receptor expression in adulthood indicates that these are expressed more strongly by the GABAergic interneurons than the excitatory neurons (Disney and Aoki, 2008; Disney et al., 2006; Erisir et al., 2001). This distribution pattern of the cholinergic receptors supports the idea that much of the cholinergic modulation of cortical plasticity is through modulation of GABAergic neurons' excitability or even of the GABA release, independent of the firing of these neurons.

2.4.4.2. Nicotinic Receptors in the Cortex during the Critical Period

The nicotinic receptors are located on in-growing TC axons. The density of nicotinic receptors in layer 4 rises from the beginning to mid-CP (first to second postnatal month for kittens) (Needler et al., 1988; Prusky et al., 1987) and persist there through adulthood (Disney et al., 2007). Since nicotinic receptors are ligand-gated and evoke rapid depolarization, they can increase the gain for detecting visual stimuli through enhanced release of glutamate from the TC axons (Disney et al., 2007). The coactivation of nicotinic receptors with glutamate receptors during visually guided behavior would be expected to favor TC synaptic transmission relative to the corticocortical synaptic transmission.

Electron microscopic immunocytochemical localization of nicotinic receptor subunits of adult cortex indicates that beta2 subunit-containing (i.e. alpha4beta2 heteromeric) nicotinic acetylcholine receptors are expressed by most TC terminals in layer 4c, and specifically at those terminals targeting the excitatory neurons, rather than the inhibitory interneurons (Disney et al., 2007). Given that the TC axons target GABAergic neurons more during the CP than after (Erisir and Dreusicke, 2005), and with all that we now know about the role played by GABAergic interneurons in ocular dominance plasticity (reviewed in Section 2.4 of this chapter), it would be interesting to reexamine the relationship between TC axons, cholinergic receptor expressions on them, and the subtypes of GABAergic interneurons targeted by the TC axons specifically during the CP for developmental plasticity.

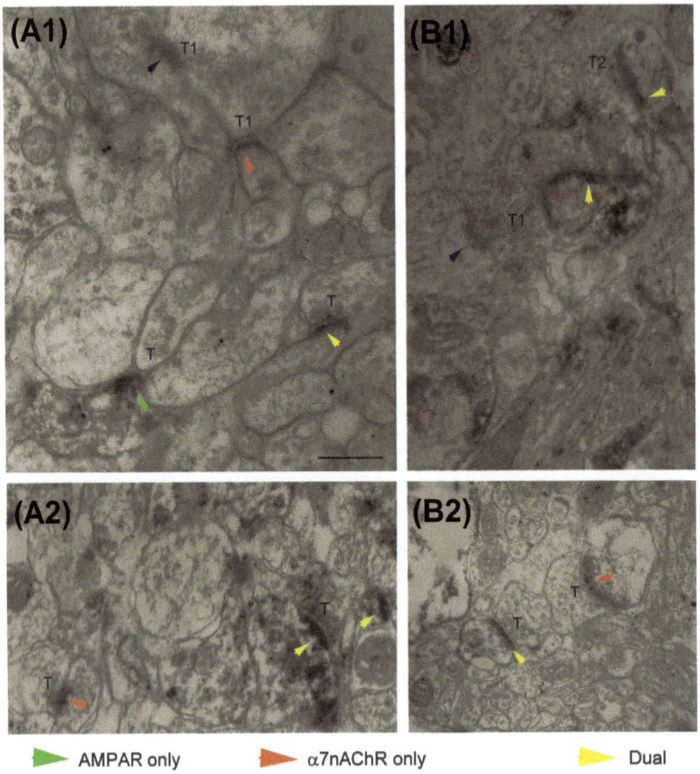

Figure 10 Excitatory synapses are heterogeneous, carrying varying levels of AMPA receptors and frequently modulated by acetylcholine via the nicotinic receptors. Nicotinic receptors containing alpha7 subunits occur at a subset of excitatory synapses at postnatal day 7 (P7, A1, and A2) and in adulthood (B1 and B2). Electron micrographs A1 and B1 were taken from layer 1, A2 was taken from layer 6, and B2 was taken from layer 2/3. The presence of alpha7 subunits of nicotinic receptors at the postsynaptic density (PSD) is indicated by the postembed colloidal gold labeling. The presence of AMPA receptor subunits (GluR1, GluR2, or GluR3) is indicated by the HRP-DAB labeling. The green arrow points to a single PSD that is detectably immunoreactive only for the AMPA receptor subunit at the PSD. The yellow arrowheads point to PSDs with detectable levels of immunoreactivity for subunits of both the AMPA and alpha7 subunit-containing nicotinic receptors. The red arrowheads point to PSDs with immunoreactivity for alpha7 subunits, only. Further details of the methods and results can be found in Levy and Aoki (2002).

The alpha7-containing (i.e. the alpha7 homomeric) nicotinic receptors occur widely within the cortex and directly over the PSD, where glutamatergic receptors are also anchored (Levy and Aoki, 2002). Since the alpha7 homomeric nicotinic receptors are also ligand-gated and are permeable to cations, there is an intriguing possibility that these nicotinic receptors may provide the depolarization necessary to unblock Mg^{2+} from the NMDARs of synapses hypothesized to be "silent", due to the absence of AMPARs (Atwood and Wojtowicz, 1999; Gomperts et al., 1998; Isaac et al., 1997; Malinow, 1998). Moreover, the alpha7-nicotinic receptors are permeable to calcium (McGehee, 1999), and are thereby able to initiate the calcium-dependent intracellular cascades more commonly ascribed to NMDARs and the GluR2-lacking AMPARs. Quantitative analysis of the adult rat neocortex reveals that approximately one-third of the asymmetric (i.e. excitatory, presumably containing NMDARs) synapses lack detectable levels of AMPAR subunits GluR1, GluR2, or GluR3 and of them, two-thirds are immune-positive for alpha7 subunits. At P7, the proportion of labeling is nearly the same: 38% of the synapses lack detectable levels of AMPAR subunits and of them, 66% are immunoreactive for alpha7 subunits (Levy and Aoki, 2002) (Figure 10). Overall, the majority of the excitatory synapses exhibit immunoreactivity for the alpha7 subunits and this finding is in agreement with that for the adult hippocampus (Fabian-Fine et al., 2001) and adult prefrontal cortex (Lubin et al., 1999).

2.4.4.3. Nicotinic Receptor-Mediated Termination of the Critical Period

The end of the CP coincides with the marked emergence of a distinct molecular pathway. The molecular brake is Lynx1, a protein that binds to nicotinic acetylcholine receptors and reduces their sensitivity to endogenous acetylcholine. The singular importance of Lynx1 blockade of nicotinic receptors can be appreciated by analysis of Lynx1 KO: such genetic deletion allows for the sensitivity to monocular deprivation to persist well beyond the CP (>P60) but not when the nicotinic receptors are antagonized. The site of action of Lynx1 may be dominated heavily by synapses associated with the PV-containing GABAergic interneurons since approximately 90% of the PV cells also express Lynx1 mRNA. However, whether the site of action of Lynx1 is the target sites of PV cells (where PV cells may release Lynx1) or autonomous (which would lead to disinhibition of PV cells) is one of many interesting questions that remain to be answered for understanding the cellular and molecular events that permit plasticity during the CP.

3. CONCLUSIONS

During early life, brain circuits are endowed with a capacity to optimize their structural connectivity in register with patterns of sensory stimuli that surround us. Several factors complicate studies that investigate the sequence of events that lead to these structural changes and the molecular players that enable these changes.

First, at any time, more than one circuit can be susceptible for experience-dependent plasticity, and these circuits may employ different molecular players, leading to different structural changes in their corresponding neuronal compartments. For example, shortly after eye opening, distinct cortical circuits that enable ocular dominance, orientation selectivity, or directional selectivity all undergo major developmental maturation and display plastic capacity, yet molecular factors that underline these capacities may not be ubiquitous. In contrast, a single molecular mechanism, such as the activation of NMDARs, can be shared by multiple plastic circuits to detect sustained changes in sensory activity, and those circuits may physically occupy the same brain regions or the same cortical layers.

Second, many molecular factors can be modulators of the same cascade; while glutamate and GABA work in concert to maintain a neuron's dynamic range, monoamines, acetylcholine, and other neuromodulators can alter the gain, and tip the balance.

Third, sustained structural changes, as observed in CP plasticity, are often the result of a cascade of many events, each of which can be detected as acute modulations of neuronal activity or transient structural changes.

Furthermore, the first step of the cascade can be necessary for the next step, yet does not, by itself, guarantee the ultimate "plastic change". Cessation of visual activity should first be detected by neurons in the deprived eye's ocular dominance column, leading to LTD and loss of dendritic spines, and this reduction of activity would trigger the step for TC axon's expansion into the deprived eye's ocular dominance column. However, similar acute physiological and structural changes in the adult do not lead to the second step of axon extension as readily as in young brains. It remains an important challenge for us to chart out the order of the cascade of events that can lead to sustained structural changes, and identify the crucial and consequential players in this process. The consequential factors in the cascade of plastic events, that is the everyday modulators of the neuronal excitability, can be as important as the crucial players, the ones that enable the flow of the plasticity cascade into the next step, or the ones that place an age-dependent brake on the cascade. In the pursuit of rendering the brain more plastic later in life, removal of the brakes of plastic processes may be detrimental for neuronal integrity while enhancing the efficiency of neuromodulatory systems on plastic neurons may prove to be more feasible.

ACKNOWLEDGMENTS

The authors wish to thank many who have read and contributed significantly to the content and form of this chapter. They include Dr Vibhakar Kotak, who has provided expert opinion on the electrophysiological and developmental aspects of the section on GABA; Dr Nicole Sabaliauskas, Dr Emma Sarro, and Dr Gauri Wable, who helped identify portions of the manuscript that were muddled, inconsistent, or grammatically incorrect; Tara Chowdhury, Mariel Rios, and Daniela Cassataro, who provided *camera lucida* drawings of Golgi-stained hippocampal pyramidal neurons and conducted the Sholl analysis.

REFERENCES

Abraham, W.C., Bear, M.F., 1996. Metaplasticity: the plasticity of synaptic plasticity. Trends Neurosci. 19, 126–130.

Aghajanian, G.K., Bloom, F.E., 1967. The formation of synaptic junctions in developing rat brain: a quantitative electron microscopic study. Brain Res. 6, 716–727.

Alldred, M.J., Mulder-Rosi, J., Lingenfelter, S.E., Chen, G., Luscher, B., 2005. Distinct gamma2 subunit domains mediate clustering and synaptic function of postsynaptic GABAA receptors and gephyrin. J. Neurosci. 25, 594–603.

Alvarez, V.A., Sabatini, B.L., 2007. Anatomical and physiological plasticity of dendritic spines. Annu. Rev. Neurosci. 30, 79–97.

Anderson, S.A., Marin, O., Horn, C., Jennings, K., Rubenstein, J.L., 2001. Distinct cortical migrations from the medial and lateral ganglionic eminences. Development 128, 353–363.
Anker, R.L., Cragg, B.G., 1974. Development of the extrinsic connections of the visual cortex in the cat. J. Comp. Neurol. 154, 29–41.
Antonini, A., Stryker, M.P., 1996. Plasticity of geniculocortical afferents following brief or prolonged monocular occlusion in the cat. J. Comp. Neurol. 369, 64–82.
Aoki, C., 1997. Postnatal changes in the laminar and subcellular distribution of NMDA-R1 subunits in the cat visual cortex as revealed by immuno-electron microscopy. Brain Res. Dev. Brain Res. 98, 41–59.
Aoki, C., Fujisawa, S., Mahadomrongkul, V., Shah, P.J., Nader, K., Erisir, A., 2003. NMDA receptor blockade in intact adult cortex increases trafficking of NR2A subunits into spines, postsynaptic densities, and axon terminals. Brain Res. 963, 139–149.
Aoki, C., Kaufman, D., Rainbow, T.C., 1986. The ontogeny of the laminar distribution of beta-adrenergic receptors in the visual cortex of cats, normally reared and dark-reared. Brain Res. 392, 109–116.
Aoki, C., Lubin, M., Fenstemaker, S., 1994a. Columnar activity regulates astrocytic beta-adrenergic receptor-like immunoreactivity in V1 of adult monkeys. Vis. Neurosci. 11, 179–187.
Aoki, C., Sabaliauskas, N., Chowdhury, T., Min, J.Y., Colacino, A.R., Laurino, K., Barbarich-Marsteller, N.C., 2012. Adolescent female rats exhibiting activity-based anorexia express elevated levels of GABA(A) receptor alpha4 and delta subunits at the plasma membrane of hippocampal CA1 spines. Synapse 66, 391–407.
Aoki, C., Venkatesan, C., Go, C.G., Mong, J.A., Dawson, T.M., 1994b. Cellular and subcellular localization of NMDA-R1 subunit immunoreactivity in the visual cortex of adult and neonatal rats. J. Neurosci. 14, 5202–5222.
Aston-Jones, G., Chiang, C., Alexinsky, T., 1991. Discharge of noradrenergic locus coeruleus neurons in behaving rats and monkeys suggests a role in vigilance. Prog. Brain Res. 88, 501–520.
Atwood, H.L., Wojtowicz, J.M., 1999. Silent synapses in neural plasticity: current evidence. Learn. Mem. 6, 542–571.
Bacci, A., Huguenard, J.R., Prince, D.A., 2005. Modulation of neocortical interneurons: extrinsic influences and exercises in self-control. Trends Neurosci. 28, 602–610.
Bache, R.J., Homans, D.C., Schwartz, J.S., Dai, X.Z., 1988. Differences in the effects of alpha-1 adrenergic blockade with prazosin and indoramin on coronary blood flow during exercise. J. Pharmacol. Exp. Ther. 245, 232–237.
Balice-Gordon, R.J., Lichtman, J.W., 1994. Long-term synapse loss induced by focal blockade of postsynaptic receptors. Nature 372, 519–524.
Bear, M.F., Carnes, K.M., Ebner, F.F., 1985. Postnatal changes in the distribution of acetylcholinesterase in kitten striate cortex. J. Comp. Neurol. 237, 519–532.
Bear, M.F., Kleinschmidt, A., Gu, Q.A., Singer, W., 1990. Disruption of experience-dependent synaptic modifications in striate cortex by infusion of an NMDA receptor antagonist. J. Neurosci. 10, 909–925.
Bear, M.F., Singer, W., 1986. Modulation of visual cortical plasticity by acetylcholine and noradrenaline. Nature 320, 172–176.
Ben-Ari, Y., Khalilov, I., Represa, A., Gozlan, H., 2004. Interneurons set the tune of developing networks. Trends Neurosci. 27, 422–427.
Bosman, L.W., Heinen, K., Spijker, S., Brussaard, A.B., 2005. Mice lacking the major adult GABAA receptor subtype have normal number of synapses, but retain juvenile IPSC kinetics until adulthood. J. Neurophysiol. 94, 338–346.
Bourgeois, J.P., Jastreboff, P.J., Rakic, P., 1989. Synaptogenesis in visual cortex of normal and preterm monkeys: evidence for intrinsic regulation of synaptic overproduction. Proc. Natl. Acad. Sci. U.S.A. 86, 4297–4301.
Bourgeois, J.P., Rakic, P., 1993. Changes of synaptic density in the primary visual cortex of the macaque monkey from fetal to adult stage. J. Neurosci. 13, 2801–2820.
Bourgeois, J.P., Rakic, P., 1996. Synaptogenesis in the occipital cortex of macaque monkey devoid of retinal input from early embryonic stages. Eur. J. Neurosci. 8, 942–950.
Breese, G.R., Knapp, D.J., Criswell, H.E., Moy, S.S., Papadeas, S.T., Blake, B.L., 2005. The neonate-6-hydroxydopamine-lesioned rat: a model for clinical neuroscience and neurobiological principles. Brain Res. Brain Res. Rev. 48, 57–73.

Burkhalter, A., Bernardo, K.L., Charles, V., 1993. Development of local circuits in human visual cortex. J. Neurosci. 13, 1916–1931.
Callaway, E.M., Katz, L.C., 1990. Emergence and refinement of clustered horizontal connections in cat striate cortex. J. Neurosci. 10, 1134–1153.
Callaway, E.M., Katz, L.C., 1992. Development of axonal arbors of layer 4 spiny neurons in cat striate cortex. J. Neurosci. 12, 570–582.
Campbell, G., Shatz, C.J., 1992. Synapses formed by identified retinogeniculate axons during the segregation of eye input. J. Neurosci. 12, 1847–1858.
Cao, Z., Lickey, M.E., Liu, L., Kirk, E., Gordon, B., 2000a. Postnatal development of NR1, NR2A and NR2B immunoreactivity in the visual cortex of the rat. Brain Res. 859, 26–37.
Cao, Z., Liu, L., Lickey, M., Gordon, B., 2000b. Development of NR1, NR2A and NR2B mRNA in NR1 immunoreactive cells of rat visual cortex. Brain Res. 868, 296–305.
Carlisle, H.J., Kennedy, M.B., 2005. Spine architecture and synaptic plasticity. Trends Neurosci. 28, 182–187.
Carmignoto, G., Vicini, S., 1992. Activity-dependent decrease in NMDA receptor responses during development of the visual cortex. Science 258, 1007–1011.
Caspary, D.M., Ling, L., Turner, J.G., Hughes, L.F., 2008. Inhibitory neurotransmission, plasticity and aging in the mammalian central auditory system. J. Exp. Biol. 211, 1781–1791.
Catalano, S.M., Robertson, R.T., Killackey, H.P., 1995. Rapid alteration of thalamocortical axon morphology follows peripheral damage in the neonatal rat. Proc. Natl. Acad. Sci. U.S.A. 92, 2549–2552.
Celio, M.R., 1993. Perineuronal nets of extracellular matrix around parvalbumin-containing neurons of the hippocampus. Hippocampus 3 Spec No. 55–60.
Celio, M.R., Chiquet-Ehrismann, R., 1993. 'Perineuronal nets' around cortical interneurons expressing parvalbumin are rich in tenascin. Neurosci. Lett. 162, 137–140.
Chapman, B., Stryker, M.P., Bonhoeffer, T., 1996. Development of orientation preference maps in ferret primary visual cortex. J. Neurosci. 16, 6443–6453.
Chattopadhyaya, B., Di Cristo, G., Wu, C.Z., Knott, G., Kuhlman, S., Fu, Y., Palmiter, R.D., Huang, Z.J., 2007. GAD67-mediated GABA synthesis and signaling regulate inhibitory synaptic innervation in the visual cortex. Neuron 54, 889–903.
Chen, L., Cooper, N.G., Mower, G.D., 2000. Developmental changes in the expression of NMDA receptor subunits (NR1, NR2A, NR2B) in the cat visual cortex and the effects of dark rearing. Brain Res. Mol. Brain Res. 78, 196–200.
Coleman, J.E., Nahmani, M., Gavornik, J.P., Haslinger, R., Heynen, A.J., Erisir, A., Bear, M.F., 2010. Rapid structural remodeling of thalamocortical synapses parallels experience-dependent functional plasticity in mouse primary visual cortex. J. Neurosci. 30, 9670–9682.
Colonnier, M., Guillery, R.W., 1964. Synaptic organization in the lateral geniculate nucleus of the monkey. Z. Zellforsch. Mikrosk. Anat. 62, 333–355.
Constantine-Paton, M., Cline, H.T., Debski, E., 1990. Patterned activity, synaptic convergence, and the NMDA receptor in developing visual pathways. Annu. Rev. Neurosci. 13, 129–154.
Corson, J., Nahmani, M., Lubarsky, K., Badr, N., Wright, C., Erisir, A., 2009. Sensory activity differentially modulates *N*-methyl-D-aspartate receptor subunits 2A and 2B in cortical layers. Neuroscience 163, 920–932.
Cragg, B.G., 1975. The development of synapses in kitten visual cortex during visual deprivation. Exp. Neurol. 46, 445–451.
Cruikshank, S.J., Lewis, T.J., Connors, B.W., 2007. Synaptic basis for intense thalamocortical activation of feedforward inhibitory cells in neocortex. Nat. Neurosci. 10, 462–468.
Cull-Candy, S.G., Leszkiewicz, D.N., 2004. Role of distinct NMDA receptor subtypes at central synapses. Sci. STKE 2004, re16.
Daw, N.W., Robertson, T.W., Rader, R.K., Videen, T.O., Coscia, C.J., 1984. Substantial reduction of cortical noradrenaline by lesions of adrenergic pathway does not prevent effects of monocular deprivation. J. Neurosci. 4, 1354–1360.
Daw, N.W., Stein, P.S., Fox, K., 1993. The role of NMDA receptors in information processing. Annu. Rev. Neurosci. 16, 207–222.
Daw, N.W., Videen, T.O., Parkinson, D., Rader, R.K., 1985. DSP-4 (*N*-(2-chloroethyl)-*N*-ethyl-2-bromobenzylamine) depletes noradrenaline in kitten visual cortex without altering the effects of monocular deprivation. J. Neurosci. 5, 1925–1933.

Disney, A.A., Aoki, C., 2008. Muscarinic acetylcholine receptors in macaque V1 are most frequently expressed by parvalbumin-immunoreactive neurons. J. Comp. Neurol. 507, 1748–1762.
Disney, A.A., Aoki, C., Hawken, M.J., 2007. Gain modulation by nicotine in macaque v1. Neuron 56, 701–713.
Disney, A.A., Domakonda, K.V., Aoki, C., 2006. Differential expression of muscarinic acetylcholine receptors across excitatory and inhibitory cells in visual cortical areas V1 and V2 of the macaque monkey. J. Comp. Neurol. 499, 49–63.
Dityatev, A., Bruckner, G., Dityateva, G., Grosche, J., Kleene, R., Schachner, M., 2007. Activity-dependent formation and functions of chondroitin sulfate-rich extracellular matrix of perineuronal nets. Dev. Neurobiol. 67, 570–588.
Dong, N., Qi, J., Chen, G., 2007. Molecular reconstitution of functional GABAergic synapses with expression of neuroligin-2 and $GABA_A$ receptors. Mol. Cell. Neurosci. 35, 14–23.
Duffy, F.H., Burchfiel, J.L., Conway, J.L., 1976. Bicuculline reversal of deprivation amblyopia in the cat. Nature 260, 256–257.
Durack, J.C., Katz, L.C., 1996. Development of horizontal projections in layer 2/3 of ferret visual cortex. Cereb. Cortex 6, 178–183.
Erisir, A., Dreusicke, M., 2005. Quantitative morphology and postsynaptic targets of thalamocortical axons in critical period and adult ferret visual cortex. J. Comp. Neurol. 485, 11–31.
Erisir, A., Harris, J.L., 2003. Decline of the critical period of visual plasticity is concurrent with the reduction of NR2B subunit of the synaptic NMDA receptor in layer 4. J. Neurosci. 23, 5208–5218.
Erisir, A., Levey, A.I., Aoki, C., 2001. Muscarinic receptor M(2) in cat visual cortex: laminar distribution, relationship to gamma-aminobutyric acidergic neurons, and effect of cingulate lesions. J. Comp. Neurol. 441, 168–185.
Fabian-Fine, R., Skehel, P., Errington, M.L., Davies, H.A., Sher, E., Stewart, M.G., Fine, A., 2001. Ultrastructural distribution of the alpha7 nicotinic acetylcholine receptor subunit in rat hippocampus. J. Neurosci. 21, 7993–8003.
Fagiolini, M., Fritschy, J.M., Low, K., Mohler, H., Rudolph, U., Hensch, T.K., 2004. Specific $GABA_A$ circuits for visual cortical plasticity. Science 303, 1681–1683.
Fagiolini, M., Hensch, T.K., 2000. Inhibitory threshold for critical-period activation in primary visual cortex. Nature 404, 183–186.
Fields, R.D., 2005. Myelination: an overlooked mechanism of synaptic plasticity? Neuroscientist 11, 528–531.
Flint, A.C., Maisch, U.S., Weishaupt, J.H., Kriegstein, A.R., Monyer, H., 1997. NR2A subunit expression shortens NMDA receptor synaptic currents in developing neocortex. J. Neurosci. 17, 2469–2476.
Fox, K., Daw, N., Sato, H., Czepita, D., 1992. The effect of visual experience on development of NMDA receptor synaptic transmission in kitten visual cortex. J. Neurosci. 12, 2672–2684.
Fox, K., Daw, N.W., 1993. Do NMDA receptors have a critical function in visual cortical plasticity? Trends Neurosci. 16, 116–122.
Fox, K., Sato, H., Daw, N., 1989. The location and function of NMDA receptors in cat and kitten visual cortex. J. Neurosci. 9, 2443–2454.
Fox, K., Sato, H., Daw, N., 1990. The effect of varying stimulus intensity on NMDA-receptor activity in cat visual cortex. J. Neurophysiol. 64, 1413–1428.
Freire, M., 1978. Effects of dark rearing on dendritic spines in layer IV of the mouse visual cortex. A quantitative electron microscopical study. J. Anat. 126, 193–201.
Fukuda, T., Aika, Y., Heizmann, C.W., Kosaka, T., 1998. GABAergic axon terminals at perisomatic and dendritic inhibitory sites show different immunoreactivities against two GAD isoforms, GAD67 and GAD65, in the mouse hippocampus: a digitized quantitative analysis. J. Comp. Neurol. 395, 177–194.
Gabbott, P.L., Somogyi, J., Stewart, M.G., Hamori, J., 1986. GABA-immunoreactive neurons in the dorsal lateral geniculate nucleus of the rat: characterisation by combined golgi-impregnation and immunocytochemistry. Exp. Brain Res. 61, 311–322.
Gabbott, P.L., Somogyi, P., 1986. Quantitative distribution of GABA-immunoreactive neurons in the visual cortex (area 17) of the cat. Exp. Brain Res. 61, 323–331.
Gandhi, S.P., Yanagawa, Y., Stryker, M.P., 2008. Delayed plasticity of inhibitory neurons in developing visual cortex. Proc. Natl. Acad. Sci. U.S.A. 105, 16797–16802.

Goldman, J.E., Abramson, B., 1990. Cyclic AMP-induced shape changes of astrocytes are accompanied by rapid depolymerization of actin. Brain Res. 528, 189–196.

Gomperts, S.N., Rao, A., Craig, A.M., Malenka, R.C., Nicoll, R.A., 1998. Postsynaptically silent synapses in single neuron cultures. Neuron 21, 1443–1451.

Gorba, T., Wahle, P., 1999. Expression of TrkB and TrkC but not BDNF mRNA in neurochemically identified interneurons in rat visual cortex in vivo and in organotypic cultures. Eur. J. Neurosci. 11, 1179–1190.

Gordon, J.A., Stryker, M.P., 1996. Experience-dependent plasticity of binocular responses in the primary visual cortex of the mouse. J. Neurosci. 16, 3274–3286.

Goto, S., Singer, W., Gu, Q., 1993. Immunocytochemical localization of calcineurin in the adult and developing primary visual cortex of cats. Exp. Brain Res. 96, 377–386.

Gu, Q.A., Bear, M.F., Singer, W., 1989. Blockade of NMDA-receptors prevents ocularity changes in kitten visual cortex after reversed monocular deprivation. Brain Res. Dev. Brain Res. 47, 281–288.

Guimaraes, A., Zaremba, S., Hockfield, S., 1990. Molecular and morphological changes in the cat lateral geniculate nucleus and visual cortex induced by visual deprivation are revealed by monoclonal antibodies Cat-304 and Cat-301. J. Neurosci. 10, 3014–3024.

Guo, Y., Kaplan, I.V., Cooper, N.G., Mower, G.D., 1997. Expression of two forms of glutamic acid decarboxylase (GAD67 and GAD65) during postnatal development of the cat visual cortex. Brain Res. Dev. Brain Res. 103, 127–141.

Hall, Z.W., Sanes, J.R., 1993. Synaptic structure and development: the neuromuscular junction. Cell 72 (Suppl.), 99–121.

Hansson, E., Ronnback, L., 1992. Adrenergic receptor regulation of amino acid neurotransmitter uptake in astrocytes. Brain Res. Bull. 29, 297–301.

Hasselmo, M.E., McGaughy, J., 2004. High acetylcholine levels set circuit dynamics for attention and encoding and low acetylcholine levels set dynamics for consolidation. Prog. Brain Res. 145, 207–231.

Hata, Y., Stryker, M.P., 1994. Control of thalamocortical afferent rearrangement by postsynaptic activity in developing visual cortex. Science 265, 1732–1735.

Heller, E., 2009. Synaptic Protein Profiling in the Mammalian Brain. The Rockefeller University, The Rockefeller University.

Hendry, S.H., Fuchs, J., deBlas, A.L., Jones, E.G., 1990. Distribution and plasticity of immunocytochemically localized GABAA receptors in adult monkey visual cortex. J. Neurosci. 10, 2438–2450.

Hendry, S.H., Jones, E.G., 1988. Activity-dependent regulation of GABA expression in the visual cortex of adult monkeys. Neuron 1, 701–712.

Hensch, T.K., 2005. Critical period plasticity in local cortical circuits. Nat. Rev. Neurosci. 6, 877–888.

Hensch, T.K., Fagiolini, M., 2005. Excitatory-inhibitory balance and critical period plasticity in developing visual cortex. Prog. Brain Res. 147, 115–124.

Hensch, T.K., Fagiolini, M., Mataga, N., Stryker, M.P., Baekkeskov, S., Kash, S.F., 1998. Local GABA circuit control of experience-dependent plasticity in developing visual cortex. Science 282, 1504–1508.

Herdon, H.J., Wilkinson, M., Madani, M.O., 1983. alpha- and beta-adrenergic binding sites in sheep cerebral cortex: characterisation, effects of photoperiod and treatment with estrogen/progesterone. Brain Res. Bull. 10, 159–161.

Hong, Y.K., Chen, C., 2011. Wiring and rewiring of the retinogeniculate synapse. Curr. Opin. Neurobiol. 21, 228–237.

Hooks, B.M., Chen, C., 2007. Critical periods in the visual system: changing views for a model of experience-dependent plasticity. Neuron 56, 312–326.

Horton, J.C., Hocking, D.R., 1997. Myelin patterns in V1 and V2 of normal and monocularly enucleated monkeys. Cereb. Cortex 7, 166–177.

Huang, Z.J., Scheiffele, P., 2008. GABA and neuroligin signaling: linking synaptic activity and adhesion in inhibitory synapse development. Curr. Opin. Neurobiol. 18, 77–83.

Hubel, D.H., Wiesel, T.N., 1970. The period of susceptibility to the physiological effects of unilateral eye closure in kittens. J. Physiol. 206, 419–436.

Hubel, D.H., Wiesel, T.N., 1979. Brain mechanisms of vision. Sci. Am. 241, 150–162.

Hubel, D.H., Wiesel, T.N., LeVay, S., 1977. Plasticity of ocular dominance columns in monkey striate cortex. Philos. Trans. R. Soc. Lond. B. Biol. Sci. 278, 377–409.

Huberman, A.D., 2007. Mechanisms of eye-specific visual circuit development. Curr. Opin. Neurobiol. 17, 73–80.

Huntsman, M.M., Isackson, P.J., Jones, E.G., 1994. Lamina-specific expression and activity-dependent regulation of seven GABAA receptor subunit mRNAs in monkey visual cortex. J. Neurosci. 14, 2236–2259.

Huttenlocher, P.R., Dabholkar, A.S., 1997. Regional differences in synaptogenesis in human cerebral cortex. J. Comp. Neurol. 387, 167–178.

Huttenlocher, P.R., de Courten, C., Garey, L.J., Van der Loos, H., 1982. Synaptogenesis in human visual cortex–evidence for synapse elimination during normal development. Neurosci. Lett. 33, 247–252.

Imamura, K., Kasamatsu, T., Shirokawa, T., Ohashi, T., 1999. Restoration of ocular dominance plasticity mediated by adenosine 3',5'-monophosphate in adult visual cortex. Proc. Biol. Sci. 266, 1507–1516.

Imamura, K., Kasamatsu, T., Tanaka, S., 2007. Neural plasticity maintained high by activation of cyclic AMP-dependent protein kinase: an age-independent, general mechanism in cat striate cortex. Neuroscience 147, 508–521.

Isaac, J.T., Crair, M.C., Nicoll, R.A., Malenka, R.C., 1997. Silent synapses during development of thalamocortical inputs. Neuron 18, 269–280.

Issa, N.P., Trachtenberg, J.T., Chapman, B., Zahs, K.R., Stryker, M.P., 1999. The critical period for ocular dominance plasticity in the Ferret's visual cortex. J. Neurosci. 19, 6965–6978.

Itakura, T., Kasamatsu, T., Pettigrew, J.D., 1981. Norepinephrine-containing terminals in kitten visual cortex: laminar distribution and ultrastructure. Neuroscience 6, 159–175.

Jiao, Y., Zhang, Z., Zhang, C., Wang, X., Sakata, K., Lu, B., Sun, Q.Q., 2011. A key mechanism underlying sensory experience-dependent maturation of neocortical GABAergic circuits in vivo. Proc. Natl. Acad. Sci. U.S.A. 108, 12131–12136.

Jonsson, G., Kasamatsu, T., 1983. Maturation of monoamine neurotransmitters and receptors in cat occipital cortex during postnatal critical period. Exp. Brain Res. 50, 449–458.

Kameyama, K., Sohya, K., Ebina, T., Fukuda, A., Yanagawa, Y., Tsumoto, T., 2010. Difference in binocularity and ocular dominance plasticity between GABAergic and excitatory cortical neurons. J. Neurosci. 30, 1551–1559.

Kaneko, T., Fujiyama, F., 2002. Complementary distribution of vesicular glutamate transporters in the central nervous system. Neurosci. Res. 42, 243–250.

Kaneko, T., Fujiyama, F., Hioki, H., 2002. Immunohistochemical localization of candidates for vesicular glutamate transporters in the rat brain. J. Comp. Neurol. 444, 39–62.

Kasamatsu, T., 1991. Adrenergic regulation of visuocortical plasticity: a role of the locus coeruleus system. Prog. Brain Res. 88, 599–616.

Kasamatsu, T., Imamura, K., Mataga, N., Hartveit, E., Heggelund, U., Heggelund, P., 1998. Roles of *N*-methyl-D-aspartate receptors in ocular dominance plasticity in developing visual cortex: re-evaluation. Neuroscience 82, 687–700.

Kasamatsu, T., Pettigrew, J.D., 1976. Depletion of brain catecholamines: failure of ocular dominance shift after monocular occlusion in kittens. Science 194, 206–209.

Kasamatsu, T., Pettigrew, J.D., 1979. Preservation of binocularity after monocular deprivation in the striate cortex of kittens treated with 6-hydroxydopamine. J. Comp. Neurol. 185, 139–161.

Kawaguchi, Y., Kubota, Y., 1997. GABAergic cell subtypes and their synaptic connections in rat frontal cortex. Cereb. Cortex 7, 476–486.

Keller, A., White, E.L., 1987. Synaptic organization of GABAergic neurons in the mouse SmI cortex. J. Comp. Neurol. 262, 1–12.

Kleinschmidt, A., Bear, M.F., Singer, W., 1987. Blockade of "NMDA" receptors disrupts experience-dependent plasticity of kitten striate cortex. Science 238, 355–358.

Klinkenberg, I., Sambeth, A., Blokland, A., 2011. Acetylcholine and attention. Behav. Brain Res. 221, 430–442.

Kohr, G., 2006. NMDA receptor function: subunit composition versus spatial distribution. Cell Tissue Res. 326, 439–446.

Kotak, V.C., Fujisawa, S., Lee, F.A., Karthikeyan, O., Aoki, C., Sanes, D.H., 2005. Hearing loss raises excitability in the auditory cortex. J. Neurosci. 25, 3908–3918.

Kotak, V.C., Takesian, A.E., Sanes, D.H., 2008. Hearing loss prevents the maturation of GABAergic transmission in the auditory cortex. Cereb. Cortex 18, 2098–2108.

Krahe, T.E., Medina, A.E., 2010. Activation of NMDA receptors is necessary for the recovery of cortical binocularity. J. Neurophysiol. 103, 2700–2706.
Kralic, J.E., Korpi, E.R., O'Buckley, T.K., Homanics, G.E., Morrow, A.L., 2002. Molecular and pharmacological characterization of GABAA receptor alpha1 subunit knockout mice. J. Pharmacol. Exp. Ther. 302, 1037–1045.
Kubota, Y., Hatada, S., Kondo, S., Karube, F., Kawaguchi, Y., 2007. Neocortical inhibitory terminals innervate dendritic spines targeted by thalamocortical afferents. J. Neurosci. 27, 1139–1150.
Lander, C., Kind, P., Maleski, M., Hockfield, S., 1997. A family of activity-dependent neuronal cell-surface chondroitin sulfate proteoglycans in cat visual cortex. J. Neurosci. 17, 1928–1939.
Lau, C.G., Zukin, R.S., 2007. NMDA receptor trafficking in synaptic plasticity and neuropsychiatric disorders. Nat. Rev. Neurosci. 8, 413–426.
Lazarus, M.S., Huang, Z.J., 2011. Distinct maturation profiles of perisomatic and dendritic targeting GABAergic interneurons in the mouse primary visual cortex during the critical period of ocular dominance plasticity. J. Neurophysiol. 106, 775–787.
Lee, H.K., Kirkwood, A., 2011. AMPA receptor regulation during synaptic plasticity in hippocampus and neocortex. Semin. Cell Dev. Biol. 22, 514–520.
LeVay, S., Wiesel, T.N., Hubel, D.H., 1980. The development of ocular dominance columns in normal and visually deprived monkeys. J. Comp. Neurol. 191, 1–51.
Levy, R.B., Aoki, C., 2002. Alpha7 nicotinic acetylcholine receptors occur at postsynaptic densities of AMPA receptor-positive and -negative excitatory synapses in rat sensory cortex. J. Neurosci. 22, 5001–5015.
Li, J.H., Wang, Y.H., Wolfe, B.B., Krueger, K.E., Corsi, L., Stocca, G., Vicini, S., 1998. Developmental changes in localization of NMDA receptor subunits in primary cultures of cortical neurons. Eur. J. Neurosci. 10, 1704–1715.
Li, Y., Fitzpatrick, D., White, L.E., 2006. The development of direction selectivity in ferret visual cortex requires early visual experience. Nat. Neurosci. 9, 676–681.
Lichtman, J.W., 1995. Synapse disassembly at the neuromuscular junction. Semin. Dev. Biol.
Lidow, M.S., Rakic, P., 1992. Scheduling of monoaminergic neurotransmitter receptor expression in the primate neocortex during postnatal development. Cereb. Cortex 2, 401–416.
Liu, X.B., Murray, K.D., Jones, E.G., 2004. Switching of NMDA receptor 2A and 2B subunits at thalamic and cortical synapses during early postnatal development. J. Neurosci. 24, 8885–8895.
Liu, Y., Shi, X., Li, Y., Zhao, K., 2012. The influences of dark rearing on the transmission characteristics of layer II/III pyramidal cells during the critical period. Brain Res.
Lubin, M., Erisir, A., Aoki, C., 1999. Ultrastructural immunolocalization of the alpha 7 nAChR subunit in guinea pig medial prefrontal cortex. Ann. N.Y. Acad. Sci. 868, 628–632.
Luhmann, H.J., Singer, W., Martinez-Millan, L., 1990. Horizontal interactions in cat striate cortex: I. Anatomical substrate and postnatal development. Eur. J. Neurosci. 2, 344–357.
Luo, L., O'Leary, D.D., 2005. Axon retraction and degeneration in development and disease. Annu. Rev. Neurosci. 28, 127–156.
Luscher, B., Fuchs, T., Kilpatrick, C.L., 2011. GABAA receptor trafficking-mediated plasticity of inhibitory synapses. Neuron 70, 385–409.
Ma, Y., Hu, H., Agmon, A., 2012. Short-term plasticity of unitary inhibitory-to-inhibitory synapses depends on the presynaptic interneuron subtype. J. Neurosci. 32, 983–988.
Malinow, R., 1998. Silent synapses in three forms of central plasticity. In: Faber, D.S., Kom, H., Redman, S.J., Thompson, S.M., Altman, J.S. (Eds.), Central Synapses: Quantal Mechanisms and Plasticity. Human Frontier Science Program, Strasbourg, France, pp. 226–334.
Malinow, R., Malenka, R.C., 2002. AMPA receptor trafficking and synaptic plasticity. Annu. Rev. Neurosci. 25, 103–126.
Marin, O., Valiente, M., Ge, X., Tsai, L.H., 2010. Guiding neuronal cell migrations. Cold Spring Harb. Perspect. Biol. 2, a001834.
Mason, C.A., 1982a. Development of terminal arbors of retino-geniculate axons in the kitten–I. Light microscopical observations. Neuroscience 7, 541–559.
Mason, C.A., 1982b. Development of terminal arbors of retino-geniculate axons in the kitten–II. Electron microscopical observations. Neuroscience 7, 561–582.

McGehee, D.S., 1999. Molecular diversity of neuronal nicotinic acetylcholine receptors. Ann. N.Y. Acad. Sci. 868, 565–577.
McLaughlin, T., Torborg, C.L., Feller, M.B., O'Leary, D.D., 2003. Retinotopic map refinement requires spontaneous retinal waves during a brief critical period of development. Neuron 40, 1147–1160.
McRae, P.A., Rocco, M.M., Kelly, G., Brumberg, J.C., Matthews, R.T., 2007. Sensory deprivation alters aggrecan and perineuronal net expression in the mouse barrel cortex. J. Neurosci. 27, 5405–5413.
Micheva, K.D., Beaulieu, C., 1995. An anatomical substrate for experience-dependent plasticity of the rat barrel field cortex. Proc. Natl. Acad. Sci. U.S.A. 92, 11834–11838.
Miller, K.D., Chapman, B., Stryker, M.P., 1989. Visual responses in adult cat visual cortex depend on *N*-methyl-D-aspartate receptors. Proc. Natl. Acad. Sci. U.S.A. 86, 5183–5187.
Miller, M.W., 1986. Maturation of rat visual cortex. III. Postnatal morphogenesis and synaptogenesis of local circuit neurons. Brain Res. 390, 271–285.
Monyer, H., Burnashev, N., Laurie, D.J., Sakmann, B., Seeburg, P.H., 1994. Developmental and regional expression in the rat brain and functional properties of four NMDA receptors. Neuron 12, 529–540.
Morales, B., Choi, S.Y., Kirkwood, A., 2002. Dark rearing alters the development of GABAergic transmission in visual cortex. J. Neurosci. 22, 8084–8090.
Movshon, J.A., 1981. Cortical effects of monocular deprivation: suppression or deafferentation? Nature 291, 284–285.
Movshon, J.A., Van Sluyters, R.C., 1981. Visual neural development. Annu. Rev. Psychol. 32, 477–522.
Mower, G.D., Berry, D., Burchfiel, J.L., Duffy, F.H., 1981. Comparison of the effects of dark rearing and binocular suture on development and plasticity of cat visual cortex. Brain Res. 220, 255–267.
Mower, G.D., Christen, W.G., Burchfiel, J.L., Duffy, F.H., 1984. Microiontophoretic bicuculline restores binocular responses to visual cortical neurons in strabismic cats. Brain Res. 309, 168–172.
Naegele, J.R., Katz, L.C., 1990. Cell surface molecules containing *N*-acetylgalactosamine are associated with basket cells and neurogliaform cells in cat visual cortex. J. Neurosci. 10, 540–557.
Nahmani, M., Erisir, A., 2005. VGluT2 immunochemistry identifies thalamocortical terminals in layer 4 of adult and developing visual cortex. J. Comp. Neurol. 484, 458–473.
Needler, M.C., Wilkinson, M., Prusky, G., Shaw, C., Cynader, M., 1988. Development of phorbol ester (protein kinase C) binding sites in cat visual cortex. Brain Res. 470, 217–227.
O'Kusky, J., Colonnier, M., 1982a. Postnatal changes in the number of astrocytes, oligodendrocytes, and microglia in the visual cortex (area 17) of the macaque monkey: a stereological analysis in normal and monocularly deprived animals. J. Comp. Neurol. 210, 307–315.
O'Kusky, J., Colonnier, M., 1982b. Postnatal changes in the number of neurons and synapses in the visual cortex (area 17) of the macaque monkey: a stereological analysis in normal and monocularly deprived animals. J. Comp. Neurol. 210, 291–306.
O'Kusky, J.R., 1985. Synapse elimination in the developing visual cortex: a morphometric analysis in normal and dark-reared cats. Brain Res. 354, 81–91.
Panzanelli, P., Gunn, B.G., Schlatter, C., Benke, D., Tyagarajan, S.K., Shcheiffele, P., Belelli, D., Lambert, J.J., Rudolph, U., Fritschy, J.-M., 2011. Distinct mechanisms regulate GABAA receptor and gephyrin clustering at perisomatic and axo-axonic synapses on CA1 pyramidal cells. J. Physiol. 589, 4959–4980.
Perez-Otano, I., Ehlers, M.D., 2005. Homeostatic plasticity and NMDA receptor trafficking. Trends Neurosci. 28, 229–238.
Pettigrew, J.D., Kasamatsu, T., 1978. Local perfusion of noradrenaline maintains visual cortical plasticity. Nature 271, 761–763.
Pinto, J.G., Hornby, K.R., Jones, D.G., Murphy, K.M., 2010. Developmental changes in GABAergic mechanisms in human visual cortex across the lifespan. Front. Cell. Neurosci. 4, 16.
Pizzorusso, T., Medini, P., Berardi, N., Chierzi, S., Fawcett, J.W., Maffei, L., 2002. Reactivation of ocular dominance plasticity in the adult visual cortex. Science 298, 1248–1251.
Portera-Cailliau, C., Price, D.L., Martin, L.J., 1996. *N*-methyl-D-aspartate receptor proteins NR2A and NR2B are differentially distributed in the developing rat central nervous system as revealed by subunit-specific antibodies. J. Neurochem. 66, 692–700.
Prusky, G.T., Shaw, C., Cynader, M.S., 1987. Nicotine receptors are located on lateral geniculate nucleus terminals in cat visual cortex. Brain Res. 412, 131–138.

Purpura, D.P., Shofer, R.J., Scarff, T., 1965. Properties of synaptic activities and spike potentials of neurons in immature neocortex. J. Neurophysiol. 28, 925–942.
Quinlan, E.M., Olstein, D.H., Bear, M.F., 1999a. Bidirectional, experience-dependent regulation of *N*-methyl-D-aspartate receptor subunit composition in the rat visual cortex during postnatal development. Proc. Natl. Acad. Sci. U.S.A. 96, 12876–12880.
Quinlan, E.M., Philpot, B.D., Huganir, R.L., Bear, M.F., 1999b. Rapid, experience-dependent expression of synaptic NMDA receptors in visual cortex in vivo. Nat. Neurosci. 2, 352–357.
Ramoa, A.S., Mower, A.F., Liao, D., Jafri, S.I., 2001. Suppression of cortical NMDA receptor function prevents development of orientation selectivity in the primary visual cortex. J. Neurosci. 21, 4299–4309.
Reiter, H.O., Stryker, M.P., 1988. Neural plasticity without postsynaptic action potentials: less-active inputs become dominant when kitten visual cortical cells are pharmacologically inhibited. Proc. Natl. Acad. Sci. U.S.A. 85, 3623–3627.
Roberts, E.B., Meredith, M.A., Ramoa, A.S., 1998. Suppression of NMDA receptor function using antisense DNA block ocular dominance plasticity while preserving visual responses. J. Neurophysiol. 80, 1021–1032.
Roberts, E.B., Ramoa, A.S., 1999. Enhanced NR2A subunit expression and decreased NMDA receptor decay time at the onset of ocular dominance plasticity in the ferret. J. Neurophysiol. 81, 2587–2591.
Ryan, T.J., Grant, S.G., 2009. The origin and evolution of synapses. Nat. Rev. Neurosci. 10, 701–712.
Rymar, V.V., Sadikot, A.F., 2007. Laminar fate of cortical GABAergic interneurons is dependent on both birth date and phenotype. J. Comp. Neurol. 501, 369–380.
Sanders, J.D., Happe, H.K., Bylund, D.B., Murrin, L.C., 2005. Development of the norepinephrine transporter in the rat CNS. Neuroscience 130, 107–117.
Sanna, E., Busonero, F., Talani, G., Carta, M., Massa, F., Peis, M., Maciocco, E., Biggio, G., 2002. Comparison of the effects of zaleplon, zolpidem, and triazolam at various GABAA receptor subtypes. Eur. J. Pharmacol. 451, 103–110.
Sarro, E.C., Kotak, V.C., Sanes, D.H., Aoki, C., 2008. Hearing loss alters the subcellular distribution of presynaptic GAD and postsynaptic GABAA receptors in the auditory cortex. Cereb. Cortex 18, 2855–2867.
Sato, H., Hata, Y., Hagihara, K., Tsumoto, T., 1987. Effects of cholinergic depletion on neuron activities in the cat visual cortex. J. Neurophysiol. 58, 781–794.
Shatz, C.J., 1983. The prenatal development of the cat's retinogeniculate pathway. J. Neurosci. 3, 482–499.
Shatz, C.J., Stryker, M.P., 1978. Ocular dominance in layer IV of the cat's visual cortex and the effects of monocular deprivation. J. Physiol. 281, 267–283.
Shatz, C.J., Stryker, M.P., 1988. Prenatal tetrodotoxin infusion blocks segregation of retinogeniculate afferents. Science 242, 87–89.
Shaw, C., Aoki, C., Wilkinson, M., Prusky, G., Cynader, M., 1987. Benzodiazepine ([^{3}H]flunitrazepam) binding in cat visual cortex: ontogenesis of normal characteristics and the effects of dark rearing. Brain Res. 465, 67–76.
Shaw, C., Needler, M.C., Wilkinson, M., Aoki, C., Cynader, M., 1985. Modification of neurotransmitter receptor sensitivity in cat visual cortex during the critical period. Brain Res. 354, 67–73.
Shen, H., Gong, Q.H., Aoki, C., Yuan, M., Ruderman, Y., Dattilo, M., Williams, K., Smith, S.S., 2007. Reversal of neurosteroid effects at alpha4beta2delta GABAA receptors triggers anxiety at puberty. Nat. Neurosci. 10, 469–477.
Shen, H., Sabaliauskas, N., Sherpa, A., Fenton, A.A., Stelzer, A., Aoki, C., Smith, S.S., 2010. A critical role for alpha4betadelta GABAA receptors in shaping learning deficits at puberty in mice. Science 327, 1515–1518.
Sheng, M., Cummings, J., Roldan, L.A., Jan, Y.N., Jan, L.Y., 1994. Changing subunit composition of heteromeric NMDA receptors during development of rat cortex. Nature 368, 144–147.
Sherman, S.M., Spear, P.D., 1982. Organization of visual pathways in normal and visually deprived cats. Physiol. Rev. 62, 738–855.
Sillito, A.M., Kemp, J.A., Patel, H., 1980. Inhibitory interactions contributing to the ocular dominance of monocularly dominated cells in the normal cat striate cortex. Exp. Brain Res. 41, 1–10.
Smith, S.S., Aoki, C., Shen, H., 2009. Puberty, steroids and GABAA receptor plasticity. Psychoneuroendocrinology 34 (Suppl. 1), S91–S103.
Sobczyk, A., Scheuss, V., Svoboda, K., 2005. NMDA receptor subunit-dependent [Ca^{2+}] signaling in individual hippocampal dendritic spines. J. Neurosci. 25, 6037–6046.

Spreafico, R., De Biasi, S., Vitellaro-Zuccarello, L., 1999. The perineuronal net: a weapon for a challenge. J. Hist. Neurosci. 8, 179–185.
Sretavan, D.W., Shatz, C.J., Stryker, M.P., 1988. Modification of retinal ganglion cell axon morphology by prenatal infusion of tetrodotoxin. Nature 336, 468–471.
Stone, E.A., Ariano, M.A., 1989. Are glial cells targets of the central noradrenergic system? A review of the evidence. Brain Res. Brain Res. Rev. 14, 297–309.
Sur, M., Frost, D.O., Hockfield, S., 1988. Expression of a surface-associated antigen on Y-cells in the cat lateral geniculate nucleus is regulated by visual experience. J. Neurosci. 8, 874–882.
Swindale, N.V., Vital-Durand, F., Blakemore, C., 1981. Recovery from monocular deprivation in the monkey. III. Reversal of anatomical effects in the visual cortex. Proc. R. Soc. Lond. B. Biol. Sci. 213, 435–450.
Taha, S.A., Stryker, M.P., 2005. Molecular substrates of plasticity in the developing visual cortex. Prog. Brain Res. 147, 103–114.
Takesian, A.E., Kotak, V.C., Sanes, D.H., 2012. Age-dependent effect of hearing loss on cortical inhibitory synapse function. J. Neurophysiol. 107, 937–947.
Tan, J., Widjaja, S., Xu, J., Shepherd, R.K., 2008. Cochlear implants stimulate activity-dependent CREB pathway in the deaf auditory cortex: implications for molecular plasticity induced by neural prosthetic devices. Cereb. Cortex 18, 1799–1813.
Tanaka, S., Tani, T., Ribot, J., O'Hashi, K., Imamura, K., 2009. A postnatal critical period for orientation plasticity in the cat visual cortex. PloS One 4, e5380.
Tretter, V., Jacob, T.C., Mukherjee, J., Fritschy, J.M., Pangalos, M.N., Moss, S.J., 2008. The clustering of GABAA receptor subtypes at inhibitory synapses is facilitated via the direct binding of receptor alpha 2 subunits to gephyrin. J. Neurosci. 28, 1356–1365.
Tsumoto, T., Hagihara, K., Sato, H., Hata, Y., 1987. NMDA receptors in the visual cortex of young kittens are more effective than those of adult cats. Nature 327, 513–514.
Turchi, J., Sarter, M., 1997. Cortical acetylcholine and processing capacity: effects of cortical cholinergic deafferentation on crossmodal divided attention in rats. Brain Res. Cogn. Brain Res. 6, 147–158.
Turrigiano, G.G., Nelson, S.B., 2004. Homeostatic plasticity in the developing nervous system. Nature reviews. Neuroscience 5, 97–107.
Valverde, F., 1971. Rate and extent of recovery from dark rearing in the visual cortex of the mouse. Brain Res. 33, 1–11.
Vos, P., Kaufmann, D., Hand, P.J., Wolfe, B.B., 1990. Beta2-adrenergic receptors are colocalized and coregulated with "whisker barrels" in rat somatosensory cortex. Proc. Natl. Acad. Sci. U.S.A. 87, 5114–5118.
Wallace, W., Bear, M.F., 2004. A morphological correlate of synaptic scaling in visual cortex. J. Neurosci. 24, 6928–6938.
Wenthold, R.J., Prybylowski, K., Standley, S., Sans, N., Petralia, R.S., 2003. Trafficking of NMDA receptors. Annu. Rev. Pharmacol. Toxicol. 43, 335–358.
White, E.L., Amitai, Y., Gutnick, M.J., 1994. A comparison of synapses onto the somata of intrinsically bursting and regular spiking neurons in layer V of rat SmI cortex. J. Comp. Neurol. 342, 1–14.
White, L.E., Coppola, D.M., Fitzpatrick, D., 2001. The contribution of sensory experience to the maturation of orientation selectivity in ferret visual cortex. Nature 411, 1049–1052.
White, L.E., Fitzpatrick, D., 2007. Vision and cortical map development. Neuron 56, 327–338.
Wiesel, T.N., Hubel, D.H., 1963. Effects of visual deprivation on morphology and physiology of cells in the cats lateral geniculate body. J. Neurophysiol. 26, 978–993.
Williams, K., Russell, S.L., Shen, Y.M., Molinoff, P.B., 1993. Developmental switch in the expression of NMDA receptors occurs in vivo and in vitro. Neuron 10, 267–278.
Winfield, D.A., 1981. The postnatal development of synapses in the visual cortex of the cat and the effects of eyelid closure. Brain Res. 206, 166–171.
Xu, H., Kotak, V.C., Sanes, D.H., 2010. Normal hearing is required for the emergence of long-lasting inhibitory potentiation in cortex. J. Neurosci. 30, 331–341.
Yazaki-Sugiyama, Y., Kang, S., Cateau, H., Fukai, T., Hensch, T.K., 2009. Bidirectional plasticity in fast-spiking GABA circuits by visual experience. Nature 462, 218–221.

CHAPTER FOURTEEN

Asynaptic and Synaptic Innervation by Acetylcholine Neurons of the Central Nervous System

Laurent Descarries[1] and Martin Parent[2]
[1]Departments of Pathology & Cell Biology and of Physiology, and Groupe de recherche sur le système nerveux central, Université de Montréal, Montreal, QC, Canada
[2]Department of Psychiatry and Neuroscience, Université Laval & Centre de recherche de l'Institut universitaire en santé mentale de Québec, Quebec City, QC, Canada

1. INTRODUCTION

1.1. Landmarks and Vocabulary

Sherrington (1897) coined the word synapse to designate sites of functional contact between distinct neuronal entities more than 50 years before synaptic junctions between central nervous system (CNS) neurons could be seen. It was only in the 1950s that electron microscopy revealed the presence of small areas of plasma membrane differentiation between neuronal elements at sites of axonal dilations where storage-like vesicles accumulated on the side of presumed transmitter release (Robertson, 1953; Palade and Palay, 1954). For the next 50 years, these junctional complexes came to be recognized as the morphological hallmark of synapses (Peters and Palay, 1996), and considered as the only site of chemical communication between CNS neurons. Based on the neuromuscular junction model, it was assumed that, in the CNS, exocytotic release of transmitter occurred alongside presynaptic thickenings and transmitter receptors were confined to postsynaptic densities.

When CNS monoamine neurons were first demonstrated to be only partly synaptic, in the sense that many of their axon varicosities or terminals[1] did not display a synaptic membrane specialization (Descarries et al., 1975, 1977), it was logical to envisage the possibility that chemical transmission from these release sites would be widespread (diffuse or volume transmission), through activation of receptors located "extrasynaptically", on a variety of cellular elements (Beaudet and Descarries, 1976). Moreover, because the existence of junctional complexes implied some fixity and permanence in the structural relationship between pre- and postsynaptic elements, the converse was suggestive of some flexibility and mobility of the releasing sites with respect to their immediate microenvironment (Descarries et al., 1975). Despite initial controversies, numerous subsequent studies confirmed the largely

[1] For lack of better terms, throughout this chapter, the words "varicosity" or "terminal" are used interchangeably to designate axon enlargements containing aggregated small vesicles and often mitochondria, whether or not they display a morphologically differentiated zone of synaptic membrane specialization (junctional complex).

The Synapse
http://dx.doi.org/10.1016/B978-0-12-418675-0.00014-6

nonjunctional (asynaptic) character of many of the monoamine innervations in CNS, but also revealed significant regional differences in their synaptic incidence, i.e. in the proportion of their axon varicosities that made morphologically differentiated synapses (for recent reviews, see Descarries and Mechawar, 2008; Descarries et al., 2010).

1.2. Scope and Purpose

It was nevertheless unexpected when, in the 1990s, the first immunoelectron microscopic investigation paying particular attention to the synaptic incidence of acetylcholine (ACh) axon terminals in brain, revealed the predominant asynaptic character of the cortical ACh innervation (Umbriaco et al., 1994). Since then, several studies have systematically examined the dual asynaptic and synaptic characters of ACh innervations in many parts of CNS and different species, as well as during postnatal development in the rat. In some of these studies, care was taken to examine the immunolabeled axon terminals in serial thin sections for electron microscopy, allowing for detailed scrutiny of most if not their entire volume and hence direct estimate of the frequency with which they make synapse. More frequently, however, the synaptic incidence of the examined varicosities was extrapolated from that observed in single thin sections for electron microscopy, using a mathematical formula devised by Beaudet and Sotelo (1981). This formula takes into account the average size of varicosity profiles, the length of their junctional complexes, and the thickness of ultrathin sections to predict the probability of seeing the junction in single sections if there is one on every varicosity. The synaptic incidence of the varicosities in question is then inferred by comparing the observed proportion in single sections to this predicted value. There is good evidence that such estimates are reliable when the single section sampling of the varicosities is sufficiently large. On several occasions, both a serial and a single section sampling of the same type of varicosities have indeed yielded very similar values of synaptic incidence (Soghomonian et al., 1989; Oleskevich et al., 1991; Descarries et al., 1992; Ridet et al., 1992; Descarries et al., 1996). Moreover, almost identical values were found by Umbriaco et al. (1994) in their large population of cortical ACh axon varicosities examined both in serial sections across their entire volume and as a randomized single section sample.

In this chapter, particular attention is paid to those studies in which the electron microscopic examination of an ACh innervation of CNS was not limited to that of morphologically differentiated synaptic profiles, and in which a sufficient number of axon terminals was visualized to draw reliable conclusions regarding synaptic incidence. Most such studies have been performed in the rat. Surprisingly, despite the increasing use of mice in neurobiology, none has yet been published using this species (but see Figure 1 for preliminary results from our ongoing experiments in the mice striatum). By way of discussion, the possible significance of the asynaptic–synaptic duality will be considered in terms of mechanisms of ACh release and mode of action of ACh in the different CNS regions examined.

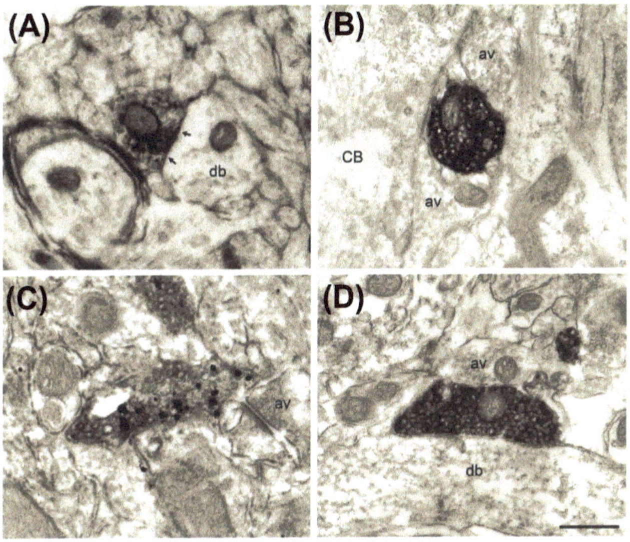

Figure 1 Examples of ACh axon varicosities (terminals) in different anatomical regions of the rat (A, B, and D) and mouse (C) brain. At the electron microscopic level, the ChAT-immunostained profiles are identified as axon varicosities by their content in aggregated synaptic vesicles, often associated with mitochondrion. (A) Cerebral cortex. This ACh varicosity from layer VI of the rat parietal cortex establishes a symmetrical synaptic contact (between arrows) with a dendritic branch (db). (B) Hippocampus. This presumed nonjunctional ACh varicosity is from the rat stratum pyramidale of CA1 and is nested between a cell body (CB) and two other axon varicosities (av). (C) Striatum. Electron micrograph of an axon varicosity immunolabeled with DAB (ChAT) and immunogold (VGLUT3) in mice. The majority of ACh axon varicosity profiles in the mouse dorsal striatum contain the VGLUT3 protein. (D) Thalamus. This ACh axon terminal observed in the parafascicular thalamic nucleus of the rat brain is directly apposed to an unlabeled dendritic branch (db) and contains many closely packed synaptic vesicles, outlined by DAB precipitate. Scale bar = 0.5 μm.

2.1. ASYNAPTIC AND SYNAPTIC ACH INNERVATIONS IN CNS

2.1.1. General Organization of the Central ACh System

Our current knowledge of the general organization of the central ACh system is largely derived from neuroanatomical studies having identified these neurons after their immunolabeling for choline acetyltransferase (ChAT), the rate-limiting enzyme for ACh synthesis. All regions of the CNS were then shown to be pervaded by relatively dense networks of ChAT-immunostained axons originating either from projection neurons located in the midbrain and basal forebrain or from local interneurons (Armstrong et al., 1983; Woolf, 1991). The major groups of ACh projection neurons are now commonly referred to as Ch1–Ch8, on the basis of their anatomical localization (Mesulam et al., 1983; Mesulam, 1988). The respective contributions of these groups to various territories of ACh innervation have been mostly defined in investigations combining tract-tracing methods and ChAT immunocytochemistry (Rye et al., 1984; Saper, 1984). Interneurons provide either small fraction (cerebral cortex), almost all (striatum), or the totality (spinal cord) of these regional ACh innervations.

The basal forebrain groups Ch1–Ch4 are the source of the rich and widespread innervations of cerebral cortex, hippocampus, olfactory bulb and amygdalar complex. Ch1 and Ch2, respectively, located in the medial septum and nucleus of the vertical limb of the diagonal band of Broca, project heavily to the hippocampal formation, whereas Ch3, in the lateral portion of the horizontal limb of the diagonal band of Broca, innervates the olfactory bulb. Together, Ch1–Ch3 contribute a limited fraction of the cortical ACh innervation, with axons restricted to limbic areas (cingulate, entorhinal, orbitofrontal, and piriform cortex). In contrast, the whole cortical mantle (including these limbic areas) and the amygdalar complex are densely innervated by Ch4 in the nucleus basalis magnocellularis of Meynert, which overlaps the substantia innominata and globus pallidus. The pontomesencephalic ACh groups Ch5 and Ch6 are the major projection pathways outside the basal forebrain, with most of their efferents reaching the thalamus and the basal forebrain. ACh neurons in the habenula (Ch7) project exclusively to the interpeduncular nucleus via the fasciculus retroflexus. Finally, the parabigeminal nucleus group Ch8 sends a majority of its axons to the superior colliculus (tectum).

2.2. CEREBRAL CORTEX

It has been estimated that, in the rat, there are 7000–9000 Ch4 neurons projecting to the cerebral cortex (Rye et al., 1984; Gritti et al., 1993). Although the great majority of ACh axon terminals in cerebral cortex originate from basal forebrain, a 20–30% proportion would be supplied by intrinsic bipolar interneurons scattered throughout layers II–VI (Johnston et al., 1981; Eckenstein and Thoenen, 1983; Eckenstein and Baughman, 1984; Levey et al., 1984).

In the frontal (Fr1), parietal (Par1), and occipital (Oc1) neocortical areas, the regional and laminar densities of ACh innervation have been estimated in number of ChAT-immunoreactive varicosities per cubic millimeter of tissue (Mechawar et al., 2000). The frontal cortex was thus found to be the most densely innervated, followed by the occipital and the parietal cortex, with respective values of 5.4, 4.6, and 3.8×10^6 varicosities/mm^3. As expected (Lysakowski et al., 1989), the laminar distribution was area-specific, and characterized by uniformly high densities throughout frontal cortex, and lower densities in layers II/III, IV and VI of the parietal cortex, as well as in layers IV and VI of the occipital cortex.

Soon after specific antibodies against ChAT were developed for immunocytochemistry, a number of electron microscopic studies described ultrastructural features of the ACh innervation of cerebral cortex. Initial reports in rat (Wainer et al., 1984; Houser et al., 1985) and cat (De Lima and Singer, 1986) demonstrated that the ChAT immunoreactive fibers observed by light microscopy indeed corresponded to unmyelinated axon segments with dilations containing aggregated small vesicles and which made morphologically defined synaptic specializations. In the cingulate, frontoparietal and visual cortex, the ChAT-immunolabeled varicosities were described as relatively small, endowed with junctional complexes that were usually symmetrical, and in synaptic contact with dendritic branches more often than spines or cell bodies (see also Clarke and Dunnett, 1986; Parnavelas et al., 1986; Beaulieu and Somoygyi, 1991; Aoki and Kabak, 1992; Garofalo et al., 1992). Although these studies obviously focused on terminals showing a synaptic junction and thus a functional target, some paid attention to the frequency with which a synaptic contact was actually made by these varicosities, as visualized in single (Garofalo et al., 1992) and/or in serial thin sections (Houser et al., 1985; De Lima and Singer, 1986; Parnavelas et al., 1986; Aoki and Kabak, 1992). The rarity of visible junctions was noted (De Lima and Singer, 1986; Aoki and Kabak, 1992), but this was attributed to not having visualized the terminals in their entirety (Houser et al., 1985; Parnavelas et al., 1986; for review, see Houser, 1990).

Taking advantage of the generous gift of a highly sensitive monoclonal antibody against whole rat ChAT (Cozzari et al., 1990), Umbriaco et al. (1994) undertook a detailed characterization of the ultrastructural morphology of the cortical ACh innervation by serial section electron microscopic examination of a large number of these axonal varicosities in every layer of adult rat parietal (Par1) cortex. In all layers of Par1, only a small fraction of ChAT immunoreactive varicosities was found to make a morphologically differentiated synaptic contact (junctional complex). The proportion of ACh varicosities making synapse (synaptic incidence) in the different cortical layers was 10%, 14%, 11%, 21%, and 14% in layers I, II/III, IV, V, and VI, respectively, for an interlayer mean of 14% and an overall average of 16%. Cortical ACh varicosities averaged 0.60 μm in diameter, and those bearing a synaptic junction were slightly but significantly larger than their nonsynaptic counterparts, i.e. 0.67 vs 0.57 μm in diameter, respectively. The junctional complexes formed by these terminals were single, occupied a small fraction of the total surface of varicosities (3%), and were almost always symmetrical (99%). The relatively few synapses made by ACh varicosities were always axodendritic, either on branches (75%) or on spines (25%). An example of such rare synaptic axon varicosity is provided in Figure 1(A). Investigations in other laboratories have confirmed that the vast majority of ACh varicosities in rat cortex are asynaptic, with reported estimates of 14% and 9% for the frontoparietal and entorhinal region, respectively (Chédotal et al., 1994; Vaucher and Hamel, 1995). The discrepant value of 66% reported by Turrini et al. (2001) for layer V of rat parietal cortex after labeling with the vesicular ACh transporter was presumably the result of a sampling bias, as suggested by the significantly larger size of the profiles examined in that particular study.

A thorough electron microscopic ChAT-immunocytochemical study of the cortical ACh innervation has also been performed in the prefrontal cortex of adult rhesus monkey (Mrzljak et al., 1995). These authors found that among 100 serially sectioned

ChAT-immunoreactive varicosities at the border of layers II and III, only 44% made synaptic contact, dendritic shafts rather than spines being again the preferred synaptic target (70%). Fifty-six percent of cortical ACh varicosities were without any visible junctional specialization, even if frequently juxtaposed to dendrites or spines receiving asymmetrical synapses. In a similar study on two samples of human anterior temporal lobe removed at surgery (Smiley et al., 1997), it was reported that 67% of 42 ACh varicosities from layers I and II of the cortex were endowed with small but identifiable synaptic specializations. Whether such variations of synaptic incidence in cortex reflect sampling biases and/or regional or species differences remains to be determined.

In rat, the ACh innervation of cerebral cortex has also been examined during postnatal development (P8, P16, and P32) (Mechawar and Descarries, 2001). In the parietal cortex (Par 1), for example, the ACh fibers grow rapidly in number from very few in the cortical subplate at P1 to a distribution and density comparable to adult in all cortical layers at P16. The number of ACh (ChAT-immunostained) varicosities increases steadily in all layers, from 2/10 μm of axon at P4 to 4/10 μm of axon at P16, the adult ratio (Mechawar and Descarries, 2001). Throughout development, these varicosities appear to be slightly smaller than in the adult and the proportion containing mitochondria increases gradually with age, from 21% at P8 to 61% in the adult (>P60) (Mechawar et al., 2002). As in the adult, only a small fraction displays a synaptic junction (13% at P8, 16% at P16, and 22% at P32), and the vast majority of synaptic ACh varicosities display small symmetrical junctional complexes exclusively made with dendritic branches (71%) and spines (29%). Since the proportion of synaptic ACh varicosities appears stable during development, it may be inferred that the number of cortical ACh synapses, amounting to 0.5 million/mm^3 of tissue and 1.0 million under 1 mm^2 of cortical surface, already reaches its adult value by the end of the second postnatal week, at least in the parietal cortex.

ACh has been proposed to act as an enhancer of functional plasticity during the postnatal development of cerebral cortex (Bear and Singer, 1986; Gu and Singer, 1993; Zhu and Waite, 1998), and to contribute to the subsequent cytodifferentiation and refinement of cortical connectivity (Höhmann et al., 1988, 1991a,b; Roerig et al., 1997; Robertson et al., 1998) and the shaping of cortical maps (Höhmann et al., 1995; Broide et al., 1996; Aramakis et al., 2000). The above data are not only compatible with such an early role for ACh in the developing cortex, but also suggest that it may depend on both diffuse and synaptic ACh transmission.

2.3. HIPPOCAMPUS

The vast majority of ACh axon terminals in hippocampus originate from the estimated 7150 ACh Ch1–Ch2 neurons in each septum (Cadete-Leite et al., 2003). Few bipolar ACh interneurons are present in rat hippocampus, most of which reside in the stratum lacunosum moleculare of CA1 (Aznavour et al., 2002). As measured in number of axon varicosities per cubic millimeter of tissue, the regional density of ACh innervation is slightly higher than in cerebral cortex, ranging from 4.9×10^6 varicosities/mm^3 in CA1 to 6.2×10^6 varicosities/mm^3 in CA3, for an average of 5.9×10^6 varicosities/mm^3 for whole hippocampus (a value 28% greater than for neocortex). In terms of laminar distribution, the densest ACh innervations are those of the stratum lacunosum moleculare of CA3, stratum pyramidale of CA1 and CA3, and stratum moleculare of the dentate gyrus, with respective values of 7.5, 8.2, 6.8, and 7.8×10^6 varicosities/mm^3 (Aznavour et al., 2002).

As in the cerebral cortex, all initial immunoelectron microscopic studies of the ACh innervation of hippocampus focused on the synaptic relationships of the observed terminals (Wainer et al., 1984; Clarke, 1985; Frotscher and Léránth, 1985, 1986; Léránth and Frotscher, 1987; Frotscher et al., 1992). The subsequent, unbiased ultrastructural analysis of this innervation by Umbriaco et al. (1995) revealed similar features as in the parietal cortex in the stratum radiatum of CA1. These hippocampal ACh varicosities averaged 0.6 μm in diameter, and only 7% were synaptic. Their rare synaptic contacts were symmetrical and made either with dendritic branches or spines. Interestingly, an even lower synaptic incidence (<3%) has been reported in a recent study in which serial electron microscopy was used to investigate the presence or absence of "any kind of synapse-like specialization" made by axon varicosities immunolabeled for the high affinity choline transporter ($n = 35$) or for the vesicular transporter of ACh ($n = 30$) in the stratum radiatum of CA1 (Yamasaki et al., 2010). In each of these two populations of ACh varicosities, only a single one was then found to make a "synapse-like" contact. It therefore seems likely that in hippocampus as well as cerebral cortex, diffuse transmission is a major determinant of the functional role of ACh.

The ACh innervation of hippocampus was also examined during the postnatal period in the rat (Aznavour et al., 2005, Figure 1(B)). At P8, varicose ACh axons already form an elaborate network throughout hippocampus. At P16, the laminar distribution of this network resembles that in the adult, but adult densities are only reached by P32. In the stratum pyramidale and the stratum radiatum of CA1, the ultrastructural features of these ACh varicosities are similar in both layers and comparable to adult at all postnatal ages, except for the increasing number exhibiting a mitochondrion (from 28% at P8 to 41% at P32). The proportion of these varicosity profiles displaying a synaptic junction is equally low at the three ages, for an average synaptic incidence of 7%, identical to that previously found in the adult. As in the cerebral cortex, these morphological data appear compatible with a permissive role of ACh, and especially diffuse ACh transmission, during hippocampal development.

2.4. NEOSTRIATUM

The striatum receives by far the strongest ACh innervation in mammalian CNS, as indicated by the high amount of cholinergic biomarkers measured throughout this region in the rat (Sorimachi and Kataoka, 1974; Cheney et al., 1975; Hoover et al., 1978;

Rea and Simon, 1981; Pascual et al., 1991). After ChAT immunolabeling, the density of this innervation appears equivalent to that of its dopamine counterpart, which has been estimated at $1.0–1.7 \times 10^8$ axon varicosities/mm^3 of tissue (Doucet et al., 1986). This profuse axonal arborization originates almost entirely from cholinergic aspiny interneurons, which account for <2% of all striatal neurons (Woolf and Butcher, 1981; Phelps et al., 1985). A minor fraction of striatal ACh axons has also been demonstrated to arise from the Ch5–Ch6 cell groups (Woolf and Butcher, 1981, 1986).

Several light and electron microscopic immunocytochemical studies carried out in rat and monkey have provided descriptions of striatal ACh axon terminals in addition to cell bodies and dendrites (Bolam et al., 1984; Wainer et al., 1984; Phelps et al., 1985; DiFiglia, 1987; Izzo and Bolam, 1988; Houser, 1990; Pickel and Chan, 1999). In all these studies, however, the emphasis had again been put on sectional profiles of ChAT-immunolabeled varicosities that exhibited a synaptic membrane specialization. Contant et al. (1996) were therefore the first to determine the actual frequency with which the ACh striatal terminals engage in synaptic relationships. They reported these ACh terminals to be generally smaller (0.43 μm in average diameter) than those in the neocortex and hippocampus, and, as previously discovered in both these regions, found them to be seldom engaged in synaptic contact. The frequency of junction was 3% in single sections, amounting to 9% when extrapolated to the whole volume of varicosities. These rare junctions were made either with synaptic branches or spines. It was noted that other axon terminals, unlabeled, were often directly apposed to these neostriatal ACh varicosities, suggesting axon terminal interactions. Occasional juxtaposition of ChAT-immunolabeled varicosities to one another was also seen.

The neostriatal ACh innervation was also characterized at both the light and electron microscopic levels during postnatal development in the rat. ACh axons pervade the entire striatum within the first 2 weeks after birth, but continue to grow at a slower rate until the fourth postnatal week (Aznavour et al., 2003). At the three postnatal ages examined (P8, P16, and P32), a significant proportion of neostriatal ACh varicosities (30% at P8, 16% at P16, and 22% at P32) appears to be endowed with a synaptic junction. These junctions are always single, more often symmetrical than asymmetrical, and mostly made with dendritic branches or spines. A few are made with cell bodies. At P8 and P16, the ACh synapses are more frequent with dendritic branches than spines, but this is no longer the case at P32. Those made with dendritic branches are always symmetrical, but those made with spines equally divided between symmetrical and asymmetrical.

Interestingly, in adult mouse as well as rat, the ACh interneurons of dorsal and ventral striatum have been shown to express vesicular glutamate transporter 3 (VGLUT3) mRNA (Gras et al., 2002; Schäfer et al., 2002), and the VGLUT3 protein has been visualized by immunofluorescence in the soma and proximal dendrites of these cells (Fremeau et al., 2002; Boulland et al., 2004) as well as a majority if not all their axon terminals (Gras et al., 2002; Schäfer et al., 2002; Boulland et al., 2004). The presence of

VGLUT3 in the majority of ACh axon varicosity and dendritic profiles of the adult mice striatum has recently been confirmed by double immunostaining at the electron microscopic level (Parent and Descarries, unpublished observations provided in Figure 1(C)). Using an array of neurobiological techniques and a Vglut3 knockout mouse model, Gras et al. (2008) have shown that VGLUT3 and the vesicular transporter of ACh (VAChT) are located in a common pool of synaptic vesicles isolated from the striatum, and that coaccumulation of glutamate results in increased ACh filling of these vesicles. They have proposed the term vesicular synergy to designate such a cooperation between two vesicular transporters, which may result in increased physiological release of the primary transmitter.

There is considerable evidence for a role of striatal interneurons in the regulation of motor behavior and reward, and a recent optogenetic study has convincingly demonstrated glutamate release from these ACh neurons (Higley et al., 2011). Determining the factors which control the release of one and/or the other transmitter from these cells, and discerning their respective contribution to the regulation of striatal function (Guzman et al., 2011) remains a fascinating challenge.

2.5. THALAMUS

Groups Ch5 in the pedunculopontine tegmental nucleus and Ch6 in the laterodorsal tegmental nucleus are the principal sources of ACh innervation of the thalamus (Paré et al., 1988; Steriade et al., 1988; Wainer and Mesulam, 1990). The ascending axons of the Ch5 and Ch6 neurons are overwhelmingly relayed in the thalamus. In all species examined to date, Ch5 and Ch6 project to virtually all thalamic nuclei. In rat, cat, and monkey, some thalamic nuclei also receive projections from the Ch4 group in the nucleus basalis (Woolf and Butcher, 1986; Hallanger et al., 1987; Levey et al., 1987; Jourdain et al., 1989; Parent et al., 1988).

There have been numerous descriptions of the ultrastructural features of ACh axon terminals in rat, cat, or monkey thalamus, as visualized by ChAT or vesicular ACh transporter immunocytochemistry (de Lima et al., 1985; Hallanger and Wainer, 1988; Isaacson and Tanaka, 1988; Raczkowski and Fitzpatrick, 1989; Hallanger et al., 1990; Schwartz and Mrzkjak, 1993; Patel and Bickford, 1997; Patel et al., 1999; Hoshino et al., 2000; Oda et al., 2003). In the lateral geniculate as well as other sensory relay or associational nuclei of thalamus, synaptic ACh terminals have been identified in contact with projection cells as well as interneurons. None of these studies has however evaluated the synaptic vs asynaptic nature of the ACh innervation in any part of thalamus.

Using immunocytochemistry with the monoclonal antibody against whole rat choline acetyltransferase produced by Cozzari et al. (1990), Parent and Descarries (2008) have characterized the ultrastructural features of thalamic ACh axon terminals in the dorsal lateral geniculate (DLG), parafascicular (PF), and reticular thalamic (Rt) nuclei of adult rat. In single thin sections, ChAT-immunostained varicosity profiles were comparable in shape and content in the three nuclei, but significantly larger in PF than in DLG and Rt (Figure 1(D)). The number of these profiles which displayed a synaptic junction was also much higher in PF than in DLG and Rt, indicating that all ChAT-immunostained varicosities in PF were synaptic, but only 39% in DLG and 33% in Rt. The hypothesis that glutamate corelease might account for the establishment and maintenance of an entirely synaptic ACh innervation in PF was refuted by the lack of colocalization of ChAT and of the VGLUT2 in PF axon varicosities after dual immunolabeling. It was concluded from this study that diffuse as well as synaptic transmission was likely to convey the modulatory effects of the ACh input from brainstem to DLG and Rt during waking, whereas the entirely synaptic ACh input to PF should allow for a direct relaying of the information from brainstem, impacting on basal ganglia function as well as perceptual awareness, including attention and pain perception.

PF was the first example of an ACh innervation of CNS formally demonstrated to be entirely synaptic. The striking difference in synaptic incidence between ACh axon varicosities from different thalamic nuclei is intriguing. There is nothing in the literature to suggest that the ACh innervation of PF and DLG would have a distinct neuronal origin. On the contrary, there is strong neuroanatomical evidence that Ch5 neurons innervate both PF and DLG (Hallanger et al., 1987). Moreover, the axons of brainstem ACh neurons are known to be highly collateralized (Woolf and Butcher, 1986; Paré et al., 1988; Steriade et al., 1988; Bolton et al., 1993; Spreafico et al., 1993). Thus, it appears that a given ACh neuron may possess axonal branches endowed with varicosities that are all synaptic and other axonal branches where the varicosities are not all junctional. This may be viewed as a further indication of the importance of local factors in determining the establishment—and perhaps even the proportion—of synaptic junctions made by a given type of transmitter-defined neuron (e.g. Daszuta et al., 1991; Descarries et al., 1992).

2.6. OTHER BRAIN REGIONS

It is beyond the scope of this chapter to list all other brain regions in which ACh terminals have been observed by ChAT immunoelectron microscopy. Worthy of note, however, is an early ChAT immunoelectron microscopic study of rat basal forebrain by Armstrong (1986), who reported that, in the amygdala, few if any ACh terminals displayed a synaptic density in single thin sections, although many were in direct contact with unlabeled neuronal elements. The VGLUT3 has recently been shown to be present in ChAT axon terminals of the basolateral amygdala arising from basal forebrain cholinergic neurons (Nickerson Poulin et al., 2006). This would also seem to preclude an important role for glutamate corelease in the establishment and maintenance of junctional synaptic specializations by axon terminals.

2.7. SPINAL CORD

As described by Barber et al. (1984), different categories of spinal ACh neurons give rise to the ACh innervation of spinal cord. The preponderant type of spinal ACh neuron is the somatic motoneuron in the ventral horn, regarded as the canonical ACh neuron since Langley's seminal work on neuromuscular activation, in the twentieth century. The only other spinal ACh neurons with axons leaving the spinal cord are the preganglionic autonomic neurons in the intermediolateral cell column at thoracolumbar and lumbosacral levels. The three other types of spinal ACh neurons are the small laminae III–V neurons of the dorsal horn, lamina VII partition neurons at the border between the dorsal and ventral horns, and lamina X neurons in the central gray surrounding the central canal. In transverse sections of adult rat spinal cord immunolabeled for ChAT, somatic motor neurons are organized in central, medial and lateral motor columns, and the medial and lateral columns can be further divided into five subcolumns (Barber et al., 1984). There are apparently widespread intra- and intercolumnar interactions, as suggested by extent of intermingling longitudinal and transverse motor dendrite bundles. Likewise, at autonomic levels, the soma and dendritic arborizations of partition cells and central canal ACh neurons are heavily mixed with those of autonomic neurons. These extensive interconnections between morphologically diverse ACh neurons have been described as the basis of a cholinergic propriospinal system (Sherriff and Henderson, 1994; Huang et al., 2000).

All layers and regions of the spinal cord, including the ependymal cell layer are pervaded by a rich ACh axon network (Phelps et al., 1984; Schäfer et al., 1998). The proportion of these varicose axons originating the different types of ACh neurons mentioned above remains undetermined. Initial reports have shown that ChAT-immunolabeled terminals contact both ACh and non-ACh elements in the spinal cord. The current prevalent view is that these terminals are mostly, if not exclusively, synaptic in nature. This view is probably biased by the fact that most studies of spinal ACh terminals have focused specifically on contacts made with the cell bodies and proximal dendrites of motoneurons, Renshaw cells, and sympathetic preganglionic neurons (e.g. Markham and Vaughn, 1990; Alvarez et al., 1999). Renshaw cells are known to be contacted by collaterals from the ACh motor neurons involved in a feedback regulation of the motor circuitry. The ACh innervation of the soma-dendrites of motoneurons, thought to arise from the canal cluster cells, consists of C-type terminals characterized by periodic post-synaptic specializations called *subsurface cisterns* (Nagy et al., 1993).

Interestingly, ChAT immunoreactive motor neurons of the rat spinal cord express Vglut1 and Vglut2 mRNA (Herzog et al., 2004), but VGLUT1 and VGLUT2 protein seem to be absent from the motor endplates of skeletal muscles (Herzog et al., 2004; Kraus et al., 2004; Nishimaru et al., 2005). The absence of VGLUT protein in axon terminals established by motor neurons that do express Vglut1 or Vglut2 transcripts has led to the suggestion that spinal motor neurons might give rise to separate sets of cholinergic terminals, with those that release glutamate possibly involved in the regulation of motor function (Nishimaru et al., 2005) (but see Gezelius et al., 2006 for conflicting results). Perhaps this represents another example of the targeting of the vesicular glutamate transporter and a coexistent transmitter to different axonal branches of the same neuron.

3. CONCLUSIONS

3.1. Functional Considerations

In view of the small proportion of ACh axon terminals which form a synaptic membrane specialization in many parts of CNS, it has been speculated that the modulatory effects of ACh on brain function are largely conveyed by diffuse (or volume) transmission in addition to synaptic transmission (Umbriaco et al., 1994; Descarries et al., 1997; Steriade and Descarries, 2006). The diffuse transmission hypothesis is also supported by all currently available data on the subcellular distribution of ACh receptors (reviewed in Volpicelli and Levey, 2004; Steriade and Descarries, 2006; see also Yamasaki et al., 2010), which indicates that the various muscarinic receptor subtypes as well as nicotinic receptor subunits are located "extrasynaptically" in every brain region where they have thus far been visualized by immunoelectron microscopy.

It has also been postulated that, in brain regions densely innervated by ACh neurons, a low ambient level of ACh is permanently maintained in the extracellular space (Descarries et al., 1995), in spite of the presence of ACh esterase (Descarries et al., 1997). Consistent with this ambient level hypothesis, resting levels of ACh in the nanomolar range have been measured by microdialysis in many CNS regions (e.g. Vinson and Justice, 1997), and such concentrations are in the range of the measured affinities for many of the muscarinic or nicotinic receptor subtypes. It has also been proposed that CNS dysregulations and/or pathological disorders may result, directly or indirectly, from alterations in the ambient level of ACh rather than lack of its release from asynaptic and/or synaptic sites. This could explain the puzzling observation that central cholinergic deficits become symptomatic only after severe, as opposed to partial, cholinergic lesions, and also the therapeutic efficacy of pharmacological agents or grafts of ACh neurons that seem capable of reverting cholinergic deficits even in the absence of restored synaptic connectivity.

Much remains to be known about the specific roles of diffuse (volume) vs synaptic ACh transmission in CNS. The ubiquity and density of the cholinergic input to cerebral cortex, hippocampus, or striatum, as well as the current knowledge on the subcellular distribution of ACh receptors in these brain regions are consistent with the presumed involvement of ACh in numerous global processes, such as control of the microcirculation, modulation of sensory information, plasticity of sensory maps, learning, memory, attention, and even conscious awareness. The largely asynaptic character of some ACh innervations throughout their postnatal development, as documented in neocortex (Mechawar et al., 2002), hippocampus (Aznavour et al., 2005), and neostriatum (Aznavour et al., 2003) must also be underscored. It suggests a hypothetical role for diffuse transmission and the ambient level of ACh in regulating the expression and/or functional state of receptors mediating trophic effects of ACh itself and/or growth factors on neurons, glial cells, and microvessels.

3.2. Mode and Mechanisms of Action

In the absence of a junctional complex on a majority of ACh terminals in many parts of brain, it is also tempting to speculate on the mechanisms of release and site of action of the transmitter contained in such varicosities. Molecules which ensure the docking and exocytosis of synaptic vesicles are known to be enriched at presynaptic densities of junctional complexes (Schoch and Gundelfinger, 2006). In the absence of junctional complexes, these may be dispersed rather evenly along the plasma membrane of asynaptic ACh terminals, allowing for release of ACh at different microlocations rather than only at predefined, electron-dense junctional zones. Along the same line of thinking, two roles have been assigned to the scaffolding and adhesion molecules located at postsynaptic densities (reviewed in Zhai and Bellen, 2004). One is to ensure the aggregation and anchoring of receptors in this particular location, and the other to ensure some permanence in the physical relationship between pre- and postsynaptic elements. A contrario, the rarity of junctions in the case of many ACh varicosities could be in keeping with the widespread, extrasynaptic location of ACh receptors on cholinoceptive elements. Moreover, it could be indicative of a lack of durable contact between the asynaptic ACh varicosities and preselected partners in their immediate microenvironment. Indeed, another striking aspect of the ACh terminals is their variable shape, which could reflect some structural instability, both in terms of overall configuration and internal content. The few 3-D reconstructions of ACh axon varicosities that have thus far been obtained from serial thin sections for electron microscopy display bumps and bulges at their surface, reflecting the presence of large organelles, and notably of mitochondria (e.g. Umbriaco et al., 1994). Moreover, all inner constituents of axon varicosities are known to travel in axons, such as the transport packets for so-called "presynaptic" proteins and active zone proteins like Bassoon and Piccolo (Ahmari et al., 2000; Shapira et al., 2003).

In this view, the dual asynaptic and synaptic characters of the ACh system in many brain regions may be viewed as a further indication of its capacity to exert modulatory and long-lasting effects on vast neuronal assemblies in addition to its well characterized, short-term, excitatory effects on synaptic partners.

ACKNOWLEDGMENTS

Supported by the Natural Sciences and Engineering Research Council of Canada (to Laurent Descarries) and by the Canadian Institutes of Health Research (MOP-115008 to Martin Parent). The authors express their sincere gratitude to Sarah Petryszyn for having performed the ChAT/VGLUT3 double immunostaining in the mice striatum and the electron microscopic analysis.

REFERENCES

Ahmari, S.E., Buchanan, J., Smith, S.J., 2000. Assembly of presynaptic active zones from cytoplasmic transport packets. Nat. Neurosci. 3, 445–451.

Alvarez, F.J., Dewey, D.E., McMillin, P., Fyffe, R.E.W., 1999. Distribution of cholinergic contacts on Renshaw cells in the rat spinal cord: a light microscopic study. J. Physiol. 515, 787–797.

Aoki, C., Kabak, S., 1992. Cholinergic terminals in the cat visual cortex: ultrastructural basis for interaction with glutamate-immunoreactive neurons and other cells. Vis. Neurosci. 8, 177–191.

Aramakis, V.B., Hsieh, C.Y., Leslie, F.M., Metherate, R., 2000. A critical period for nicotine-induced disruption of synaptic development in rat auditory cortex. J. Neurosci. 20, 6106–6116.

Armstrong, D.M., 1986. Ultrastructural characterization of choline acetyltransferase-containing neurons in the basal forebrain of rat: evidence for a cholinergic innervation of intracerebral blood vessels. J. Comp. Neurol. 250, 81–92.

Armstrong, D.M., Saper, C.B., Levey, A.I., Wainer, B.H., Terry, R.D., 1983. Distribution of cholinergic neurons in rat brain: demonstrated by the immunocytochemical localization of choline acetyltransferase. J. Comp. Neurol. 216, 53–68.

Aznavour, N., Mechawar, N., Descarries, L., 2002. Comparative analysis of cholinergic innervation in the dorsal hippocampus of adult mouse and rat: a quantitative immunocytochemical study. Hippocampus 12, 206–217.

Aznavour, N., Mechawar, N., Watkins, K.C., Descarries, L., 2003. Fine structural features of the acetylcholine innervation in the developing neostriatum of rat. J. Comp. Neurol. 460, 280–291.

Aznavour, N., Watkins, K.C., Descarries, L., 2005. Postnatal development of the cholinergic innervation in the dorsal hippocampus of rat: quantitative light and electron microscopic immunocytochemical study. J. Comp. Neurol. 486, 61–75.

Barber, R.P., Phelps, P.E., Houser, C.R., Crawford, G.D., Salvaterra, P.M., Vaughn, J.E., 1984. The morphology and distribution of neurons containing choline acetyltransferase in the adult rat spinal cord: an immunocytochemical study. J. Comp. Neurol. 229, 329–346.

Bear, M.F., Singer, W., 1986. Modulation of visual cortical plasticity by acetylcholine and noradrenaline. Nature 320, 172–176.

Beaudet, A., Descarries, L., 1976. Quantitative data on serotonin nerve terminals in adult rat neocortex. Brain Res. 111, 301–309.

Beaudet, A., Sotelo, C., 1981. Synaptic remodeling of serotonin axon terminals in rat agranular cerebellum. Brain Res. 206, 305–329.

Beaulieu, C., Somogyi, P., 1991. Enrichment of cholinergic synaptic terminals on GABAergic neurons and coexistence of immunoreactive GABA and choline acetyltransferase in the same synaptic terminals in the striate cortex of the cat. J. Comp. Neurol. 304, 666–680.

Bolam, J.P., Wainer, B.H., Smith, A.D., 1984. Characterization of cholinergic neurons in the rat neostriatum. A combination of choline acetyltransferase immunocytochemistry, Golgi-impregnation and electron microscopy. Neuroscience 12, 711–718.

Bolton, R.F., Cornwall, J., Phillipson, O.T., 1993. Collateral axons of cholinergic pontine neurones projecting to midline, mediodorsal and parafascicular thalamic nuclei in the rat. J. Chem. Neuroanat. 6, 101–114.

Boulland, J.L., Qureschi, T., Seal, R.P., Rafiki, A., Gundersen, V., Bergersen, L.H., Fremeau Jr., R.T., Edwards, R.H., Storm-Mathisen, J., Chaudhry, F.A., 2004. Expression of the vesicular glutamate transporters during development indicates the widespread co-release of multiple neurotransmitters. J. Comp. Neurol. 480, 264–280.

Broide, R.S., Robertson, R.T., Leslie, F.M., 1996. Regulation of a7 nicotinic acetylcholine receptors in the developing rat somatosensory cortex by thalamocortical afferents. J. Neurosci. 16, 2956–2971.

Cadete-Leite, A., Pereira, P.A., Madeira, M.D., Paula-Barbosa, M.M., 2003. Nerve growth factor prevents cell death and induces hypertrophy of basal forebrain cholinergic neurons in rats withdrawn from prolonged ethanol intake. Neuroscience 119, 1055–1069.

Chédotal, A., Umbriaco, D., Descarries, L., Hartman, B.K., Hamel, E., 1994. Light and electron microscopic immunocytochemical analysis of the neurovascular relationships of choline acetyltransferase and vasoactive intestinal polypeptide nerve terminals in the rat cerebral cortex. J. Comp. Neurol. 343, 57–71.

Cheney, D.L., LeFevre, H.F., Racagni, G., 1975. Choline acetyltransferase activity and mass fragmentographic measurement of acetylcholine in specific nuclei and tracts of rat brain. Neuropharmacology 14, 801–809.

Clarke, D.J., 1985. Cholinergic innervation of the rat dentate gyrus. An immunocytochemical and electron microscopic study. Brain Res. 360, 349–354.

Clarke, D.J., Dunnett, S.B., 1986. Ultrastructural organization of choline-acetyltransferase-immunoreactive fibres innervating the neocortex from embryonic ventral forebrain grafts. J. Comp. Neurol. 250, 192–206.

Contant, C., Umbriaco, U., Garcia, S., Watkins, K.C., Descarries, L., 1996. Ultrastructural characterization of the acetylcholine innervation in adult rat neostriatum. Neuroscience 71, 933–947.

Cozzari, C., Howard, J., Hartman, B.K., 1990. Analysis of Epitopes on Choline Acetyltransferase (ChAT) Using Monoclonal Antibodies (Mabs). Program No 200. Abstract Viewer/Itinerary Planner. Society for Neuroscience, Washington, DC.

Daszuta, A., Chazal, G., Garcia, S., Oleskevich, S., Descarries, L., 1991. Ultrastructural features of serotonin neurons grafted to adult rat hippocampus. An immunocytochemical analysis of their cell bodies and axon terminals. Neuroscience 42, 793–811.

De Lima, A.D., Singer, W., 1986. Cholinergic innervation of the cat striate cortex. A choline acetyltransferase immunocytochemical analysis. J. Comp. Neurol. 250, 324–338.

De Lima, A.D., Montero, V.M., Singer, W., 1985. The cholinergic innervation of the visual thalamus: an EM immunocytochemical study. Exp. Brain Res. 59, 206–212.

Descarries, L., Mechawar, N., 2008. Structural organization of monoamine and acetylcholine neuron systems in the rat CNS. In: Lajtha, A. (Ed.), Handbook of Neurochemistry and Molecular Neurobiology, third ed. In: Vizi, E.S. (vol. Ed.), Neurotransmitter Systems, Springer, New York, pp. 1–20.

Descarries, L., Beaudet, A., Watkins, K.C., 1975. Serotonin nerve terminals in adult rat neocortex. Brain Res. 100, 563–588.

Descarries, L., Watkins, K.C., Lapierre, Y., 1977. Noradrenergic axon terminals in the cerebral cortex of rat. III. Topometric ultrastructural analysis. Brain Res. 133, 197–222.

Descarries, L., Soghomonian, J.-J., Garcia, S., Doucet, G., Bruno, J.P., 1992. Ultrastructural analysis of the serotonin hyperinnervation in adult rat neostriatum following neonatal dopamine denervation with 6-hydroxydopamine. Brain Res. 569, 1–13.

Descarries, L., Umbriaco, D., Contant, C., Watkins, K.C., 1995. A new hypothesis of acetylcholine (ACh) function in densely ACh-innervated areas of brain. IVth IBRO World Congr. Neurosci. Abstr. A3.15, 101.

Descarries, L., Watkins, K.C., Garcia, S., Bosler, O., Doucet, G., 1996. Dual character, asynaptic and synaptic, of the dopamine innervation in adult rat neostriatum. A quantitative autoradiographic and immunocytochemical analysis. J. Comp. Neurol. 375, 167–186.

Descarries, L., Gisiger, V., Steriade, M., 1997. Diffuse transmission by acetylcholine in the CNS. Prog. Neurobiol. 53, 603–625.

Descarries, L., Riad, M., Parent, M., 2010. Ultrastructure of the serotonin innervation in the mammalian central nervous system. In: Müller, C.P., Jacobs, B.L. (Eds.), Handbook of the Behavioural Neurobiology of Serotonin. Elsevier, Amsterdam, pp. 65–101.

DiFiglia, M., 1987. Synaptic organization of cholinergic neurons in the monkey neostriatum. J. Comp. Neurol. 255, 245–258.

Doucet, G., Descarries, L., Garcia, S., 1986. Quantification of the dopamine innervation in adult rat neostriatum. Neuroscience 19, 427–445.

Eckenstein, F., Thoenen, H., 1983. Cholinergic neurons in the rat cerebral cortex demonstrated by immunohistochemical localization of choline acetyltransferase. Neurosci. Lett. 36, 211–215.

Eckenstein, F., Baughman, R.W., 1984. Two types of cholinergic innervation in cortex, one co-localized with vasoactive intestinal polypeptide. Nature 309, 153–155.
Fremeau Jr., R.T., Burman, J., Qureshi, T., Tran, C.H., Proctor, J., Johnson, J., Zhang, H., Sulzer, D., Copenhagen, D.R., Storm-Mathisen, J., Reimer, R.J., Chaudhry, F.A., Edwards, R.H., 2002. The identification of vesicular glutamate transporter 3 suggests novel modes of signaling by glutamate. Proc. Natl. Acad. Sci. U.S.A. 99, 14488–14493.
Frotscher, M., Léránth, C., 1985. Cholinergic innervation of the rat hippocampus as revealed by choline acetyltransferase immunocytochemistry: a combined light and electron microscopic study. J. Comp. Neurol. 239, 237–246.
Frotscher, M., Léránth, C., 1986. The cholinergic innervation of the rat fascia dentate: identification of target structures on granule cells by combining choline acetyltransferase immunocytochemistry and Golgi impregnation. J. Comp. Neurol. 243, 58–70.
Frotscher, M., Soriano, E., Léránth, C., 1992. Cholinergic and GABAergic neurotransmission in the fascia dentate: electron microscopic immunocytochemical studies in rodents and primates. Epilepsy Res. Suppl. 7, 65–78.
Garofalo, J., Ribeiro-Da-Silva, A., Cuello, A.C., 1992. Nerve growth factor-induced synaptogenesis and hypertrophy of cortical cholinergic terminals. Proc. Natl. Acad. Sci. U.S.A. 89, 2639–2645.
Gezelius, H., Wallén-Mackenzie, Å., Enjin, A., Lagerström, M., Kullander, K., 2006. Role of glutamate in locomotor rhythm generating neuronal circuitry. J. Physiol. Paris 100, 297–303.
Gras, C., Herzog, E., Bellenchi, G.C., Bernard, V., Ravassard, P., Pohl, M., Gasnier, B., Giros, B., El Mestikawy, S., 2002. A third vesicular glutamate transporter expressed by cholinergic and serotoninergic neurons. J. Neurosci. 22, 5442–5451.
Gras, C., Amilhon, B., Lepicard, E.M., Poirel, O., Vinatier, J., Herbin, M., Dumas, S., Tzavara, E.T., Wade, M.R., Nomikos, G.G., Hanoun, N., Saurini, F., Kemel, M.L., Gasnier, B., Giros, B., El Mestikawy, S., 2008. The vesicular glutamate transporter VGLUT3 synergizes striatal acetylcholine tone. Nat. Neurosci. 11, 292–300.
Gritti, I., Mainville, L., Jones, B.E., 1993. Codistribution of GABA- with acetylcholine-synthesizing neurons in the basal forebrain of the rat. J. Comp. Neurol. 329, 438–457.
Gu, Q., Singer, W., 1993. Effects of intracortical infusion of anticholinergic drugs on neuronal plasticity in kitten striate cortex. Eur. J. Neurosci. 5, 475–485.
Guzman, M.S., De Jaeger, X., Raulic, S., Souza, I.A., Li, A.X., Schmid, S., Menon, R.S., Gainetdinov, R.R., Caron, M.G., Bartha, R., Prado, V.F., Prado, M.A., 2011. Elimination of the vesicular acetylcholine transporter in the striatum reveals regulation of behaviour by cholinergic-glutamatergic co-transmission. PLoS Biol. 9, e1001194.
Hallanger, A.E., Wainer, B.H., 1988. Ultrastructure of ChAT-immunoreactive synaptic terminals in the thalamic reticular nucleus of the rat. J. Comp. Neurol. 278, 486–497.
Hallanger, A.E., Levey, A.I., Lee, H.J., Rye, D.B., Wainer, B.H., 1987. The origins of cholinergic and other subcortical afferents to the thalamus in the rat. J. Comp. Neurol. 262, 105–124.
Hallanger, A.E., Price, S.D., Lee, H.J., Steininger, T.L., Wainer, B.H., 1990. Ultrastructure of cholinergic synaptic terminals in the thalamic anteroventral, ventroposterior, and dorsal lateral geniculate nuclei of the rat. J. Comp. Neurol. 299, 482–492.
Herzog, E., Landry, M., Buhler, E., Bouali-Benazzouz, R., Legay, C., Henderson, C.E., Nagy, F., Dreyfus, P., Giros, B., El Mestikawy, S., 2004. Expression of vesicular glutamate transporters, VGLUT1 and VGLUT2, in cholinergic spinal motoneurons. Eur. J. Neurosci. 20, 1752–1760.
Higley, M.J., Gittis, A.H., Oldenburg, I.A., Balthasar, N., Seal, R.P., Edwards, R.H., Lowell, B.B., Kreitzer, A.C., Sabatini, B.L., 2011. Cholinergic interneurons mediate fast VGluT3-dependent glutamatergic transmission in the striatum. PLoS One 6, e19155.
Höhmann, C.F., Brooks, A.R., Coyle, J.T., 1988. Neonatal lesions of the basal forebrain cholinergic neurons result in abnormal cortical development. Dev. Brain Res. 42, 253–264.
Höhmann, C.F., Kwiterovich, K.K., Oster-Granite, M.L., Coyle, J.T., 1991a. Newborn basal forebrain lesions disrupt cortical cytodifferentiation as visualized by rapid golgi staining. Cereb. Cortex 1, 143–157.
Höhmann, C.F., Wilson, W., Coyle, J.T., 1991b. Efferent and afferent connections of mouse sensory-motor cortex following cholinergic deafferentation at birth. Cereb. Cortex 1, 158–172.

Höhmann, C.F., Potter, E.D., Levey, A.I., 1995. Development of muscarinic receptor subtypes in the forebrain of the mouse. J. Comp. Neurol. 358, 88–101.
Hoover, D.B., Muth, E.A., Jacobowitz, D.M., 1978. A mapping of the distribution of acetylcholine, choline acetyltransferase and acetylcholinesterase in discrete areas of rat brain. Brain Res. 153, 295–306.
Hoshino, K., Hicks, T.P., Hirano, S., Norita, M., 2000. Ultrastructural organization of transmitters in the cat lateralis medialis-suprageniculate nucleus of the thalamus: an immunohistochemical study. J. Comp. Neurol. 419, 257–270.
Houser, C.R., 1990. Cholinergic synapses in the central nervous system. Studies of the immunocytochemical localization of choline acetyltransferase. J. Electron Microsc. Tech. 15, 2–19.
Houser, C.R., Crawford, G.D., Salvaterra, P.M., Vaughn, J.E., 1985. Immunocytochemical localization of choline acetyltransferase in rat cerebral cortex: a study of cholinergic neurons and synapses. J. Comp. Neurol. 234, 17–34.
Huang, A., Noga, B.R., Carr, P.A., Fedirchuk, B., Jordan, L.M., 2000. Spinal cholinergic neurons activated during locomotion: localization and electrophysiological characterization. J. Neurophysiol. 83, 3537–3547.
Isaacson, L.G., Tanaka Jr., D., 1988. Cholinergic innervation of canine thalamostriatal projection neurons: an ultrastructural study combining choline acetyltransferase immunocytochemistry and WGA-HRP retrograde labeling. J. Comp. Neurol. 277, 529–540.
Izzo, P.N., Bolam, J.P., 1988. Cholinergic synaptic input to different parts of spiny striatonigral neurons in the rat. 269, 219–234.
Johnston, M.V., McKinney, M., Coyle, J.T., 1981. Neocortical cholinergic innervation: a description of extrinsic and intrinsic components in the rat. Exp. Brain Res. 43, 159–172.
Jourdain, A., Semba, K., Fibiger, H.C., 1989. Basal forebrain and mesopontine tegmental projections to the reticular thalamic nucleus: an axonal collateralization and immunohistochemical study in the rat. Brain Res. 505, 55–65.
Kraus, T., Neuhuber, W.L., Raab, M., 2004. Vesicular glutamate transporter 1 immunoreactivity in motor endplates of striated esophageal but not skeletal muscles in the mouse. Neurosci. Lett. 360, 53–56.
Léránth, C., Frotscher, M., 1987. Cholinergic innervation of hippocampal GAD- and somatostatin-immunoreactive commissural neurons. J. Comp. Neurol. 261, 33–47.
Levey, A.I., Wainer, B.H., Rye, D.B., Mufson, E.J., Mesulam, M.-M., 1984. Choline acetyltransferase-immunoreactive neurons intrinsic to rodent cortex and distinction from acetylcholinesterase neurons. Neuroscience 13, 341–353.
Levey, A.I., Hallanger, A.E., Wainer, B.H., 1987. Choline acetyltransferase immunoreactivity in the rat thalamus. J. Comp. Neurol. 257, 317–332.
Lysakowski, A., Wainer, B.H., Bruce, G., Hersh, L.B., 1989. An atlas of the regional and laminar distribution of choline acetyltransferase immunoreactivity in rat cerebral cortex. Neuroscience 28, 291–336.
Markham, J.A., Vaughn, J.E., 1990. Ultrastructural analysis of choline acetyltransferase-immunoreactive sympathetic preganglionic neurons and their dendritic bundles in rat thoracic spinal cord. Synapse 5, 299–312.
Mechawar, N., Descarries, L., 2001. The cholinergic innervation develops early and rapidly in the rat cerebral cortex: a quantitative immunocytochemical study. Neuroscience 108, 555–567.
Mechawar, N., Cozzari, C., Descarries, L., 2000. Cholinergic innervation of adult rat cerebral cortex: a quantitative immunocytochemical description. J. Comp. Neurol. 428, 305–318.
Mechawar, N., Watkins, K.C., Descarries, L., 2002. Ultrastructural features of the acetylcholine innervation in the developing parietal cortex of rat. J. Comp. Neurol. 443, 250–258.
Mesulam, M.-M., 1988. Central cholinergic pathways: neuroanatomy and some behavioral implications. In: Avoli, M., Reader, T.A., Gloor, D., Dykes, R.W. (Eds.), Neurotransmitters and Cortical Function: From Molecules to Mind. Plenum Press, New York, pp. 237–260.
Mesulam, M.-M., Mufson, E.J., Wainer, B.H., Levey, A.I., 1983. Central cholinergic pathways in the rat: an overview based on an alternative nomenclature (Ch1–Ch6). Neuroscience 10, 1185–1201.
Mrzljak, L., Pappy, M., Léránth, C., Goldman-Rakic, P.S., 1995. Cholinergic synaptic circuitry in the macaque pre-frontal cortex. J. Comp. Neurol. 357, 603–617.
Nagy, J.I., Yamamoto, T., Jordan, L.M., 1993. Evidence for the cholinergic nature of c-terminals associated with subsurface cisterns in alpha-motoneurons of rat. Synapse 15, 17–32.

Nickerson Poulin, A., Guerci, A., El Mestikawy, S., Semba, K., 2006. Vesicular glutamate transporter 3 immunoreactivity is present in cholinergic basal forebrain neurons projecting to the basolateral amygdala in rat. J. Comp. Neurol. 498, 690–711.

Nishimaru, H., Restrepo, C.E., Ryge, J., Yanagawa, Y., Kiehn, O., 2005. Mammalian motor neurons co-release glutamate and acetylcholine at central synapses. Proc. Natl. Acad. Sci. USA. 102, 5245–5249.

Oda, S., Kuroda, M., Kakuta, S., Tanihata, S., Ishikawa, Y., Kishi, K., 2003. Ultrastructure of ascending cholinergic terminals in the anteroventral thalamic nucleus of the rat: a comparison with the mammillothalamic terminals. Brain Res. Bull. 59, 473–483.

Oleskevich, S., Descarries, L., Watkins, K.C., Séguéla, P., Daszuta, A., 1991. Ultrastructural features of the serotonin innervation in adult rat hippocampus: an immunocytochemical description in single and serial thin sections. Neuroscience 42, 777–791.

Palade, G.E., Palay, S.L., 1954. Electron microscope observations of interneuronal and neuromuscular synapses. Anat. Rec. 118, 335–336.

Paré, D., Smith, Y., Parent, A., Steriade, M., 1988. Projections of brainstem core cholinergic and non-cholinergic neurons of cat to intralaminar and reticular thalamic nuclei. Neuroscience 25, 69–86.

Parent, A., Paré, D., Smith, Y., Steriade, M., 1988. Basal forebrain cholinergic and noncholinergic projections to the thalamus and brainstem in cats and monkeys. J. Comp. Neurol. 277, 281–301.

Parent, M., Descarries, L., 2008. The acetylcholine innervation of the adult rat thalamus: distribution and ultrastructural features in dorsolateral geniculate, parafascicular and reticular thalamic nuclei. J. Comp. Neurol. 511, 678–691.

Parnavelas, J.G., Kelly, W., Franke, E., Eckenstein, F., 1986. Cholinergic neurons and fibres in the rat visual cortex. J. Neurocytol. 15, 329–336.

Pascual, J., Gonzalez, A.M., Pazos, A., 1991. Further studies on the biochemical characterization and autoradiographic distribution of [^{3}H]hemicholinium-3 binding sites in rat brain: a presynaptic cholinergic marker. Pharmac. Res. 24, 345–355.

Patel, N.C., Bickford, M.E., 1997. Synaptic targets of cholinergic terminals in the pulvinar nucleus of the cat. J. Comp. Neurol. 387, 266–278.

Patel, N.C., Carden, W.B., Bickford, M.E., 1999. Synaptic targets of cholinergic terminals in the cat lateral posterior nucleus. J. Comp. Neurol. 410, 31–41.

Peters, A., Palay, S.L., 1996. The morphology of synapses. J. Neurocytol. 25, 687–700.

Phelps, P.E., Barber, R.P., Houser, C.R., Crawford, G.D., Salvaterra, P.M., Vaughn, J.E., 1984. Postnatal development of neurons containing choline acetyltransferase in rat spinal cord: an immunocytochemical study. J. Comp. Neurol. 229, 347–361.

Phelps, P.E., Houser, C.R., Vaughn, J.E., 1985. Immunocytochemical localization of choline acetyltransferase within the rat neostriatum: a correlated light and electron microscopic study of cholinergic neurons and synapses. J. Comp. Neurol. 238, 286–307.

Pickel, V.M., Chan, J., 1999. Ultrastructural localization of the serotonin transporter in limbic and motor compartments of the nucleus accumbens. J. Neurosci. 19, 7356–7366.

Raczkowski, D., Fitzpatrick, D., 1989. Organization of cholinergic synapses in the cat's dorsal lateral geniculate and perigeniculate nuclei. J. Comp. Neurol. 288, 676–690.

Rea, M.A., Simon, J.R., 1981. Regional distribution of cholinergic parameters within the rat striatum. Brain Res. 219, 317–326.

Ridet, J.-L., Sandillon, F., Rajaofetra, N., Geffard, M., Privat, A., 1992. Spinal dopaminergic system of the rat: light and electron microscopic study using an antiserum against dopamine, with particular emphasis on synaptic incidence. Brain Res. 598, 233–241.

Robertson, J.D., 1953. Ultrastructure of two invertebrate synapses. Proc. Soc. Exp. Biol. Med. 82, 219–223.

Robertson, R.T., Gallardo, K.A., Claytor, K.J., Ha, D.H., Ku, K.-H., Yu, B.P., Lauterborn, J.C., Wiley, R.G., Yu, J., Gall, C.M., Leslie, F.M., 1998. Neonatal treatment with 192 IgG-saporin produces long-term forebrain cholinergic deficits and reduces dendritic branching and spine density of neocortical pyramidal neurons. Cereb. Cortex 8, 142–155.

Roerig, B., Nelson, D.A., Katz, L.C., 1997. Fast synaptic signaling by nicotinic acetylcholine and serotonin 5-HT_3 receptors in developing visual cortex. J. Neurosci. 17, 8353–8362.

Rye, D.B., Wainer, B.H., Mesulam, M.-M., Mufson, E.J., Saper, C.B., 1984. Cortical projections arising from the basal forebrain: a study of cholinergic and noncholinergic components employing combined retrograde tracing and immunohistochemical localization of choline acetyltransferase. Neuroscience 13, 627–643.
Saper, C.B., 1984. Organization of cerebral cortical afferent systems in the rat. II. Magnocellular basal nucleus. J. Comp. Neurol. 222, 313–342.
Schäfer, M.K., Eiden, L.E., Weihe, E., 1998. Cholinergic neurons and terminal fields revealed by immunohistochemistry for the vesicular acetylcholine transporter. I. Central nervous system. Neuroscience 84, 331–359.
Schäfer, M.K.-H., Varoqui, H., Defamie, N., Weihe, E., Erickson, J.D., 2002. Molecular cloning and functional identification of mouse vesicular glutamate transporter 3 and its expression in subsets of novel excitatory neurons. J. Biol. Chem. 277, 50734–50748.
Schoch, S., Gundelfinger, E.D., 2006. Molecular organization of the presynaptic active zone. Cell Tissue Res. 326, 379–391.
Schwartz, M.L., Mrzljak, L., 1993. Cholinergic innervation of the mediodorsal thalamic nucleus in the monkey: ultrastructural evidence supportive of functional diversity. J. Comp. Neurol. 327, 48–62.
Shapira, M., Zhai, R.G., Dresbach, T., Bresler, T., Torres, V.I., Gundelfinger, E.D., Ziv, N.E., Garner, C.C., 2003. Unitary assembly of presynaptic active zones from Piccolo-Bassoon transport vesicles. Neuron 38, 237–252.
Sherriff, F.E., Henderson, Z., 1994. A cholinergic propriospinal innervation of the rat spinal cord. Brain Res. 634, 150–154.
Sherrington, C.S., 1897. The central nervous system. In: Foster, M., Sherrington, C.S. (Eds.), A Textbook of Physiology. Macmillan, London, pp. 915–1000.
Smiley, J.F., Morrell, F., Mesulam, M.-M., 1997. Cholinergic synapses in human cerebral cortex: an ultrastructural study in serial sections. Exp. Neurol. 144, 361–368.
Soghomonian, J.-J., Descarries, L., Watkins, K.C., 1989. Serotonin innervation in adult rat neostriatum. II. Ultrastructural features: a radioautographic and immunocytochemical study. Brain Res. 481, 67–86.
Sorimachi, M., Kataoka, K., 1974. Choline uptake by nerve terminals: a sensitive and a specific marker of cholinergic innervation. Brain Res. 72, 350–353.
Spreafico, R., Amadeo, A., Angoscini, P., Panzica, F., Battaglia, G., 1993. Branching projections from mesopontine nuclei to the nucleus reticularis and related thalamic nuclei: a double labelling study in the rat. J. Comp. Neurol. 336, 481–492.
Steriade, M., Descarries, L., 2006. Cholinergic modulation of cortical activity. In: Giacobini, E., Pepeu, G. (Eds.), The Brain Cholinergic System in Health and Disease. Taylor and Francis Books, Oxforshire, UK, pp. 191–207.
Steriade, M., Paré, D., Parent, A., Smith, Y., 1988. Projections of cholinergic and non-cholinergic neurons of the brainstem core to relay and associational thalamic nuclei in the cat and macaque monkey. Neuroscience 25, 47–67.
Turrini, P., Casu, M.A., Wong, T.P., De Koninck, Y., Ribeiro-Da-Silva, A., Cuello, A.C., 2001. Cholinergic nerve terminals establish classical synapses in the rat cerebral cortex: synaptic pattern and age-related atrophy. Neuroscience 105, 277–285.
Umbriaco, D., Watkins, K.C., Descarries, L., Cozzari, C., Hartman, B.K., 1994. Ultrastructural and morphometric features of the acetylcholine innervation in adult rat parietal cortex: an electron microscopic study in serial sections. J. Comp. Neurol. 348, 351–373.
Umbriaco, D., Garcia, S., Beaulieu, C., Descarries, L., 1995. Relational features of acetylcholine, noradrenaline, serotonin and GABA axonal varicosities in the stratum radiatum of adult rat hippocampus (CA1). A comparative electron microscopic analysis. Hippocampus 5, 605–620.
Vaucher, E., Hamel, E., 1995. Cholinergic basal forebrain neurons project to cortical microvessels in the rat: electron microscopic study with anterogradely transported *Phaseolus vulgaris* leucoagglutinin and choline acetyltransferase immunocytochemistry. J. Neurosci. 15, 7427–7441.
Vinson, P.N., Justice, J.B., 1997. Effect of neostigmine on concentration and extraction fraction of acetylcholine using quantitative microdialysis. J. Neurosci. Methods 73, 61–67.
Volpicelli, L.A., Levey, A.I., 2004. Muscarinic acetylcholine receptor subtypes in cerebral cortex and hippocampus. In: Descarries, L., Krnjevic, K., Steriade, M. (Eds.), Acetylcholine in the Cerebral Cortex. Progress in Brain Research, vol. 145. Elsevier, Amsterdam, pp. 59–66.

Wainer, B.H., Bolam, J.P., Freund, T.F., Henderson, Z., Totterdell, S., Smith, A.G., 1984. Cholinergic synapses in the rat brain: a correlated light and electron microscopic immunohistochemical study employing a monoclonal antibody against choline acetyltransferase. Brain Res. 308, 69–76.

Wainer, B.H., Mesulam, M.M., 1990. Ascending cholinergic pathways in the rat brain. In: Steriade, M., Biesold, D. (Eds.), Brain Cholinergic Systems. Oxford University Press, Oxford, pp. 65–119.

Woolf, N.J., 1991. Cholinergic systems in mammalian brain and spinal cord. Prog. Neurobiol. 37, 475–524.

Woolf, N.J., Butcher, L.L., 1981. Cholinergic neurons in the caudate-putamen complex proper are intrinsically organized: a combined Evans blue and acetylcholinesterase analysis. Brain Res. Bull. 7, 487–507.

Woolf, N.J., Butcher, L.L., 1986. Cholinergic systems in the rat brain. III. Projections from the pontomesencephalic tegmentum to the thalamus, tectum, basal ganglia, and basal forebrain. Brain Res. Bull. 16, 603–637.

Yamasaki, M., Matsui, M., Watanabe, M., 2010. Preferential localization of muscarinic M_1 receptor on dendritic shaft and spine of cortical pyramidal cells and its anatomical evidence for volume transmission. J. Neurosci. 30, 4408–4418.

Zhai, R.G., Bellen, H.J., 2004. The architecture of the active zone in the presynaptic nerve terminal. Physiology (Bethesda) 19, 262–270.

Zhu, X.O., Waite, P.M.E., 1998. Cholinergic depletion reduces plasticity of barrel field cortex. Cereb. Cortex 8, 63–72.

CHAPTER FIFTEEN

Prefrontal Cortical Dopamine Transmission: Ultrastructural Studies and Their Functional Implications

Susan R. Sesack
Departments of Neuroscience and Psychiatry, University of Pittsburgh, Pittsburgh, PA, USA

1. INTRODUCTION

1.1. Functions of the PFC and Its DA Innervation

The prefrontal cortex (PFC) encompasses regions in the frontal lobe that perform executive functions requiring temporal ordering of goal-directed behavior (Fuster, 2001; Kesner and Churchwell, 2011; Miller, 2000; Miller and Cohen, 2001; Wise, 2008). Such functions include working memory, future planning, rule learning, decision making, and other cognitive abilities (Kesner and Churchwell, 2011; Miller, 2000). The PFC reaches its greatest extent in humans but is represented to some extent in many mammalian species (Fuster, 1997; Preuss, 1995; Reep, 1984; Uylings et al., 2003). Although not all cytoarchitectonic divisions have homologous representation in lower species (Preuss, 1995; Wise, 2008), even rodents require temporal organization of complex behaviors and so have some rudiments of PFC structure and function residing within the frontal lobe (Kesner and Churchwell, 2011; Uylings et al., 2003).

Given the important cognitive functions carried out by the PFC, it is perhaps not surprising that malfunction of this region has been implicated in so many mental disorders (Miller, 2000). These include schizophrenia (Beneyto and Lewis, 2011; Tan et al., 2009), major depressive disorder (Mayberg, 1997; Murray et al., 2011), posttraumatic stress disorder (Koenigs and Grafman, 2009; Shin et al., 2006), attention-deficit hyperactivity disorder (Brennan and Arnsten, 2008; Liston et al., 2011), and substance abuse (Goldstein and Volkow, 2011). For many of these conditions, either pathophysiological or treatment models suggest that disruptions in the dopamine (DA) innervation to the PFC may contribute to symptom states (Abi-Dargham et al., 2002; Akil et al., 1999; Arnsten and Pliszka, 2011; Durstewitz and Seamans, 2008; Gamo and Arnsten, 2011; Rajkowska, 2000; Steketee, 2003). Indeed, lesions of the DA input in animals lead to altered PFC function that is sometimes as profound as occurs with damage to the PFC itself (Brozoski et al., 1979; Cohen and Servan-Schreiber, 1992; Collins et al., 1998). Structural alterations are also observed in this region following DA denervation (Wang and Deutch, 2008). Such observations are consistent with the critical modulatory role that DA mediates in regulating PFC cognitive

The Synapse
http://dx.doi.org/10.1016/B978-0-12-418675-0.00015-8

functions such as working memory, set-shifting, and decision making (Chudasama and Robbins, 2004; Cohen et al., 2002; Floresco et al., 2006; Goldman-Rakic, 1995; Goldman-Rakic et al., 2000; Hitchcott et al., 2007; Sawaguchi and Goldman-Rakic, 1994; Seamans and Yang, 2004; Vijayraghavan et al., 2007).

These considerations highlight the importance of understanding the detailed anatomical features of the DA innervation to the PFC at multiple levels. In this chapter, we will review what is known about this specific monoamine input from studies using light and especially electron microscopy. Where most relevant, information from neurochemical and electrophysiological experiments is added to provide functional significance to the anatomical characterizations. The principal focus will be on studies from rodents and, unless otherwise specified, will mainly include the prelimbic and infralimbic cortices of Krettek and Price (1977). These correspond roughly to Brodmann's areas 32 and 25, respectively. Information from studies of nonhuman primates is included primarily where it provides greater depth of knowledge than is available from rodent analyses. A complete species comparison of all anatomical features of the midbrain DA system is beyond the scope of this chapter.

2.1. CORTICAL DA LABELING METHODS, CELLS OF ORIGIN, AND TERMINATIONS

2.1.1. Labeling Methodology

The anatomical characterization of DA axons within the PFC is based primarily on studies using immunoreactivity for the synthetic enzyme tyrosine hydroxylase (TH). That this enzyme also contributes to the synthesis of norepinephrine raises questions as to its selectivity for labeling DA processes. Nevertheless, over three decades of accumulated evidence indicate that TH levels are abundant in DA axons but remarkably lower, even undetectable, in many noradrenergic fibers within the PFC and certain other cortical regions (Asan, 1998; Lewis et al., 1987; Miner et al., 2003; Noack and Lewis, 1989; Pickel et al., 1975; Schmidt and Bhatnagar, 1979). The reason for these low levels is not known, but may include the fact that the norepinephrine transporter (NET) preferentially transports DA (Morón et al., 2002; Yamamoto and Novotney, 1998), providing a ready source of norepinephrine precursor that does not require TH synthesis (Miner and Sesack, 2007; Miner et al., 2006; Valentini et al., 2004) (Section 4). Another possible interpretation is that noradrenergic transmission is low in naive animals under standard laboratory conditions and that TH levels are correspondingly reduced as a result. In support of this idea, we have found increased TH detection in norepinephrine fibers following a chronic stress that is known to activate the PFC noradrenergic projection in rats (Miner and Sesack, 2007; Miner et al., 2006).

In addition to examinations of TH, some studies of the PFC are based on immunostaining for DA itself. Although useful for confirming the relative selectivity of TH immunocytochemistry for labeling DA axons, these investigations are limited by the fixative required (Chagnaud et al., 1987; Séguéla et al., 1988), which prohibits detection of most other proteins of interest. Immunolabeling for the DA transporter (DAT) should, in theory, prove to be the most reliable marker for DA axons in the PFC, given that only DA neurons are capable of making this protein. Unfortunately, most of the DA projection to the PFC is remarkably devoid of DAT (Section 4). Hence, TH immunoreactivity continues to be relied upon for studies examining the anatomical properties of the mesoprefrontal projection, with the caveat that a small contribution by cortical noradrenergic axons cannot be ruled out.

2.1.2. DA Cells Innervating the PFC and Their Axon Terminations

The basic anatomy of the mesoprefrontal DA pathway has been amply described since its discovery in 1973 (Beckstead et al., 1979; Berger et al., 1991, 1974; Björklund and Dunnett, 2007; Björklund and Lindvall, 1984; Descarries et al., 1987; Fallon and Loughlin, 1987; Lammel et al., 2008; Lewis and Sesack, 1997; Manitt et al., 2011; Mundorf et al., 2001; Swanson, 1982; Thierry et al., 1973; Williams and Goldman-Rakic, 1998). Hence, only the key features need to be covered here. The DA neurons innervating the PFC reside in the ventral midbrain, including the ventral tegmental area (VTA), substantia nigra (SN), and, to a lesser extent, the retrorubral field. These cells are intermixed with non-DA neurons that contribute to the mesoprefrontal projection, including both GABAergic (Carr and Sesack, 2000b) and glutamatergic (Gorelova et al., 2012; Hur and Zaborszky, 2005; Yamaguchi et al., 2011) phenotypes.

The axons of all these cells travel by way of the medial forebrain bundle and terminate in multiple cytoarchitectonic divisions of the frontal cortex. In rodents, the greatest DA innervation density occurs within the medial PFC, i.e. prelimbic and infralimbic cortices, with a somewhat lower density in the anterior cingulate region. Within the PFC, DA axons are densest in the deep layers V and VI and less substantial in more superficial layers, I–III. Layer IV is absent in the rat PFC (Krettek and Price, 1977). Retrograde tracing studies, in combination with lesions and biochemical characterizations, indicate that the deep layer DA axons are primarily derived from the VTA, whereas those in superficial layers, such as seen in the anterior cingulate cortex, arise mainly from the SN (Berger et al., 1991; Björklund and Dunnett, 2007). Hence, DA cells targeting different cortical areas represent nonbranching, distinct neuronal populations. On the whole, DA cells projecting to the PFC also show only sparse collateralization to subcortical targets (Björklund and Dunnett, 2007; Fallon and Loughlin, 1982; Loughlin and Fallon, 1984; Swanson, 1982).

In the primate, the DA innervation to the frontal lobe is considerably denser than in rodent (Berger et al., 1991; Björklund and Dunnett, 2007). It reaches its greatest density in the motor cortex and overall demonstrates a more uniform cytoarchitectonic and laminar pattern (Levitt et al., 1984; Lewis et al., 1988; Lewis and Sesack, 1997; Rosenberg and Lewis, 1995; Williams and Goldman-Rakic, 1993). Moreover, the DA input to the frontal lobe, including the dorsolateral PFC, derives more extensively from cells in the SN as well as the retrorubral field than is true in the rat (Berger et al., 1991; Gaspar et al., 1992; Williams and Goldman-Rakic, 1998). DA cells projecting to the prelimbic and infralimbic regions, however, are similar in monkeys and rats, deriving primarily from the parabrachial and midline subdivisions of the VTA (Williams and Goldman-Rakic, 1998). As seen in the rat, a substantial portion of non-DA neurons contributes to the PFC projections (Williams and Goldman-Rakic, 1998), although the phenotype of these cells has not been examined. The issue of collateralization also has not been well studied in the primate although some DA neurons innervating the PFC are known to also project to motor or supplementary motor cortices within the frontal lobes (Gaspar et al., 1992). Whether these cells also send branched projections to subcortical structures has not been investigated.

2.2. ULTRASTRUCTURAL FEATURES, SYNAPTIC INCIDENCE, AND TARGETS OF CORTICAL DA AXONS

2.2.1. General Ultrastructural Features

The basic ultrastructure of the DA innervation to the PFC is well known, having been thoroughly described by many laboratories in the 1980s and 1990s (Descarries et al., 1987; Goldman-Rakic et al., 1989; Krimer et al., 1997; Séguéla et al., 1988; Sesack et al., 1995b; Smiley and Goldman-Rakic, 1993; Smiley et al., 1992; Van Eden et al., 1987; Verney et al., 1990). The axons, regardless of species and layer, are relatively small and unmyelinated and give rise to varicosities en passant, some of which form synaptic contacts with dendritic shafts and spines. Somata are rarely, if ever, contacted. As illustrated in Figure 1, the majority of synapses are of the symmetric type, suggestive of an inhibitory or modulatory influence (Carlin et al., 1980; Gray, 1959; Mori, 1966), consistent with the physiological actions of DA (Section 5.3). Many DA synapses are relatively small in length and difficult to observe without serial section analysis (Descarries and Umbriaco, 1995; Séguéla et al., 1988; Smiley and Goldman-Rakic, 1993). Approximately 10–20% of DA axons form asymmetric synapses in the PFC of rat (Séguéla et al., 1988; Yokofujita et al., 2008) and human (Smiley et al., 1992).

2.2.2. Synaptic Incidence

An important issue for characterizing the ultrastructure of DA axons within different target regions is determining the extent to which individual varicosities form classical synaptic junctions, i.e. the synaptic incidence. For example, extensive serial reconstruction of DA fibers in the striatum of the rat indicates that only about 35% of the varicosities form recognizable synapses (Descarries et al., 1996). This does not necessarily preclude the release of DA from varicosities without synaptic membrane specializations or from intervaricose axon segments (Fuxe and Agnati, 1991; Zoli et al., 1998). Nevertheless, the detection of synaptic complexes increases confidence that traditional exocytosis occurs at those sites. In the PFC, a major species difference distinguishes the DA innervation in this regard. In the primate dorsolateral PFC, 39% of serially sectioned DA varicosities make detectable synapses (Smiley and Goldman-Rakic, 1993), a proportion similar to the rat striatum. Conversely, in the rat mPFC, as many as 93% of DA axon varicosities have been observed making synaptic contacts, by far the highest reported in any forebrain target region (Séguéla et al., 1988).

The significance of this species and regional distinction in synaptic incidence is unclear. In keeping with the relative constancy of synaptic density across cortical evolution (Harrison et al., 2002), one would have hypothesized the rat PFC to have at least an equivalent synaptic incidence to the primate although invariant scaling has mainly been reported for corticocortical connections. Methodological differences are more likely to be at the heart of discrepant findings for DA synapse number, particularly given their

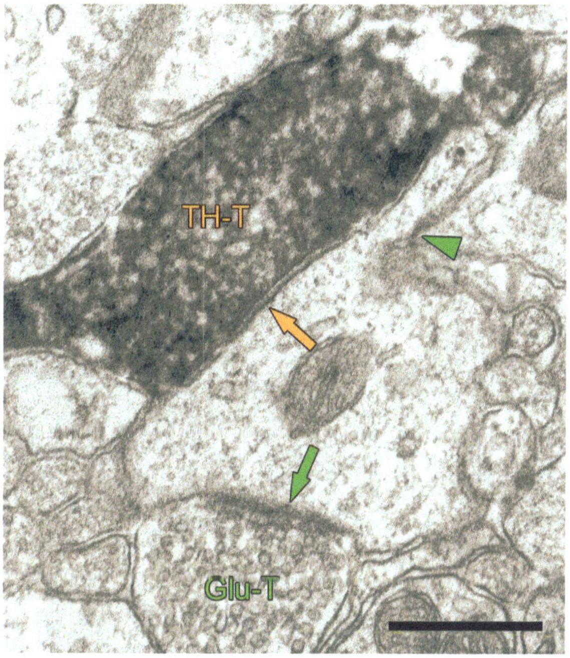

Figure 1 Electron micrograph from the rat prelimbic PFC showing a typical DA axon labeled by immunoperoxidase for the synthetic enzyme tyrosine hydroxylase (TH-T) and forming a symmetric synapse onto a distal dendrite (gold arrow). The dendrite receives additional synaptic input from a presumed glutamate axon terminal (Glu-T) associated with an asymmetric specialization (green arrow). Near the top of the image, a spine neck emanates from the dendrite (green arrowhead) indicating that this structure likely represents a spiny pyramidal neuron. It should be noted that the tissue was dually labeled by immunogold for the norepinephrine transporter and the absence of immunogold particles in the TH-T verifies its identity as a DA axon varicosity. Scale bar, 0.5 μm.

small size and weak morphological features (Figure 1). These features make them particularly difficult to detect (Séguéla et al., 1988; Sesack, 2002; Smiley, 1996), such that the application of overly zealous criteria may lead to the rejection of probable synapses from some samples, or the opposite may occur. Regarding increasing cortical complexity in the course of evolution (Harrison et al., 2002), it should be noted that in the rat PFC, the majority of DA axons synapse onto dendritic shafts as opposed to spines (Séguéla et al., 1988), whereas this proportion is reversed in human and nonhuman primates (Smiley and Goldman-Rakic, 1993; Smiley et al., 1992). These findings may suggest an increase in the specificity of DA modulation for particular sources of glutamate afferents that synapse onto the same spines (Carr et al., 1999; Goldman-Rakic et al., 1989; Séguéla et al., 1988) (Section 3.4).

2.2.3. Pyramidal Cell Target

The observation of synapses onto spines suggests that spiny pyramidal neurons are among the cellular targets of DA axons within the PFC (Figure 2). In the rat, the pyramidal nature of some postsynaptic targets has been verified by retrograde transport of tracer from terminal regions. More specifically, DA axons have been shown to synapse onto pyramidal neurons with cortical projections (Carr and Sesack, 2000a) as well as those with subcortical projections (Carr et al., 1999). These are likely to represent two separate cell populations, as indicated by dual retrograde tract-tracing studies (Pinto et al., 1997).

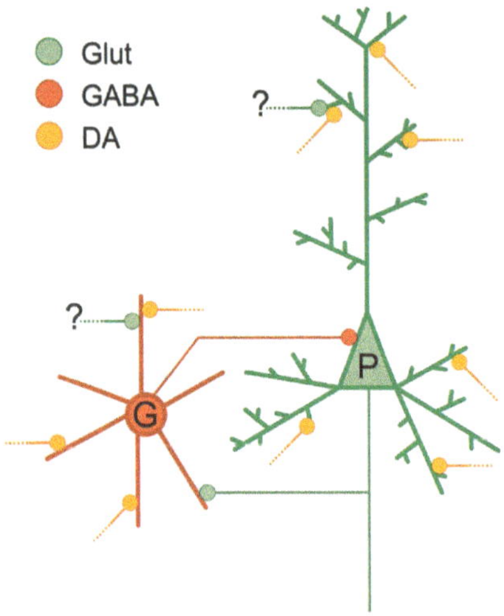

Figure 2 Schematic drawing illustrating the main synaptic targets of DA axon terminals in the PFC. These include spiny pyramidal neurons that are glutamatergic (P) and aspiny nonpyramidal neurons containing GABA (G). The latter are estimated to receive fewer total contacts from DA axons (Krimer et al., 1997). Along distal dendrites and spines, DA synapses are often in close proximity to those formed by glutamate axons, but the source of these convergent inputs remains to be determined, as represented by the question marks.

In the primate, the identification of pyramidal neurons among the synaptic targets of PFC DA axons has been verified by careful integration of light and electron microscopic analysis (Goldman-Rakic et al., 1989; Krimer et al., 1997; Smiley and Goldman-Rakic, 1993). In particular, the detailed investigation of Krimer and associates revealed that dendrites of pyramidal neurons are contacted by DA axons at secondary and higher branches. Approximately one DA axon varicosity contacts each dendrite, with some minor variations across layers (range 0.7–1.4 contacts per 100 μm dendritic length) (Krimer et al., 1997). Some of these contacts have been verified as synaptic junctions by electron microscopy. The findings of Krimer et al. suggest that DA axonal arbors exhibit an intrinsic arrangement in the PFC similar to that designed to exert influence at regular intervals on target neurons in the striatum (Moss and Bolam, 2008). The observations further indicate that DA transmission may modulate glutamate signaling from all possible sources arriving within the distal dendritic trees of pyramidal cells (Krimer et al., 1997) (Section 3.4).

Synaptic inputs to spines constitute one of the more important aspects of DA synaptology in the PFC. Spines serve as biochemically isolated compartments along the dendritic tree, with all spines receiving an asymmetric synapse onto the head that is presumed, and in most cases verified, to derive from glutamate inputs (Koch and Poggio, 1983; Koch et al., 1992). A proportion of these spines receive secondary input along the head, neck, or base (Koch and Poggio, 1983), in many cases from DA axons (Goldman-Rakic et al., 1989; Séguéla et al., 1988). The three-part arrangement of spine, glutamate input, and DA synapse has been referred to as an Atriad@ (Goldman-Rakic et al., 1989), which has been speculated to serve as a temporal coincidence detector. According to theoretical models, activation of the secondary input can modulate transmission at that specific glutamate synapse, but only within a narrow time frame (Koch and Poggio, 1983; Yao et al., 2008).

2.2.4. Convergence with Glutamate Afferents

In theory, DA synapses onto spines can modulate quite specific sources of glutamate drive. In the striatal complex, approximately 10–40% of spines receive such dual input (Arbuthnott and Wickens, 2007; Smith and Bolam, 1990), and DA axons have been found to converge onto common spines with every source of cortical and thalamic afferent investigated to date (Freund et al., 1984; Moss and Bolam, 2008; Pinto et al., 2003; Sesack and Pickel, 1990, 1992; Smith and Bolam, 1990; Smith et al., 1994; Totterdell and Smith, 1989). These observations from the basal ganglia suggest a relative lack of selectivity of DA modulation for particular glutamate inputs (Moss and Bolam, 2008). Indeed, a rather uniform modulatory influence of DA nigrostriatal axons has been suggested based on their formation of a regular lattice structure, with varicosities relatively evenly spaced from each other and from cortical and thalamic afferents (Moss and Bolam, 2008). Additional uniformity derives from the temporality of DA release events

(Arbuthnott and Wickens, 2007). The probability of DA axons forming a uniform spatial arrangement within the PFC seems less likely given the more heterogeneous neuropil and greater number and variety of input sources. Perhaps a lattice-type array might be observed within the dense DA innervation to the deep layers of the rat PFC or within the overall more uniform distribution of DA axons in the primate PFC (Williams and Goldman-Rakic, 1993). On the other hand, the localization of DA receptors in both rat and monkey PFC argues for the existence of specific microenvironments of DA transmission (Sections 5.1.4 and 5.2.2) and therefore for a nonregular arrangement of modulatory sites (Paspalas and Goldman-Rakic, 2004; Santana et al., 2009).

In the PFC, no known source of glutamate input has yet been shown to converge synaptically onto common spines with DA axons (Figure 2). Indeed, our laboratory has endeavored to identify specific glutamate inputs to the PFC that engage in this triadic arrangement, but to date we have obtained negative results for the hippocampus (Carr and Sesack, 1996), callosal PFC axons (Carr and Sesack, 1998), the paraventricular thalamus (Pinto et al., 2003), and the basolateral amygdala (Pinto and Sesack, 1999). A few instances of potential convergence with the local collaterals of pyramidal neurons were detected (Sesack and Miner, 1997), although these could not be replicated in more extended studies (Sesack, unpublished observations). One laboratory did report that projections from the VTA to the PFC converged onto the apical dendrites of layer V pyramidal cells with axons from the mediodorsal nucleus of the thalamus (Kuroda et al., 1996). Specific examples of triadic arrangements involving spines, however, were not reported. Moreover, the study did not examine DA axons specifically, which is an important omission given that the bulk of the mesocortical projection is carried by fibers from GABA and glutamate neurons (Section 2.2).

The findings outlined above could be interpreted as indicating that DA axons in the PFC converge onto spines with only one or a few specific sources of glutamate, and these remain to be identified. Alternatively, DA axons may synapse onto spines receiving convergent input from all glutamate afferent sources, as observed for the striatal complex, but at rates too low for the limits of detection afforded by current ultrastructural methods. Certainly the DA innervation density of the PFC is lower than that seen in the striatum. Moreover, DA axons in the rat PFC more frequently target dendritic shafts as opposed to spines (Section 3.2). Still, it should be noted that in the investigation of hippocampal inputs to the PFC, not a single spine postsynaptic to hippocampal axons was found to receive a secondary synapse, even from unlabeled axons (Carr and Sesack, 1996). Similarly, in the study of callosal axons from the contralateral PFC, no examples of convergence onto spines with DA axons were detected, but triadic arrangements involving GABA axons were found (Carr and Sesack, 1998). These observations argue for specificity in the modulation by DA of particular sources of glutamate at the level of dendritic spines. Further studies are needed in order to verify this hypothesis. Regardless, it needs to be kept in mind that DA's ability to modulate glutamate transmission in the immediate sense may not be its primary function. Rather, it may be the long-term plasticity changes evoked by DA receptor activation (Section 5.3), and how this plasticity is modified throughout learning, that is most relevant to understanding the capacity of this essential signaling agent for modulating appropriate cognitive functions of the PFC (Miller, 2000).

2.2.5. Nonpyramidal Cell Targets

In addition to spiny pyramidal neurons, DA axons in the PFC of rats and monkeys synapse onto nonpyramidal local circuit interneurons as illustrated schematically in Figure 2 (Krimer et al., 1997; Sesack et al., 1995b; Smiley and Goldman-Rakic, 1993). These contacts appear to be less numerous than those onto pyramidal cells, with only 0.1–0.8 contacts per 100 μm length of dendrite (Krimer et al., 1997). In addition, these DA synapses are almost exclusively onto dendritic shafts, as expected given the relative absence of spines on nonpyramidal neurons. Whether there is selectivity for particular subpopulations of interneurons (Condé et al., 1994; Gabbott and Bacon, 1996) has not yet been fully investigated. In the monkey PFC, ultrastructural analyses have reported DA synaptic inputs onto local circuit neurons expressing the calcium-binding protein parvalbumin (Sesack et al., 1997) but not onto those expressing calretinin (Sesack et al., 1995a). Calbindin-containing cells are difficult to assess as this calcium-binding protein is also found in a population of pyramidal cells. Furthermore, similar studies have not been conducted in the rat to examine whether DA axons in rodent PFC target specific classes of interneurons. To some extent, even partial selectivity in contacting the parvalbumin class of local circuit neurons is not surprising, given that this population receives more total synaptic input compared to other interneuron types (Glausier et al., 2009; Gulyás et al., 1999; Melchitzky and Lewis, 2003). Once again, it remains to be determined whether this represents true synaptic specificity for DA axons innervating the PFC or whether the contacts of this neuromodulator are in fact diffuse and nonspecific as originally speculated (Parnavelas and Papadopoulos, 1989).

2.3. DA TRANSPORTER LOCALIZATION

One of the most surprising characteristics of the mesocortical innervation is the sparse presence of specific plasma membrane transporter for dopamine (DAT) as illustrated schematically in Figure 3. Early light microscopic views of DAT in the rat mPFC clearly indicated a relative absence of this protein, especially when compared to other markers for DA axons in adjacent sections (Ciliax et al., 1995; Sesack et al., 1998). Similar observations were made in primates and humans (Ciliax et al., 1999; Lewis et al., 2001). Electron microscopic investigation in the rat and monkey PFC also revealed low levels of this protein and a predominant localization to intervaricose segments of axons (Lewis et al., 2001; Sesack et al., 1998). These observations differ markedly from what

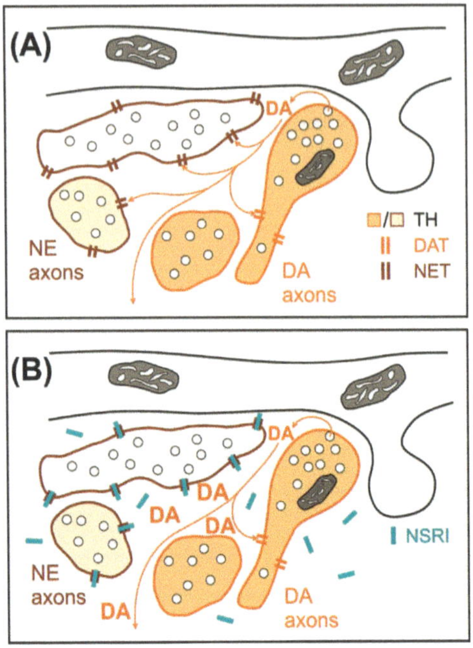

Figure 3 Schematic drawings representing the neuropil of the PFC. (A) DA axons contain readily detectable TH but sparse or no labeling for DAT. DAT that is present tends to occur mainly along intervaricose segments of axons. Norepinephrine axons express weakly detectable or no TH but contain abundant NET. As NET has a higher affinity for DA than for norepinephrine, it represents a principal means for clearing extracellular DA in addition to diffusion. (B) Administration of a norepinephrine-selective reuptake inhibitor (NSRI) will block NET and therefore cause a rise in extracellular levels of DA as well as norepinephrine (not shown).

is observed for DA axons in the striatal complex, which express abundant DAT within varicosities and at perisynaptic locations (Hersch et al, 1997; Nirenberg et al., 1997, 1996). Recently, a transgenic mouse line was created in which DAT was overexpressed (Salahpour et al., 2008). Despite finding threefold higher total and membrane-associated DAT in the striatum, no increase in DAT-immunoreactive axons was detected in the mPFC of these animals, suggesting that the normally low level of DAT in this region is regulated by posttranslational mechanisms.

The relative absence of DAT from DA axons within the mPFC is largely predictable based on the broader literature. First, not all DA neurons within the ventral midbrain express robust levels of DAT mRNA. DAT message is especially low or absent from so-called dorsal tier cells that express calbindin and are relatively resistant to neurotoxic lesions (Björklund and Dunnett, 2007; Blanchard et al., 1994; Haber et al., 1995; Sanghera et al., 1994). Whether this population specifically includes neurons projecting to the PFC has not yet been investigated. Second, neurochemical studies indicate that DA is transported more slowly and less efficiently in the cortex versus the striatum (Cass and

Gerhardt, 1995; Garris et al., 1993; Garris and Wightman, 1994; Moghaddam et al., 1990; Mundorf et al., 2001; Wayment et al., 2001). Third, the major metabolite of DA in the PFC is 3-methoxytyramine (3-MT) (Karoum et al., 1994), a product of the postsynaptic enzyme catechol o methyl transferase (COMT) (Matsumoto et al., 2003); whereas less DA is converted to 3,4 dihydroxyphenylacetic acid (DOPAC) via presynaptic monoamine oxidase (MAO) (Karoum et al., 1994). Additional observations from COMT knock-out mice indicate that O-methylation of DA rather than uptake through DAT serves as a major means for clearing extracellular DA in the PFC (Gogos et al., 1998; Käenmäki et al., 2010). The findings have important implications for understanding the contribution of COMT polymorphisms in humans to the working memory functions of the PFC and the risk for developing schizophrenia (Weinberger et al., 2001).

In addition to breakdown by COMT, a substantial literature indicates that clearance of extracellular DA in the PFC occurs via uptake through the NET, which has considerable affinity for DA (Bymaster et al., 2002; Carboni et al., 1990; Cass and Gerhardt, 1995; Gresch et al, 1995; Morón et al., 2002; Valentini et al., 2004; Yamamoto and Novotney, 1998). The fact that DA can be readily converted to NE within NE axons suggests one potential explanation for why these fibers expend the energy necessary to transport DA (Miner and Sesack, 2007). This idea may further explain the difficulty in detecting the synthetic enzyme TH within cortical NE axons (Section 2.1). Furthermore, the findings have important implications for understanding how the ability of NET-selective blockers to raise DA levels specifically in the PFC might contribute to their efficacy in the treatment of depression and attention deficit hyperactivity disorder (ADHD) (Bymaster et al., 2002; Carboni and Silvagni, 2004; Garnock-Jones and Keating, 2010; Swanson et al., 2006; Tanda et al., 1994).

2.4. DA RECEPTORS AND PHYSIOLOGY

2.4.1. Overview

All five major subtypes of DA receptors are present in the PFC to variable extent. The literature as it pertains to receptor expression detected by light microscopic immunocytochemical and in situ hybridization methods has been reviewed recently (Sesack, 2010). Although that review focused primarily on the D2 receptor, some information regarding expression of the other subtypes was also included. Additional reviews on DA receptor localization are also available (Beaulieu and Gainetdinov, 2011; De Almeida et al., 2008; Yao et al., 2008). The focus here will be on ultrastructural observations within the PFC based on immunocytochemistry; where possible, an added emphasis will be on the subcellular localization of these proteins. The bulk of this literature comes from studies of the primate PFC and represents the venerable expertise of the late Patricia Goldman-Rakic and her associates. Observations from the rodent are, therefore, only where specifically indicated below.

2.4.2. D1 Family

2.4.2.1. General Observations

The D1 receptor family consists of D1 and D5 subtypes, which are pharmacologically indistinguishable, save for the 10-fold higher affinity of the D5 receptor for DA (Missale et al., 1998; Sunahara et al., 1991). The D1 class of receptors appears to be of greater importance to the working memory functions of the PFC than the D2 receptor family (Abi-Dargham et al., 2002; Goldman-Rakic et al., 2000; Sawaguchi and Goldman-Rakic, 1994; Vijayraghavan et al., 2007; Zahrt et al., 1997). Within the PFC, the D1 receptor is found in greater abundance than the D5 subtype. In fact, the D1 receptor has the densest expression in the PFC of all the DA receptor subtypes (Lidow et al., 1991; Meador-Woodruff et al., 1996), although this density is notably lower than in the basal ganglia. Similar observations have been made in the rat (Santana et al., 2009).

2.4.2.2. Postsynaptic D1 Receptor: Cellular and Subcellular Localization

Within the perikarya of pyramidal neurons, the D1 receptor is found mainly in relationship to the Golgi apparatus and not in association with the plasmalemma (Paspalas and Goldman-Rakic, 2004). As shown schematically in Figure 4(A), D1 receptor protein that is associated with the plasma membrane is observed principally in distal compartments, specifically spines and small diameter dendritic shafts (Bergson et al., 1995; Bordelon-Glausier et al., 2008; Muly et al., 2001; Paspalas and Goldman-Rakic, 2005; Smiley et al., 1994). Within spines, the D1 receptor is often colocalized with the signaling agent protein phosphatase-1, specifically the alpha and gamma-1 subtypes (Muly et al., 2001). D1 receptor labeling is frequently found in close proximity to the asymmetric synapses formed by glutamate nerve terminals in distal dendrites and spines (see Figure 4(A)), and this relationship has been interpreted as an anatomical substrate for D1 modulation of glutamate transmission (Smiley et al., 1994). This supposition might then explain the functional significance of the triadic arrangement involving DA and glutamate axons forming convergent synapses onto common spines (Goldman-Rakic et al., 2000) (Section 3.4). However, as Figure 4(A) depicts, the D1 receptor protein typically is not found immediately postsynaptic to DA axons (Smiley et al., 1994). The later finding indicates a dominant extrasynaptic placement for D1 receptors despite their distal position within the pyramidal cell dendritic tree.

The D1 receptor subtype has also been reported in GABA local circuit interneurons in the PFC (Glausier et al., 2009; Muly et al., 1998; Paspalas and Goldman-Rakic, 2005) (Figure 4(A)). Dual immunofluorescence studies indicate enrichment in parvalbumin-containing as opposed to calretinin-containing cells. Within GABA neurons, the D1 receptor is observed mainly in association with the Golgi apparatus in addition to the endoplasmic reticulum and vesicular membranes. D1 receptor immunoreactivity located in proximity to the plasma membrane is observed mainly within distal dendrites (Bordelon-Glausier et al., 2008; Muly et al., 1998).

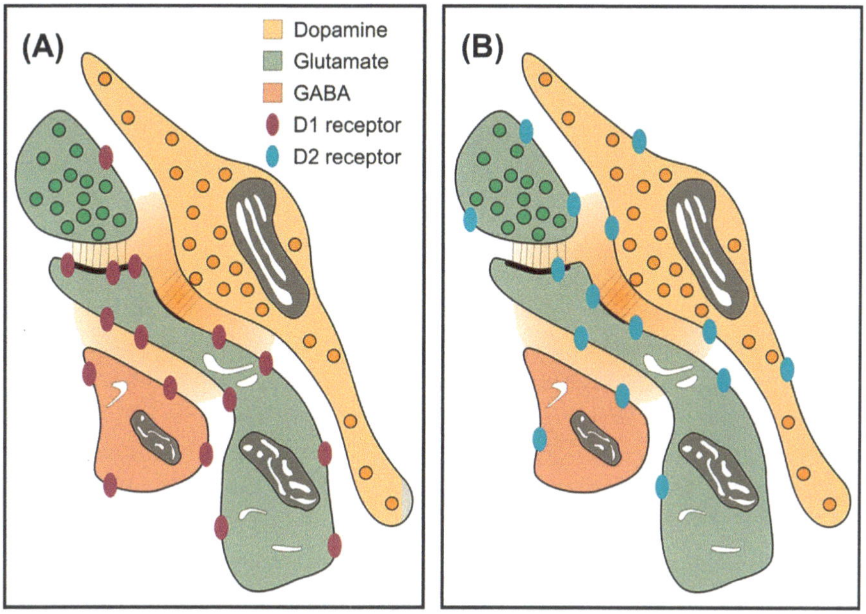

Figure 4 Schematic depiction of DA receptor plasmalemmal distribution in distal dendritic and axonal compartments of the PFC. The principal pre- and postsynaptic sites of D1 (panel A) and D2 (panel B) receptor membrane localization are illustrated. See text for detailed description.

2.4.2.3. Presynaptic D1 Receptor

Evidence suggests that GABA cells synthesizing the D1 receptor transport it to their intrinsic axons, given that presynaptic profiles forming symmetric synapses onto somata and dendrites contain immunolabeling for this receptor (Bordelon-Glausier et al., 2008; Muly et al., 1998). The morphology of these axons and their content of parvalbumin are consistent with their derivation from GABA interneurons (Muly et al., 1998). Other axon varicosities labeled for the D1 receptor have excitatory-type, presumed glutamate morphology (Bergson et al., 1995; Bordelon-Glausier et al., 2008; Paspalas and Goldman-Rakic, 2005; Smiley et al., 1994) (Figure 4(A)). In the deep layers of the PFC, these D1 receptor-immunoreactive axons form asymmetric synapses onto spines, most likely of pyramidal neurons, and do not synapse onto GABA interneurons labeled for parvalbumin (Paspalas and Goldman-Rakic, 2005). Hence, in theory at least, this localization pattern creates target specificity in the presynaptic actions of DA on glutamate transmission via local collaterals, i.e. presynaptic D1 receptors represent a substrate for DA modulation of pyramidal to pyramidal cell transmission but not pyramidal to nonpyramidal cell communication (Gao and Goldman-Rakic, 2003; Gao et al., 2001; Paspalas and Goldman-Rakic, 2005). Anatomical proof of this concept awaits further experimentation.

2.4.2.4. Postsynaptic D5 Receptor: Cellular and Subcellular Localization

The D5 receptor, which is closely related to D1 receptors in structure and pharmacological specificity, has also been localized within the PFC. Most D5-immunoreactive neurons appear to be pyramidal cells, and these frequently coexpress the D1 receptor (Bergson et al., 1995). By ultrastructural examination, the D5 receptor is observed in more proximal locations relative to the D1 receptor, i.e. the D5 protein is less commonly found in spines and more often seen in dendritic shafts and in perikarya (Bergson et al., 1995; Paspalas and Goldman-Rakic, 2004). There is still some degree of overlap of the two subtypes, however, in that a population of spines expressing D1 receptor immunolabeling is also immunoreactive for the D5 subtype (Bordelon-Glausier et al., 2008). The functional advantage of having spines with both D1 and D5 receptors is as yet unclear although such spines will have a greater dynamic range of responses to DA.

The overall more proximal location of the D5 receptor is contrary to the relatively distal placement of DA axon synapses (Sections 3.1 and 3.2). This observation suggests that the D5 receptor, like the D1 subtype, is activated primarily via extrasynaptic means, albeit following DA diffusion over a greater distance. The relatively proximal location of the D5 receptor further suggests that DA acting at this site serves to modulate the excitability of pyramidal neurons (Section 5.3.2) regardless of which specific source(s) of glutamate transmission produced depolarization further out in the distal dendrites (Goldman-Rakic et al., 2000; Paspalas and Goldman-Rakic, 2004). Together, these observations are consistent with findings that DA acting through a D1-type receptor modulates specific ionic conductances that operate along proximal compartments of pyramidal cells (Gorelova and Yang, 2000; Henze et al., 2000; Yang and Seamans, 1996).

Regarding the D5 receptor, an important paper examining its localization within the somata of PFC neurons demonstrated a specific relationship with the subsurface cisterns of the perisomatic plasma membrane (Paspalas and Goldman-Rakic, 2004). This particular site of D5 receptor localization, termed an Amicrodomain@, occurs only in pyramidal neurons and is specific for D5 versus D1 or D2 receptors. The known function of subsurface cisterns in calcium sequestration and release (Park et al., 2000) suggests that this microdomain may underlie the ability of D1-type receptors to regulate intracellular levels of calcium (Missale et al., 1998). Further support of this interpretation lies in the extensive colocalization of the D5 receptor with the inositol trisphosphate (IP3) receptor in the same microdomains (Paspalas and Goldman-Rakic, 2004); the binding of IP3 typically increases intracellular calcium.

The D5 microdomain is important because it represents a highly specific functional compartment that is not activated by synaptically released DA (Paspalas and Goldman-Rakic, 2004). Indeed, axosomatic synapses of DA axons are rare in the PFC (Smiley and Goldman-Rakic, 1993), and axons with the morphological features of DA varicosities have not been seen in juxtaposition to the D5 receptor microdomains. Instead, the regions of D5 receptor expression along subsurface cisterns are typically juxtaposed by axons forming asymmetric synapses onto structures in the neuropil adjacent to the somata (Paspalas and Goldman-Rakic, 2004). Hence, DA can only reach the perisomatic D5 receptor microdomain via a considerable degree of diffusion in the extracellular space. The presence of this mechanism within the PFC may offer some explanation for the low levels of DA transporter in this region, which facilitates diffusion (Section 4).

2.4.2.5. Presynaptic D5 Receptor

In addition to somata and dendrites, the D5 receptor has also been localized to the axon initial segments of pyramidal neurons (Bergson et al., 1995) (but see Paspalas and Goldman-Rakic, 2004), and D5 receptor protein has been found in axon varicosities with morphological features of a glutamate phenotype. The majority of these boutons have been reported to coexpress the D1 receptor (Bordelon-Glausier et al., 2008). D5 immunoreactive axons of nonglutamate morphology have also been observed (Bergson et al., 1995). This finding is consistent with D5 receptor localization to local circuit interneurons, particularly those that express calretinin as opposed to parvalbumin, the opposite expression pattern from the D1 receptor (Glausier et al., 2009). Of course, the D5 receptor-labeled axons with nonglutamate morphology may also be of extrinsic origin (Bergson et al., 1995).

2.4.3. D2 Family

2.4.3.1. General Observations

The D2 receptor family for DA transduction consists of D2, D3, and D4 subtypes. Interest in this class of receptors is fueled in part by the known dependence of antipsychotic drug efficacy on blockade of D2-type receptors (Strange, 2001; Wilson et al., 1998). While perhaps not exhaustively tested, evidence from rodent studies suggests that D1 and D2 receptors are mostly if not exclusively expressed by separate populations of PFC neurons, the only potential exception being layer V where these two subtypes show the greatest overlap (Santana et al., 2009). The one study where colocalization was specifically investigated, however, found no evidence to support coexpression by single classes of cells in the rat (Gaspar et al., 1995).

2.4.3.2. Postsynaptic D2 Receptor: Cellular and Subcellular Localization

D2 receptors are expressed by both pyramidal and nonpyramidal cells in the PFC of rodents, monkeys, and humans (de Almeida and Mengod, 2010; Fitzgerald et al., 2012; Khan et al., 1998a, 1998b, 2001; Santana et al., 2009). In rats, the D2 receptor is more commonly observed in parvalbumin- than calbindin-containing local circuit interneurons (Le Moine and Gaspar, 1998) (see also Khan et al., 2001). The D2 receptor is also expressed by astrocytes and their processes in this region across all species examined to date (Fitzgerald et al., 2012; Khan et al., 2001; Negyessy and Goldman-Rakic, 2005; Paspalas et al., 2006).

Within PFC neurons of primate, including pyramidal cells, the D2 receptor protein is seldom observed in the perikarya; when weakly present, it is typically in association with the Golgi apparatus, endoplasmic reticulum, endosomes, and rarely with the plasma membrane (Negyessy and Goldman-Rakic, 2005; Paspalas and Goldman-Rakic, 2004; Paspalas et al., 2006). In a recent study of the mouse prelimbic PFC, however, moderate to intense D2 receptor labeling within somata was detected (Fitzgerald et al., 2012). The subcellular distribution of this protein included endomembranes and the plasmalemma. Whether these observations reflect true species differences or the characteristics of the different D2 receptor antibodies employed remains to be determined.

In primate PFC neurons, stronger D2 receptor immunoreactivity is more typically detected in small diameter dendrites and spines (Negyessy and Goldman-Rakic, 2005; Paspalas and Goldman-Rakic, 2004; Paspalas et al., 2006) as shown schematically in Figure 4(B). D2 receptor protein is often found in association with cytoplasmic organelles in these regions, including the spine apparatus and other vacuoles (Negyessy and Goldman-Rakic, 2005; Paspalas et al., 2006). Furthermore, the D2 receptor protein is located at both nonsynaptic (Negyessy and Goldman-Rakic, 2005; Paspalas et al., 2006) and perisynaptic portions of the plasmalemma (Negyessy and Goldman-Rakic, 2005) (Figure 4(B)). The perisynaptic distribution includes the asymmetric synapses formed by presumed glutamate nerve terminals and therefore resembles the localization of the D1 receptor (Smiley et al., 1994).

D2 receptor labeling in distal compartments is often colocalized with immunostaining for the interacting protein neuronal calcium sensor-1 (NCS-1) (Negyessy and Goldman-Rakic, 2005). Furthermore, the DA D2 receptor has been shown to undergo endocytosis via a clathrin-mediated process solely along distal dendritic membranes (Paspalas et al., 2006). The dendrites exhibiting D2 receptor immunoreactive clathrin-associated structures typically do not receive synaptic input, although the labeled portion of the membrane is apposed by nonsynapsing axons. The phenotype of these axons has not yet been investigated, although they have morphological features suggestive of DA fibers, in which case the D2 receptors may be endocytosed upon binding DA after short-range diffusion; this would constitute an additional microdomain for DA transmission (Paspalas et al., 2006).

2.4.3.3. Presynaptic D2 Receptor

Cortical neurons expressing D2 receptors might transport this protein into local collaterals, as D2 receptor-labeled axon terminals with both glutamate-type (shown in Figure 4(B)) and non-glutamate-type morphology have been observed within the PFC of monkeys (Negyessy and Goldman-Rakic, 2005) and mice (Chiu et al., 2010; Fitzgerald et al., 2012). Some of the terminals without glutamate morphology contain parvalbumin immunoreactivity, indicating a GABAergic phenotype (Fitzgerald et al., 2012). That D2 receptor-labeled axons might also derive from extrinsic sources is of course a possibility. Specifically, some of the immunoreactive profiles forming symmetric synapses or failing to exhibit synaptic specializations might represent D2 autoreceptor presentation in DA axons (Figure 4(B)). The relatively infrequent observation of such axon varicosities, however, is consistent with biochemical and electrophysiological evidence for relatively low autoreceptor control in the mesocortical DA cell population (Tzschentke, 2001). Conversely, the relatively frequent occurrence of D2 receptor-labeled axon terminals forming asymmetric synapses in both rodent and primate (Fitzgerald et al., 2012; Negyessy and Goldman-Rakic, 2005) indicates that presynaptic regulation of glutamate transmission is an important function of D2 receptors in the PFC (illustrated in Figure 4(B)). In addition, PFC pyramidal cells that express the D2 receptor clearly transport this protein into their extrinsically directed axons, where they are observed in striatal and amygdala target regions, at least in the rat (Pinto and Sesack, 2008; Wang and Pickel, 2002).

2.4.4. D3 and D4 Receptors

The other variants of the D2 receptor family, D3 and D4 subtypes, are expressed in the PFC although less is known about their detailed subcellular localization. The presence of the D3 receptor has been described in layer III–V nonpyramidal, and possibly pyramidal cells, in rat and human cortex (Khan et al., 1998a). To date, no ultrastructural studies have investigated the expression and subcellular distribution of this receptor subtype in the PFC, perhaps due to concerns regarding antibody specificity.

The D4 receptor is reportedly the most abundant of the D2 family of DA receptors expressed in the rat cortex (Khan et al., 1998a); estimates differ about whether this is also true in the primate (de Almeida and Mengod, 2010; Khan et al., 1998a). This observation is of particular interest given the known involvement of the PFC in the pathophysiology of schizophrenia (Beneyto and Lewis, 2011; Tan et al., 2009) and the fact that clozapine, an antipsychotic medication with few motoric side effects, has higher affinity for D4 than for other DA receptors (Seeman, 1992).

Within cortical regions in both rats and primates, pyramidal and nonpyramidal neurons express immunoreactivity for this protein (Khan et al., 1998a; Mrzljak et al., 1996; Rivera et al., 2008; Wedzony et al., 2000). By light microscopy, D4 receptor immunoreactivity has been shown to extend from somata into the proximal apical and basilar dendritic trees of presumed pyramidal cells and also has been detected within axon initial segments (Rivera et al., 2008; Wedzony et al., 2000). The D4 receptor appears to be robustly localized to parvalbumin-containing local circuit interneurons, and some calbindin neurons expressing D4 receptor have been detected as well (de Almeida and Mengod, 2010; Mrzljak et al., 1996; Wedzony et al., 2000). At the subcellular level, D4 receptor immunolabeling in the cytoplasm was observed in association with the endoplasmic reticulum; immunoreactivity was also seen along the plasma membrane (Mrzljak et al., 1996). Although DA axons were not identified in this study, there did not appear to be axon varicosities in contact with the membrane at sites of D4 receptor protein expression, indicating that this receptor subtype is dominantly extrasynaptic.

2.5. PHYSIOLOGY

2.5.1. Conceptual and Methodological Complexities

The issue of DA physiology is a complicated one for several reasons, as described in greater detail by many others (Cohen et al., 2002; Dreher and Burnod, 2002; Durstewitz and Seamans, 2008; O'Donnell, 2003; Seamans and Yang, 2004). First, DA is capable of issuing a range of subtle and complex actions on PFC neurons that depend in part on the levels of DA and the subtype of receptor that is activated; the D1 receptor in particular is associated with a well-documented U-shaped dose response curve (Goldman-Rakic et al., 2000; Vijayraghavan et al., 2007). Second, other important factors contribute to the observed actions of DA, including the specific layer of cortex examined, the ion channels and signaling pathways acted upon, the activity state and synaptic input of the target neurons, direct versus indirect effects produced via pyramidal versus nonpyramidal local circuit neurons, respectively, as well as the developmental age of the animals (O'Donnell, 2010; Tseng and O'Donnell, 2007b). Third, in addition to the particulars of neuronal conditions surrounding the application or release of DA, there are a myriad of methodological issues that can alter the physiological response characteristics reported for DA on PFC neurons. These include whether the recording is performed in vivo versus in vitro, from anesthetized or awake animals, using sharp intracellular or patch electrodes, and whether DA is applied systemically, focally, or via bath perfusion. When recorded in vivo, the neuronal responses to DA would further depend on the oscillatory patterns of PFC cells and, hence, on task demands (Benchenane et al., 2011; Towers and Hestrin, 2008).

2.5.2. DA Actions on Cell Excitability

Despite these complexities, there is some relative consensus regarding the effects of DA that appear to be most consistent. DA acting on D1-type receptors primarily tends to increase neuronal excitability of pyramidal cells (Chen et al., 2007; Seamans and Yang, 2004; Sun et al., 2005; Tseng and O'Donnell, 2004; Tseng and O'Donnell, 2005; Zheng et al., 1999), although exceptions are reported (Orozco-Cabal et al., 2008). Consistent with increased excitability, D1-type receptors tend to enhance long-term potentiation of glutamate signaling (Chen et al., 2007; Gurden et al., 2000; Sun et al., 2005). Conversely, DA effects mediated via D2-type receptors lead mainly to reductions of excitability and of synaptic responses to glutamate (Gulledge and Jaffe, 1998; Sun et al., 2005; Tseng and O'Donnell, 2004; Tseng and O'Donnell, 2007a; Zheng et al., 1999). D2-type receptor actions also tend to favor long-term depression of glutamate transmission (Otani et al.,

1998). There are some reports of D2 receptors increasing excitability in pyramidal cells (Gee et al., 2012) and local circuit interneurons (Tseng and O'Donnell, 2007a).

As a result of these observations, it has been proposed that DA acting on D1-type receptors serves a gating function by inducing a combination of increased membrane potential but decreased firing activity (Bernardi et al., 1982; O'Donnell, 2003) that may restrict responsivity of pyramidal cells only to strong inputs (Durstewitz and Seamans, 2008; O'Donnell, 2003; Seamans and Yang, 2004). In this way, DA would support neuronal ensembles with synchronous firing so that they can maintain information Ain mind@ for cognitive computations (Durstewitz and Seamans, 2008). Conversely, acting via a D2-type receptor mechanism, DA would reduce the strength of neuronal networks, allowing them greater flexibility for switching between states (Durstewitz and Seamans, 2008).

2.5.3. DA Actions on Neuronal Plasticity

In addition to these acute actions, several studies indicate that DA also contributes to long-term plasticity in target neurons via structural changes involving dendritic spines. In brief, DA appears to exert an Atrophic@ influence that enhances spine stability. Much of this work has focused on the basal ganglia, where DA acting via a D2 receptor and subsequent regulation of a Cav1.3 L-type calcium channel serves as a principal mechanism for spine stabilization (Day et al., 2006). Lesion-induced loss of DA signal, such as occurs in animal models of Parkinson's disease, results in the retraction of dendritic spines of medium spiny neurons and loss of the glutamate synapses they receive (Ingham et al., 1998). This effect is specific to cells that express the D2 receptor and does not occur in D1 receptor-expressing neurons (Day et al., 2006). In addition to spine loss, alterations in dendritic length and branching have also been observed (Day et al., 2006). A substantial body of research has also demonstrated a more complex set of changes in spine density following psychostimulant use and/or withdrawal; such alterations depend on the ability of these drugs to affect DA transmission (Burchett, 2005; Castner and Williams, 2007; Dietz et al., 2009; Robinson and Kolb, 2004).

Psychostimulant exposure can also modify spine density in pyramidal neurons of the PFC (Castner and Williams, 2007; Robinson and Kolb, 2004). The mechanisms underlying these drug-induced effects are difficult to establish, given the variable treatment and withdrawal procedures used by different laboratories. Arguably, a more simplified approach is the lesion-induced loss of DA, which until recently had not been examined for the PFC. Studies have now determined that loss of DA after axonal denervation does in fact lead to a reduction of spine density in layer V pyramidal neurons in the rat PFC (Wang and Deutch, 2008). Basal but not apical dendritic length is also reduced, and structural alterations are specific for the PFC versus the motor cortex, which in the rat receives only sparse DA input. Perhaps paradoxically, the loss of spines is reversed by the atypical antipsychotic olanzapine; the typical antipsychotic drug haloperidol does not produce this effect (Wang and Deutch, 2008). Also surprising is the fact that both D1 and D2 receptors appear to be involved in spine stabilization in the PFC, which is in contradistinction to the selective influence of the D2 receptor in the striatum. This may reflect, at least in part, the fact that the study was conducted in mutant mouse lines in which spines of layer V pyramidal cells were reduced in D1 and D2 receptors knock-out mice, but not in D4 receptor knock-out animals (Wang et al., 2009).

2.6. COLOCALIZATION OF DA AND GLUTAMATE

2.6.1. Potential Excitatory Actions of the Mesocortical Pathway

Early electrophysiological studies using intracellular recordings suggested that DA mediated an excitatory influence on mPFC neurons (Bernardi et al., 1982). Conversely, most experiments utilizing extracellular physiology reported a dominant inhibitory action of DA in the PFC (Bunney and Aghajanian, 1976; Ferron et al., 1984; Pirot et al., 1992; Sesack and Bunney, 1989; Thierry et al., 1986). When electrical stimulation of the VTA instead of local DA application was examined, short latency excitatory postsynaptic potentials were observed (Mercuri et al., 1985). Based on the estimated conduction velocity, however, it was determined that the excitatory responses to VTA stimulation occurred with a latency too fast for the unmyelinated and slowly conducting DA axons (Mercuri et al., 1985; Wang, 1981). Interpretations of the original findings therefore had to encompass the possibility of electrical stimulation of fibers of passage and/or the induction of antidromic action potentials that subsequently produced orthodromic excitation of neighboring cells through local collaterals.

The third idea that was not considered until much later was the presence within the VTA of a population of glutamate-containing cells that had faster conduction speeds than DA neurons. Even after its conception, considerable time passed before that particular theory was adequately tested. Electrophysiological studies mindful of this new interpretation again reported excitatory actions of VTA stimulation on mPFC neurons. In this case, however, lesions of DA neurons by the selective neurotoxin 6-hydroxydopamine suggested that the excitatory effects were in fact mediated by glutamate released from DA axons (Lapish et al., 2007; Lavin et al., 2005). Nevertheless, a more recent consideration of this work indicates that nonspecific damage associated with the lesion was actually responsible for loss of the evoked excitatory responses (Gorelova et al., 2012).

2.6.2. A Viable Cellular Marker for Glutamate Neurons

The main impediment with anatomical identification of transmitter glutamate was the difficulty in proving that this amino acid was not present within cells as a metabolic constituent. Hence, a major advance in identifying transmitter glutamate came with the discovery of proteins responsible for transporting glutamate into synaptic vesicles

(Fremeau et al., 2001), a necessary preliminary step in the vesicular exocytotic mechanism. A series of studies then followed in which the presence within the VTA of mRNA for the vesicular transporter type 2 (vGlut2) protein was identified (Hur and Zaborszky, 2005; Kawano et al., 2006; Yamaguchi et al., 2007). The first conclusive evidence that these cells had projections to the mPFC was offered by Hur and Zaborszky who combined retrograde tract-tracing with in situ hybridization for vGlut2 mRNA (Hur and Zaborszky, 2005). Despite this clear demonstration of a mesocortical glutamate projection, two major questions remained, namely whether any portion of the VTA glutamate cell population also contained DA as a cotransmitter and whether such neurons projected to the mPFC. A complete consideration of the essential issue of DA-glutamate colocalization is beyond the scope of this chapter, which will therefore be confined to certain key observations.

2.6.3. Glutamate in Midbrain DA Neurons

DA neurons isolated in primary cultures clearly express glutamate characteristics and form excitatory autapses (Dal Bo et al., 2004; Sulzer et al., 1998). These observations may largely indicate regression to an earlier stage of development in which many midbrain DA cells express the glutamate phenotype (Berube-Carriere et al., 2009; Dal Bo et al., 2008; Descarries et al., 2008). Such high levels of coexpression are also detected in adults following neurotoxic injury to the DA system (Berube-Carriere et al., 2009; Descarries et al., 2008). In the healthy adult brain, however, colocalization of markers for DA and glutamate phenotype occurs at a frequency much lower than during early development or following injury. Estimates suggest that only ~3% of VTA neurons in total contain mRNA for vGlut2 (Nair-Roberts et al., 2008), although there is an obvious gradient increasing lateral to medial, and certain subdivisions of the VTA contain a higher proportion of these cells (Kawano et al., 1996; Yamaguchi et al., 2007, 2011). The proportion of DA neurons that coexpress vGlut2 mRNA also varies by a similar gradient and subdivision dependence (Kawano et al., 1996; Yamaguchi et al., 2007) (Figure 5).

As illustrated in Figure 5, the SN appears to be devoid of a glutamatergic cell population in adult rats (Yamaguchi et al., 2007). This observation is consistent with the virtual absence of vGlut proteins from DA axons throughout the striatal complex, including the accumbens, in adult animals (Berube-Carriere et al., 2009; Descarries et al., 2008; Gorelova et al., 2012; Moss et al., 2011). The later observations are inconsistent with one other anatomical study (Yamaguchi et al., 2011) and several electrophysiological experiments using optogenetic approaches and transgenic mice indicating that the DA innervation to the accumbens does corelease glutamate (Chuhma et al., 2004; Hnasko et al., 2010; Stuber et al., 2010; Tecuapetla et al., 2010) (Figure 5). Discrepancies between anatomical and physiological observations are not uncommon in the DA literature, and as with others before, this particular difference will no doubt be resolved with further experimentation.

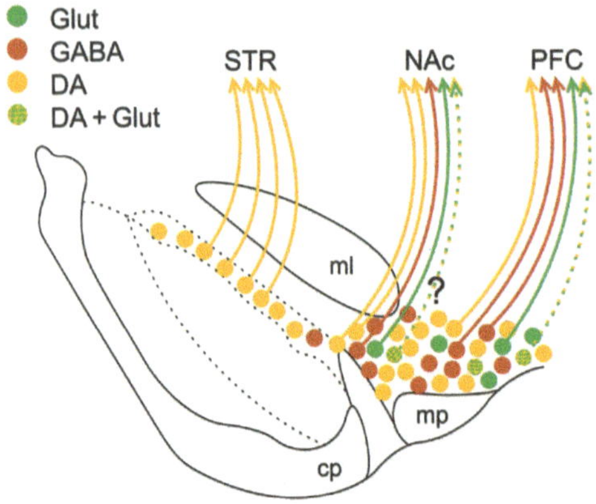

Figure 5 Schematic depiction of the rat midbrain illustrating the approximate distribution of DA, GABA, and glutamate (Glut) neurons and their main projections to forebrain targets including the dorsal striatum (STR), nucleus accumbens (NAc), and PFC. The projections to the NAc and PFC are shown as spatially distinct for ease of illustration; in reality, they are more intermixed. The contribution of DA cells to these projections decreases along a lateral to medial gradient with GABA and Glut neurons making up an increasing proportion of the pathways. DA neurons that colocalize Glut (DA + Glut), as well as Glut only cells, have been verified to project to the PFC (Gorelova et al., 2012; Yamaguchi et al., 2011). The characterization of DA + Glut projections to the NAc has yet to reach consensus, as represented by the question mark (see text for more details).

2.6.4. Glutamate in Mesocortical DA Cells

Returning to the PFC, two recent papers have advanced this important line of investigation by demonstrating that both glutamate-only and glutamate-DA coexpressing neurons contribute to the mesoprefrontal projection, especially from the medial cell groups located in the rostrolinear, rostral VTA, and parabrachial nuclei (Gorelova et al., 2012; Yamaguchi et al., 2011). The innervation has been demonstrated by retrograde tract-tracing with dual immunocytochemistry and in situ hybridization as well as by dual immunofluorescence and confocal microscopy. These observations are depicted schematically in Figure 5. The findings bring greater understanding to the mesocortical pathway and the possibility that some part of its functional purpose may be to produce glutamatergic excitation of mPFC neurons in concert with DA neuromodulatory effects. Of course the synaptic targets of the glutamate-only projections are not known, nor is it clear whether the axons colocalizing DA and glutamate form a particular subset of the overall DA synaptic population. It is interesting to speculate that the synapses exhibiting asymmetric character in the rat mPFC (Séguéla et al., 1988) (Section 3.1) might be those that contain both transmitters. Furthering this speculation, one might add that vGlut2 mRNA is absent from the SN and that DA axons in the dorsal striatum never form asymmetric synapses (Descarries et al., 1996; Hanley and Bolam, 1997; Moss and Bolam, 2008; Pickel et al., 1981), nor do they contain vGlut2 protein (Moss et al., 2011). Hence, there is at least a rough correlation between the presence/absence of vGlut2 and the expression of synapse morphology with asymmetric character.

The presence of glutamate as a cotransmitter in some mesoprefrontal DA axons raises obvious questions as to the functional significance of this dual input. If a major purpose of DA is to modulate glutamate transmission, how does the presence of glutamate within some DA axons make sense within this construct? One as yet unexplored issue that might shed light on this question is the identity of the synaptic targets innervated by this specific population of afferents. If one can assume that asymmetric synapses equate to the presence of glutamate presynaptically, then it appears that the targets are no different from those innervated by DA axons forming symmetric synapses (Séguéla et al., 1988). More advanced studies are needed, however, to verify whether this is actually the case. One speculation that might be offered regarding corelease of glutamate deals with the fact that DA seems to be capable of altering the up and down states that are observed for PFC and other forebrain neurons in anesthetized animals (Lewis and O'Donnell, 2000; O'Donnell, 2003; Peters et al., 2004; Tseng and O'Donnell, 2005). Indeed, the function of forebrain DA may be the facilitation of neuronal ensemble formation rather than increasing or decreasing neuronal excitability per se (O'Donnell, 2003) (Section 5.3.2). In this case, having a low level of glutamate released along with DA from a subset of axons within the PFC may, by some as yet poorly understood mechanism, facilitate the creation and/or maintenance of stable ensembles representing cognitive information held in mind (O'Donnell, 2003).

3. CONCLUSIONS

The ultrastructural features of DA axons within the rodent and primate PFC indicate that this critical modulatory substance communicates via both synaptic and extrasynaptic means. The later mode of transmission is facilitated by the relative absence of DAT and the demonstrated presence of DA receptors at extrasynaptic sites on somata, dendrites, and axons. Both pyramidal and nonpyramidal cells are targeted by the PFC DA innervation, and DA axons converge onto common spines and distal dendrites with glutamate inputs. The origin of glutamate afferents participating in this convergent "triadic" arrangement has yet to be identified, however, with the negative findings to date suggesting that DA axons may selectively modulate glutamate transmission from one or only a few sources.

In contrast to this synaptic selectivity, a consideration of DA receptor localization and physiology creates an impression of relatively diffuse and nonspecific effects of this modulator, given that most receptor sites occur at a distance from DA varicosities and that DA alters ionic conductances across multiple somatodendritic compartments. Nevertheless, the identification of specific "microdomains" of distribution within both D1 and D2 receptor families argues that DA may have selective modulatory actions following release in concentrations high enough to permit long-range diffusion (Chergui et al., 1994). Whether such high release events correspond to the bursts by midbrain DA neurons that signal behaviorally relevant events or stress (Anstrom and Woodward, 2005; Roesch et al, 2007; Schultz, 1998) remains to be determined. The fact that the PFC itself can trigger such bursts (Gariano and Groves, 1988; Overton et al., 1996; Tong et al., 1996), combined with circuitry indicating that the PFC innervates mesoprefrontal DA cells (Carr and Sesack, 2000c), suggests a potential circuit mechanism whereby the PFC can respond to strong afferent drive by increasing its own modulatory influence by DA.

This chapter has detailed findings regarding DA synaptology and receptor and transporter localization in the PFC that have important implications for understanding the pathophysiology and treatment of mental disorders. For example, reduced modulatory influence by DA in the PFC of schizophrenics (Akil et al., 1999) will alter the excitability and plasticity functions of PFC neurons in a manner that contributes to cognitive deficits. The fact that DA axons synapse onto spines and can affect spine stability may, at least in part, help to explain the reduced spine density observed for pyramidal cell basilar dendrites in this disorder (Glantz and Lewis, 2000). Similarly, understanding how patients with depression or ADHD can respond positively to drugs that selectively block the NET (Arnsten and Pliszka, 2011; Gamo and Arnsten, 2011) is advanced by knowledge that DAT levels are low in the PFC and that the NET clears extracellular DA as well as norepinephrine in this region. Important investigations remain, including a determination of DA concentration following synaptic versus nonsynaptic release, estimation of the precise distances between DA release sites and receptors, and delineation of the functional significance of glutamate corelease with DA. Such studies will further our understanding of DA transmission in the PFC and the impact of this modulator on proper cognitive function as well as on mental disorders and their treatment.

REFERENCES

Abi-Dargham, A., Mawlawi, O., Lombardo, I., Gil, R., Martinez, D., Huang, Y., Hwang, D.R., Keilp, J., Kochan, L., Van Heertum, R., Gorman, J.M., Laruelle, M., 2002. Prefrontal dopamine D1 receptors and working memory in schizophrenia. J. Neurosci. 22, 3708–3719.

Akil, M., Pierri, J.N., Whitehead, R.E., Edgar, C.L., Mohila, C., Sampson, A.R., Lewis, D.A., 1999. Lamina-specific alterations in the dopamine innervation of the prefrontal cortex in schizophrenic subjects. Am. J. Psychiatry 156, 1580–1589.

Anstrom, K.K., Woodward, D.J., 2005. Restraint increases dopaminergic burst firing in awake rats. Neuropsychopharmacology 30, 1832–1840.

Arbuthnott, G.W., Wickens, J., 2007. Space, time and dopamine. Trends Neurosci. 30, 62–69.

Arnsten, A.F., Pliszka, S.R., 2011. Catecholamine influences on prefrontal cortical function: relevance to treatment of attention deficit/hyperactivity disorder and related disorders. Pharmacol. Biochem. Behav. 99, 211–216.

Asan, E., 1998. The catecholaminergic innervation of the rat amygdala. Adv. Anat. Embryol. Cell Biol. 142, 1–118.

Beaulieu, J.M., Gainetdinov, R.R., 2011. The physiology, signaling, and pharmacology of dopamine receptors. Pharmacol. Rev. 63, 182–217.

Beckstead, R.M., Domesick, V.B., Nauta, W.J.H., 1979. Efferent connections of the substantia nigra and ventral tegmental area in the rat. Brain Res. 175, 191–217.

Benchenane, K., Tiesinga, P.H., Battaglia, F.P., 2011. Oscillations in the prefrontal cortex: a gateway to memory and attention. Curr. Opin. Neurobiol. 21, 475–485.

Beneyto, M., Lewis, D.A., 2011. Insights into the neurodevelopmental origin of schizophrenia from postmortem studies of prefrontal cortical circuitry. Int. J. Dev. Neurosci. 29, 295–304.
Berger, B., Gaspar, P., Verney, C., 1991. Dopaminergic innervation of the cerebral cortex: unexpected differences between rodents and primates. Trends Neurosci. 14, 21–27.
Berger, B., Tassin, J.P., Blanc, G., Moyne, M.A., Thierry, A.M., 1974. Histochemical confirmation for dopaminergic innervation of the rat cerebral cortex after destruction of the noradrenergic ascending pathways. Brain Res. 81, 332–337.
Bergson, C., Mrzljak, L., Smiley, J.F., Pappy, M., Levenson, R., Goldman-Rakic, P.S., 1995. Regional, cellular, and subcellular variations in the distribution of D1 and D5 receptors in primate brain. J. Neurosci. 15, 7821–7836.
Bernardi, G., Cherubini, E., Marciani, M.G., Mercuri, N., Stanzione, P., 1982. Responses of intracellularly recorded cortical neurons to the iontophoretic application of dopamine. Brain Res. 245, 267–274.
Berube-Carriere, N., Riad, M., Dal Bo, G., Levesque, D., Trudeau, L.E., Descarries, L., 2009. The dual dopamine-glutamate phenotype of growing mesencephalic neurons regresses in mature rat brain. J. Comp. Neurol. 517, 873–891.
Björklund, A., Dunnett, S.B., 2007. Dopamine neuron systems in the brain: an update. Trends Neurosci. 30, 194–202.
Björklund, A., Lindvall, O., 1984. Dopamine-containing systems in the CNS. In: Björklund, A., Hökfelt, T. (Eds.), Handbook of Chemical Neuroanatomy. Classical Neurotransmitters in the CNS, Part 1, vol. 2. Elsevier Science Publishers, Amsterdam, pp. 55–122.
Blanchard, V., Raisman-Vozari, R., Vyas, S., Michel, P.P., Javoy-Agid, F., Uhl, G.R., Agid, Y., 1994. Differential expression of tyrosine hydroxylase and membrane dopamine transporter genes in subpopulations of dopaminergic neurons of the rat mesencephalon. Mol. Brain Res. 22, 29–40.
Bordelon-Glausier, J.R., Khan, Z.U., Muly, E.C., 2008. Quantification of D1 and D5 dopamine receptor localization in layers I, III, and V of *Macaca mulatta* prefrontal cortical area 9: coexpression in dendritic spines and axon terminals. J. Comp. Neurol. 508, 893–905.
Brennan, A.R., Arnsten, A.F., 2008. Neuronal mechanisms underlying attention deficit hyperactivity disorder: the influence of arousal on prefrontal cortical function. Ann. N.Y. Acad. Sci. 1129, 236–245.
Brozoski, T.J., Brown, R.M., Rosvold, H.E., Goldman, P.S., 1979. Cognitive deficit caused by regional depletion of dopamine in prefrontal cortex of rhesus monkey. Science 205, 929–932.
Bunney, B.S., Aghajanian, G.K., 1976. Dopamine and norepinephrine innervated cells in the rat prefrontal cortex: pharmacological differentiation using microiontophoretic techniques. Life Sci. 19, 1783–1792.
Burchett, S.A., 2005. Psychostimulants, madness, memory… and RGS proteins? Neuromolecular Med. 7, 101–127.
Bymaster, F.P., Katner, J.S., Nelson, D.L., Hemrick-Luecke, S.K., Threlkeld, P.G., Heiligenstein, J.H., Morin, S.M., Gehlert, D.R., Perry, K.W., 2002. Atomoxetine increases extracellular levels of norepinephrine and dopamine in prefrontal cortex of rat: a potential mechanism for efficacy in attention deficit/hyperactivity disorder. Neuropsychopharmacology 27, 699–711.
Carboni, E., Silvagni, A., 2004. Dopamine reuptake by norepinephrine neurons: exception or rule? Crit. Rev. Neurobiol. 16, 121–128.
Carboni, E., Tanda, G.L., Frau, R., Di Chiara, G., 1990. Blockade of the noradrenaline carrier increases extracellular dopamine concentrations in the prefrontal cortex: evidence that dopamine is taken up in vivo by noradrenergic terminals. J. Neurochem. 55, 1067–1070.
Carlin, R.K., Grab, D.J., Cohen, R.S., Siekevitz, P., 1980. Isolation and characterization of postsynaptic densities from various brain regions: enrichment of different types of postsynaptic densities. J. Cell Biol. 86, 831–845.
Carr, D.B., O'Donnell, P., Card, J.P., Sesack, S.R., 1999. Dopamine terminals in the rat prefrontal cortex synapse on pyramidal cells that project to the nucleus accumbens. J. Neurosci. 19, 11049–11060.
Carr, D.B., Sesack, S.R., 1996. Hippocampal afferents to the rat prefrontal cortex: synaptic targets and relation to dopamine terminals. J. Comp. Neurol. 369, 1–15.
Carr, D.B., Sesack, S.R., 1998. Callosal terminals in the rat prefrontal cortex: synaptic targets and association with GABA-immunoreactive structures. Synapse 29, 193–205.
Carr, D.B., Sesack, S.R., 2000a. Dopamine terminals synapse on callosal projection neurons in the rat prefrontal cortex. J. Comp. Neurol. 425, 275–283.

Carr, D.B., Sesack, S.R., 2000b. GABA-containing neurons in the rat ventral tegmental area project to the prefrontal cortex. Synapse 38, 114–123.
Carr, D.B., Sesack, S.R., 2000c. Projections from the rat prefrontal cortex to the ventral tegmental area: target specificity in the synaptic associations with mesoaccumbens and mesocortical neurons. J. Neurosci. 20, 3864–3873.
Cass, W.A., Gerhardt, G.A., 1995. In vivo assessment of dopamine uptake in rat medial prefrontal cortex: comparison with dorsal striatum and nucleus accumbens. J. Neurochem. 65, 201–207.
Castner, S.A., Williams, G.V., 2007. From vice to virtue: insights from sensitization in the nonhuman primate. Prog. Neuro-Psychopharmacol. Biol. Psychiatry 31, 1572–1592.
Chagnaud, J.L., Mons, N., Tuffet, S., Grandier-Vazeilles, X., Geffard, M., 1987. Monoclonal antibodies against glutaraldehyde-conjugated dopamine. J. Neurochem. 49, 487–494.
Chen, L., Bohanick, J.D., Nishihara, M., Seamans, J.K., Yang, C.R., 2007. Dopamine D1/5 receptor-mediated long-term potentiation of intrinsic excitability in rat prefrontal cortical neurons: Ca^{2+}-dependent intracellular signaling. J. Neurophysiol. 97, 2448–2464.
Chergui, K., Suaud-Chagny, M.F., Gonon, F., 1994. Nonlinear relationship between impulse flow, dopamine release and dopamine elimination in the rat brain in vivo. Neuroscience 62, 641–645.
Chiu, C.Q., Puente, N., Grandes, P., Castillo, P.E., 2010. Dopaminergic modulation of endocannabinoid-mediated plasticity at GABAergic synapses in the prefrontal cortex. J. Neurosci. 30, 7236–7248.
Chudasama, Y., Robbins, T.W., 2004. Dopaminergic modulation of visual attention and working memory in the rodent prefrontal cortex. Neuropsychopharmacology 29, 1628–1636.
Chuhma, N., Zhang, H., Masson, J., Zhuang, X., Sulzer, D., Hen, R., Rayport, S., 2004. Dopamine neurons mediate a fast excitatory signal via their glutamatergic synapses. J. Neurosci. 24, 972–981.
Ciliax, B.J., Drash, G.W., Staley, J.K., Haber, S., Mobley, C.J., Miller, G.W., Mufson, E.J., Mash, D.C., Levey, A.I., 1999. Immunocytochemical localization of the dopamine transporter in human brain. J. Comp. Neurol. 409, 38–56.
Ciliax, B.J., Heilman, C., Demchyshyn, L.L., Pristupa, Z.B., Ince, E., Hersch, S.M., Niznik, H.B., Levey, A.I., 1995. The dopamine transporter: immunocytochemical characterization and localization in brain. J. Neurosci. 15, 1714–1723.
Cohen, J.D., Braver, T.S., Brown, J.W., 2002. Computational perspectives on dopamine function in prefrontal cortex. Curr. Opin. Neurobiol. 12, 223–229.
Cohen, J.D., Servan-Schreiber, D., 1992. Context, cortex, and dopamine: a connectionist approach to behavior and biology in schizophrenia. Psychol. Rev. 99, 45–77.
Collins, P., Roberts, A.C., Dias, R., Everitt, B.J., Robbins, T.W., 1998. Perseveration and strategy in a novel spatial self-ordered sequencing task for nonhuman primates: effects of excitotoxic lesions and dopamine depletions of the prefrontal cortex. J. Cognit. Neurosci. 10, 332–354.
Condé, F., Lund, J.S., Jacobowitz, D.M., Baimbridge, K.G., Lewis, D.A., 1994. Local circuit neurons immunoreactive for calretinin, calbindin D-28k or parvalbumin in monkey prefrontal cortex: distribution and morphology. J. Comp. Neurol. 341, 95–116.
Dal Bo, G., Berube-Carriere, N., Mendez, J.A., Leo, D., Riad, M., Descarries, L., Levesque, D., Trudeau, L.E., 2008. Enhanced glutamatergic phenotype of mesencephalic dopamine neurons after neonatal 6-hydroxydopamine lesion. Neuroscience 156, 59–70.
Dal Bo, G., St-Gelais, F., Danik, M., Williams, S., Cotton, M., Trudeau, L.E., 2004. Dopamine neurons in culture express VGLUT2 explaining their capacity to release glutamate at synapses in addition to dopamine. J. Neurochem. 88, 1398–1405.
Day, M., Wang, Z., Ding, J., An, X., Ingham, C.A., Shering, A.F., Wokosin, D., Ilijic, E., Sun, Z., Sampson, A.R., Mugnaini, E., Deutch, A.Y., Sesack, S.R., Arbuthnott, G.W., Surmeier, D.J., 2006. Selective elimination of glutamatergic synapses on striatopallidal neurons in Parkinson disease models. Nat. Neurosci. 9, 251–259.
de Almeida, J., Mengod, G., 2010. D2 and D4 dopamine receptor mRNA distribution in pyramidal neurons and GABAergic subpopulations in monkey prefrontal cortex: implications for schizophrenia treatment. Neuroscience 170, 1133–1139.
De Almeida, J., Palacios, J.M., Mengod, G., 2008. Distribution of 5-HT and DA receptors in primate prefrontal cortex: implications for pathophysiology and treatment. Prog. Brain Res. 172, 101–115.
Descarries, L., Berube-Carriere, N., Riad, M., Bo, G.D., Mendez, J.A., Trudeau, L.E., 2008. Glutamate in dopamine neurons: synaptic versus diffuse transmission. Brain Res. Rev. 58, 290–302.

Descarries, L., Lemay, B., Doucet, G., Berger, B., 1987. Regional and laminar density of the dopamine innervation in adult rat cerebral cortex. Neuroscience 21, 807–824.
Descarries, L., Umbriaco, D., 1995. Ultrastructural basis of monoamine and acetylcholine function in CNS. Semin. Neurosci. 7, 309–318.
Descarries, L., Watkins, K.C., Garcia, S., Bosler, O., Doucet, G., 1996. Dual character, asynaptic and synaptic, of the dopamine innervation in adult rat neostriatum: a quantitative autoradiographic and immunocytochemical analysis. J. Comp. Neurol. 375, 167–186.
Dietz, D.M., Dietz, K.C., Nestler, E.J., Russo, S.J., 2009. Molecular mechanisms of psychostimulant-induced structural plasticity. Pharmacopsychiatry 42 (Suppl. 1), S69–S78.
Dreher, J.C., Burnod, Y., 2002. An integrative theory of the phasic and tonic modes of dopamine modulation in the prefrontal cortex. Neural Netw. 15, 583–602.
Durstewitz, D., Seamans, J.K., 2008. The dual-state theory of prefrontal cortex dopamine function with relevance to catechol-o-methyltransferase genotypes and schizophrenia. Biol. Psychiatry 64, 739–749.
Fallon, J.H., Loughlin, S.E., 1982. Monoamine innervation of the forebrain: collateralization. Brain Res. Bull. 9, 295–307.
Fallon, J.H., Loughlin, S.E., 1987. Monoamine innervation of cerebral cortex and a theory of the role of monoamines in cerebral cortex and basal ganglia. In: Peters, A., Jones, E.G. (Eds.), Cerebral Cortex, Further Aspects of Cortical Function, Including Hippocampus. Plenum Press, New York, pp. 41–127.
Ferron, A., Thierry, A.M., Le Douarin, C., Glowinski, J., 1984. Inhibitory influence of the mesocortical dopaminergic system on spontaneous activity or excitatory response induced from the thalamic mediodorsal nucleus in the rat medial prefrontal cortex. Brain Res. 302, 257–265.
Fitzgerald, M.L., Chan, J., Mackie, K., Lupica, C.R., Pickel, V.M., 2012. Altered dendritic distribution of dopamine D2 receptors and reduction in mitochondrial number in parvalbumin-containing interneurons in the medial prefrontal cortex of cannabinoid-1 (CB1) receptor knockout mice. J. Comp. Neurol. (Epub ahead of print).
Floresco, S.B., Magyar, O., Ghods-Sharifi, S., Vexelman, C., Tse, M.T., 2006. Multiple dopamine receptor subtypes in the medial prefrontal cortex of the rat regulate set-shifting. Neuropsychopharmacology 31, 297–309.
Fremeau Jr, R.T., Troyer, M.D., Pahner, I., Nygaard, G.O., Tran, C.H., Reimer, R.J., Bellocchio, E.E., Fortin, D., Storm-Mathisen, J., Edwards, R.H., 2001. The expression of vesicular glutamate transporters defines two classes of excitatory synapse. Neuron 31, 247–260.
Freund, T.F., Powell, J.F., Smith, A.D., 1984. Tyrosine hydroxylase-immunoreactive boutons in synaptic contact with identified striatonigral neurons, with particular reference to dendritic spines. Neuroscience 13, 1189–1215.
Fuster, J.M., 1997. The Prefrontal Cortex: Anatomy, Physiology, and Neuropsychology of the Frontal Lobe. Lippincott-Raven Publishers, Philadelphia.
Fuster, J.M., 2001. The prefrontal cortex—an update: time is of the essence. Neuron 30, 319–333.
Fuxe, K., Agnati, L.F., 1991. Two principal modes of electrochemical communication in the brain: volume versus wiring transmission. In: Fuxe, K., Agnati, L.F. (Eds.), Volume Transmission in the Brain: Novel Mechanisms for Neural Transmission. Raven Press, Ltd., New York, pp. 1–9.
Gabbott, P.L.A., Bacon, S.J., 1996. Local circuit neurons in the medial prefrontal cortex (areas 24a, b, c, 25 and 32) in the monkey: I. cell morphology and morphometrics. J. Comp. Neurol. 364, 567–608.
Gamo, N.J., Arnsten, A.F., 2011. Molecular modulation of prefrontal cortex: rational development of treatments for psychiatric disorders. Behav. Neurosci. 125, 282–296.
Gao, W.J., Goldman-Rakic, P.S., 2003. Selective modulation of excitatory and inhibitory microcircuits by dopamine. Proc. Natl. Acad. Sci. USA 100, 2836–2841.
Gao, W.J., Krimer, L.S., Goldman-Rakic, P.S., 2001. Presynaptic regulation of recurrent excitation by D1 receptors in prefrontal circuits. Proc. Natl. Acad. Sci. USA 98, 295–300.
Gariano, R.F., Groves, P.M., 1988. Burst firing induced in midbrain dopamine neurons by stimulation of the medial prefrontal and anterior cingulate cortices. Brain Res. 462, 194–198.
Garnock-Jones, K.P., Keating, G.M., 2010. Spotlight on atomoxetine in attention-deficit hyperactivity disorder in children and adolescents. CNS Drugs 24, 85–88.
Garris, P.A., Collins, L.B., Jones, S.R., Wightman, R.M., 1993. Evoked extracellular dopamine in vivo in the medial prefrontal cortex. J. Neurochem. 61, 637–647.

Garris, P.A., Wightman, R.M., 1994. Different kinetics govern dopaminergic transmission in the amygdala, prefrontal cortex, and striatum: an in vivo voltammetric study. J. Neurosci. 14, 442–450.
Gaspar, P., Bloch, B., Le Moine, C., 1995. D1 and D2 receptor gene expression in the rat frontal cortex: cellular localization in different classes of efferent neurons. Eur. J. Neurosci. 7, 1050–1063.
Gaspar, P., Stepniewska, I., Kaas, J.H., 1992. Topography and collateralization of the dopaminergic projections to motor and lateral prefrontal cortex in owl monkeys. J. Comp. Neurol. 325, 1–21.
Gee, S., Ellwood, I., Patel, T., Luongo, F., Deisseroth, K., Sohal, V.S., 2012. Synaptic activity unmasks dopamine D2 receptor modulation of a specific class of layer V pyramidal neurons in prefrontal cortex. J. Neurosci. 32, 4959–4971.
Glantz, L.A., Lewis, D.A., 2000. Decreased dendritic spine density on prefrontal cortical pyramidal neurons in schizophrenia. Arch. Gen. Psychiatry 57, 65–73.
Glausier, J.R., Khan, Z.U., Muly, E.C., 2009. Dopamine D1 and D5 receptors are localized to discrete populations of interneurons in primate prefrontal cortex. Cereb. Cortex 19, 1820–1834.
Gogos, J.A., Morgan, M., Luine, V., Santha, M., Ogawa, S., Pfaff, D., Karayiorgou, M., 1998. Catechol-O-methyltransferase-deficient mice exhibit sexually dimorphic changes in catecholamine levels and behavior. Proc. Natl. Acad. Sci. USA 95, 9991–9996.
Goldman-Rakic, P.S., 1995. Cellular basis of working memory. Neuron 14, 477–485.
Goldman-Rakic, P.S., Leranth, C., Williams, S.M., Mons, N., Geffard, M., 1989. Dopamine synaptic complex with pyramidal neurons in primate cerebral cortex. Proc. Natl. Acad. Sci. USA 86, 9015–9019.
Goldman-Rakic, P.S., Mully, E.C., Williams, G.V., 2000. D1 receptors in prefrontal cells and circuits. Brain Res. Rev. 31, 295–301.
Goldstein, R.Z., Volkow, N.D., 2011. Dysfunction of the prefrontal cortex in addiction: neuroimaging findings and clinical implications. Nat. Rev. Neurosci. 12, 652–669.
Gorelova, N., Mulholland, P.J., Chandler, L.J., Seamans, J.K., 2012. The glutamatergic component of the mesocortical pathway emanating from different subregions of the ventral midbrain. Cereb. Cortex 22, 327–336.
Gorelova, N.A., Yang, C.R., 2000. Dopamine D1/D5 receptor activation modulates a persistent sodium current in rat prefrontal cortical neurons in vitro. J. Neurophysiol. 84, 75–87.
Gray, E.G., 1959. Axo-somatic and axo-dendritic synapses of the cerebral cortex: an electron microscope study. J. Anat. 93, 420–433.
Gresch, P.J., Sved, A.F., Zigmond, M.J., Finlay, J.M., 1995. Local influence of endogenous norepinephrine on extracellular dopamine in rat medial prefrontal cortex. J. Neurochem. 65, 111–116.
Gulledge, A.T., Jaffe, D.B., 1998. Dopamine decreases the excitability of layer V pyramidal cells in the rat prefrontal cortex. J. Neurosci. 18, 9139–9151.
Gulyás, A.I., Megías, M., Emri, Z., Freund, T.F., 1999. Total number and ratio of excitatory and inhibitory synapses converging onto single interneurons of different types in the CA1 area of the rat hippocampus. J. Neurosci. 19, 10082–10097.
Gurden, H., Takita, M., Jay, T.M., 2000. Essential role of D1 but not D2 receptors in the NMDA receptor-dependent long-term potentiation at hippocampal-prefrontal cortex synapses in vivo. RC106 , J. Neurosci. 20, 1–5.
Haber, S.N., Ryoo, H., Cox, C., Lu, W., 1995. Subsets of midbrain dopaminergic neurons in monkeys are distinguished by different levels of mRNA for the dopamine transporter: comparison with the mRNA for the D2 receptor, tyrosine hydroxylase and calbindin immunoreactivity. J. Comp. Neurol. 362, 400–410.
Hanley, J.J., Bolam, J.P., 1997. Synaptology of the nigrostriatal projection in relation to the compartmental organization of the neostriatum in the rat. Neuroscience 81, 353–370.
Harrison, K.H., Hof, P.R., Wang, S.S., 2002. Scaling laws in the mammalian neocortex: does form provide clues to function? J. Neurocytol. 31, 289–298.
Henze, D.A., Gonzalez-Burgos, G.R., Urban, N.N., Lewis, D.A., Barrionuevo, G., 2000. Dopamine increases excitability of pyramidal neurons in primate prefrontal cortex. J. Neurophysiol. 84, 2799–2809.
Hersch, S.M., Yi, H., Heilman, C.J., Edwards, R.H., Levey, A.I., 1997. Subcellular localization and molecular topology of the dopamine transporter in the striatum and substantia nigra. J. Comp. Neurol. 388, 211–227.
Hitchcott, P.K., Quinn, J.J., Taylor, J.R., 2007. Bidirectional modulation of goal-directed actions by prefrontal cortical dopamine. Cereb. Cortex 17, 2820–2827.

Hnasko, T.S., Chuhma, N., Zhang, H., Goh, G.Y., Sulzer, D., Palmiter, R.D., Rayport, S., Edwards, R.H., 2010.Vesicular glutamate transport promotes dopamine storage and glutamate corelease in vivo. Neuron 65, 643–656.

Hur, E.E., Zaborszky, L., 2005.Vglut2 afferents to the medial prefrontal and primary somatosensory cortices: a combined retrograde tracing in situ hybridization. J. Comp. Neurol. 483, 351–373.

Ingham, C.A., Hood, S.H., Taggart, P., Arbuthnott, G.W., 1998. Plasticity of synapses in the rat neostriatum after unilateral lesion of the nigrostriatal dopaminergic pathway. J. Neurosci. 18, 4732–4743.

Käenmäki, M., Tammimäki, A., Myöhänen, T., Pakarinen, K., Amberg, C., Karayiorgou, M., Gogos, J.A., Männistö, P.T., 2010. Quantitative role of COMT in dopamine clearance in the prefrontal cortex of freely moving mice. J. Neurochem. 114, 1745–1755.

Karoum, F., Chrapusta, S.J., Egan, M.F., 1994. 3-Methoxytyramine is the major metabolite of released dopamine in the rat frontal cortex: reassessment of the effects of antipsychotics on the dynamics of dopamine release and metabolism in the frontal cortex, nucleus accumbens, and striatum by a simple two pool model. J. Neurochem. 63, 972–979.

Kawano, H., Decker, K., Reuss, S., 1996. Is there a direct retina-raphe-suprachiasmatic nucleus pathway in the rat? Neurosci. Lett. 212, 143–146.

Kawano, M., Kawasaki, A., Sakata-Haga, H., Fukui, Y., Kawano, H., Nogami, H., Hisano, S., 2006. Particular subpopulations of midbrain and hypothalamic dopamine neurons express vesicular glutamate transporter 2 in the rat brain. J. Comp. Neurol. 498, 581–592.

Kesner, R.P., Churchwell, J.C., 2011. An analysis of rat prefrontal cortex in mediating executive function. Neurobiol. Learn. Mem. 96, 417–431.

Khan, Z.U., Gutierrez, A., Martin, R., Penafiel, A., Rivera, A., De La Calle, A., 1998a. Differential regional and cellular distribution of dopamine D2-like receptors: an immunocytochemical study of subtype-specific antibodies in rat and human brain. J. Comp. Neurol. 402, 353–371.

Khan, Z.U., Koulen, P., Rubinstein, M., Grandy, D.K., Goldman-Rakic, P.S., 2001. An astroglia-linked dopamine D2-receptor action in prefrontal cortex. Proc. Natl. Acad. Sci. USA 98, 1964–1969.

Khan, Z.U., Mrzljak, L., Gutierrez, A., de la Calle, A., Goldman-Rakic, P.S., 1998b. Prominence of the dopamine D2 short isoform in dopaminergic pathways. Proc. Natl. Acad. Sci. USA 95, 7731–7736.

Koch, C., Poggio, T., 1983. A theoretical analysis of electrical properties of spines. Proc. R. Soc. London 218, 455–477.

Koch, C., Zador, A., Brown, T.H., 1992. Dendritic spines: convergence of theory and experiment. Science 256, 973–974.

Koenigs, M., Grafman, J., 2009. Posttraumatic stress disorder: the role of medial prefrontal cortex and amygdala. Neuroscientist 15, 540–548.

Krettek, J.E., Price, J.L., 1977. The cortical projections of the mediodorsal nucleus and adjacent thalamic nuclei in the rat. J. Comp. Neurol. 171, 157–192.

Krimer, L.S., Jakab, R.L., Goldman-Rakic, P.S., 1997. Quantitative three-dimensional analysis of the catecholaminergic innervation of identified neurons in the macaque prefrontal cortex. J. Neurosci. 17, 7450–7461.

Kuroda, M., Murakami, K., Igarashi, H., Okada, A., 1996. The convergence of axon terminals from the mediodorsal thalamic nucleus and ventral tegmental area on pyramidal cells in layer V of the rat prelimbic cortex. Eur. J. Neurosci. 8, 1340–1349.

Lammel, S., Hetzel, A., Hackel, O., Jones, I., Liss, B., Roeper, J., 2008. Unique properties of mesoprefrontal neurons within a dual mesocorticolimbic dopamine system. Neuron 57, 760–773.

Lapish, C.C., Kroener, S., Durstewitz, D., Lavin, A., Seamans, J.K., 2007. The ability of the mesocortical dopamine system to operate in distinct temporal modes. Psychopharmacology 191, 609–625.

Lavin, A., Nogueira, L., Lapish, C.C., Wightman, R.M., Phillips, P.E., Seamans, J.K., 2005. Mesocortical dopamine neurons operate in distinct temporal domains using multimodal signaling. J. Neurosci. 25, 5013–5023.

Le Moine, C., Gaspar, P., 1998. Subpopulations of cortical GABAergic interneurons differ by their expression of D1 and D2 dopamine receptor subtypes. Mol. Brain Res. 58, 231–236.

Levitt, P., Rakic, P., Goldman-Rakic, P., 1984. Region-specific distribution of catecholamine afferents in primate cerebral cortex: a fluorescence histochemical analysis. J. Comp. Neurol. 227, 23–36.

Lewis, B.L., O'Donnell, P., 2000.Ventral tegmental area afferents to the prefrontal cortex maintain membrane potential 'up' states in pyramidal neurons via D(1) dopamine receptors. Cereb. Cortex 10, 1168–1175.

Lewis, D.A., Campbell, M.J., Foote, S.L., Goldstein, M., Morrison, J.H., 1987. The distribution of tyrosine hydroxylase-immunoreactive fibers in primate neocortex is widespread but regionally specific. J. Neurosci. 7, 279–290.
Lewis, D.A., Foote, S.L., Goldstein, M., Morrison, J.H., 1988. The dopaminergic innervation of monkey prefrontal cortex: a tyrosine hydroxylase immunohistochemical study. Brain Res. 449, 225–243.
Lewis, D.A., Melchitzky, D.S., Sesack, S.R., Whitehead, R.E., Auh, S., Sampson, A., 2001. Dopamine transporter immunoreactivity in monkey cerebral cortex: regional, laminar and ultrastructural localization. J. Comp. Neurol. 432, 119–136.
Lewis, D.A., Sesack, S.R., 1997. Dopamine systems in the primate brain. In: Bloom, F.E., Björklund, A., Hökfelt, T. (Eds.), Handbook of Chemical Neuroanatomy, the Primate Nervous System, Part I. Elsevier Science Publishers, Amsterdam, pp. 261–373.
Lidow, M.S., Goldman-Rakic, P.S., Gallager, D.W., Rakic, P., 1991. Distribution of dopaminergic receptors in the primate cerebral cortex: quantitative autoradiographic analysis using [3H]raclopride, [3H]spiperone and [3H]SCH23390. Neuroscience 40, 657–671.
Liston, C., Malter Cohen, M., Teslovich, T., Levenson, D., Casey, B.J., 2011. Atypical prefrontal connectivity in attention-deficit/hyperactivity disorder: pathway to disease or pathological end point? Biol. Psychiatry 69, 1168–1177.
Loughlin, S.E., Fallon, J.H., 1984. Substantia nigra and ventral tegmental area projections to cortex: topography and collateralization. Neuroscience 11, 425–435.
Manitt, C., Mimee, A., Eng, C., Pokinko, M., Stroh, T., Cooper, H.M., Kolb, B., Flores, C., 2011. The netrin receptor DCC is required in the pubertal organization of mesocortical dopamine circuitry. J. Neurosci. 31, 8381–8394.
Matsumoto, M., Weickert, C.S., Akil, M., Lipska, B.K., Hyde, T.M., Herman, M.M., Kleinman, J.E., Weinberger, D.R., 2003. Catechol O-methyltransferase mRNA expression in human and rat brain: evidence for a role in cortical neuronal function. Neuroscience 116, 127–137.
Mayberg, H.S., 1997. Limbic-cortical dysregulation: a proposed model of depression. J. Neuropsychiatry Clin. Neurosci. 9, 471–481.
Meador-Woodruff, J.H., Damask, S.P., Wang, J.C., Haroutunian, V., Davis, K.L., Watson, S.J., 1996. Dopamine receptor mRNA expression in human striatum and neocortex. Neuropsychopharmacology 15, 17–29.
Melchitzky, D.S., Lewis, D.A., 2003. Pyramidal neuron local axon terminals in monkey prefrontal cortex: differential targeting of subclasses of GABA neurons. Cereb. Cortex 13, 452–460.
Mercuri, N., Calabresi, P., Stanzione, P., Bernardi, G., 1985. Electrical stimulation of mesencephalic cell groups (A9-A10) produces monosynaptic excitatory potentials in rat frontal cortex. Brain Res. 338, 192–195.
Miller, E.K., 2000. The prefrontal cortex and cognitive control. Nat. Rev. Neurosci. 1, 59–65.
Miller, E.K., Cohen, J.D., 2001. An integrative theory of prefrontal cortex function. Ann. Rev. Neurosci. 24, 167–202.
Miner, L.A.H., Schroeter, S., Blakely, R.D., Sesack, S.R., 2003. Ultrastructural localization of the norepinephrine transporter in superficial and deep layers of the rat prelimbic prefrontal cortex and its spatial relationship to probable dopamine terminals. J. Comp. Neurol. 466, 478–494.
Miner, L.A.H., Sesack, S.R., 2007. Anatomical characteristics of norepinephrine axons in the prefrontal cortex: unexpected findings that may indicate low activity state in naïve animals. In: Tseng, K.Y., Atorzi, M. (Eds.), Monoaminergic Modulation of Cortical Excitability. Springer, New York, NY, pp. 35–65.
Miner, L.H., Jedema, H.P., Moore, F.W., Blakely, R.D., Grace, A.A., Sesack, S.R., 2006. Chronic stress increases the plasmalemmal distribution of the norepinephrine transporter and the coexpression of tyrosine hydroxylase in norepinephrine axons in the prefrontal cortex. J. Neurosci. 26, 1571–1578.
Missale, C., Nash, S.R., Robinson, S.W., Jaber, M., Caron, M.G., 1998. Dopamine receptors: from structure to function. Physiol. Rev. 78, 189–225.
Moghaddam, B., Roth, R.H., Bunney, B.S., 1990. Characterization of dopamine release in the rat medial prefrontal cortex as assessed by in vivo microdialysis: comparison to the striatum. Neuroscience 36, 669–676.
Mori, S., 1966. Some observations on the fine structure of the corpus striatum of the rat brain. Z. Zellforsch Mikrosk Anat. 70, 461–488.

Morón, J.A., Brockington, A., Wise, R.A., Rocha, B.A., Hope, B.T., 2002. Dopamine uptake through the norepinephrine transporter in brain regions with low levels of the dopamine transporter: evidence from knock-out mouse lines. J. Neurosci. 22, 389–395.
Moss, J., Bolam, J.P., 2008. A dopaminergic axon lattice in the striatum and its relationship with cortical and thalamic terminals. J. Neurosci. 28, 11221–11230.
Moss, J., Ungless, M.A., Bolam, J.P., 2011. Dopaminergic axons in different divisions of the adult rat striatal complex do not express vesicular glutamate transporters. Eur. J. Neurosci. 33, 1205–1211.
Mrzljak, L., Bergson, C., Pappy, M., Huff, R., Levenson, R., Goldman-Rakic, P.S., 1996. Localization of dopamine D4 receptors in GABAergic neurons of the primate brain. Nature 381, 245–248.
Muly, E.C., Greengard, P., Goldman-Rakic, P.S., 2001. Distribution of protein phosphatases-1 alpha and -1 gamma 1 and the D(1) dopamine receptor in primate prefrontal cortex: evidence for discrete populations of spines. J. Comp. Neurol. 440, 261–270.
Muly, E.C.I., Szigeti, K., Goldman-Rakic, P.S., 1998. D1 receptor in interneurons of macaque prefrontal cortex: distribution and subcellular localization. J. Neurosci. 18, 10553–10565.
Mundorf, M.L., Joseph, J.D., Austin, C.M., Caron, M.G., Wightman, R.M., 2001. Catecholamine release and uptake in the mouse prefrontal cortex. J. Neurochem. 79, 130–142.
Murray, E.A., Wise, S.P., Drevets, W.C., 2011. Localization of dysfunction in major depressive disorder: prefrontal cortex and amygdala. Biol. Psychiatry 69, e43–54.
Nair-Roberts, R.G., Chatelain-Badie, S.D., Benson, E., White-Cooper, H., Bolam, J.P., Ungless, M.A., 2008. Stereological estimates of dopaminergic, GABAergic and glutamatergic neurons in the ventral tegmental area, substantia nigra and retrorubral field in the rat. Neuroscience 152, 1024–1031.
Negyessy, L., Goldman-Rakic, P.S., 2005. Subcellular localization of the dopamine D2 receptor and coexistence with the calcium-binding protein neuronal calcium sensor-1 in the primate prefrontal cortex. J. Comp. Neurol. 488, 464–475.
Nirenberg, M.J., Chan, J., Pohorille, A., Vaughan, R.A., Uhl, G.R., Kuhar, M.J., Pickel, V.M., 1997. The dopamine transporter: comparative ultrastructure of dopaminergic axons in limbic and motor compartments of the nucleus accumbens. J. Neurosci. 17, 6899–6907.
Nirenberg, M.J., Vaughan, R.A., Uhl, G.R., Kuhar, M.J., Pickel, V.M., 1996. The dopamine transporter is localized to dendritic and axonal plasma membranes of nigrostriatal dopaminergic neurons. J. Neurosci. 16, 436–447.
Noack, H.J., Lewis, D.A., 1989. Antibodies directed against tyrosine hydroxylase differentially recognize noradrenergic axons in monkey neocortex. Brain Res. 500, 313–324.
O'Donnell, P., 2003. Dopamine gating of forebrain neural ensembles. Eur. J. Neurosci. 17, 429–435.
O'Donnell, P., 2010. Adolescent maturation of cortical dopamine. Neurotoxicity Res. 18, 306–312.
Orozco-Cabal, L., Liu, J., Pollandt, S., Schmidt, K., Shinnick-Gallagher, P., Gallagher, J.P., 2008. Dopamine and corticotropin-releasing factor synergistically alter basolateral amygdala-to-medial prefrontal cortex synaptic transmission: functional switch after chronic cocaine administration. J. Neurosci. 28, 529–542.
Otani, S., Blond, O., Desce, J.M., Crepel, F., 1998. Dopamine facilitates long-term depression of glutamatergic transmission in rat prefrontal cortex. Neuroscience 85, 669–676.
Overton, P.G., Tong, Z.Y., Clark, D., 1996. A pharmacological analysis of the burst events induced in midbrain dopaminergic neurons by electrical stimulation of the prefrontal cortex in the rat. J. Neural Transm. 103, 523–540.
Park, M.K., Petersen, O.H., Tepikin, A.V., 2000. The endoplasmic reticulum as one continuous $Ca^{(2+)}$ pool: visualization of rapid $Ca^{(2+)}$ movements and equilibration. Embo J. 19, 5729–5739.
Parnavelas, J.G., Papadopoulos, G.C., 1989. The monoaminergic innervation of the cerebral cortex is not diffuse and nonspecific. Trends Neurosci. 12, 315–319.
Paspalas, C.D., Goldman-Rakic, P.S., 2004. Microdomains for dopamine volume neurotransmission in primate prefrontal cortex. J. Neurosci. 24, 5292–5300.
Paspalas, C.D., Goldman-Rakic, P.S., 2005. Presynaptic D1 dopamine receptors in primate prefrontal cortex: target-specific expression in the glutamatergic synapse. J. Neurosci. 25, 1260–1267.
Paspalas, C.D., Rakic, P., Goldman-Rakic, P.S., 2006. Internalization of D2 dopamine receptors is clathrin-dependent and select to dendro-axonic appositions in primate prefrontal cortex. Eur. J. Neurosci. 24, 1395–1403.

Peters, Y., Barnhardt, N.E., O'Donnell, P., 2004. Prefrontal cortical up states are synchronized with ventral tegmental area activity. Synapse 52, 143–152.
Pickel, V.M., Beckley, S.C., Joh, T.H., Reis, D.J., 1981. Ultrastructural immunocytochemical localization of tyrosine hydroxylase in the neostriatum. Brain Res. 225, 373–385.
Pickel, V.M., Joh, T.H., Field, P.M., Becker, C.G., Reis, D.J., 1975. Cellular localization of tyrosine hydroxylase by immunohistochemistry. J. Histochem. Cytochem. 23, 1–12.
Pinto, A., Jankowski, M., Sesack, S.R., 2003. Projections from the paraventricular nucleus of the thalamus to the rat prefrontal cortex and nucleus accumbens shell: ultrastructural characteristics and spatial relationships with dopamine afferents. J. Comp. Neurol. 459, 142–155.
Pinto, A., Miner, L.A.H., Sesack, S.R., 1997. Limited collateralization of neurons in the rat prefrontal cortex that project to the nucleus accumbens. Neuroscience 97, 635–642.
Pinto, A., Sesack, S.R., 1999. Basolateral amygdala afferents to the rat prefrontal cortex: ultrastructure and relation to dopamine afferents. Soc. Neurosci. Abstr. 25, 1216.
Pinto, A., Sesack, S.R., 2008. Ultrastructural analysis of prefrontal cortical inputs to the rat amygdala: spatial relationships to presumed dopamine axons and D1 and D2 receptors. Brain Struct. Funct. 213, 159–175.
Pirot, S., Godbout, R., Mantz, J., Tassin, J.-P., Glowinski, J., Thierry, A.-M., 1992. Inhibitory effects of ventral tegmental area stimulation on the activity of prefrontal cortical neurons: evidence for the involvement of both dopaminergic and GABAergic components. Neuroscience 49, 857–865.
Preuss, T.M., 1995. Do rats have a prefrontal cortex? The Rose-Woolsey-Akert program reconsidered. J. Cognit. Neurosci. 7, 1–24.
Rajkowska, G., 2000. Histopathology of the prefrontal cortex in major depression: what does it tell us about dysfunctional monoaminergic circuits? Prog. Brain Res. 126, 397–412.
Reep, R., 1984. Relationship between prefrontal and limbic cortex: a comparative anatomical review. Brain Behav. Evol. 25, 5–80.
Rivera, A., Peñafiel, A., Megías, M., Agnati, L.F., López-Téllez, J.F., Gago, B., Gutiérrez, A., de la Calle, A., Fuxe, K., 2008. Cellular localization and distribution of dopamine D(4) receptors in the rat cerebral cortex and their relationship with the cortical dopaminergic and noradrenergic nerve terminal networks. Neuroscience 155, 997–1010.
Robinson, T.E., Kolb, B., 2004. Structural plasticity associated with exposure to drugs of abuse. Neuropharmacology 47 (Suppl. 1), 33–46.
Roesch, M.R., Calu, D.J., Schoenbaum, G., 2007. Dopamine neurons encode the better option in rats deciding between differently delayed or sized rewards. Nat. Neurosci. 10, 1615–1624.
Rosenberg, D.R., Lewis, D.A., 1995. Postnatal maturation of the dopaminergic innervation of monkey prefrontal cortices: a tyrosine hydroxylase immunohistochemical analysis. J. Comp. Neurol. 358, 383–400.
Salahpour, A., Ramsey, A.J., Medvedev, I.O., Kile, B., Sotnikova, T.D., Holmstrand, E., Ghisi, V., Nicholls, P.J., Wong, L., Murphy, K., Sesack, S.R., Wightman, R.M., Gainetdinov, R.R., Caron, M.G., 2008. Increased amphetamine-induced hyperactivity and reward in mice overexpressing the dopamine transporter. Proc. Natl. Acad. Sci. USA 105, 4405–4410.
Sanghera, M.K., Manaye, K.F., Liang, C.L., Lacopino, A.M., Bannon, M.J., German, D.C., 1994. Low dopamine transporter mRNA levels in midbrain regions containing calbindin. Neuroreport 5, 1641–1644.
Santana, N., Mengod, G., Artigas, F., 2009. Quantitative analysis of the expression of dopamine D1 and D2 receptors in pyramidal and GABAergic neurons of the rat prefrontal cortex. Cereb. Cortex 19, 849–860.
Sawaguchi, T., Goldman-Rakic, P.S., 1994. The role of D1-dopamine receptor in working memory: local injections of dopamine antagonists into the prefrontal cortex of rhesus monkeys performing an oculomotor delayed-response task. J. Neurophysiol. 71, 515–528.
Schmidt, R.H., Bhatnagar, R.K., 1979. Assessment of the effects of neonatal subcutaneous 6-hydroxydopamine on noradrenergic and dopaminergic innervation of the cerebral cortex. Brain Res. 166, 309–319.
Schultz, W., 1998. Predictive reward signal of dopamine neurons. J. Neurophysiol. 80, 1–27.
Seamans, J.K., Yang, C.R., 2004. The principal features and mechanisms of dopamine modulation in the prefrontal cortex. Prog. Neurobiol. 74, 1–58.
Seeman, P., 1992. Dopamine receptor sequences. Therapeutic levels of neuroleptics occupy D2 receptors, clozapine occupies D4. Neuropsychopharmacology 7, 261–284.
Séguéla, P., Watkins, K.C., Descarries, L., 1988. Ultrastructural features of dopamine axon terminals in the anteromedial and the suprarhinal cortex of adult rat. Brain Res. 442, 11–22.

Sesack, S.R., 2002. Synaptology of dopamine neurons. In: Di Chiara, G. (Ed.), Handbook of Experimental Pharmacology, Dopamine in the CNS I. Springer-Verlag, Berlin Heidelberg, pp. 63–119.
Sesack, S.R., 2010. Functional implications of dopamine D2 receptor localization in relation to glutamate neurons. In: Bjorklund, A., Dunnett, S., Iversen, L., Iversen, S. (Eds.), Dopamine Handbook. Oxford University Press, New York, NY.
Sesack, S.R., Bressler, C.N., Lewis, D.A., 1995a. Ultrastructural associations between dopamine terminals and local circuit neurons in the monkey prefrontal cortex: a study of calretinin-immunoreactive cells. Neurosci. Lett. 200, 9–12.
Sesack, S.R., Bunney, B.S., 1989. Pharmacological characterization of the receptor mediating electrophysiological effects of dopamine in the rat medial prefrontal cortex: a microiontophoretic study. J. Pharmacol. Exp. Ther. 248, 1323–1333.
Sesack, S.R., Hawrylak, V.A., Guido, M.A., Levey, A.I., 1998. Dopamine axon varicosities in the rat prefrontal cortex exhibit sparse immunoreactivity for the dopamine transporter. J. Neurosci. 18, 2697–2708.
Sesack, S.R., Hawrylak, V.A., Melchitzky, D.S., Lewis, D.A., 1997. Dopamine innervation of a subclass of local circuit neurons in monkey prefrontal cortex: ultrastructural analysis of tyrosine hydroxylase and parvalbumin immunoreactive structures. Cereb. Cortex 8, 614–622.
Sesack, S.R., Miner, L.A.H., 1997. In the rat prefrontal cortex, dopamine terminals converge with local axon collaterals onto common dendritic processes. Soc. Neurosci. Abstr. 23, 1213.
Sesack, S.R., Pickel, V.M., 1990. In the rat medial nucleus accumbens, hippocampal and catecholaminergic terminals converge on spiny neurons and are in apposition to each other. Brain Res. 527, 266–279.
Sesack, S.R., Pickel, V.M., 1992. Prefrontal cortical efferents in the rat synapse on unlabeled neuronal targets of catecholamine terminals in the nucleus accumbens septi and on dopamine neurons in the ventral tegmental area. J. Comp. Neurol. 320, 145–160.
Sesack, S.R., Snyder, C.L., Lewis, D.A., 1995b. Axon terminals immunolabeled for dopamine or tyrosine hydroxylase synapse on GABA-immunoreactive dendrites in rat and monkey cortex. J. Comp. Neurol. 363, 264–280.
Shin, L.M., Rauch, S.L., Pitman, R.K., 2006. Amygdala, medial prefrontal cortex, and hippocampal function in PTSD. Ann. N.Y. Acad. Sci. 1071, 67–79.
Smiley, J.F., 1996. Monoamines and acetylcholine in primate cerebral cortex: what anatomy tells us about function. Rev. Bras Biol. 56 (Suppl. 1 Pt 1), 153–164.
Smiley, J.F., Goldman-Rakic, P.S., 1993. Heterogeneous targets of dopamine synapses in monkey prefrontal cortex demonstrated by serial section electron microscopy: a laminar analysis using the silver-enhanced diaminobenzidine sulfide (SEDS) immunolabeling technique. Cereb. Cortex 3, 223–238.
Smiley, J.F., Levey, A.I., Ciliax, B.J., Goldman-Rakic, P.S., 1994. D1 dopamine receptor immunoreactivity in human and monkey cerebral cortex: predominant and extrasynaptic localization in dendritic spines. Proc. Natl. Acad. Sci. USA 91, 5720–5724.
Smiley, J.F., Williams, S.M., Szigeti, K., Goldman-Rakic, P.S., 1992. Light and electron microscopic characterization of dopamine-immunoreactive axons in human cerebral cortex. J. Comp. Neurol. 321, 325–335.
Smith, A.D., Bolam, J.P., 1990. The neural network of the basal ganglia as revealed by the study of synaptic connections of identified neurones. Trends Neurosci. 13, 259–265.
Smith, Y., Bennett, B.D., Bolam, J.P., Parent, A., Sadikot, A.F., 1994. Synaptic relationship between dopaminergic afferents and cortical or thalamic input in the sensorimotor territory of the striatum in monkey. J. Comp. Neurol. 344, 1–19.
Steketee, J.D., 2003. Neurotransmitter systems of the medial prefrontal cortex: potential role in sensitization to psychostimulants. Brain Res. Rev. 41, 203–228.
Strange, P.G., 2001. Antipsychotic drugs: importance of dopamine receptors for mechanisms of therapeutic actions and side effects. Pharmacol. Rev. 53, 119–133.
Stuber, G.D., Hnasko, T.S., Britt, J.P., Edwards, R.H., Bonci, A., 2010. Dopaminergic terminals in the nucleus accumbens but not the dorsal striatum corelease glutamate. J. Neurosci. 30, 8229–8233.
Sulzer, D., Joyce, M.P., Lin, L., Geldwert, D., Haber, S.N., Hattori, T., Rayport, S., 1998. Dopamine neurons make glutamatergic synapses in vitro. J. Neurosci. 18, 4588–4602.
Sun, X., Zhao, Y., Wolf, M.E., 2005. Dopamine receptor stimulation modulates AMPA receptor synaptic insertion in prefrontal cortex neurons. J. Neurosci. 25, 7342–7351.

Sunahara, R.K., Guan, H.C., O'Dowd, B.F., Seeman, P., Laurier, L.G., Ng, G., George, S.R., Torchia, J., Van Tol, H.H., Niznik, H.B., 1991. Cloning of the gene for a human dopamine D5 receptor with higher affinity for dopamine than D1. Nature 350, 614–619.

Swanson, C.J., Perry, K.W., Koch-Krueger, S., Katner, J., Svensson, K.A., Bymaster, F.P., 2006. Effect of the attention deficit/hyperactivity disorder drug atomoxetine on extracellular concentrations of norepinephrine and dopamine in several brain regions of the rat. Neuropharmacology 50, 755–760.

Swanson, L.W., 1982. The projections of the ventral tegmental area and adjacent regions: a combined fluorescent retrograde tracer and immunofluorescence study in the rat. Brain Res. Bull. 9, 321–353.

Tan, H.Y., Callicott, J.H., Weinberger, D.R., 2009. Prefrontal cognitive systems in schizophrenia: towards human genetic brain mechanisms. Cognit. Neuropsychiatry 14, 277–298.

Tanda, G., Carboni, E., Frau, R., Di Chiara, G., 1994. Increase of extracellular dopamine in the prefrontal cortex: a trait of drugs with antidepressant potential? Psychopharmacology 115, 285–288.

Tecuapetla, F., Patel, J.C., Xenias, H., English, D., Tadros, I., Shah, F., Berlin, J., Deisseroth, K., Rice, M.E., Tepper, J.M., Koos, T., 2010. Glutamatergic signaling by mesolimbic dopamine neurons in the nucleus accumbens. J. Neurosci. 30, 7105–7110.

Thierry, A.M., Blanc, G., Sobel, A., Stinus, L., Glowinski, J., 1973. Dopaminergic terminals in the rat cortex. Science 182, 499–501.

Thierry, A.M., Le Douarin, C., Penit, J., Ferron, A., Glowinski, J., 1986. Variation in the ability of neuroleptics to block the inhibitory influence of dopaminergic neurons on the activity of cells in the rat prefrontal cortex. Brain Res. Bull. 16, 155–160.

Tong, Z.-Y., Overton, P.G., Clark, D., 1996. Stimulation of the prefrontal cortex in the rat induces patterns of activity in midbrain dopaminergic neurons which resemble natural burst events. Synapse 22, 195–208.

Totterdell, S., Smith, A.D., 1989. Convergence of hippocampal and dopaminergic input onto identified neurons in the nucleus accumbens of the rat. J. Chem. Neuroanat. 2, 285–298.

Towers, S.K., Hestrin, S., 2008. D1-like dopamine receptor activation modulates GABAergic inhibition but not electrical coupling between neocortical fast-spiking interneurons. J. Neurosci. 28, 2633–2641.

Tseng, K.Y., O'Donnell, P., 2004. Dopamine-glutamate interactions controlling prefrontal cortical pyramidal cell excitability involve multiple signaling mechanisms. J. Neurosci. 24, 5131–5139.

Tseng, K.Y., O'Donnell, P., 2005. Post-pubertal emergence of prefrontal cortical up states induced by D1-NMDA co-activation. Cereb. Cortex 15, 49–57.

Tseng, K.Y., O'Donnell, P., 2007a. D2 dopamine receptors recruit a GABA component for their attenuation of excitatory synaptic transmission in the adult rat prefrontal cortex. Synapse 61, 843–850.

Tseng, K.Y., O'Donnell, P., 2007b. Dopamine modulation of prefrontal cortical interneurons changes during adolescence. Cereb. Cortex 17, 1235–1240.

Tzschentke, T.M., 2001. Pharmacology and behavioral pharmacology of the mesocortical dopamine system. Prog. Neurobiol. 63, 241–320.

Uylings, H.B.M., Groenewegen, H.J., Kolb, B., 2003. Do rats have a prefrontal cortex? Behav. Brain Res. 146, 3–17.

Valentini, V., Frau, R., Di Chiara, G., 2004. Noradrenaline transporter blockers raise extracellular dopamine in medial prefrontal but not parietal and occipital cortex: differences with mianserin and clozapine. J. Neurochem. 88, 917–927.

Van Eden, C.G., Hoorneman, E.M.D., Buijs, R.M., Matthijssen, M.A.H., Geffard, M., Uylings, H.B.M., 1987. Immunocytochemical localization of dopamine in the prefrontal cortex of the rat at the light and electron microscopic level. Neuroscience 22, 849–862.

Verney, C., Alvarez, C., Geffard, M., Berger, B., 1990. Ultrastructural double-labelling study of dopamine terminals and GABA-containing neurons in rat anteromedial cerebral cortex. Eur. J. Neurosci. 2, 960–972.

Vijayraghavan, S., Wang, M., Birnbaum, S.G., Williams, G.V., Arnsten, A.F., 2007. Inverted-U dopamine D1 receptor actions on prefrontal neurons engaged in working memory. Nat. Neurosci. 10, 376–384.

Wang, H., Pickel, V.M., 2002. Dopamine D2 receptors are present in prefrontal cortical afferents and their targets in patches of the rat caudate-putamen nucleus. J. Comp. Neurol. 442, 392–404.

Wang, H.D., Deutch, A.Y., 2008. Dopamine depletion of the prefrontal cortex induces dendritic spine loss: reversal by atypical antipsychotic drug treatment. Neuropsychopharmacology 33, 1276–1286.

Wang, H.D., Stanwood, G.D., Grandy, D.K., Deutch, A.Y., 2009. Dystrophic dendrites in prefrontal cortical pyramidal cells of dopamine D1 and D2 but not D4 receptor knockout mice. Brain Res. 1300, 58–64.

Wang, R.Y., 1981. Dopaminergic neurons in the rat ventral tegmental area. I. Identification and characterization. Brain Res. Rev. 3, 123–140.

Wayment, H.K., Schenk, J.O., Sorg, B.A., 2001. Characterization of extracellular dopamine clearance in the medial prefrontal cortex: role of monoamine uptake and monoamine oxidase inhibition. J. Neurosci. 21, 35–44.

Wedzony, K., Chocyk, A., Mackowiak, M., Fijal, K., Czyrak, A., 2000. Cortical localization of dopamine D4 receptors in the rat brain–immunocytochemical study. J. Physiol. Pharmacol. 51, 205–221.

Weinberger, D.R., Egan, M.F., Bertolino, A., Callicott, J.H., Mattay, V.S., Lipska, B.K., Berman, K.F., Goldberg, T.E., 2001. Prefrontal neurons and the genetics of schizophrenia. Biol. Psychiatry 50, 825–844.

Williams, S.M., Goldman-Rakic, P.S., 1993. Characterization of the dopaminergic innervation of the primate frontal cortex using a dopamine-specific antibody. Cereb. Cortex 3, 199–222.

Williams, S.M., Goldman-Rakic, P.S., 1998. Widespread origin of the primate mesofrontal dopamine system. Cereb. Cortex 8, 321–345.

Wilson, J.M., Sanyal, S., Van Tol, H.H., 1998. Dopamine D2 and D4 receptor ligands: relation to antipsychotic action. Eur. J. Pharmacol. 351, 273–286.

Wise, S.P., 2008. Forward frontal fields: phylogeny and fundamental function. Trends Neurosci. 31, 599–608.

Yamaguchi, T., Sheen, W., Morales, M., 2007. Glutamatergic neurons are present in the rat ventral tegmental area. Eur. J. Neurosci. 25, 106–118.

Yamaguchi, T., Wang, H.L., Li, X., Ng, T.H., Morales, M., 2011. Mesocorticolimbic glutamatergic pathway. J. Neurosci. 31, 8476–8490.

Yamamoto, B.K., Novotney, S., 1998. Regulation of extracellular dopamine by the norepinephrine transporter. J. Neurochem. 71, 274–280.

Yang, C.R., Seamans, J.K., 1996. Dopamine D1 receptor actions in layers V–VI rat prefrontal cortex neurons in vitro: modulation of dendritic-somatic signal integration. J. Neurosci. 16, 1922–1935.

Yao, W.D., Spealman, R.D., Zhang, J., 2008. Dopaminergic signaling in dendritic spines. Biochem. Pharmacol. 75, 2055–2069.

Yokofujita, J., Oda, S., Igarashi, H., Sato, F., Kuroda, M., 2008. Synaptic characteristics between cortical cells in the rat prefrontal cortex and axon terminals from the ventral tegmental area that utilize different neurotransmitters. Int. J. Neurosci. 118, 1443–1459.

Zahrt, J., Taylor, J.R., Mathew, R.G., Arnsten, A.F., 1997. Supranormal stimulation of D1 dopamine receptors in the rodent prefrontal cortex impairs spatial working memory performance. J. Neurosci. 17, 8528–8535.

Zheng, P., Zhang, X.X., Bunney, B.S., Shi, W.X., 1999. Opposite modulation of cortical N-methyl-D-aspartate receptor-mediated responses by low and high concentrations of dopamine. Neuroscience 91, 527–535.

Zoli, M., Torri, C., Ferrari, R., Jansson, A., Zini, I., Fuxe, K., Agnati, L.F., 1998. The emergence of the volume transmission concept. Brain Res. Rev. 26, 136–147.

INDEX

Note: Page numbers followed by "f" denote figures; "t" tables.

B

C

D

P

T

V

FSC
www.fsc.org
MIX
Papier | Fördert gute Waldnutzung
FSC® C083411

Kolibri 360 GmbH
Ferdinand-Jühlke-Straße 7
99095 Erfurt, Deutschland
produktsicherheit@kolibri360.de